Customizing AutoCAD

Release 14

Online Companion

The Online Companion™ is your link to AutoCAD on the Internet. We have compiled supporting resources with links to a variety of sites. Not only can you find out about training and education, industry sites, and the online community, we also point to valuable archives compiled for AutoCAD® users from various Web sites. In addition, there are pages specifically for users of Customizing AutoCAD. These include an owner's page with updates, a swap bank where you can share your drawings with other AutoCAD students, and a page where you can send us your comments. You can find the Online Companion at:

http://www.autodeskpress.com/onlinecompanion.html

When you reach the Online Companion page, click on the title Customizing AutoCAD.

Customizing AutoCAD

Sham Tickoo

Professor
Department of Manufacturing Engineering Technologies
and Supervision
Purdue University Calumet
Hammond, Indiana

Contributing Author

Gregory Neff

Associate Professor
Department of Manufacturing Engineering Technologies
and Supervision
Purdue University Calumet
Hammond, Indiana

I(T)P® International Thomson Publishing

Albany • Bonn • Boston • Cincinnati • Detroit • London • Madrid
Melbourne • Mexico City • New York • Pacific Grove • Paris • San Francisco
Singapore • Tokyo • Toronto • Washington

NOTICE TO THE READER

Publisher does not warrant or guarantee any of the products described herein or perform any independent analysis in connection with any of the product information contained herein. Publisher does not assume, and expressly disclaims, any obligation to obtain and include information other than that provided to it by the manufacturer.

The reader is expressly warned to consider and adopt all safety precautions that might be indicated by the activities herein and to avoid all potential hazards. By following the instructions contained herein, the reader willingly assumes all risks in connection with such instructions.

The publisher makes no representation or warranties of any kind, including but not limited to, the warranties of fitness for particular purpose or merchantability, nor are any such representations implied with respect to the material set forth herein, and the publisher takes no responsibility with respect to such material. The publisher shall not be liable for any special, consequential, or exemplary damages resulting, in whole or part, from the readers' use of, or reliance upon, this material. Autodesk does not guarantee the performance of the software and Autodesk assumes no responsibility or liability for the performance of the software or for errors in this manual.

Trademarks: Autodesk, the Autodesk logo, and AutoCAD are registered trademarks of Autodesk, Inc. Microsoft and Windows 95 are registered trademarks of the Microsoft Corporation. Windows NT is a trademark of Microsoft Corp. Online Companion is a trademark of International Thomson Publishing. All other product names are acknowledged as trademarks of their respective owners.

Cover: AutoCAD image © 1998 Autodesk, Inc.

Staff
Publisher: Alar Elken
Acquisitions Editor: Sandy Clark
Production Coordinator: Jennifer Gaines
Art & Design Coordinator: Mary Beth Vought
Editorial Assistant: Christopher Leonard

COPYRIGHT © 1998 Delmar Publishers Inc. Autodesk Press imprint.

For more information, contact:

Delmar Publishers/Autodesk Press
3 Columbia Circle, Box 15-015
Albany, New York USA 12212-5015

International Thomson Publishing Europe
Berkshire House 168-173
High Holborn
London, WC1V 7AA
United Kingdom

Thomas Nelson Australia
102 Dodds Street
South Melbourne, 3205
Victoria, Australia

Nelson Canada
1120 Birchmont Road
Scarborough, Ontario
Canada, M1K 5G4

International Thomson Publishing
Southern Africa
Building 18, Constantia Park
240 Old Pretoria Road
P.O. Box 2459
Halfway House, 1685 South Africa

International Thomson Editores
Campos Eliseos 385, Piso 7
Colonia Polanco
11560 Mexico D. F. Mexico

International Thomson Publishing GmbH
Konigswinterer Strasse 418
53227 Bonn Germany

International Thomson Publishing France
Tour Maine-Montparnasse
33, Avenue du Maine
75755 Paris Cedex 15, France

International Thomson Publishing -Japan
Hirakawacho Kyowa Building, 3F
2-2-1 Hirakawa-cho Chiyoda-ku
Tokyo 102 Japan

International Thomson Publishing Asia
221 Henderson Road
#05-10 Henderson Building
Singapore 0315

All rights reserved. No part of this work covered by the copyright hereon may be reproduced or used in any form or by any means – graphic, electronic, or mechanical, including photocopying, recording, taping, or information storage and retrieval systems without the written permission of the publisher.

1 2 3 4 5 6 7 8 9 10 XXX 03 02 01 00 99 98 97

Library of Congress Cataloging-in-Publication Data

Tickoo, Sham.
    Customizing AutoCAD release 14 / Sham Tickoo
       p.   cm.
    Includes index.
    ISBN 0-7668-0364-3
    1. Computer graphics. 2. AutoCAD (Computer file) I. Title.
T385.T5292 1998
620'.0042'02855369--dc21
                                        97-51556
                                           CIP

Table of Contents

Preface — xvii

Dedication — xxi

Chapter 1: Template Drawings
- Creating Template Drawings — 1-1
- The Standard Template Drawings — 1-1
- Loading a Template Drawing — 1-5
  - Using the Dialog Box — 1-5
  - Using the NEW Command — 1-6
- Customizing Drawings with Layers and Dimensioning Specifications — 1-6
- Customizing Drawings According to Plot Size and Drawing Scale — 1-10
- Customizing Drawings with Viewports — 1-14
- Customizing a Drawing with Paper Space — 1-17
- Using Drawing Files as Template Files — 1-20

Chapter 2: Script Files and Slide Shows
- What are Script Files? — 2-1
- SCRIPT Command — 2-3
- RSCRIPT Command — 2-8
- DELAY Command — 2-9
- RESUME Command — 2-10
- Command Line Switches — 2-10
- Invoking a Script File When Loading AutoCAD — 2-11
- What is a Slide Show? — 2-18
- What are Slides? — 2-18
- MSLIDE Command — 2-18
- VSLIDE Command — 2-19
- Preloading Slides — 2-21
- Slide Libraries — 2-23

Chapter 3: Creating Linetypes and Hatch Patterns
- Standard Linetypes — 3-1
- Linetype Definition — 3-1
  - Header Line — 3-2
  - Pattern Line — 3-2
- Elements of Linetype Specification — 3-3
- Creating Linetypes — 3-3
  - Using the AutoCAD Linetype Command — 3-4
  - Using a Text Editor — 3-6
- Creating Linetype Files — 3-8
- Alignment Specification — 3-8
- LTSCALE Command — 3-9

LTSCALE Factor for Plotting	3-11
Alternate Linetypes	3-12
Modifying Linetypes	3-13
Current Linetype Scaling (CELTSCALE)	3-17
Complex Linetypes	3-17
Creating a String Complex Linetype	3-18
Creating a Shape Complex Linetype	3-22
Hatch Pattern Definition	3-25
Header Line	3-25
Hatch Descriptors	3-25
Hatch Angle	3-26
How Hatch Works	3-27
Simple Hatch Pattern	3-28
Effect of Angle and Scale Factor on Hatch	3-29
Hatch Pattern with Dashes and Dots	3-30
Hatch with Multiple Descriptors	3-31
Saving Hatch Patterns in a Separate File	3-36
Custom Hatch Pattern File	3-36
Adding Hatch Pattern Slides to AutoCAD Slide Library	3-37

Chapter 4: Customizing the ACAD.PGP File

What is the ACAD.PGP File?	4-1
Sections of the ACAD.PGP File	4-6
Comments	4-6
External Command	4-6
Command Aliases	4-7
REINIT Command	4-9

Chapter 5: Pull-down, Cascading, Cursor, and Partial Menus and Customizing Toolbars

AutoCAD Menu	5-1
Standard Pull-down Menus	5-2
Writing a Pull-down Menu	5-3
Loading Menus	5-10
Restrictions	5-12
Cascading Submenus in Pull-down Menus	5-13
Cursor Menus	5-18
Submenus	5-22
Submenu Definition	5-22
Submenu Reference	5-23
Displaying a Submenu	5-23
Loading Menus	5-23
Loading Screen Menus	5-23
Loading an Image Tile Menu	5-24
Partial Menus	5-27
Menu Section Labels	5-27
Writing Partial Menus	5-28
Accelerator Keys	5-32
Toolbars	5-33
Menu-Specific Help	5-38

Customizing the Toolbars	5-38
Creating a New Image and Tooltip for an Icon	5-39
Deleting the Icons from a Toolbar	5-41
Deleting a Toolbar	5-41
Copying a Tool Icon	5-41
Creating Custom Toolbars with Flyout Icons	5-42

Chapter 6: Tablet Menus

Standard Tablet Menus	6-2
Advantages of a Tablet Menu	6-3
Customizing a Tablet Menu	6-3
Writing a Tablet Menu	6-4
Tablet Configuration	6-7
Loading Menus	6-8
Tablet Menus with Different Block Sizes	6-9
Assigning Commands to a Tablet	6-13
Automatic Menu Swapping	6-14

Chapter 7: Image Tile Menus

Image Tile Menus	7-1
Submenus	7-2
Submenu Definition	7-2
Submenu Reference	7-3
Displaying a Submenu	7-3
Writing an Image Tile Menu	7-3
Slides for Image Tile Menus	7-8
Loading Menus	7-9
Restrictions	7-10
Image Tile Menu Item Labels	7-11
Menu Item Label Formats	7-11

Chapter 8: Button and Auxiliary Menus

Button Menus	8-1
Auxiliary Menus	8-2
Writing Buttons and Auxiliary Menus	8-2
Special Handling for Button Menus	8-5
Submenus	8-8
Submenu Definition	8-8
Submenu Reference	8-8
Loading Menus	8-9
Loading Screen Menus	8-9
Loading a Pull-down Menu	8-9
Loading an Image Menu	8-9

Chapter 9: Screen Menus

Screen Menu	9-1
Loading Menus	9-6
Submenus	9-7

Submenu Definition	9-8
Submenu Reference	9-8
Nested Submenus	9-9
Multiple Submenus	9-16
Long Menu Definitions	9-31
Menu Command Repetition	9-33
Automatic Menu Swapping	9-35
MENUECHO System Variable	9-35
Menus for Foreign Languages	9-35
Use of Control Characters in Menu Items	9-36
Special Characters	9-37
Command Definition without Enter or Space	9-38
Menu Items with Single Object Selection Mode	9-40
Use of AutoLISP in Menus	9-40
Diesel Expression in Menus	9-41

Chapter 10: Customizing the Standard AutoCAD Menu

The Standard AutoCAD Menu	10-1
Submenus	10-14
Submenu Definition	10-15
Submenu Reference	10-15
Loading Screen Menus	10-16
Loading Pull-down Menus	10-16
Loading Image Tile Menus	10-16
Customizing Tablet Area-1	10-17
Submenus	10-22
Customizing Tablet Area-2	10-26
Customizing Tablet Area-3	10-27
Customizing Tablet Area-4	10-29
Customizing Buttons and Auxiliary Menus	10-31
Customizing Pull-down and Cursor Menus	10-36
Cascading Submenus in Pull-down Menus	10-37
Cursor Menus	10-41
Submenus	10-42
Swapping Pull-down Menus	10-42
Customizing IMAGE TILE Menus	10-42
Image Tile Menu Item Labels	10-44
Customizing the Screen Menu	10-47
Submenus	10-47
Nested Submenus	10-48

Chapter 11: Shapes and Text Fonts

Shape Files	11-1
Shape Description	11-1
Header	11-1
Shape Specification	11-2
Vector Length and Direction Encoding	11-2
Compiling and Loading Shape/Font Files	11-3
Header Line	11-3
Shape Specification	11-3

Special Codes	11-6
Standard Codes	11-6
Code 000: End of Shape Definition	11-6
Code 001: Activate Draw Mode	11-6
Code 002: Deactivate Draw Mode	11-6
Code 003: Divide Vector Lengths	11-8
Code 004: Multiply Vector Lengths	11-9
Codes 005 and 006: Location Save/Restore	11-9
Code 007: Subshape	11-10
Code 008: X-Y Displacement	11-10
Code 009: Multiple X-Y Displacements	11-11
Code 00A or 10: Octant Arc	11-11
Code 00B or 11: Fractional Arc	11-12
Code 00C or 12: Arc Definition by Displacement and Bulge	11-13
Code 00D or 13: Multiple Bulge-Specified Arc	11-14
Code 00E or 14: Flag Vertical Text	11-14
Text Font Files	11-18
Text Font Description	11-18
Line Feed	11-19
Shape Definition	11-19

Chapter 12: AutoLISP

About AutoLISP	12-1
Mathematical Operations	12-2
Addition	12-2
Subtraction	12-3
Multiplication	12-3
Division	12-3
Incremented, Decremented, and Absolute Numbers	12-4
Incremented Number	12-4
Decremented Number	12-4
Absolute Number	12-4
Trigonometrical Functions	12-5
sin	12-5
cos	12-5
atan	12-5
angtos	12-6
Relational Statements	12-6
Equal to	12-7
Not equal to	12-7
Less than	12-7
Less than or equal to	12-7
Greater than	12-8
Greater than or equal to	12-8
defun, setq, getpoint, and Command Functions	12-8
defun	12-8
setq	12-9
getpoint	12-10
Command	12-11
Loading an AutoLISP Program	12-14
getcorner, getdist, and setvar Functions	12-15
getcorner	12-15

getdist	12-16
setvar	12-16
list Function	12-19
car, cdr, and cadr Functions	12-19
car	12-19
cdr	12-20
cadr	12-20
graphscr, textscr, princ, and terpri Functions	12-21
graphscr	12-21
textscr	12-21
princ	12-21
terpri	12-21
getangle and getorient Functions	12-25
getangle	12-25
getorient	12-26
getint, getreal, getstring, and getvar Functions	12-28
getint	12-28
getreal	12-28
getstring	12-28
getvar	12-29
polar and sqrt Functions	12-29
polar	12-29
sqrt	12-30
itoa, rtos, strcase, and prompt Functions	12-33
itoa	12-33
rtos	12-33
strcase	12-34
prompt	12-34
Flowcharts	12-38
Conditional Functions	12-38
if	12-38
progn	12-42
while	12-43
repeat	12-46
Persistent AutoLISP	12-47

Chapter 13: AutoLISP: Editing the Drawing Database

Editing the Drawing Database	13-1
ssget	13-1
ssget "X"	13-3
Group Codes for ssget "X"	13-4
sslength	13-5
ssname	13-5
entget	13-5
assoc	13-6
cons	13-6
subst	13-6
entmod	13-7
How the Database is Retrieved and Edited	13-8

Chapter 14: Programmable Dialog Boxes Using Dialog Control Language

Dialog Control Language	14-1
Dialog Box	14-2
Dialog Box Components	14-2
Button and Text Tiles	14-4
Button Tile	14-4
Text Tile	14-4
Tile Attributes	14-5
Types of Attribute Values	14-5
Predefined Attributes	14-6
key, label, and is_default Attributes	14-6
key Attribute	14-6
label Attribute	14-6
is_default Attribute	14-7
fixed_width and alignment Attributes	14-8
fixed_width Attribute	14-8
alignment Attribute	14-8
Loading a DCL File	14-10
Displaying a New Dialog Box	14-10
Using AutoLISP Function to Load a DCL File	14-12
Use of Standard Button Subassemblies	14-12
AutoLISP Functions	14-13
load_dialog	14-13
unload_dialog	14-13
new_dialog	14-13
start_dialog	14-14
done_dialog	14-14
action_tile	14-14
Managing Dialog Boxes with AutoLISP	14-14
Row and Boxed Row Tiles	14-16
Row Tile	14-16
Boxed Row Tile	14-16
Column, Boxed Column, and Toggle Tiles	14-17
Column Tile	14-17
Boxed Column Tile	14-17
Toggle Tile	14-17
Mnemonic Attribute	14-17
AutoLISP Functions	14-21
logand and logior	14-21
atof and rtos Functions	14-22
get_tile and set_tile Functions	14-22
Predefined Radio Button, Radio Column, Boxed Radio Column, and Radio Row Tiles	14-25
Predefined Radio Button Tile	14-25
Predefined Radio Column Tile	14-25
Predefined Boxed Radio Column Tile	14-25
Predefined Radio Row Tile	14-26
Edit Box Tile	14-31
width and edit width Attributes	14-31
width Attribute	14-31
edit Width Attribute	14-32

Slider and Image Tiles	14-35
Slider Tile	14-35
Image Tile	14-35
min_value, max_value, small_increment, and big_increment Attributes	14-35
min_value and max_value Attributes	14-35
small_increment and big_increment Attributes	14-36
aspect_ratio and color Attributes	14-36
aspect_ratio Attribute	14-36
color Attribute	14-36
AutoLISP Functions	14-39
dimx_tile and dimy_tile	14-39
vector_image	14-40
fill_image	14-40
start_image	14-40
end_image	14-41
$value	14-41

Chapter 15: DIESEL: A String Expression Language

DIESEL	15-1
Status Line	15-1
Modemacro System Variable	15-2
Customizing the Status Line	15-3
Macro Expressions Using DIESEL	15-4
Using AutoLISP with Modemacro	15-6
DIESEL Expressions in Menus	15-8
Macrotrace System Variable	15-9
DIESEL String Functions	15-10
Addition	15-10
Subtraction	15-10
Multiplication	15-11
Division	15-11
Relational Statements	15-11

Chapter 16: Visual Basic

About Visual Basic	16-1
Installing VBA	16-2
Objects 16-2	
Add Method	16-3
AddCircle	16-3
AddLine	16-3
AddArc	16-4
AddText	16-4
Finding Help on Methods and Properties	16-4
Loading and Saving VBA Projects	16-5
GetPoint, GetDistance, and GetAngle Methods	16-10
GetPoint Method	16-10
GetDist Method	16-10
GetAngle Method	16-10
PolarPoint and AngleFromXAxis Methods	16-14

PolarPoint Method	16-14
AngleFromXAxis	16-15
Additional VBA Examples	16-20

Chapter 17: Accessing External Databases AutoCAD SQL Extension (ASE)

AutoCAD SQL2 Environment (ASE)	17-1
Environment	17-2
Catalog	17-2
Schema	17-2
Session	17-2
Transaction	17-2
Understanding Databases	17-2
Database	17-2
Database Management System	17-3
Relational Database	17-3
Components of a Table	17-4
Defining Keys	17-5
Isolation Levels	17-5
Dirty Read Transaction	17-5
Nonrepeatable Read Transaction	17-6
Phantom Read Transaction	17-6
Establishing the Database Environment	17-6
Accessing Data in External Databases	17-9
Selection of Rows	17-9
Editing Data in a Table	17-10
Linking a Database with a Drawing	17-13
Editing Links (ASELINKS)	17-14
Deleting Links	17-15
Creating Displayable Attributes	17-16
Editing Rows	17-18
Forming Selection Sets	17-19
ASESELECT Command Options	17-21
Using SQL Statements (ASESQLED)	17-21
Generating Reports from Exported Data	17-26

Chapter 18: Defining Block Attributes

Attributes	18-1
Defining Attributes	18-2
DDATTDEF Command	18-2
ATTDEF Command	18-4
Editing Attribute Tags	18-7
Using the DDEDIT Command	18-7
Using the Change Command	18-7
Inserting Blocks with Attributes	18-7
Using the Dialog Box	18-7
Using the Command Line	18-8
Extracting Attributes	18-11
Using the Dialog Box (DDATTEXT)	18-11
Using the Command Line (ATTEXT)	18-14

xiv　Customizing AutoCAD

Controlling Attribute Visibility (ATTDISP Command)	18-15
Editing Attributes (DDATTE Command)	18-16
Editing Attributes (ATTEDIT Command)	18-17
Global Editing of Attributes	18-17
Editing Visible Attributes Only	18-18
Editing All Attributes	18-18
Editing Specific Attributes	18-18
Editing Attributes with Specific Attribute Tag Names	18-18
Editing Attributes with Specific Attribute Value	18-19
Individual Editing of Attributes	18-22
Inserting Text Files in the Drawing	18-25
Using MTEXT Command	18-25

Chapter 19: Rendering

Rendering	19-1
Determining which Sides are to be Rendered in a Model	19-1
Points to be Remembered while Defining a Model	19-2
Loading and Unloading AutoCAD Render	19-2
Elementary Rendering	19-3
Selecting Different Properties for Rendering	19-5
Rendering Type	19-6
Rendering Options	19-6
Rendering Procedures	19-7
Destination	19-7
Sub Sampling	19-8
Background	19-8
Fog/Depth Cue	19-9
AutoCAD Render Light Source	19-9
Ambient Light	19-9
Point Light	19-9
Spotlight	19-9
Attenuation	19-10
Distant Light	19-11
Inserting and Modifying Lights	19-11
Inserting Distant Light	19-11
Modifying Distant Light	19-13
Inserting Point Light	19-15
Defining and Rendering a Scene	19-15
Modifying a Scene	19-18
Obtaining Rendering Information	19-18
Attaching Materials	19-19
Assigning Materials to an Object	19-19
Assigning Materials to the AutoCAD Color Index (ACI)	19-21
Assigning Materials to Layers	19-22
Detaching Materials	19-23
Changing the Parameters of a Material	19-23
Attributes Area	19-24
Defining New Materials	19-25
Exporting a Material from Drawing to the Library of Materials	19-26
Saving a Rendering	19-26
Saving a Rendering to a File	19-26
Saving a Viewport Rendering	19-17

        Saving a Render-Window Rendered Image   19-27
    Replaying a Rendered Image   19-28
        Replaying a Rendered Image to a Viewport   19-28
        Replaying a Rendered Image to the Windows Render Window   19-28

Chapter 20: AutoCAD on the Internet

    AutoCAD on the Internet*   20-1
        Launching the Web Browser (BROWSER Command)   20-1
        The Uniform Resource Locator   20-3
        How URLs are Used in AutoCAD   20-4
        Attaching URLs to Objects (ATTACHURL Command)   20-6
        Selecting a URL (SELECTURL Command)   20-7
        Listing a URL (LISTURL Command)   20-7
        Removing a URL (DETACHURL Command)   20-8
    The Drawing Web Format   20-11
        Creating a DWF File (DWFOUT Command)   20-11
        Viewing DWF Files   20-12
        DWF Plug-in Commands   20-14
        Drag and Drop   20-15
        Embedding a DWF File   20-16
        Accessing a Drawing on the Internet (OPENURL Command)   20-17
        Inserting a Block from the Internet (INSERTURL Command)   20-20
        Saving a Drawing to the Internet (SAVEURL Command)   20-20

Chapter 21: Data Exchange, Object Linking and Embedding, Multilines, and Digitizing

    Data Exchange in AutoCAD   21-1
    DXF File Format (Data Interchange File)   21-2
        Creating a Data Interchange File (DXFOUT Command)   21-2
        Information in a DXF File   21-3
        Converting DXF Files into the Drawing Editor DXB File Format   21-4
        Importing Scanned Files into the Drawing Editor DXB File Format   21-4
    Data Interchange Through Raster Files   21-5
        Exporting the Raster Files (SAVEIMG Command)   21-5
    Raster Images*   21-8
        Attaching Raster Images   21-8
        Attach Image Dialog Box   21-10
    Editing Raster Image Files*   21-10
        Clipping Raster Images   21-10
        Adjusting Raster Image   21-11
        Image Quality   21-11
        Transparency   21-12
        Image Frame   21-12
        Other Editing Commands   21-12
        Scaling Raster Images   21-12
    Postscript Files   21-12
        PSOUT Command   21-13
        Prolog Section Name   21-14
        What to Plot   21-14
        Preview   21-14

Size Units	21-15
Scale	21-15
Paper Size	21-15
PSIN Command	21-15
PostScript Fill Patterns (PSFILL Command)	21-16
Object Linking and Embedding	21-18
Clipboard	21-18
Object Embedding	21-19
Linking Objects	21-22
Creating Multilines	21-24
Defining Multiline Style (MLSTYLE Command)	21-24
Multiline Style Area	21-24
Element Properties	21-25
Multiline Properties	21-26
Drawing Multilines (MLINE Command)	21-27
Justification Option	21-28
Scale Option	21-28
STyle Option	21-29
Editing Multilines (Using GRIPS)	21-29
Editing Multilines (Using the MLEDIT Command)	21-29
Cross Section	21-30
Tee Intersection	21-31
Corner Joint	21-31
Adding and Deleting Vertices	21-31
Cutting and Welding Multilines	21-32
-MLEDIT Command	21-33
System Variables for MLINE	21-33
Digitizing Drawings	21-34
TABLET Command	21-35
Floating Screen Pointing Area	21-36

Appendices

Appendix A: System Requirements
Appendix B: Bonus Tools
Appendix C: AutoCAD Linetypes
Appendix D: AutoCAD Hatch Patterns
Appendix E: AutoCAD Text Fonts
Appendix F: Dialog Boxes
Appendix G: Pull-down and Cascading Menus
Appendix H: Toolbars
Appendix I: AutoCAD Commands
Appendix J: AutoCAD System Variables

Index

Preface

AutoCAD, developed by Autodesk Inc., is the most popular PC-CAD system available in the market. Nearly 1.8 million people in 80 countries around the world are using AutoCAD to generate various kinds of drawings. In 1997 the market share of AutoCAD grew to 78 percent, making it the worldwide standard for generating drawings. Also, AutoCAD's open architecture has allowed third-party developers to write application software that has significantly added to its popularity. For example, the author of this book has developed a software package "SMLayout" for sheet metal products that generates flat layout of various geometrical shapes such as transitions, intersections, cones, elbows, and tank heads. Several companies in Canada and the United States are using this software package with AutoCAD to design and manufacture various products. AutoCAD has also provided facilities that allow users to customize AutoCAD to make it more efficient and therefore increase their productivity.

The purpose of this book is to unravel the customizing power of AutoCAD and explain it in a way that is easy to understand. Every customizing technique is thoroughly explained with examples and illustrations that make it easy to comprehend the customizing concepts of AutoCAD. When you are done reading this book, you will be able to generate a Template drawing, write script files, edit existing menus, write your own menus, write shape and text files, create new linetypes and hatch patterns, define new commands, write programs in the AutoLISP programming language, edit the existing drawing database, create your own dialog boxes using DCL, customize the status line using DIESEL, and edit the Program Parameter file (ACAD.PGP). In the process, you will discover some new applications of AutoCAD that are unique and might have a significant effect on your drawings. You will also get a better idea of why AutoCAD has become such a popular software package and an international standard in PC-CAD.

To use this book, you do not need to be an AutoCAD expert or a programmer. If you know the basic AutoCAD commands, you will have no problem in understanding the material presented in this book. The book contains a detailed description of various customizing techniques that you can use to customize your system. Every chapter has several examples that illustrate some possible applications of these customizing techniques. At the end of each chapter are some exercises that provide a challenge to the user to solve the problems on his/her own. In a class situation, these exercises can be assigned to students to test their understanding of the material explained in the chapter. The chapters on AutoLISP programming assume that the user has no programming background and therefore all commands have been thoroughly explained in a way that makes programming easy to understand and interesting to learn. **All chapters, except Slide Shows and Editing the Drawing Database, are independent and can be read in any order and used without reading the rest of the book.** The user needs only to read the chapters on Script Files before the chapter on Slide Shows and the chapter on AutoLISP before the chapter on Editing the Drawing Database. However, in order to get a good understanding of customizing techniques, it is recommended to start from Chapter 1 and then progress through the chapters. AutoCAD Release 14 features are indicated by asterisk (*) at the end of the feature. The following is a summary of each chapter.

Chapter 1: Template Drawings
This chapter explains how to create a Template drawing and how to standardize the information that is common to all drawings. It also describes how to create a Template drawing with paper space and predefined viewports.

Chapter 2: Script Files and Slide Shows
This chapter introduces the user to script files and how to utilize them to group AutoCAD commands in a predetermined sequence to perform a given operation. This chapter also explains how to use script files to create a slide show that can be used for product presentation.

Chapter 3: Creating Linetypes and Hatch Patterns
This chapter explains how to create a new linetype and how to edit the linetype file, ACAD.LIN. This chapter also describes the techniques of creating a new hatch pattern and the effect of hatch scale and hatch angle on hatch.

Chapter 4: Customizing the ACAD.PGP File
This chapter explains the use of AutoCAD's Program Parameter file (ACAD.PGP) to define aliases for the operating system commands and some of the AutoCAD commands.

Chapter 5: Pull-down and Cursor Menus
This chapter explains how to write a pull-down menu and how to load screen and Image tile menus from the pull-down menus.

Chapter 6: Tablet Menus
This chapter explains how to write a tablet menu, and how to load other menus from the tablet menu. Advantages of the tablet menu, design of the tablet menu, and how AutoCAD assigns commands to different blocks of the tablet menu are also discussed.

Chapter 7: Image Tile Menus
This chapter explains the Image tile menus and how to write an Image tile menu. It also discusses submenus and how to make slides for the Image tile menu.

Chapter 8: Buttons and Auxiliary Menus
This chapter deals with buttons and auxiliary menus and how to assign AutoCAD commands to different buttons of a multi-button pointing device.

Chapter 9: Screen Menus
This chapter describes the procedure to write a screen menu with multiple submenus and how to load Image tile or pull-down menus from the screen menu.

Chapter 10: Customizing the Standard AutoCAD Menu
This chapter describes how to edit and change various menu sections of the standard AutoCAD menu, ACAD.MNU. It also contains information about submenus and how to load different submenus.

Chapter 11: Shapes and Text Fonts
This chapter explains what shapes are and how to create shape and text fonts. It also contains a detailed description of special codes and their application to creating shapes and text fonts.

Chapter 12: AutoLISP
This chapter explains different AutoLISP functions and how to use these functions to write a program. It also introduces the user to basic programming techniques and use of relational and conditional statements in a program.

Chapter 13: AutoLISP: Editing the Drawing Database
This chapter describes those AutoLISP functions that allow a user to edit the drawing database.

Chapter 14: Programmable Dialog Boxes Using Dialog Control Language
This chapter is an introduction to Dialog Control Language and its applications in customizing the existing dialog boxes and writing new dialog boxes. It also explains the use of AutoLISP in controlling the dialog boxes.

Chapter 15: DIESEL: A String Expression Language
This chapter describes the DIESEL string expression language and its application in customizing the status line by altering the value of AutoCAD system variable MODEMACRO.

Chapter 16: Visual Basic
This chapter describes how to install the AutoCAD preview VBA, load and run sample VBA projects, utilize the visual basic editor, use AutoCAD objects and object properties, and apply and use AutoCAD methods.

Chapter 17: Accessing External Database, AutoCAD SQL Extension (ASE)
This chapter describes how to use SQL to access and manipulate the data that is stored in the external database and link data from the database to objects in a drawing. The chapter also explains how to edit the database.

Chapter 18: Defining Block Attributes
This chapter describes how to extract the attributes that have been assigned to blocks by using DDATTEXT or ATTEXT command. The chapter also explains how to create an attribute definition by using the DDATTDEF or ATTDEF commands and how to edit attributes.

Chapter 19: Rendering
This chapter describes how to render images that makes it easier to visualize the shape and size of a three-dimensional (3D) object, compared to a wireframe image or a shaded image. The chapter also explains how to control the appearance of the object by defining the surface material and reflective quality of the surface and by adding lights to get the desired effects.

Chapter 20: AutoCAD on the Internet
This chapter explains how to use Internet to exchange digital information around the world by bringing together text, graphics, audio, and movies in an easy to use format. The chapter also describes the other uses of the Internet including FTP (file transfer protocol for effortless binary file transfer), Gopher (presents data in a structured, subdirectory-like format), and USENET.

Chapter 21: Data Exchange, Object Linking and Embedding, Multilines, and Digitizing
This chapter describes how to use various data exchange formats that enable transfer (translation) of data from one data processing software to another. The chapter also explains how to work with different Windows-based applications by transferring information between them by creating links between the different applications and then updating those links, which in turn updates or modifies the information in the corresponding applications. It also describes how to create, edit, and use multiline and how to digitize drawings.

In the end, it is the author's sincere hope that the material provided in this book will prove valuable in mastering the customizing techniques and advanced features of AutoCAD.

DEDICATION

*To teachers, who make it possible to disseminate knowledge
to enlighten the young and curious minds
of our future generations*

*To students, who are dedicated to learning new technologies
and making the world a better place to live*

Thanks

*To the faculty and students of the METS department of Purdue
University Calumet for their cooperation*

Chapter 1

Template Drawings

Learning objectives

After completing this chapter, you will be able to:
- *Create template drawings.*
- *Load template drawings using dialog boxes and the command line.*
- *Do initial drawing setup.*
- *Customize drawings with layers and dimensioning specifications.*
- *Customize drawings with viewports and paper space.*

CREATING TEMPLATE DRAWINGS

One way to customize AutoCAD is to create template drawings that contain initial drawing setup information and, if desired, visible objects and text. When the user starts a new drawing, the settings associated with the template drawing are automatically loaded. If you start a new drawing from scratch, AutoCAD loads default setup values. For example, the default limits are (0.0,0.0), (12.0,9.0) and the default layer is 0 with white color and continuous linetype. Generally, these default parameters need to be reset before generating a drawing on the computer using AutoCAD. A considerable amount of time is required to set up the layers, colors, linetypes, limits, snaps, units, text height, dimensioning variables, and other parameters. Sometimes, border lines and a title block may also be needed.

In production drawings, most of the drawing setup values remain the same. For example, the company title block, border, layers, linetypes, dimension variables, text height, LTSCALE, and other drawing setup values do not change. You will save considerable time if you save these values and reload them when starting a new drawing. You can do this by making template drawings, which can contain the initial drawing setup information, set according to company specifications. They can also contain a border, title block, tolerance table, block definitions, floating viewports in the paper space, and perhaps some notes and instructions that are common to all drawings.

THE STANDARD TEMPLATE DRAWINGS

The AutoCAD software package comes with standard template drawings like acad.dwt, acadiso.dwt, ansi_a.dwt, din_a.dwt, iso_a4.dwt, jis_a3.dwt. The ansi, din, and iso template drawings are based on the drawing standards developed by ANSI (American National Standards Institute), DIN (German), and ISO (International Organization for Standardization). When you

1-2 Customizing AutoCAD

start a new drawing, AutoCAD displays the **Create New Drawing** dialog box on the screen. To load the template drawing, select the **Use a Template** button and AutoCAD will display the list of standard template drawings. From this list you can select any template drawing according to your requirements. If you want to start a drawing with default setting, select the **Start from Scratch** button in the **Create New Drawing** dialog box. The following are some of the system variables, with the default values that are assigned to new drawing:

System Variable Name	Default Value
BASE	0.0000,0.0000,0.0000
BLIPMODE	Off
CHAMFERA	0.5000
CHAMFERB	0.5000
COLOR	Bylayer
DIMALT	Off
DIMALTD	2
DIMALTF	25.4
DIMPOST	None
DIMASO	On
DIMASZ	0.18
DRAGMODE	Auto
ELEVATION	0
FILLMODE	On
FILLETRAD	0.5000
GRID	0.5000
GRIDMODE	0
ISOPLANE	Left
LIMMIN	0.0000,0.0000
LIMMAX	12.0000,9.0000
LTSCALE	1.0
MIRRTEXT	1 (Text mirrored like other objects)
ORTHOMODE	0 (Off)
TILEMODE	1 (On)
TRACEWID	0.0500

Example 1

Create a template drawing with the following specifications. The name of the template drawing is **PROTO1**.

Limits	18.0,12.0
Snap	0.25
Grid	0.50
Text height	0.125
Units	Decimal
	2-digits to the right of decimal point
	Decimal degrees

2-digits to the right of decimal point
0 angle along positive X axis (east)
Angle positive if measured counterclockwise

Start AutoCAD and select the **Start from Scratch** button in the **Create New Drawing** dialog box (Figure 1-1). You can also invoke the **Create New Drawing** dialog box by selecting **New** in the **File** pull-down menu or entering **NEW** at AutoCAD Command: prompt.

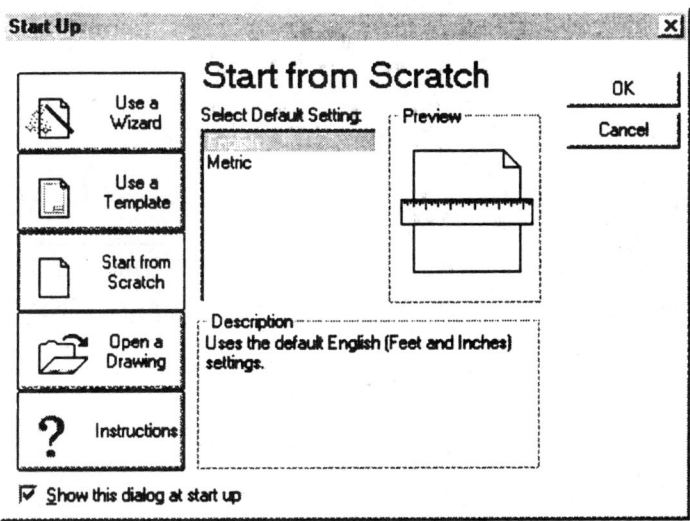

Figure 1-1 Start Up dialog box

Once you are in the Drawing Window, use the following AutoCAD commands to set up the values as given in Example 1.

    Command: **LIMITS**
    ON/OFF/<Lower left corner> <0.00,0.00>: **0,0**
    Upper right corner<12.0,9.0>: **18.0,12.0**

    Command: **SNAP**
    Snap spacing or ON/OFF/Aspect/Rotate/Style<0.5000>: **0.25**

    Command: **GRID**
    Grid spacing(X) or ON/OFF/Snap/Aspect<0.5000>: **0.50**

    Command: **SETVAR**
    Variable name or ?: **TEXTSIZE**
    New value for textsize<0.2000>: **0.125**

Using the Dialog Box. You can use the **Unit Control** dialog box (Figure 1-2) to set the units. To invoke the Unit Control dialog box, enter **DDUNITS** at AutoCAD Command: prompt or select Units in the Format pull-down menu.

1-4 Customizing AutoCAD

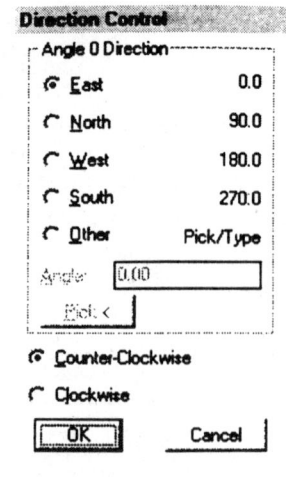

Figure 1-2 Unit Control dialog box

Figure 1-3 Direction Control dialog box

Using UNITS Command. You can also use the UNITS command to set the units.

Command: **UNITS**

Report formats:	Examples:
1. Scientific	1.55E+01
2. Decimal	15.50
3. Engineering	1'-3.50"
4. Architectural	1'-3 1/2"
5. Fractional	15 1/2

With the exception of the Engineering and Architectural formats, these formats can be used with any basic units of measurement. For example, Decimal mode is perfect for metric units as well as decimal English units.

Enter choice, 1 to 5 <2>: **2**
Number of digits to right of decimal point (0 to 8) <4>: **2**

Systems of angle measure:	Examples:
1. Decimal degrees	45.0000
2. Degrees/minutes/seconds	45d0'0"
3. Grads	50.0000g
4. Radians	0.7854r
5 Surveyor's units	N 45d0'0"

Enter choice, 1 to 5 <1>: **1**
Number of fractional places for display of angles (0 to 8) <0>: **2**
Direction for angle 0.00:

East	3 o'clock	= 0.00
North	12 o'clock	= 90.00
West	9 o'clock	= 180.00

South   6 o'clock   = 270.00

Enter direction for angle 0.00<0.00>: **0**
Do you want angles measured clockwise?<N>: **N**

Now, save the drawing as **PROTO1.DWT** using AutoCAD's SAVE or SAVEAS command. You must select template (*DWT) from the list box in the dialog box. This drawing is now saved as PROTO1.DWT on the default drive. You can also save this drawing on a floppy diskette in drives A or B.

Command: **SAVE**
Save Drawing As <Drawing.dwg>: **A:PROTO1.DWT**

LOADING A TEMPLATE DRAWING
Using the Dialog Box

You can use the template drawing any time you want to start a new drawing. To use the preset values of the template drawing, start AutoCAD or enter the NEW command (Command: **NEW**) to start a new drawing. AutoCAD displays the **Create New Drawing** dialog box (Figure 1-4).

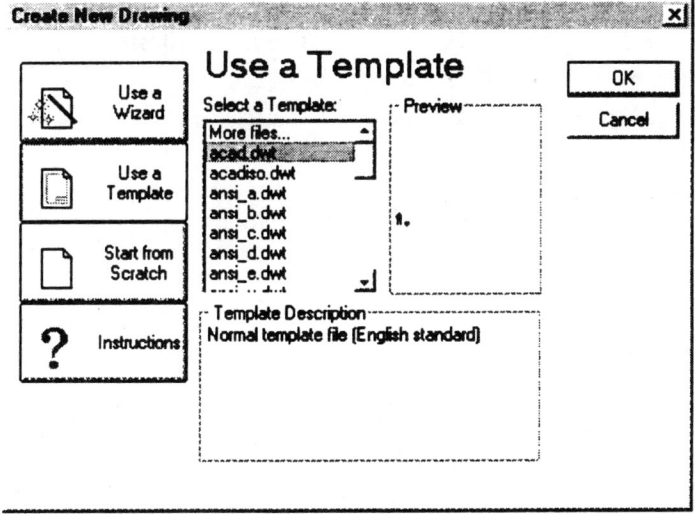

Figure 1-4 Create New Drawing dialog box

You can also start a new drawing by selecting the **File** pull-down menu and then selecting the **New** option from this menu. To load the template drawing, select the **Use a Template** button; AutoCAD will display the list of template drawings. Select PROTO1 template drawing and then select the OK button to exit the dialog box. AutoCAD will start a new drawing that will have the same setup as that of template drawing, **PROTO1**.

1-6 Customizing AutoCAD

You can have several template drawings, each with a different setup. For example, **PROTOB** for a 18" by 12" drawing, **PROTOC** for a 24" by 18" drawing, **PROTOD** for a 36" by 24" drawing, and **PROTOE** for a 48" by 36" drawing. Each template drawing can be created according to user-defined specifications. You can then load any of these template drawings as discussed previously.

Using the NEW Command

You can create a new drawing without using the dialog boxes by assigning a value of 0 to the AutoCAD system variable **FILEDIA**. To start a new drawing, enter NEW at the Command: prompt (Command: NEW) and AutoCAD will prompt you to enter the name of the drawing. The format for entering the name of the template drawing is:

Command: **NEW**
Enter template file (or . for none) <current>: *Enter template drawing name.*

CUSTOMIZING DRAWINGS WITH LAYERS AND DIMENSIONING SPECIFICATIONS

Most production drawings need multiple layers for different groups of objects. In addition to layers, it is good practice to assign different colors to different layers to control the line width at the time of plotting. You can generate a template drawing that contains the desired number of layers with linetypes and colors according to your company specifications. You can then use this template drawing to make a new drawing. The next example illustrates the procedure used for customizing a drawing with layers, linetypes, and colors.

Example 2

You want to create a template drawing (**PROTO2**) that has a border and the company's title block, as shown in Figure 1-5. In addition to this, you want the following initial drawing setup:

Limits	48.0,36.0
Snap	1.0
Grid	4.00
Text height	0.25
PLINE width	0.02
Ltscale	4.0
DIMENSIONS	
DIMSCALE	4.0
DIMTAD	ON
DIMTIX	ON
DIMTOH	OFF
DIMTIH	OFF
DIMSCALE	25

(Use DIMSTYLE command to save these values in MYDIM1 dimension style file.)

LAYERS

Layer Names	Line Type	Color
0	Continuous	White
OBJ	Continuous	Red
CEN	Center	Yellow
HID	Hidden	Blue
DIM	Continuous	Green
BOR	Continuous	Magenta

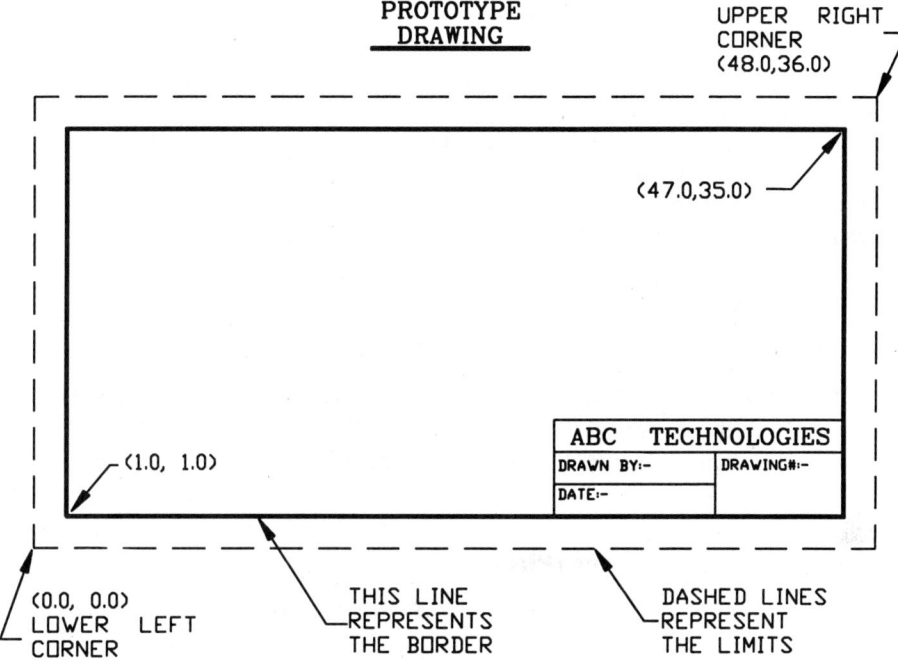

Figure 1-5 Template drawing for Example 2

Start a new drawing with default parameters. You can do this by selecting the Start from Scratch button in the Create New Drawing dialog box. Once you are in the drawing editor, use the AutoCAD commands to set up the values as given for this example. Also, draw a border and a title block as shown in Figure 1-5. In this figure the hidden lines indicate the drawing limits. The border lines are 0.5 units inside the drawing limits. For the border lines, use a polyline of width 0.02 units. Use the following command sequence to produce the prototype drawing for Example 2:

Command: **LIMITS**
ON/OFF/<Lower left corner> <0.00,0.00>:**0,0**
Upper right corner<12.0,9.0>: **48.0,36.0**
Command: **SNAP**
Snap spacing or ON/OFF/Aspect/Rotate/Style<1.0>: **1.0**

Command: **GRID**
Grid spacing(X) or ON/OFF/Snap/Aspect <1.0>: **4.00**

Command: **SETVAR**
Variable name or ?: **TEXTSIZE**
New value for textsize <0.180>: **0.25**

Command: **PLINEWID**
New value for PLINEWID: **0.02**

Command: **PLINE**
From point: **1.0,1.0**
Current line-width is **0.02**
Arc/Close/Halfwidth/Length/Undo/Width/<Endpoint of line>:**47,1**
Arc/Close/Halfwidth/Length/Undo/Width/<Endpoint of line>:**47,35**
Arc/Close/Halfwidth/Length/Undo/Width/<Endpoint of line>:**1,35**
Arc/Close/Halfwidth/Length/Undo/Width/<Endpoint of line>:**C**

Command: **LTSCALE**
New scale factor <1.0>: **4.0**

Using the Dimension Styles Dialog Box. You can use the **Dimension Styles** dialog box (Figure 1-6) to set the dimension variables. To invoke the **Dimension Styles** dialog box, enter **DDIM** at AutoCAD Command: prompt or select **Style** in the **Dimension** pull-down menu.

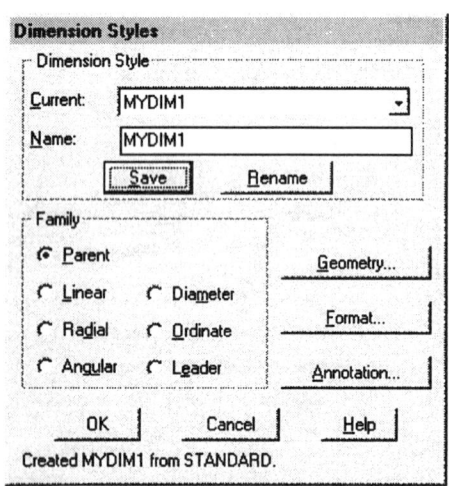

Figure 1-6 Dimension Styles dialog box

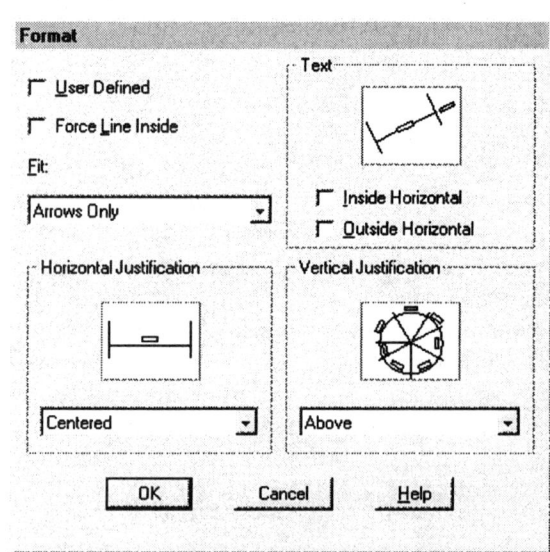

Figure 1-7 Format dialog box

Using the DIM Command. You can also use the DIM command to set the dimensions. The following is the command prompt sequence for setting the dimension variables and saving the dimension style file.

```
Command: DIM
Dim: DIMSCALE
Current value<1.00> New value: 4.0
Dim: DIMTIX
Current value<Off> New value: ON
Dim: Dimtad
Current value<Off> New value: ON
Dim: Dimtoh
Current value<On> New value: OFF
Dim: Dimtih
Current value<On> New value: OFF
Dim: Style
New text style <STANDARD>: MYDIM1
Dim: Esc        (Press the Escape key)
```

Using the Layer & Linetype Properties Dialog Box. You can use the Layer & Linetype Properties dialog box (Figure 1-8) to set the layers and linetypes. To invoke this dialog box, enter LAYER at AutoCAD Command: prompt or select Layer in the Format pull-down menu.

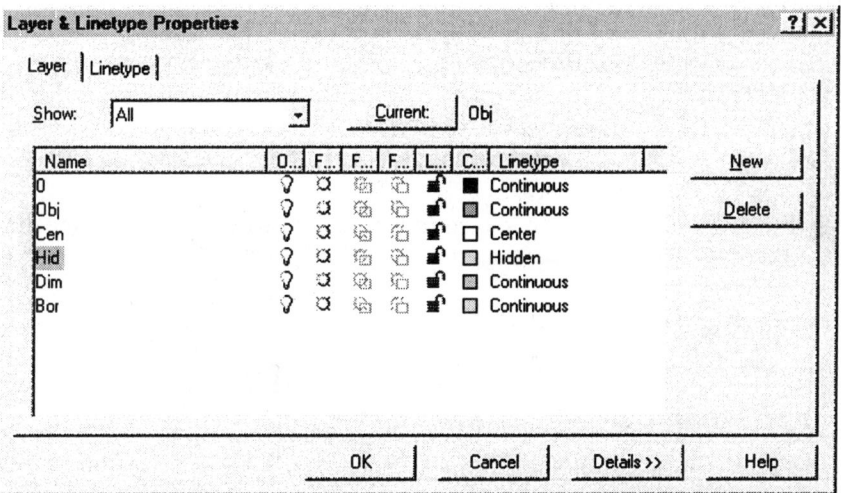

Figure 1-8 Layer and Linetype Properties dialog box

Using the LAYER Command. You can also use the -LAYER command to set the layers and linetypes. The following is the command prompt sequence for setting the layers and linetypes.

```
Command: -LAYER
?/Make/Set/New/ON/OFF/Color/Ltype/Freeze/Thaw/LOck/Unlock: N
New layer name(s): OBJ,CEN,HID,DIM,BOR

?/Make/Set/New/ON/OFF/Color/Ltype/Freeze/Thaw/LOck/Unlock: L
Linetype (or ?)<CONTINUOUS>: HIDDEN
Layer name(s) for linetype HIDDEN<0>: HID
```

?/Make/Set/New/ON/OFF/Color/Ltype/Freeze/Thaw/LOck/Unlock: **L**
Linetype (or ?)<CONTINUOUS>: **CENTER**
Layer name(s) for linetype CENTER<0>: **CEN**

?/Make/Set/New/ON/OFF/Color/Ltype/Freeze/Thaw/LOck/Unlock: **C**
Color: **RED**
Layer name(s) for color 1 (red)<0>: **OBJ**

?/Make/Set/New/ON/OFF/Color/Ltype/Freeze/Thaw/LOck/Unlock: **C**
Color: **YELLOW**
Layer name(s) for color 2 (yellow)<0>: **CEN**

?/Make/Set/New/ON/OFF/Color/Ltype/Freeze/Thaw/LOck/Unlock: **C**
Color: **BLUE**
Layer name(s) for color 5 (blue)<0>: **HID**

?/Make/Set/New/ON/OFF/Color/Ltype/Freeze/Thaw/LOck/Unlock: **C**
Color: **GREEN**
Layer name(s) for color 3 (green)<0>: **DIM**

?/Make/Set/New/ON/OFF/Color/Ltype/Freeze/Thaw/LOck/Unlock: **C**
Color: **MAGENTA**
Layer name(s) for color 6 (magenta)<0>: **BOR**
?/Make/Set/New/ON/OFF/Color/Ltype/Freeze/Thaw/LOck/Unlock: (**RETURN**)

Next, add the title block and the text as shown in Figure 1-5. After completing the drawing, save it as PROTO2.DWT. You have created a template drawing (PROTO2) that contains all of the information given in Example 2.

CUSTOMIZING DRAWINGS ACCORDING TO PLOT SIZE AND DRAWING SCALE

You can generate a template drawing according to plot size and scale. For example, if the scale is 1/16" = 1' and the drawing is to be plotted on a 36" by 24" area, you can calculate drawing parameters like limits, DIMSCALE, and LTSCALE and save them in a template drawing. This will save considerable time in the initial drawing setup and provide uniformity in the drawings. The next example explains the procedure involved in customizing a drawing according to a certain plot size and scale. (Note, you can also use the paper space to specify the paper size and scale.)

Example 3

Generate a template drawing (**PROTO3**) with the following specifications:

Plotted sheet size	36" by 24" (Figure 1-9)
Scale	1/8" = 1.0'
Snap	3'

Grid	6'
Text height	1/4" on plotted drawing
Ltscale	Calculate
Dimscale	Calculate
Units	Architectural
	16-denominator of smallest fraction
	Angle in degrees/minutes/seconds
	4-number of fractional places for display of angles
	0 angle along positive X axis
	Angle positive if measured counterclockwise
Border	Border should be 1" inside the edges of the plotted drawing sheet, using PLINE 1/32" wide when plotted (Figure 1-9)

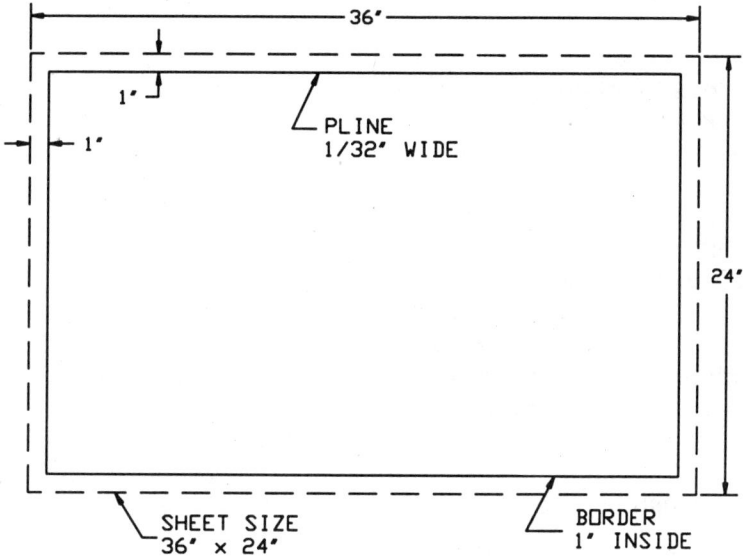

Figure 1-9 Border of prototype drawing

In this example, you need to calculate some values before you set the parameters. For example, the limits of the drawing depend on the plotted size of the drawing and the scale of the drawing. Similarly, **LTSCALE** and **DIMSCALE** depend on the limits of the drawing. The following calculations explain the procedure for finding the values of limits, ltscale, dimscale, and text height.

Limits

Given:
Sheet size   36" x 24"
Scale   1/8" = 1'
              or 1" = 8'

Calculate:
XLimit
YLimit
Since sheet size is 36" x 24" and scale is 1/8"=1'
Therefore, XLimit = 36 x 8' = 288'
           YLimit = 24 x 8' = 192'

Text height

Given:
Text height when plotted = 1/4"
Sheet size   36" x 24"
Scale   1/8" = 1'

Calculate:
Text height
Since scale is 1/8" = 1'
         or 1/8" = 12"
         or   1" = 96"
Therefore, scale factor = 96
           Text height = 1/4" x 96
                        = 24" = 2'

Ltscale and Dimscale

Known:
Since scale is 1/8" = 1'
         or 1/8" = 12"
         or   1" = 96"

Calculate:
Ltscale and Dimscale
Since scale factor = 96
Therefore, LTSCALE = Scale factor = 96
Similarly, DIMSCALE = 96
(All dimension variables, like DIMTXT and DIMASZ, will be multiplied by 96.)

Pline Width

Given:
Scale is 1/8" = 1'

Calculate:
PLINE width
```
Since scale is 1/8" = 1'
         or    1"  = 8'
         or    1"  = 96"
```

Therefore,
```
PLINE width = 1/32 x 96
            = 3"
```

After calculating the parameters, use the following AutoCAD commands to set up the drawing, then save the drawing as **PROTO3.DWT**.

Command: **UNITS**

Report formats:		**Examples:**
1.	Scientific	1.55E+01
2.	Decimal	15.50
3.	Engineering	1'-3.50"
4.	Architectural	1'-3 1/2"
5.	Fractional	15 1/2

With the exception of Engineering and Architectural formats, these formats can be used with any basic unit of measurement. For example, decimal mode is perfect for metric units as well as decimal English units.

Enter choice, 1 to 5<2>: **4**
Denominator of smallest fraction to display
1, 2, 4, 8, 16, 32, or 64 <16>: **16**

Systems of angle measure:		**Examples:**
1.	Decimal degrees	45.0000
2.	Degrees/minutes/seconds	45d0'0"
3.	Grads	50.0000g
4.	Radians	0.7854r
5.	Surveyor's units	N 45d0'0" E

Enter choice, 1 to 5<1>: **2**
Number of fractional places for display of angles (0 to 8)<0>: **4**

Direction for angle 0.00:
East	3 o'clock	= 0d0'0'
North	12 o'clock	= 90d0'0"
West	9 o'clock	= 180d0'0"
South	6 o'clock	= 270d0'0"

Enter direction for angle 0d0'0" <0d0'0">: **RETURN**

Do you want angles measured clockwise? <N> : **N**

Command: **LIMITS**
ON/OFF/<Lower left corner> <0'-0",0'-0"> :**0,0**
Upper right corner <1'-0",0'-9"> : **288',192'**

Command: **SNAP**
Snap spacing or ON/OFF/Aspect/Rotate/Style <0'-1"> : **3'**

Command: **GRID**
Grid spacing(X) or ON/OFF/Snap/Aspect <0'-0"> : **6'**

Command: **SETVAR**
Variable name or ?: **TEXTSIZE**
New value for textsize <0'-0 3/16"> : **2'**

Command: **LTSCALE**
New scale factor <1.0> : **96**

Command: **DIM**
Dim: **Dimscale**
Current value <1.00> New value: **96**

Dim: **Style**
New text style <STANDARD> : **MYDIM2**

Command: **PLINE**
From point: **8',8'**
Current line-width is **0.00**
Arc/Close/Halfwidth/Length/Undo/Width/<Endpoint of line> :**W**
Starting width <0.00> : **3**
Ending width <0'-3"> : **RETURN**
Arc/Close/Halfwidth/Length/Undo/Width/<Endpoint of line> :**280',8'**
Arc/Close/Halfwidth/Length/Undo/Width/<Endpoint of line> :**280',184'**
Arc/Close/Halfwidth/Length/Undo/Width/<Endpoint of line> :**8',184'**
Arc/Close/Halfwidth/Length/Undo/Width/<Endpoint of line> :**C**

CUSTOMIZING DRAWINGS WITH VIEWPORTS

In certain applications you might need a standard viewport configuration to display different views of an object. It involves setting up the desired viewports and then changing the viewpoint for different viewports. You can generate a prototype drawing that contains a required number of viewports and the viewpoint information. Now, if you insert a 3D object in one of the viewports of the prototype drawing, you will automatically get different views of the object without setting viewports or viewpoints. The following example illustrates the procedure for creating a prototype drawing with a standard number (four) of viewports and viewpoints.

Example 4

Generate a prototype drawing with four viewports, as shown in Figure 1-10. The viewports should have the following viewpoints (vpoints):

Viewports	Vpoint	View
Top right	1,-1,1	3D view
Top left	0,0,1	Top view
Lower right	1,0,0	Right side view
Lower left	0,-1,0	Front view

Start AutoCAD and create a new drawing, PROTO5. Use the following commands to set the viewports and vpoints.

Command: **VPORTS**
Save/Restore/Delete/Join/SIngle/?/2/<3>/4: **4**
Regenerating drawing

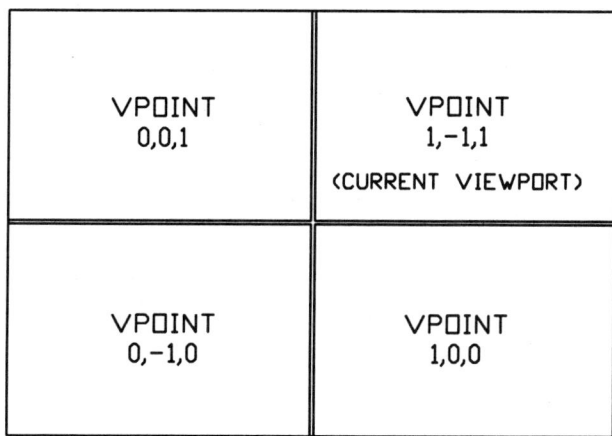

Figure 1-10 Viewports with different viewpoints

Make the top right viewport current and use the following command to set up the vpoint:

Command: **VPOINT**
Rotate/<View point>:<0.00,0.00,1.00>: **1,-1,1**
Regenerating drawing

Make the top left viewport current and use the following command to set up the vpoint:

Command: **VPOINT**
Rotate/<View point>:<0.00,0.00,1.00>: **0,0,1**
Regenerating drawing

1-16 Customizing AutoCAD

Make the lower right viewport current and use the following command to set up the vpoint:

Command: **VPOINT**
Rotate/<View point>:<0.00,0.00,1.00>: **1,0,0**
Regenerating drawing

Make the lower left viewport current and use the following command to set up the vpoint:

Command: **VPOINT**
Rotate/<View point>:<0.00,0.00,1.00>: **0,-1,0**
Regenerating drawing

Command: **SAVE**
Save Drawing As <C:\acadr14\Drawing.dwg>: **PROTO5**

The drawing will be saved under the filename, PROTO5. Now, start a new drawing, TBLOCK. Draw the 3D tapered block shown in Figure 1-11 and save it as TBLOCK. (Assume proportionate dimensions for the tapered block.)

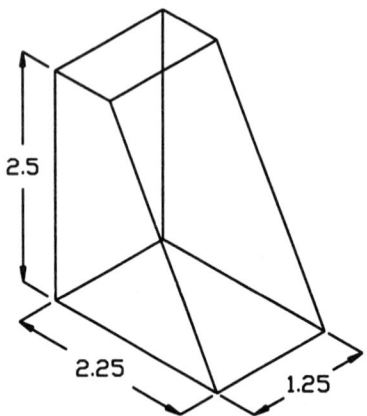

Figure 1-11 3D tapered block

Again, start a new drawing, TEST, using the prototype drawing PROTO5. Make the top right viewport current and insert the drawing TBLOCK. Four different views will be automatically displayed on the screen as shown in Figure 1-12.

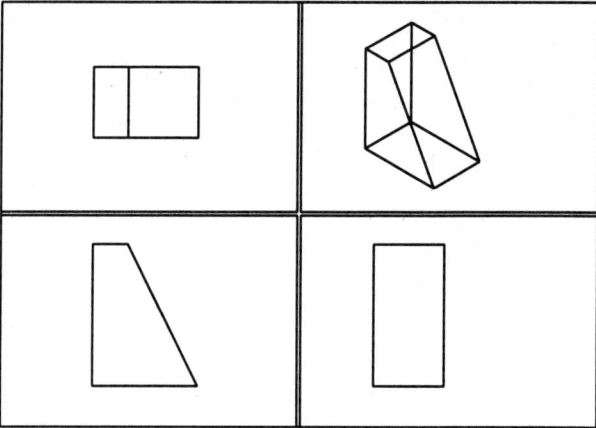

Figure 1-12 Different views of 3D tapered block

CUSTOMIZING A DRAWING WITH PAPER SPACE

Paper space provides a convenient way to plot multiple views of a 3D drawing or multiple views of a regular 2D drawing. It takes quite some time to set up the viewports in model space with different vpoints and scale factors. You can create prototype drawings that contain predefined viewport settings, with vpoint and other desired information. Now if you create a new drawing, or insert a drawing, the views are automatically generated. The following example illustrates the procedure for generating a prototype drawing with paper space and model space viewports.

Example 5

Generate a prototype of the drawing in Figure 1-13 with four views in paper space that display front, top, side, and 3D views of the object. The plot size is 9 by 6 inches. The plot scale of the model space is 0.5 or 1/2" = 1". The model space viewports should have the following vpoint setting:

MODEL SPACE

Viewports	Vpoint	View
Top right	1,-1,1	3D view
Top left	0,0,1	Top view
Lower right	1,0,0	Right side view
Lower left	0,-1,0	Front view

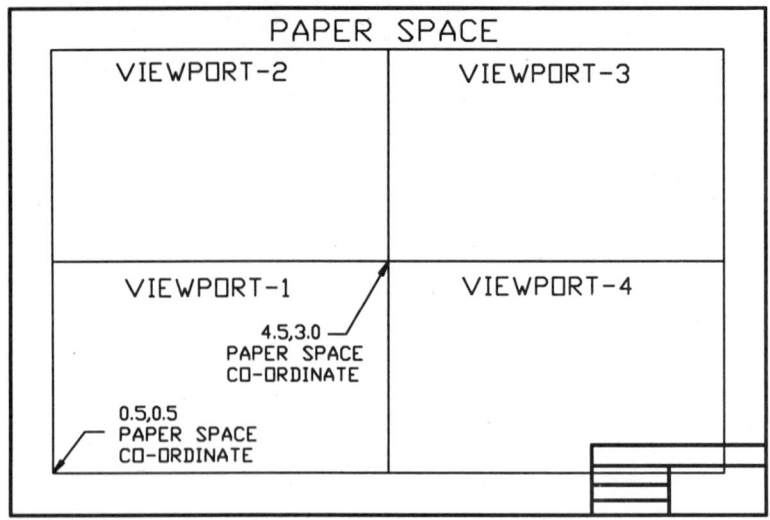

Figure 1-13 Paper space with four viewports

Start AutoCAD and create a new drawing, PROTO6. Use the following commands to set up various parameters.

The first step is to change the TILEMODE to zero and set up the limits for the paper space. At the time of plotting, if you want to use a plot scale factor of 1=1, set up the paper space limits equal to plot size. In this example, the plot size is 9 by 6; therefore, the limits are 0,0 and 9,6. After setting the limits, use AutoCAD's ZOOM command with All option to display the new limits.

```
Command: TILEMODE
New value for TILEMODE<1>: 0

Command: LIMITS
Reset paper space limits:
ON/OFF/<Lower left corner> <0.00,0.00>: Press Enter (↲).
Upper right corner<12.00,9.00>: 9.0,6.0

Command: ZOOM
All/Center/Dynamic/Extents/Previous/Scale(X/XP)/Window/<Realtime>: ALL
```

The second step is to set up a layer (VIEW) for viewports and assign it a color (green).
```
Command: -Layer
?/Make/Set/New/ON/OFF/Color/Ltype/Freezw/Thaw/LOck/Unlock: M
New current layer<0>: VIEW
?/Make/Set/New/ON/OFF/Color/Ltype/Freezw/Thaw/LOck/Unlock: C
Color: GREEN
Layer name(s) for color 3 (green)<VIEW>: Press Enter (↲).
```

Template Drawings 1-19

?/Make/Set/New/ON/OFF/Color/Ltype/Freezw/Thaw/LOck/Unlock: ←⏎

Now, use AutoCAD's MVIEW command to set up a viewport, then switch to model space to zoom the display to half the size.

Command: **MVIEW**
ON/OFF/Hideplot/Fit/2/3/4/Restore/<First Point>:**0.5,0.5**
Other corner: **4.5,3.0**

Command: **MSPACE**   (or **MS**)
Command: **ZOOM**
All/Center/Dynamic/Extents/Previous/Scale(X/XP)/Window/<Realtime>: **0.5XP**

Use AutoCAD's PSPACE command to change to paper space and make four copies of the viewport as shown in Figure 1-13.

Use AutoCAD's MSPACE command to change to model space, then change the vpoints of different model space viewports by using the VPOINT command. The vpoint values for different viewports are shown in Example 5.

Use the PSPACE command to change to paper space and set a new layer, PBORDER, with the color yellow. Make the PBORDER layer current, draw a border, and if needed, a title block using the PLINE command.

Command: **PSPACE**
Command: **PLINE**

From point: **0,0**
Current line-width is 0.00
Arc/Close/Halfwidth/Length/Undo/Width/<Endpoint of line>: **9.0,0**
Arc/Close/Halfwidth/Length/Undo/Width/<Endpoint of line>: **9.0,6.0**
Arc/Close/Halfwidth/Length/Undo/Width/<Endpoint of line>: **0,6.0**
Arc/Close/Halfwidth/Length/Undo/Width/<Endpoint of line>: **C**

The last step is to change the TILEMODE to 1 and save the prototype drawing. To test the paper space that you just created, insert the TBLOCK drawing that you created in Example 5. If you change TILEMODE to zero, you will find four different views of TBLOCK (Figure 1-14). You can freeze the layer VIEW so that the viewports do not appear on the drawing. Now you can plot this drawing with a plot scale factor of 1=1 and the size of the plot will be exactly 9 by 6 inches.

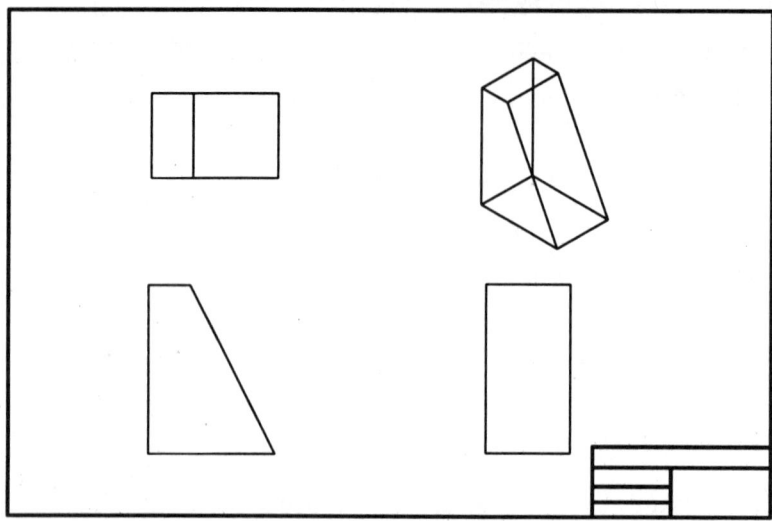

Figure 1-14 Four views of TBLOCK in paper space

USING DRAWING FILES AS TEMPLATE FILES

You can also use the regular drawings as prototype (template) drawings. For example, let us assume that you have a drawing PROJ1.DWG that you want to use as a template drawing. Start a new drawing and select the **Use a Template** button in the **Create New Drawing** dialog box. In the template file list box, select **More files**; the **Select template** dialog box (Figure 1-15) is displayed. In the **File of type** drop-down list, select **Drawing (.dwg)** file type; AutoCAD will display the drawing files. Select PROJ1 file and then select the **Open** button to open the file. You can also copy PROJ1.DWG to PROJ1.DWT and use this file as a template file.

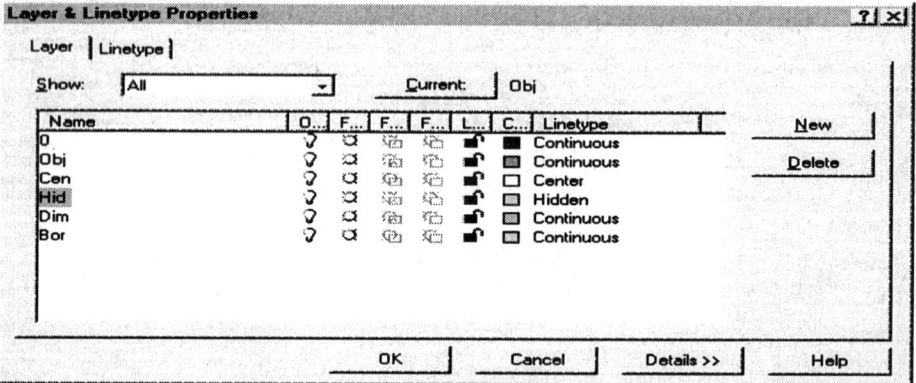

Figure 1-15 Select template dialog box

REVIEW QUESTIONS

1. The three standard template drawings that come with AutoCAD software are _____ , _____ , and _____ .

2. To use a template file, select the _____ button in the Create New Drawing dialog box.

3. To start a drawing with default setup, select the _____ button in the Create New Drawing dialog box.

4. The default value of **DIMSCALE** is _____ .

5. The default value for **DIMTXT** is _____ .

6. The default value for **SNAP** is _____ .

7. Architectural units can be selected by using AutoCAD's _____ or _____ commands.

8. If plot size is 36" x 24", and the scale is 1/2" = 1', then XLimit = _____ and YLimit = _____ .

9. If the plot size is 24" x 18", and the scale is 1 = 20, the XLimit = _____ and YLimit = _____ .

10. If the plot size is 200 x 150 and limits are (0.00,0.00) and (600.00,450.00), the **LTSCALE** factor = _____ .

11. _____ provides a convenient way to plot multiple views of a 3D drawing or multiple views of a regular 2D drawing.

12. You can use AutoCAD's _____ command to set up a viewport in paper space.

13. You can use AutoCAD's _____ command to change to paper space.

14. You can use AutoCAD's _____ command to change to model space.

15. The values that can be assigned to TILEMODE are _____ and _____ .

16. In the model space, if you want to reduce the display size by half, the scale factor you enter in ZOOM-Scale command is _____ .

EXERCISES

Exercise 1

Generate a template drawing (**PROTOE1**) with the following specifications:

Limits	36.0,24.0
Snap	0.5
Grid	1.0
Text height	0.25
Units	Decimal
	2-number of digits to right of decimal point
	Decimal degrees
	0-number of fractional places for display of angles
	0-angle along positive X axis
	Angle positive if measured counterclockwise

Exercise 2

Generate a template drawing **PROTOE2** with the following specifications:

Limits	48.0,36.0
Snap	0.5
Grid	2.0
Text height	0.25
PLINE width	0.03
Ltscale	Calculate
Dimscale	Calculate
Plot size	10.5 x 8

LAYERS

Layer Names	Line Type	Color
0	Continuous	White
OBJECT	Continuous	Green
CENTER	Center	Magenta
HIDDEN	Hidden	Blue
DIM	Continuous	Red
BORDER	Continuous	Cyan

Exercise 3

Generate a template drawing (**PROTOE3**) with following specifications:

Plotted sheet size	36" x 24" (Figure 1-16)
Scale	1/2" = 1.0'
Text height	1/4" on plotted drawing
Ltscale	Calculate
Dimscale	Calculate
Units	Architectural
	32-denominator of smallest fraction to display
	Angle in degrees/minutes/seconds
	4-number of fractional places for display of angles
	0d0'0"-direction for angle
	Angle positive if measured counterclockwise
Border	Border is 1-1/2" inside the edges of the plotted drawing sheet, using PLINE 1/32" wide when plotted (Figure 1-16)

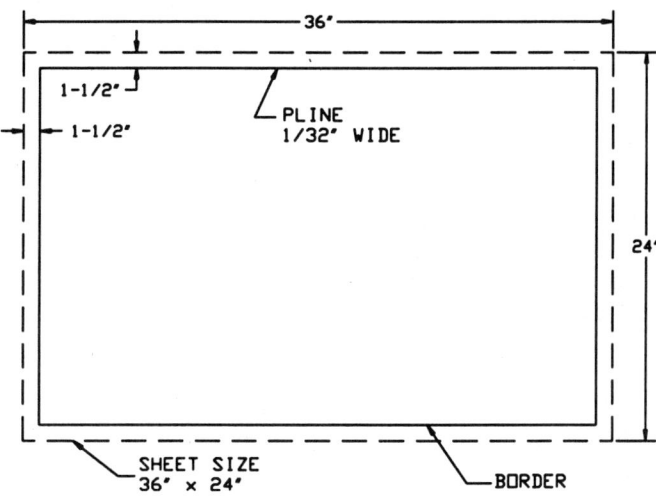

Figure 1-16 Drawing for Exercise 3

Exercise 4

Generate a prototype drawing with the following specifications (the name of the drawing is PROTOE4):

Plotted sheet size	24" x 18" (Figure 1-17)
Scale	1 = 50
Border	The border is 1" inside the edges of the plotted drawing sheet, using PLINE 0.05" wide when plotted (Figure 1-17)
DIMTAD	ON
DIMTIX	ON

DIMTOH	OFF
DIMTIH	OFF
DIMSCALE	Calculate
DIMALT	ON
DIMASO	OFF
DIMTOFL	ON

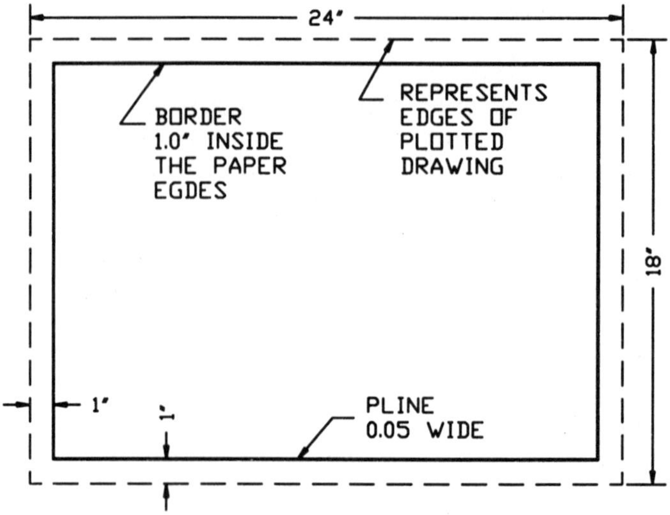

Figure 1-17 Prototype drawing

Exercise 5

You want to set up a prototype drawing with the following specifications (the name of the drawing is PROTOE5):

Limits	36.0,24,0
Border	35.0,23.0
Grid	1.0
Snap	0.5
Text height	0.15
Dimscale	2.0
Units	Decimal (up to 2 places)
Ltscale	3
Current layer	Object

LAYERS

Layer Name	Linetype	Color
0	Continuous	White
Object	Continuous	Red
Hidden	Hidden	Yellow

Center	Center	Green
Dim	Continuous	Blue
Border	Continuous	Magenta
Notes	Continuous	White

This prototype drawing should also have a border line and a title block as shown in Figure 1-18.

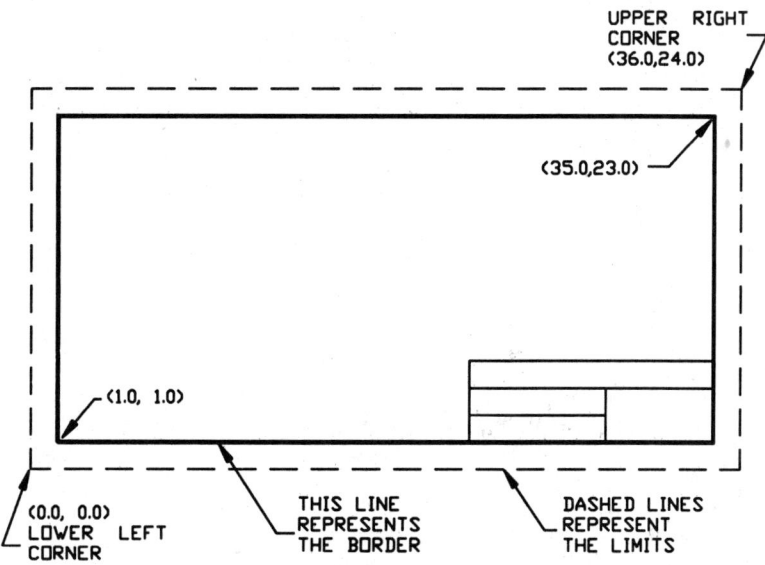

Figure 1-18 Prototype drawing

Chapter 2

Script Files and Slide Shows

Learning objectives

After completing this chapter, you will be able to:
- *Write script files and use the SCRIPT command to run script files.*
- *Use the RSCRIPT and DELAY commands in script files.*
- *Invoke script files when loading AutoCAD.*
- *Create a slide show.*
- *Preload slides when running a slide show.*

WHAT ARE SCRIPT FILES?

AutoCAD has provided a facility called **script files** that allows you to combine different AutoCAD commands and execute them in a predetermined sequence. The commands can be written as a text file using any text editor like Notepad or AutoCAD's EDIT command (if the **ACAD.PGP** file is present and EDIT is defined in the file). These files, generally known as script files, have extension **.SCR** (example: **PLOT1.SCR**). A script file is executed with the AutoCAD SCRIPT command.

Script files can be used to generate a slide show, do the initial drawing setup, or plot a drawing to a predefined specification. They can also be used to automate certain command sequences that are used frequently in generating, editing, or viewing a drawing. Scripts cannot access dialog boxes or menus. When commands that open the file and plot dialog boxes are issued from a script file, AutoCAD runs the command line version of the command instead of opening the dialog box.

Example 1

Write a script file that will perform the following initial setup for a drawing (file name **SCRIPT1.SCR**):

Ortho	On	Zoom	All
Grid	2.0	Text height	0.125
Grid	Off	Ltscale	4.0

Snap	0.5	Dimscale	4.0
Limits	0,0		
	48.0,36.0		

Before writing a script file, you need to know the AutoCAD commands and the entries required in response to the command prompts. To find out the sequence of the prompt entries, you can type the command at the keyboard and then respond to different prompts. The following is a list of AutoCAD commands and prompt entries for Example 1:

Command: **ORTHO**
ON/OFF<Off>: **ON**

Command: **GRID**
Grid spacing(X) or ON/OFF/Snap/Aspect<1.0>: **2.0**

Command: **GRID**
Grid spacing(X) or ON/OFF/Snap/Aspect<1.0>: **OFF**

Command: **SNAP**
Snap spacing or ON/OFF/Aspect/Rotate/Style<1.0>: **0.5**

Command: **LIMITS**
ON/OFF/<Lower left corner> <0.00,0.00>:**0,0**
Upper right corner<12.0,9.0>: **48.0,36.0**

Command: **ZOOM**
All/Center/Dynamic/Extents/Left/Previous/Vmax/Window/<Scale(X/XP)>: **A**

Command: **SETVAR**
Variable name or ?: **TEXTSIZE**
New value for textsize<0.02>: **0.125**

Command: **LTSCALE**
New scale factor<1.0000>: **4.0**

Command: **SETVAR**
Variable name or ?: **DIMSCALE**
New value for dimscale<1.0000>: **4.0**

Once you know the AutoCAD commands and the required prompt entries, you can write the script file using the AutoCAD EDIT command or any text editor. The following file is a listing of the script file for Example 1:

ORTHO
ON
GRID

2.0
GRID
OFF
SNAP
0.5
LIMITS
0,0
48.0,36.0
ZOOM
ALL
SETVAR
TEXTSIZE
0.125
LTSCALE
4.0
SETVAR
DIMSCALE 4.0

Notice that the commands and the prompt entries in this file are in the same sequence as mentioned before. You can also combine several statements in one line, as shown in the following list:

ORTHO ON
GRID 2.0 GRID OFF
SNAP 0.5 SNAP ON
LIMITS 0,0 48.0,36.0 ZOOM ALL
SETVAR TEXTSIZE 0.125
LTSCALE 4.0
SETVAR DIMSCALE 4.0

Note

In the script file, a space is used to terminate a command or a prompt entry. Therefore, spaces are very important in these files. Make sure there are no extra spaces, unless they are required to press Enter more than once.

After you change the limits, it is good practice to use the ZOOM command with the All option to display the new limits on the screen.

AutoCAD ignores and does not process any lines that begin with a semicolon (;). This allows you to put comments in the file.

SCRIPT COMMAND

The AutoCAD SCRIPT command allows you to run a script file while you are in the drawing editor. To execute the script file, type the SCRIPT command and press Enter. AutoCAD will prompt you to enter the name of the script file. You can accept the default file name or enter a new file name. The default script file name is the same as the drawing name. If you want to enter a new file name, type the name of the script file **without** the file extension (**.SCR**). (The file extension is assumed and need not be included with the file name.)

To run the script file of Example 1, type the SCRIPT command and press Enter; AutoCAD will display the **Select Script File** dialog box (Figure 2-1). In the dialog box, select the file SCRIPT1 and then choose the Open button. You will see the changes taking place on the screen as the script file commands are executed. The format of the SCRIPT command is:

Command: **SCRIPT**

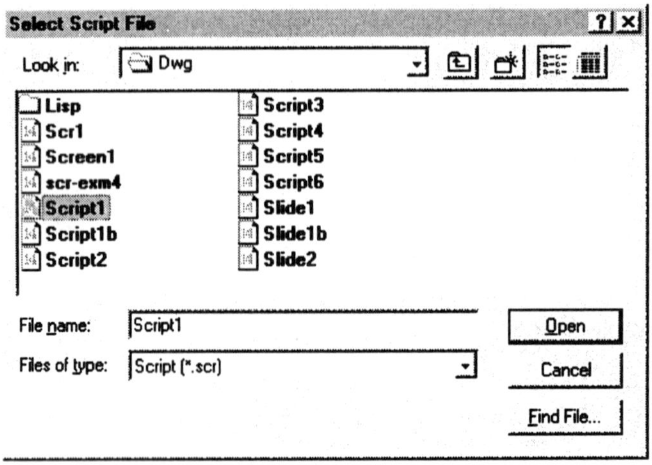

Figure 2-1 Select Script File dialog box

You can also enter the name of the script file at the Command: prompt by setting FILEDIA=0. The format of the SCRIPT command is:

Command: **FILEDIA**
New value for FILEDIA <1>: **0**
Command: **SCRIPT**
Script file<default>: *Script file name.*

For example:
Command: **SCRIPT**
```
Script file <CUSTOM>: SCRIPT1
                │         └─ Name of the script file
                └─ Default drawing file name
```

Example 2

Write a script file that will set up the following layers with the given colors and linetypes (file name **SCRIPT2.SCR**).

Layer Names	Color	Linetype
Object	Red	Continuous
Center	Yellow	Center
Hidden	Blue	Hidden
Dimension	Green	Continuous
Border	Magenta	Continuous
Hatch	Cyan	Continuous

As mentioned earlier, you need to know the AutoCAD commands and the required prompt entries before writing a script file. For Example 2, you need the following commands to create the layers with the given colors and linetypes:

Command: **LAYER**
?/Make/Set/New/ON/OFF/Color/Ltype/Freeze/Thaw: **N**
New layer name(s):**OBJECT,CENTER,HIDDEN,DIM,BORDER,HATCH**

?/Make/Set/New/ON/OFF/Color/Ltype/Freeze/Thaw: **L**
Linetype (or ?)<CONTINUOUS>: **CENTER**
Layer name(s) for linetype CENTER<0>: **CENTER**

?/Make/Set/New/ON/OFF/Color/Ltype/Freeze/Thaw: **L**
Linetype (or ?)<CONTINUOUS>: **HIDDEN**
Layer name(s) for linetype HIDDEN<0>: **HIDDEN**

?/Make/Set/New/ON/OFF/Color/Ltype/Freeze/Thaw: **C**
Color: **RED**
Layer name(s) for color 1 (red)<0>: **OBJECT**

?/Make/Set/New/ON/OFF/Color/Ltype/Freeze/Thaw: **C**
Color: **YELLOW**
Layer name(s) for color 2 (yellow)<0>: **CENTER**

?/Make/Set/New/ON/OFF/Color/Ltype/Freeze/Thaw: **C**
Color: **BLUE**
Layer name(s) for color 5 (blue)<0>: **HIDDEN**

?/Make/Set/New/ON/OFF/Color/Ltype/Freeze/Thaw: **C**
Color: **GREEN**
Layer name(s) for color 3 (green)<0>: **DIM**

?/Make/Set/New/ON/OFF/Color/Ltype/Freeze/Thaw: **C**
Color: **MAGENTA**
Layer name(s) for color 6 (magenta)<0>: **BORDER**

?/Make/Set/New/ON/OFF/Color/Ltype/Freeze/Thaw: **C**
Color: **CYAN**

Layer name(s) for color 4 (cyan)<0>: **HATCH**
?/Make/Set/New/ON/OFF/Color/Ltype/Freeze/Thaw: **(RETURN)**

The following file is a listing of the script file that creates different layers and assigns the given colors and linetypes to these layers:

```
;This script file will create new layers and
;assign different colors and linetypes to layers
LAYER
NEW
OBJECT,CENTER,HIDDEN,DIM,BORDER,HATCH
L
CENTER
CENTER
L
HIDDEN
HIDDEN
C
RED
OBJECT
C
YELLOW
CENTER
C
BLUE
HIDDEN
C
GREEN
DIM
C
MAGENTA
BORDER
C
CYAN
HATCH

;(This is a blank line to terminate the LAYER command.)
;End of script file
```

Example 3

Write a script file that will rotate the circle and the line, as shown in Figure 2-2, around the lower endpoint of the line through 45-degree increments. The script file should be able to produce a continuous rotation of the given objects with a delay of two seconds after every 45-degree rotation (file name **SCRIPT3.SCR**).

Script Files and Slide Shows 2-7

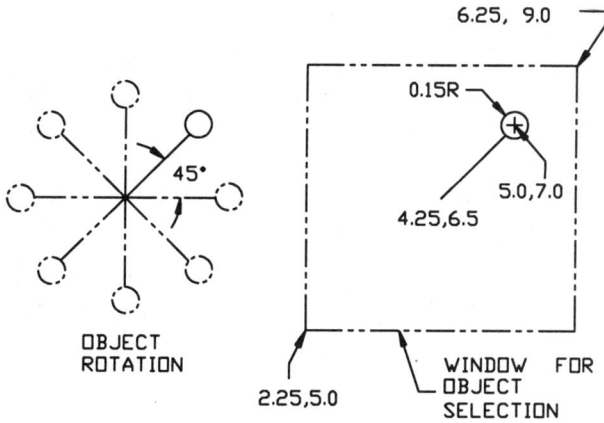

Figure 2-2 Line and circle rotated through 45-degree increments

Before writing the script file, enter the required commands and the prompt entries at the keyboard. Write down the exact sequence of the entries in which they have been entered to perform the given operations. The following is a listing of the AutoCAD command sequence needed to rotate the circle and the line around the lower endpoint of the line:

    Command: **ROTATE**   *(Enter ROTATE command.)*
    Select objects: **W**   *(Window option to select objects.)*
    First corner: **2.25,5.0**
    Other corner: **6.25,9.0**
    Select objects: **<RETURN>**
    Base point: **4.25,6.5**
    <Rotation angle>/Reference: **45**

Once the AutoCAD commands, command options, and their sequences are known, you can write a script file. As mentioned earlier, you can use any text editor to write a script file. The following file is a listing of the script file that will create the required rotation of the circle and the line of Example 3.

    ROTATE   1
    W   2
    2.25,5.0   3
    6.25,9.0   4
                  (Blank line for Return.)   5
    4.25,6.5   6
    45   7

Line 1
ROTATE
In this line, ROTATE is an AutoCAD command that rotates the objects.

2-8 Customizing AutoCAD

Line 2
W
In this line, W is the Window option for selecting the objects that need to be edited.

Line 3
2.25,5.0
In this line, 2.25 defines the X coordinate and 5.0 defines the Y coordinate of the lower left corner of the object selection window.

Line 4
6.25,9.0
In this line, 6.25 defines the X coordinate and 9.0 defines the Y coordinate of the upper right corner of the object selection window.

Line 5
Line 5 is a blank line that terminates the object selection process.

Line 6
4.25,6.5
In this line, 4.25 defines the X coordinate and 6.5 defines the Y coordinate of the base point for rotation.

Line 7
45
In this line, 45 is the incremental angle for rotation.

Note
One of the limitations of the script files is that all the information has to be contained within the file. These files do not let you enter information. For instance, in Example 3, if you want to use the Window option to select the objects, the Window option (W) and the two points that define this window must be contained within the script file. The same is true for the base point and all other information that goes in a script file. There is no way that a script file can prompt you to enter a particular piece of information and then resume the script file, unless you embed AutoLISP commands to prompt for user input.

RSCRIPT COMMAND

The AutoCAD RSCRIPT command allows the user to execute the script file indefinitely until canceled. It is a very desirable feature when the user wants to run the same file continuously. For example, in the case of a slide show for a product demonstration, the RSCRIPT command can be used to run the script file again and again until it is terminated by pressing the Esc (Escape) key from the keyboard. Similarly, in Example 3, the rotation command needs to be repeated indefinitely to create a continuous rotation of the objects. This can be accomplished by adding RSCRIPT at the end of the file, as in the following file:

ROTATE
W
2.25,5.0
6.25,9.0
(Blank line for Return.)
4.25,6.5
45
RSCRIPT

The RSCRIPT command on line 8 will repeat the commands from line 1 to line 7, and thus set the script file in an indefinite loop. The script file can be stopped by pressing the Esc or the Backspace key.

Note
You cannot provide conditional statements in a script file to terminate the file when a particular condition is satisfied, unless you use the AutoLISP functions in the script file.

DELAY COMMAND

In the script files, some of the operations happen very quickly and make it difficult to see the operations taking place on the screen. It might be necessary to intentionally introduce a pause between certain operations in a script file. For example, in a slide show for a product demonstration, there must be a time delay between different slides so that the audience has enough time to see them. This is accomplished by using the AutoCAD DELAY command, which introduces a delay before the next command is executed. The general format of the DELAY command is:

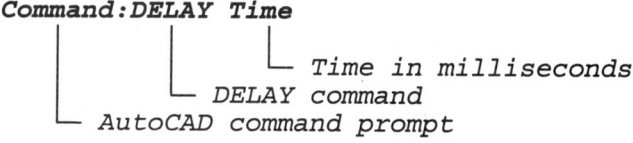

The DELAY command is to be followed by the delay time in milliseconds. For example, a delay of 2,000 milliseconds means that AutoCAD will pause for approximately two seconds before executing the next command. It is approximately two seconds because computer processing speeds vary. The maximum time delay you can enter is 32,767 milliseconds (about 33 seconds). In Example 3, a two-second delay can be introduced by inserting a DELAY command line between line 7 and line 8, as in the following file listing:

ROTATE
W
2.25,5.0
6.25,9.0
(Blank line for Return.)
4.25,6.5
45

DELAY 2000
RSCRIPT

The first seven lines of this file rotate the objects through a 45-degree angle. Before the RSCRIPT command on line 8 is executed, there is a delay of 2,000 milliseconds (about two seconds). The RSCRIPT command will repeat the script file that rotates the objects through another 45-degree angle. Thus, a slide show is created with a time delay of two seconds after every 45-degree increment.

RESUME COMMAND

If you cancel a script file and then want to continue it, you can do so by using the AutoCAD RESUME command.

Command: **RESUME**

The RESUME command can also be used if the script file has encountered an error that causes it to be suspended. The RESUME command will skip the command that caused the error and continue with the rest of the script file. If the error occurred when the command was in progress, use a leading apostrophe with the RESUME command ('RESUME) to invoke the RESUME command in transparent mode.

Command: **'RESUME**

COMMAND LINE SWITCHES

The command line switches can be used as arguments to the acad.exe file that launches AutoCAD. You can also use the **Preferences** dialog box to set the environment or by adding a set of environment variables in the autoexec.bat file. The command line switches and environment variables override the values set in Preferences dialog box for the current session only. These switches do not alter the system registry. The following is the list of the command line switches:

Switch	Function
/c	Controls where AutoCAD stores and searches for the hardware configuration file (acad14.cfg)
/s	Specifies which directories to search for support files if they are not in the current directory
/d	Specifies which directories to search for ADI drivers
/b	Designates a script to run after AutoCAD starts
/t	Specifies a template to use when creating a new drawing
/nologo	Starts AutoCAD without first displaying the logo screen
/v	Designates a particular view of the drawing to be displayed upon start-up of AutoCAD
/r	Reconfigures AutoCAD with the default device configuration settings
/p	Specifies the profile to use on start-up

INVOKING A SCRIPT FILE WHEN LOADING AUTOCAD

The script files can also be run when loading AutoCAD, without getting into the drawing editor. The format of the command for running a script file when loading AutoCAD is:

Drive > ACADR14 [existing-drawing] [/t template] [/v view] /b script-file

In the following example, AutoCAD will open the existing drawing (Mydwg1) and then run the script file (Setup).

Example
```
C:\ACADR14>ACAD Mydwg1 /b Setup
                              └─ Name of the script file
                     └─ Existing drawing file name
            └─ ACAD command to start AutoCAD
  └─ ACAD subdirectory containing
     AutoCAD system files
```

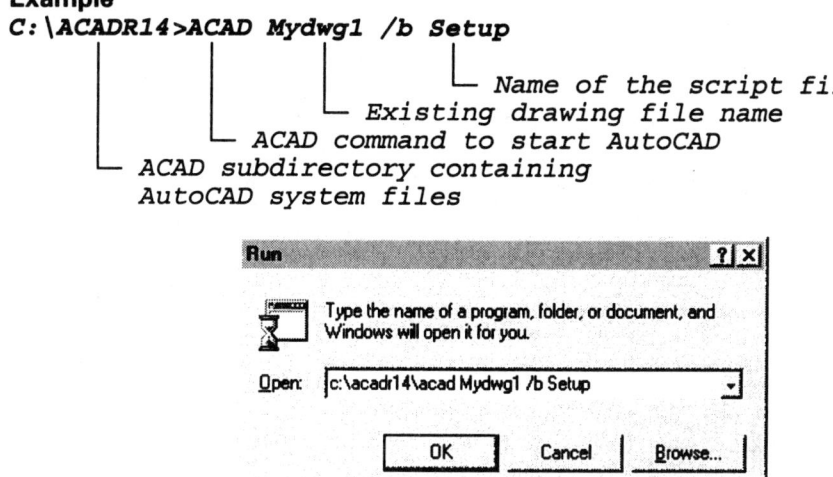

Figure 2-3 Invoking script file when loading AutoCAD using the Run dialog box

In the following example, AutoCAD will start a new drawing with the default name (Drawing), using the template file **temp1**, and then run the script file (Setup).

Example
```
C:\ACADR14\ACAD /t temp1 /b Setup
                             └─ Name of the script file
                     └─ Existing template file name
```
or
```
C:\ACADR14\ACAD /t c:\acadr14\mytemp\temp1 /b Setup
                    └─ Path name
```

In the following example, AutoCAD will start a new drawing with the default name (Drawing), and then run the script file (Setup).

2-12 Customizing AutoCAD

Example
`C:\ACADR14\ACAD /b Setup`
　　　　　　　　　　└─ *Name of the script file*

Here, it is assumed that the AutoCAD system files are loaded in the ACADR14 directory.

> **Note**
> *For invoking a script file when loading AutoCAD, the drawing file or the template file specified in the command must exist in the search path. You cannot start a new drawing with a given name.*
>
> *You should avoid abbreviations to prevent any confusion. For example, a C can be used as a close option when you are drawing lines. It can also be used as a command alias for drawing a circle. If you use both of these in a script file, it might be confusing.*

Example 4

Write a script file that can be invoked when loading AutoCAD and create a drawing with the following setup (filename SCRIPT4.SCR):

Grid	3.0
Snap	0.5
Limits	0,0
	36.0,24.0
Zoom	All
Text height	0.25
Ltscale	3.0
Dimscale	3.0

Layers

Name	Color	Linetype
Obj	Red	Continuous
Cen	Yellow	Center
Hid	Blue	Hidden
Dim	Green	Continuous

First, write a script file and save the file under the name SCRIPT4.SCR. The following file is a listing of this script file that does the initial setup for a drawing:

```
GRID 2.0
SNAP 0.5
LIMITS 0,0 36.0,24.0 ZOOM ALL
SETVAR TEXTSIZE 0.25
LTSCALE 3
SETVAR DIMSCALE 3.0
LAYER NEW
```

Script Files and Slide Shows 2-13

```
OBJ,CEN,HID,DIM
L CENTER CEN
L HIDDEN HID
C RED OBJ
C YELLOW CEN
C BLUE HID
C GREEN DIM
```
(Blank line for Enter.)

After you have written and saved the file, quit the drawing editor. To run the script file, SCRIPT4, select Start, Run, and then enter the following command line:

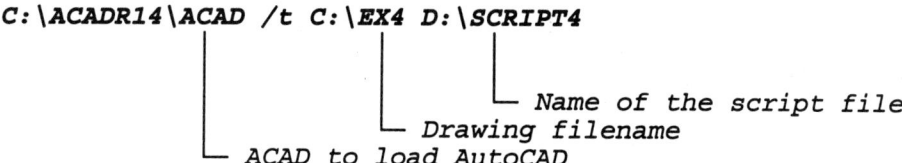

C:\ACADR14\ACAD /t C:\EX4 D:\SCRIPT4

– Name of the script file
– Drawing filename
– ACAD to load AutoCAD

Here it is assumed that the template file (EX4) is on C drive and the script file (SCRIPT4) is on D drive.

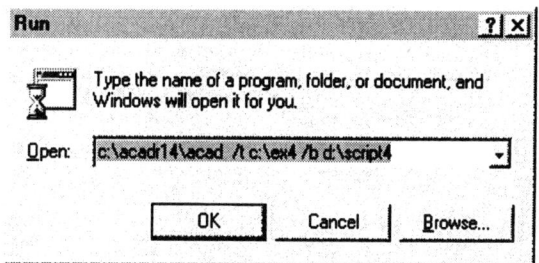

Figure 2-4 Invoking script file when loading AutoCAD using the Run dialog box

When you enter this line, AutoCAD is loaded and the file EX4.DWT is opened. The script file, SCRIPT4, is then automatically loaded and the commands defined in the file are executed.

In the following example, AutoCAD will start a new drawing with the default name (Drawing), and then run the script file (SCRIPT4).

Example
C:\ACADR14\ACAD /b SCRIPT4

– Name of the script file

Here, it is assumed that the AutoCAD system files are loaded in the ACADR14 directory.

Example 5

Write a script file that will plot a 36" by 24" to maximum plot size, using your system printer/plotter. Use the Window option to select the drawing to be plotted.

Before writing a script file to plot a drawing, find out the plotter specifications that must be entered in the script file to obtain the desired output. To determine the prompt entries and their sequences to set up the plotter specifications, enter the AutoCAD PLOT command at the keyboard. **Make sure the system variable CMDDIA is set to zero**; otherwise, AutoCAD will display the Plot dialog box. However, when you run a script file, AutoCAD ignores the CMDDIA setting; therefore, you can leave the CMDDIA setting as 1. Note the entries you make and their sequence (the entries for your printer or plotter will probably be different). The following is a listing of the plotter specification with the new entries:

Command: **CMDDIA**
New value for CMDDIA <1>: **0**
Command: **PLOT**
What to plot -- Display, Extents, Limits, View, or Window <D>: **W**
First corner: Other corner:
Device Name: HP LaserJet 5P
Output Port: LPT1:
Driver Version: 4.0

Use the Control Panel to make permanent changes to
a printer's configuration.

Plot device is System Printer ADI 4.3 - by Autodesk, Inc.
Description: Default System Printer
Plot optimization level = 0
Plot will NOT be written to a selected file
Sizes are in Inches and the style is landscape
Plot origin is at (0.00,0.00)
Plotting area is 7.93 wide by 10.48 high (MAX size)
Plot is NOT rotated
Area fill will NOT be adjusted for pen width
Hidden lines will NOT be removed
Plot will be scaled to fit available area
0. No changes, proceed to Plot
1. Merge partial configuration from .pcp file
2. Replace configuration from .pc2 file
3. Save partial configuration as .pcp file
4. Save configuration as .pc2 file
5. Detailed plot configuration

Enter choice, 0-5 <0>: **5**

Pen widths are in inches.

Object Color	Pen No.	Line-type	Pen Speed	Object Color	Pen No.	Line-type	Pen Speed
1 (red)	1	0	36	9	1	0	36
2 (yellow)	1	0	36	10	1	0	36
3 (green)	1	0	36	11	1	0	36
4 (cyan)	1	0	36	12	1	0	36
5 (blue)	1	0	36	13	1	0	36
6 (magenta)	1	0	36	14	1	0	36
7 (white)	1	0	36	15	1	0	36
8	1	0	36				

0. Solid Line
1. - - - - - - - - - - - - - - -
2. ....
3. -.-.-.-.-.-.-.-.-.-.-.-.-.-
4. -..-..-..-..-..-..-..-..-.

Do you want to change any of the above parameters? <N>: N
Write the plot to a file? <N> N
Size units (Inches or Millimeters) <I>: I
Plot origin in Inches <0.00,0.00>: 0,0
Rotate plot clockwise 0/90/180/270 degrees <0>: 0
Adjust area fill boundaries for pen width? <N> N
Remove hidden lines? <N> N

Specify scale by entering:
Plotted Inches=Drawing Units or Fit or ? <F>: 1=4
Device Name: HP LaserJet 5P
Output Port: LPT1:
Driver Version: 4.0

Use the Control Panel to make permanent changes to
a printer's configuration.

Plot device is System Printer ADI 4.3 - by Autodesk, Inc.
Description: Default System Printer
Plot optimization level = 0
Plot will NOT be written to a selected file
Sizes are in Inches and the style is landscape
Plot origin is at (0.00,0.00)
Plotting area is 7.93 wide by 10.48 high (MAX size)
Plot is NOT rotated
Area fill will NOT be adjusted for pen width
Hidden lines will NOT be removed
Plot will be scaled to fit available area
0. No changes, proceed to Plot

2-16 Customizing AutoCAD

1. Merge partial configuration from .pcp file
2. Replace configuration from .pc2 file
3. Save partial configuration as .pcp file
4. Save configuration as .pc2 file
5. Detailed plot configuration

Enter choice, 0-5 <0>: **0**
Effective plotting area: 7.93 wide by 6.17 high

Plot complete.

Now you can write the script file by entering the responses to these prompts in the file. The following file is a listing of the script file that will plot a 36" by 24" drawing on 9" by 6" paper after making the necessary changes in the plot specifications. The comments on the right are **not** a part of the file.

```
PLOT
W                       (Window option.)
0,0 36,24               (First corner, other corner.)
5                       (Detail plot configuration.)
N                       (Do you want to change plotter parameters.)
N
I
0,0
0                       (Rotation angle.)
N
N
1=4
0
```

Note
You can use a blank line to accept the default value for a prompt. A blank line in the script file will cause a Return. However, you must not accept the default plot specifications because the file might have been altered by another user or by another script file. Therefore, always enter the actual values in the file so that when you run a script file, it does not take the default values.

Example 6

Write a script file that will plot a 288' by 192' drawing on a 36" x 24" sheet of paper. The drawing scale is 1/8" = 1'. The following table lists the assignment of pens for different colors and speeds. (The filename is SCRIPT6.SCR. In this example assume that AutoCAD is configured for the HPGL plotter and the plotter description is HPGL-Plotter.)

Object Color	Pen No.	Line-type	Pen Speed
1 (red)	1	0	25

2 (yellow)	2	0	30
3 (green)	1	0	36
4 (cyan)	1	0	36
5 (blue)	2	0	36
6 (magenta)	2	6	36

The following file is a listing of the script file for plotting a 144' by 96' drawing on a 36" by 24" size paper after making the necessary changes in the plot specifications:

<pre>
PLOT
L
Y   (Do you want to change anything?)
Y   (Do you want to change plotter?)
HPGL-Plotter
L
Y   (Do you want to change anything?)
N   (Do you want to change plotter?)
20   (How many seconds should we wait?)
Y   (Do you want to change any of the above
                           parameters?)
1 0 25 0.010   (Pen No., Linetype, Pen Speed, Pen width.)
2 0 30 0.010
1 0 36 0.010
1 0 36 0.010
2 0 36 0.010
2 0 36 0.010
X   (X for exit.)
N
I
0,0
9,6
0
N
N
1=96   (Plotted inches = drawing units.)
                        (Blank line for Enter.)
</pre>

Note

You can use a blank line to accept the default value for a prompt. A blank line in the script file will cause Enter (⏎). However, you must not accept the default plot specifications, because the file might have been altered by another user or by another script file. Therefore, always enter the actual values in the file so that when you run a script file, it does not take the default values.

WHAT IS A SLIDE SHOW?

AutoCAD provides a facility using script files to combine the slides in a text file and display them in a predetermined sequence. In this way, you can generate a slide show for a slide presentation. You can also introduce a time delay in the display so that the viewer has enough time to view a slide.

A drawing or parts of a drawing can also be displayed by using the AutoCAD display commands. For example, you can use ZOOM, PAN, or other commands to display the details you want to show. If the drawing is very complicated, it takes quite some time to display the desired information, and it may not be possible to get the desired views in the right sequence. However, with slide shows you can arrange the slides in any order and present them in a definite sequence. In addition to saving time, this will also help to minimize the distraction that might be caused by constantly changing the drawing display. Also, some drawings are confidential in nature and you may not want to display some portions or views of them. By making slides, you can restrict the information that is presented through them. You can send a slide show to a client without losing control of the drawings and the information that is contained in them.

WHAT ARE SLIDES?

A **slide** is the snapshot of a screen display; it is like taking a picture of a display with a camera. The slides do not contain any vector information like AutoCAD drawings, which means that the entities do not have any information associated with them. For example, the slides do not retain any information about the layers, colors, linetypes, start point, or endpoint of a line or viewpoint. Therefore, slides cannot be edited like drawings. If you want to make any changes in the slide, you need to edit the drawing and then make a new slide from the edited drawing.

MSLIDE COMMAND

Slides are created by using the AutoCAD MSLIDE command. If **FILEDIA** is set to 0, the command will prompt you to enter the slide file name.

```
Command: MSLIDE
Slide file <Default>: Slide file name.
```

Example
Command: **MSLIDE**
Slide File: <NEWDWG> SLIDE1
　　　　　　　　　　｜　　　　｜
　　　　　　　　　　｜　　　　└─Slide file name
　　　　　　　　　　└─Default slide file name

In the preceding example, AutoCAD will save the slide file as **SLIDE1.SLD**. If **FILEDIA** is set to 1, the MSLIDE command displays the **Create Slide File** dialog (Figure 2-5) box on the screen. You can enter the slide file name in this dialog box.

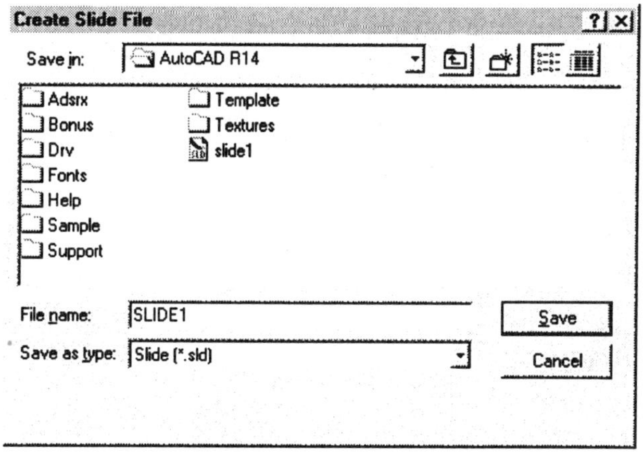

Figure 2-5 Create Slide File dialog box

> **Note**
> *In model space, you can use the MSLIDE command to make a slide of the existing display in the current viewport.*
>
> *If you are in the paper space viewport, you can make a slide of the display in the paper space that includes any floating viewports.*
>
> *When the viewports are not active, the MSLIDE command will make a slide of the current screen display.*

VSLIDE COMMAND

To view a slide, use the VSLIDE command. AutoCAD will then prompt you to enter the slide file name. Enter the name of the slide you want to view and press Enter. Do **not** enter the extension after the slide file name. AutoCAD automatically assumes the extension **.SLD**.

Command: **VSLIDE**
Slide file <Default>: *Name*.

Example
Command: **VSLIDE**
Slide file <NEWDWG>: SLIDE1
                         └─ Name of slide file
           └─ Default slide file name

You can also use the **Select Slide File** dialog box (Figure 2-6) to view a slide.

2-20 Customizing AutoCAD

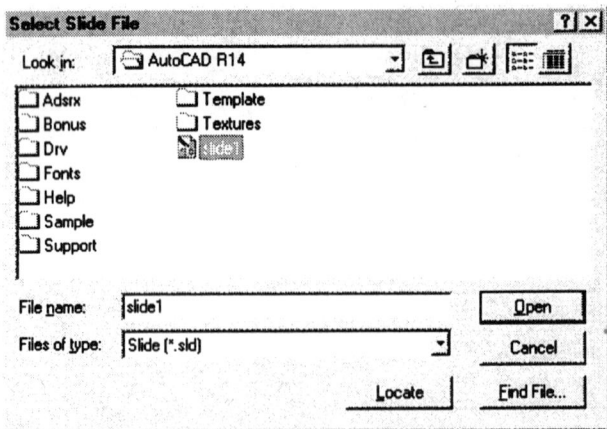

Figure 2-6 Select Slide File dialog box

Note

After viewing a slide, you can use the AutoCAD REDRAW command to remove the slide display and return to the existing drawing on the screen.

Any command that is automatically followed by a redraw will also display the existing drawing. For example, AutoCAD GRID, ZOOM ALL, and REGEN commands will automatically return to the existing drawing on the screen.

You can view the slides on high-resolution or low-resolution monitors. Depending on the resolution of the monitor, AutoCAD automatically adjusts the image. However, if you are using a high-resolution monitor, it is better to make the slides on the same monitor to take full advantage of that monitor.

Example 7

Write a script file that will generate a slide show of the following slide files, with a time delay of 15 seconds after every slide:

SLIDE1, SLIDE2, SLIDE3, SLIDE4

The first step in a slide show is to create the slides. Figure 2-7 shows the drawings that have been saved as slide files **SLIDE1, SLIDE2, SLIDE3,** and **SLIDE4**. The second step is to find out the sequence in which you want these slides to be displayed, with the necessary time delay, if any, between slides. Then you can use any text editor or the AutoCAD EDIT command (provided the **ACAD.PGP** file is present and EDIT is defined in the file) to write the script file with the extension **.SCR**.

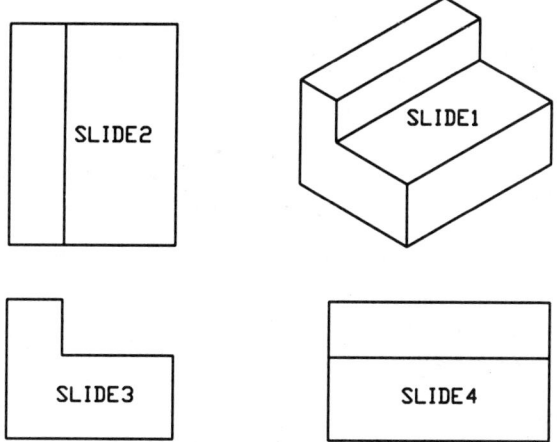

Figure 2-7 Slides for slide show

The following file is a listing of the script file that will create a slide show of the slides in Figure 2-7. The name of the script file is **SLDSHOW1**.

    VSLIDE SLIDE1
    DELAY 15000
    VSLIDE SLIDE2
    DELAY 15000
    VSLIDE SLIDE3
    DELAY 15000
    VSLIDE SLIDE4
    DELAY 15000

To run this slide show, type SCRIPT in response to the AutoCAD Command: prompt. Now, type the name of the script file (**SLDSHOW1**) and press Enter. The slides will be displayed on the screen, with an approximate time delay of 15 seconds between them.

PRELOADING SLIDES

In the script file of Example 7, VSLIDE SLIDE1 in line 1 loads the slide file, **SLIDE1**, and displays it on screen. After a pause of 15,000 milliseconds, it starts loading the second slide file, **SLIDE2**. Depending on the computer and the disk access time, you will notice that it takes some time to load the second slide file; the same is true for the other slides. To avoid the delay in loading the slide files, AutoCAD has provided a facility to preload a slide while viewing the previous slide. This is accomplished by placing an asterisk (*) in front of the slide file name.

    VSLIDE SLIDE1   *(View slide, SLIDE1.)*
    VSLIDE *SLIDE2   *(Preload slide, SLIDE2.)*
    DELAY 15000   *(Delay of 15 seconds.)*
    VSLIDE   *(Display slide, SLIDE2.)*
    VSLIDE *SLIDE3   *(Preload slide, SLIDE3.)*

DELAY 15000   (Delay of 15 seconds.)
VSLIDE   (Display slide, SLIDE3.)
VSLIDE *SLIDE4
DELAY 15000
VSLIDE
DELAY 15000
RSCRIPT   (Restart the script file.)

Example 8

Write a script file to generate a continuous slide show of the following slide files, with a time delay of two seconds between slides:

SLD1, SLD2, SLD3

The slide files are located in different subdirectories, as shown in Figure 2-8. The subdirectory **SUBDIR1** is the current subdirectory.

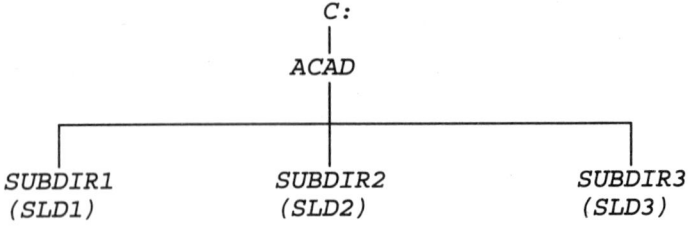

Figure 2-8 Subdirectories of the C drive

Where:
C: (Root directory.)
ACAD   (Subdirectory where the AutoCAD files are loaded.)
SUBDIR1   (Drawing subdirectory.)
SUBDIR2   (Drawing subdirectory.)
SUBDIR3   (Drawing subdirectory.)
SLD1   (Slide file in SUBDIR1 subdirectory.)
SLD2   (Slide file in SUBDIR2 subdirectory.)
SLD3   (Slide file in SUBDIR3 subdirectory.)

The following file is the listing of the script files that will generate a slide show for the slides in Example 8:

VSLIDE SLD1
DELAY 2000
VSLIDE C:\ACAD\SUBDIR2\SLD2
DELAY 2000
VSLIDE C:\ACAD\SUBDIR3\SLD3

DELAY 2000
RSCRIPT

Line 1
VSLIDE SLD1
In this line, the AutoCAD command VSLIDE loads the slide file **SLD1**. Since, in this example, we are assuming you are in the subdirectory **SUBDIR1** and the first slide file, **SLD1**, is located in the same subdirectory, it does not require any path definition.

Line 2
DELAY 2000
This line uses the AutoCAD DELAY command to create a pause of approximately two seconds before the next slide is loaded.

Line 3
VSLIDE C:\ACAD\SUBDIR2\SLD2
In this line, the AutoCAD command VSLIDE loads the slide file **SLD2**, located in the subdirectory **SUBDIR2**. If the slide file is located in a different subdirectory, you need to define the path with the slide file.

Line 5
VSLIDE C:\ACAD\SUBDIR3\SLD3
In this line, the VSLIDE command loads the slide file **SLD3**, located in the subdirectory **SUBDIR3**.

Line 7
RSCRIPT
In this line, the RSCRIPT command executes the script file again and displays the slides on the screen. This process continues indefinitely until the script file is canceled by pressing the Esc key or the Backspace key.

SLIDE LIBRARIES

AutoCAD provides a utility, SLIDELIB, which constructs a library of the slide files. The format of the SLIDELIB utility command is:

C:\> **SLIDELIB (Library filename) < (Slide list filename)**

Example
```
C:\>SLIDELIB SLDLIB <SLDLIST
      |        |        |
      |        |        └─ List of slide filenames
      |        └─ Slide library filename
      └─ AutoCAD's SLIDELIB utility
```

2-24　Customizing AutoCAD

The SLIDELIB utility is supplied with the AutoCAD software package. You can find this utility (SLIDELIB.EXE) in the support subdirectory. The slide file list is a list of the slide filenames that you want in a slide show. It is a text file that can be written by using any text editor or AutoCAD's EDIT command (provided ACAD.PGP file is present and EDIT is defined in the file). The slide files in the slide file list should not contain any file extension (.SLD). However, if you want to add a file extension it should be .SLD.

The slide file list can be also created by using the following command, if you have DOS version 5.0 or above:

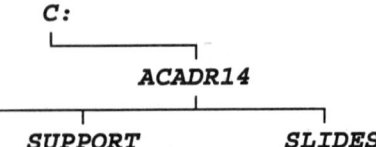

C:\ACAD\SLIDES > **DIR *.SLD/B > SLDLIST**

In this example assume that the name of the slide file list is **SLDLIST** and all slide files are in the SLIDES subdirectory. To use this command to create a slide file list, all slide files must be in the same directory.

When you use the SLIDELIB utility, it reads the slide filenames from the file that is specified in the slide list and the file is then written to the file specified by library. In the Example 9, the SLIDELIB utility reads the slide filenames from the file SLDLIST and writes them to the library file SLDLIB:

C:\ > **SLIDELIB SLDLIB < SLDLIST**

Note
*You **cannot** edit a slide library file. If you want to change anything, you have to create a new list of the slide files and then use the SLIDELIB utility to create a new slide library.*

If you edit a slide while the slide is displayed on the screen, the slide is not edited. Instead, the current drawing that is behind the slide will be edited. Therefore, do not use any editing commands while you are viewing a slide. Use the VSLIDE and DELAY commands only when viewing a slide.

The path name is not saved in the slide library; therefore, if you have more than one slide with the same name, even though they are in different subdirectories, only one slide will be saved in the slide library.

If you are using DOS, use a smaller library file. If the library file contains a large number of files, it might exceed the computer's memory. In that case, the SLIDELIB utility will display a warning message.

Example 9

Use AutoCAD's SLIDELIB utility to generate a continuous slide show of the following slide files with a time delay of 2.5 seconds between the slides. (The filenames are: SLDLIST for slide list file, SLDSHOW1 for slide library, SHOW1 for script file.)

FRONT, TOP, RSIDE, STAIRS, 3DVIEW, LROOM, FROOM, BROOM

The slide files are located in different subdirectories as shown in Figure 2-9.

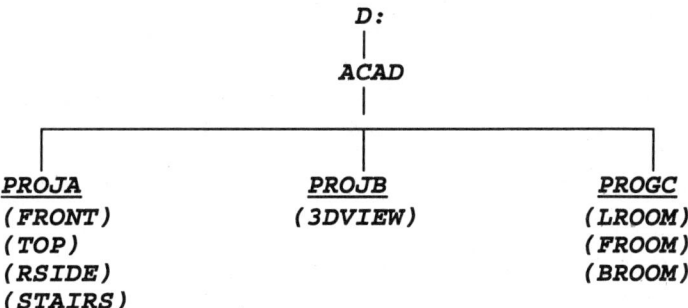

Figure 2-9 Subdirectories of D drive

Where
D   *(D drive.)*
ACAD   *(Subdirectory where the AutoCAD files are loaded.)*
PROJA   *(Drawing subdirectory.)*
PROJB   *(Drawing subdirectory.)*
PROJC   *(Drawing subdirectory.)*

Step 1
The first step is to create a list of the slide filenames with the drive and the directory information. Assume that you are in the ACAD subdirectory. You can use a text editor or AutoCAD's EDIT function to create a list of the slide files that you want to include in the slide show. These files do not need a file extension. However, if you choose to give them a file extension, it should be .SLD. The following file is a listing of the file SLDLIST for Example 9:

    D:\ACADR14\PROJA\FRONT
    D:\ACADR14\PROJA\TOP
    D:\ACADR14\PROJA\RSIDE
    D:\ACADR14\PROJA\STAIRS
    D:\ACADR14\PROJB\3DVIEW
    D:\ACADR14\PROJC\LROOM
    D:\ACADR14\PROJC\FROOM
    D:\ACADR14\PROJC\BROOM

Step 2
The second step is to use AutoCAD's SLIDELIB utility program to create the slide library. The name of the slide library is assumed to be SLDSHOW1 for this example.

2-26 Customizing AutoCAD

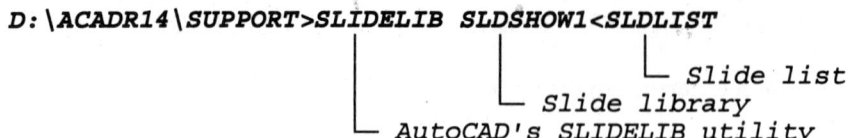

```
D:\ACADR14\SUPPORT>SLIDELIB SLDSHOW1<SLDLIST
```
- Slide list
- Slide library
- AutoCAD's SLIDELIB utility

Step 3
Now you can write a script file for the slide show that will use the slides in the slide library. The name of the script file for this example is assumed to be SHOW1.

```
VSLIDE SLDSHOW1(FRONT)
DELAY 2500
VSLIDE SLDSHOW1(TOP)
DELAY 2500
VSLIDE SLDSHOW1(RSIDE)
DELAY 2500
VSLIDE SLDSHOW1(STAIRS)
DELAY 2500
VSLIDE SLDSHOW1(3DVIEW)
DELAY 2500
VSLIDE SLDSHOW1(LROOM)
DELAY 2500
VSLIDE SLDSHOW1(RROOM)
DELAY 2500
VSLIDE SLDSHOW1(BROOM)
DELAY 2500
RSCRIPT
```

Step 4
Start AutoCAD, if you are not in AutoCAD. With AutoCAD's SCRIPT command run the script file, SHOW1, and you will see the slides displayed on the screen.

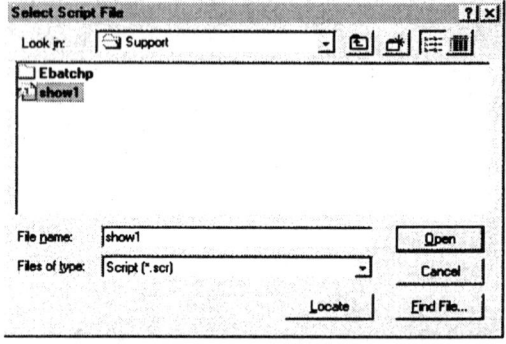

Figure 2-10 Invoking script file

Script Files and Slide Shows 2-27

You can also invoke the SCRIPT command from the Command prompt after setting the system variable FILEDIA to 0.

Command: **SCRIPT**
Slide file <default>:**SHOW1**

REVIEW QUESTIONS

SCRIPT FILES

1. AutoCAD has provided a facility of _____ that allows you to combine different AutoCAD commands and execute them in a predetermined sequence.

2. The _____ files can be used to generate a slide show, do the initial drawing setup, or plot a drawing to a predefined specification.

3. Before writing a script file, you need to know the AutoCAD _____ and the _____ required in response to the command prompts.

4. In a script file, you can _____ several statements in one line.

5. In a script file, the _____ are used to terminate a command or a prompt entry.

6. The AutoCAD _____ command is used to run a script file.

7. When you run a script file, the default script file name is the same as the _____ name.

8. When you run a script file, type the name of the script file without the file _____.

9. One of the limitations of script files is that all the information has to be contained _____ the file.

10. The AutoCAD _____ command allows you to re-execute a script file indefinitely until the command is canceled.

11. You cannot provide a _____ statement in a script file to terminate the file when a particular condition is satisfied.

12. The AutoCAD _____ command introduces a delay before the next command is executed.

13. The DELAY command is to be followed by _____ in milliseconds.

14. If the script file was canceled and you want to continue the script file, you can do so by using the AutoCAD _____ command.

SLIDE SHOWS

15. AutoCAD provides a facility through _____ files to combine the slides in a text file and display them in a predetermined sequence.

16. A _____ can also be introduced in the script file so that the viewer has enough time to view a slide.

17. Slides are the _____ of a screen display.

18. Slides do not contain any _____ information, which means that the entities do not have any information associated with them.

19. Slides _____ be edited like a drawing.

20. Slides can be created using the AutoCAD _____ command.

21. Slide file names can be up to _____ characters long.

22. In model space, you can use the MSLIDE command to make a slide of the _____ display in the _____ viewport.

23. If you are in paper space, you can make a slide of the display in paper space that _____ any floating viewports.

24. To view a slide, use the AutoCAD _____ command.

25. If you want to make any change in the slide, you need to _____ the drawing, then make a new slide from the edited drawing.

26. If the slide is in the slide library and you want to view it, the slide library name has to be _____ with the slide filename.

27. AutoCAD provides a utility that constructs a library of the slide files. This is done with AutoCAD's utility program called _____.

28. You cannot _____ a slide library file. If you want to change anything, you have to create a new list of the slide files and then use the _____ utility to create a new slide library.

29. The path name _____ be saved in the slide library. Therefore, if you have more than one slide with the same name, although with different subdirectories, only one slide will be saved in the slide library.

EXERCISES

SCRIPT FILES

Exercise 1

Write a script file that will do the following initial setup for a drawing:

Grid	2.0
Snap	0.5
Limits	0,0
	18.0,12.0
Zoom	All
Text height	0.25
Ltscale	2.0
Dimscale	2.0
Dimtix	On
Dimtoh	Off
Dimtih	Off
Dimtad	1
Dimcen	0.75

Exercise 2

Write a script file that will set up the following layers with the given colors and linetypes (filename SCRIPTE2.SCR):

Contour	Red	Continuous
SPipes	Yellow	Center
WPipes	Blue	Hidden
Power	Green	Continuous
Manholes	Magenta	Continuous
Trees	Cyan	Continuous

Exercise 3

Write a script file that will do the following initial setup for a new drawing:

Limits	0,0 24,18
Grid	1.0
Snap	0.25
Ortho	On
Snap	On
Zoom	All
Pline width	0.02
PLine	0,0 24,0 24,18 0,18 0,0
Units	Decimal units

2-30 Customizing AutoCAD

                              Number of decimal digits (2)
                              Decimal degrees
                              Number of decimal digits (2)
                              Direction of 0 angle (3 o'clock)
                              Angle measured counterclockwise
Ltscale   1.5

Layers

Name	Color	Linetype
Obj	Red	Continuous
Cen	Yellow	Center
Hid	Blue	Hidden
Dim	Green	Continuous

Exercise 4

Write a script file that will PLOT a given drawing according to the following specifications. (Use the plotter for which your system is configured and adjust the values accordingly.)

    Plot, using the Window option
    Window size (0,0 24,18)
    Do not write the plot to file
    Size in Inch units
    Plot origin (0.0,0.0)
    Maximum plot size (8.0,10.5)
    90 degree plot rotation
    No removal of hidden lines
    Plotting scale (Fit)

Exercise 5

Write a script file that will continuously rotate a line in 10 degree increments around its mid point. The time delay between increments is one second.

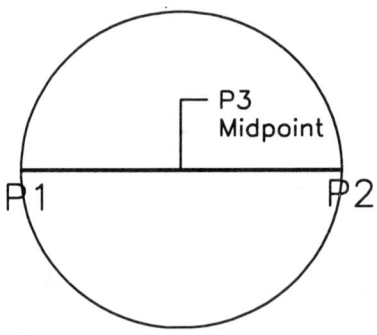

Figure 2-11 Drawing for Exercise 5

SLIDE SHOWS

Exercise 6

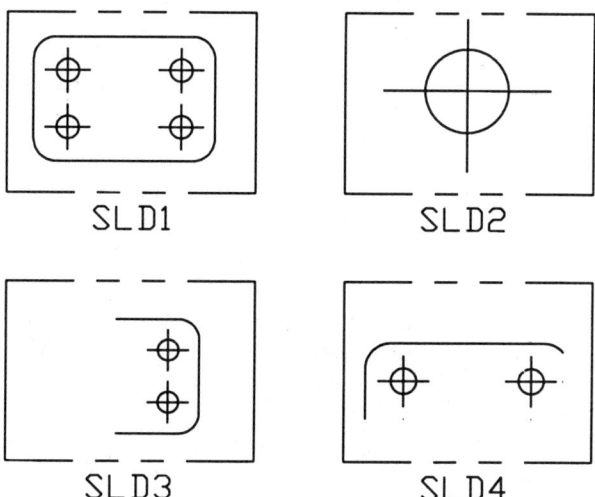

Figure 2-12 Slides for slide show

Make the slides shown in Figure 2-12 and write a script file for a continuous slide show. Provide a time delay of 10 seconds after every slide. (You do not have to use the slides shown in Figure 2-12; you can use any slides of your choice.)

Exercise 7

From Exercise 6, list the slides in a file SLDLIST2 and create a slide library file SLDLIB2. Then write a script file SHOW2 using the slide library with a time delay of five seconds after every slide.

Exercise 8

Write a script file to generate a continuous slide show of the following slide files with a time delay of three seconds between the slides.

    SLDX1, SLDX2, SLDX3
    SLDY1, SLDY2, SLDY3
    SLDZ1, SLDZ2, SLDZ3

Assume that the slide files are located in different subdirectories as shown in Figure 2-13 and ACAD is the current subdirectory.

2-32 Customizing AutoCAD

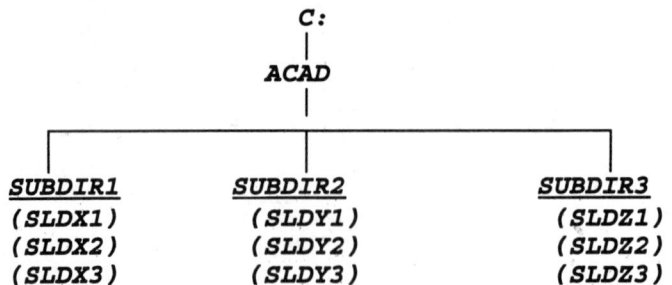

Figure 2-13 Subdirectories of C drive

Where
C:	*(C drive.)*
ACAD	*(Subdirectory where the AutoCAD files are loaded.)*
SUBDIR1	*(Drawing subdirectory.)*
SUBDIR2	*(Drawing subdirectory.)*
SUBDIR3	*(Drawing subdirectory.)*

Chapter 3

Creating Linetypes and Hatch Patterns

Learning objectives

After completing this chapter, you will be able to:
Create Linetypes:
- *Write linetype definitions.*
- *Create different linetypes.*
- *Create linetype files.*
- *Determine LTSCALE for plotting the drawing to given specifications.*
- *Define alternate linetypes and modify existing linetypes.*
- *Create string and shape complex linetypes.*

Create Hatch Patterns:
- *Understand hatch pattern definition.*
- *Create new hatch patterns.*
- *Determine the effect of angle and scale factor on hatch.*
- *Create hatch patterns with multiple descriptors.*
- *Save hatch patterns in a separate file.*
- *Define custom hatch pattern file.*
- *Add hatch pattern slides to AutoCAD slide library.*

STANDARD LINETYPES

The AutoCAD software package comes with a library of standard linetypes that has 38 different linetypes, including ISO linetypes. These linetypes are saved in the **ACAD.LIN** file. You can modify existing linetypes or create new ones.

LINETYPE DEFINITION

All linetype definitions consist of two parts: **header line and pattern line.**

Header Line

The **header line** consists of an asterisk (*) followed by the name of the linetype and the linetype description. The name and the linetype description should be separated by a comma. If there is no description, the comma that separates the linetype name and the description is not required.

The format of the header line is:

* Linetype Name, Description

Example

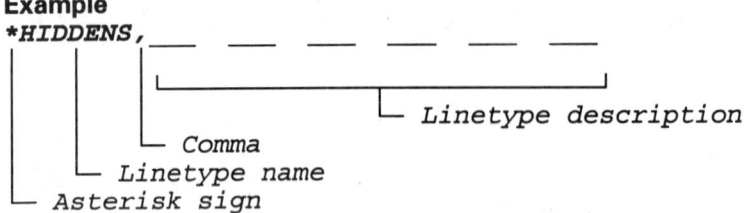

All linetype definitions require a linetype name. When you want to load a linetype or assign a linetype to an object, AutoCAD recognizes the linetype by the name you have assigned to the linetype definition. The names of the linetype definition should be selected to help the user recognize the linetype by its name. For example, a linetype name LINEFCX does not give the user any idea about the type of line. However, a linetype name like DASHDOT gives a better idea about the type of line that a user can expect.

The linetype description is a textual representation of the line. This representation can be generated by using dashes, dots, and spaces at the keyboard. The graphic is used by AutoCAD when you want to display the linetypes on the screen by using the AutoCAD LINETYPE command with the ? option or by using the dialog box. The linetype description cannot exceed 47 characters.

Pattern Line

The **pattern line** contains the definition of the line pattern. The definition of the line pattern consists of the alignment field specification and the linetype specification. The alignment field specification and the linetype specification are separated by a comma.

The format of the pattern line is:

Alignment Field Specification, Linetype Specification

Example

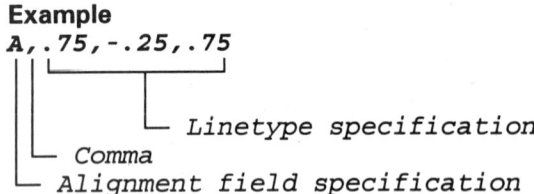

Creating Linetypes and Hatch Patterns 3-3

The letter used for alignment field specification is A. This is the only alignment field supported by AutoCAD; therefore, the pattern line will always start with the letter A. The linetype specification defines the configuration of the dash-dot pattern to generate a line. The maximum number for dash length specification in the linetype is 12, provided the linetype pattern definition fits on one 80-character line.

ELEMENTS OF LINETYPE SPECIFICATION

All linetypes are created by combining the basic elements in a desired configuration. There are three basic elements that can be used to define a linetype specification.

   Dash   (Pen down)
   Dot   (Pen down, 0 length)
   Space   (Pen up)

Example

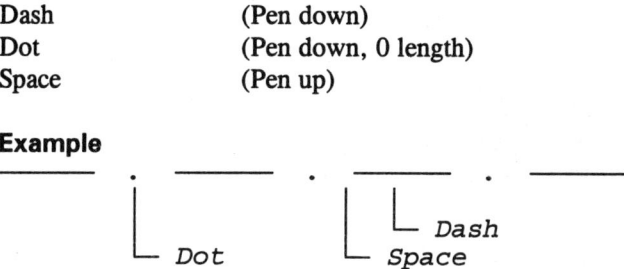

The dashes are generated by defining a positive number. For example, .5 will generate a dash 0.5 units long. Similarly, spaces are generated by defining a negative number. For example, -.2 will generate a space 0.2 units long. The dot is generated by defining a 0 length.

Example

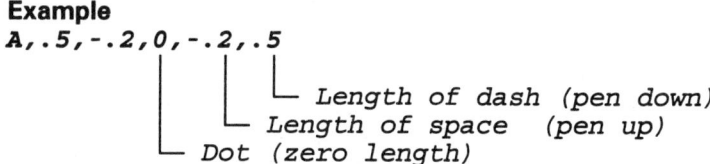

CREATING LINETYPES

Before creating a linetype, you need to decide the type of line you want to generate. Draw the line on a piece of paper and measure the length of each element that constitutes the line. You need to define only one segment of the line, because the pattern is repeated when you draw a line. Linetypes can be created or modified by one of the following methods:

   Using the AutoCAD LINETYPE command
   Using a text editor (such as Notepad).

Consider the following example, which creates a new linetype, first using the AutoCAD LINETYPE command and then using a text editor.

Example 1

Using the AutoCAD LINETYPE command, create linetype DASH3DOT (Figure 3-1) with the following specifications:

Length of the first dash 0.5
Blank space 0.125
Dot
Blank space 0.125
Dot
Blank space 0.125
Dot
Blank space 0.125

Using the AutoCAD Linetype Command

To create a linetype using the AutoCAD LINETYPE command, first make sure that you are in the drawing editor. Then enter the LINETYPE command and select the Create option to create a linetype.

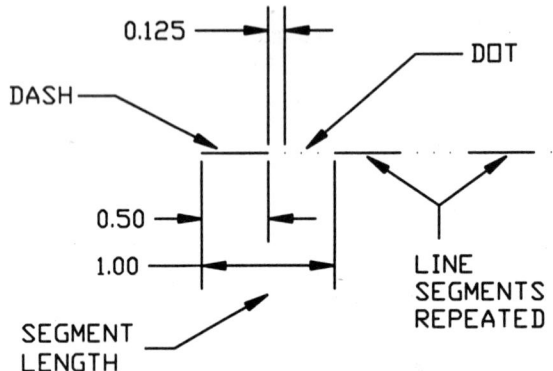

Figure 3-1 Linetype specifications of DASH3DOT

Command: -LINETYPE
?/Create/Load/Set: C

Enter the name of the linetype and the name of the library file in which you want to store the definition of the new linetype.

Name of linctype to create: **DASH3DOT**

If FILEDIA=1, the **Create or Append Linetype File** dialog box (Figure 3-2) will appear on the screen. If FILEDIA=0, AutoCAD will prompt you to enter the name of the file.

File for storage of linetype <default>: **Acad**

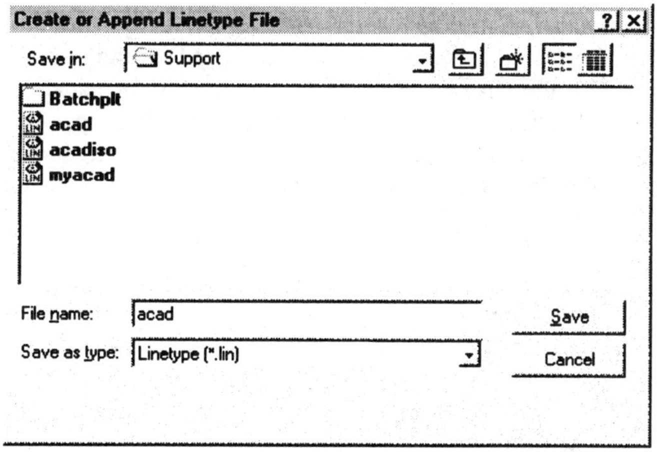

Figure 3-2 Create or Append Linetype File dialog box

If the linetype already exists, the following message will be displayed on the screen:

(Name) already exists in this file.
Current definition is:
*Linetype name [,description]
alignment, dash-1, dash-2,__.
Overwrite? <N>

If you want to redefine the existing line style, enter Y; otherwise, type N or press Return to choose the default value of N. You can then repeat the process with a different name of the linetype.

After entering the name of the linetype and the library file name, AutoCAD will prompt you to enter the descriptive text and the pattern of the line.

Descriptive text: **DASH3DOT,____ . . . ____ . . . ____
Enter pattern (on next line):
A,.5,-.125,0,-.125,0,-.125,0,-.125

Descriptive Text
***DASH3DOT,____ . . . ____ . . . ____**

For the descriptive text, you have to type an asterisk (*) followed by the name of the linetype. For Example 1, the name of the linetype is DASH3DOT. The name *DASH3DOT can be followed by the description of the linetype; the length of this description cannot exceed 47 characters. In this example, the description is dashes and dots ____ . . . ____. It could be any text or alphanumeric string. The description is displayed on the screen when you list the linetypes.

Pattern
A,.5,-.125,0,-.125,0,-.125,0,-.125

The line pattern should start with an alignment definition. Currently, AutoCAD supports only one type of alignment--A. Therefore, it is automatically displayed on the screen when you select the **LINETYPE** command with the Create option. After entering A for pattern alignment, you must define the pen position. A positive number (.5 or 0.5) indicates a "pen-down" position, and a negative number (-.25 or -0.25) indicates a "pen-up" position. The length of the dash or the space is designated by the magnitude of the number. For example, 0.5 will draw a dash 0.5 units long, and -0.25 will leave a blank space of 0.25 units. A dash length of 0 will draw a dot (.). Here are the pattern definition elements for Example 1:

.5	pen down	0.5 units long dash
-.125	pen up	.125 units blank space
0	pen down	dot
-.125	pen up	.125 units blank space
0	pen down	dot
-.125	pen up	.125 units blank space
0	pen down	dot
-.125	pen up	.125 units blank space

After you enter the pattern definition, the linetype (DASH3DOT) is automatically saved in the **ACAD.LIN** file. The linetype (DASH3DOT) can be loaded using the AutoCAD **LINETYPE** command and selecting the **Load** option.

> **Note**
> *The name and the description must be separated by a comma (,). The description is optional. If you decide not to give one, omit the comma after the linetype name DASH3DOT.*

Using a Text Editor

You can also use a text editor (like Notepad) to create a new linetype. Using the text editor, load the file and insert the lines that define the new linetype. The following file is a partial listing of the **ACAD.LIN** file after adding a new linetype to the file:

```
*BORDER,__ __ . __ __ . __ __ . __ __ . __ __ . __ __ .
A,.5,-.25,.5,-.25,0,-.25
*BORDER2,__ . __ . __ . __ . __ . __ . __ . __ . __ . __ .
A,.25,-.125,.25,-.125,0,-.125
*BORDERX2,____ ____ . ____ ____ . ____ ____ .
A,1.0,-.5,1.0,-.5,0,-.5

*CENTER,____ _ ____ _ ____ _ ____ _ ____ _ ____
A,1.25,-.25,.25,-.25
*CENTER2,___ _ ___ _ ___ _ ___ _ ___ _ ___
A,.75,-.125,.125,-.125
*CENTERX2,_____ __ _____ __ _____ __ __
A,2.5,-.5,.5,-.5
*DASHDOT,__ . __ . __ . __ . __ . __ . __ . __ .
A,.5,-.25,0,-.25
```

*DOTX2,. . . . .
A,0,-.5
*HIDDEN,__ __ __ __ __ __ __ __ __ __
A,.25,-.125
*HIDDEN2,_ _ _ _ _ _ _ _ _ _ _ _ _ _ _
A,.125,-.0625
*HIDDENX2,___ ___ ___ ___ ___ ___ ___
A,.5,-.25
*PHANTOM,_____ __ __ _____ __ __ _____ __ __
A,1.25,-.25,.25,-.25,.25,-.25
*PHANTOM2,___ _ _ ___ _ _ ___ _ _ ___
A,.625,-.125,.125,-.125,.125,-.125
*PHANTOMX2,_____ ____ ____ _____
A,2.5,-.5,.5,-.5,.5,-.5
DASH3DOT,___ . . . ___ . . . ___
A,.5,-.125,0,-.125,0,-.125,0,-.125

The last two lines of this file define the new linetype, DASH3DOT. The first line contains the name DASH3DOT and the description of the line (___ . . . ___). The second line contains the alignment and the pattern definition. Save the file and then load the linetype using the AutoCAD LINETYPE command with the Load option. The lines and polylines that this linetype will generate are shown in Figure 3-3.

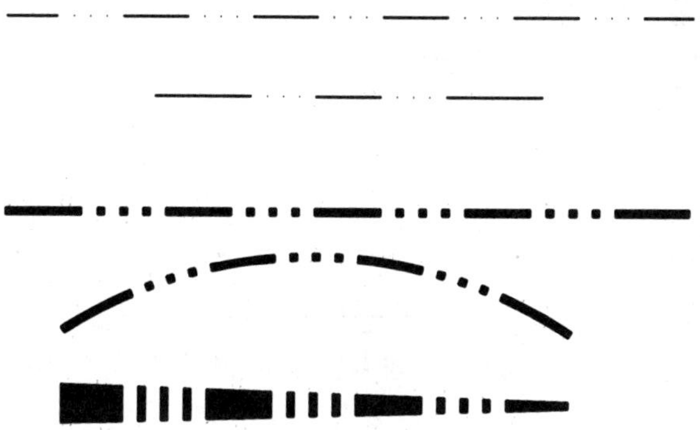

Figure 3-3 Lines created by linetype DASH3DOT

> **Note**
> *If you change the LTSCALE factor, all lines in the drawing are affected by the new ratio.*

CREATING LINETYPE FILES

You can start a new linetype file and then add the line definitions to this file. Use any text editor like Notepad to start a new file (NEWLT.LIN) and then add the following two lines to the file to define the DASH3DOT linetype.

*DASH3DOT,___ . . . ___ . . . ___
A,.5,-.125,0,-.125,0,-.125,0,-.125

You can load the DASH3DOT linetype from this file using AutoCAD's LINETYPE command and then selecting the LOAD option.

Command: **-LINETYPE**
?/Create/Load/Set: **L**
Linetype(s) load: **DASH3DOT**
File to search <default>: **NEWLT**

You can also load the linetype by using Layer & Linetype Properties dialog box (Figure 3-4) that can be invoked by entering LINETYPE at the Command: prompt, selecting Linetype from the Format pull-down menu, or selecting Layer from the Format pull-down menu and then selecting the Linetype tab.

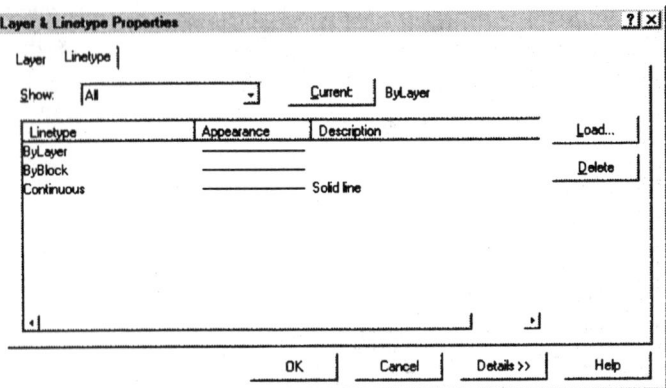

Figure 3-4 Loading linetypes using Layer & Linetype Properties dialog box

ALIGNMENT SPECIFICATION

The alignment specifies the pattern alignment at the start and the end of the line, circle, or arc. In other words, the line would always start and end with the dash (___). The alignment definition "A" requires that the first element be a dash or dot (pen down), followed by a negative (pen up) segment. The minimum number of dash segments for alignment A is two. If there is not enough space for the line AutoCAD will draw a continuous line.

Creating Linetypes and Hatch Patterns 3-9

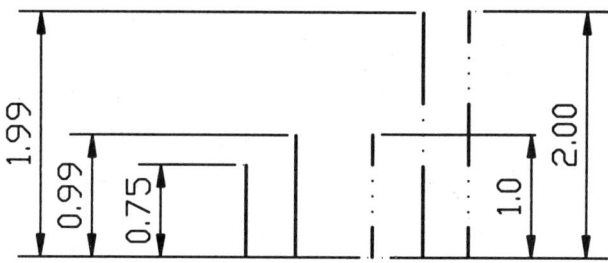

Figure 3-5 Alignment of linetype DASH3DOT

For example, in the linetype DASH3DOT of Example 1, the length of each line segment is 1.0 (.5 + .125 + .125 + .125 + .125 = 1.0). If the length of the line drawn is less than 1.00, the line will be drawn as a continuous line (Figure 3-5). If the length of the line is 1.00 or greater, the line will be drawn according to DASH3DOT linetype. AutoCAD automatically adjusts the length of the dashes and the line will always start and end with a dash. The length of the starting and ending dashes will be at least half the length of the dash as specified in the file. If the length of the dash as specified in the file is 0.5, the length of the starting and ending dashes will be at least 0.25. To fit a line that starts and ends with a dash, the length of these dashes can also increase as shown in Figure 3-5.

LTSCALE COMMAND

As we mentioned previously, the length of each line segment in the DASH3DOT linetype is 1.0 (.5 + .125 + .125 + .125 + .125 = 1.0). If you draw a line that is less than 1.0 units long, AutoCAD will draw a single dash that looks like a continuous line (Figure 3-6). This problem can be rectified by changing the linetype scale factor variable LTSCALE to a smaller value. This can be accomplished by using AutoCAD's LTSCALE command:

   Command: **LTSCALE**
   New scale factor <default>: *New value.*

The default value of the LTSCALE variable is 1.0. If the LTSCALE is changed to 0.75, the length of each segment is reduced by 0.75 (1.0 x 0.75 = 0.75). Then, if you draw a line 0.75 units or longer, it will be drawn according to the definition of DASH3DOT (___ . . . ___) (Figures 3-6 and 3-7).

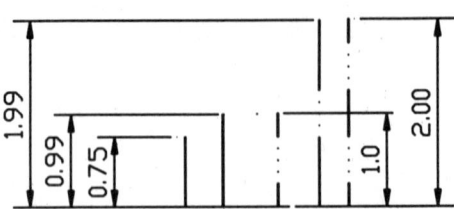

Figure 3-6 Alignment when Ltscale = 1

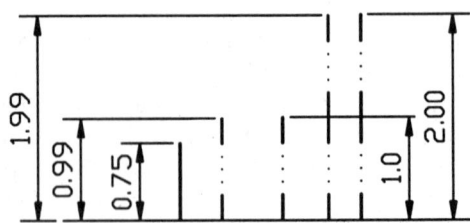

Figure 3-7 Alignment when Ltscale = 0.99

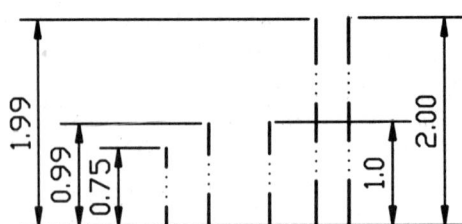

Figure 3-8 Alignment when Ltscale = 0.75

The appearance of the lines is also affected by the limits of the drawing. Most of the AutoCAD linetypes work fine for a drawing that has the limits 12,9. Figure 3-9 shows a line of linetype DASH3DOT that is four units long and the limits of the drawing are 12,9. If you increase the limits to 48,36 the lines will appear as continuous lines. If you want the line to appear the same as before **on the screen**, the LTSCALE should be changed. Since the limits of the drawing have increased four times, the LTSCALE should also increase by the same amount. If you change the scale factor to four, the line segments will also increase by a factor of four. As shown in Figure 3-9, the length of the starting and the ending dash has increased to one unit.

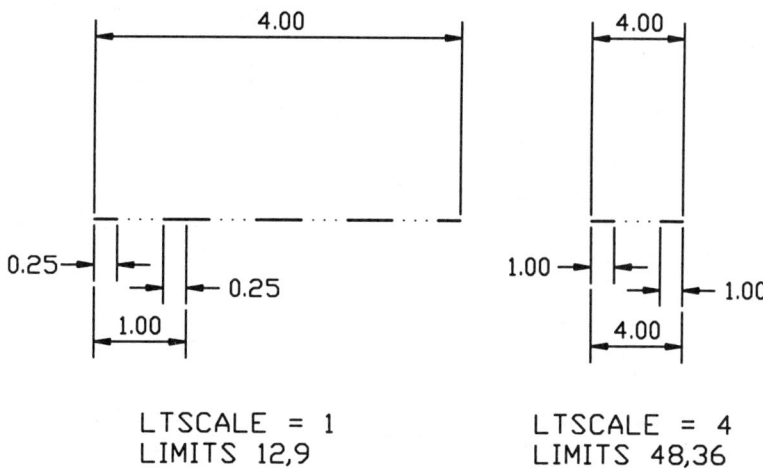

Figure 3-9 Linetype DASH3DOT before and after changing LTSCALE factor

In general, the approximate LTSCALE factor for **screen display** can be obtained by dividing the X-limit of the drawing by the default X-limit (12.00).

LTSCALE factor for SCREEN DISPLAY = X-limits of the drawing/12.00

Example
  Drawing limits are 48,36
  LTSCALE factor for screen display = 48/12 = 4

  Drawing sheet size is 36,24 and scale is 1/4" = 1'
  LTSCALE factor for screen display = 12 x 4 x (36 / 12) = 144

LTSCALE FACTOR FOR PLOTTING

The LTSCALE factor for plotting depends on the size of the sheet you are using to plot the drawing. For example, if the limits are 48 by 36, the drawing scale is 1:1, and you want to plot the drawing on a 48" by 36" size sheet, the LTSCALE factor is 1. If you check the specification of a hidden line in the ACAD.LIN file, the length of each dash is 0.25. Therefore, when you plot a drawing with 1:1 scale, the length of each dash in a hidden line is 0.25.

However, if the drawing scale is 1/8" = 1' and you want to plot the drawing on a 48" by 36" paper, the LTSCALE factor must be 96 (8 x 12 = 96). The length of each dash in the hidden line will increase by a factor of 96 because the LTSCALE factor is 96. Therefore, the length of each dash will be 24 units (0.25 x 96 = 24). At the time of plotting, the scale factor for plotting must be 1:96 to plot the 384' by 288' drawing on a 48" by 36" size paper. Each dash of the hidden line that was 24" long on the drawing will be 0.25 (24/96 = 0.25) inch long when plotted. Similarly, if the desired text size on the paper is 1/8", the text height in the drawing must be 12" (1/8 x 96 = 12").

> **Ltscale Factor for PLOTTING = Drawing Scale**

Sometimes your plotter may not be able to plot a 48" by 36" drawing or you might like to decrease the size of the plot so that the drawing fits within a specified area. To get the correct dash lengths for hidden, center, or other lines, you must adjust the LTSCALE factor. For example, if you want to plot the above mentioned drawing in 45" by 34" area, the reduction factor is:

Reduction factor	= 48/45
	= 1.0666
New LTSCALE factor	= LTSCALE factor x Reduction factor
	= 96 x 1.0666
	= 102.4

> **New Ltscale Factor for PLOTTING = Drawing Scale x Reduction Factor**

Note
If you change the LTSCALE factor, all lines in the drawing are affected by the new ratio.

ALTERNATE LINETYPES

One of the problems with the LTSCALE factor is that it affects all the lines in the drawing. As shown in Figure 3-10A, the length of each segment in all DASH3DOT type lines is approximately equal, no matter how long the lines. You might want to have a small segment length if the lines are small and a longer segment length if the lines are long. You can accomplish this by using CELTSCALE (discussed later in this chapter) or by defining an alternate linetype with a different segment length. For example, you can define a linetype DASH3DOT and DASH3DOTX with different line pattern specifications.

```
*DASH3DOT,____ . . . ____ . . . ____ . . . ____
A,0.5,-.125,0,-.125,0,-.125,0,-.125
*DASH3DOTX,_____ . . . _____
A,1.0,-.25,0,-.25,0,-.25,0,-.25
```

In DASH3DOT linetype the segment length is one unit, whereas in DASH3DOTX linetype the segment length is two units. You can have several alternate linetypes to produce the lines with different segment lengths. Figure 3-10B shows the lines generated by DASH3DOT and DASH3DOTX.

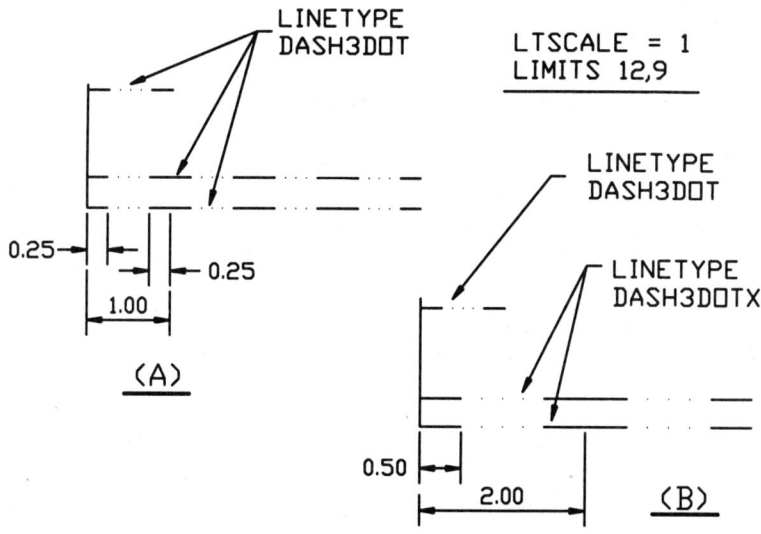

Figure 3-10 Linetypes generated by DASH3DOT and DASH3DOTX

Note
Although you might have used different linetypes with different segment lengths, the lines will be affected equally when you change the LTSCALE factor. For example, if the LTSCALE factor is 0.5, the segment length of DASH3DOT line will be 0.5 and the segment length of DASH3DOTX will be 1.0 units.

MODIFYING LINETYPES

You can also modify the linetypes that are defined in the ACAD.LIN file. You need a text editor, such as Notepad, to modify the linetype. You can also use the EDIT function of DOS, or AutoCAD's EDIT command (provided the ACAD.PGP file is present and EDIT is defined in the file). For example, if you want to change the dash length of the border linetype from 0.5 to 0.75, load the file, then edit the pattern line of the border linetype. The following file is a partial listing of the ACAD.LIN file after changing the border and centerx2 linetypes.

```
;;  AutoCAD Linetype Definition file,  Version 2.0
;;  Copyright 1991, 1992, 1993, 1994, 1996 by Autodesk, Inc.


;;
*BORDER,__ __ . __ __ . __ __ . __ __ . __ __ . __ __ . __ __ . __ __ .
A,.75,-.25,.75,-.25,0,-.25
*BORDER2,Border (.5x) __.__.__.__.__.__.__.__.__.
A,.25,-.125,.25,-.125,0,-.125
*BORDERX2,Border (2x) ____  ____  .  ____  ____  .  ____
A,1.0,-.5,1.0,-.5,0,-.5

*CENTER,Center ____  _  ____  _  ____  _  ____  _  ____
```

A,1.25,-.25,.25,-.25
*CENTER2,Center (.5x) ___ _ ___ _ ___ ___ _ ___ _ ___
A,.75,-.125,.125,-.125
CENTERX2,Center (2x) _____ __ _____ __ ____
A,3.5,-.5,.5,-.5

*DASHDOT,Dash dot __ . __ . __ . __ . __ . __ . __
A,.5,-.25,0,-.25
*DASHDOT2,Dash dot (.5x) _._._._._._._._._._._._.
A,.25,-.125,0,-.125
*DASHDOTX2,Dash dot (2x) ____ . ____ . ____ . ___
A,1.0,-.5,0,-.5

*DASHED,Dashed __ __ __ __ __ __ __ __ __ __ __
A,.5,-.25
*DASHED2,Dashed (.5x) _ _ _ _ _ _ _ _ _ _ _ _ _ _ _
A,.25,-.125
*DASHEDX2,Dashed (2x) ___ ___ ___ ___ ___ ___
A,1.0,-.5

*DIVIDE,Divide ____ . . ____ . . ____ . . ____ . . ____
A,.5,-.25,0,-.25,0,-.25
*DIVIDE2,Divide (.5x) __._._._._._._._._._._._._._
A,.25,-.125,0,-.125,0,-.125
*DIVIDEX2,Divide (2x) _____ . . _____ . . _
A,1.0,-.5,0,-.5,0,-.5

*DOT,Dot . . . .
A,0,-.25
*DOT2,Dot (.5x) ……………………………………
A,0,-.125
*DOTX2,Dot (2x) . . . .
A,0,-.5

*HIDDEN,Hidden __ __ __ __ __ __ __ __ __ __ __
A,.25,-.125
*HIDDEN2,Hidden (.5x) _ _ _ _ _ _ _ _ _ _ _ _ _ _ _
A,.125,-.0625
*HIDDENX2,Hidden (2x) ___ ___ ___ ___ ___ ___
A,.5,-.25

*PHANTOM,Phantom _____ __ __ ____ __ __ ____
A,1.25,-.25,.25,-.25,.25,-.25
*PHANTOM2,Phantom (.5x) __ _ __ __ __ __ __ __
A,.625,-.125,.125,-.125,.125,-.125
*PHANTOMX2,Phantom (2x) _____ ___ ___ _

A,2.5,-.5,.5,-.5,.5,-.5
;; ISO 128 (ISO/DIS 12011) linetypes
;; The size of the line segments for each defined ISO line, is
;; defined for an usage with a pen width of 1 mm. To use them with
;; the other ISO predefined pen widths, the line has to be scaled
;; with the appropriate value (e.g. pen width 0,5 mm -> ltscale 0.5).
;;
*ACAD_ISO02W100,ISO dash __ __ __ __ __ __ __ __ __ __ __
A,12,-3
*ACAD_ISO03W100,ISO dash space __   __   __   __   __   __
A,12,-18
*ACAD_ISO04W100,ISO long-dash dot ____ . ____ . ____ . ____ . _
A,24,-3,.5,-3
*ACAD_ISO05W100,ISO long-dash double-dot ____ .. ____ .. ____ .
A,24,-3,.5,-3,.5,-3
*ACAD_ISO06W100,ISO long-dash triple-dot ____ ... ____ ... ____
A,24,-3,.5,-3,.5,-3,.5,-3
*ACAD_ISO07W100,ISO dot . . . .
A,.5,-3
*ACAD_ISO08W100,ISO long-dash short-dash ____ __ ____ __ ____ _
A,24,-3,6,-3
*ACAD_ISO09W100,ISO long-dash double-short-dash ____ __ __ ____
A,24,-3,6,-3,6,-3
*ACAD_ISO10W100,ISO dash dot __ . __ . __ . __ . __ . __ . __ .
A,12,-3,.5,-3
*ACAD_ISO11W100,ISO double-dash dot __ __ . __ __ . __ __ . __
A,12,-3,12,-3,.5,-3
*ACAD_ISO12W100,ISO dash double-dot __ . . __ . . __ . . __ . .
A,12,-3,.5,-3,.5,-3
*ACAD_ISO13W100,ISO double-dash double-dot __ __ . . __ __ . . _
A,12,-3,12,-3,.5,-3,.5,-3
*ACAD_ISO14W100,ISO dash triple-dot __ . . . __ . . . __ . . . _
A,12,-3,.5,-3,.5,-3,.5,-3
*ACAD_ISO15W100,ISO double-dash triple-dot __ __ . . . __ __ . .
A,12,-3,12,-3,.5,-3,.5,-3,.5,-3

;; Complex linetypes
;;
;; Complex linetypes have been added to this file.
;; These linetypes were defined in LTYPESHP.LIN in
;; Release 13, and are incorporated in ACAD.LIN in
;; Release 14.
;;
;; These linetype definitions use LTYPESHP.SHX.
;;
*FENCELINE1,Fenceline circle ----0-----0----0-----0----0-----0--

```
A,.25,-.1,[CIRC1,ltypeshp.shx,x=-.1,s=.1],-.1,1
*FENCELINE2,Fenceline square ----[]-----[]----[]-----[]----[]---
A,.25,-.1,[BOX,ltypeshp.shx,x=-.1,s=.1],-.1,1
*TRACKS,Tracks -|-|-|-|-|-|-|-|-|-|-|-|-|-|-|-|-|-|-|-|-|-|-
A,.15,[TRACK1,ltypeshp.shx,s=.25],.15
*BATTING,Batting SSSSSSSSSSSSSSSSSSSSSSSSSSSSSSSSSSSSSSSSSSSSSS
A,.0001,-.1,[BAT,ltypeshp.shx,x=-.1,s=.1],-.2,[BAT,ltypeshp.shx,r=180,x=.1,s=.1],-.1
*HOT_WATER_SUPPLY,Hot water supply ---- HW ---- HW ---- HW ----
A,.5,-.2,["HW",STANDARD,S=.1,R=0.0,X=-0.1,Y=-.05],-.2
*GAS_LINE,Gas line ----GAS----GAS----GAS----GAS----GAS----GAS--
A,.5,-.2,["GAS",STANDARD,S=.1,R=0.0,X=-0.1,Y=-.05],-.25
*ZIGZAG,Zig zag /\/\/\/\/\/\/\/\/\/\/\/\/\/\/\/\/\/\/\/\
A,.0001,-.2,[ZIG,ltypeshp.shx,x=-.2,s=.2],-.4,[ZIG,ltypeshp.shx,r=180,x=.2,s=.2],-.2
```

Example 2

Create a new file, **NEWLINET.LIN**, and define a linetype, **VARDASH**, with the following specifications:

Length of first dash 1.0
Blank space 0.25
Length of second dash 0.75
Blank space 0.25
Length of third dash 0.5
Blank space 0.25
Dot
Blank space 0.25
Length of next dash 0.5
Blank space 0.25
Length of next dash 0.75

Use a text editor and insert the following lines that define the new linetype VARDASH. Save the file as NEWLINET.LIN.

```
*VARDASH,-------- ---- -- . -- ---- --------
A,1,-.25,.75,-.25,.5,-.25,0,-.25,.5,-.25,.75,-.25
```

The type of lines that this linetype will generate are shown in Figure 3-11.

Figure 3-11 Lines generated by linetype VARDASH

CURRENT LINETYPE SCALING (CELTSCALE)

Like LTSCALE, the CELTSCALE system variable controls the linetype scaling. The difference is that CELTSCALE determines the current linetype scaling. For example, if you set the CELTSCALE to 0.5, all lines drawn after setting the new value for CELTSCALE will have the linetype scaling factor of 0.5. The value is retained in the CELTSCLAE system variable. The first line (a) in Figure 3-12 is drawn with the CELTSCALE factor of 1 and the second line (b) is drawn with the CELTSCALE factor of 0.5. The length of the dashed is reduced by a factor of 0.5 when the CELTSCALE is 0.5.

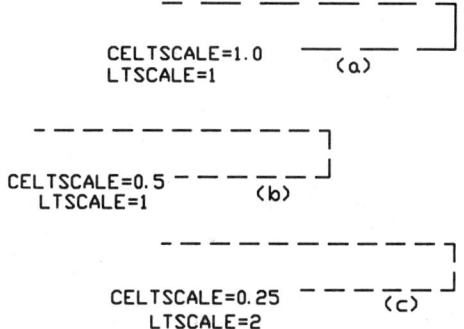

Figure 3-12 Using CELTSCALE to control current linetype scaling

The LTSCALE system variable controls the global scale factor. For example, if LTSCALE is set to 2, all lines in the drawing will be affected by a factor of 2. The net scale factor is equal to the product of CELTSCALE and LTSCALE. Figure 3-12(c) shows a line that is drawn with LTSCALE of 2 and CELTSCALE of 0.25. The net scale factor is = LTSCALE x CELTSCALE = 2 x 0.25 = 0.5.

COMPLEX LINETYPES

AutoCAD has provided a facility to create complex linetypes. The complex linetypes can be classified into two groups: string complex linetype and shape complex linetype. The difference between the two is that the string complex linetype has a text string inserted in the line, whereas the shape complex linetype has a shape inserted in the line. The facility of creating complex linetypes increases the functionality of lines. For example, if you want to draw a line around a building that indicates the fence line, you can do it by defining a complex linetype that will automatically give you the desired line with the text string (Fence). Similarly, you can define a complex linetype that will insert a shape (symbol) at predefined distances along the line.

Creating a String Complex Linetype

When writing the definition of a string complex linetype, the actual text and its attributes must be included in the linetype definition. The format of the string complex linetype is:

["String", Text Style, Text Height, Rotation, X-Offset, Y-Offset]

String. It is the actual text that you want to insert along the line. The text string must be enclosed in quotation marks (" ").

Text Style. This is the name of the text style file that you want to use for generating the text string. The text style must be predefined.

Text Height. This is the actual height of the text, if the text height defined in the text style is 0. Otherwise, it acts as a scale factor for the text height specified in the text style. In Figure 3-13, the height of the text is 0.1 units.

Rotation. This is the rotation of the text string with respect to the positive X axis. The angle is always measured with respect to positive X axis, no matter what AutoCAD's direction setting. The angle can be specified in radians (r), grads (g), or degrees (d). The default is degrees.

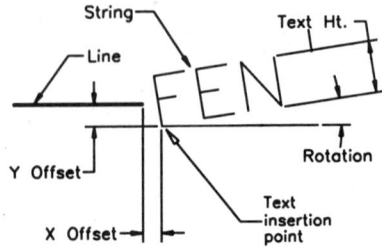

Figure 3-13 The attributes of a string complex linetype

X-Offset. This is the distance of the lower left corner of the text string from the endpoint of the line segment measured along the line. If the line is horizontal, then the X-Offset distance is measured along the X axis. In Figure 3-13, the X-Offset distance is 0.05.

Y-Offset. This is the distance of the lower left corner of the text string from the endpoint of the line segment measured perpendicular to the line. If the line is horizontal, then the Y-Offset distance is measured along the Y axis. In Figure 3-13, the Y-Offset distance is -0.05. The distance is negative because the start point of the text string is 0.05 units **below** the endpoint of the first line segment.

Example 3

In the following example, you will write the definition of a string complex linetype that consists of the text string "Fence" and line segments. The length of each line segment is 0.75. The height of the text string is 0.1 units, and the space between the end of the text string and the following line segment is 0.05 (Figure 3-14).

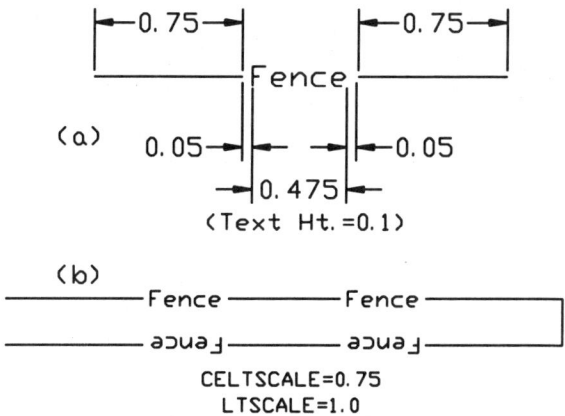

Figure 3-14 The attributes of the string complex linetype and line specifications for Example 3

Step 1
Before writing the definition of a new linetype, it is important to determine the line specification. One of the ways this can be done is to actually draw the lines and the text the way you want them to appear in the drawing. Once you have drawn the line and the text to your satisfaction, measure the distances needed to define the string complex linetype. In this example, the values are given as follows:

Text string=	Fence
Text style=	Standard
Text height=	0.1
Text rotation=	0
X-Offset=	0.05
Y-Offset=	-0.05
Length of the first line segment=	0.75
Distance between the line segments=	0.575

Step 2
Use a text editor to write the definition of the string complex linetype. You can add the definition to the AutoCAD **ACAD.LIN** file or create a separate file. The extension of the file must be **.LIN**. The following file is the listing of the **FENCE.LIN** file for Example 3. The name of the linetype is NEWFence.

```
*NEWFence1,New fence boundary line
A,0.75,["Fence",Standard,S=0.1,A=0,X=0.05,Y=-0.05],-0.575
or
A,0.75,-0.05["Fence",Standard,S=0.1,A=0,X=0,Y=-0.05],-0.525
```

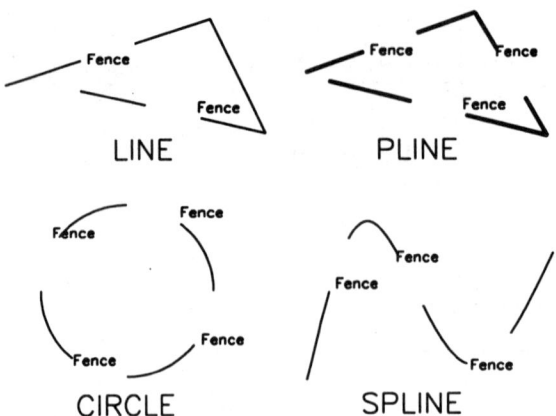

Figure 3-15 Using String Complex Linetype with angle A=0

Step 3
To test the linetype, load the linetype using the **LINETYPE** command with the Load option, and assign it to a layer. Draw a line or any object to check if the line is drawn to the given specifications. Notice that the text is always drawn along X axis. Also, when you draw a line at an angle, polyline, circle, or spline, the text string does not align with the object (Figure 3-15).

Step 4
In the NEWFence linetype definition, the specified angle is 0 degrees (Absolute angle A = 0). Therefore, when you use the NEWFence linetype to draw a line, circle, polyline, or spline, the text string (Fence) will be at zero degrees. If you want the text string (Fence) to align with the polyline (Figure 3-16), spline, or circle, specify the angle as relative angle (R = 0) in the NEWFence linetype definition. The following is the linetype definition for NEWFence linetype with relative angle R = 0:

*NEWFence2,New fence boundary line
A,0.75,["Fence",Standard,S=0.1,R=0,X=0.05,Y=-0.05],-0.575

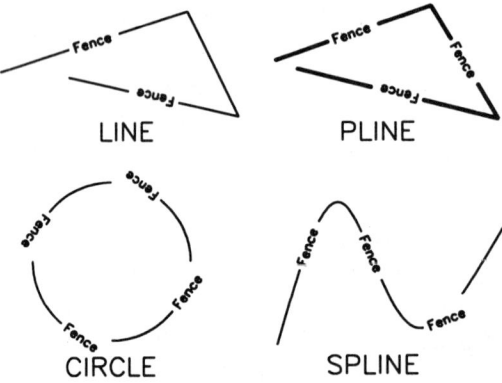

Figure 3-16 Using string complex linetype with angle R = 0

Step 5

In Figure 3-16, you might notice that the text string is not properly aligned with the circumference of the circle. This is because AutoCAD draws the text string in a direction that is tangent to the circle at the text insertion point. To resolve this problem, you must define the middle point of the text string as the insertion point. Also, the line specifications should be measured accordingly. Figure 3-17 gives the measurements of the NEWFence linetype with the middle point of the text as the insertion point.

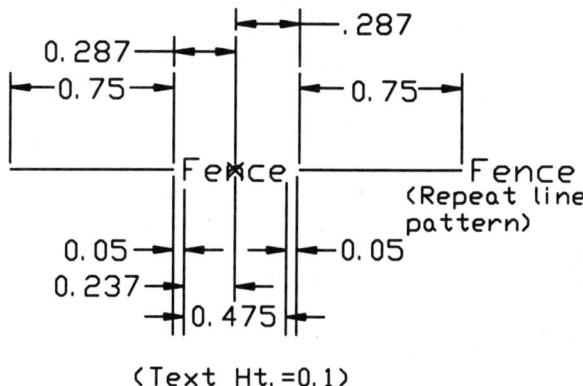

Figure 3-17 Specifications of string complex linetype with the middle point of the text string as the text insertion point

The following is the linetype definition for NEWFence linetype:

*NEWFence3,New fence boundary line
A,0.75,-0.287,["FENCE",Standard,S=0.1,X=-0.237,Y=-0.05],-0.287

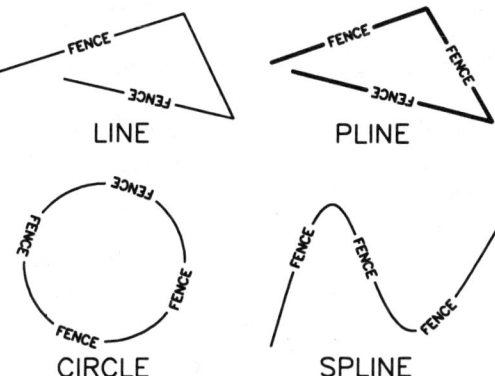

Figure 3-18 Using string complex linetype with the middle point of the text string as the text insertion point

3-22 Customizing AutoCAD

> **Note**
> *If no angle is defined in the line definition, it defaults to angle R = 0.*

Creating a Shape Complex Linetype

As with the string complex linetype, when you write the definition of a shape complex linetype, the name of the shape, the name of the shape file, and other shape attributes, like rotation, scale, X-Offset, and Y-Offset, must be included in the linetype definition. The format of the shape complex linetype is:

[Shape Name, Shape File, Scale, Rotation, X-Offset, Y-Offset]

Shape Complex Linetype Attributes Assigned to the Shape (Figure 3-19).

Shape Name. This is the name of the shape that you want to insert along the line. The shape name must exist; otherwise, no shape will be generated along the line.

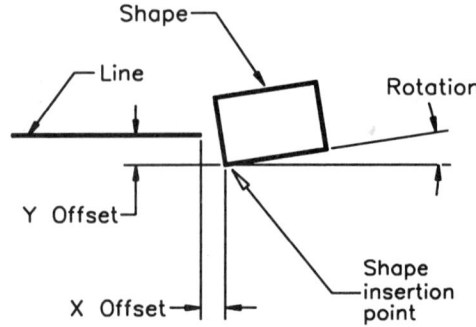

Figure 3-19 The attributes of a shape complex linetype

Shape File. This is the name of the **compiled** shape file (**.SHX**) that contains the definition of the shape being inserted in the line. The name of the subdirectory where the shape file is located must be in the ACAD search path. The shape files (**.SHP**) must be compiled before using the SHAPE command to load the shape.

Scale. This is the scale factor by which the defined shape size is to be scaled. If the scale is 1, the size of the shape will be same as defined in the shape definition (**.SHP** file).

Rotation. This is the rotation of the shape with respect to the positive X axis.

X-Offset. This is the distance of the shape insertion point from the endpoint of the line segment measured along the line. If the line is horizontal, then the X-Offset distance is measured along the X axis. In Figure 3-19, the X-Offset distance is 0.2.

Y-Offset. This is the distance of the shape insertion point from the endpoint of the line segment measured perpendicular to the line. If the line is horizontal, then the Y-Offset distance is measured along the Y axis. In Figure 3-19, the Y-Offset distance is 0.

Example 4

In the following example, you will write the definition of a shape complex linetype that consists of the shape (Manhole; the name of the shape is MH) and a line. The scale of the shape is 0.1, the length of each line segment is 0.75, and the space between line segments is 0.2.

Step 1

Before writing the definition of a new linetype, it is important to determine the line specifications. One of the ways this can be done is to actually draw the lines and the shape the way you want them to appear in the drawing (Figure 3-20). Once you have drawn the line and the shape to your satisfaction, measure the distances needed to define the shape complex linetype. In this example, the values are as follows:

Shape name   MH
Shape file name   MHOLE.SHX   (Name of the compiled shape file.)
Scale   0.1
Rotation   0
X-Offset   0.2
Y-Offset   0
Length of the first line segment = 0.75
Distance between the line segments = 0.2

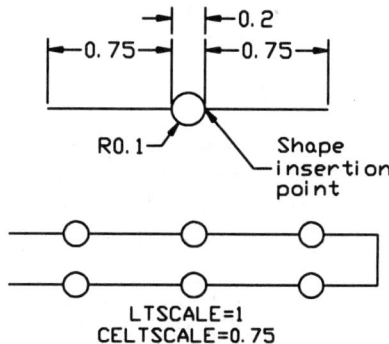

Figure 3-20 The attributes of the shape complex linetype and line specifications for Example 4

Step 2

Use a text editor to write the definition of the shape file. The extension of the file must be **.SHP**. The following file is the listing of the **MHOLE.SHP** file for Example 4. The name of the shape is MH. (For details, see Chapter 29.)

*215,9,MH
001,10,(1,007),
001,10,(1,071),0

Step 3

Use the COMPILE command to compile the shape file (.SHP file). When you use this command, AutoCAD will prompt you to enter the name of the shape file (Figure 3-21). For this example, the name is **MHOLE.SHP**. The following is the command sequence for compiling the shape file:

Command: **COMPILE**
Enter NAME of shape file: **MHOLE**

3-24 Customizing AutoCAD

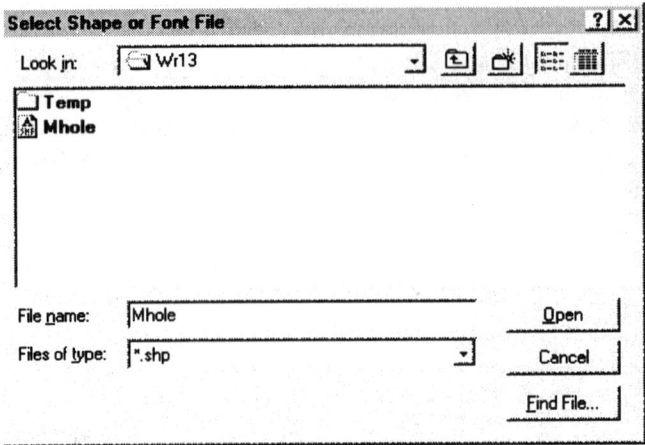

Figure 3-21 Select Shape or Font File dialog box

You can also compile the shape from the command line by setting FILEDIA=0 and then using the COMPILE command.

Step 4
Use a text editor to write the definition of the shape complex linetype. You can add the definition to the AutoCAD **ACAD.LIN** file or create a separate file. The extension of the file must be **.LIN**. The following file is the listing of the **MHOLE.LIN** file for Example 4. The name of the linetype is MHOLE.

   *MHOLE,Line with Manholes
   A,0.75,[MH,MHOLE.SHX,S=0.10,X=0.2,Y=0],-0.2

Step 5
To test the linetype, load the linetype using the **LINETYPE** command with the Load option and assign it to a layer. Draw a line or any object to check if the line is drawn to the given specifications. The shape is drawn upside down when you draw a line from right to left.

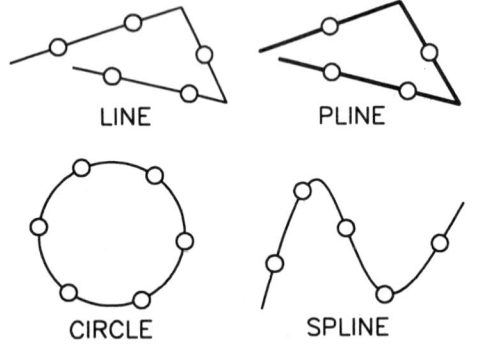

Figure 3-22 Using shape complex linetype

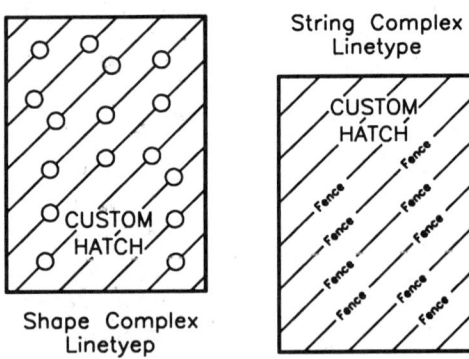

Figure 3-23 Using shape and string complex linetypes to create custom hatch

HATCH PATTERN DEFINITION

The AutoCAD software comes with a hatch pattern library file, **ACAD.PAT**, that contains 53 hatch patterns. These hatch patterns are sufficient for general drafting work. However, if you need a different hatch pattern, AutoCAD lets you create your own. There is no limit to the number of hatch patterns you can define.

The hatch patterns you define can be added to the hatch pattern library file, **ACAD.PAT**. You can also create a new hatch pattern library file, provided the file contains only one hatch pattern definition, and the name of the hatch is the same as the name of the file. The hatch pattern definition consists of the following two parts: **header line and hatch descriptors.**

Header Line

The **header line** consists of an asterisk (*) followed by the name of the hatch pattern. The hatch name is the name used in the hatch command to hatch an area. After the name, you can give the hatch description, which is separated from the hatch name by a comma (,). The general format of the header line is:

```
*HATCH Name [, Hatch Description]
   |       |            |
   |       |            └─ Description of hatch pattern
   |       └─ Name of hatch pattern
   └─ Asterisk
```

The description can be any text that describes the hatch pattern. It can also be omitted, in which case, a comma should **not** follow the hatch pattern name.

Example

```
*DASH45,Dashed lines at 45 degrees
   |           |
   |           └─ Hatch description
   └─ Hatch name
```

Hatch Descriptors

The **hatch descriptors** consist of one or more lines that contain the definition of the hatch lines. The general format of the hatch descriptor is:

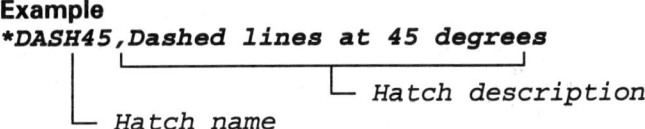

```
          ┌─ Angle of hatch lines
          |      ┌─ X coordinate of hatch line
          |      |      ┌─ Y coordinate of hatch line
        Angle, X-origin, Y-origin, D1, D2 [,Dash Length.....]
        Displacement of second line ┘   |              |
        (Delta-X)                       |              |
           Distance between hatch lines ┘              |
           (Delta-Y)                                   |
                      Length of dashes and spaces ─────┘
                      (Pattern line definition)
```

Example

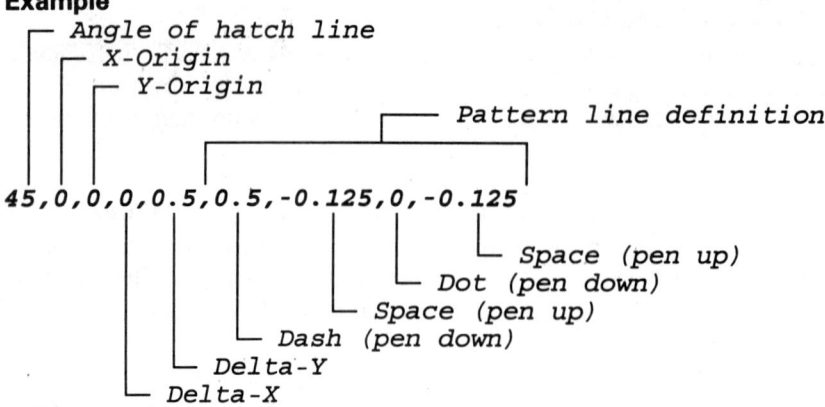

```
          ┌─ Angle of hatch line
          │  ┌─ X-Origin
          │  │  ┌─ Y-Origin
          │  │  │        ┌─ Pattern line definition
          │  │  │        │
         45,0,0,0.5,0.5,-0.125,0,-0.125
                    │   │   │    │   │
                    │   │   │    │   └─ Space (pen up)
                    │   │   │    └─ Dot (pen down)
                    │   │   └─ Space (pen up)
                    │   └─ Dash (pen down)
                    └─ Delta-Y
                  └─ Delta-X
```

Hatch Angle

X-origin and Y-origin. The hatch angle is the angle that the hatch lines make with the positive X axis. The angle is positive if measured counterclockwise (Figure 3-24), and negative if the angle is measured clockwise. When you draw a hatch pattern, the first hatch line starts from the point defined by X-origin and Y-origin. The remaining lines are generated by offsetting the first hatch line by a distance specified by delta-X and delta-Y. In Figure 3-25(a), the first hatch line starts from the point with the coordinates X = 0 and Y = 0. In Figure 3-25(b) the first line of hatch starts from a point with the coordinates X = 0 and Y = 0.25.

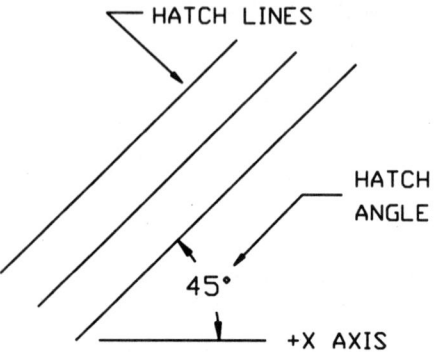

Figure 3-24 Hatch angle

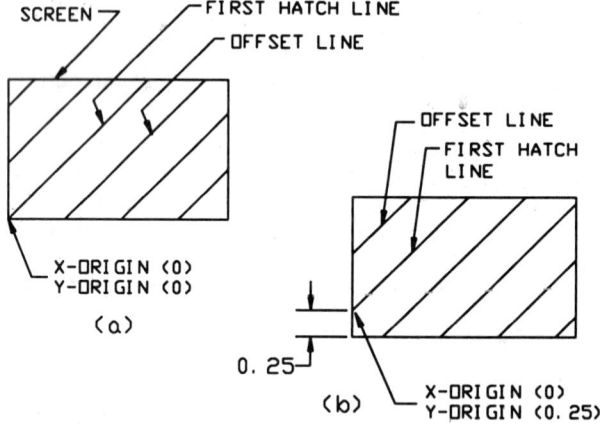

Figure 3-25 X-origin and Y-origin of hatch lines

Delta-X and Delta-Y. Delta-X is the displacement of the offset line in the direction in which the hatch lines are generated. For example, if the lines are drawn at a 0-degree angle and delta-X = 0.5, the offset line will be displaced by a distance delta-X (0.5) along the 0-angle direction. Similarly, if the hatch lines are drawn at a 45-degree angle, the offset line will be displaced by a distance delta-X (0.5) along a 45-degree direction (Figure 3-26).

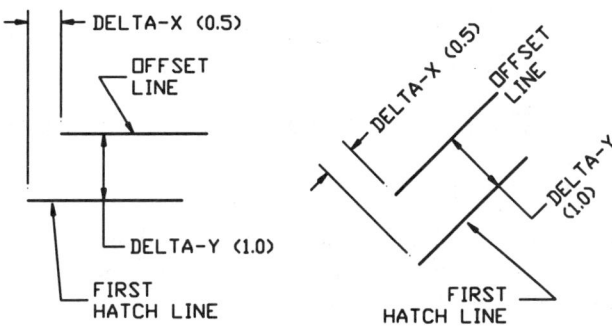

Figure 3-26 Delta-X and delta-Y of hatch lines

Delta-Y is the displacement of the offset lines measured perpendicular to the hatch lines. For example, if delta-Y = 1.0, the space between any two hatch lines will be 1.0 (Figure 3-26).

HOW HATCH WORKS

When you hatch an area, AutoCAD generates an infinite number of hatch lines of infinite length. The first hatch line always passes through the point specified by the X-origin and Y-origin. The remaining lines are generated by offsetting the first hatch line in both directions. The offset distance is determined by delta-X and delta-Y. All selected entities that form the boundary of the hatch area are then checked for intersection with these lines. Any hatch lines found within the defined hatch boundaries are turned on, and the hatch lines outside the hatch boundary are turned off, as shown in Figure 3-27. Since the hatch lines are generated by offsetting, the hatch lines in different areas of the drawing are automatically aligned relative to the drawing's snap origin. Figure 3-27(a) shows the hatch lines as computed by AutoCAD. These lines are **not** drawn on the screen; they are shown here for illustration only. Figure 3-27(b) shows the hatch lines generated in the circle that was defined as the hatch boundary.

3-28 Customizing AutoCAD

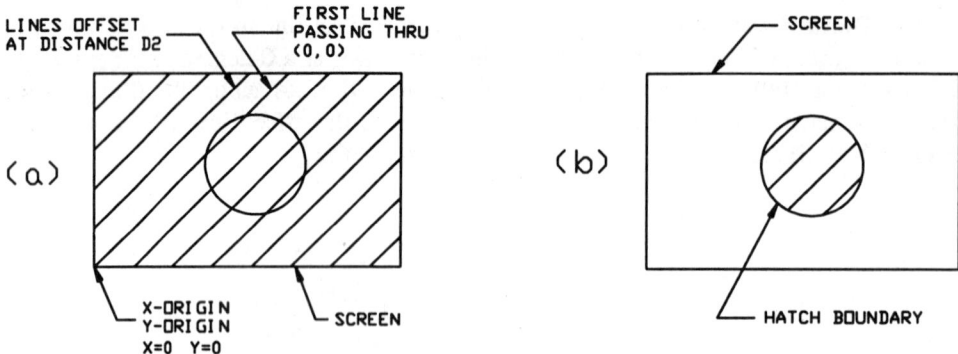

Figure 3-27 Hatch lines outside the hatch boundary are turned off

SIMPLE HATCH PATTERN

It is good practice to develop the hatch pattern specification before writing a hatch pattern definition. For simple hatch patterns it may not be that important, but for more complicated hatch patterns you should know the detailed specifications. Example 5 illustrates the procedure for developing a simple hatch pattern.

Example 5

Write a hatch pattern definition for the hatch pattern shown in Figure 3-28, with the following specifications:

Name of the hatch pattern =	HATCH1
X-Origin =	0
Y-Origin =	0
Distance between hatch lines =	0.5
Displacement of hatch lines =	0
Hatch line pattern =	Continuous

This hatch pattern definition can be added to the existing **ACAD.PAT** hatch file. You can use any text editor (like Notepad) to write the file. Load the **ACAD.PAT** file that is located in **ACADR14\SUPPORT** directory and insert the following two lines at the end of the file.

```
*HATCH1,Hatch Pattern for Example 5
45,0,0,0,.5
 | | | | |
 | | | | └─ Distance between hatch lines
 | | | └─── Displacement of second hatch line
 | | └───── Y-origin
 | └─────── X-origin
 └───────── Hatch angle
```

The first field of hatch descriptors contains the angle of the hatch lines. That angle is 45 degrees with respect to the positive X axis. The second and third fields describe the X and Y coordinates of the first hatch line origin. The first line of the hatch pattern will pass through this point. If the values of the X-origin and Y-origin were 0.5 and 1.0, respectively, then the first line would pass through the point with the X coordinate of 0.5 and the Y coordinate of 1.0, with respect to the drawing origin 0,0. The remaining lines are generated by offsetting the first line by a distance 0.5 on both sides of the line, as shown in Figure 3-28.

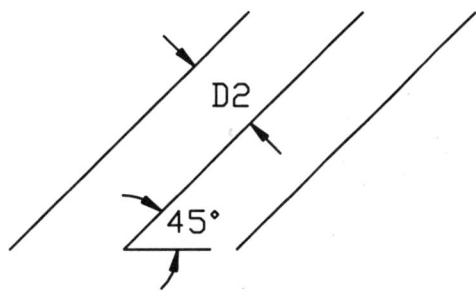

Figure 3-28 Hatch pattern angle and offset distance

EFFECT OF ANGLE AND SCALE FACTOR ON HATCH

When you hatch an area, you can alter the angle and displacement of hatch lines you have specified in the hatch pattern definition to get a desired hatch spacing. You can do this by entering an appropriate value for angle and scale factor in the AutoCAD HATCH command.

    Command: **HATCH**
    Pattern(? or name/U,style)<default>: **Hatch1**
    Scale for pattern<default>: **1**
    Angle of pattern<default>: **0**

To understand how the angle and the displacement can be changed, hatch an area with the hatch pattern HATCH1 of Example 5. You will notice that the hatch lines have been generated according to the definition of hatch pattern HATCH1. Notice the effect of hatch angle and scale factor on the hatch. Figure 3-29(a) shows a hatch that is generated by the AutoCAD HATCH command with a 0-degree angle and a scale factor of 1.0. If the angle is 0, the hatch will be generated with the same angle as defined in the hatch pattern definition (45 degrees in Example 5). Similarly, if the scale factor is 1.0, the distance between the hatch lines will be same as defined in the hatch pattern definition (0.5 in Example 5). Figure 3-29(b) shows a hatch that is generated when the hatch scale factor is 0.5. If you measure the distance between the successive hatch lines, it will be 0.5 x 0.5 = 0.25. Figures 3-29(c) and (d) show the hatch when the angle is 45 degrees and the scale factors are 1.0 and 0.5, respectively. You can enter any value in response to the HATCH command prompts to generate hatch lines at any angle and with any line spacing.

3-30 Customizing AutoCAD

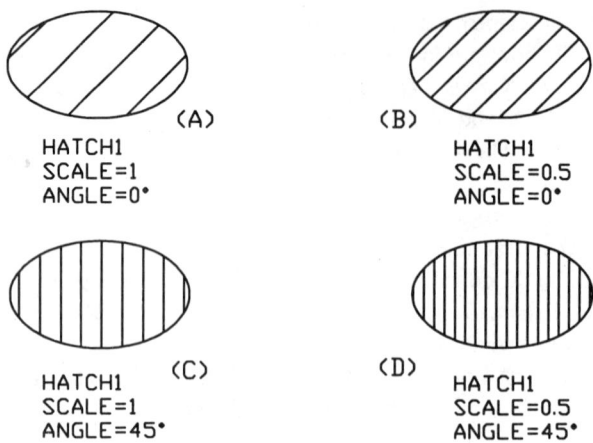

Figure 3-29 Effect of angle and scale factor on hatch

HATCH PATTERN WITH DASHES AND DOTS

The lines you can use in a hatch pattern definition are not restricted to continuous lines. You can define any line pattern to generate a hatch pattern. The lines can be a combination of dashes, dots, and spaces in any configuration. However, the maximum number of dashes you can specify in the line pattern definition of a hatch pattern is six. Example 6 uses a dash-dot line to create a hatch pattern.

Example 6

Write a hatch pattern definition for the hatch pattern shown in Figure 3-30, with the following specifications:

Name of the hatch pattern	HATCH2
Hatch angle =	0
X-origin =	0
Y-origin =	0
Displacement of lines (D1) =	0.25
Distance between lines (D2) =	0.25
Length of each dash =	0.5
Space between dashes and dots =	0.125
Space between dots =	0.125

You can use the AutoCAD EDIT command to edit the **ACAD.PAT** file. The general format of the header line and the hatch descriptors is:

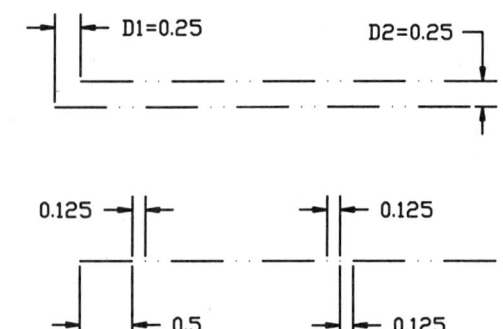

Figure 3-30 Hatch lines made of dashes and dots

*HATCH NAME, Hatch Description
Angle, X-Origin, Y-Origin, D1, D2 [,Dash Length.....]

Substitute the values from Example 6 in the corresponding fields of header line and field descriptor:

```
*HATCH2,Hatch with dashes and dots
0,0,0,0.25,0.25,0.5,-0.125,0,-0.125,0,-0.125
```

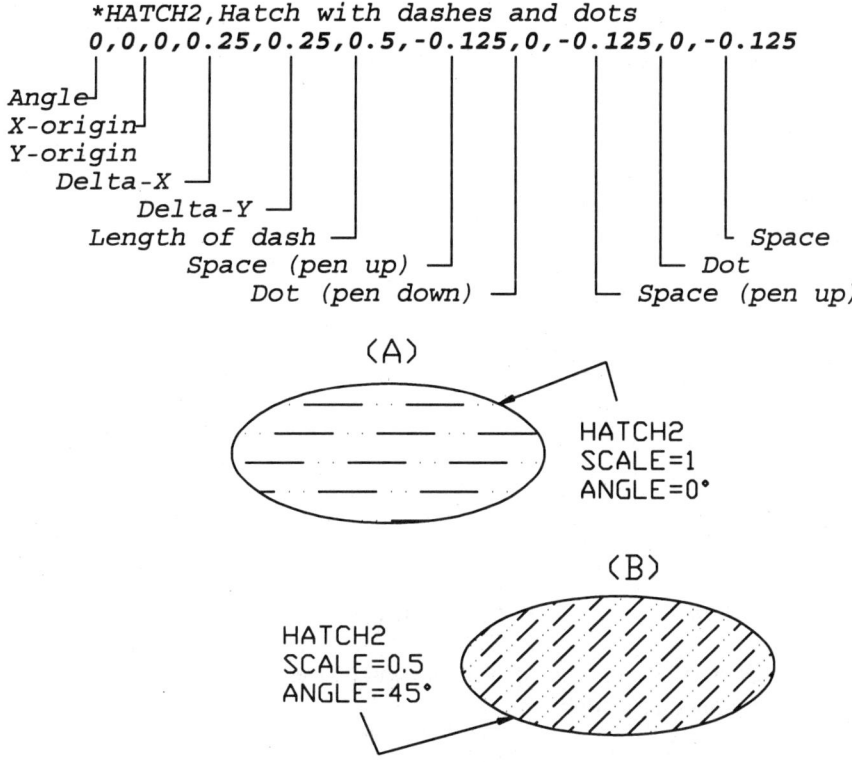

Angle
X-origin
Y-origin
Delta-X
Delta-Y
Length of dash
Space (pen up)
Dot (pen down)
Space
Dot
Space (pen up)

Figure 3-31 Hatch pattern at different angles and scales

The hatch pattern this hatch definition will generate is shown in Figure 3-31. Figure 3-31(a) shows the hatch with 0-degree angle and a scale factor of 1.0. Figure 3-31(b) shows the hatch with a 45-degree angle and a scale factor of 0.5.

HATCH WITH MULTIPLE DESCRIPTORS

Some hatch patterns require multiple lines to generate a shape. For example, if you want to create a hatch pattern of a brick wall, you need a hatch pattern that has four hatch descriptors to generate a rectangular shape. You can have any number of hatch descriptor lines in a hatch pattern definition. It is up to the user to combine them in any conceivable order. However, there are some shapes you cannot generate. A shape that has a non-linear element, like an arc, cannot be generated by hatch pattern definition. However, you can simulate an arc by defining short line segments because you can use only straight lines to generate a hatch pattern. Example 7 uses three lines to define a triangular hatch pattern.

3-32 Customizing AutoCAD

Example 7

Write a hatch pattern definition for the hatch pattern shown in Figure 3-32, with the following specifications:

Name of the hatch pattern =	HATCH3
Vertical height of the triangle =	0.5
Horizontal length of the triangle =	0.5
Vertical distance between the triangles =	0.5
Horizontal distance between the triangles =	0.5

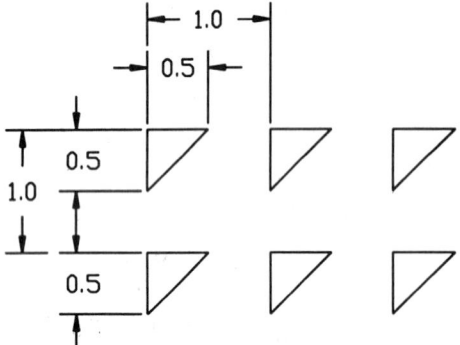

Figure 3-32 Triangle hatch pattern

Figure 3-33 Vertical line

Each triangle in this hatch pattern consists of the following three elements: a vertical line, a horizontal line, and a line inclined at 45 degrees.

Vertical Line. For the vertical line, the specifications are:

Hatch angle =	90 degrees
X-origin =	0
Y-origin =	0
Delta-X (D1) =	0
Delta-Y (D2) =	1.0
Dash length =	0.5
Space =	0.5

Substitute the values from the vertical line specification in various fields of the hatch descriptor to get the following line:

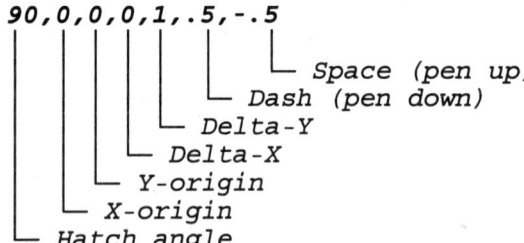

Horizontal Line. For the horizontal line (Figure 3-34), the specifications are:

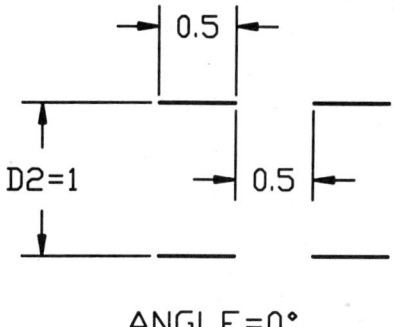

Hatch angle =	0 degrees
X-origin =	0
Y-origin =	0.5
Delta-X (D1) =	0
Delta-Y (D2) =	1.0
Dash length =	0.5
Space =	0.5

Figure 3-34 Horizontal line

The only difference between the vertical line and the horizontal line is the angle. For the horizontal line, the angle is 0 degrees, whereas for the vertical line, the angle is 90 degrees. Substitute the values from the vertical line specification to obtain the following line:

Line Inclined at 45 Degrees. This line is at an angle; therefore, you need to calculate the distances delta-X (D1) and delta-Y (D2), the length of the dashed line, and the length of space. Figure 3-35 shows the calculations to find these values.

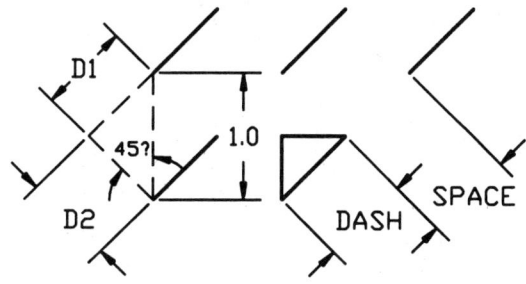

$$D1 = 1.0 \times COS\ 45 \qquad D2 = 1.0 \times SIN\ 45$$
$$D1 = 0.7071 \qquad D2 = 0.7071$$

$$DASH = SQRT(0.5**2 + 0.5**2)$$
$$= .7071$$
$$SPACE = DASH = .7071$$

Figure 3-35 Line inclined at 45 degrees

3-34 Customizing AutoCAD

Hatch angle =	45 degrees
X-Origin =	0
Y-Origin =	0
Delta-X (D1) =	0.7071
Delta-Y (D2) =	0.7071
Dash length =	0.7071
Space =	0.7071

After substituting the values in the general format of the hatch descriptor, you will obtain the following line:

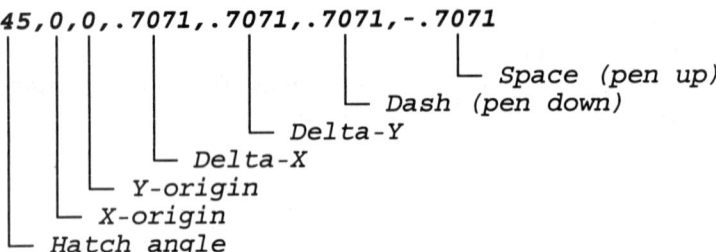

Now you can combine these three lines and insert them at the end of the **ACAD.PAT** file. You can also use the AutoCAD EDIT command to edit the file and insert the lines.

Figure 3-36 shows the hatch pattern that will be generated by this hatch pattern (HATCH3). In Figure 3-36(a) the hatch pattern is at a 0-degree angle and the scale factor is 0.5. In Figure 3-36(b) the hatch pattern is at a -45-degree angle and the scale factor is 0.5.

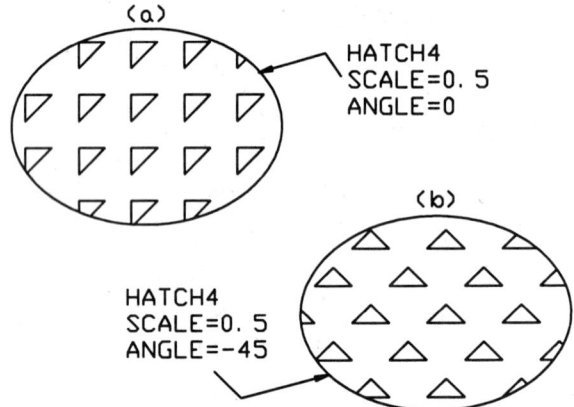

Figure 3-36 Hatch generated by HATCH3 pattern

The following file is a partial listing of the **ACAD.PAT** file, after adding the hatch pattern definitions from Examples 5, 6, and 7.

*angle,Angle steel
0, 0,0, 0,.275, .2,-.075
90, 0,0, 0,.275, .2,-.075
*ansi31,ANSI Iron, Brick, Stone masonry
45, 0,0, 0,.125
*ansi32,ANSI Steel
45, 0,0, 0,.375
45, .176776695,0, 0,.375
*ansi33,ANSI Bronze, Brass, Copper
45, 0,0, 0,.25
45, .176776695,0, 0,.25, .125,-.0625
*ansi34,ANSI Plastic, Rubber
45, 0,0, 0,.75
45, .176776695,0, 0,.75
45, .353553391,0, 0,.75
45, .530330086,0, 0,.75
*ansi35,ANSI Fire brick, Refractory material
45, 0,0, 0,.25
45, .176776695,0, 0,.25, .3125,-.0625,0,-.0625
*ansi36,ANSI Marble, Slate, Glass
45, 0,0, .21875,.125, .3125,-.0625,0,-.0625
*ansi37,ANSI Lead, Zinc, Magnesium, Sound/Heat/Elec Insulation
45, 0,0, 0,.125
135, 0,0, 0,.125

*steel,Steel material
45, 0,0, 0,.125
45, 0,.0625, 0,.125
*swamp,Swampy area
0, 0,0, .5,.866025403, .125,-.875
90, .0625,0, .866025403,.5, .0625,-1.669550806
90, .078125,0, .866025403,.5, .05,-1.682050806
90, .046875,0, .866025403,.5, .05,-1.682050806
60, .09375,0, .5,.866025403, .04,-.96
120, .03125,0, .5,.866025403, .04,-.96
*trans,Heat transfer material
0, 0,0, 0,.25
0, 0,.125, 0,.25, .125,-.125
*triang,Equilateral triangles
60, 0,0, .1875,.324759526, .1875,-.1875
120, 0,0, .1875,.324759526, .1875,-.1875
0, -.09375,.162379763, .1875,.324759526, .1875,-.1875
*zigzag,Staircase effect
0, 0,0, .125,.125, .125,-.125
90, .125,0, .125,.125, .125,-.125

*HATCH1,Hatch at 45 Degree Angle
45,0,0,0,.5
*HATCH2,Hatch with Dashes & Dots:
0,0,0,.25,.25,0.5,-.125,0,-.125,0,-.125
*HATCH3,Triangle Hatch:
90,0,0,0,1,.5,-.5
0,0,0.5,0,1,.5,-.5
45,0,0,.7071,.7071,.7071,-.7071

SAVING HATCH PATTERNS IN A SEPARATE FILE

When you load a certain hatch pattern, AutoCAD looks for that definition in the ACAD.PAT file; therefore, the hatch pattern definitions must be in that file. However, you can add the new pattern definition to a different file and then copy that file to ACAD.PAT. Be sure to make a copy of the original ACAD.PAT file so that you can copy that file back when needed. Assume the name of the file that contains your custom hatch pattern definitions is CUSTOMH.PAT.

1. Copy ACAD.PAT file to ACADORG.PAT
2. Copy CUSTOMH.PAT to ACAD.PAT

If you want to use the original hatch pattern file, copy the ACADORG.PAT file to ACAD.PAT.

CUSTOM HATCH PATTERN FILE

As mentioned earlier, you can add the new hatch pattern definitions to the **ACAD.PAT** file. There is no limit to the number of hatch pattern definitions you can add to this file. However, if you have only one hatch pattern definition, you can define a separate file. It has the following three requirements:

1. The name of the file has to be the same as the hatch pattern name.
2. The file can contain only one hatch pattern definition.
3. The hatch pattern name--and, therefore, the hatch file name--should be unique.
4. If you want to save the hatch pattern on the A drive, then the drive letter (A:) should precede the hatch name. For example, if the hatch name is HATCH3, the header line will be ***A:HATCH3, Triangle Hatch:** and the file name **HATCH3.PAT**.

*HATCH3,Triangle Hatch:
90,0,0,0,1,.5,-.5
0,0,0.5,0,1,.5,-.5
45,0,0,.7071,.7071,.7071,-.7071

Note
The hatch lines can be edited after exploding the hatch with the AutoCAD EXPLODE command. After exploding, each hatch line becomes a separate object.

It is good practice not to explode a hatch because it increases the size of the drawing database. For example, if a hatch consists of 100 lines, save it as a single object. However, after you explode the hatch, every line becomes a separate object and you have 99 additional objects in the drawing.

Keep the hatch lines in a separate layer to facilitate editing of the hatch lines.

Assign a unique color to hatch lines so that you can control the width of the hatch lines at the time of plotting.

You can also add hatch pattern slides to AutoCAD slide library.

ADDING HATCH PATTERN SLIDES TO AUTOCAD SLIDE LIBRARY

The hatch patterns that you create can be added to AutoCAD's slide library so that the hatch patterns are displayed in the hatch icons. If you enter the **BHATCH** command, AutoCAD displays the **Boundary Hatch** dialogue box on the screen. If you select the **Pattern...** box, AutoCAD displays the hatch icons with the hatch names. Once you have created the new hatch patterns, you can make slides of these hatch patterns and add them to AutoCAD's slide library.

The first step is to make slides of the hatch patterns that you have created (refer to Chapters 2 and 8). **Make sure that the names of the slide files are same as the names of the hatch pattern.** For example, if the name of the hatch pattern is HATCH1, the name of the slide file must be HATCH1.SLD. When the slide name in the ACAD.SLB file matches with the hatch pattern name in the ACAD.PAT file, the slide is displayed in the **Hatch pattern palette** dialog box.

Before you make any changes in the standard slide library file (ACAD.SLB) supplied with AutoCAD, make a copy of this file. You can find this file in the support subdirectory. The original slide files that were used to create the slide file list are not supplied with the software. However, there are some programs available from third-party software developers that can be used to extract slides from the ACAD.SLB file.

Note
One such shareware program (Slide Manager, SLIDEMGR) that will allow you to extract the slides from AutoCAD's ACAD.SLB file is available from JMIcro, 335 Washington Street, Suite 178, Woburn, MA 01801.

After you make the new slides and extract the slides from the ACAD.SLB file, move the slide files to separate directory on the hard disk of your computer. For example, you can make a subdirectory, SLIDES, which will contain all the slide files that you are using in your applications. Next, use any text editor to write a slide file list (ACADSLD.LST). This file lists all the slide filenames that are used in the ACAD.SLB file. Add the slide filenames of your hatch patterns to this file.

The slide file list can be also created by using the following command, if you have DOS version 5.0 or above:

C:\ACADR14\SLIDES > **DIR *.SLD/B > ACADSLD.SLD**

3-38 Customizing AutoCAD

In this example, assume that the name of the slide file list is **ACADSLD.SLD** (the file extension .SLD is optional) and all slide files are in the SLIDES subdirectory. (To use this command to create a slide file list, all slide files must be in the same directory.)

AutoCAD provides a SLIDELIB utility that constructs a library of the slide files. When you use the SLIDELIB utility, it reads the slide filenames from the slide file list (ACADSLD.SLD) and the file is then written to the slide library file (ACAD.SLB). Use the following command to create the slide library, ACAD.SLB:

> C:\ACADR14\SLIDES> **SLIDELIB ACAD<ACADSLD.SLD**
> or
> C:\ACADR14\SLIDES> **SLIDELIB ACAD<ACADSLD.LST**

The new hatch patterns that you create have been added to the slide library ACAD.SLB and these hatch patterns will be displayed in the hatch icons.

Note
The SLIDELIB utility that constructs the slide library file is in the SUPPORT subdirectory. If this subdirectory is not in the path, use the following command to create the slide library file, ACAD.SLB:
C:\ACADR14\SLIDES>C:\ACAD\SUPPORT\SLIDELIB ACAD<ACADSLD.SLD

REVIEW QUESTIONS

CREATING LINETYPES

1. The AutoCAD _____ command can be used to create a new linetype.

2. The AutoCAD _____ command can be used to load a linetype.

3. The AutoCAD _____ command can be used to change the linetype scale factor.

4. In AutoCAD, the linetypes are saved in the _____ file.

5. The linetype description should not be more than _____ characters long.

6. A positive number denotes a pen _____ segment.

7. The segment length _____ generates a dot.

8. AutoCAD supports only _____ alignment field specification.

9. A line pattern definition always starts with _____.

10. A header line definition always starts with _____.

CREATING HATCH PATTERNS

11. The **ACAD.PAT** file contains _____ number of hatch pattern definitions.

12. The header line consists of an asterisk, the pattern name, and _____.

13. The first hatch line passes through a point whose coordinates are specified by _____ and _____.

14. The perpendicular distance between the hatch lines in a hatch pattern definition is specified by _____.

15. The displacement of the second hatch line in a hatch pattern definition is specified by _____.

16. The maximum number of dash lengths that can be specified in the line pattern definition of a hatch pattern is _____.

17. The hatch lines in different areas of the drawing will automatically _____ since the hatch lines are generated by offsetting.

18. The hatch angle as defined in the hatch pattern definition can be changed further when you use the AutoCAD _____ command.

19. When you load a hatch pattern, AutoCAD looks for that hatch pattern in the _____ file.

20. The hatch lines can be edited after _____ the hatch by using the AutoCAD _____ command.

EXERCISES

Creating Linetypes

Exercise 1

Using the AutoCAD LINETYPE command, create a new linetype "DASH3DASH" with the following specifications:

Length of the first dash 0.75
Blank space 0.125
Dash length 0.25
Blank space 0.125
Dash length 0.25
Blank space 0.125

Dash length 0.25
Blank space 0.125

Exercise 2

Use a text editor to create a new file, **NEWLT2.LIN**, and a new linetype, **DASH2DASH**, with the following specifications:

Length of the first dash 0.5
Blank space 0.1
Dash length 0.2
Blank space 0.1
Dash length 0.2
Blank space 0.1

Exercise 3

Using AutoCAD's LINETYPE command, create a linetype DASH3DOT with the following specifications:

Length of the first dash 0.75
Blank space 0.25
Dot
Blank space 0.25
Dot
Blank space 0.25
Dot
Blank space 0.25

Exercise 4

Using AutoCAD's EDIT command, create a new file, NEWLINET.LIN, and define a linetype VARDASHX with the following specifications:

Length of first dash 1.5
Blank space 0.25
Length of second dash 0.75
Blank space 0.25
Dot
Blank space 0.25
Length of third dash 0.75

Exercise 5

a. Write the definition of a string complex linetype (Hot water line) as shown in Figure 3-37(a).

b. Write the definition of a string complex linetype (Gas line) as shown in Figure 3-37(b).

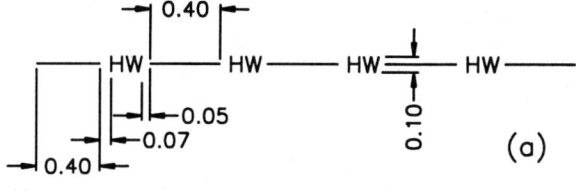

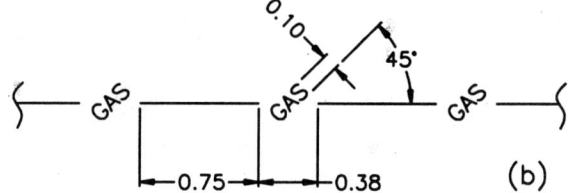

Figure 3-37 Specifications for string complex linetype

Exercise 6

Write the shape file for the shape shown in Figure 3-38(a). Compile the shape and use it in defining the shape complex linetype so that you can draw a fence line as shown in Figure 3-38(b).

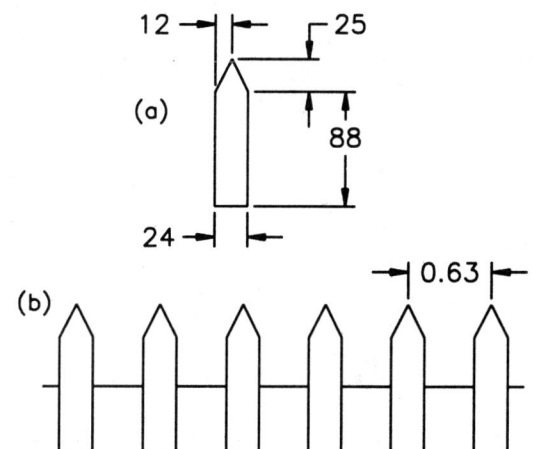

Figure 3-38 Specifications for shape complex linetype

Creating Hatch Patterns

Exercise 7

Determine the hatch specifications and write a hatch pattern definition for the hatch pattern in Figure 3-39.

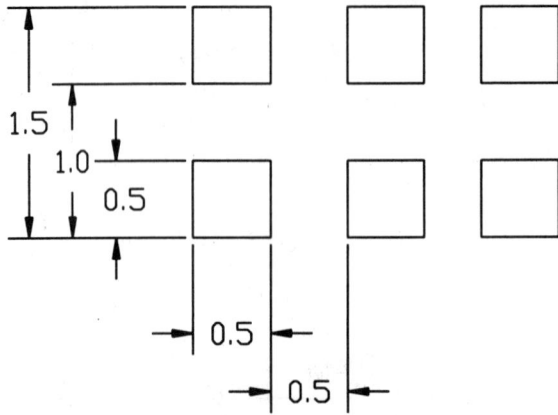

Figure 3-39 Square hatch pattern

Exercise 8

Determine the hatch pattern specifications and write a hatch pattern definition for the hatch pattern in Figure 3-40.

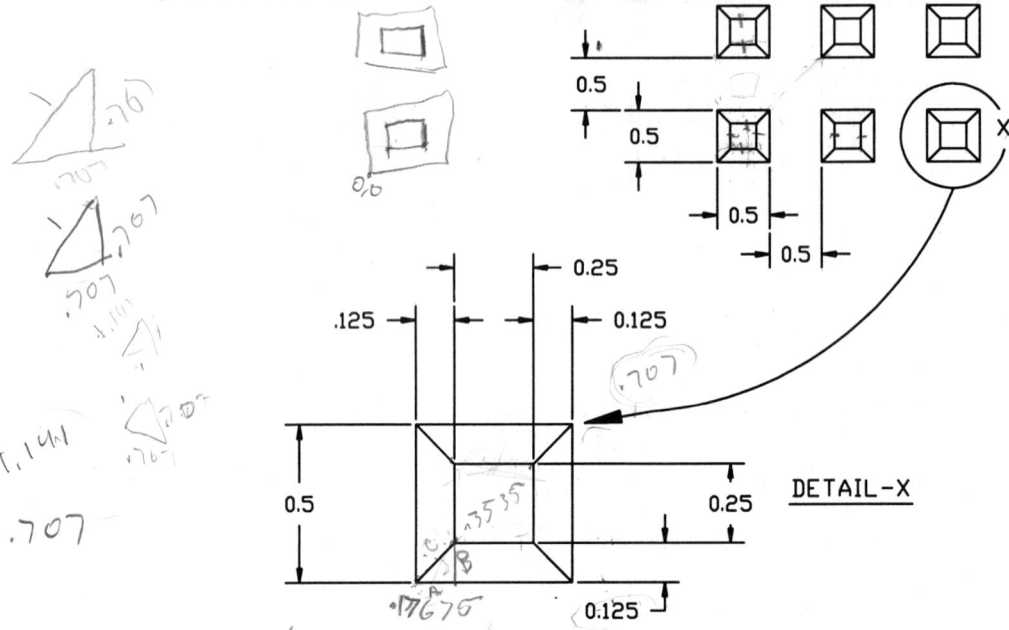

Figure 3-40 Hatch pattern for Exercise 8

Exercise 9

Determine the hatch specifications and write a hatch pattern definition for the hatch pattern as shown in Figure 3-41. Use this hatch to hatch a circle or rectangle.

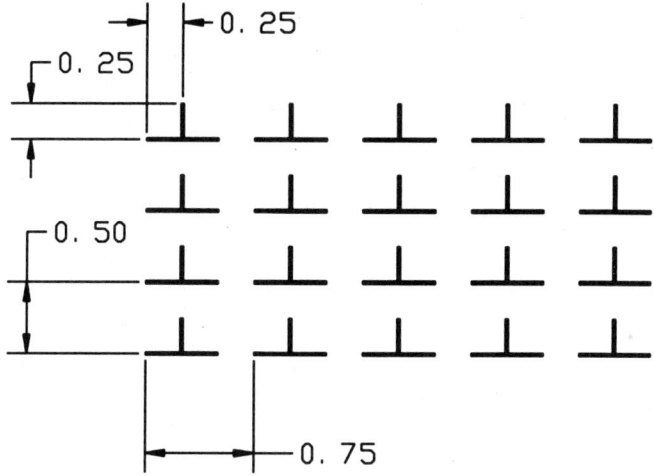

Figure 3-41 Hatch pattern for Exercise 9

Exercise 10

Determine the hatch specifications and write a hatch pattern definition for the hatch pattern as shown in Figure 3-42. Use this hatch to hatch a circle or rectangle.

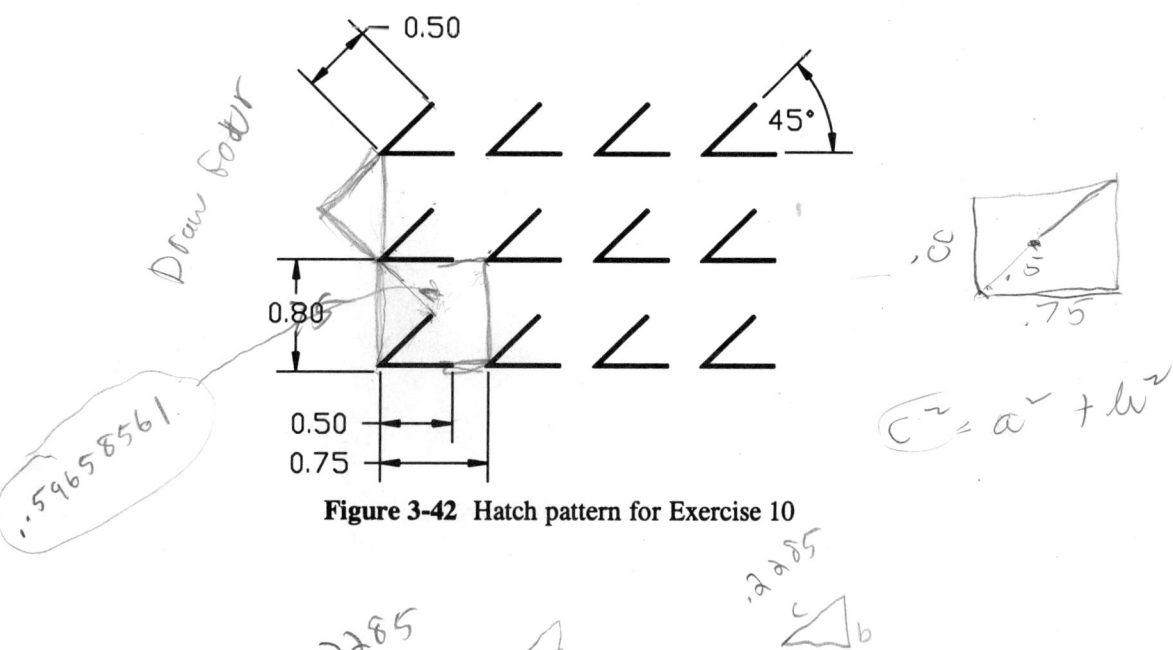

Figure 3-42 Hatch pattern for Exercise 10

Chapter 4

Customizing the ACAD.PGP File

Learning objectives

After completing this chapter, you will be able to:
- *Customize the ACAD.PGP file.*
- *Edit different sections of the ACAD.PGP file.*
- *Abbreviate commands by defining command aliases.*
- *Use the REINIT command to reinitialize the PGP file.*

WHAT IS THE ACAD.PGP FILE?

AutoCAD software comes with the program parameters file **ACAD.PGP**, which defines aliases for the operating system commands and some of the AutoCAD commands. When you install AutoCAD, this file is automatically copied on the **ACADR14\SUPPORT** subdirectory of the hard drive. The **ACAD.PGP** file lets you access the operating system commands from the drawing editor. For example, if you want to delete a file, all you need to do is enter DEL at the Command: prompt (**Command: DEL**), and then AutoCAD will prompt you to enter the name of the file you want to delete.

The file also contains command aliases of some frequently used AutoCAD commands. For example, the command alias for the LINE command is L. If you enter L at the Command: prompt (**Command: L**), AutoCAD will treat it as the LINE command. The **ACAD.PGP** file also contains comment lines that give you some information about different sections of the file.

The following file is a partial listing of the standard **ACAD.PGP** file. Some of the lines have been deleted to make the file shorter.

```
;AutoCAD Program Parameters File For AutoCAD Release 14
;External Command and Command Alias Definitions
;Copyright (C) 1997 by Autodesk, Inc.
;Each time you open a new or existing drawing, AutoCAD searches
;the support path and reads the first acad.pgp file that it finds.
```

```
;While AutoCAD is running, you can invoke other programs or utilities,
;such Windows system commands, utilities, and applications.
;You define external commands by specifying a command name to be used
;from the AutoCAD command prompt and an executable command string
;that is passed to the operating system.
;You can abbreviate frequently used AutoCAD commands by defining
;aliases for them in the command alias section of acad.pgp.
;You can create a command alias for any AutoCAD command,
;device driver command, or external command.
;Recommendation: back up this file before editing it.
;There is a bonus application for editing command aliases as well as
;a sample acad.pgp file with many more command aliases.
;See the bonus\cadtools folder for more details.

;External command format:
; <Command name>,[<DOS request>],<Bit flag>,[*]<Prompt>,
;The bits of the bit flag have the following meanings:
;First bit (1): if set, don't wait for the application to finish
;Second bit (2): if set, run the application minimized
;Third bit (4): if set, run the application "hidden"
;Bits 2 and 4 are mutually exclusive; if both are specified only the 2 bit is used.
;The most useful values are likely to be 0 (start the application and wait
;for it to finish), 1 (start the application and don't wait), 3 (minimize and don't
;wait), and 5 (hide and don't wait). Values of 2 and 4 should normally be avoided,
;as they make AutoCAD unavailable until the application has completed.
;Examples of external commands for command windows

; External Command format:
; <Command name>,[<DOS request>],<Memory reserve>,[*]<Prompt>
; Example of External Commands for DOS
; For Windows NT Add START before all items to prevent locking AutoCAD
CATALOG, DIR /W          0,File specification:,
DEL,     DEL,            0,File to delete: ,
DIR,     DIR,            0,File specification: ,         ⎯ External
EDIT,    START EDIT,     0,File to edit: ,                 commands
SH,      ,               0,*OS Command: ,                  Section
SHELL,   ,               0,*OS Command: ,
TYPE,    TYPE,           0,File to list: ,
o
 o
  o
; Examples of external commands for Windows
; See also the STARTAPP AutoLISP function for an alternative method.

EXPLORER, START EXPLORER, 1,,
NOTEPAD,  START NOTEPAD,  1,*File to edit: ,
PBRUSH,   START PBRUSH,   1,,
;
```

```
;The following are guidelines for creating new command aliases.
;1. Try the first character of the command, then try the first two,
;then the first three.
;2. Ignore "DD" at the beginning of a command.
;3. Abbreviate the following prefixes:
;    Examples: 3 for 3D, A for ASE, D for Dim, I for Image, R for render.
;4. Once an alias is defined, add suffixes for related aliases:
;    Examples: R for Redraw, RA for Redrawall, L for Line, LT for Linetype.
;5. An alias should reduce a command by at least two characters.
;6. Commands with a control key equivalent, status bar button, or function key
;    do not require a command alias.
;  Examples: Use Control-N, -O, -P, and -S for New, Open, Print, and Save
;7. Use a hyphen to differentiate between command line and dialog box commands.
;8. Exceptions to the rules include AA for Area, T for Mtext, X for Explode.
;Sample aliases for AutoCAD commands
;These examples include most frequently used commands.
;
3A,      *3DARRAY
3F,      *3DFACE
3P,      *3DPOLY
A,       *ARC
AA,      *AREA                         ───────── Command alias section
AL,      *ALIGN
AP,      *APPLOAD
AR,      *ARRAY
AAD,     *ASEADMIN

C,       *CIRCLE
CH,      *DDCHPROP
-CH,     *CHANGE
CHA,     *CHAMFER
COL,     *DDCOLOR
CO,      *COPY
D,       *DDIM
DAL,     *DIMALIGNED
DAN,     *DIMANGULAR
DBA,     *DIMBASELINE
DCE,     *DIMCENTER
DCO,     *DIMCONTINUE
DDI,     *DIMDIAMETER
DED,     *DIMEDIT
DI,      *DIST
DIV,     *DIVIDE
DLI,     *DIMLINEAR
```

DO,	*DONUT
DOR,	*DIMORDINATE
DOV,	*DIMOVERRIDE
DR,	*DRAWORDER
DRA,	*DIMRADIUS
DST,	*DIMSTYLE
DT,	*DTEXT
DV,	*DVIEW
E,	*ERASE
ED,	*DDEDIT
EL,	*ELLIPSE
EX,	*EXTEND
EXIT,	*QUIT
EXP,	*EXPORT
EXT,	*EXTRUDE
F,	*FILLET
LI,	*LIST
LS,	*LIST
LT,	*LINETYPE
-LT,	*-LINETYPE
LTS,	*LTSCALE
M,	*MOVE
MA,	*MATCHPROP
ME,	*MEASURE
MI,	*MIRROR
ML,	*MLINE
MO,	*DDMODIFY
MS,	*MSPACE
MT,	*MTEXT
MV,	*MVIEW
O,	*OFFSET
OS,	*DDOSNAP
-OS,	*-OSNAP
P,	*PAN
-P,	*-PAN
PA,	*PASTESPEC
PE,	*PEDIT
PL,	*PLINE
PO,	*POINT
POL,	*POLYGON
PR,	*PREFERENCES
PRE,	*PREVIEW
PRINT,	*PLOT

```
SC,       *SCALE
SCR,      *SCRIPT
SE,       *DDSELECT
SEC,      *SECTION
SET,      *SETVAR
SHA,      *SHADE
SL,       *SLICE
SN,       *SNAP
SO,       *SOLID
SP,       *SPELL
SPL,      *SPLINE
SPE,      *SPLINEDIT
ST,       *STYLE
SU,       *SUBTRACT
T,        *MTEXT
-T,       *-MTEXT
TA,       *TABLET
TH,       *THICKNESS
TI,       *TILEMODE
TO,       *TOOLBAR
```

```
;
; The following are alternative aliases and aliases as supplied in AutoCAD Release 14.
;
AV,        *DSVIEWER
CP,        *COPY
DIMALI,    *DIMALIGNED
DIMANG,    *DIMANGULAR
DIMBASE,   *DIMBASELINE
DIMCONT,   *DIMCONTINUE
DIMDIA,    *DIMDIAMETER
DIMED,     *DIMEDIT
DIMTED,    *DIMTEDIT
DIMLIN,    *DIMLINEAR
DIMORD,    *DIMORDINATE
DIMRAD,    *DIMRADIUS
DIMSTY,    *DIMSTYLE
DIMOVER,   *DIMOVERRIDE
LEAD,      *LEADER
TM,        *TILEMODE
```

SECTIONS OF THE ACAD.PGP FILE

The contents of the AutoCAD program parameters file (**ACAD.PGP**) can be categorized into three sections. These sections merely classify the information that is defined in the **ACAD.PGP** file. They do not have to appear in any definite order in the file, and they have no section headings. For example, the comment lines can be entered anywhere in the file; the same is true with external commands and AutoCAD command aliases. The **ACAD.PGP** file can be divided into these three sections: **comments, external commands,** and **command aliases.**

Comments

The comments of **ACAD.PGP** file can contain any number of comment lines and can occur anywhere in the file. Every comment line must start with a semicolon (; This is a comment line). Any line that is preceded by a semicolon is ignored by AutoCAD. You should use the comment line to give some relevant information about the file that will help other AutoCAD users to understand, edit, or update the file.

External Command

In the external command section you can define any valid external command that is supported by your system. The information must be entered in the following format:

<Command name>, [OS Command name], <Memory reserve>,
<Command prompt>,

Command Name. This is the name you want to use to activate the external command from the AutoCAD drawing editor. For example, you can use **goword** as a command name to load the word program (**Command: goword**). The command name must not be an AutoCAD command name or an AutoCAD system variable name. If the name is an AutoCAD command name, the command name in the **PGP** file will be ignored. Also, if the name is an AutoCAD system variable name, the system variable will be ignored. You should use the command names that reflect the expected result of the external commands. (For example, **hallo** is not a good command name for a directory file.) The command names can be uppercase or lowercase.

OS Command Name. The OS Command name is the name of a valid system command that is supported by your operating system. For example, in DOS the command to delete files is DEL; therefore, the OS Command name used in the **ACAD.PGP** file must be DEL. The following is a list of the types of commands that can be used in the PGP file:

OS Commands (del, dir, type, copy, rename, edlin, etc.)
Commands for starting a word processor, or text editors (word, shell, etc.)
Name of the user-defined programs and batch files

Memory Reserve. This field must contain a number, preferably zero. The memory reserve field has no effect on the PGP file. In older releases of AutoCAD, the memory reserve field was used to specify the number of bytes AutoCAD must release to execute an external command. In AutoCAD Release 14, this field keeps the PGP files compatible with the older releases of AutoCAD.

Command Prompt. The command prompt field of the command line contains the prompt you want to display on the screen. It is an optional field that must be replaced by a comma if there is no prompt. If the operating system (OS) command that you want to use contains spaces, the prompt must be preceded by an asterisk (*). For example, the DOS command EDLIN NEW.PGP contains a space between EDLIN and NEW; therefore, the prompt used in this command line must be preceded by an asterisk. The command can be terminated by pressing Enter. If the OS command consists of a single word (DIR, DEL, TYPE), the preceding asterisk must be omitted. In this case you can terminate the command by pressing the Spacebar or Enter.

Command Aliases

It is time-consuming to enter AutoCAD commands at the keyboard because it requires typing the complete command name before pressing Enter. AutoCAD provides a facility that can be used to abbreviate the commands by defining aliases for the AutoCAD commands. This is made possible by the AutoCAD program parameters file (**ACAD.PGP** file). Each command alias line consists of two fields (**L, *LINE**). The first field (**L**) defines the alias of the command; the second field (***LINE**) consists of the AutoCAD command. The AutoCAD command must be preceded by an asterisk for AutoCAD to recognize the command line as a command alias. The two fields must be separated by a comma. The blank lines and the spaces between the two fields are ignored. In addition to AutoCAD commands, you can also use aliases for AutoLISP command names, provided the programs that contain the definition of these commands are loaded.

Example 1

Add the following external commands and AutoCAD command aliases to the AutoCAD program parameters file (**ACAD.PGP**).

External Commands

Abbreviation	Command Description
GOWORD	This command loads the word processor (word) program from the C:\MIS\WORD directory.
RN	This command executes the rename command of DOS.
COP	This command executes the copy command of DOS.

Command Aliases Section

Abbreviation	Command	Abbreviation	Command
EL	Ellipse	T	Trim
CO	Copy	CH	Chamfer
O	Offset	ST	Stretch
S	Scale	MI	Mirror

The **ACAD.PGP** file is an ASCII text file. To edit this file you can use the AutoCAD EDIT command (provided the EDIT command is defined in the **ACAD.PGP** file), or any text editor. The following is a partial listing of the **ACAD.PGP** file after insertion of the lines for the command aliases of Example 5. **The line numbers are not a part of the file; they are shown here for reference only**. The lines that have been added to the file are highlighted in bold.

4-8 Customizing AutoCAD

```
DEL,DEL,                    0,File to delete: ,                           1
DIR,DIR,                    0,File specification ,                        2
EDIT, START EDIT,           0,File to edit: ,                             3
SH,,                        0,*OS Command: ,                              4
SHELL,,                     0,*OS Command: ,                              5
START,                      START,                                        1,Appl
ication to start: ,         6
goword, word,               0,,                                           8
rn, rename,                 0, *[drive](File name) [drive](File name): ,  9
COP, copy,                  0, *[drive](File name) [drive](File name): ,  10

DIMLIN      *DIMLINEAR                                                    11
DIMORD,     *DIMORDINATE                                                  12
DIMRAD,     *DIMRADIUS                                                    13
DIMSTY,     *DIMSTYLE                                                     14
DIMOVER,    *DIMOVERRIDE                                                  15
LEAD,       *LEADER                                                       16
TM,         *TILEMODE                                                     17
EL,         *ELLIPSE                                                      18
CO,         *COPY                                                         19
O,          *OFFSET                                                       20
S,          *SCALE                                                        21
MI,         *MIRROR                                                       22
ST,         *STRETCH                                                      23
```

Explanation

Lines 8

goword, word, 0,,

In line 8, **goword** loads the word processor program **(word)**. For these aliases to work, the subdirectory where these programs reside must be in the search path.

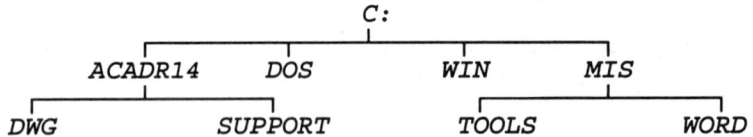

For example, if the word program is in the **word** subdirectory, the **word** subdirectory must be in the search path.

PATH= C:\;C:\DOS;C:\BAT;C:\MIS\WORD;

If the subdirectory that contains the file is not in the search path, the programs can also be loaded by defining a batch file. For example, the following batch file will load the word program:

C:\MIS\WORD\Word

Customizing the ACAD.PGP File 4-9

Assume here that the name of the batch file is **WORD.BAT** and that it is in the **BAT** subdirectory that is defined in the search path. When you enter the GOWORD command, AutoCAD will search for the **WORD.BAT** file and then execute the commands defined in the file.

Lines 9 and 10
rn, rename, 0, *[drive](File name) [drive](File name): ,
COP, copy, 0, *[drive](File name) [drive](File name): ,
Line 9 defines the alias for the DOS command **RENAME**, and the next line defines the alias for the DOS command **COPY**. The 0 after **rename** has no function, and the command prompt ***[drive](File name) [drive](File name):** is automatically displayed to let you know the format and the type of information that is expected.

Lines 18 and 19
EL, *ELLIPSE
CO, *COPY
Line 18 defines the alias (**EL**) for the AutoCAD command **ELLIPSE**, and the next line defines the alias (**CO**) for the **COPY** command. The AutoCAD commands must be preceded by an asterisk. You can put any number of spaces between the alias abbreviation and the AutoCAD command.

REINIT COMMAND

When you make any changes in the **ACAD.PGP** file, there are two ways to reinitialize the **ACAD.PGP** file. One is to quit AutoCAD and then re-enter it. When you start AutoCAD, the **ACAD.PGP** file is automatically loaded.

You can also reinitialize the **ACAD.PGP** file by using the AutoCAD REINIT command. The REINIT command lets you reinitialize the I/O ports, digitizer, display, and AutoCAD program parameters file, **ACAD.PGP**. When you enter the REINIT command, AutoCAD will display a dialog box (Figure 4-1). To reinitialize the **ACAD.PGP** file, select the corresponding toggle box, and then select OK. AutoCAD will reinitialize the program parameters file (**ACAD.PGP**), and then you can use the command aliases defined in the file.

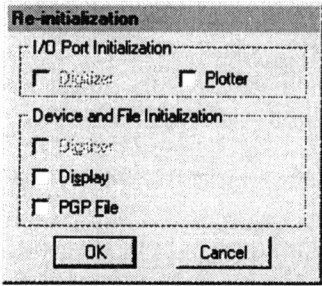

Figure 4-1 Re-initialization dialog box

REVIEW QUESTIONS

Indicate whether the following statements are true or false.

1. The comment section can contain any number of lines. (T/F)

2. AutoCAD ignores any line that is preceded by a semicolon. (T/F)

3. The command alias must not be an AutoCAD command. (T/F)

4. The memory reserve field must contain a zero. (T/F)

5. In the command alias section, the command alias must be preceded by a semicolon. (T/F)

6. You cannot use aliases for AutoLISP commands. (T/F)

7. If the bit code in the return code field is 2, AutoCAD will return to the text screen. (T/F)

8. The ACAD.PGP file does not come with AutoCAD software. (T/F)

9. The ACAD.PGP file is an ASCII file. (T/F)

EXERCISES

Exercise 1

Add the following external commands and AutoCAD command aliases to the AutoCAD program parameters file (**ACAD.PGP**).

External Command Section

Abbreviation	Command description
DBASE	This command loads the Dbase program that resides in the C:\DBASE directory.
LOTUS	This command loads the spreadsheet program that resides in the C:\LOTUS directory.
CD	This command executes the CHKDSK command of DOS.
FORMAT	This command executes the FORMAT command of DOS.

Command Aliases Section

Abbreviation	Command	Abbreviation	Command
BL	BLOCK	LT	LTSCALE
INS	INSERT	EX	EXPLODE
DIS	DISTANCE	G	GRID
TIME	T		

Chapter 5

Pull-down, Cascading, Cursor, and Partial Menus and Customizing Toolbars

Learning objectives

After completing this chapter, you will be able to:
- Write pull-down menus.
- Load menus.
- Write cascading submenus in pull-down menus.
- Write cursor menus.
- Swap pull-down menus.
- Write partial menus.
- Define accelerator keys.
- Write toolbar definitions.
- Write menus to access online help.
- Customize toolbars.

AUTOCAD MENU

The AutoCAD menu provides a powerful tool to customize AutoCAD. The AutoCAD software package comes with a standard menu file named **ACAD.MNU**. When you start AutoCAD, the menu file **ACAD.MNU** is automatically loaded. The AutoCAD menu file contains AutoCAD commands, separated under different headings for easy identification. For example, all draw commands are under Draw and all editing commands are under Modify. The headings are named and arranged to make it easier for you to locate and access the commands. However, there are some commands that you may never use. Also, some users might like to regroup and rearrange the commands so that it is easier to access those most frequently used.

AutoCAD lets the user eliminate rarely used commands from the menu file and define new ones. This is made possible by editing the existing **ACAD.MNU** file or writing a new menu file. There is no limit to the number of files you can write. You can have a separate menu file for each application. For example, you can have separate menu files for mechanical, electrical, and architectural drawings. You can load these menu files any time by using the AutoCAD MENU command. The menu files are text files with the extension .MNU. These files can be written by using any text editor like Notepad.

The menu file can be divided into six sections, each section identified by a section label. AutoCAD uses the following labels to identify different sections of the AutoCAD menu file:

    ***SCREEN
    ***TABLET(n)   n is from 1 to 4
    ***IMAGE
    ***POP(n)   n is from 0 to 16
    ***BUTTONS(n)   n is from 1 to 4
    ***AUX(n)   n is from 1 to 4
    ***MENUGROUP
    ***TOOLBARS
    ***HELPSTRING
    ***ACCELERATORS

The tablet menu can have up to four different sections. The POP menu (pull-down and cursor menu) can have up to 16 sections, and auxiliary and buttons menus can have up to four sections.

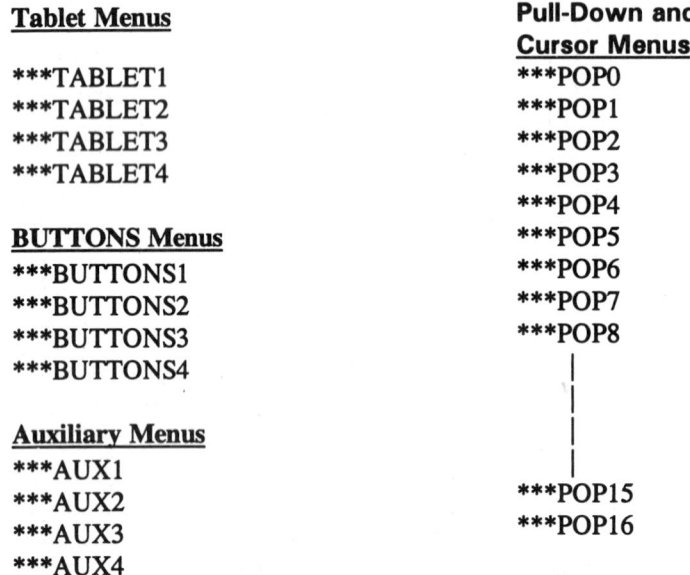

STANDARD PULL-DOWN MENUS

The pull-down menu is a part of the AutoCAD standard menu file, **ACAD.MNU**. The **ACAD.MNU** file is automatically loaded when you start AutoCAD, provided the standard configuration of AutoCAD has not been changed.

The pull-down menus can be selected by moving the crosshairs to the top of the screen, into the menu bar area. If you move the pointing device sideways, different menu bar titles are highlighted and you can select the desired item by pressing the pick button on your pointing device. Once the item is selected, the corresponding pull-down menu is displayed directly under the title (Figure 5-1). The pull-down menu can have 16 sections, named as POP1, POP2, POP3, . . ., POP16.

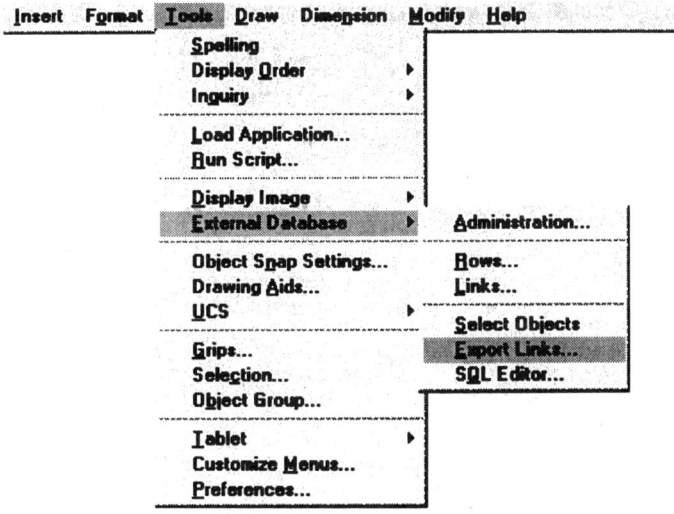

Figure 5-1 Pull-down and cascading menus

WRITING A PULL-DOWN MENU

Before you write a menu, you need to design a menu and to know the exact sequence of AutoCAD commands and the prompts associated with particular commands. To design a menu, you should select and arrange the commands in a way that provides easy access to the most frequently used commands. A careful design will save a lot of time in the long run. Therefore, consider several possible designs with different command combinations, and then select the one best suited for the job. Suggestions from other CAD operators can prove very valuable.

The second important thing in developing a menu is to know the exact sequence of the commands and the prompts associated with each command. To better determine the prompt entries required in a command, you should enter all the commands and the prompt entries at the keyboard. The following is a description of some of the commands and the prompt entries required for Example 1.

LINE Command
Command: **LINE**

Notice the command and input sequence:

LINE
<RETURN>

CIRCLE (C,R) Command
Command: **CIRCLE**
3P/2P/TTR/<Center point>: *Specify center point*
Diameter/<Radius>: *Enter radius*

Notice the command and input sequence:

CIRCLE
<RETURN>
Center point
<RETURN>
Radius
<RETURN>

CIRCLE (C,D) Command
Command: **CIRCLE**
3P/2P/TTR/<Center point>: *Specify center point*
Diameter/<Radius>: **D**
Diameter: *Enter diameter*

Notice the command and input sequence:

CIRCLE
<RETURN>
Center Point
<RETURN>
D
<RETURN>
Diameter
<RETURN>

CIRCLE (2P) Command
Command: **CIRCLE**
3P/2P/TTR/<Center Point>: **2P**
First point on diameter: *Specify first point*
Second point on diameter: *Specify second point*

Notice the command and input sequence:

CIRCLE
<RETURN>
2P
<RETURN>
Select first point on diameter
<RETURN>

Select second point on diameter
<RETURN>

ERASE Command
Command: **ERASE**

Notice the command and input sequence:

ERASE
<RETURN>

MOVE Command
Command: **MOVE**

Notice the command and prompt entry sequence:

MOVE
<RETURN>

The difference between the Center-Radius and Center-Diameter options of the CIRCLE command is that in the first one the RADIUS is the default, whereas in the second one you need to enter D to use the diameter option. This difference, although minor, is very important when writing a menu file. Similarly, the 2P (two-point) option of the CIRCLE command is different from the other two options. Therefore, it is important to know both the correct sequence of the AutoCAD commands and the entries made in response to the prompts associated with those commands.

You can use any text editor (like Notepad) to write the file. You can also use AutoCAD's EDIT command to write the menu file. If you use the EDIT command, AutoCAD will prompt you to enter the file name you want to edit. The file name can be up to eight characters long, and the file extension must be **.MNU**. If the file name exists, it will be automatically loaded; otherwise a new file will be created. To understand the process of developing a pull-down menu, consider the following example.

Note
*If AutoCAD's EDIT command does not work, check the **ACAD.PGP** file and make sure that this command is defined in the file.*

Example 1

Write a pull-down menu for the following AutoCAD commands:

LINE	ERASE	REDRAW	SAVE
PLINE	MOVE	REGEN	QUIT
CIRCLE C,R	COPY	ZOOM ALL	PLOT
CIRCLE C,D	STRETCH	ZOOM WIN	
CIRCLE 2P	EXTEND	ZOOM PRE	
CIRCLE 3P	OFFSET		

The first step in writing any menu is to design the menu so that the commands are arranged in the desired configuration. Figure 5-2 shows one of the possible designs of this menu.

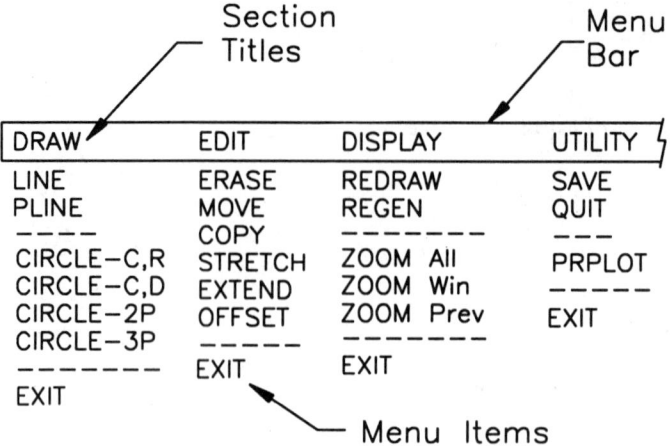

Figure 5-2 Design of pull-down menu

This menu has four different groups of commands; therefore, it will have four sections: POP1, POP2, POP3, and POP4, and each section will have a section label. The following file is a listing of the pull-down menu file for Example 1. **The line numbers are not a part of the file; they are shown here for reference only.**

```
***POP1                         1
[DRAW]                          2
[LINE]*^C^CLINE                 3
[PLINE]^C^CPLINE                4
[--]                            5
[CIR-C,R]^C^CCIRCLE             6
[CIR-C,D]^C^CCIRCLE \D          7
[CIR-2P]^C^CCIRCLE 2P           8
[CIR-3P]^C^CCIRCLE 3P           9
[--]                           10
[Exit]^C                       11
***POP2                        12
[EDIT]                         13
[ERASE]*^C^CERASE              14
[MOVE]^C^CMOVE                 15
[COPY]^C^CCOPY                 16
[STRETCH]^C^CSTRETCH;C         17
[OFFSET]^C^COFFSET             18
[EXTEND]^C^CEXTEND             19
[--]                           20
[Exit]^C                       21
***POP3                        22
```

[DISPLAY]	23
[REDRAW]'REDRAW	24
[REGEN]^C^CREGEN	25
[--]	26
[ZOOM-All]^C^C'ZOOM A	27
[ZOOM-Window]'ZOOM W	28
[ZOOM-Prev]'ZOOM PREV	29
[--]	30
[~Exit]^C	31
***POP4	32
[UTILITY]	33
[SAVE]^C^CSAVE;	34
[QUIT]^C^CQUIT	35
[----]	36
[PLOT]^C^CPLOT	37
[--]	38
[Exit]^C	39

Explanation

Line 1
***POP1
POP1 is the section label for the first pull-down menu. All section labels in the AutoCAD menu begin with three asterisks (***), followed by the section label name, such as POP1.

Line 2
[DRAW]
In this menu item DRAW is the menu bar title displayed when the cursor is moved in the menu bar area. The title names should be chosen so you can identify the type of commands you expect in that particular pull-down menu. In this example, all the draw commands are under the title DRAW, all edit commands are under EDIT, and so on for other groups of items. The menu bar

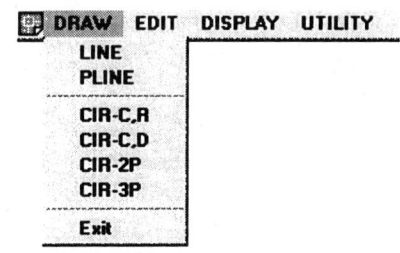

title can be of any length. However, it is recommended to keep them short to accommodate other menu items. Some of the display devices provide a maximum of 80 characters. To have 16 sections in a single row in the menu bar, the length of each title should not exceed five characters. If the length of the menu bar items exceeds 80 characters, AutoCAD will wrap the items that cannot be accommodated in 80 characters space, and display them in the next line. This will result in two menu item lines in the menu bar.

If the first line in a pull-down menu section is blank, the title of that section is not displayed in the menu bar area. Since the menu bar title is not displayed, you **cannot** access that pull-down menu. This allows you to turn off the pull-down menu section. For example, if you replace [DRAW] with a blank line, the DRAW section (POP1) of the pull-down menu will be disabled; the second section (POP2) will be displayed in its place.

Example
```
***POP1                    Section label
                           Blank line (turns off POP1)
[LINE:]^CLINE              Menu item
[PLINE:]^CPLINE
[CIRCLE:]^CCIRCLE
```

The menu bar titles are left-justified. If the first title is not displayed, the rest of the menu titles will be shifted to the left. In Example 1, if the DRAW title is not displayed in the menu bar area, then the EDIT, DISPLAY, and FILE sections of the pull-down menu will move to the left.

Line 3
*^C^CLINE
In this menu item, the command definition starts with an asterisk (*). This feature allows the command to be repeated automatically until it is canceled by entering CTRL C or by selecting another menu command. ^C^C cancels the existing command twice; LINE is an AutoCAD command that generates lines.

```
*^C^CLINE
 │ │   └─ AutoCAD's LINE command
 │ └─ Cancels existing command twice
 └─ Repeats the menu item (command)
```

Line 5
[--]
To separate two groups of commands in any section, you can use a menu item that consists of two or more hyphens (--). This line automatically expands to fill the **entire width** of the pull-down menu. **You cannot use a blank line in a pull-down menu.** If any section of a pull-down menu (**POP section) has a blank line, the items beyond the blank line are not displayed.

Line 11
[Exit]^C
In this menu item, ^C command definition has been used to cancel the pull-down menu. This item provides you with one more option for canceling the pull-down menu. This is especially useful for new AutoCAD users who are not familiar with all AutoCAD features. The pull-down menu can also be canceled by any of the following actions:

1. Selecting a point.
2. Selecting an item in the screen menu area.
3. Selecting or typing another command.
4. Pressing Esc at the keyboard.
5. Selecting any menu title in the menu bar.

Line 28
[ZOOM-Window]'ZOOM W

In this menu item the single quote (') preceding the ZOOM command makes the ZOOM Window command transparent. When a command is transparent, the existing command is not canceled. After the ZOOM Window command, AutoCAD will automatically resume the current operation.

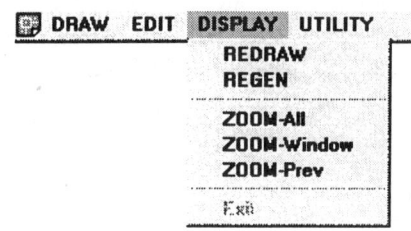

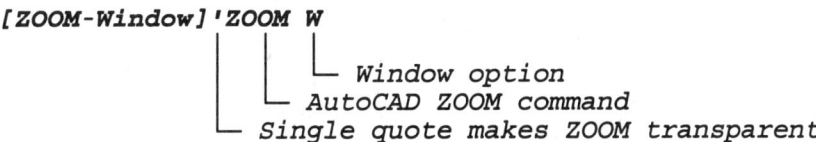

Line 31
[~Exit]^C

This menu item is similar to the menu item in line 11, except for the tilde (~). Since this menu item has a tilde (~), the menu item is not available (displayed grayed out), and if you select this item it will not cancel the pull-down menu. You can use this feature to disable a menu item or to indicate that the item is not a valid selection. If there is an instruction associated with the item, the instruction will not be executed when you select the item. For example, [~OSNAPS]^C^C$S=OSNAPS will not load the OSNAPS submenu on the screen.

Line 34
[SAVE]^C^CSAVE;

In this menu item, the semicolon (;) that follows the SAVE command enters Return. The semicolon is not required; the command will also work without a semicolon.

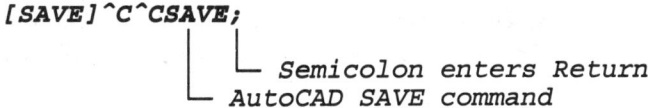

Line 36
[----]

This menu item has four hyphens; therefore, the line will not extend. Only when there are two hyphens (--) does the line extend the entire width of the pull-down menu.

> **Note**
> *For all pull-down menus, the menu items are displayed directly beneath the menu title and are left-justified. If any pull-down menu [for example, the rightmost pull-down menu (POP16)] does not have enough space to display the entire menu item, the pull-down menu will expand to the left to accommodate the entire length of the longest menu item.*
>
> *You can use // (two forward slashes) for comment lines. AutoCAD ignores the lines that start with //.*

From this example it is clear that every statement in the menu is based on the AutoCAD commands and the information that is needed to complete those commands. This forms the basis for creating a menu file and should be given consideration. Following is a summary of the AutoCAD commands used in Example 1 and their equivalents in the menu file.

AutoCAD Commands	Menu File
Command: **LINE**	[LINE]^C^CLINE
Command: **CIRCLE** 3P/2P/TTR/<Center point>: Diameter/<Radius>:	[CIR-C,R]^C^CCIRCLE
Command: **CIRCLE** 3P/2P/TTR/<Center point>: Diameter/<Radius>: D Diameter:	[CIR-C,D]^C^CCIRCLE;\D
Command: **CIRCLE** 3P/2P/TTR/<Center point>: 2P First point on diameter: Second point on diameter:	[CIR- 2P]^C^CCIRCLE;2P
Command: **ERASE**	[ERASE]^C^CERASE
Command: **MOVE**	[MOVE]^C^CMOVE

LOADING MENUS

AutoCAD automatically loads the **ACAD.MNU** file when you get into the AutoCAD drawing editor. However, you can also load a different menu file by using the AutoCAD MENU command.

Command: **Menu**

When you enter the Menu command, AutoCAD displays the **Select Menu File** dialog box (Figure 5-3) on the screen. Select the menu file that you want to load and then select the Open button.

Pull-down, Cascading, Cursor, and Partial Menus and Customizing Toolbars 5-11

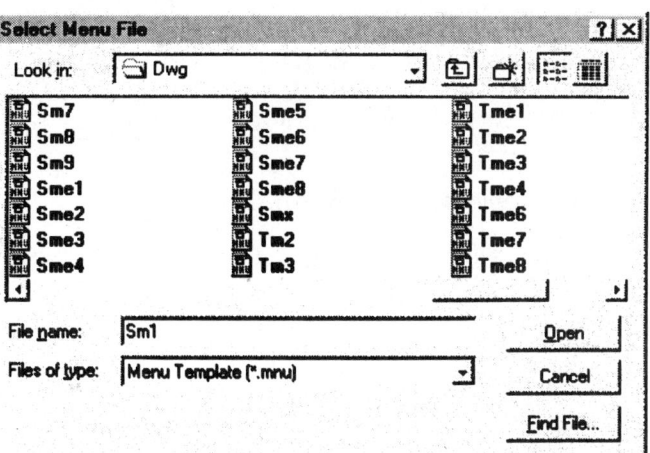

Figure 5-3 Select Menu File dialog box

You can also load the menu file from the command line.

    Command: **FILEDIA**
    New value for FILEDIA <1>: **0**
    Command: **Menu**
    Menu filename <ACAD>: PDM1

                              Name of menu file
                      Default menu file

After you enter the MENU command, AutoCAD will prompt for the file name. Enter the name of the menu file without the file extension (.MNU), since AutoCAD assumes the extension .MNU. AutoCAD will automatically compile the menu file into MNC and MNR files. When you load a menu file in windows, AutoCAD creates the following files:

    .mnc and .mnr files   When you load a menu file (.mnu), AutoCAD compiles the menu file and creates .mnc and .mnr files. The .mnc file is a compiled menu file. The .mnr file contains the bitmaps used by the menu.

    .mns file   When you load the menu file, AutoCAD also creates an .mns file. This is an ASCII file that is same as the .mnu file when you initially load the menu file. Each time you make a change in the contents of the file, AutoCAD changes the .mns file.

Note
1. After you load the new menu, you cannot use the screen menu, buttons menu, or digitizer because the original menu, ACAD.MNU, is not present and the new menu does not contain these menu areas.

2. To activate the original menu again, load the menu file by using the MENU command:

5-12   Customizing AutoCAD

Command: **Menu**
Menu file name or . for none <SM1>: **ACAD**

3. If you need to use input from a keyboard or a pointing device, use the backslash (\). The system will pause for you to enter data.

4. There should be no space after the backslash (\).

5. The menu items, menu labels, and command definition can be uppercase, lowercase, or mixed.

6. You can introduce spaces between the menu items to improve the readability of the menu file.

7. If there are more items in the menu than the number of spaces available, the excess items are not displayed on the screen. For example, if the display device limits the number of items to 21, items in excess of 21 will not be displayed on the screen and are therefore inaccessible.

8. If you are using a high-resolution graphics board, you can increase the number of lines that can be displayed on the screen. On some devices this is 80 lines.

RESTRICTIONS

The pull-down menus are easy to use and provide a quick access to frequently used AutoCAD commands. However, the menu bar and the pull-down menus are disabled during the following commands:

DTEXT Command
After you assign the text height and the rotation angle to a DTEXT command, the pull-down menu is automatically disabled.

SKETCH Command
The pull-down menus are disabled after you set the record increment in the SKETCH command.

VPOINT Command
The pull-down menus are disabled while the axis tripod and compass are displayed on the screen.

ZOOM and DVIEW Commands
The pull-down menus are disabled when the dynamic zoom or dynamic view is in progress.

Exercise 1

Write a pull-down menu for the following AutoCAD commands.

DRAW	**EDIT**	**DISP/TEXT**	**UTILITY**
LINE	FILLET0	DTEXT,C	SAVE
PLINE	FILLET	DTEXT,L	QUIT

ELLIPSE	CHAMFER	DTEXT,R	END
POLYGON	STRETCH	ZOOM WIN	DIR
DONUT	EXTEND	ZOOM PRE	PLOT
	OFFSET		

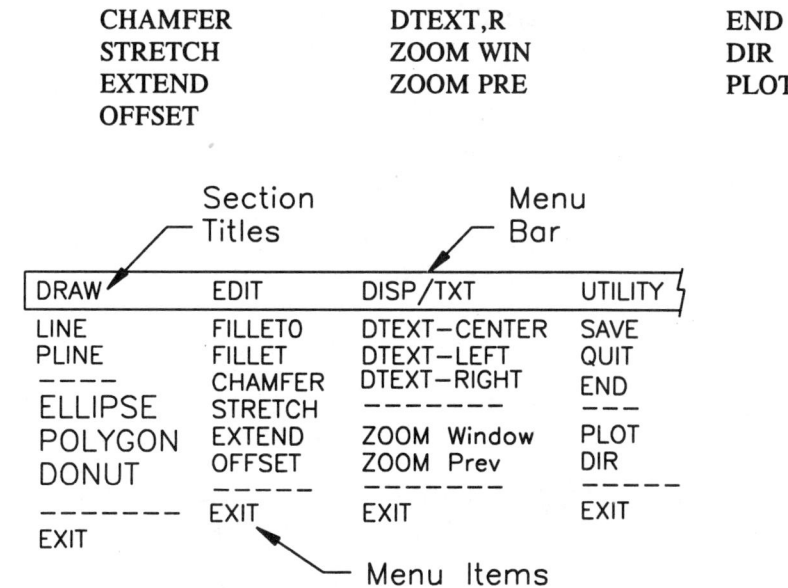

Figure 5-4 Pull-down menu display for Exercise 1

CASCADING SUBMENUS IN PULL-DOWN MENUS

The number of items in a pull-down menu or cursor menu can be very large, and sometimes they cannot all be accommodated on one screen. For example, the maximum number of items that can be displayed on some display devices is 21. If the pull-down menu or the cursor menu has more items than can be displayed, the excess menu items are not displayed on the screen and cannot be accessed. You can overcome this problem by using cascading menus that let you define smaller groups of items within a menu section. When an item is selected, it loads the cascading menu and displays the items, defined in the cascading menu, on the screen.

The cascading feature of AutoCAD allows pull-down and cursor menus to be displayed in a hierarchical order that makes it easier to select submenus. To use the cascading feature in pull-down and cursor menus, AutoCAD has provided some special characters. For example, -> defines a cascaded submenu and <- designates the last item in the pull-down menu. The following table lists some of the characters that can be used with the pull-down or cursor menus.

Character	Character Description
--	The item label consisting of two hyphens automatically expands to fill the entire width of the pull-down menu. Example: [--]
+	Used to continue the menu item to the next line. This character has to be the last character of the menu item.

Example: [Triang:]^C^CLine;1,1;+3,1;2,2;

->   This label character defines a cascaded submenu; it must precede the name of the submenu.
Example: [->Draw]

<-   This label character designates the last item of the cascaded pull-down or cursor menu. The character must precede the label item.
Example: [<-CIRCLE 3P]^C^CCIRCLE;3P

<-<-...   This label character designates the last item of the pull-down or cursor menu and also terminates the parent menu. The character must precede the label item.
Example: [<-<-Center Mark]^C^C_dim;_center

$(This label character can be used with the pull-down and cursor menus to evaluate a DIESEL expression. The character must precede the label item.
Example: $(if,$(getvar,orthomode),Ortho)

~   This item indicates the label item is not available (displayed grayed out); the character must precede the item.
Example: [~Application not available]

Some display devices provide space for a maximum of 80 characters. Therefore, if there are 10 pull-down menus, the length of each menu title should average eight characters. If the combined length of all menu bar titles exceeds 80 characters, AutoCAD automatically wraps the excess menu items and displays them on the next line in the menu bar. The following is a list of some additional features of the pull-down menu.

1. The section labels of the pull-down menus are ***POP1 through ***POP16. The menu bar titles are displayed in the menu bar.

2. The pull-down menus can be accessed by selecting the menu title from the menu bar at the top of the screen.

3. A maximum of 999 menu items can be defined in the pull-down menu. This includes the items that are defined in the pull-down submenus. The menu items in excess of 999 are ignored.

4. The number of menu items that can be displayed depends on the display device you are using. If the cursor or the pull-down menu contains more items than can be accommodated on the screen, the excess items are truncated. For example, if your system can display only 21 menu items, the menu items in excess of 21 are automatically truncated.

Example 2

Write a pull-down menu for the commands shown in Figure 5-5. The pull-down menu must use the AutoCAD cascading feature.

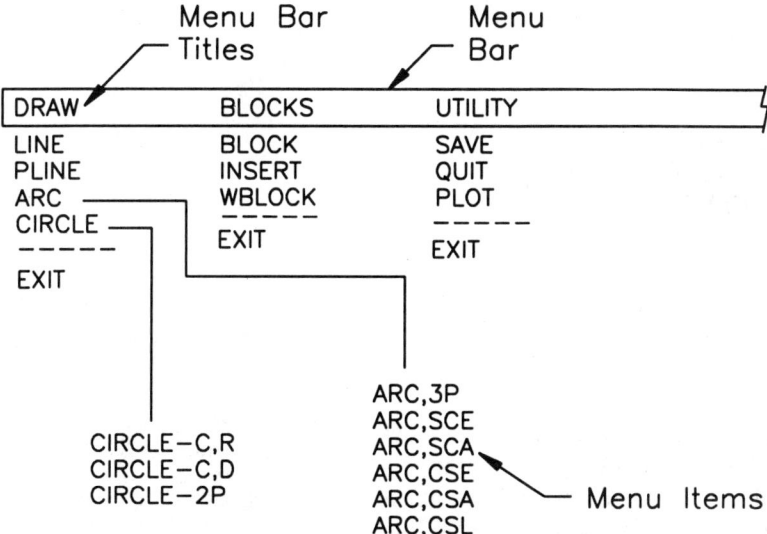

Figure 5-5 Pull-down menu structure for Example 2

The following file is a listing of the pull-down menu for Example 2. **The line numbers are not a part of the menu; they are shown here for reference only.**

```
***POP1                                  1
[DRAW]                                   2
[LINE]^C^CLINE                           3
[PLINE]^C^CPLINE                         4
[->ARC]                                  5
  [ARC]^C^CARC                           6
  [ARC,3P]^C^CARC;\\DRAG                 7
  [ARC,SCE]^C^CARC;\C;\DRAG              8
  [ARC,SCA]^C^CARC;\C;\A;DRAG            9
  [ARC,CSE]^C^CARC;C;\\DRAG             10
  [ARC,CSA]^C^CARC;C;\\A;DRAG           11
  [<-ARC,CSL]^C^CARC;C;\\L;DRAG         12
[->CIRCLE]                              13
  [CIRCLE C,R]^C^CCIRCLE                14
  [CIRCLE C,D]^C^CCIRCLE;\D             15
  [CIRCLE 2P]^C^CCIRCLE;2P              16
  [<-CIRCLE 3P]^C^CCIRCLE;3P            17
[--]                                    18
[Exit]^C                                19
***POP2                                 20
```

```
    [BLOCKS]                                              21
    [BLOCK]$S=X $S=BLKX ^C^CBLOCK                         22
    [INSERT]$S=X $S=BLK *^C^CINSERT                       23
    [WBLOCK]$S=X $S=WBLK ^C^CWBLOCK                       24
    [--]                                                  25
    [Exit]^C                                              26
    ***POP3                                               27
    [UTILITY]                                             28
    [SAVE]^C^CSAVE                                        29
    [QUIT]^C^CQUIT                                        30
    [PLOT]^C^CPLOT                                        31
    [--]                                                  32
    [Exit]^C                                              33
```

Explanation

Line 5
[->ARC]
In this menu item, **ARC** is the menu item label that is preceded by the special label character ->. This special character indicates that the menu item has a submenu. The menu items that follow it (lines 6-12) are the submenu items.

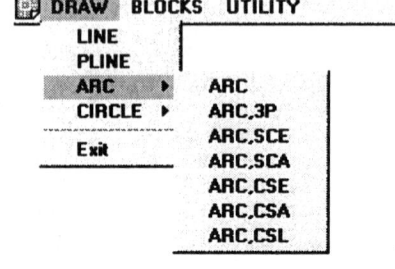

Line 12
[<-ARC,CSL]^C^CARC;C;\\L;DRAG
In this line, the menu item label **ARC,CSL** is preceded by another special label character, **<-**, which indicates the end of the submenu. The item that contains this character must be the last menu item of the submenu.

Lines 13 and 17
[->CIRCLE]
[<-CIRCLE 3P]^C^CCIRCLE;3P
The special character **->** in front of **CIRCLE** indicates that the menu item has a submenu; the character **<-** in front of **CIRCLE 3P** indicates that this item is the last menu item in the submenu. When you select the menu item **CIRCLE** from the pull-down menu, it will automatically display the submenu on the side.

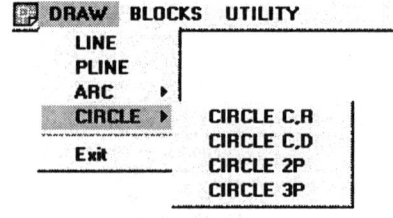

Example 3

Write a pull-down menu that has the cascading submenus for the commands shown in Figure 5-6.

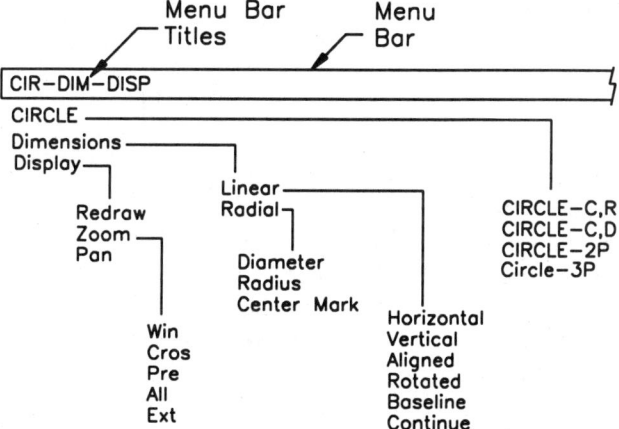

Figure 5-6 Pull-down menu structure for Example 3

The following file is a listing of the pull-down menu for Example 3. **The line numbers are not a part of the menu; they are shown here for reference only.**

***POP1	1
[DRAW]	2
[->CIRCLE]	3
[CIRCLE C,R]^C^C_CIRCLE	4
[CIRCLE C,D]^C^C_CIRCLE;_D	5
[CIRCLE 2P]^C^C_CIRCLE;_2P	6
[<-CIRCLE 3P]^C^C_CIRCLE;_3P	7
[->Dimensions]	8
[->Linear]	9
[Horizontal]^C^C_dim;_horizontal	10
[Vertical]^C^C_dim;_vertical	11
[Aligned]^C^C_dim;_aligned	12
[Rotated]^C^C_dim;_rotated	13
[Baseline]^C^C_dim;_baseline	14
[<-Continue]^C^C_dim;_continue	15
[->Radial]	16
[Diameter]^C^C_dim;_diameter	17
[Radius]^C^C_dim;_radius	18
[<-<-Center Mark]^C^C_dim;_center	19
[->DISPLAY]	20
[REDRAW]^C^CREDRAW	21
[->ZOOM]	22
[...Win]^C^C_ZOOM;_W	23
[...Cros]^C^C_ZOOM;_C	24

```
        [...Pre]^C^C_ZOOM;_P                                   25
        [...All]^C^C_ZOOM;_A                                   26
        [<-...Ext]^C^C_ZOOM;_E                                 27
        [<-PAN]^C^C_Pan                                        28
```

Explanation
Lines 8 and 9
[->Dimensions]
[->Linear]

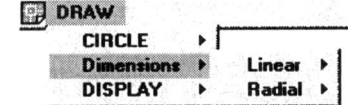

The special label character -> in front of the menu item **Dimensions** indicates that it has a submenu, and the character -> in front of **Linear** indicates that there is another submenu. The second submenu **Linear** is within the first submenu **Dimensions**. The menu items on lines 10 to 15 are defined in the **Linear** submenu, and the menu items **Linear** and **Radial** are defined in the submenu **Dimensions**.

Line 16
[->Radial]
This menu item defines another submenu; the menu items on line numbers 17, 18, and 19 are part of this submenu.

Line 19
[<-<-Center Mark]^C^C_dim;_center
In this menu item the special label character <-<- terminates the **Radial** and **Dimensions** (parent submenu) submenus.

Lines 27 and 28
[<-...Ext]^C^C_ZOOM;_E
[<-PAN]^C^C_Pan

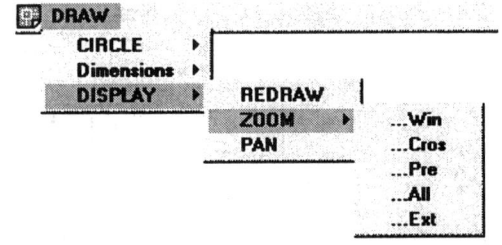

The special character <- in front of the menu item ...Ext terminates the ZOOM submenu; the special character in front of the menu item PAN terminates the DISPLAY submenu.

CURSOR MENUS
The cursor menus are similar to the pull-down menus, except the cursor menu can contain only 499 items compared with 999 items in the pull-down. The section label of the cursor menu must be ***POP0. The cursor menus are displayed near or at the cursor location. Therefore, they can be used to provide convenient and quick access to some of the frequently used commands. The following is a list of some of the features of the cursor menu.

1. The section label of the cursor menu is ***POP0. The menu bar title defined under this section label is not displayed in the menu bar.

2. On most systems, the menu bar title is not displayed at the top of the cursor menu. However, for compatibility reasons you should give a dummy menu bar title.

Pull-down, Cascading, Cursor, and Partial Menus and Customizing Toolbars 5-19

3. The cursor menu can be displayed through the **$P0=*** menu command only. This command can be issued by a menu item in another menu, such as the buttons menu, auxiliary menu, or screen menu. The command can also be issued from an AutoLISP or ADS program.

4. A maximum of 499 menu items can be defined in the cursor menu. This includes the items that are defined in the cursor submenus. The menu items in excess of 499 are ignored.

5. The number of menu items that can be displayed on the screen depends on the system you are using. If the cursor or pull-down menu contains more items than your screen can accommodate, the excess items are truncated. For example, if your system displays 21 menu items, the menu items in excess of 21 are automatically truncated.

Example 4

Write a cursor menu for the following AutoCAD commands using cascading submenus. The menu should be compatible with foreign language versions of AutoCAD. Use the third button of the BUTTONS menu to display the cursor menu.

<u>Osnaps</u>	<u>Draw</u>	<u>**DISPLAY**</u>
Center	Line	REDRAW
Endpoint	PLINE	ZOOM
Intersection	CIR C,R	...Win
Midpoint	CIR 2P	...Cen
Nearest	ARC SCE	...Prev
Perpendicular	ARC CSE	...All
Quadrant		...Ext
Tangent		PAN
None		

The following file is a listing of the menu file for Example 4. **The line numbers are not a part of the file; they are for reference only.**

```
***AUX1                                          1
;                                                2
$P0=*                                            3
***POP0                                          4
[Osnaps]                                         5
[Center]_Center                                  6
[End point]_Endp                                 7
[Intersection]_Int                               8
[Midpoint]_Mid                                   9
[Nearest]_Nea                                    10
[Perpendicular]_Per                              11
[Quadrant]_Qua                                   12
[Tangent]_Tan                                    13
[None]_Non                                       14
```

```
    [--]                                   15
    [->Draw]                               16
     [Line]^C^C_Line                       17
     [PLINE]^C^C_Pline                     18
     [CIR C,R]^C^C_Circle                  19
     [CIR 2P]^C^C_Circle;_2P               20
     [ARC SCE]^C^C_ARC;\C                  21
     [<-ARC CSE]^C^C_Arc;C                 22
    [--]                                   23
    [->DISPLAY]                            24
     [REDRAW]^C^_REDRAW                    25
     [->ZOOM]                              26
      [...Win]^C^C_ZOOM;_W                 27
      [...Cen]^C^C_ZOOM;_C                 28
      [...Prev]^C^C_ZOOM;_P                29
      [...All]^C^C_ZOOM;_A                 30
      [<-...Ext]^C^C_ZOOM;_E               31
     [<-PAN]^C^C_Pan                       32
  ***POP1                                  33
  [Draw]                                   34
```

Explanation

Line 1
***AUX1

AUX1 is the section label for the first auxiliary menu; ******* designates the menu section. The menu items that follow it, until the second section label, are a part of this buttons menu.

Lines 2 and 3

;
$P0=*

The semicolon (;) is assigned to the second button of the pointing device (the first button of the pointing device is the pick button); the special command **$P0=*** is assigned to the third button of the pointing device.

Lines 4 and 5
***POP0
[Osnaps]

The menu label **POP0** is the menu section label for the cursor menu; **Osnaps** is the menu bar title. The menu bar title is not displayed, but is required. Otherwise, the first item will be interpreted as a title and will be disabled.

Line 6
[Center]_Center
In this menu item, _Center is the center object Snap mode. The menu files can be used with foreign language versions of AutoCAD, if AutoCAD commands and the command options are preceded by the underscore (_) character.

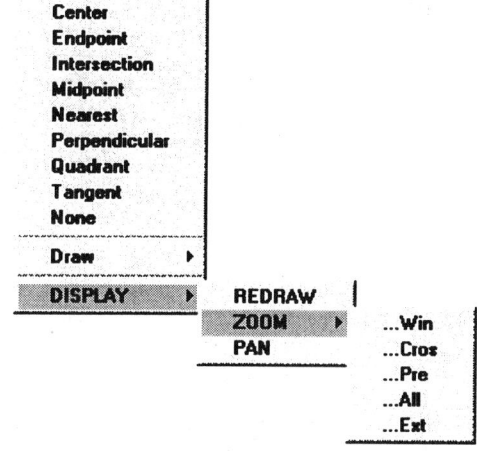

After loading the menu, if you press the third button of your pointing device, the cursor menu will be displayed at the cursor (screen crosshairs) location. If the cursor is close to the edges of the screen, the cursor menu will be displayed at a location that is closest to the cursor position. When you select a submenu, the items contained in the submenu will be displayed, even if the cursor menu is touching the edges of the screen display area.

Lines 33 and 34
***POP1
[Draw]
***POP1 defines the first pull-down menu. If no POPn sections are defined or the status line is turned off, the cursor menu is automatically disabled.

Exercise 2

Write a pull-down menu for the following AutoCAD commands. Use a cascading menu for the LINE command options in the pull-down menu. (The layout of the pull-down menu is shown in Figures 5-7.)

LINE	ZOOM All	TIME
Continue	ZOOM Win	LIST
Close	ZOOM Pre	DISTANCE
Undo	PAN	AREA
.X	DBLIST	
.Y	STATUS	
.Z		
CIRCLE		
ELLIPSE		

```
       DRAW          DISPLAY      INQUIRY
       LINE          ZOOM-A:      TIME:
       CIRCLE:       ZOOM-W:      LIST:
       ELLIPSE:      ZOOM-P:      DISTANCE:
               LINE: PAN:         AREA:
                 Cont.            DBLIST:
                 Close            STATUS:
                 Undo
                 Filters
                   .X
                   .Y             PULL-DOWN MENU
                   .Z
```

Figure 5-7 Design of pull-down menu for Exercise 2

SUBMENUS

The number of items in a pull-down menu or cursor menu can be very large and sometimes they cannot all be accommodated on one screen. For example, the maximum number of items that can be displayed on most of the display devices is 21. If the pull-down menu or the cursor menu has more than 21 items, the menu excess items are not displayed on the screen and cannot be accessed. You can overcome this problem by using submenus that let you define smaller groups of items within a menu section. When a submenu is selected, it loads the submenu items and displays them on the screen.

The pull-down menus that use AutoCAD's cascading feature are the most efficient and easy to write. The submenus follow a logical pattern that are easy to load and use without causing any confusion. It is strongly recommended to use the cascading menus whenever you need to write the pull-down or the cursor menus. However, AutoCAD provides the option to swap the submenus in the pull-down menus. These menus can sometimes cause distraction because the original pull-down menu is completely replaced by the submenu when swapping the menus.

Submenu Definition

A submenu definition consists of two asterisk signs (**) followed by the name of the submenu. A menu can have any number of submenus and every submenu should have a unique name. The items that follow a submenu, up to the next section label or submenu label, belong to that submenu. Following is the format of a submenu label:

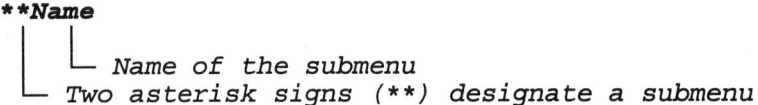

Note
The submenu name can be up to 31 characters long.

The submenu name can consist of letters, digits, and the special characters like: $ (dollar), - (hyphen), and _ (underscore).

> *The submenu name cannot have any embedded blanks.*
>
> *The submenu names should be unique in a menu file.*

Submenu Reference

The submenu reference is used to reference or load a submenu. It consists of a "$" sign followed by a letter that specifies the menu section. The letter that specifies a pull-down menu section is Pn, where n designates the number of the pull-down menu section. The menu section is followed by "=" sign and the name of the submenu that the user wants to activate. The submenu name should be without "**". Following is the format of a submenu reference:

```
$Section=Submenu
 │   │       │
 │   │       └─ Name of submenu
 │   └─ "=" sign
 └─ Menu section specifier
└─ "$" sign
```

Example
```
$P1=P1A
    │  │
    │  └─ Name of submenu
    └─ P1-Specifies pull-down menu section 1
```

Displaying a Submenu

When you load a submenu in a pull-down menu, the submenu items are not automatically displayed on the screen. For example, when you load a submenu P1A that has DRAW-ARC as the first item, the current title of POP1 will be replaced by the DRAW-ARC. But the items that are defined under DRAW-ARC are not displayed on the screen. To force the display of the new items on the screen AutoCAD uses a special command $Pn=*.

```
$Pn=*
 │ │ │
 │ │ └─ Asterisk sign (*)
 │ └─ Pull-down menu section number (1 to 10)
 └─ P for pull-down menu
```

LOADING MENUS

From the pull-down menu, you can load any menu that is defined in the screen or image tile menu sections by using the appropriate load commands. It may not be needed in most of the applications, but if you want to, you can load the menus that are defined in other menu sections.

Loading Screen Menus

You can load any menu that is defined in the screen menu section from the pull-down menu by using the following load command:

```
$S=X $S=LINE
     │    └─ Submenu name defined in screen menu section
     └─ Submenu name defined in screen menu section
  └─ S specifies screen menu
```

The first load command ($S=X) loads the submenu X that has been defined in the screen menu section of the menu file. The X submenu can contain 21 blank lines, so that when it is loaded it clears the screen menu. The second load command ($S=LINE) loads the submenu LINE that has also been defined in the screen menu section of the menu file.

Loading an Image Tile Menu

You can also load an image tile menu from the pull-down menu by using the following load command:

```
$I=IMAGE1 $I=*
    │       └─ To display the dialog box
    └─ Load the submenu IMAGE1
```

This menu item consists of two load commands. The first load command $I=IMAGE1 loads the image tile submenu IMAGE1 that has been defined in the image tile menu section of the file. The second load command $I=* displays the new dialog box on the screen.

Example 5

Write a pull-down menu for the following AutoCAD commands. Use submenus for the ARC and CIRCLE commands.

```
LINE                BLOCK              QUIT
PLINE               INSERT             SAVE
ARC                 WBLOCK             ----
    ARC 3P                             PLOT
    ARC SCE
    ARC SCA
    ARC CSE
    ARC CSA
    ARC CSL
CIRCLE
    CIRCLE C,R
    CIRCLE C,D
    CIRCLE 2P
```

The layout shown in Figure 5-8 is one of the possible designs for this pull-down menu. The ARC and CIRCLE commands are in separate groups that will be defined as submenus in the menu file.

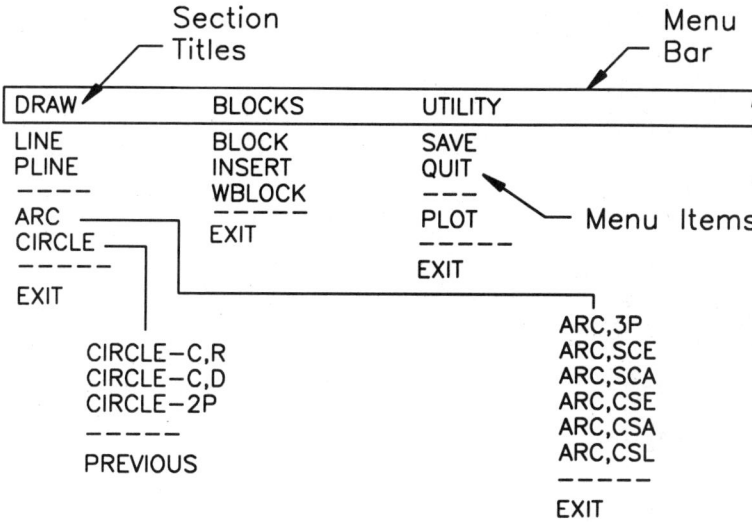

Figure 5-8 Design of pull-down menu for Example 5

The following file is a listing of the pull-down menu for Example 5. The line numbers are not a part of the menu file. They are given here for reference only.

```
***POP1                                         1
**P1A                                           2
[DRAW]                                          3
[LINE]^C^CLINE                                  4
[PLINE]^C^CPLINE                                5
[--]                                            6
[ARC]^C^C$P1=P1B $P1=*                          7
[CIRCLE]^C^C$P1=P1C $P1=*                       8
[--]                                            9
[Exit]^C                                       10
                                               11
**P1B                                          12
[ARC]                                          13
[ARC,3P]^C^CARC \\DRAG                         14
[ARC,SCE]^C^CARC \C \DRAG                      15
[ARC,SCA]^C^CARC \C \A DRAG                    16
[ARC,CSE]^C^CARC C \\DRAG                      17
[ARC,CSA]^C^CARC C \\A DRAG                    18
[ARC,CSL]^C^CARC C \\L DRAG                    19
[--]                                           20
[Exit]^C                                       21
                                               22
**P1C                                          23
```

```
            [CIRCLE]                                          24
            [CIRCLE C,R]^C^CCIRCLE                            25
            [CIRCLE C,D]^C^CCIRCLE \D                         26
            [CIRCLE 2P]^C^CCIRCLE 2P                          27
            [--]                                              28
            [PREVIOUS]$P1=P1A $P1=*                           29
                                                              30
            ***POP2                                           31
            [BLOCKS]                                          32
            [BLOCK]^C^CBLOCK                                  33
            [INSERT]*^C^CINSERT                               34
            [WBLOCK]^C^CWBLOCK                                35
            [--]                                              36
            [Exit]$P1=P1A $P1=*                               37
                                                              38
            ***POP3                                           39
            [UTILITY]                                         40
            [SAVE]^C^CSAVE                                    41
            [QUIT]^C^CQUIT                                    42
            [~--]                                             43
            [PLOT]^C^CPLOT                                    44
            [~--]                                             45
            [Exit]^C                                          46
                                                              47
```

Line 2
**P1A

**P1A defines the submenu P1A. All the submenus have two asterisk signs (**) followed by the name of the submenu. The submenu can have any valid name. In this example, P1A has been chosen because it is easy to identify the location of the submenu. P indicates that it is a pull-down menu, 1 indicates that it is in the first pull-down menu (POP1), and A indicates that it is the first submenu in that section.

Line 6
[--]

The two hyphens enclosed in the brackets will automatically expand to fill the entire width of the pull-down menu. This menu item cannot be used to define a command. If it does contain a command definition, the command is ignored. For example, if the menu item is [--]^C^CLINE, the command ^C^CLINE will be ignored.

Line 7
[ARC]^C^C$P1=P1B $P1=*

In this menu item, $P1=P1B loads the submenu P1B and assigns it to the first menu section (POP1), but the new pull-down menu is not displayed on the screen. $P1=* forces the display of the new pull-down menu on the screen.

For example, if you select CIRCLE from the first pull-down menu (POP1), the menu bar title DRAW will be replaced by CIRCLE, but the new menu is not displayed on the screen. Now, if you select CIRCLE from the menu bar, the command defined under the CIRCLE submenu will be displayed in the pull-down menu.

To force the display of the menu that is currently assigned to POP1, you can use AutoCAD's special command **$P1=***. If you select CIRCLE from the first pull-down menu (POP1), the CIRCLE submenu will be loaded and automatically displayed on the screen.

Line 21
[EXIT]^C
When you select this menu item the current pull-down menu will be canceled. It will not return to the previous submenu (DRAW). If you check the menu bar, it will display ARC as the section title, not DRAW. Therefore, it is not a good practice to cancel a submenu. It is better to return to the first menu before canceling it or define a command that automatically loads the previous menu and then cancels the pull-down menu.

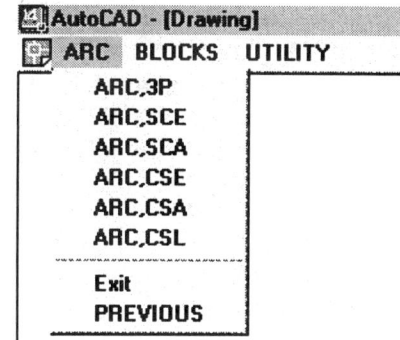

Example
[EXIT]$P1=P1A $P1=* ^C^C

Line 29
[PREVIOUS]$P1=P1A $P1=*
In this menu item $P1=P1A loads the submenu P1A, which happens to be the previous menu in this case. You can also use $P1= to load the previous menu. $P1=* forces the display of the submenu P1A.

PARTIAL MENUS
AutoCAD has provided a facility that allows the user to write their own menus and then load them in the menu bar. For example, in Windows you can write partial menus, toolbars, and definitions for accelerator keys. After you write the menu, AutoCAD lets you load the menu and use it with the standard menu. For example, you could load a partial menu and use it like a pull-down menu. You can also unload the menus that you do not want to use. These features make it convenient to use the menus that have been developed by AutoCAD users and developers.

Menu Section Labels
The following is a list of the additional menu section labels for Windows.

Section label	Description
***MENUGROUP	Menu file group name
***TOOLBARS	Toolbar definition
***HELPSTRING	Online help
***ACCELERATORS	Accelerator key definitions

Writing Partial Menus

The following example illustrates the procedure for writing a partial menu for Windows:

Example 6

In this example you will write a partial menu for Windows. The menu file has two pull-down menus, POP1 (MyDraw) and POP2 (MyEdit), as shown in Figure 5-9.

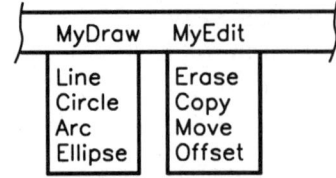

Figure 5-9 Pull-down menus

Step 1
Use a text editor to write the following menu file. The name of the file is assumed to be **MYMENU1.MNU**. The following is the listing of the menu file for this example:

```
***MENUGROUP=Menu1                1
***POP1                           2
[/MMyDraw]                        3
[/LLine]^C^CLine                  4
[/CCircle]^C^CCircle              5
[/AArc]^C^CArc                    6
[/EEllipse]^C^CEllipse            7
***POP2                           8
[/EMyEdit]                        9
[/EErase]^C^CErase               10
[/CCopy]^C^CCopy                 11
[/MMove]^C^CMove                 12
[/OOffset]^C^COffset             13
```

Explanation
Line 1
***MENUGROUP=Menu1
MENUGROUP is the section label and the Menu1 is the name tag for the menu group. The MENUGROUP label must precede all menu section definitions. The name of the MENUGROUP (Menu1) can be up to 32 characters long (alphanumeric), excluding spaces and punctuation marks. There is only one MENUGROUP in a menu file. All section labels must be preceded by *** (***MENUGROUP).

Line 2
***POP1
POP1 is the pull-down menu section label. The items on line numbers 3 through 7 belong to this section. Similarly, the items on line numbers 9 through 13 belong to the pull-down menu section **POP2**.

Pull-down, Cascading, Cursor, and Partial Menus and Customizing Toolbars 5-29

Line 3
[/MMyDraw]
/M defines the mnemonic key you can use to activate the menu item. For example, /M will display an underline under the letter M in the text string that follows it. If you enter the letter M, AutoCAD will execute the command defined in that menu item. MyDraw is the menu item label. The text string inside the brackets [ ], except /M, has no function. It is used for displaying the function name so that the user can recognize the command that will be executed by selecting that item.

Line 4
[/LLine]^C^CLine
In this line, the /L defines the mnemonic key, and the Line that is inside the brackets is the menu item label. ^C^C cancels the command twice, and the Line is the AutoCAD LINE command. The part of the menu item statement that is outside the brackets is executed when you select an item from the menu. When you select Line 4, AutoCAD will execute the LINE command.

Step 2
Save the file and then load the partial menu file using the AutoCAD MENULOAD command.

Command: **MENULOAD**

When you enter the **MENULOAD** command, AutoCAD displays the **Menu Customization** dialog box [Figure 5-10(a)]. To load the menu file, enter the name of the menu file, **MYMENU1.MNU**, in the **File Name:** edit box. You can also use the Browse... option to invoke the **Select Menu File** dialog box. Select the name of the file, and then use the OK button to return to the **Menu Customization** dialog box. To load the selected menu file, select the LOAD button. The name of the menu group (MENU1) will be displayed in the Menu Groups list box.

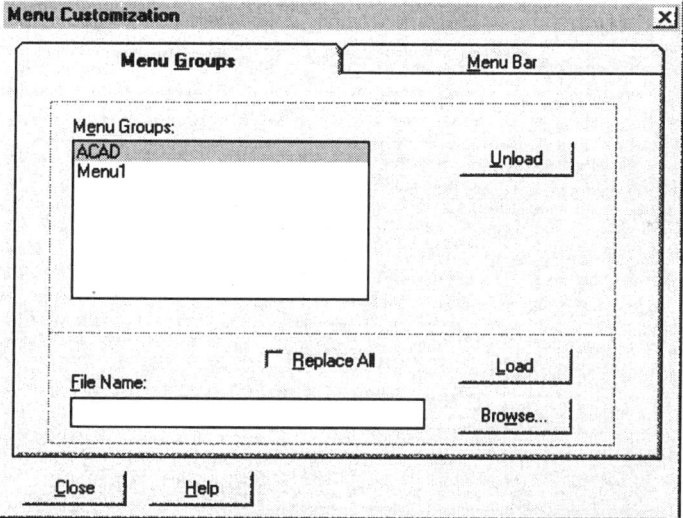

Figure 5-10(a) Menu Customization dialog box (Menu Groups tab)

5-30 Customizing AutoCAD

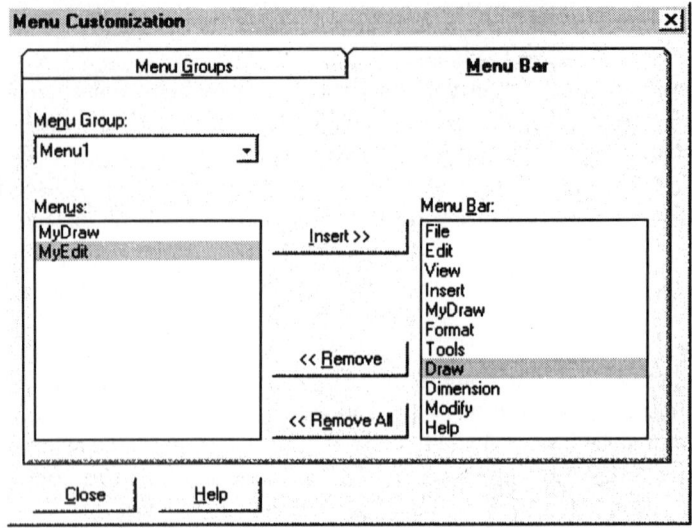

Figure 5-10(b) Menu Customization dialog box (Menu Bar tab)

You can also load the menu file from the command line, as follows:

Command: **FILEDIA**
New value for FILEDIA <1>: **0**   *(Disables the file dialog boxes.)*
Command: **MENULOAD**
Enter name of menu file to load: **MYMENU1.MNU**

You can also use the AutoLISP functions to set the **FILEDIA** system variable to 0 and then load the partial menu.

Command: **(SETVAR "FILEDIA" 0)**
Command: **(Command "MENULOAD" "MYMENU1")**

Step 3

In the Menu Customization dialog box, select the Menu Bar tab to display the menu bar options [Figure 5-10(b)]. In the Menu Groups list box select Menu1; the menus defined in the menu group (Menu1) will be displayed in the Menus list box. In the Menus list box select the menu (MyDraw) that you want to insert in the menu bar. In the Menu Bar list box select the position where you want to insert the new menu. For example, if you want to insert the new menu in front of Format, select the Format menu in the Menu Bar list box. Select the Insert button to insert the selected menu (MyDraw) in the menu bar. The menu (MyDraw1) is displayed in the menu bar located at the top of your screen.

You can also load the menu from the Command line. Once the menu is loaded, use the MENUCMD command (AutoLISP function) to display the partial menus.

Command: **(MENUCMD "P5=+Menu1.POP1")**
Command: **(MENUCMD "P6=+Menu1.POP2")**

After you enter these commands, AutoCAD will display the pull-down menu titles in the menu bar, as shown in Figure 5-11. If you select MyDraw, the corresponding pull-down menu as defined in the menu file will be displayed on the screen. Similarly, selecting MyEdit will display the corresponding edit pull-down menu.

Figure 5-11 MENUCMD command places the menu titles in the menu bar

MENUCMD is an AutoLISP function, and P5 determines where the POP1 menu will be displayed. In this example the POP1 (MyDraw) pull-down menu will be displayed as the fifth pull-down menu. Menu1 is the MENUGROUP name as defined in the menu file, and POP1 is the pull-down menu section label. The MENUGROUP name and the menu section label must be separated by a period (.).

Step 4
If you want to unload the menu (MyDraw), select the Menu Bar tab of the Menu Customization dialog box. In the Menu Bar list box select the menu item that you want to unload. Select the Remove button to remove the selected item.

Step 5
You can also unload the menu groups (Menu1) using Menu Customization dialog box. Enter the MENULOAD or MENUUNLOAD command to invoke the **Menu Customization** dialog box [Figure 5-10(a)]. AutoCAD will display the names of the menu files in the Menu Groups: list box. Select the **Menu1** menu, and then select the **Unload** button. AutoCAD will unload the menu group. Select the **Close** button to exit the dialog box. You can also unload the menu file from the command line, as follows:

    Command: **FILEDIA**
    New value for FILEDIA <1>: **0**
    Command: **MENUUNLOAD**
    Enter the name of the MENUGROUP to unload: **MENU1.MNU**

The **MENUUNLOAD** command unloads the entire menu group. You can also unload an individual pull-down menu without unloading the entire menu group by using the following command:

    Command: **(MENUCMD "P5=-)**

This command will unload the pull-down menu at position five (P5, MyDraw pull-down menu). The menu group is still loaded, but the P5 pull-down menu is not visible. The menu can also be reinitialized by using the MENU command to load the base menu **ACAD.MNU**. This will remove all partial menus and the tag definitions associated with the partial menus.

Accelerator Keys

AutoCAD for Windows also supports user-defined accelerator keys. For example, if you enter C at the Command: prompt, AutoCAD draws a circle because it is a command alias for circle as defined in the ACAD.PGP file. You cannot use the C key to enter the COPY command. To use the C key for entering the COPY command, you can define the accelerator keys. You can combine the Shift key with C in the menu file so that when you hold down the Shift key and then press the C key, AutoCAD will execute the COPY command. The following example illustrates the use of accelerator keys.

Example 7

In this example you will add the following accelerator keys to the partial menu of Example 6.

    CONTROL+"E"   to draw an ellipse (ELLIPSE command)
    SHIFT+"C"   to copy (COPY command)
    [CONTROL"Q"]   to quit (QUIT command)

The following file is the listing of the partial menu file that uses the accelerator keys of Example 7.

```
***MENUGROUP=Menu1
***POP1
**Alias
[/MMyDraw]
[/LLine]^C^CLine
[/CCircle]^C^CCircle
[/AArc]^C^CArc
ID_Ellipse [/EEllipse]^C^CEllipse

***POP2
[/EMyEdit]
[/EErase]^C^CErase
ID_Copy [/CCopy]^C^CCopy
[/OOffset]^C^COffset
[/VMove]^C^CMov

***ACCELERATORS
ID_Ellipse [CONTROL+"E"]
ID_Copy [SHIFT+"C"]
[CONTROL"Q"]^C^CQuit
```

This menu file defines three accelerator keys. The **ID_Copy [SHIFT+"C"]** accelerator key consists of two parts. The ID_Copy is the name tag, which must be the same as used earlier in the menu item definition. The SHIFT+"C" is the label that contains the modifier (SHIFT) and the keyname (C). The keyname or the string, such as "Escape", must be enclosed in quotation marks.

Pull-down, Cascading, Cursor, and Partial Menus and Customizing Toolbars 5-33

After you load the file, Shift+C will enter the COPY command and Ctrl+E will draw an ellipse. Similarly, Ctrl+Q will cancel the existing command and enter the QUIT command.

The accelerator keys can be defined in two ways. One way is to give the name tag followed by the label containing the modifier. The modifier is followed by a single character or a special virtual key enclosed in quotation marks [CONTROL+"E"] or ["ESCAPE"]. You can also use the plus sign (+) to concatenate the modifiers [SHIFT + CONTROL + "L"]. The other way of defining an accelerator key is to give the modifier and the key string, followed by a command sequence [CONTROL "Q"]^C^CQuit.

Special Virtual Keys

The following are the special virtual keys. These keys must be enclosed in quotation marks when used in the menu file.

String	Description	String	Description
"F1"	F1 key	"NUMBERPAD0"	0 key
"F2"	F2 key	"NUMBERPAD1"	1 key
"F3"	F3 key	"NUMBERPAD2"	2 key
"F4"	F4 key	"NUMBERPAD3"	3 key
"F5"	F5 key	"NUMBERPAD4"	4 key
"F6"	F6 key	"NUMBERPAD5"	5 key
"F7"	F7 key	"NUMBERPAD6"	6 key
"F8"	F8 key	"NUMBERPAD7"	7 key
"F9"	F9 key	"NUMBERPAD8"	8 key
"F10"	F10 key	"NUMBERPAD9"	9 key
"F11"	F11 key	"UP"	Up-arrow key
"F12"	F12 key	"DOWN"	Down-arrow key
"HOME"	Home key	"LEFT"	Left-arrow key
"END"	End key	"RIGHT"	Right-arrow key
"INSERT"	Ins key	"ESCAPE"	Esc key
"DELETE"	Del key		

Valid Modifiers. The following are the valid modifiers:

String	Description
CONTROL	The control key on the keyboard
SHIFT	The Shift key (left or right)
COMMAND	The Apple key on Macintosh keyboards
META	The meta key on UNIX keyboards

Toolbars

The contents of the toolbar and its default layout can be specified in the Toolbar section (***TOOLBARS) of the menu file. Each toolbar must be defined in a separate submenu.

Toolbar Definition. The following is the general format of the toolbar definition:

    ***TOOLBARS
    **MYTOOLS1

TAG1 [Toolbar ("tbarname", orient, visible, xval, yval, rows)]
TAG2 [Button ("btnname", id_small, id_large)]macro
TAG3 [Flyout ("flyname", id_small, id_large, icon, alias)]macro
TAG4 [control (element)]
[--]

***TOOLBARS** is the section label of the toolbar, and **MYTOOLS1** is the name of the submenu that contains the definition of a toolbar. Each toolbar can have five distinct items that control different elements of the toolbar: TAG1, TAG2, TAG3, TAG4, and separator ([--]).

The first line of the toolbar (TAG1) defines the characteristics of the toolbar. In this line, **Toolbar** is the keyword, and it is followed by a series of options enclosed in parentheses. The following describes the available options.

tbarname	This is a text string that names the toolbar. The tbarname text string must consist of alphanumeric characters with no punctuation other than a dash (-) or an underscore (_).
orient	This determines the orientation of the toolbar. The acceptable values are Floating, Top, Bottom, Left, and Right. These values are not case-sensitive.
visible	This determines the visibility of the toolbar. The acceptable values are Show and Hide. These values are not case-sensitive.
xval	This is a numeric value that specifies the X ordinate in pixels. The X ordinate is measured from the left edge of the screen to the right side of the toolbar.
yval	This is a numeric value that specifies the Y ordinate in pixels. The Y ordinate is measured from the top edge of the screen to the top of the toolbar.
rows	This is a numeric value that specifies the number of rows.

The second line of the toolbar (TAG2) defines the button. In this line the **Button** is the key word and it is followed by a series of options enclosed in parentheses. The following is the description of the available options.

btnname	This is a text string that names the button. The text string must consist of alphanumeric characters with no punctuation other than a dash (-) or an underscore (_). This text string is displayed as ToolTip when you place the cursor over the button.
id_small	This is a text string that names the ID string of the small image resource (16 by 16 bitmap). The text string must consist of alphanumeric characters with no punctuation other than a dash (-) or an underscore (_). The id_small text string can also specify a user-defined bitmap (Example: ICON_16_CIRCLE).

id_big   This is a text string that names the ID string of the large image resource (32 by 32 bitmap). The text string must consist of alphanumeric characters with no punctuation other than a dash (-) or an underscore (_). The id_big text string can also specify a user-defined bitmap (Example: ICON_32_CIRCLE).

macro   The second line (TAG2), which defines a button, is followed by a command string (macro). For example, the macro can consist of ^C^CLine. It follows the same syntax as that of any standard menu item definition.

The third line of the toolbar (TAG3) defines the flyout control. In this line the **Flyout** is the key word, and it is followed by a series of options enclosed in parentheses. The following describes the available options.

flyname   This is a text string that names the flyout. The text string must consist of alphanumeric characters with no punctuation other than a dash (-) or an underscore (_). This text string is displayed as ToolTip when you place the cursor over the flyout button.

id_small   This is a text string that names the ID string of the small image resource (16 by 16 bitmap). The text string must consist of alphanumeric characters with no punctuation other than a dash (-) or an underscore (_). The id_small text string can also specify a user-defined bitmap.

id_big   This is a text string that names the ID string of the large image resource (32 by 32 bitmap). The text string must consist of alphanumeric characters with no punctuation other than a dash (-) or an underscore (_). The id_big text string can also specify a user-defined bitmap.

icon   This is a Boolean key word that determines whether the button displays its own icon or the last icon selected. The acceptable values are **ownicon** and **othericon**. These values are not case-sensitive.

alias   The alias specifies the name of the toolbar submenu that is defined with the standard ****aliasname** syntax.

macro   The third line (TAG3), which defines a flyout control, is followed by a command string (macro). For example, the macro can consist of ^C^CCircle. It follows the same syntax as that of any standard menu item definition.

The fourth line of the toolbar (TAG4) defines a special control element. In this line the **Control** is the key word, and it is followed by the type of control element enclosed in parentheses. The following describes the available control element types.

element   This parameter can have one of the following three values:
Layer: This specifies the layer control element.
Linetype: This specifies the linetype control element.
Color: This specifies the color control element.

The fifth line ([--]) defines a separator.

Example 8

In this example you will write a menu file for a toolbar for the LINE, PLINE, CIRCLE, ELLIPSE, and ARC commands. The name of the toolbar is MyDraw1 (Figure 5-12).

The following is the listing of the menu file:
***TOOLBARS
**TB_MyDraw1
ID_MyDraw1[_Toolbar("MyDraw1", _Floating, _Hide, 10, 200, 1)]
ID_Line [_Button("Line", ICON_16_LINE, ICON_32_LINE)]^C^C_line
ID_Pline [_Button("Pline", ICON_16_PLine, ICON_32_PLine)]^C^C_PLine
ID_Circle[_Button("Circle", ICON_16_CirRAD, ICON_32_CirRAD)]^C^C_Circle
ID_ELLIPSE[_Button("Ellipse", ICON_16_EllCEN, ICON_32_EllCEN)]^C^C_ELLIPSE
ID_Arc[_Button("Arc 3Point", ICON_16_Arc3Pt, ICON_32_Arc3Pt)]^C^C_Arc

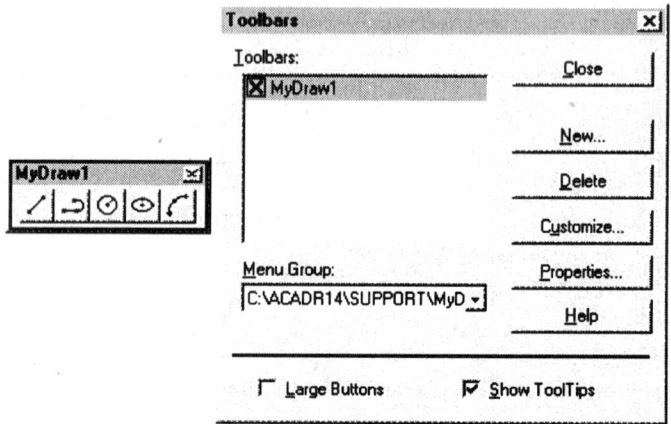

Figure 5-12 MyDraw1 toolbar for Example 8 and Toolbars dialog box

Use the MENULOAD command to load the MyDraw1 menu group as discussed earlier. To display the new toolbar (MyDraw1) on the screen, select Toolbars from the View pull-down menu and then select MyDraw1 in the Menu Groups list box. Turn the MyDraw1 toolbar on in the Toolbars dialog box; MyDraw1 toolbar is displayed on the screen.

You can also load the new toolbar from the command line. After using the MENULOAD command to load the MyDraw1 menu group, use the -TOOLBAR command to display the MyDraw1 toolbar.

Command: **-TOOLBAR**
Toolbar name (or ALL): **MYDRAW1**
Show/Hide/Left/Right/Top/Bottom/Float: <Show>: **S**

Example 9

In this example you will write a menu file for a toolbar with a flyout. The name of the toolbar is MyDraw2 (Figure 5-13), and it contains two icons, Circle and Arc. When you select the Circle icon, it should display a flyout with radius, diameter, 2P, and 3P icons (Figure 5-14). Similarly, when you select the Arc icon, it should display the 3Point, SCE, and SCA icons.

The following is the listing of the menu file:
***Menugroup=M2
***TOOLBARS
**TB_MyDraw2
ID_MyDraw2[_Toolbar("MyDraw2", _Floating, _Show, 10, 100, 1)]
ID_TbCircle[_Flyout("Circle", ICON_16_Circle, ICON_32_Circle, _OtherIcon, M2.TB_Circle)]
ID_TbArc[_Flyout("Arc", ICON_16_Arc, ICON_32_Arc, _OtherIcon, M2.TB_Arc)]

**TB_Circle
ID_TbCircle[_Toolbar("Circle", _Floating, _Hide, 10, 150, 1)]
ID_CirRAD[_Button("Circle C,R", ICON_16_CirRAD, ICON_32_CirRAD)]^C^C_Circle
ID_CirDIA[_Button("Circle C,D", ICON_16_CirDIA, ICON_32_CirDIA)]^C^C_Circle;\D
ID_Cir2Pt[_Button("Circle 2Pts", ICON_16_Cir2Pt, ICON_32_Cir2Pt)]^C^C_Circle;2P
ID_Cir3Pt[_Button("Circle 3Pts", ICON_16_Cir3Pt, ICON_32_Cir3Pt)]^C^C_Circle;3P

**TB_Arc
ID_TbArc[_Toolbar("Arc", _Floating, _Hide, 10, 150, 1)]
ID_Arc3PT[_Button("Arc,3Pts", ICON_16_Arc3PT, ICON_32_Arc3PT)]^C^C_Arc
ID_ArcSCE[_Button("Arc,SCE", ICON_16_ArcSCE, ICON_32_ArcSCE)]^C^C_Arc;\C
ID_ArcSCA[_Button("Arc,SCA", ICON_16_ArcSCA, ICON_32_ArcSCA)]^C^C_Arc;\C;\A

Explanation

ID_TbCircle[_Flyout("Circle", ICON_16_Circle, ICON_32_Circle, _OtherIcon, **M2.TB_Circle)]**

In this line M2 is the MENUGROUP name (***MENUGROUP=M2) and TB_Circle is the name of the toolbar submenu. **M2.TB_Circle** will load the submenu TB_Circle that has been defined in the M2 menugroup. If M2 is missing, AutoCAD will not display the flyout when you select the Circle icon.

ID_CirDIA[_Button("Circle C,D", ICON_16_**CirDIA**, ICON_32_**CirDIA**)]^C^C_Circle;\D

CirDIA is a user-defined bitmap that displays the Circle-diameter icon. If you use any other name, AutoCAD will not display the desired icon.

To load the toolbar, use the MENULOAD command to load the menu file. The MyDraw1 toolbar will be displayed on the screen.

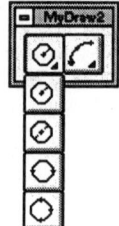

Figure 5-13 Toolbar for Example 9   **Figure 5-14** Toolbar with flyout

Menu-Specific Help

AutoCAD for Windows allows access to online help. For example, if you want to define a helpstring for the CIRCLE and ARC commands, the syntax is as follows:

    ***HELPSTRING
    ID_Copy *(This command will copy the selected object.)*
    ID_Ellipse *(This command will draw an ellipse.)*

The ***HELPSTRING** is the section label for the helpstring menu section. The lines defined in this section start with a name tag (ID_Copy) and are followed by the label enclosed in square brackets. If you press the F1 key when the menu item is highlighted, the help engine gets activated and AutoCAD displays the helpstring defined with the name tag.

CUSTOMIZING THE TOOLBARS

AutoCAD has provided several toolbars that should be sufficient for general use. However, sometimes you may need to customize the toolbars so that the commands that you use frequently are grouped in one toolbar. This saves time in selecting commands. It also saves the drawing space because you do not need to have several toolbars on the screen. The following example explains the process involved in creating and editing the toolbars.

Example 10

In this example you will create a new toolbar (MyToolbar1) that has Line, Polyline, Circle (Center, Radius option), Arc (Center, Start, End option), Spline, and Paragraph Text (MText) commands. You will also change the image and tooltip of Line icon and perform other operations like deleting toolbars and icons and copying icons between toolbars.

1. Select the toolbars in the View pull-down menu. Toolbars dialog box will appear on the screen. You can also invoke this dialog box by entering TBCONFIG at AutoCAD's Command: prompt.

2. Select New button in the Toolbars dialog box; AutoCAD will display the New Toolbar dialog box at the top of the existing (Toolbars) dialog box.

3. Enter the name of the toolbar in the Toolbar Name edit box (Example MyToolbar1) and then select the OK button to exit the dialog box. The name of the new toolbar (MyToolbar1) is displayed in the Toolbars dialog box (Figure 5-15).

4. Select the new toolbar, if it is not already selected, by selecting the box that is located just to the left of the toolbar name (MyToolbar1). The new toolbar (MyToolbar1) appears on the screen.

5. Select the Customize button in the Toolbars dialog box (Figure 5-16). The Customize Toolbars dialog box is displayed on the screen.

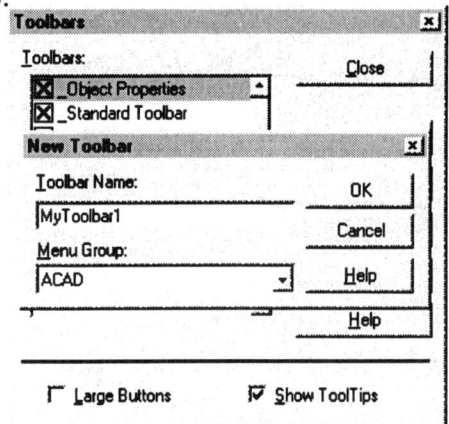

6. In the Categories edit box select the down arrow to display the list of toolbars and select the draw toolbar. (You can also type D to display the toolbars that start with the letter D). AutoCAD displays the draw tool icons.

Figure 5-15 Toolbars dialog box

7. Select and drag the Line icon and position it in the MyToolbar1 toolbox. Repeat the same for Polyline, Circle (Center, Radius option), Arc (Center, Start, End option), Spline and Text (Paragraph text) icons.

8. Now, select the Dimensioning category and then select and drag some of the dimensioning commands to MyToolbar1.

9. Similarly, you can open any toolbar category and add commands to the new toolbar.

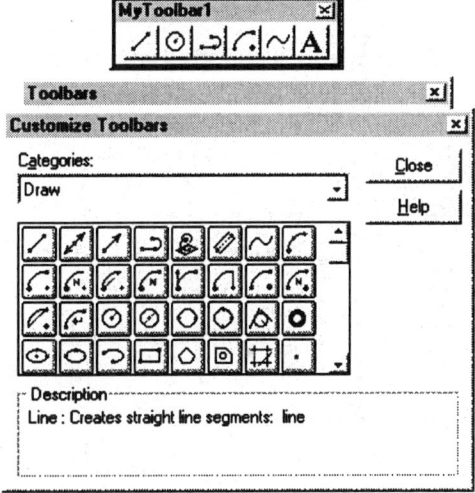

10. When you are done defining the commands for the new toolbar, select the Close button in the Customize Toolbars dialog box to return to the Toolbars dialog box. Select the Close button again to return to AutoCAD screen.

11. Test the icons in the new toolbar (MyToolbar1).

Figure 5-16 MyToolbar1 toolbar and Customize Toolbars dialog box

Creating a New Image and Tooltip for an Icon

1. Make sure the icon with the image you want to edit is displayed on the screen. In this example we want to edit the image of the Line icon of MyToolbar1.

5-40 Customizing AutoCAD

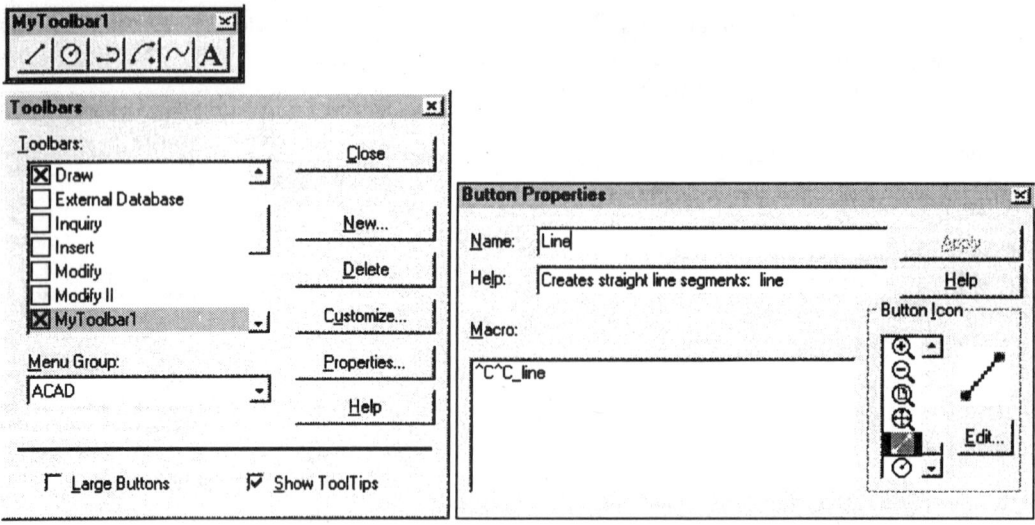

Figure 5-17 MyToolbar1 toolbar, Toolbars and Button Properties dialog box

2. Right-click on the Line icon of MyToolbar1; the Toolbars dialog box appears on the screen. Again, right-click on the Line icon; the Button Properties dialog box is displayed with the image of the existing Line icon (Figure 5-17).

3. To edit the shape of the image, select the edit button to access the Button Editor. Select the Grid to display the grid lines.

   You can edit the shape by using different tools in the Button Editor. You can draw a Line by selecting the Line button and specifying two points. You can draw a circle or an ellipse by using the Circle button. The Erase button can be used to erase the image.

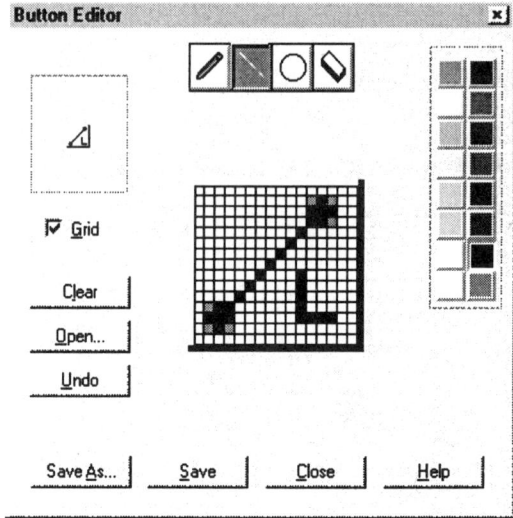

Figure 5-18 Button Editor dialog box

4. In this example, we want to change the color of Line image. To accomplish this, erase the existing line, and then select the color and draw a line. Also, create the shape L in the lower right corner (Figure 5-18).

5. Select the SaveAs button and save the image as MyLine in Tutorial directory. Select the Close button to exit the Button Editor.

6. In the Button Properties dialog box enter MyLine in the Name edit box. This changes the tooltip of this icon. Now, select the Apply button to apply the changes to the image. Close the dialog boxes to return to AutoCAD screen. Notice the change in the icon image and tooltip (Figure 5-19).

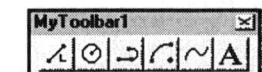

Figure 5-19 MyToollbar1 Toolbar

Deleting the Icons from a Toolbar

1. Right-click on the icon that you want to remove from the toolbox. The Toolbars dialog box is displayed. In this example, we want to remove the spline icon from the MyToolbar1 toolbox.

2. Select the customize button in the Toolbars dialog box. The Customize Toolbars dialog box is displayed.

3. Click and drag out the Spline icon from the MyToolbar1 toolbox. The icon will be deleted from the toolbox.

4. Repeat the above step to delete other icons, if needed.

5. Close the dialog boxes to return to screen.

Deleting a Toolbar

1. Select Toolbars from the View pull-down menu.

2. Select the Toolbar that you want to delete and then select the delete button. The toolbar you selected is deleted.

Copying a Tool Icon

In this example we want to copy the ordinate dimensioning icon from Dimensioning toolbar to MyToolbar1.

1. Select the Toolbars from the View pull-down menu or right-click on any icon in any toolbar.

2. In the Toolbars dialog box select the check box for Dimensioning and MyToolbar1. The Dimensioning and MyToolbar1 toolbars are displayed on the screen.

3. In the Toolbars dialog box select the Customizing button; the Customize Toolbars dialog box appears.

4. Hold the Ctrl key down and then select and drag the Ordinate dimensioning icon from the Dimensioning toolbar to MyToolbar1. The Ordinate dimensioning toolbar is copied to MyToolbar1.

Note
If you do not hold down the Ctrl key, the icon you selected will be moved from the Dimensioning toolbar to MyToolbar1.

Any changes made in the toolbars are saved in the ACAD.MNS and ACAD.MNR files. The following is the partial listing of ACAD.MNS file:

**MYTOOLBAR1
ID_MyToolbar1_0 [_Toolbar("MyToolbar1", _Floating, _Show, 512, 177, 1)]
ID_Line_0   [_Button("MyLine", "ICON.bmp", "ICON_24_LINE")]^C^C_line
ID_CircleCenterRadius_0 [_Button("Circle Center Radius", "ICON_16_CIRRAD", "ICON_24_CIRRAD")]^C^C_circle
ID_Polyline_0 [_Button("Polyline", "ICON_16_PLINE", "ICON_24_PLINE")]^C^C_pline
ID_ArcCenterStartEnd_0 [_Button("Arc Center Start End", "ICON_16_ARCCSE", "ICON_24_ARCCSE")]^C^C_arc _c
ID_Text_0   [_Button("Text", "ICON_16_MTEXT", "ICON_24_MTEXT")]^C^C_mtext

Creating Custom Toolbars with Flyout Icons

In this example you will create a custom toolbar with flyout icons.

1. Right click on any toolbar, the **Toolbars** dialog box is displayed on the screen.

2. In the **Toolbars** dialog box, select the New button to display the New Toolbars dialog box. Enter the name of the new toolbar (For example, MYToolbar2) and then select the OK button to exit the box.

3. In the Toolbars dialog, select the new toolbar name (MYToolbar2) and then select the customize button to display the **Customize Toolbars** dialog box. From the pop-down list select custom and then drag the flyout icon to the new toolbar.

4. Right click twice on the flyout icon of the new toolbar. The **Flyout Properties** dialog box (Figure 5-20) is displayed on the screen. To associate it with a predefined toolbar, select the toolbar name (For example, ACAD.Draw) from the list.

5. Select the Apply button and then close the dialog boxes. If you click on the flyout icon of the new toolbar, the corressponding flyout icons are displayed.

6. Now, you can add new icons to the toolbar or delete the ones you do not want in this toolbar.

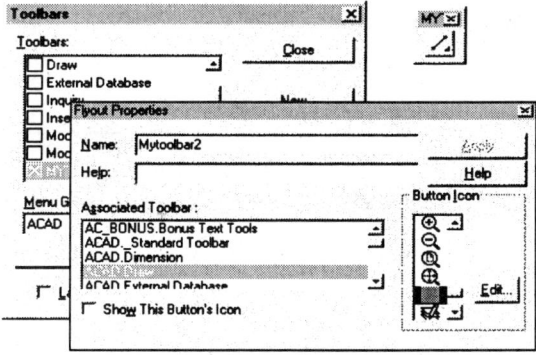

Figure 5-20 Flyout Properties dialog box and the new toolbar (MYToolbar2)

REVIEW QUESTIONS

1. A pull-down menu can have _____ sections.

2. The length of the section title should not exceed _____ characters.

3. The section titles in a pull-down menu are _____ justified.

4. In a pull-down menu, a line consisting of two hyphens ([--]) _____ automatically to fill the _____ of the pull-down menu.

5. If the menu item begins with a tilde (~), the items will be _____ .

6. Pull-down menus are _____ when the dynamic zoom is in progress.

7. Every cascading menu in the menu file should have a _____ name.

8. The cascading menu name can be _____ characters long.

9. The cascading menu names should not have any _____ blanks.

10. In Windows you can write partial menus, toolbars, and accelerator key definitions. (T/F)

11. A menu file can contain only one MENUGROUP. (T/F)

12. You can load the partial menu file by using the AutoCAD _____ command.

13. Once the menu is loaded, you can use the _____ (AutoLISP function) to display the partial menus.

EXERCISES

Exercise 3

Write a pull-down menu for the following AutoCAD commands. (The layout of the pull-down menu is shown in Figure 5-21.)

LINE	DIM HORZ	DTEXT LEFT
CIRCLE C,R	DIM VERT	DTEXT RIGHT
CIRCLE C,D	DIM RADIUS	DTEXT CENTER
ARC 3P	DIM DIAMETER	DTEXT ALIGNED
ARC SCE	DIM ANGULAR	DTEXT MIDDLE
ARC CSE	DIM LEADER	DTEXT FIT

```
                PULL-DOWN MENU

    ┌─────────────────────────────────────────┐
    │  DRAW        DIM            DTEXT       │
    └─────────────────────────────────────────┘
       LINE         DIM-HORZ       DTEXT-LEFT
       CIRCLE C,R   DIM-VERT       DTEXT-RIGHT
       CIRCLE C,D   DIM-RADIUS     DTEXT-CENTER
       ARC 3P       DIM-DIAMETER   DTEXT-ALIGNED
       ARC SCE      DIM-ANGULAR    DTEXT-MIDDLE
       ARC CSE      DIM-LEADER     DTEXT-FIT
```

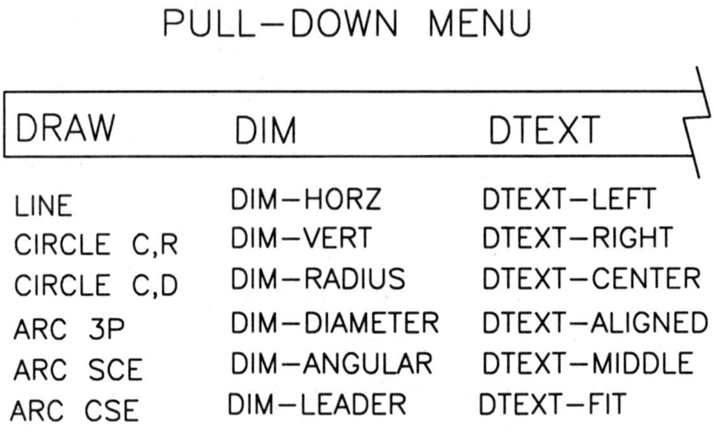

Figure 5-21 Layout of pull-down menu

Exercise 4

Write a pull-down menu for the following AutoCAD commands.

LINE	BLOCK
PLINE	WBLOCK
CIRCLE C,R	INSERT
CIRCLE C,D	BLOCK LIST
ELLIPSE AXIS ENDPOINT	DDATTDEF
ELLIPSE CENTER	DDATTE

Exercise 5

Write a partial menu for Windows. The menu file should have two pull-down menus, POP1 (MyArc) and POP2 (MyDraw). The MyArc pull-down menu should contain all Arc options and must be displayed at the sixth position. Similarly, the MyDraw pull-down menu should contain Line, Circle, Pline, Trace, Dtext, and Mtext commands and should occupy the ninth position.

Exercise 6

Write a menu file for a toolbar with a flyout. The name of the toolbar is MyDrawX1, and it contains two icons, Polygon and Ellipse. When you select the Polygon icon, it should display a flyout with rectangle and polygon icons. Similarly, when you select the Ellipse icon, it should display the Ellipse-Center Option and Ellipse-Edge Option icons.

Exercise 7

Write a pull-down for the following AutoCAD commands. (The layouts of the pull-down menu is shown in Figure 5-22.)

LAYER NEW	SNAP 0.25	UCS WORLD
LAYER MAKE	SNAP 0.5	UCS PREVIOUS
LAYER SET	GRID 1.0	VPORTS 2
LAYER LIST	DRID 10.0	VPORTS 4
LAYER ON	APERTURE 5	VPORTS SING.
LAYER OFF	PICKBOX 5	

PULL-DOWN MENU

LAYER	SETTINGS	UCS-PORT
LAYER-New	SNAP 0.25	UCS-World
LAYER-Make	SNAP 0.5	UCS-Pre
LAYER-Set	GRID 1.0	VPORTS-2
LAYER-List	GRID 10.0	VPORTS-4
LAYER-On	APERTURE 5	VPORTS-1
LAYER-Off	PICKBOX 5	

Figure 5-22 Design of pull-down menu for Exercise 7

Chapter 6

Tablet Menus

Learning objectives

After completing this chapter, you will be able to:
- *Understand the advantages of tablet menus.*
- *Write and customize tablet menus.*
- *Load menus and configure tablet menus.*
- *Write tablet menus with different block sizes.*
- *Assign commands to tablet overlays.*

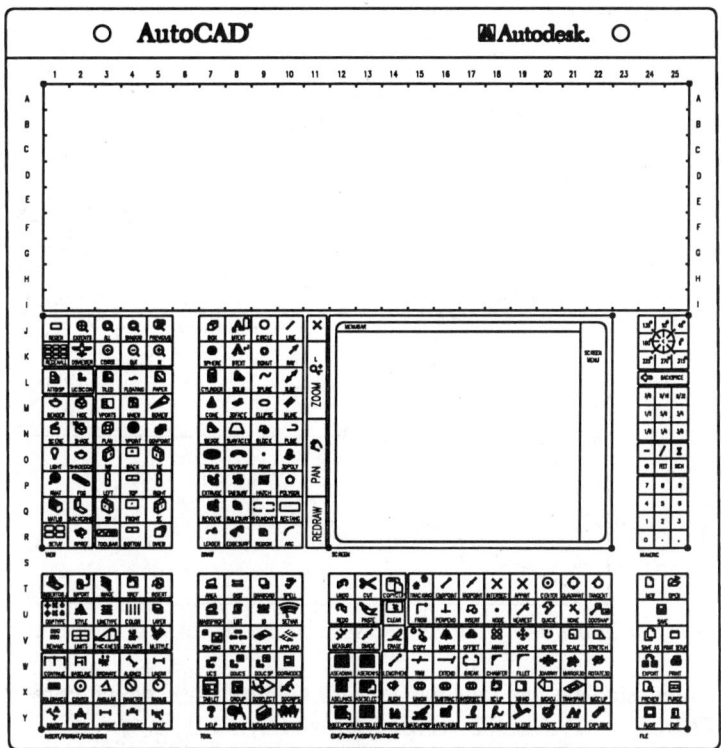

Figure 6-1 Sample tablet template

STANDARD TABLET MENU

The tablet menu provides a powerful alternative for entering commands. In the tablet menu, the commands are selected from the template that is secured on the surface of a digitizing tablet. To use the tablet menu you need a digitizing tablet and a pointing device. You also need a tablet template (Figure 6-1) that contains AutoCAD commands arranged in various groups for easy identification.

The standard AutoCAD menu file has four tablet menu sections: TABLET1, TABLET2, TABLET3, and TABLET4. When you start AutoCAD and get into the drawing editor, the tablet menu sections TABLET1, TABLET2, TABLET3, and TABLET4 are automatically loaded. The commands defined in these four sections are then assigned to different blocks of the template.

The first tablet menu section (TABLET1) has 225 blank blocks that can be used to assign up to 225 menu items. The remaining tablet menu sections contain AutoCAD commands, arranged in functional groups that make it easier to identify and access the commands. The commands contained in the TABLET2 section include RENDER, SOLID MODELING, DISPLAY, INQUIRY, DRAW, ZOOM, and PAPER SPACE commands. The TABLET3 section contains numbers, fractions, and angles. The commands contained in the TABLET4 section include TEXT, DIMENSIONING, OBJECT SNAPS, EDIT, UTILITY, XREF, and SETTINGS.

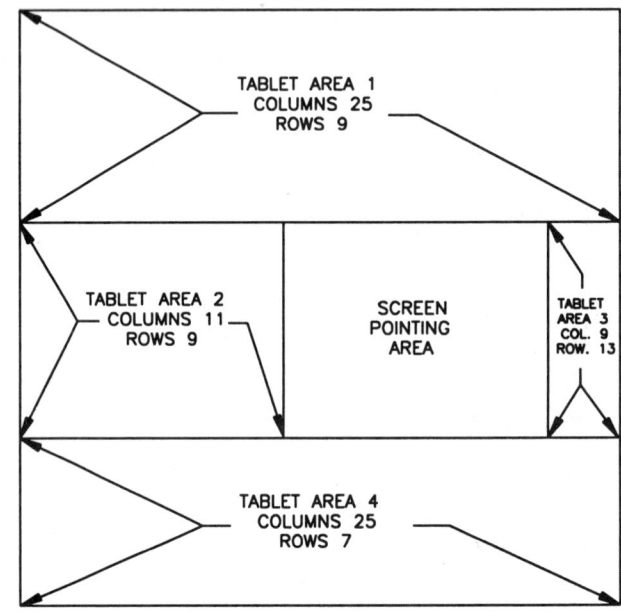

Figure 6-2 Four tablet areas of the AutoCAD tablet template

The AutoCAD tablet template has four tablet areas (Figure 6-2), which correspond to four tablet menu sections, TABLET1, TABLET2, TABLET3, and TABLET4.

ADVANTAGES OF A TABLET MENU

A tablet menu has the following advantages over the screen menu, pull-down menu, image menu, or keyboard.

1. In the tablet menu the commands can be arranged so that the most frequently used commands can be accessed directly. This can save considerable time in entering AutoCAD commands. In the screen menu or the pull-down menu some of the commands cannot be accessed directly. For example, to generate a horizontal dimension you have to go through several steps. You first select Dimension from the root menu and then Linear. In the tablet menu you can select the Linear dimensioning command directly from the digitizer. This saves time and eliminates the distraction that takes place as you page through different screens.

2. You can have the graphical symbols of the AutoCAD commands drawn on the tablet template. This makes it much easier to recognize and select commands. For example, if you are not an expert in AutoCAD dimensioning, you may find Baseline and Continue dimensioning confusing. But if the command is supported by the graphical symbol illustrating what a command does, the chances of selecting a wrong command are minimized.

3. You can assign any number of commands to the tablet overlay. The number of commands you can assign to a tablet is limited only by the size of the digitizer and the size of the rectangular blocks.

CUSTOMIZING A TABLET MENU

As with a screen menu, you can customize the AutoCAD tablet menu. It is a powerful customizing tool to make AutoCAD more efficient.

The tablet menu can contain a maximum of four sections: TABLET1, TABLET2, TABLET3, and TABLET4. Each section represents a rectangular area on the digitizing tablet. These rectangular areas can be further divided into any number of rectangular blocks. The size of each block depends on the number of commands that are assigned to the tablet area. Also, the rectangular tablet areas can be located anywhere on the digitizer and can be arranged in any order. The AutoCAD TABLET command configures the tablet. The MENU command loads and assigns the commands to the rectangular blocks on the tablet template.

Before writing a tablet menu file, it is very important to design the layout of the tablet template. A well-thought-out design can save a lot of time in the long run. The following points should be considered when designing a tablet template:

1. Learn the AutoCAD commands that you use in your profession.

2. Group the commands based on their function, use, or relationship with other commands.

3. Draw a rectangle representing a template so that it is easy for you to move the pointing device around. The size of this area should be appropriate to your application. It should not be too large or too small. Also, the size of the template depends on the active area of the digitizer.

6-4 Customizing AutoCAD

4. Divide the remaining area into four different rectangular tablet areas for TABLET1, TABLET2, TABLET3, and TABLET4. It is not necessary to use all four areas; you can have fewer tablet areas, but four is the maximum.

5. Determine the number of commands you need to assign to a particular tablet area; then determine the number of rows and columns you need to generate in each area. The size of the blocks does not need to be the same in every tablet area.

6. Use the TEXT command to print the commands on the tablet overlay, and draw the symbols of the command, if possible.

7. Plot the tablet overlay on good-quality paper or a sheet of Mylar. If you want the plotted side of the template to face the digitizer board, you can create a mirror image of the tablet overlay and then plot the mirror image.

WRITING A TABLET MENU

When writing a tablet menu you must understand the AutoCAD commands and the prompt entries required for each command. Equally important is the design of the tablet template and the placement of various commands on it. Give considerable thought to the design and layout of the template, and, if possible, invite suggestions from AutoCAD users in your trade. To understand the process involved in developing and writing a tablet menu, consider Example 1.

Example 1

Write a tablet menu for the following AutoCAD commands. The commands are to be arranged as shown in Figure 6-3. Make a tablet menu template for configuration and command selection (filename **TM1.MNU**).

LINE	CIRCLE	PLINE
CIRCLE C,D	ERASE	CIRCLE 2P

Figure 6-3 represents one of the possible template designs, where the AutoCAD commands are in one row at the top of the template, and the screen pointing area is in the center. There is only one area in this template; therefore, you can place all these commands under the section label TABLET1. To write a menu file you can use any text editor like Notepad.

The name of the file is **TM1**, and the extension of the file is **.MNU**. **The line numbers are not part of the file. They are shown here for reference only.**

```
***TABLET1                        1
^C^CLINE                          2
^C^CPLINE                         3
^C^CCIRCLE                        4
^C^CCIRCLE \D                     5
^C^CCIRCLE 2P                     6
^C^CERASE                         7
```

Explanation

Line 1
***TABLET1
TABLET1 is the section label of the first tablet area. All the section labels are preceded by three asterisks (***).

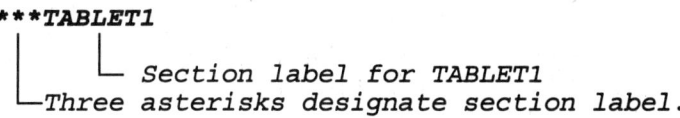

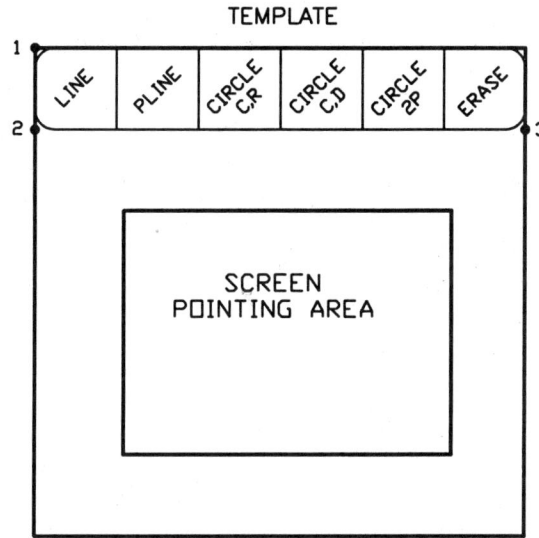

Figure 6-3 Design of tablet template

Line 2
^C^CLINE
^C^C cancels the existing command twice; LINE is an AutoCAD command. There is no space between the second ^C and LINE.

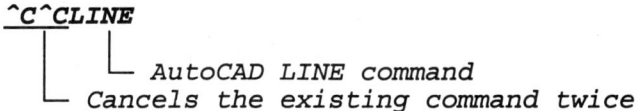

Line 3
^C^CPLINE
^C^C cancels the existing command twice; PLINE is an AutoCAD command.

Line 4
^C^CCIRCLE

6-6 Customizing AutoCAD

^C^C cancels the existing command twice; CIRCLE is an AutoCAD command. The default input for the CIRCLE command is the center and the radius of the circle; therefore, no additional input is required for this line.

Line 5
^C^CCIRCLE \D
^C^C cancels the existing command twice; CIRCLE is an AutoCAD command like the previous line. However, this command definition requires the diameter option of the circle command. This is accomplished by using \D in the command definition. There should be no space between the backslash (\) and the D, but there should always be a space **before** the backslash (\). The backslash (\) lets the user enter a point, and in this case it is the center point of the circle. After you enter the center point, the diameter option is selected by the letter D, which follows the backslash (\).

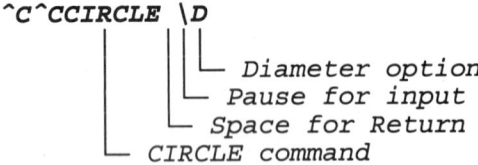

Line 6
^C^CCIRCLE 2P
^C^C cancels the existing command twice; CIRCLE is an AutoCAD command. The 2P selects the two-point option of the CIRCLE command.

Line 7
^C^CERASE
^C^C cancels the existing command twice; ERASE is an AutoCAD command that erases the selected objects.

Note
In the tablet menu, the part of the menu item that is enclosed in the brackets is used for screen display only. For example, in the following menu item, T1-6 will be ignored and will have no effect on the command definition.

```
                  ┌─ For reference only and has no effect
                  │  on the command definition
                  │
          [T1-6]^C^CCIRCLE 2P
            │       │
            │       └─ Item number 6
            └─ Tablet area 1
```

The reference information can be used to designate the tablet area and the line number.

Before you can use the commands from the new tablet menu, you need to configure the tablet and load the tablet menu.

TABLET CONFIGURATION

To use the new template to select the commands, you need to configure the tablet so that AutoCAD knows the location of the tablet template and the position of the commands assigned to each block. This is accomplished by using the AutoCAD TABLET command. Secure the tablet template (Figure 6-3) on the digitizer with the edges of the overlay approximately parallel to the edges of the digitizer. Enter the AutoCAD TABLET command, select the Configure option, and respond to the following prompts. Figure 6-4 shows the points you need to select to configure the tablet.

Command: **TABLET**
Options (ON/OFF/CAL/CFG):**CFG**
Enter number of tablet menus desired (0-4): **1**
Do you want to realign tablet menu areas? <N>: **Y**
Digitize upper left corner of menu area 1; **P1**
Digitize lower left corner of menu area 1; **P2**
Digitize lower right corner of menu area 1; **P3**
Enter the number of columns for menu area 1: **6**
Enter the number of rows for menu area 1: **1**
Do you want to respecify the Fixed Screen Pointing Area? <N>: **Y**
Digitize lower left corner of Fixed Screen pointing area: **P4**
Digitize upper right corner of Fixed Screen pointing area: **P5**
Do you want to specify the Floating Screen Pointing Area? <N>: **N**

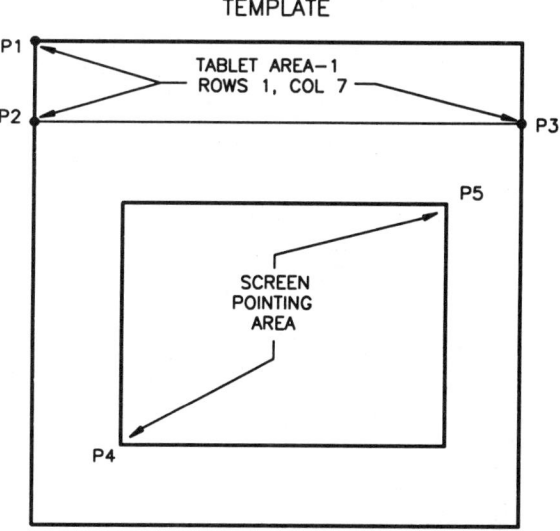

Figure 6-4 Points that need to be selected to configure the tablet

Note
The three points P1, P2, and P3 should form a 90-degree angle. If the selected points do not form a 90-degree angle, AutoCAD will prompt you to enter the points again until they do.

6-8 Customizing AutoCAD

> *The tablet areas should not overlap the screen pointing area.*
>
> *The screen pointing area can be any size and located anywhere on the tablet as long as it is within the active area of the digitizer. The screen pointing area should not overlap other tablet areas. The screen pointing area you select will correspond to the monitor screen area. Therefore, the length-to-width ratio of the screen pointing area should be the same as that of the monitor, unless you are using the screen pointing area to digitize a drawing.*

LOADING MENUS

AutoCAD automatically loads the **ACAD.MNU** file when you get into the AutoCAD drawing editor. However, you can also load a different menu file by using the AutoCAD MENU command.

Command: **Menu**

When you enter the Menu command, AutoCAD displays the **Select Menu File** dialog box on the screen (Figure 6-5). Select the menu file that you want to load and then select the Open button.

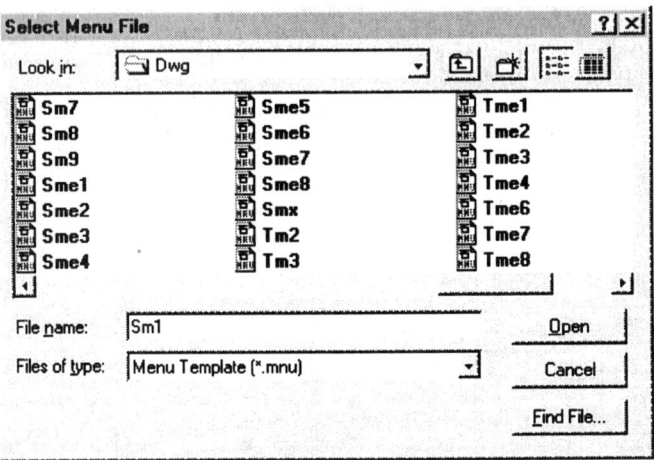

Figure 6-5 Select Menu File dialog box

You can also load the menu file from the command line.

    Command: **FILEDIA**
    New value for FILEDIA <1>: **0**
    Command: **Menu**
    `Menu filename <ACAD>: PDM1`
                                      Name of menu file
                        Default menu file

After you enter the MENU command, AutoCAD will prompt for the file name. Enter the name of the menu file without the file extension (.MNU), since AutoCAD assumes the extension .MNU. AutoCAD will automatically compile the menu file into **MNC** and **MNR** files.

Exercise 1

Write a tablet menu for the following AutoCAD commands. Make a tablet menu template for configuration and command selection (filename **TME1.MNU**).

    LINE   TEXT-Center
    CIRCLE C,R   TEXT-Left
    ARC C.S.E   TEXT-Right
    ELLIPSE   TEXT-Aligned
    DONUT

Use the template in Figure 6-6 to arrange the commands. The draw and text commands should be placed in two separate tablet areas.

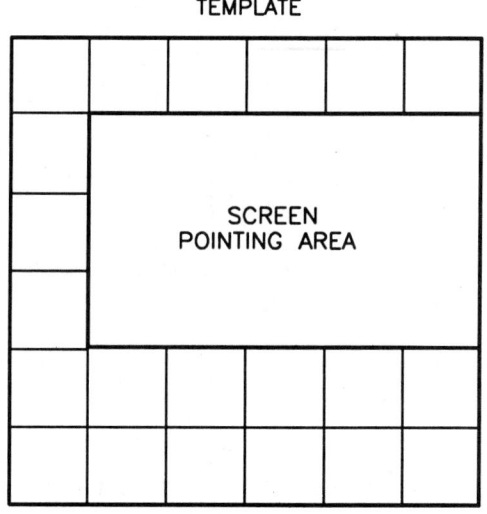

Figure 6-6 Template for Exercise 1

TABLET MENUS WITH DIFFERENT BLOCK SIZES

As mentioned earlier, the size of each tablet area can be different. The size of the blocks in these tablet areas can also be different. But the size of every block in a particular tablet area must be the same. This provides you with a lot of flexibility in designing a template. For example, you may prefer to have smaller blocks for numbers, fractions, or letters, and larger blocks for draw commands. You can also arrange these tablet areas to design a template layout with different shapes, such as L-shape and T-shape.

6-10 Customizing AutoCAD

Tablet Area-1

Tablet Area-2

The following example illustrates the use of multiple tablet areas with different block sizes.

Example 2

Write a tablet menu for the tablet overlay shown in Figure 6-7(a). Figure 6-7(b) shows the number of rows and columns in different tablet areas (filename **TM2.MNU**).

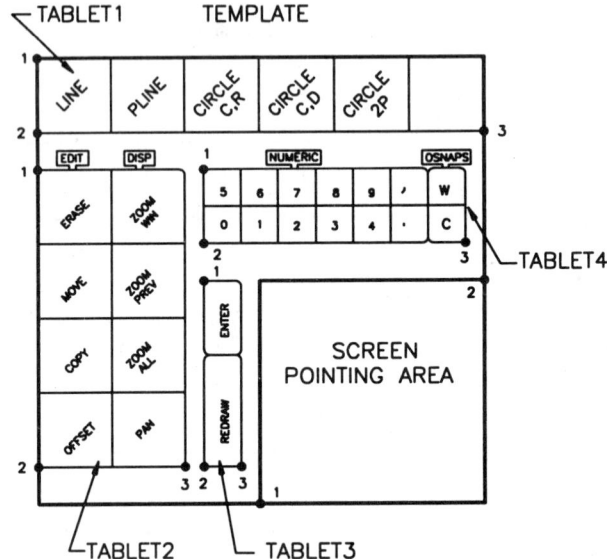

Figure 6-7(a) Tablet overlay for Example 2

Notice that this tablet template has four different sections in addition to the screen pointing area. Therefore, this menu will have four section labels: TABLET1, TABLET2, TABLET3, and TABLET4. You can use any text editor (Notepad) to write the file. The following file is a listing of the tablet menu of Example 2.

```
***TABLET1                                                    1
^C^CLINE                                                      2
^C^CPLINE                                                     3
^C^CCIRCLE                                                    4
^C^CCIRCLE \D                                                 5
^C^CCIRCLE 2P                                                 6
***TABLET2                                                    7
```

^C^CERASE	8
^C^CZOOM W	9
^C^CMOVE	10
^C^CZOOM P	11
^C^CCCOPY	12
^C^CZOOM A	13
^C^COFFSET	14
^C^CPAN	15
***TABLET3	16
;	17
;	18
'REDRAW	19
'REDRAW	20
'REDRAW	21
***TABLET4	22
5\	23
6\	24
7\	25
8\	26
9\	27
,\	28
WINDOW	29
0\	30
1\	31
2\	32
3\	33
4\	34
.\	35
CROSSING	36

Explanation

Lines 1-6
The first six lines are identical to the first six lines of the tablet menu in Example 1.

Line 9
^C^CZOOM W
ZOOM is an AutoCAD command; W is the window option of the ZOOM command.

```
^C^CZOOM W
       │ │└─ Window option of ZOOM command
       │ └── Space for Return
       └──── AutoCAD ZOOM command
```

This menu item could also be written as:

```
^C^CZOOM;W
        └─ Semicolon for Return
```

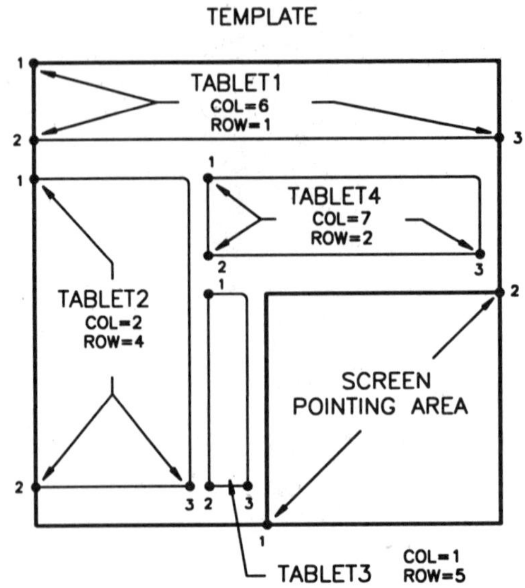

TEMPLATE

Figure 6-7(b) Number of rows and columns in different tablet areas

Lines 17 and 18
The semicolon (;) is for Return. It has the same effect as entering Return at the keyboard.

Lines 19-21
'REDRAW
REDRAW is an AutoCAD command that redraws the screen. Notice that there is no ^C^C in front of the REDRAW command. If it had ^C^C, the existing command would be canceled before redrawing the screen. This may not be desirable in most applications because you might want to redraw the screen without canceling the existing command. The apostrophe (') in front of REDRAW makes the REDRAW command transparent.

Line 23
5
The backslash (\) is used to introduce a pause for user input. Without the backslash you cannot enter another number or a character, because after you select the digit 5 it will automatically be followed by Return. For example, without the backslash (\), you will not be able to enter a number like 5.6. Therefore, you need the backslash to enable you to enter decimal numbers or any characters. To terminate the input, enter Return at the keyboard or select Return from the digitizer.

```
5\
 └─ Backslash for user input
```

ASSIGNING COMMANDS TO A TABLET

After loading the menu by means of the AutoCAD MENU command, you must configure the tablet. At the time of configuration AutoCAD actually generates and stores the information about the rectangular blocks on the tablet template. When you load the menu, the commands defined in the tablet menu are assigned to various blocks. For example, when you select the three points for tablet area 4, [Figure 6-7(a)], and enter the number of rows and columns, AutoCAD generates a grid of seven columns and two rows, as shown in the following diagram.

After Configuration

When you load the new menu, AutoCAD takes the commands under the section label TABLET4 and starts filling the blocks from left to right. That means "5", "6", "7", "8", "9", "," and "Window" will be placed in the top row. The next seven commands will be assigned to the next row, starting from the left, as shown in the diagram on page 6-14.

After Loading Tablet Menu

5	6	7	8	9	,	Window
0	1	2	3	4	.	Cross

Similarly, tablet area 3 has been divided into five rows and one column. At first, it appears that this tablet area has only two rows and one column, [Figure 6-7(a)]. When you configure this tablet area by specifying the three points and entering the number of rows and columns, AutoCAD divides the area into one column and five rows, as shown in the following diagrams:

After loading the menu, AutoCAD takes the commands in the TABLET3 section of the tablet menu and assigns them to the blocks. The first command (;) is placed in the first block. Since there are no more blocks in the first row, the next command (;) is placed in the second row. Similarly, the three REDRAW commands are placed in the next three rows. If you pick a point in the first two rows, you will select the ENTER command. Similarly, if you pick a point in the next three blocks, you will select the REDRAW command.

After Configuration

After Loading Menu

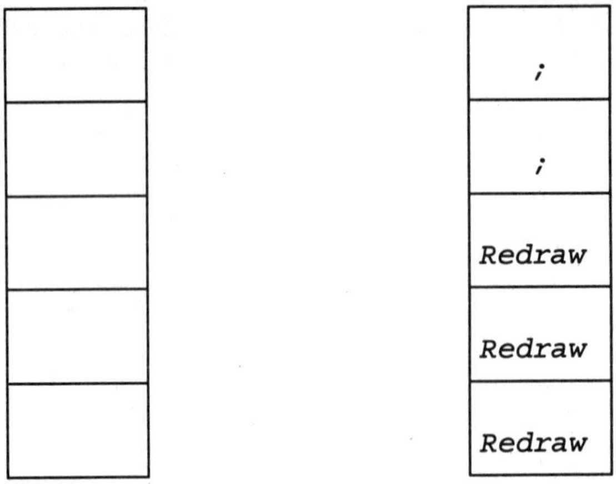

;
;
Redraw
Redraw
Redraw

This process is carried out for all the of tablet areas, and the information is stored in the AutoCAD configuration file, **ACAD.CFG**. If for any reason the configuration is not right, the tablet menu may not perform the desired function.

AUTOMATIC MENU SWAPPING

The screen menus can be automatically swapped by using the system variable **MENUCTL**. When the system variable **MENUCTL** is set to 1, AutoCAD automatically issues **$S=CMDNAME** command, where CMDNAME is the name of the command that loads the submenu. For example, if you select the LINE command from the digitizer, pull-down menu, or enter the LINE command from the keyboard, CMDNAME command will load the LINE submenu and display it in the screen menu area. To utilize this feature of AutoCAD, the command name and the submenu name must be the same. For example, if the name of the arc submenu is ARC and you select the ARC command, the arc submenu will be automatically loaded on the screen. However, if the submenu name is different (for example MYARC), then AutoCAD will not load the arc submenu on the screen. The default value of MENUCTL system variable is 1. If you set the MENUCTL variable to 0, AutoCAD will not utilize the $S=CMDNAME command feature to load the submenus.

REVIEW QUESTIONS

1. The maximum number of tablet menu sections is _____.

2. A tablet menu area is _____ in shape.

3. The blocks in any tablet menu area are _____ in shape.

4. A tablet menu area can have _____ number of rectangular blocks.

5. You _____ assign the same command to more than one block on the tablet menu template.

6. The AutoCAD _____ command is used to configure the tablet menu template.

7. The AutoCAD _____ command is used to load a new menu.

EXERCISES

Exercise 2

Design the template and write a tablet menu to insert the following user-defined blocks:

BX1	BX5	BX9
BX2	BX6	BX10
BX3	BX7	BX11
BX4	BX8	BX12

Exercise 3

Design the template and the screen menu for the following AutoCAD commands:

LINE	ZOOM-Win	DIM-Horz
PLINE	ZOOM-Dyn	DIM-Vert
ARC	ZOOM-All	DIM-Alig
CIRCLE	ZOOM-Pre	DIM-Ang
ELLIPSE	ZOOM-Ext	DIM-Rad
POLYGON	ZOOM-Scl	DIM-Cen

Exercise 4

Write a tablet menu for the commands shown in the tablet menu template of Figure 6-8. Make a tablet menu template for configuration and command selection.

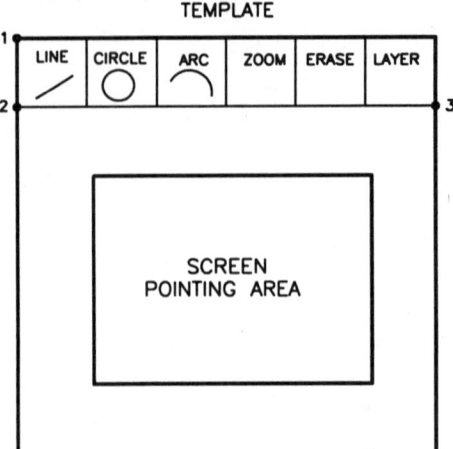

Figure 6-8 Tablet menu template for Exercise 4

Exercise 5

Write a tablet menu for the AutoCAD commands shown in Figure 6-9. Configure the tablet and then load the new menu. Make a tablet menu template for configuration and command selection.

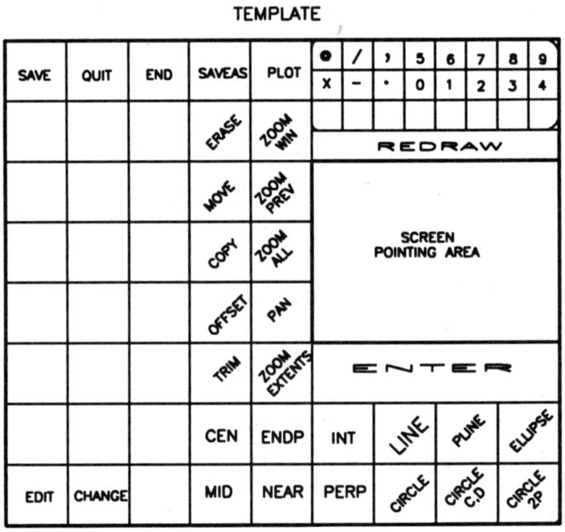

Figure 6-9 Tablet overlay for Exercise 5

Tablet Menus   6-17

Exercise 6

Write a tablet menu file for the AutoCAD commands shown in the template of Figure 6-10. Make a tablet menu template for configuration and command selection.

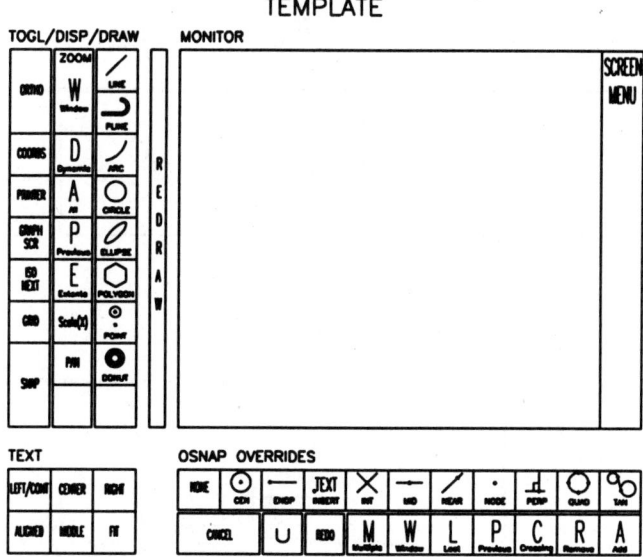

Figure 6-10 Tablet menu template for Exercise 6

Exercise 7

Write a combined pull-down and tablet menu file for the commands shown in the tablet menu template in Figure 6-11. Make a tablet menu template for configuration and command selection.

TEMPLATE

LINE	CIRCLE	ARC			
ERASE					ZOOM ALL
MOVE		SCREEN POINTING AREA			ZOOM WIN.
COPY					LIST
FILLET					AREA
TRIM					HELP

Figure 6-11 Tablet menu template for Exercise 7

Template for Example 1

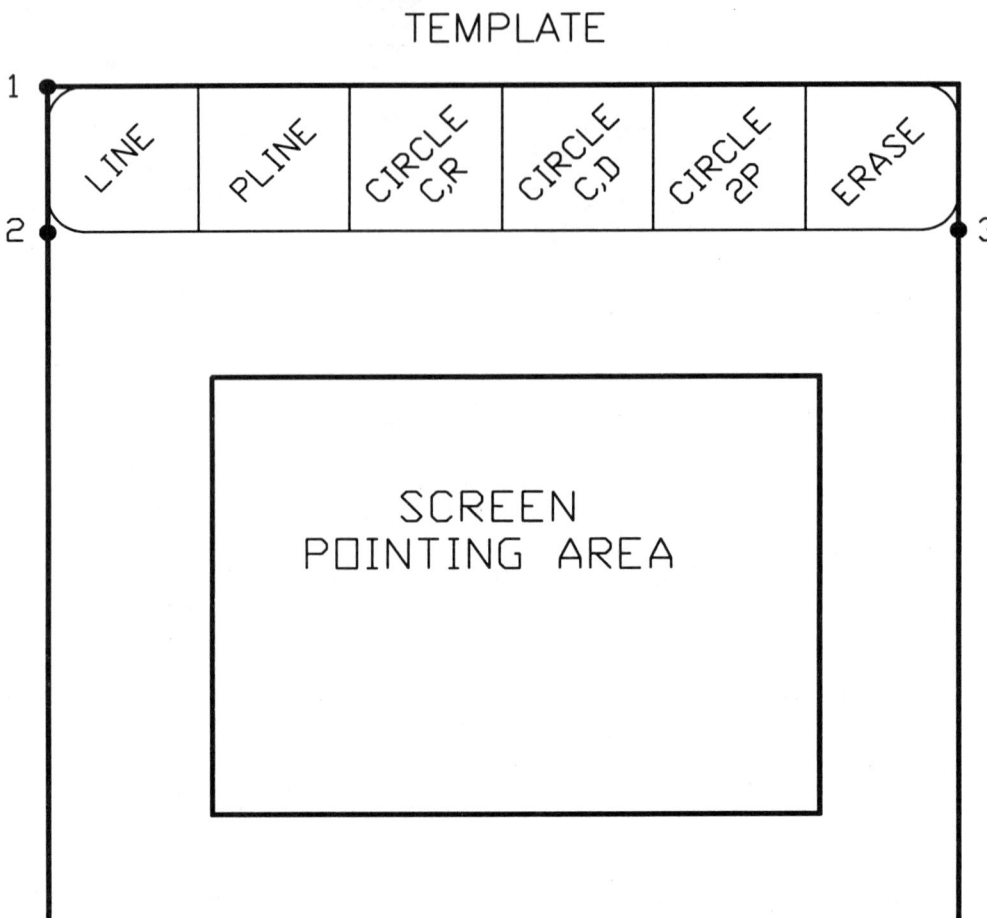

Note
This template is for tablet configuration. You may make a copy of this page and then secure it on the digitizer surface for configuration.

Tablet Menus 6-19

Template for Exercise 1

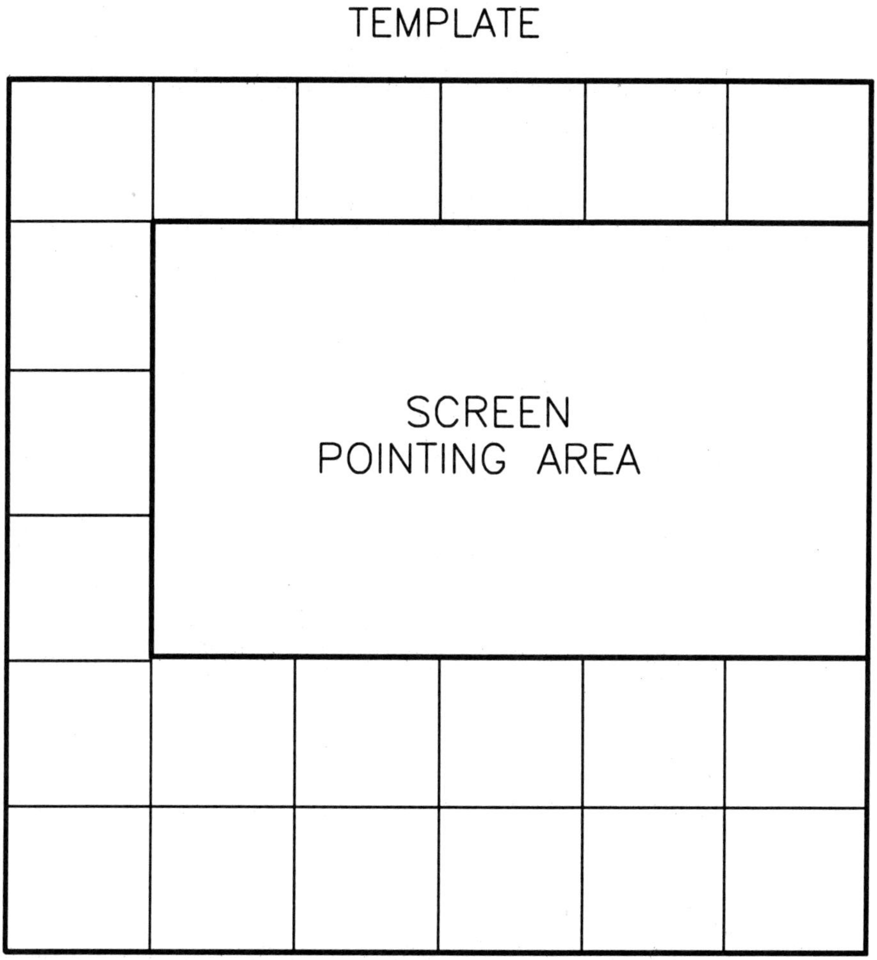

TEMPLATE

SCREEN POINTING AREA

Note
This template is for tablet configuration. You may make a copy of this page and then secure it on the digitizer surface for configuration.

Template for Example 2

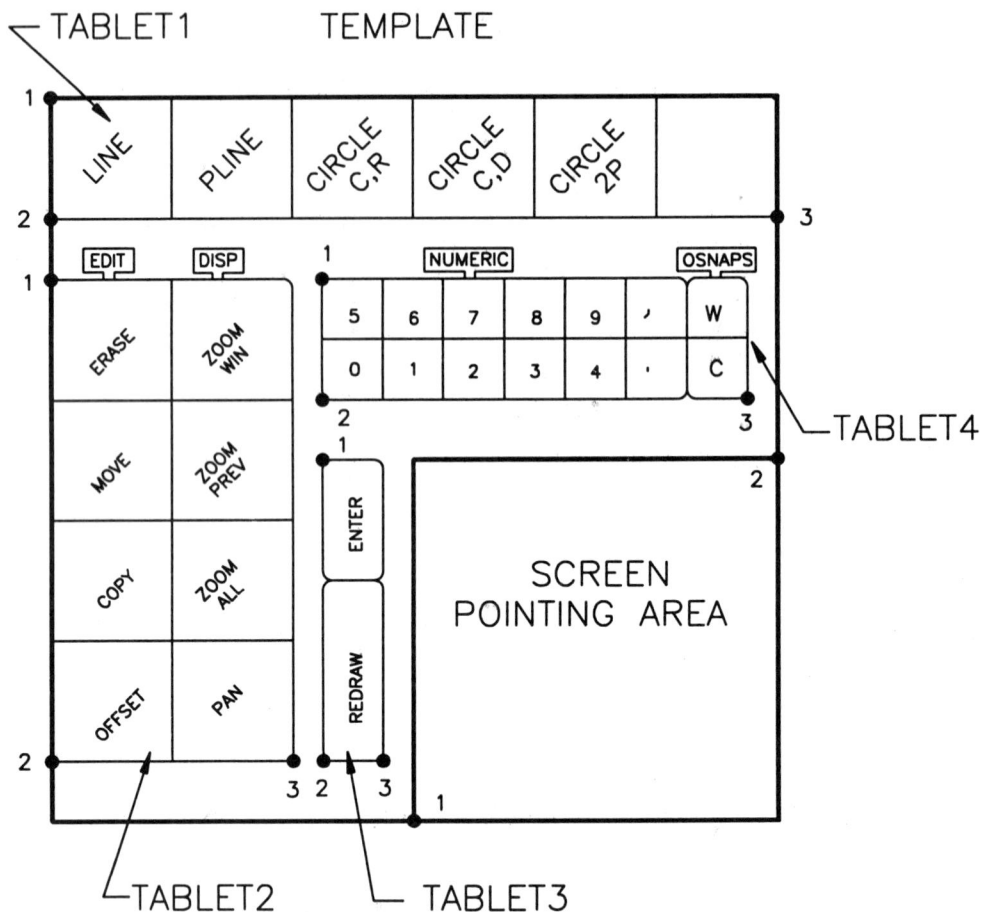

Note
This template is for tablet configuration. You may make a copy of this page and then secure it on the digitizer surface for configuration.

Template for Example 2

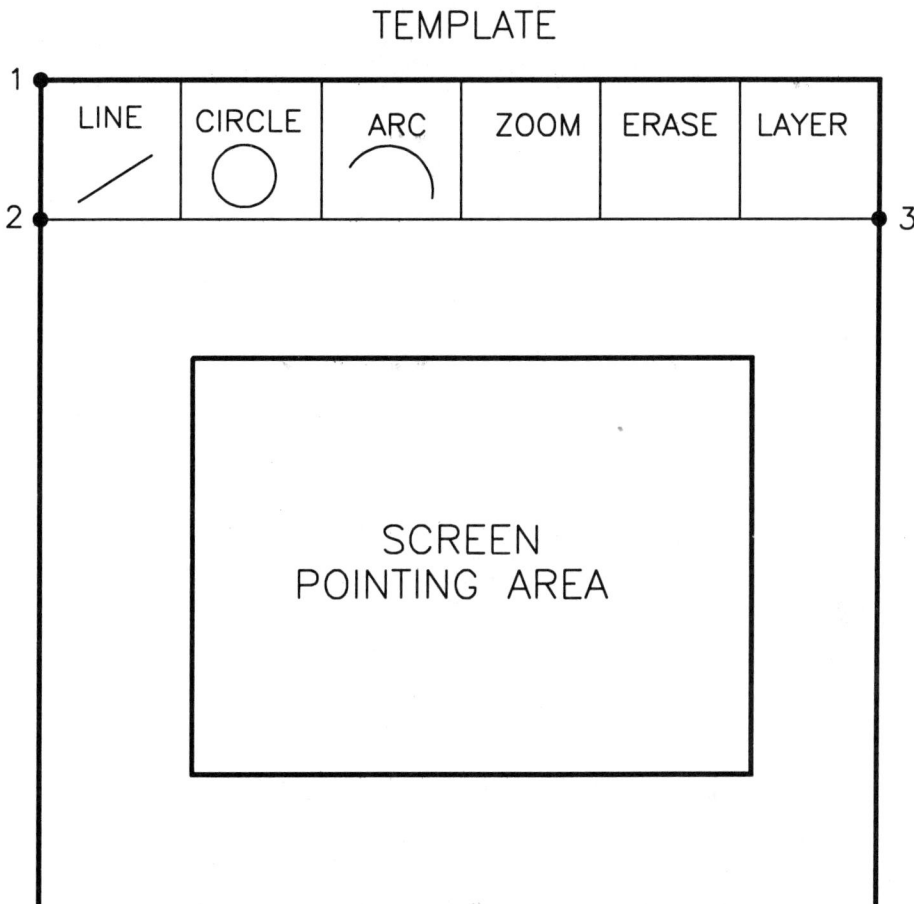

Note
This template is for tablet configuration. You may make a copy of this page and then secure it on the digitizer surface for configuration.

Template for Exercise 5

TEMPLATE

SAVE	QUIT	END	SAVEAS	PLOT	@ / , 5 6 7 8 9						
					X	–	.	0	1	2	3	4
			ERASE	ZOOM WIN	REDRAW								
			MOVE	ZOOM PREV	SCREEN POINTING AREA								
			COPY	ZOOM ALL	^								
			OFFSET	PAN	^								
			TRIM	ZOOM EXTENTS	ENTER								
			CEN	ENDP	INT	LINE		PLINE		ELLIPSE		
EDIT	CHANGE		MID	NEAR	PERP	CIRCLE		CIRCLE C,D		CIRCLE 2P		

> **Note**
> *This template is for tablet configuration. You may make a copy of this page and then secure it on the digitizer surface for configuration.*

Template for Exercise 6

TEMPLATE

TOGL/DISP/DRAW			MONITOR									SCREEN MENU
ORTHO	ZOOM **W** Window	LINE										
		PLINE	R E D R A W									
COORDS	**D** Dynamic	ARC										
PRINTER	**A** All	CIRCLE										
GRAPH SCR	**P** Previous	ELLIPSE										
ISO NEXT	**E** Extents	POLYGON										
GRID	Scale(X)	POINT										
SNAP	PAN	DONUT										

TEXT

LEFT/CONT	CENTER	RIGHT
ALIGNED	MIDDLE	FIT

OSNAP OVERRIDES

NONE	CEN	ENDP	TEXT INSERT	INT	MID	NEAR	NODE	PERP	QUAD	TAN
CANCEL	U	REDO	M Multiple	W Window	L Last	P Previous	C Crossing	R Remove	A Add	

Note
This template is for tablet configuration. You may make a copy of this page and then secure it on the digitizer surface for configuration.

Template for Exercise 7

TEMPLATE

LINE /	CIRCLE ○	ARC ⌒				
ERASE						ZOOM ALL
MOVE		SCREEN POINTING AREA				ZOOM WIN.
COPY						LIST
FILLET						AREA
TRIM						HELP

Note
This template is for tablet configuration. You may make a copy of this page and then secure it on the digitizer surface for configuration.

Chapter 7

Image Tile Menus

Learning objectives

After completing this chapter, you will be able to:
♦ Write image tile menus.
♦ Reference and display submenus.
♦ Make slides for image tile menus.

IMAGE TILE MENUS

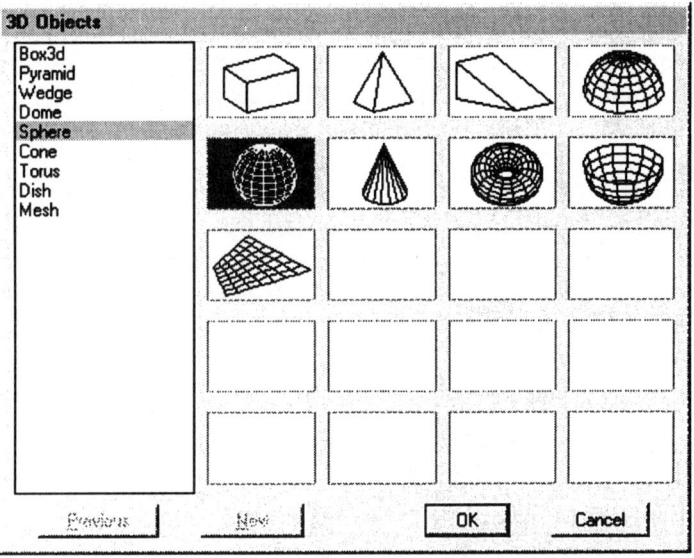

Figure 7-1 Sample image tile menu display

The image tile menus, also known as **icon menus**, are extremely useful for inserting a block, selecting a text font, or drawing a 3D object. You can also use the image tile menus to load an AutoLISP routine or a predefined macro. Thus, the image tile menu is a powerful tool for customizing AutoCAD.

The image tile menus can be accessed from the pull-down, tablet, button, or screen menu. However, the image tile menus **cannot** be loaded by entering the command from the keyboard. When you select an image tile, a dialog box that contains **20 image tiles** is displayed on the screen (Figure 7-1). The names of the slide files associated with image tiles appear on the left side of the dialog box with a scrolling bar that can be used to scroll the file names. The title of the image tile menu is displayed at the top of the dialog box (Figure 7-1). When you activate the image tile menu, an arrow that can be moved to select any image tile appears on the screen. You can select an image tile by selecting the slide file name from the dialog box and then selecting the OK button from the dialog box or double-clicking on the slide file name.

When you select the slide file, AutoCAD highlights the corresponding image tile by drawing a rectangle around the image tile (Figure 7-1). You can also select an image tile by moving the arrow to the desired image tile, and then pressing the pick button of the pointing device. The corresponding slide file name will be automatically highlighted. And if you select the OK button or double-click on the image tile, the command associated with that menu item will be executed. You can cancel an image tile menu by pressing Escape on the keyboard, or selecting an image tile from the dialog box.

SUBMENUS

You can define an unlimited number of menu items in the image tile menu, but only 20 image tiles will be displayed at a time. If the number of items exceeds 20, you can use the **Next** and **Previous** buttons of the dialog box to page through different pages of image tiles. You can also define submenus that let you define smaller groups of items within an image tile menu section. When you select a submenu, the submenu items are loaded and displayed on the screen.

Submenu Definition

A submenu label consists of two asterisks (**) followed by the name of the submenu. The image tile menu can have any number of submenus, and every submenu should have a unique name. The items that follow a submenu, up to the next section label or submenu label, belong to that submenu. The format of a submenu label is:

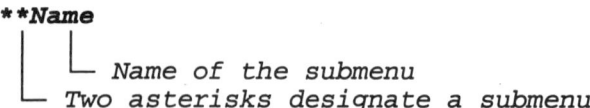

> **Note**
>
> *The submenu name can be up to 31 characters long.*
>
> *The submenu name can consist of letters, digits, and special characters, like $ (dollar), - (hyphen), and _ (underscore).*
>
> *The submenu name should not have any embedded blanks.*
>
> *Submenu names should be unique in a menu file.*

Image Tile Menus 7-3

Submenu Reference

The submenu reference is used to reference or load a submenu. It consists of a $ sign followed by a letter that specifies the menu section. The letter that specifies an image tile menu section is I. The menu section is followed by an = sign and the name of the submenu you want to activate. The submenu name should be without the **. Following is the format of a submenu reference:

```
$Section=Submenu
 │       │   │
 │       │   └─ Name of submenu
 │       └─ "=" sign
 │     └─ Menu section specifier
 └─ "$" sign
```

```
$I=IMAGE1
 │    │
 │    └─ Name of submenu
 └─ I specifies image tile menu section
```

Displaying a Submenu

When you load a submenu, the new dialog box and the image tiles are not automatically displayed. For example, if you load submenu IMAGE1, the items contained in this submenu will not be displayed. To force the display of the new image tile menu on the screen, AutoCAD uses the special command $I=*:

```
$I=*
 │  │
 │  └─ Asterisk (*)
 └─ I for image tile menu
```

WRITING AN IMAGE TILE MENU

The image tile menu consists of the section label ***IMAGE followed by image tiles or image tile submenus. The menu file can contain only one image tile menu section (***IMAGE); therefore, all image tiles must be defined in this section.

```
***IMAGE
 │   │
 │   └─ Section label for an image tile
 └─ Three asterisks designate a section label
```

You can define any number of submenus in the image tile menu. All submenus have two asterisks followed by the name of the submenu (**PARTS or **IMAGE1):

```
**IMAGE1
 │   │
 │   └─ Name of submenu
 └─ Two asterisks designate a submenu
```

7-4 Customizing AutoCAD

The first item in the image tile menu is the title of the image tile menu, which is also displayed at the top of the dialog box. The image tile dialog box title has to be enclosed in brackets ([PLC-SYMBOLS]) and should not contain any command definition. If it does contain a command definition, AutoCAD ignores the definition. The remaining items in the image tile menu file contain slide names in the brackets and the command definition outside the brackets.

```
***IMAGE                    Image tile menu section
**BOLTS                     Image tile submenu (BOLTS)
[HEX-HEAD BOLTS]            Image tile title
[BOLT1]^C^CINSERT;B1        BOLT1 is slide file name;
                            B1 is block name
```

Example 1

Write an image tile menu that will enable you to insert the block shapes from Figure 7-2 in a drawing by selecting the corresponding image tile from the dialog box. Use the pull-down menu to load the image tile menu.

PLC SYMBOLS
NO (NORMALLY OPEN)
NC (NORMALLY CLOSED)
COIL

ELECTRIC SYMBOLS
RESIS (RESISTANCE)
DIODE
GROUND

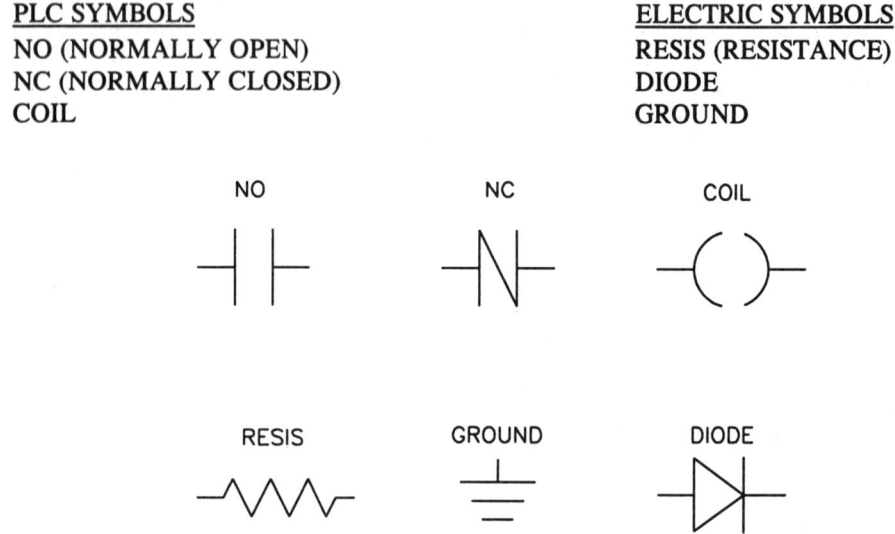

Figure 7-2 Block shapes for image tile menu

As mentioned in earlier chapters, the first step in writing a menu is to design the menu so that the commands are arranged in a desired configuration. Figure 7-3 shows one possible design for the pull-down menu and the image tile menu for Example 1.

You can use the AutoCAD EDIT command to write the file. **The line numbers in the following file are for reference only and are not a part of the menu file.**

```
***POP1                                    1
[ELECTRIC]                                 2
[PLC-SYMBOLS]$I=IMAGE1 $I=*                3
[ELEC-SYMBOLS]$I=IMAGE2 $I=*               4
***IMAGE                                   5
**IMAGE1                                   6
[PLC-SYMBOLS]                              7
[NO]^C^CINSERT;NO;\1.0;1.0;0               8
[NC]^C^CINSERT;NC;\1.0;1.0;0;              9
[COIL]^C^CINSERT;COIL                     10
[ No-Image]                               11
[blank]                                   12
**IMAGE2                                  13
[ELECTRICAL SYMBOLS]                      14
[RESIS]^C^CINSERT;RESIS;\\\\              15
[DIODE]^C^CINSERT;DIODE;\1.0;1.0;\        16
[GROUND]^C^CINSERT;GRD;\1.5;1.5;0;;       17
```

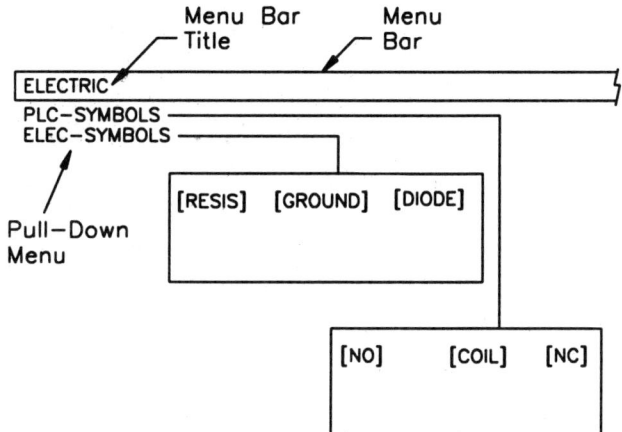

Figure 7-3 Design of pull-down menu and image tile menu for Example 1

Explanation

Line 1
***POP1
In this menu item, ***POP1 is the section label and defines the first section of the pull-down menu.

Line 2
[ELECTRIC]
In this menu item [ELECTRIC] is the menu bar label for the POP1 pull-down menu. It will be displayed in the menu bar.

Line 3
[PLC-SYMBOLS]$I=IMAGE1 $I=*
In this menu item, $I=IMAGE1 loads the submenu IMAGE1; $I=* displays the current image tile menu on the screen.

```
[PLC-SYMBOLS]$I=IMAGE1 $I=*
                |        |
                |        └─ Forces display of current menu
                └─ Loads submenu IMAGE1
```

Line 5
***IMAGE
In this menu item, ***IMAGE is the section label of the image tile menu. All the image tile menus have to be defined within this section; otherwise, AutoCAD cannot locate them.

Line 6
**IMAGE1
In this menu item, **IMAGE1 is the name of the image tile submenu.

Line 7
[PLC-SYMBOLS]
When you select line 3 ([PLC-SYMBOLS]$I=IMAGE1 $I=*), AutoCAD loads the submenu IMAGE1 and displays the title of the image tile at the top of the dialog box (Figure 7-4). This title is defined in line 7. If this line is missing, the next line will be displayed at the top of the dialog box. Image tile titles can be any length, as long as they fit the length of the dialog box.

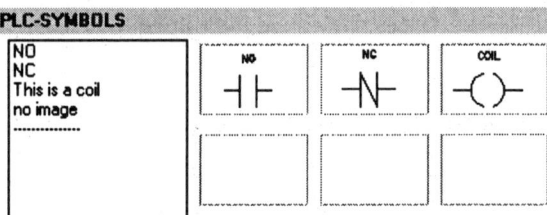

Figure 7-4 Image tile box for PLC-SYMBOLS

Line 8
[NO]^C^CINSERT;NO;\1.0;1.0;0
In this menu item, the first NO is the name of the slide and has to be enclosed within brackets. The name should not have any trailing or leading blank

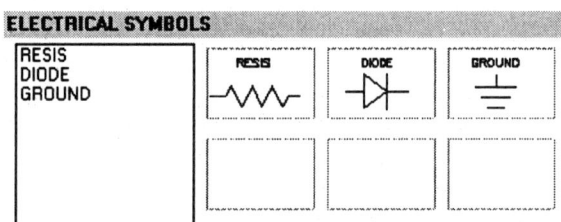

Figure 7-5 Image tile box for ELECTRICAL-SYMBOLS

spaces. If the slides are not present, AutoCAD will not display any graphical symbols in the image tiles. However, the menu items will be loaded, and if you select this item, the command associated with the image tile will be executed. The second NO is the name of the block that is to be inserted. The backslash (\) pauses for user input; in this case it is the block insertion point. The first 1.0 defines the Xscale factor. the second 1.0 defines the Yscale factor, and the following 0 defines the rotation.

Image Tile Menus 7-7

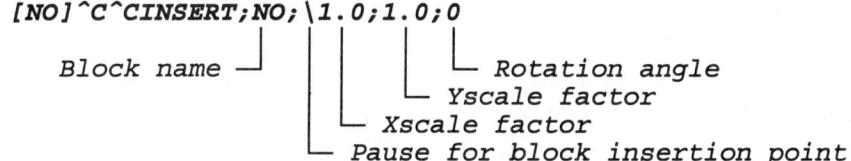
```
[NO]^C^CINSERT;NO;\1.0;1.0;0
```
- Block name
- Pause for block insertion point
- Xscale factor
- Yscale factor
- Rotation angle

When you select this item, it will automatically enter all the prompts of the INSERT command and insert the NO block at the given location. The only input you need to enter is the insertion point of the block.

Line 10
[COIL]^C^CINSERT;COIL
In this menu item the block name is given, but you need to define other parameters when inserting this block.

Line 11
[No-Image]
Notice the **blank space before No-Image**. If there is a space following the open bracket, AutoCAD does not look for a slide. AutoCAD instead displays the text, enclosed within the brackets, in the **slide file list** box of the dialog box.

Line 12
[blank]
Line 12 consists of **blank**; this displays a separator line in the list box and a blank image (no image) in the image box.

Line 15
[RESIS]^C^CINSERT;RESIS;
This menu item inserts the block RESIS. The first backslash (\) is for the block insertion point. The second and third backslashes are for the Xscale and Yscale factors. The fourth backslash is for the rotation angle. This menu item could also be written as:

    [RESIS]^C^CINSERT;RESIS;
          or
    [RESIS]^C^CINSERT;RESIS

Line 16
[DIODE]^C^CINSERT;DIODE;\1.0;1.0;
If you select this menu item, AutoCAD will prompt you to enter the block insertion point and the rotation angle. The first backslash is for the block insertion point; the second backslash is for the rotation angle.

```
[DIODE]^C^CINSERT;DIODE;\1.0;1.0;\
```
- Pause for insertion point
- Pause for rotation angle

7-8 Customizing AutoCAD

Line 17
[GROUND]^C^CINSERT;GRD;\1.5;1.5;0;;
This menu item has two semicolons (;) at the end. The first semicolon after 0 is for Return and completes the block insertion process. The second semicolon enters a Return and repeats the INSERT command. However, when the command is repeated you will have to respond to all of the prompts. It does not accept the values defined in the menu item.

> **Note**
>
> *The ***IMAGE section label replaces the ***ICON section used in previous releases. The ***ICON is still valid for AutoCAD Release14. It may be dropped in future releases.*
>
> *The menu item repetition feature cannot be used with the image tile menus. For example, if the command definition starts with an asterisk ([GROUND]*^C^CINSERT;GRD;\1.5;1.5;0;;), the command is not automatically repeated, as is the case with a pull-down menu.*
>
> *A blank line in an image tile menu terminates the menu and clears the image tiles.*
>
> *The menu command $I=*, which displays the current menu, cannot be entered at the keyboard.*
>
> *If you want to cancel or exit an image tile menu, press the Esc (Escape) key on the keyboard. AutoCAD ignores all other entries from the keyboard.*
>
> *You can define any number of image tile menus and submenus in the image tile menu section of the menu file.*

SLIDES FOR IMAGE TILE MENUS

The idea behind creating slides for the image tile menus is to display graphical symbols in the image tiles. This symbol makes it easier for you to identify the operation that the image tile will perform. Any slide can be used for the image tile. However, the following guidelines should be kept in mind when creating slides for the image tile menu:

1. When you make a slide for an image tile menu, draw the object so that it fills the entire screen. The MSLIDE command makes a slide of the existing screen display. If the object is small, the picture in the image tile menu will be small. Use ZOOM Extents or ZOOM Window to display the object before making a slide.

2. When you use the image tile menu, it takes some time to load the slides for display in the image tiles. The more complex the slides, the more time it will take to load them. Therefore, the slides should be kept as simple as possible and at the same time give enough information about the object.

3. Do not fill the object, because it takes a long time to load and display a solid object. If there is a solid area in the slide, AutoCAD does not display the solid area in the image tile.

4. If the objects are too long or too wide, it is better to center the image with the AutoCAD PAN command before making a slide.

Image Tile Menus 7-9

5. The space available on the screen for image tile display is limited. Make the best use of this small area by giving only the relevant information in the form of a slide.

6. The image tiles that are displayed in the dialog box have the length-to-width ratio (aspect ratio) of 1.5:1. For example, if the length of the image tile is 1.5 units, the width is 1 unit. If the drawing area of your screen has an aspect ratio of 1.5 and the slide drawing is centered in the drawing area, the slide in the image tile will also be centered.

LOADING MENUS

AutoCAD automatically loads the **ACAD.MNU** file when you get into the AutoCAD drawing editor. However, you can also load a different menu file by using the AutoCAD MENU command.

Command: **Menu**

When you enter the Menu command, AutoCAD displays the **Select Menu File** dialog box (Figure 7-6) on the screen. Select the menu file that you want to load and then select the Open button.

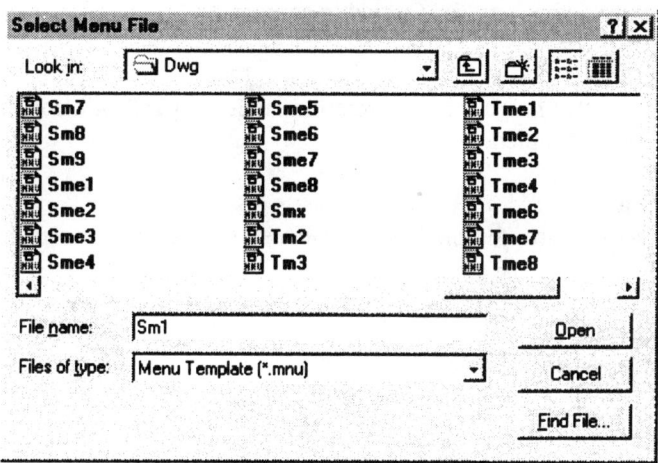

Figure 7-6 Select Menu File dialog box

You can also load the menu file from the command line.

```
Command: FILEDIA
New value for FILEDIA <1>: 0
Command: Menu
Menu filename <ACAD>: PDM1
                       │        └─ Name of menu file
                       └─ Default menu file
```

7-10 Customizing AutoCAD

After you enter the MENU command, AutoCAD will prompt for the file name. Enter the name of the menu file without the file extension (.MNU), since AutoCAD assumes the extension .MNU. AutoCAD will automatically compile the menu file into **MNC** and **MNR** files.

> **Note**
>
> *When you load the image tile menu, some of the commands will be displayed on the screen menu area. This happens because the screen menu area is empty. If the menu file contains a screen menu also, the pull-down menu items will not be displayed in the screen menu area.*
>
> *When you load a new menu the original menu you had on the screen prior to loading the new menu is disabled. You cannot select commands from the screen menu, pull-down menu, pointing device, or the digitizer, unless these sections are defined in the new menu file.*
>
> *To load the original menu, use the MENU command again and enter the name of the menu file.*

RESTRICTIONS

The pull-down and image tile menus are very easy to use and provide a quick access to some of the frequently used AutoCAD commands. However, the menu bar, the pull-down menus, and the image tile menus are disabled during the following commands:

DTEXT Command
After you assign the text height and the rotation angle to a DTEXT command, the pull-down menu is automatically disabled.

SKETCH Command
The pull-down menus are disabled after you set the record increment in the SKETCH command.

VPOINT Command
The pull-down menus are disabled while the axis tripod and compass are displayed on the screen.

ZOOM Command
The pull-down menus are disabled when the dynamic zoom is in progress.

DVIEW
The pull-down menus are disabled when the dynamic view is in progress.

Exercise 1

Write an image tile menu for inserting the blocks shown in Figure 7-7. Arrange the blocks in two groups so that you have two submenus in the image tile menu.

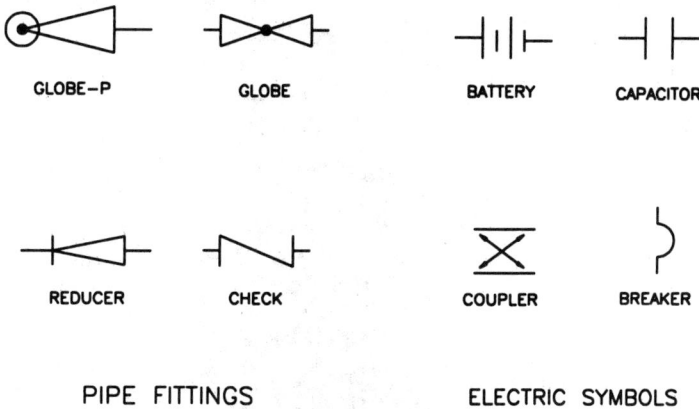

Figure 7-7 Block shapes for Exercise 1

PIPE FITTINGS	ELECTRIC SYMBOLS
GLOBE-P	BATTERY
GLOBE	CAPACITOR
REDUCER	COUPLER
CHECK	BREAKER

IMAGE TILE MENU ITEM LABELS

As with screen and pull-down menus, you can use menu item labels in the image menus. However, the menu item labels in the image tile menus use different formats, and each format performs a particular function in the image tile menu. The menu item labels appear in the slide list box of the dialog box. The maximum number of characters that can be displayed in this box is 23. The characters in excess of 23 are not displayed in the list box. However, this does not affect the command that is defined with the menu item.

Menu Item Label Formats

[slidename]. In this menu item label format, **slidename** is the name of the slide displayed in the image tile. This name (slidename) is also displayed in the list box of the corresponding dialog box.

[slidename,label]. In this menu item label format, **slidename** is the name of the slide displayed in the image tile. However, unlike the previous format, the **slidename** is **not** displayed in the list box. The **label** text is displayed in the list box. For example, if the menu item label is **[BOLT1,1/2-24UNC-3LG]**, **BOLT1** is the name of the slide and **1/2-24UNC-3LG** is the label that will be displayed in the list box.

[slidelib(slidename)]. In this menu item label format, **slidename** is the name of the slide in the slide library file **slidelib**. The slide (slidename) is displayed in the image tile, and the slide file name (slidename) is also displayed in the list box of the corresponding dialog box.

[slidelib(slidename,label)]. In this menu item label format, **slidename** is the name of the slide in the slide library file **slidelib**. The slide (slidename) is displayed in the image tile, and the **label** text is displayed in the list box of the corresponding dialog box.

[blank]. This menu item will draw a line that extends through the width of the list box. It also displays a blank image tile in the dialog box.

[label]. If the **label** text is preceded by a space, AutoCAD does not look for a slide. The **label** text is displayed in the list box only. For example, if the menu item label is [EXIT]^C, the label text (EXIT) will be displayed in the list box. If you select this item, the cancel command (^C) defined with the item will be executed. The **label** text is **not** displayed in the image tile of the dialog box.

Example 2

Write the pull-down and image tile menus for inserting the following commands. B1 to B15 are the block names.

BLOCK
WBLOCK
ATTDEF
LIST
INSERT
    BL1   BL6   BL11
    BL2   BL7   BL12
    BL3   BL8   BL13
    BL4   BL9   BL14
    BL5   BL10   BL15

The first step in writing a menu is to design the menu. Figure 7-8 shows the design of the pull-down menu. If you select Insert from the pull-down menu, the image tiles and block names will be displayed in the dialog box.

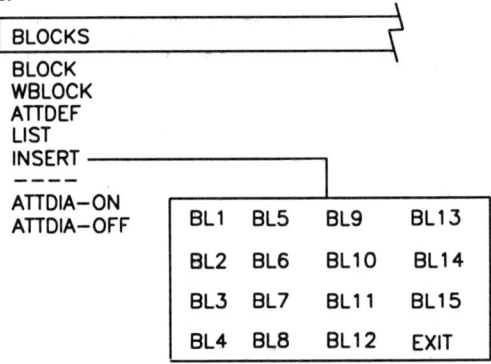

Figure 7-8 Design of screen and pull-down menus for Example 2

The following file is a listing of the menu file for Example 2. The file contains the screen, tablet, pull-down, and image tile menu sections. **The line numbers are not a part of the menu file; they are given for reference only.**

```
***POP1                                              1
[INSERT]                                             2
[BLOCK]^C^CBLOCK                                     3
[WBLOCK]^C^CWBLOCK                                   4
[ATTRIBUTE DEFINITION]^C^CATTDEF                     5
[LIST BLOCK NAMES]^C^CINSERT;?                       6
[INSERT]^C^C$I=IMAGE1 $I=*                           7
[--]                                                 8
[ATTDIA-ON]^C^CSETVAR ATTDIA 1                       9
[ATTDIA-OFF]^C^CSETVAR ATTDIA 0                     10
                                                    11
***IMAGE                                            12
**IMAGE1                                            13
[BLOCK INSERTION FOR EXAMPLE-2]                     14
[BL1]^C^C$S=INSERT1 INSERT;BL1;\1.0;1.0;\           15
[BL2]^C^C$S=INSERT1 INSERT;BL2;\1.0;1.0;0           16
[BL3]^C^C$S=INSERT1 INSERT;BL3;\;;\                 17
[BL4]^C^C$S=INSERT1 INSERT;BL4;\;;;                 18
[BL5]^C^C$S=INSERT1 INSERT;*BL5;\1.75               19
[BL6]^C^C$S=INSERT1 INSERT;BL6;\XYZ                 20
[BL7]^C^C$S=INSERT1 INSERT;BL7;\XYZ;;;\0            21
[BL8]^C^C$S=INSERT1 INSERT;BL8;\XYZ;;;;             22
[BL9]^C^C$S=INSERT1 INSERT;BL9;\XYZ;;;;\            23
[BL10]^C^C$S=INSERT1 INSERT;*BL10;\XYZ;\            24
[BL11]^C^C$S=INSERT2 INSERT;BL11;\XYZ;1;1.5;2;45    25
[BL12]^C^C$S=INSERT2 INSERT;BL12;\XYZ;\\;;          26
[BL13]^C^C$S=INSERT2 INSERT;*BL13;\\45              27
[BL14]^C^C$S=INSERT2 INSERT;BL14;\C;@1.0,1.0;0      28
[BL15]^C^C$S=INSERT2 INSERT;BL15;\C;@1.0,2.0;\      29
```

Explanation
Line 1
***POP1
This is the section label of the first pull-down menu. The menu items defined on lines 2 to 10 are defined in this section.

Line 12
***IMAGE
This is the section label of the image tile menu.

7-14 Customizing AutoCAD

Line 13
**IMAGE1
IMAGE1 is the name of the submenu; the items on lines 14 to 29 are defined in this submenu.

Line 15
[BL1]^C^C$S=INSERT1 INSERT;BL1;\1.0;1.0;
In this menu item, BL1 is the name of the slide and $S=INSERT1 loads the submenu INSERT1 on the screen.

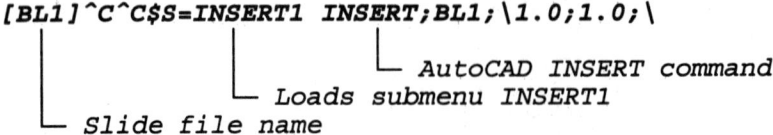

Exercise 2
Write a tablet and image tile menu for inserting the following blocks. (The template of the tablet menu is shown in Figure 7-9.)

B1	B4	B7
B2	B5	B8
B3	B6	B9

TEMPLATE

B1	B2	B3				
B4	B5	B6				
B7	B8	B9				

Figure 7-9 Tablet template for Exercise 2

REVIEW QUESTIONS

1. The image tiles are displayed in the _____ box.

2. An image tile menu can be canceled by entering _____ at the keyboard.

3. The image title dialog box can contain a maximum of _____ image tiles.

4. A blank line in an image tile menu _____ the image tile menu.

5. The menu item repetition feature _____ be used with the image tile menu.

6. The drawing for a slide should be _____ on the entire screen before making a slide.

7. You _____ fill a solid area in a slide for image tile menu.

8. An image tile menu _____ be accessed from a tablet menu.

EXERCISES

Exercise 3
Write an image tile menu for the following commands. Make the slides that will graphically illustrate the function of the command.

    LINE   CIRCLE C,R
    PLINE   CIRCLE C,D
                        CIRCLE 2P

Exercise 4
Write a tablet and image tile menu for inserting the following blocks. The layout of the template for the tablet menu is shown in Figure 7-10.

    B1 B2 B3   C1 C2 C3
    B4 B5 B6   C4 C5 C6
    B7 B8 B9   C7 C8 C9

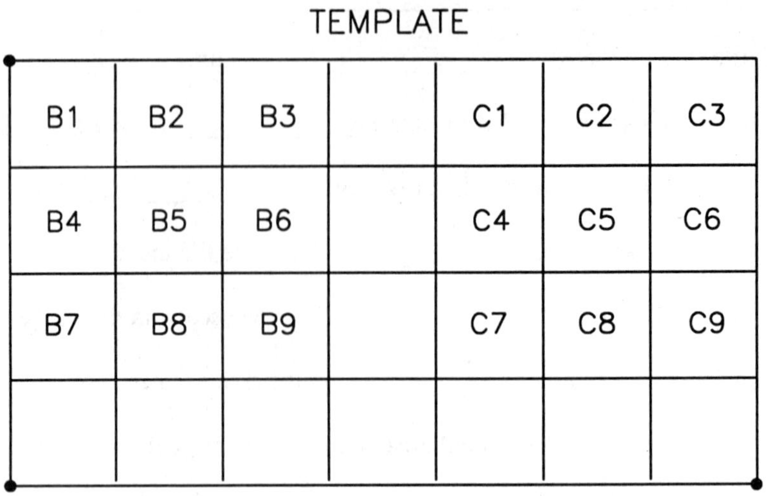

TEMPLATE

B1	B2	B3		C1	C2	C3
B4	B5	B6		C4	C5	C6
B7	B8	B9		C7	C8	C9

Figure 7-10 Tablet template for Exercise 4

Chapter 8

Button and Auxiliary Menus

Learning objectives

After completing this chapter, you will be able to:
◆ Write Buttons menus.
◆ Learn special handling for button menus.
◆ Define and load submenus.

BUTTON MENUS

You can use a multibutton pointing device to specify points, select objects, and execute commands. These pointing devices come with different numbers of buttons, but four-button and twelve-button pointing devices are very common. In addition to selecting points and objects, the multibutton pointing devices can be used to provide easy access to frequently used AutoCAD commands. The commands are selected by pressing the desired button; AutoCAD automatically executes the command or the macro that is assigned to that button. Figure 8-1 shows one such pointing device with 12 buttons.

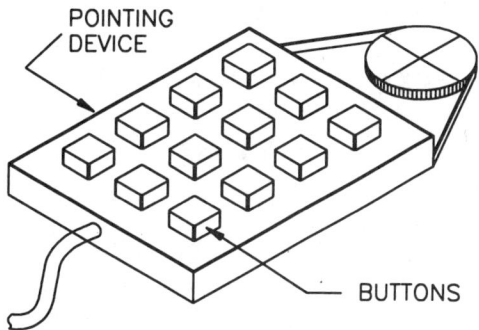

Figure 8-1 Pointing device with 12 buttons

The AutoCAD software package comes with a standard button menu that is part of the **ACAD.MNU** file. The standard menu is automatically loaded when you start AutoCAD and enter the drawing editor. You can write your own button menu and assign the desired commands or macros to various buttons of the pointing device.

AUXILIARY MENUS

In a menu file, you can have up to four auxiliary menu sections (AUX1 through AUX4). The auxiliary menu sections (***AUXn) are identical to the button menu sections. The difference is in the hardware. If your computer uses a system mouse, it will automatically use the auxiliary menu.

WRITING BUTTON AND AUXILIARY MENUS

In a menu file, you can have up to four button menus (BUTTONS1 through BUTTONS4) and four auxiliary menus (AUX1 through AUX4). The buttons and the auxiliary menus are identical. However, they are OS dependent. For example, in R14 the system mouse uses auxiliary menus. If your system has a pointing device (digitizer puck), AutoCAD automatically assigns the commands defined in the BUTTONS sections of the menu file to the buttons of the pointing device. When you load the menu file, the commands defined in the BUTTONS1 section of the menu file are assigned to the pointing device (digitizer puck) and if your computer has a system mouse, the mouse will use the auxiliary menus. You can also access other button menus (BUTTONS2 through BUTTONS4) by using the following keyboard-and-button (buttons of the pointing device-digitizer puck) combinations.

Aux Menu	Buttons Menu	Keyboard + Button Sequence
AUX1	**BUTTONS1**	Press the button of the pointing device.
AUX2	**BUTTONS2**	Hold down the **Shift** key and press the button of the pointing device.
AUX3	**BUTTONS3**	Hold down the **Ctrl** key and press the button of the pointing device.
AUX4	**BUTTONS4**	Hold down the **Shift** and **Ctrl** keys and press the button of the pointing device.

One of the buttons, generally the first, is used as a pick button to specify the coordinates of the screen crosshairs and send that information to AutoCAD. This button can also be used to select commands from various other menus, such as tablet menu, screen menu, pull-down menu, and image tile menu. This button cannot be used to enter a command, but AutoCAD commands can be assigned to other buttons of the pointing device. Before writing a button menu, you should decide the commands and options you want to assign to different buttons, and know the prompts associated with those commands. The following example illustrates the working of the button menu and the procedure for assigning commands to different buttons.

Example 1

Write a buttons menu for the following AutoCAD commands. The pointing device has 12 buttons (Figure 8-2), and button number 1 is used as a pick button (filename **BM1.MNU**).

Button	Function	Button	Function
2	RETURN	3	CANCEL
4	CURSOR MENU	5	SNAP

6	ORTHO	7	AUTO
8	INT,END	9	LINE
10	CIRCLE	11	ZOOM Win
12	ZOOM Prev		

You can use the AutoCAD EDIT command or any other text editor to write the menu file. The following file is a listing of the button menu for Example 1. **The line numbers are for reference only and are not a part of the menu file.**

```
***BUTTONS1                                      1
;                                                2
^C^C                                             3
$P0=*                                            4
^B                                               5
^O                                               6
AUTO                                             7
INT,ENDP                                         8
^C^CLINE                                         9
^C^CCIRCLE                                      10
'ZOOM;Win                                       11
'ZOOM;Prev                                      12
```

Explanation

Line 1
***BUTTONS
***BUTTONS1 is the section label for the first button menu. When the menu is loaded, AutoCAD compiles the menu file and assigns the commands to the buttons of the pointing device.

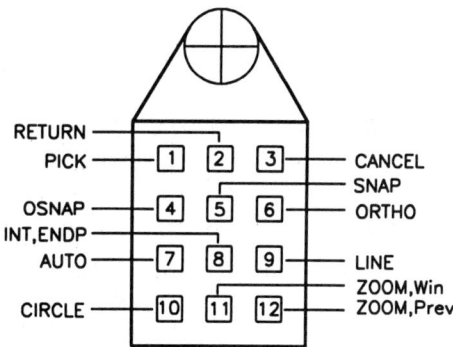

Figure 8-2 Pointing device

Line 2
;
This menu item assigns a semicolon (;) to button number 2. When you specify the second button on the pointing device, it enters a Return. It is like entering Return at the keyboard or the digitizer.

Line 3
^C^C
This menu item cancels the existing command twice (^C^C). This command is assigned to button number 3 of the pointing device. When you pick the third button on the pointing device, it cancels the existing command twice.

Line 4
$P0=*

8-4 Customizing AutoCAD

This menu item loads and displays the cursor menu POP0, which contains various object snap modes. It is assumed that the POP0 pull-down menu has been defined in the menu file. This command is assigned to button number 4 of the pointing device. If you press this button, it will load and display the cursor menu on the screen near the crosshairs location.

Line 5
^B
This menu item changes the Snap mode; it is assigned to button number 5 of the pointing device. When you pick the fifth button on the pointing device, it turns the Snap mode on or off. It is like holding down the CTRL key and then pressing the B key.

Line 6
^O
This menu item changes the ORTHO mode; it is assigned to button number 6. When you pick the sixth button on the pointing device, it turns the ORTHO mode on or off.

Line 7
AUTO
This menu item selects the AUTO option for creating a selection set; this command is assigned to button number 7 on the pointing device.

Line 8
INT,ENDP
In this menu item, INT is for the intersection osnap, and ENDP is for the endpoint osnap. This command is assigned to button number 8 on the pointing device. When you pick this button, AutoCAD looks for the intersection point. If it cannot find an intersection point, it then starts looking for the endpoint of the object that is within the pick box.

```
INT,ENDP
 |   L Endpoint object snap
 L Intersection object snap
```

Line 9
^C^CLINE
This menu item defines the LINE command; it is assigned to button number 9. When you select this button, AutoCAD cancels the existing command, and then selects the LINE command.

Line 10
^C^CCIRCLE
This menu item defines the CIRCLE command; it is assigned to button number 10. When you pick this button, AutoCAD automatically selects the CIRCLE command and prompts for the user input.

Line 11
'ZOOM;Win
This menu item defines a transparent ZOOM command with Window option; it is assigned to button number 11 of the pointing device.

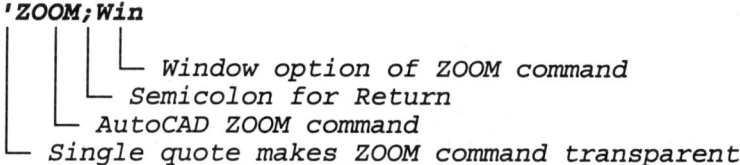

```
'ZOOM;Win
     │ │ └─ Window option of ZOOM command
     │ └─ Semicolon for Return
     └─ AutoCAD ZOOM command
 └─ Single quote makes ZOOM command transparent
```

Line 12
'ZOOM;Prev
This menu item defines a transparent ZOOM command with previous option; it is assigned to button number 12 of the pointing device.

> **Note**
> *If the button menu has more menu items than the number of buttons on the pointing device, the menu items in excess of the number of buttons are ignored. This does not include the pick button. For example, if a pointing device has three buttons in addition to the pick button, the first three menu items will be assigned to the three buttons (buttons 2, 3, and 4). The remaining lines of the button menu are ignored.*
>
> *The commands are assigned to the buttons in the same order in which they appear in the file. For example, the menu item that is defined on line 3 will automatically be assigned to the fourth button of the pointing device. Similarly, the menu item that is on line 4 will be assigned to the fifth button of the pointing device. The same is true of other menu items and buttons.*

SPECIAL HANDLING FOR BUTTON MENUS

When you press any button on the multi-button pointing device (Figure 8-3), AutoCAD receives the following information:

1. **The button number**
2. **The coordinates of the screen cross-hairs**

You can write a button menu that uses one or both pieces of information. The following example uses only the button number and ignores the coordinates of the screen cross-hairs:

Example
^C^CLINE

If this command is assigned to the second button of the pointing device and you select this button, AutoCAD will receive the button number and the coordinates of the screen cross-hairs. AutoCAD will execute the command that is assigned to the second button, but it will ignore the coordinates of the cross-hairs. The following example uses both button number and the coordinates of the screen cross hairs:

Example
^C^CLINE;\

8-6 Customizing AutoCAD

In this menu item the LINE command is followed by a semicolon (;) and a backslash (\). The semicolon inputs an Enter and the backslash normally causes a pause for user input. However, in the buttons menu AutoCAD will not pause for the user input. The backslash (\) in this menu item will use the coordinates of the screen cross-hairs supplied by the pointing device as the starting point (From point) of the line. AutoCAD will then prompt for the second point of the line (To point).

Example 2

Write a button menu for the following AutoCAD commands. The menu items should use the information about the coordinate points of the screen cross-hairs, where applicable (filename BM2.MNU).

Button	Function
2	Enter
3	ERASE (with SI and NEAR options)
4	INT,END
5	LINE
6	PLINE
7	CIRCLE

The following file is a listing of the button menu for Example 2. The line numbers are not a part of the file; they are for reference only.

```
***BUTTONS1                    1
;                              2
^C^CERASE;SI;NEAR;\            3
INT,ENDP;\                     4
LINE;\                         5
PLINE;\                        6
CIRCLE;\                       7
```

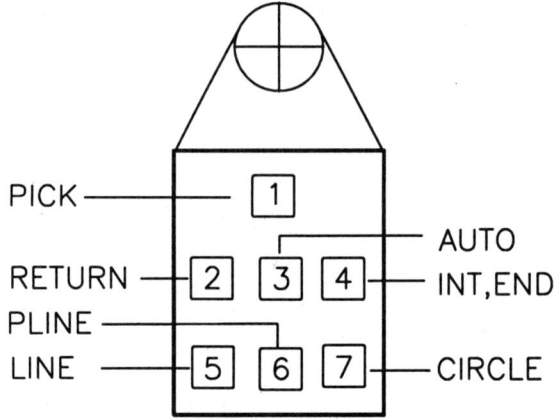

Figure 8-3 Pointing device with seven buttons

Button and Auxiliary Menus 8-7

Line 3
^C^CERASE;SI;NEAR;
This menu item defines an ERASE command with single object selection option (SI) and near object snap (NEAR). The backslash is used to accept the coordinates supplied by the screen cross-hairs. This command is assigned to button number 3 of the pointing device. If you point to an object and press the third button on the pointing device, the object will be erased and AutoCAD will automatically return to the command prompt.

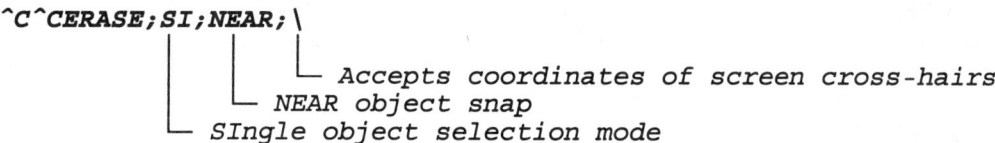

Line 4
INT,ENDP;
In this menu item INT is for the intersection snap and ENDP is for the endpoint snap. This command is assigned to button number 8 on the pointing device. When you pick this button, AutoCAD looks for the intersection point. If it cannot find an intersection point, it starts looking for the endpoint of the object that is within the pick box. The backslash (\) is used here to accept the coordinates of the screen cross-hairs. If the cross-hairs are near an object within the pick box AutoCAD will snap to the intersection's point, or the endpoint of the object when you press button number 4 on the pointing device.

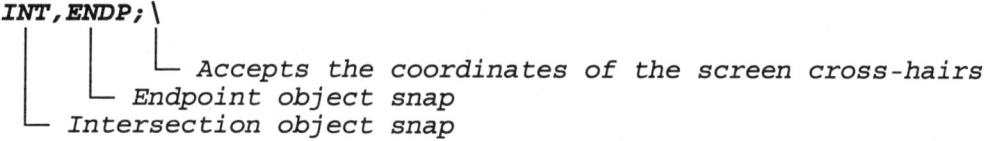

Line 7
CIRCLE;
This menu item generates a circle. The backslash (\) is used here to accept the coordinates of the screen cross-hairs as the center of the circle. If you press button number 7 of the pointing device to draw a circle, you do not need to enter the center of the circle because the current position of the screen cross-hairs automatically becomes the center of the circle. You only have to define the radius to generate the circle.

Note
The coordinate information associated with the button can be used with the first backslash only.

If a menu item in a button menu has more than one backslash (\), the remaining backslashes are ignored. For example, in the following menu item the first backslash uses the coordinates of screen cross-hairs as the insertion point and the remaining backslashes are ignored and do not cause a pause as in other menus.

8-8 Customizing AutoCAD

Example
```
INSERT;B1\\\0
         ││││
         │││└─ Rotation
         ││└── Y-scale factor
         │└─── X-scale factor
         └──── Insertion point
```

SUBMENUS

The facility to define submenus is not limited to screen, pull-down, and image menus only. You can also define submenus in the button menu.

Submenu Definition

A submenu label consists of two asterisk signs (**) followed by the name of the submenu. The buttons menu can have any number of submenus and every submenu should have a unique name. The items that follow a submenu, up to the next section label or submenu label, belong to that submenu. The format of a submenu label is:

```
**Name
│ │
│ └─ Name of the submenu
└─── Two asterisk signs (**) designate a submenu
```

> **Note**
> *The submenu name can be up to 31 characters long.*
>
> *The submenu name can consist of letters, digits, and the special characters like: $ (dollar), - (hyphen), and _ (underscore).*
>
> *The submenu name should not have any embedded blanks.*
>
> *The submenu names should be unique in a menu file.*

Submenu Reference

The submenu reference is used to reference or load a submenu. It consists of a "$" sign followed by a letter that specifies the menu section. The letter that specifies a button menu section is B. The menu section is followed by "=" sign and the name of the submenu that the user wants to activate. The submenu name should be without "**". Following is the format of a submenu reference:

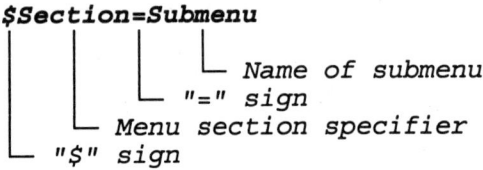

```
$Section=Submenu
│   │     │
│   │     └─ Name of submenu
│   │    └── "=" sign
│   └─────── Menu section specifier
└─────────── "$" sign
```

Button and Auxiliary Menus 8-9

Example
```
$B=BUTTON1
     │    └─ Name of submenu
     └─ B specifies buttons menu section
```

LOADING MENUS

From the buttons menu, you can load any menu that is defined in the screen, pull-down, or image menu sections by using the appropriate load commands. It may not be needed in most of the applications, but if you want to, you can load the menus that are defined in other menu sections.

Loading Screen Menus

You can load any menu that is defined in the screen menu section from the button menu by using the following load commands:

```
$S=X  $S=INSERT
   │        └─ Submenu name defined in screen menu section
   └─ Submenu name defined in screen menu section
 └─ S specifies screen menu
```

The first load command ($S=X) loads the submenu X that has been defined in the screen menu section of the menu file. The X submenu can contain 21 blank lines, so that when it is loaded it clears the screen menu. The second load command ($S=INSERT) loads the submenu INSERT that has been defined in the screen menu section of the menu file.

Loading a Pull-down Menu

You can load a pull-down menu from the button menu by using the following command:

```
$P1=P1A  $P1=*
     │        └─ Forces the display of new menu items
     └─ Loads the submenu P1A
```

The first load command $P1=P1A loads the submenu P1A that has been defined in the POP1 section of the menu file. The second load command $P1=* forces the new menu items to be displayed on the screen.

Loading an Image Menu

You can also load an image menu from the button menu by using the following load command:

```
$I=IMAGE1  $I=*
       │        └─ To display the dialog box
       └─ Load the submenu IMAGE1
```

8-10 Customizing AutoCAD

This menu item consists of two load commands. The first load command $I=IMAGE1 loads the image submenu IMAGE1 that has been defined in the image menu section of the file. The second load command $I=* displays the new dialog box on the screen.

Example 3

Write a button menu for a pointing device that has six buttons. The functions assigned to different buttons are shown in the following table (filename BM3.MNU):

SUBMENU B1	SUBMENU B2
1. PICK	1. PICK
2. Enter	2. Enter
3. LOAD OSNAPS SUBMENU	3. LOAD IMAGE1 SUBMENU
4. LOAD ZOOM1 SUBMENU	4. EXPLODE
5. LOAD BUTTON SUBMENU B1	5. LOAD BUTTON SUBMENU B1
6. LOAD BUTTON SUBMENU B2	6. LOAD BUTTON SUBMENU B2

Note

OSNAPS submenu is defined in the POP1 section of the pull-down menu.

ZOOM1 submenu is defined in the screen section of the menu file.

IMAGE1 submenu is defined in the IMAGE section of the menu file. The IMAGE1 submenu contains four images for inserting the blocks.

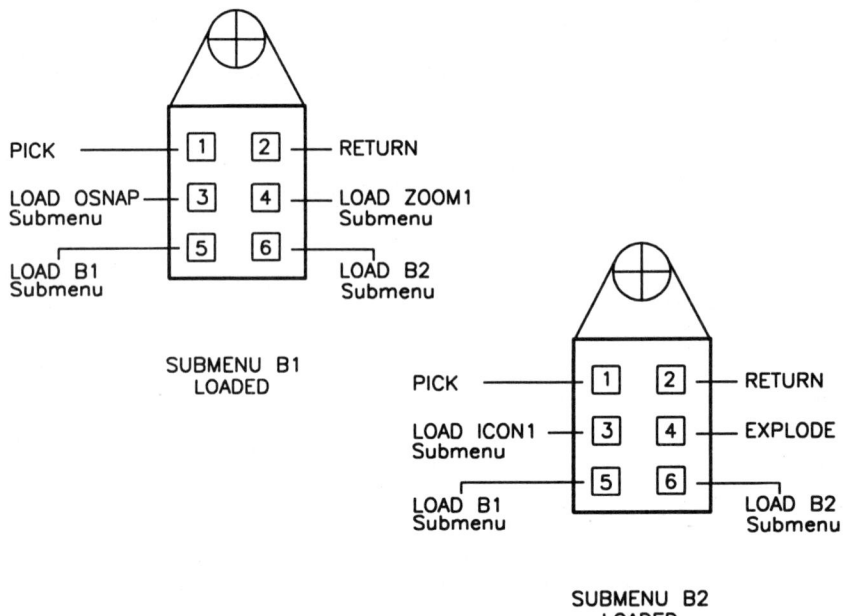

Figure 8-4 Commands assigned to different buttons of pointing device

Button and Auxiliary Menus 8-11

In this example there are two submenus. The submenus are loaded by picking button 5 for submenu B1 and button 6 for submenu B2. When the submenu B1 is loaded, AutoCAD assigns the commands that are defined under the submenu B1 to the buttons of the pointing device. Similarly, when the submenu B2 is loaded, AutoCAD assigns the commands that are defined under the submenu B2 to the buttons of the pointing device. Figure 8-4 shows the commands that are assigned to the buttons after loading submenu B1, and submenu B2

You can use AutoCAD's EDIT command to write the file. The following file is the listing of the buttons menu for Example 3. The line numbers are not a part of the file. They are shown here for reference only.

```
***BUTTON                                        1
**B1                                             2
;                                                3
$P1=*                                            4
$S=X $S=ZOOM1 'ZOOM;Win                          5
$B=B1                                            6
$B=B2                                            7
**B2                                             8
;                                                9
^C^C$I=IMAGE1 $I=*                              10
EXPLODE;\                                       11
$B=B1                                           12
$B=B2                                           13
```

Line 2
**B1
This menu item defines a submenu. The name of the submenu is B1.

Line 4
$P1=*
This menu item loads and displays the pull-down menu that has been defined in the POP1 section of the pull-down menu.

Line 5
$S=X $S=ZOOM1 'ZOOM;Win
This menu item will load and display the ZOOM1 submenu on the screen and enter a transparent ZOOM Window command. $S=X loads the submenu X that has been defined in the screen menu section of the menu file. $S=ZOOM1 loads the submenu ZOOM1 that has also been defined in the screen menu section. 'ZOOM;Win is a transparent zoom command with the window option.

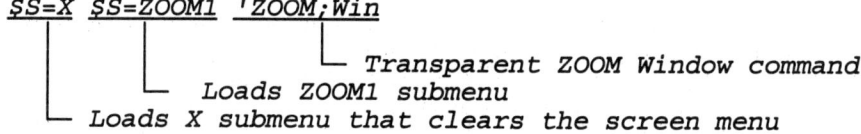

8-12 Customizing AutoCAD

Line 8
**B2
This menu item defines the submenu B2.

Line 10
^C^C$I=IMAGE1 $I=*
This menu item cancels the existing command twice and then loads the submenu IMAGE1 that has been defined in the IMAGE menu. $I=* displays the current dialog box on the screen.

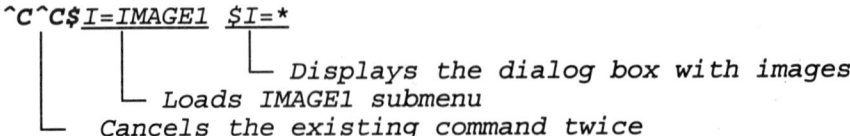

Line 11
EXPLODE;
This menu item will explode an object. It utilizes the special feature of the pointing device buttons that supply the coordinates of the screen cross-hair. When you select this button, it will explode the object where the screen cross-hair is located.

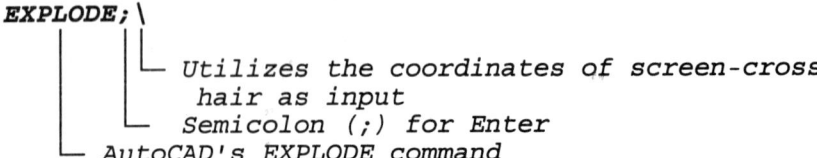

Line 12
$B=B1
This menu item loads the submenu B1 and assigns the functions defined under this submenu to different buttons of the pointing device.

Line 13
$B=B2
This menu item loads the submenu B2 and assigns the functions defined under this submenu to different buttons of the pointing device.

REVIEW QUESTIONS

1. A multibutton pointing device can be used to specify _____, or select _____, or enter AutoCAD _____.

2. AutoCAD receives the button _____ and _____ of screen crosshairs when a button is activated on the pointing device.

3. If the number of menu items in the button menu is more than the number of buttons on the pointing device, the excess lines are _____.

4. Commands are assigned to the buttons of the pointing device in the _____ order in which they appear in the buttons menu.

5. The format of referencing or loading a submenu that has been defined in the image menu is _____.

6. The format of displaying a loaded submenu that has been defined in the image menu is _____.

7. The format of the LOAD command for loading a submenu that has been defined in the image menu is _____.

EXERCISES

Exercise 1

Write a button menu for the following AutoCAD commands. The pointing device has 10 buttons (Figure 8-5), and button number 1 is used for specifying the points. The blocks are to be inserted with a scale factor of 1.00 and a rotation of 0 degrees (file name **BME1.MNU**).

1. PICK BUTTON
2. RETURN
3. CANCEL
4. OSNAPS
5. INSERT B1
6. INSERT B2
7. INSERT B3
8. ZOOM Window
9. ZOOM All
10. ZOOM Previous

8-14 Customizing AutoCAD

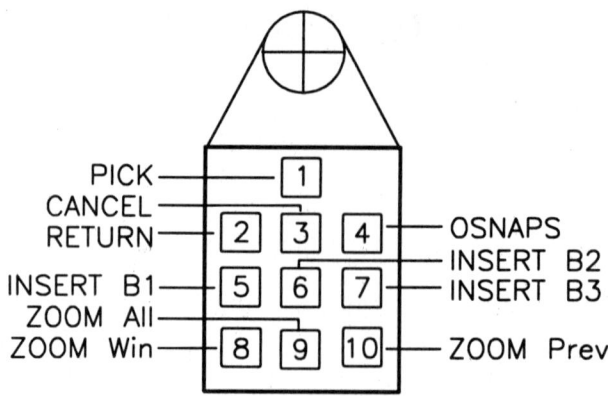

Figure 8-5 Pointing device with 10 buttons

1. B1, B2, B3 are the names of the blocks or Wblocks that have already been created.
2. Assume that the Osnap submenu has already been defined in the screen menu section of the menu file.
3. Use the transparent ZOOM command for ZOOM Previous and ZOOM Window.

Exercise 2

Write a button menu for a pointing device that has 10 buttons. The functions assigned to different buttons are shown in the following table and in Figure 8-6 (filename BME2.MNU):

SUBMENU B1		SUBMENU B2	
1.	PICK	1.	PICK
2.	Enter	2.	Enter
3.	LINE	3.	LOAD IMAGE1
4.	CIRCLE	4.	LOAD IMAGE2
5.	LOAD OSNAPS	5.	LOAD P2 (Pull-down)
6.	ZOOM Win	6.	LOAD P3 (Pull-down)
7.	ZOOM Prev	7.	INSERT
8.	ERASE	8.	EXPLODE
9.	LOAD BUTTON MENU B1	9.	LOAD BUTTON MENU B1
10.	LOAD BUTTON MENU B2	10.	LOAD BUTTON MENU B2

Button and Auxiliary Menus 8-15

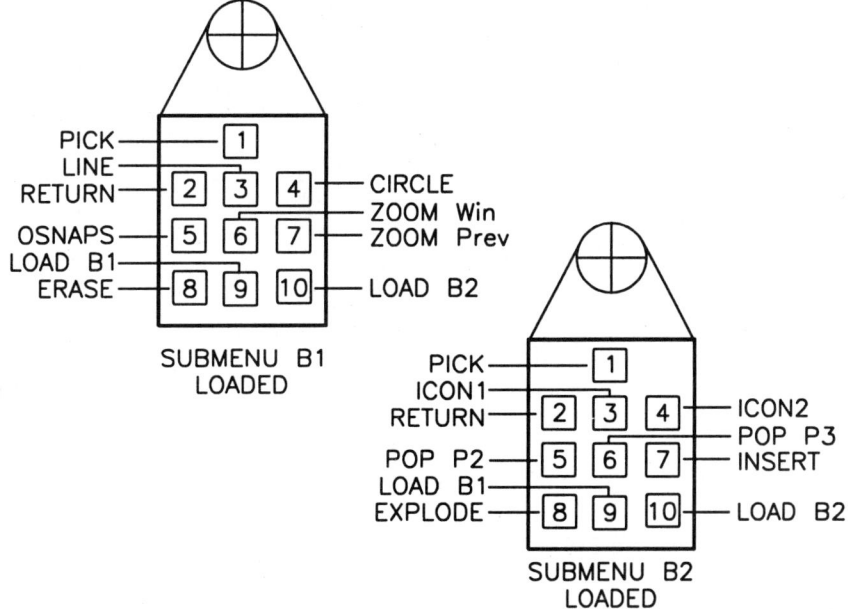

Figure 8-6 Commands assigned to different buttons of pointing device

Assume that:
1. OSNAPS submenu is defined in the POP1 or POP0 (Cursor menu) section of the pull-down menu.
2. Pull-down menus P2 and P3 are defined in the POP2 and POP3 sections of the pull-down menu.
3. IMAGE1 and IMAGE2 submenus are defined in the IMAGE section of the menu file. The IMAGE1 and IMAGE2 submenus contain four images each for inserting the blocks.

Chapter 9

Screen Menus

Learning objectives

After completing this chapter, you will be able to:
- *Write screen menus.*
- *Load screen menus.*
- *Write submenus and referencing submenus.*
- *Write menu files with multiple submenus.*
- *Use menu command repetition in menus.*
- *Write menus for foreign languages.*
- *Use control and special characters in menu items.*
- *Use command definition without return or space.*
- *Use menu items with single object selection mode.*
- *Use AutoLISP and DIESEL expressions in menus.*

SCREEN MENU

When you are in the AutoCAD drawing editor the screen menu is displayed on the right of the screen. The AutoCAD screen menu displays AutoCAD at the top, followed by asterisk signs (* * * *) and a list of commands (Figure 9-1).

Depending on the scope of the menu, the size of the menu file can vary from a few lines to several hundred. A menu file consists of section labels, submenus, and menu items. A menu item consists of an item label and a command definition. The menu item label is enclosed in brackets, and the command definition (menu macro) is outside the brackets.

The menu item label that is enclosed in the brackets is displayed in the screen menu area of the monitor and is not a part of the command definition. The command definition, the part of the menu item outside the bracket, is the executable part of the menu item. To understand the process of developing and writing a screen menu, consider the following example.

9-2 Customizing AutoCAD

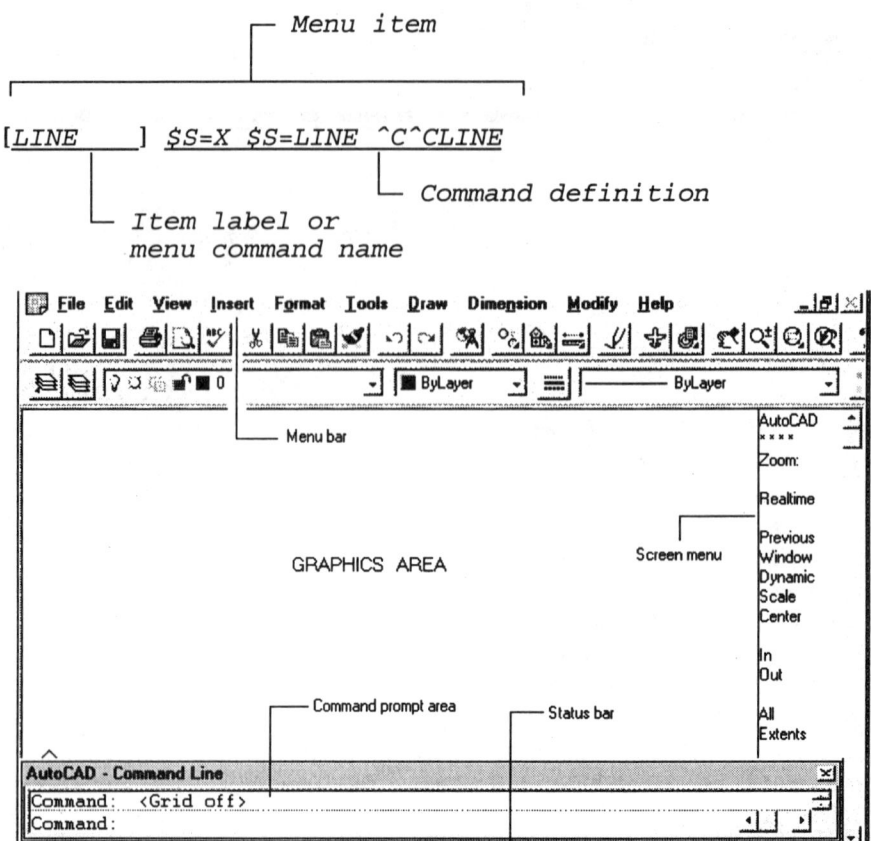

Figure 9-1 Screen display

Example 1

Write a screen menu for the following AutoCAD commands (file name **SM1.MNU**):
   Line
   Circle C,R
   Circle C,D
   Circle 2P
   Erase
   Move

The layout of these commands is shown in Figure 9-2(a). This menu is named MENU-1 and it should be displayed at the top of the screen menu. It lets you know the menu you are using. You can use any text editor to write the file. The file name can be up to eight characters long, and the file extension must be **.MNU**. For Example 1 the file name is **SM1.MNU**. **SM1** is the name of the screen menu file, and **.MNU** is the extension of this file. All menu files have the extension **.MNU**.

Screen Menus 9-3

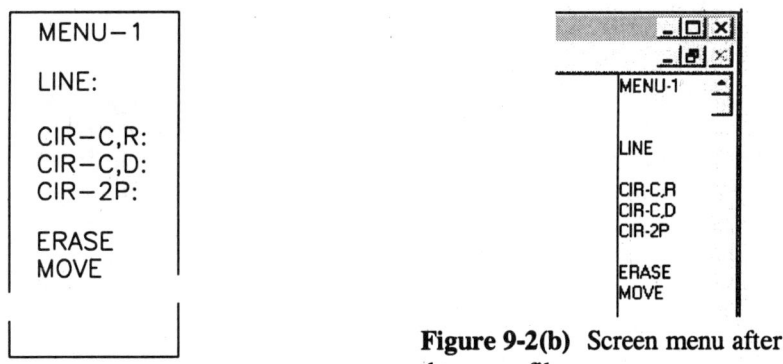

Figure 9-2(a) Layout of screen menu

Figure 9-2(b) Screen menu after loading the menu file

The following file is a listing of the screen menu for Example 1. **The line numbers on the right are not a part of the file. They are shown here for reference only.**

```
***SCREEN                                           1
[ MENU-1     ]                                      2
[            ]                                      3
[            ]                                      4
[LINE        ]^C^CLINE                              5
[            ]                                      6
[CIR-C,R     ]^C^CCIRCLE                            7
[CIR-C,D     ]^C^CCIRCLE;\D                         8
[CIR- 2P     ]^C^CCIRCLE;2P                         9
[            ]                                     10
[ERASE       ]^C^CERASE                            11
[MOVE        ]^C^CMOVE                             12
```

Explanation

Line 1
***SCREEN
***SCREEN is the section label for the screen menu. The lines that follow the screen menu are treated as a part of this menu. The screen menu definition will be terminated by another section label, such as ***TABLET1 or ***POP1.

Line 2
[MENU-1]
This menu item displays MENU-1 on the screen. Anything that is inside the brackets is for display only and has no effect on the command. The maximum number of characters or spaces that can be displayed inside these brackets is eight, because the width of the screen menu column on the screen is eight characters. If the number of characters is more than eight, the remaining characters are not displayed on the screen and can be used for comments. The part of the menu item that is outside the brackets is executed even if the number of characters inside the bracket is more than eight.

9-4 Customizing AutoCAD

Lines 3 and 4

[   ]

These menu items print a blank line on the screen menu. There are eight blank spaces inside the bracket. You could also use the brackets with no spaces between the brackets ([]). When the blank spaces are printed on the screen menu, it displays a blank line. This line does not contain anything outside the bracket; therefore, no command is executed. To provide the space in the menu you can also leave a blank space in the menu file or have two brackets ([ ]). The next line, line 4, also prints a blank line.

Line 5

[LINE   ]^C^CLINE

This menu item displays LINE on the screen. The first ^C (caret C) cancels the existing command, and the **second** ^C cancels the command again. The two CANCEL (^C^C) commands are required to make sure that the existing commands are canceled before a new command is executed. Most AutoCAD commands can be canceled with just one CANCEL command. However, some commands, like dimensioning and pedit, need to be canceled twice to get out of the command. **LINE** will prompt the user to enter the points to draw a line. Since there is nothing after the LINE, the spaces after the line automatically enters a Return.

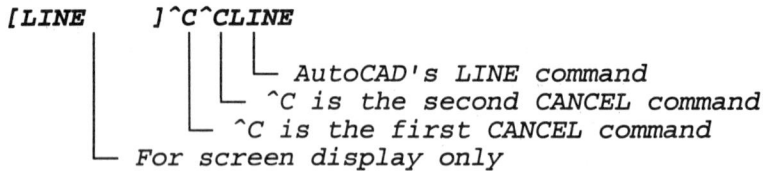

Line 7

[CIR-C,R]^C^CCIRCLE

The part of the menu item that is enclosed within the brackets is for screen display only. The part of the menu item that is outside the brackets is executed when this line is selected. ^C^C (caret C) cancels the existing command twice. CIRCLE is an AutoCAD command that generates a circle. **The space after CIRCLE automatically causes a Return. The space acts like pressing the spacebar on the keyboard.**

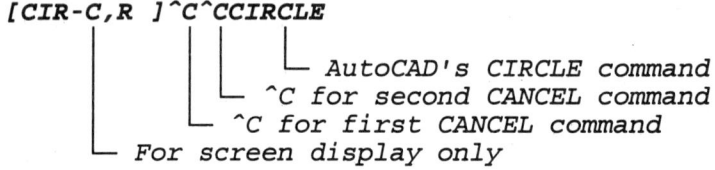

Line 8

[CIR-C,D]^C^CCIRCLE;\D

The part of the menu item that is enclosed in the brackets is for screen display only, and the part that is outside the brackets is the executable part. ^C^C will cancel the existing command twice.

Screen Menus 9-5

```
[CIR-C,D ]^C^CCIRCLE;\D
                     │││ │
                     │││ └─ Diameter option
                     ││└── AutoCAD pauses for user input
                     │└─── Semicolon (;) for Return
                     └──── CIRCLE command
          └───────────── ^C^C cancels the existing command
```

The CIRCLE command is followed by a semicolon, a backslash (\), and a D for diameter option. **The semicolon (;) after the CIRCLE command causes a Return, which has the same effect as entering Return at the keyboard. The backslash (\) pauses for user input.** In this case it is the center point of the circle. D is for the diameter option and it is automatically followed by Return. The semicolon in this example can also be replaced by a blank space, as shown in the following line. However, the semicolon is easier to spot.

```
[CIR-C,D ]^C^CCIRCLE \D
                     │
                     └─ Blank space for Return
```

Line 9
[CIR- 2P]^C^CCIRCLE;2P
In this menu item ^C^C cancels the existing command twice. The semicolon after CIRCLE enters a Return. 2P is for the two-point option, followed by the blank space that causes a Return. You will notice that the sequence of the command and input is the same as discussed earlier. Therefore, it is essential to know the exact sequence of the command and input; otherwise, the screen menu is not going to work.

```
[CIR- 2P ]^C^CCIRCLE;2P
                    │ │
                    │ └─ 2-Point option for CIRCLE
                    └─── Semicolon (;) for Return
           └──────────── CIRCLE command
           └──────────── ^C^C Cancels existing command twice
```

The semicolon after the CIRCLE command can be replaced by a blank space, as shown in the following line. The blank space causes a Return, like a semicolon (;).

```
[CIR- 2P ]^C^CCIRCLE 2P
                    │
                    └─ Blank space or semicolon for Return
```

Line 11
[ERASE]^C^CERASE
In this menu item ^C^C cancels the existing command twice, and ERASE is an AutoCAD command that erases the selected objects.

```
[ERASE    ]^C^CERASE
               │
               └─ AutoCAD ERASE command
```

Line 12
[MOVE]^C^CMOVE
In this menu item ^C^C cancels the existing command twice, and the MOVE command will move the selected objects.

```
[MOVE    ]^C^CMOVE
                 └─ AutoCAD MOVE command
```

LOADING MENUS

AutoCAD automatically loads the **ACAD.MNU** file when you get into the AutoCAD drawing editor. However, you can also load a different menu file by using the AutoCAD MENU command.

Command: **Menu**

When you enter the Menu command, AutoCAD displays the **Select Menu File** dialog box (Figure 9-3) on the screen. Select the menu file that you want to load and then select the Open button.

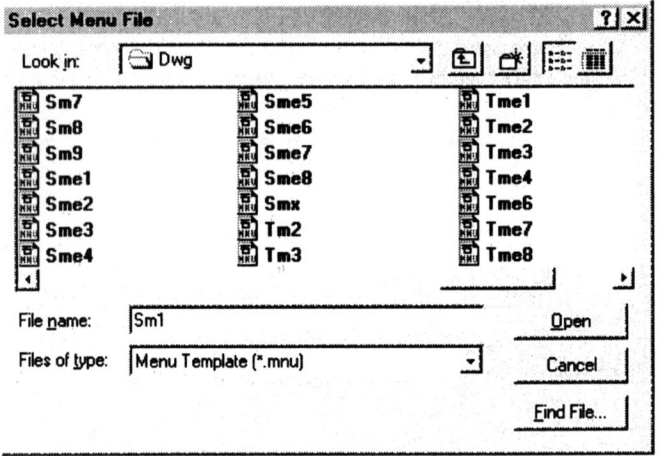

Figure 9-3 Select Menu File dialog box

You can also load the menu file from the command line.

```
Command: FILEDIA
New value for FILEDIA <1>: 0
Command: Menu
Menu filename <ACAD>: PDM1
                │       └─ Name of menu file
                └─ Default menu file
```

After you enter the MENU command, AutoCAD will prompt for the file name. Enter the name of the menu file without the file extension (.MNU), since AutoCAD assumes the extension .MNU. AutoCAD will automatically compile the menu file into **MNC** and **MNR** files.

> **Note**
>
> *1. After you load the new menu, you cannot use the screen menu, buttons menu, or digitizer because the original menu, **ACAD.MNU**, is not present and the new menu does not contain these menu areas.*
>
> *2. To activate the original menu again, load the menu file by using the MENU command:*
>
> Command: **Menu**
> Menu file name or . for none <SM1>: **ACAD**
>
> *3. If you need to use input from a keyboard or a pointing device, use the backslash (\). The system will pause for you to enter data.*
>
> *4. There should be no space after the backslash (\).*
>
> *5. The menu items, menu labels, and command definition can be uppercase, lowercase, or mixed.*
>
> *6. You can introduce spaces between the menu items to improve the readability of the menu file.*
>
> *7. If there are more items in the menu than the number of spaces available, the excess items are not displayed on the screen. For example, if the display device limits the number of items to 21, items in excess of 21 will not be displayed on the screen and are therefore inaccessible.*
>
> *8. If you are using a high-resolution graphics board, you can increase the number of lines that can be displayed on the screen. On some devices this is 80 lines.*

Exercise 1

Design and write a screen menu for the following AutoCAD commands (file name **SME1.MNU**):

PLINE
ELLIPSE (Center)
ELLIPSE (Axis endpoint)
ROTATE
OFFSET
SCALE

SUBMENUS

The screen menu file can be so large that all items cannot be accommodated on one screen. For example, the maximum number of items that can be displayed on most screens is 21. If the screen menu has more than 21 items, the menu items in excess of 21 are not displayed on the screen and therefore cannot be accessed. You can overcome this problem by using submenus that enable the user to define smaller groups of items within a menu section. When a submenu is selected, it loads the submenu items and displays them on screen. However, depending on the resolution of the monitor and the graphics card, the number of items you can display can be higher and you may not need submenus.

Submenu Definition

A submenu definition consists of two asterisks (**) followed by the name of the submenu. A menu can have any number of submenus, and every submenu should have a unique name. The items that follow a submenu, up to the next section label, or submenu label belong to that submenu. The format of a submenu definition is:

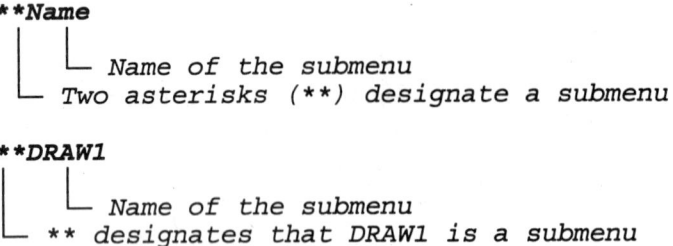

> **Note**
> The submenu name can be up to 31 characters long.
>
> The submenu name can consist of letters, digits, and such special characters as $ (dollar sign), - (hyphen), and _ (underscore).
>
> The submenu name should not have any embedded blanks (spaces).
>
> The submenu names should be unique in a menu file.

Submenu Reference

The submenu reference is used in a menu item to reference or load a submenu. It consists of a "$" sign followed by a letter that specifies the menu section. The letter that specifies a screen menu section is S. The section is followed by an "=" sign and the name of the submenu the user wants to activate. The submenu name should be without the **. The following is the format of a submenu reference:

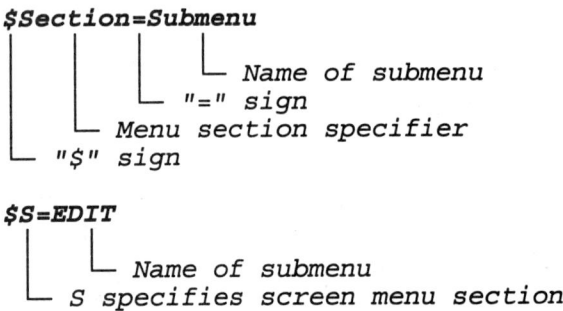

Screen Menus 9-9

> **Note**
>
> $ *A special character code used to load a submenu in a menu file.*
> $M=   *This is used to load a DIESEL macro from a menu item.*
> *The following are the section specifiers:*
> S   *Specifies the SCREEN menu.*
> P0 - P16   *Specifies the POP menus, POP0 through POP16.*
> I   *Specifies the ICON menu.*
> B1 - B4   *Specifies the BUTTONS menu, B1 through B4.*
> T1 - T4   *Specifies the TABLET menus, T1 through T4.*
> A1 - A4   *Specifies the AUX menus A1 through A4.*

Nested Submenus

When a submenu is activated, the current menu is copied to a stack. If you select another submenu, the submenu that was current will be copied or pushed to the top of the stack. The maximum number of menus that can be stacked is eight. If the stack size increases to more than eight, the menu at the bottom of the stack is removed and forgotten. You can call the previous submenu by using the nested submenu call. The format of this call is:

```
$S=
│ │ └── "=" sign
│ └──── Screen menu specifier
└────── "$" sign
```

The maximum number of nested submenu calls is eight. Each time you call a submenu (issue $S=), this pops the last item from the stack and reactivates it.

Example 2

Design a menu layout and write a screen menu for the following commands.

    LINE   ERASE
    PLINE   MOVE
    ELLIPSE-C   ROTATE
    ELLIPSE-E   OFFSET
    CIR-C,R   COPY
    CIR-C,D   SCALE
    CIR- 2P

As mentioned earlier, the first and the most important part of writing a menu is the design of the menu and knowing the commands and the prompts associated with those commands. You should know the way you want the menu to look and the way you want to arrange the commands for maximum efficiency. Write out the menu on a piece of paper, and check it thoroughly to make sure you have arranged the commands the way you want them. Use submenus to group the commands based on their use, function, and relationship with other submenus. Make provisions to access other frequently used commands without going through the root menu.

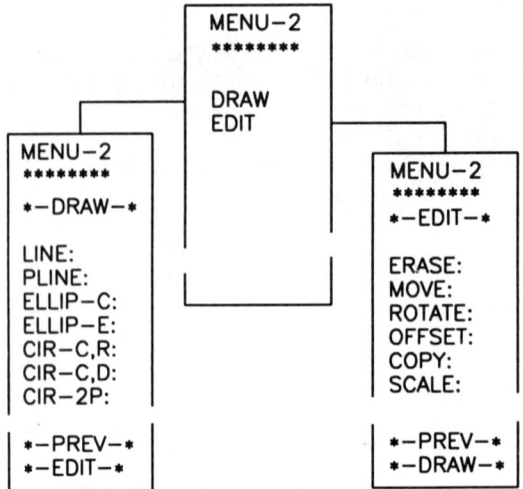

Figure 9-4 Screen menu design

Figure 9-4 shows one of the possible arrangements of the commands and the design of the screen menu. It has one main menu and two submenus. One of the submenus is for draw commands, the other is for edit commands. The colon (:) at the end of the commands is not required. It is used here to distinguish the commands from those items that are not used as commands. For example, DRAW in the root menu is not a command; therefore, it has no colon at the end. On the other hand, if you select ERASE from the EDIT menu, it executes the ERASE command, so it has a colon (:) at the end of the command.

The following file is a listing of the menu file of Example 1. **The line numbers on the right are not a part of the file; they are given here for reference only.**

```
***SCREEN                                   1
[ MENU-2      ]                             2
[*******      ]                             3
[             ]                             4
[             ]                             5
[             ]                             6
[             ]                             7
[DRAW         ]^C^C$S=DRAW                  8
[EDIT         ]^C^C$S=EDIT                  9
                                           10
**DRAW                                     11
[ MENU-2      ]^C^C$S=SCREEN               12
[*******      ]                            13
[                                          14
[*-DRAW-*     ]                            15
[             ]                            16
[LINE:        ]^C^CLINE                    17
[PLINE:       ]^C^CPLINE;\W;0.1;0.1        18
```

[ELLIP-C:	]^C^CELLIPSE;C	19
[ELLIP-E:	]^C^CELLIPSE	20
[CIR-C,R:	]^C^CCIRCLE	21
[CIR-C,D:	]^C^CCIRCLE;\D	22
[CIR-2P:	]^C^CCIRCLE;2P	23
[	]	24
[	]	25
[	]	26
[	]	27
[	]	28
[	]	29
[*-PREV-*	]^C^C$S=	30
[*-EDIT-*	]^C^C$S=EDIT	31
		32
**EDIT		33
[MENU-2	]^C^C$S=SCREEN	34
[********	]	35
[	]	36
[*-EDIT-*	]	37
[	]	38
[ERASE:	]^C^CERASE	39
[MOVE:	]^C^CMOVE	40
[ROTATE:	]^C^CROTATE	41
[OFFSET:	]^C^COFFSET	42
[COPY:	]^C^CCOPY	43
[SCALE:	]^C^CSCALE	44
[	]	45
[	]	46
[	]	47
[	]	48
[	]	49
[	]	50
[	]	51
[*-PREV-*	]^C^C$S=	52
[*-DRAW-*	]^C^C$S=DRAW	53

Explanation

Line 1
***SCREEN
***SCREEN is the section label for the screen menu.

Line 2
[MENU-2]
This menu item displays MENU-2 at the top of the screen menu.

9-12 Customizing AutoCAD

Line 3
[********]
This menu item prints eight asterisk signs (********) on the screen menu.

Lines 4-7
[   ]
These menu items print four blank lines on the screen menu. The brackets are not required. They could be just four blank lines without brackets.

Line 8
[DRAW   ]^C^C$S=DRAW
[DRAW   ] displays DRAW on the screen, letting you know that by selecting this function you can access the draw commands. ^C^C cancels the existing command, and $S=DRAW loads the DRAW submenu on the screen.

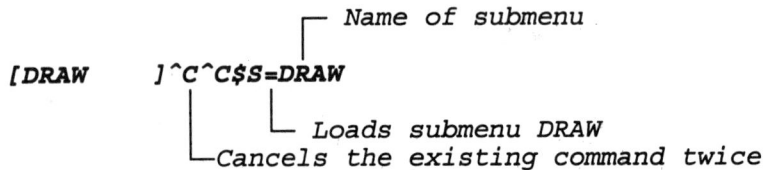

Line 9
[EDIT   ]^C^C$S=EDIT
[EDIT   ] displays EDIT on the screen. ^C^C cancels the current command, and $S=EDIT loads the submenu **EDIT**.

Line 10
The blank lines between the submenus or menu items are not required. It just makes it easier to read the file.

Line 11
**DRAW
**DRAW is the name of the submenu; lines 12 through 31 are defined under this submenu.

Line 15
[*-DRAW-*]
This prints *-DRAW-* as a heading on the screen menu to let the user know that the commands listed on the menu are draw commands.

Line 18
[PLINE:]^C^CPLINE;\W;0.1;0.1
[PLINE:] displays PLINE: on the screen. ^C^C cancels the command, and PLINE is the AutoCAD polyline command. The semicolons are for Return. The semicolons can be replaced by a blank space. The backslash (\) is for user input. In this case it is the start point of the polyline. W selects the width option of the polyline. The first 0.1 is the starting width of the polyline, and the second 0.1 is the ending width. This command will draw a polyline of 0.1 width.

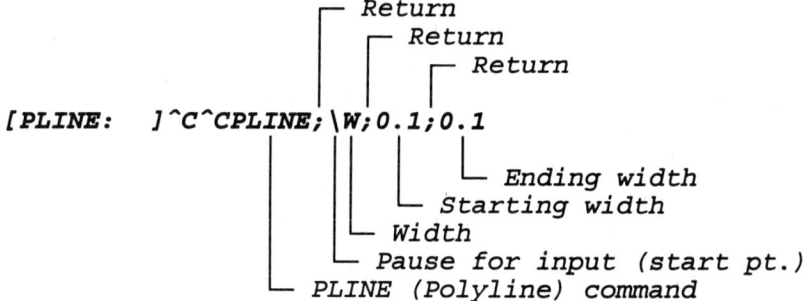

Line 19
[ELLIP-C:]^C^CELLIPSE;C
[ELLIP-C:] displays ELLIP-C: on the screen menu. ^C^C cancels the existing command, and ELLIPSE is an AutoCAD command to generate an ellipse. The semicolon is for Return, and C selects the center option of the ELLIPSE command.

```
[ELLIP-C:]^C^CELLIPSE;C
                   │ │└─ Center option for ellipse
                   │ └── Return
                   └──── ELLIPSE command
```

Line 20
[ELLIP-E:]^C^CELLIPSE
[ELLIP-E:] displays ELLIP-E: on the screen menu. ^C^C cancels the existing command, and ELLIPSE is an AutoCAD command. Here the ELLIPSE command uses the default option, axis endpoint, instead of center.

Line 30
[*-PREV-*]^C^C$S=
[*-PREV-*] displays *-PREV-* on the screen menu, and ^C^C cancels the existing command. $S= restores the previous menu that was displayed on the screen before loading the current menu.

```
[*-PREV-*]^C^C$S=
                └─ Restores the previous screen menu
```

Line 31
[*-EDIT-*]^C^C$S=EDIT
[*-EDIT-*] displays *-EDIT-* on the screen menu, and ^C^C cancels the existing command. $S=EDIT loads the submenu EDIT on the screen. This lets you access the EDIT commands without going back to the root menu and selecting EDIT from there.

```
                          ┌─ Name of the submenu
                          │
        [*-EDIT-*]^C^C$S=EDIT
                               └─ Loads the submenu EDIT
```

Line 33
**EDIT
**EDIT is the name of the submenu, and lines 34 to 53 are defined under this submenu.

Line 34
[MENU-2]^C^C$S=SCREEN
[MENU-2] displays MENU-2 on the screen menu, and ^C^C cancels the existing command. $S=SCREEN loads the root menu SCREEN on the screen menu.

Line 39
[ERASE:]^C^CERASE
[ERASE:] displays ERASE: on the screen menu, and ^C^C cancels the existing command. ERASE is an AutoCAD command for erasing the selected objects.

Line 53
[*-DRAW-*]^C^C$S=DRAW
$S=DRAW loads the DRAW submenu on the screen. It lets you load the DRAW menu without going through the root menu. The root menu is the first menu that appears on the screen when you load a menu or start AutoCAD.

When you select DRAW from the root menu, the submenu DRAW is loaded on the screen. The menu items in the draw submenu completely replace the menu items of the root menu. If you select MENU-2 from the screen menu now, the root menu will be loaded on the screen, but some of the items are not cleared from the screen menu (Figure 9-5). This is because the root menu does not have enough menu items to completely replace the menu items of the DRAW submenu.

One of the ways to clear the screen is to define a submenu that has 21 blank lines. When this submenu is loaded it will clear the screen. If you then load another submenu, there will be no overlap, because the screen menu has already been cleared and there are no menu items left on the screen (Figure 9-6). Example 3 illustrates the use of such a submenu.

Screen Menus 9-15

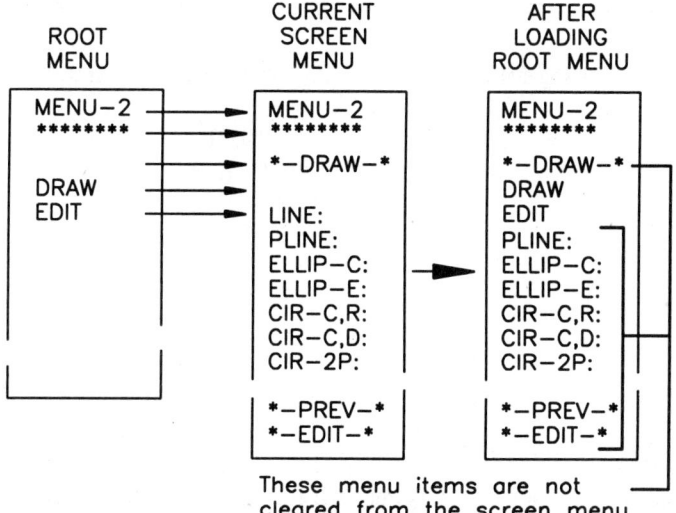

Figure 9-5 Screen menu display after loading the root menu

Another way to avoid overlapping the menu items is to define every submenu so that they all have the same number of menu items. The disadvantage with this approach is that the menu file will be long, because every submenu will have 21 lines.

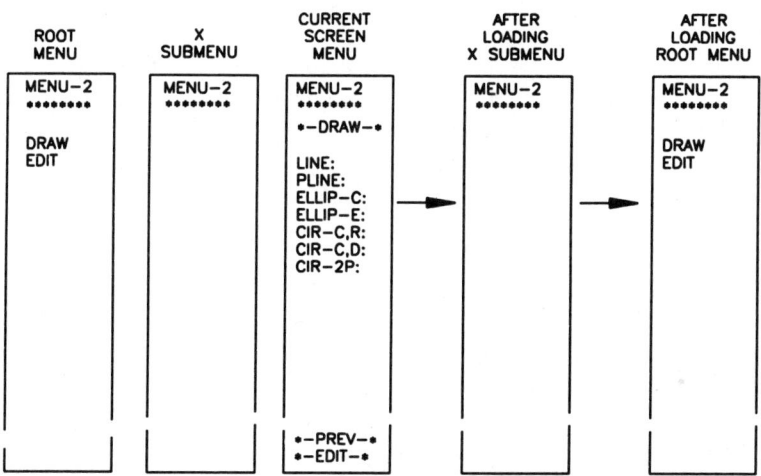

Figure 9-6 Screen menu display after loading the root menu

Exercise 2

Write a screen menu file for the following AutoCAD commands (file name **SM2.MNU**).

ARC	MIRROR
-3P	BREAK-F
-SCE	BREAK-@
-SCA	EXTEND
-SCL	STRETCH
-SEA	FILLET-0
POLYGON-C	FILLET
POLYGON-E	CHAMFER

MULTIPLE SUBMENUS

A menu file can have any number of submenus. All the submenu names have two asterisks (**) in front of them, even if there is a submenu within a submenu. Also, several submenus can be loaded in a single statement. If there is more than one submenu in a menu item, they should be separated from one another by a space. Example 3 illustrates the use of multiple submenus.

Example 3

Design the layout of the menu, and then write the screen menu for the following AutoCAD commands.

Draw	**ARC**	**Edit**	**Display**
LINE	3Point	EXTEND	ZOOM
Continue	SCE	STRETCH	REGEN
Close	SCA	FILLET	SCALE
Undo	CSE		PAN
.X	CSA		
.Y	CSL		
.Z			
.XY			
.XZ			
.YZ			

There are several different ways of arranging the commands, depending on your requirements. The following layout is one of the possible screen menu designs for the given AutoCAD commands (Figure 9-7).

Screen Menus 9-17

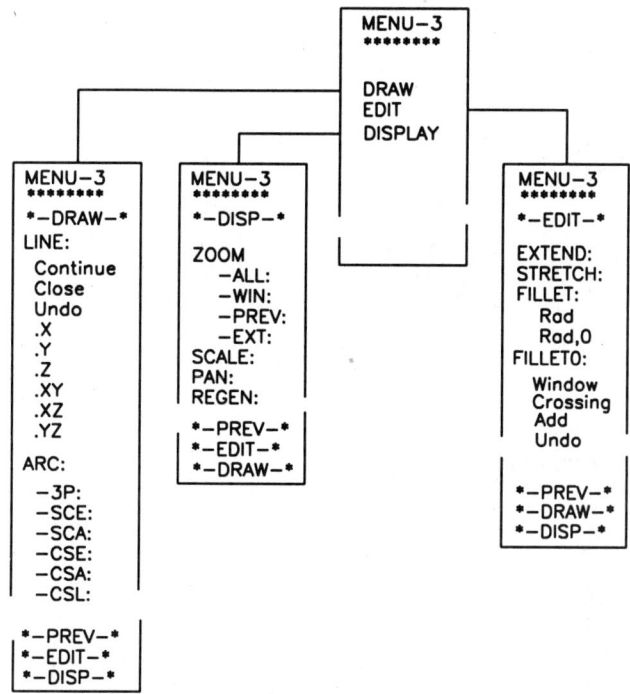

Figure 9-7 Screen menu design with submenus

The name of the following menu file is **SM3.MNU**. You can use a text editor to write the file. The following file is the listing of the menu file **SM3.MNU**. **The line numbers on the right are not a part of the file. They are shown here for reference only**.

```
***SCREEN                                                           1
**S                                                                 2
[ MENU-3      ]^C^C$S=X $S=S                                        3
[********     ]$S=OSNAP                                             4
[             ]                                                     5
[DRAW         ]^C^C$S=X $S=DRAW                                     6
[EDIT         ]^C^C$S=X $S=EDIT                                     7
[DISPLAY      ]^C^C$S=X $S=DISP                                     8
**DRAW 3                                                            9
[*-DRAW-*     ]                                                    10
[             ]                                                    11
[LINE:        ]^C^CLINE                                            12
[ Continue    ]^C^CLINE;;                                          13
[ Close       ]CLOSE                                               14
[ Undo        ]UNDO                                                15
[ .X          ].X                                                  16
[ .Y          ].Y                                                  17
[ .Z          ].Z                                                  18
```

```
[ .XY            ].XY                                  19
[ .XZ            ].XZ                                  20
[ .YZ            ].YZ                                  21
[                ]                                     22
[ARC:            ]^c^cArc                              23
[  -3P:          ]^C^CARC \\DRAG                       24
[  -SCE:         ]^C^CARC \C \DRAG                     25
[  -SCA:         ]^C^CARC \C \A DRAG                   26
[  -CSE:         ]^C^CARC C \\DRAG                     27
[  -CSA:         ]^C^CARC C \\A DRAG                   28
[  -CSL:         ]^C^CARC C \\L DRAG                   29
[                ]                                     30
[*-PREV-*        ]$S= $S=                              31
[*-EDIT-*        ]^C^C$S=X $S=EDIT                     32
[*-DISP-*        ]$S=X $S=DISP                         33
**EDIT 3                                               34
[*-EDIT-*        ]                                     35
[                ]                                     36
[EXTEND:         ]^C^CEXTEND                           37
[STRETCH:        ]^C^CSTRETCH;C                        38
[FILLET:         ]^C^CFILLET                           39
[ Rad            ]R;\Fillet                            40
[ Rad 0          ]R;0;Fillet                           41
[FILLET0:        ]^C^CFillet;R;0::                     42
  Win.                                                 43
  Cross.                                               44
  Add                                                  45
  Undo                                                 46
[                ]                                     47
[*-PREV-*        ]$S= $S=                              48
[*-DRAW-*        ]^C^C$S=X $S=DRAW                     49
[*-DISP-*        ]$S=X $S=DISP                         50
**DISP 3                                               51
[*-DISP-*        ]                                     52
[                ]                                     53
[ZOOM:           ]'ZOOM                                54
[ -ALL           ]A                                    55
[ -WIN           ]W                                    56
[ -PREV          ]P                                    57
[ -EXT           ]E                                    58
[                ]                                     59
[SCALE:          ]'ZOOM                                60
[PAN:            ]'PAN                                 61
[REGEN:          ]^C^CREGEN                            62
[                ]                                     63
[*-PREV-*        ]$S= $S=                              64
```

[*-EDIT-*	]^C^C$S=X $S=EDIT	65
[*-DRAW-*	]$S=X $S=DRAW	66
**X 3		67
[	]	68
[	]	69
[	]	70
[	]	71
[	]	72
[	]	73
[	]	74
[	]	75
[	]	76
[	]	77
[	]	78
[	]	79
[	]	80
[	]	81
[	]	82
[	]	83
[	]	84
[	]	85
[	]	86
**OSNAP 2		87
[-OSNAPS-	]	88
[	]	89
[Center	]CEN $S=	90
[Endpoint	]END $S=	91
[Insert	]INS $S=	92
[Intersec	]INT $S=	93
[Midpoint	]MID $S=	94
[Nearest	]NEA $S=	95
[Node	]NOD $S=	96
[Perpend	]PER $S=	97
[Quadrant	]QUA $S=	98
[Tangent	]TAN $S=	99
[None	]NONE $S=	100
[	]	101
[	]	102
[	]	103
[	]	104
[	]	105
[*-PREV-*	]$S=	106

Explanation
Line 3
[MENU-3]^C^C$S=X $S=S

9-20 Customizing AutoCAD

In this menu item, [MENU-3] displays MENU-3 at the top of the screen menu. ^C^C cancels the command, and $S=X loads submenu X. Similarly, $S=S loads submenu S. Submenu X, defined on line 171, consists of 18 blank lines. Therefore, when this submenu is loaded it prints blank lines on the screen and clears the screen menu area. After loading submenu X, the system loads submenu S, and the items that are defined in this submenu are displayed on the screen menu.

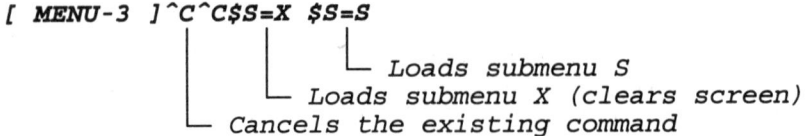

Line 4
[********]$S=OSNAP
This menu item prints eight asterisks on the screen menu, and $S=OSNAP loads submenu OSNAP. The OSNAP submenu, defined on line 191, consists of object snap modes.

Line 6
[DRAW   ]^C^C$S=X $S=DRAW
This menu item displays DRAW on the screen menu area, cancels the existing command, and then loads submenu X and submenu DRAW. Submenu X clears the screen menu area, and submenu DRAW, as defined on line 11, loads the items defined under the DRAW submenu.

Line 9
**DRAW 3
This line is the submenu label for DRAW; 3 indicates that the first line of the DRAW submenu will be printed on line 3. Nothing will be printed on line 1 or line 2. The first line of the DRAW submenu will be on line 3, followed by the rest of the menu. All the submenus except submenus S and OSNAP have a 3 at the end of the line. Therefore, the first two lines (MENU-3) and (********) will never be cleared and will be displayed on the screen all the time. If you select MENU-3 in any menu, it will load submenu S. If you select ********, it will load submenu OSNAP.

Line 12
[LINE:]^C^CLINE
^C^C cancels the existing command, and LINE is the AutoCAD LINE command.

Line 13
[Continue]^C^CLINE;;
In this menu item, the LINE command is followed by two semicolons that continue the LINE command. To understand it, look at the LINE command to see how a continue option is used in this command:

Command: **LINE**
From point: **RETURN**   **(Continue)**
To point:

Following is the command and prompt entry sequence for the LINE command, if you want to start a line from the last point:

LINE
RETURN
RETURN
SELECT A POINT

Therefore, in the screen menu file the LINE command has to be followed by two Returns to continue from the previous point.

Line 16
[.X   ].X
In this menu item, .X extracts the X coordinate from a point. The bracket, .X, and spaces inside the brackets ([.X   ])are not needed. The line could consist of .X only. The same is true for lines 39 thru 43.

Example

[.X	].X	can also be written (without brackets) as	.X
[.Y	].Y		.Y
[.Z	].Z		.Z

Line 31
[*-PREV-*]$S= $S=
This menu item recalls the previous menu twice. AutoCAD keeps track of the submenus that were loaded on the screen. The first $S= will load the previously loaded menu, and the second $S= will load the submenu that was loaded before that. For example, in line 16 ([ARC:]$S=X $S=ARC), two submenus have been loaded: first X, and then ARC. Submenu X is stacked before loading submenu ARC, which is current and is not stacked yet. The first $S= will recall the previous menu, in this case submenu X. The second $S= will load the menu that was on the screen before selecting the item in line 16.

```
[*-PREV-*]$S=  $S=
               |    └─ Loads next to last menu
               └─ Loads last menu
```

Line 33
[*-DISP-*]$S=X $S=DISP
In this menu item, $S=X loads submenu X, and $S=DISP loads submenu DISP. Notice that there is no CANCEL (^C^C) command in the line. There are some menus you might like to load without canceling the existing command. For example, if you are drawing a line, you might want to zoom without canceling the existing command. You can select [*-DISP-*], select the

9-22 Customizing AutoCAD

appropriate zoom option, and then continue with the line command. However, if the line has a CANCEL command ([*-DISP-*]^C^C$S=X $S=DISP), you cannot continue with the LINE command, because the existing command will be canceled when you select *-DISP-* from the screen. In line 28 ([*-EDIT-*]^C^C$S=X $S=EDIT), ^C^C cancels the existing command, because you cannot use any editing command unless you have canceled the existing command.

Line 38
[STRETCH:]^C^CSTRETCH;C

^C^C cancels the existing command, and STRETCH is an AutoCAD command. The C is for the crossing option that prompts you to enter two points to select the objects.

Line 40
[RAD]R;\FILLET

This menu item selects the radius option of the FILLET command, and then waits for you to enter the radius. After entering the radius, execute the FILLET command again to generate the desired fillet between the two selected objects.

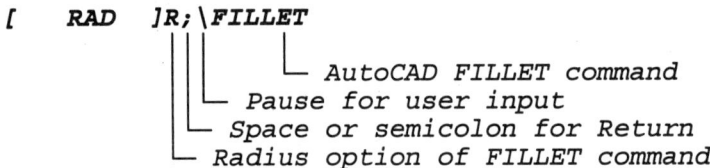

Line 41
[RAD 0]R;0;FILLET

This menu item selects the radius option of the FILLET command and assigns it a 0 value. Then it executes the FILLET command to generate a 0 radius fillet between the two selected objects.

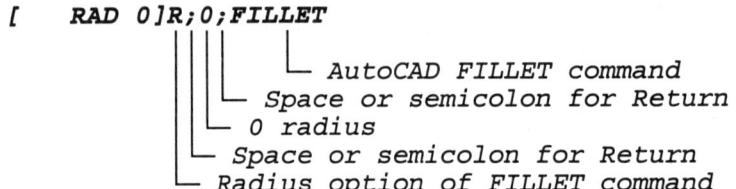

Line 42
[FILLET0:]^C^CFILLET;R 0;;

This menu item defines a FILLET command with 0 radius, and then generates 0 fillet between the two selected objects.

Screen Menus 9-23

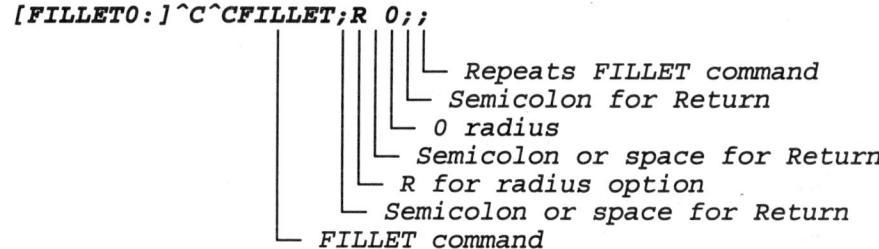

```
[FILLET0:]^C^CFILLET;R 0;;
```

Line 54
[ZOOM:]'ZOOM
This menu item defines a transparent ZOOM command.

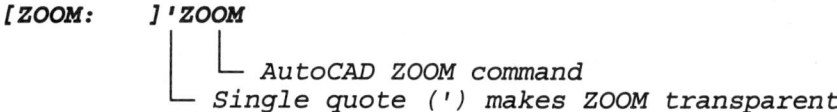

Line 90
[Center]CEN $X=
In this menu item CEN is for center object snap, and $X= automatically recalls the previous screen menu after selecting the object.

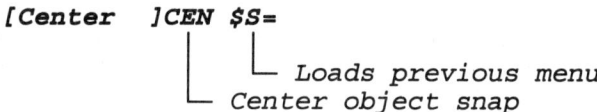

> **Note**
> *If any menu item on a screen menu includes more than one load command, the commands must be separated by a space.*
>
> **Example**
> *[LINE:]$S=X $S=LINE*
>
>   └─ *Blank space*
>
> *Similarly, if a menu item includes both a load command and an AutoCAD command, they should be separated by a space.*
>
> **Example**
> *[LINE:]$S=LINE ^C^CLINE*
>
>   └─ *Blank space*

Exercise 3

Write a screen menu for the AutoCAD commands shown in Figure 9-8 (file name **SME3.MNU**). The ******** are to access the object snap submenu.

9-24 Customizing AutoCAD

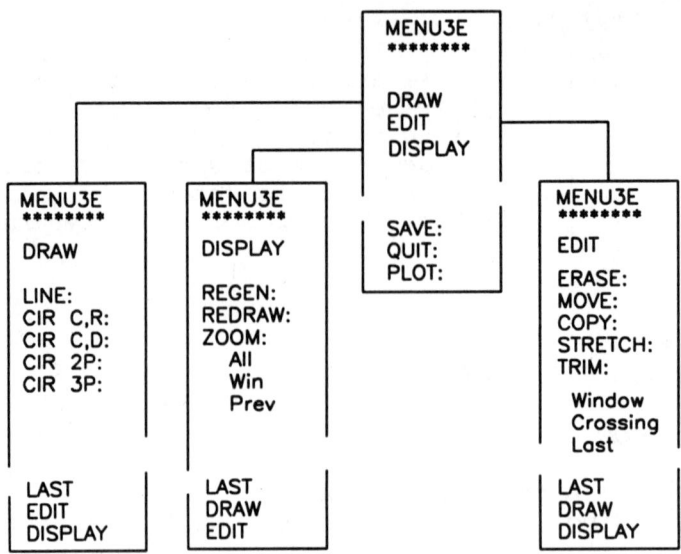

Figure 9-8 Screen menu design with submenus

Example 4

Write a combined screen and tablet menu for the commands shown in the tablet menu template (Figure 9-9A) and the screen menu (Figure 9-9B). When the user selects a command from the digitizer template, it should automatically load the corresponding screen menu on the screen (filename TM3.MNU).

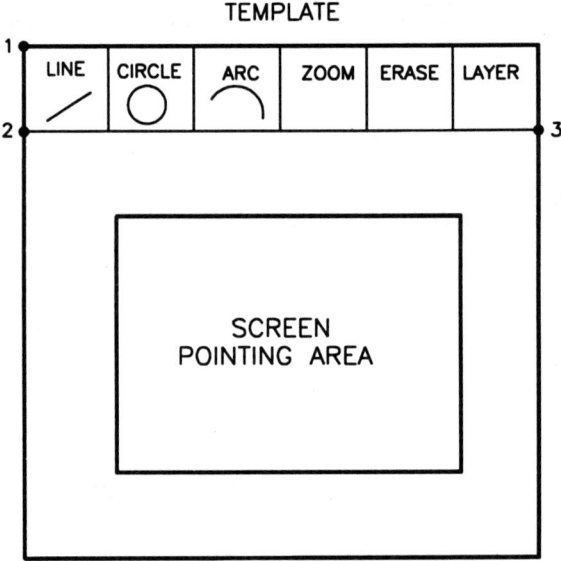

Figure 9-9A Tablet menu template for Example 4

Screen Menus 9-25

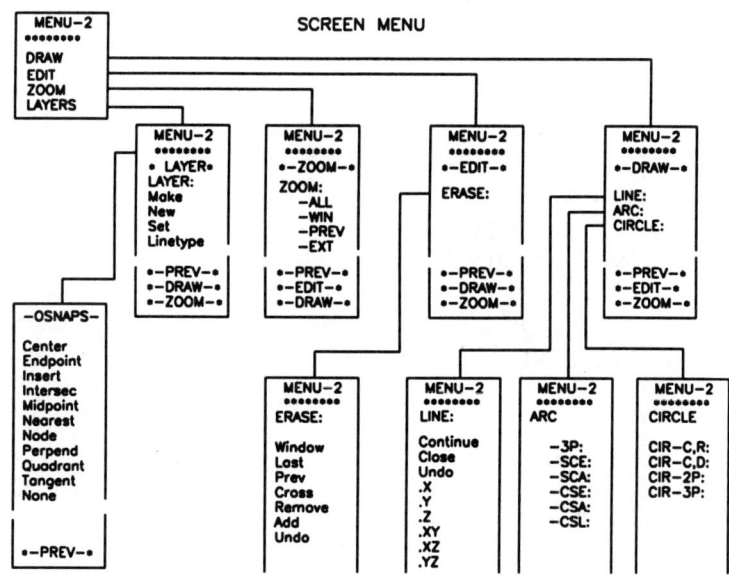

Figure 9-9B Screen menu displays

The following file is a listing of the combined menu for Example 4. The line numbers are not a part of the file; they are shown here for reference only.

```
    ***SCREEN                                             1
    **S                                                   2
    [ MENU-3       ]^C^C$S=X $S=S                         3
    [*******       ]$S=OSNAP                              4
    [              ]                                      5
    [              ]                                      6
    [DRAW          ]^C^C$S=X $S=DRAW                      7
    [EDIT          ]^C^C$S=X $S=EDIT                      8
    [ZOOM          ]^C^C$S=X $S=ZOOM                      9
    [LAYER         ]^C^C$S=X $S=LAYER                    10
                                                         11
    **DRAW 3                                             12
    [*-DRAW-*      ]                                     13
    [              ]                                     14
    [              ]                                     15
    [LINE:         ]$S=X $S=LINE ^C^CLINE                16
    [ARC:          ]$S=X $S=ARC                          17
    [CIRCLE:       ]$S=X $S=CIRCLE ^C^CCIRCLE            18
    [              ]                                     19
    [              ]                                     20
    [              ]                                     21
    [              ]                                     22
```

```
[                    ]                                  23
[                    ]                                  24
[                    ]                                  25
[                    ]                                  26
[                    ]                                  27
[*-PREV-*            ]$S= $S=                           28
[*-EDIT-*            ]^C^C$S=X $S=EDIT                  29
[*-ZOOM-*            ]$S=X $S=ZOOM                      30
                                                        31
**LINE 3                                                32
[LINE:               ]^C^CLINE                          33
[                    ]                                  34
[                    ]                                  35
[Continue            ]^C^CLINE;;                        36
[Close               ]CLOSE                             37
[Undo                ]UNDO                              38
[.X                  ].X                                39
[.Y                  ].Y                                40
[.Z                  ].Z                                41
[.XY                 ].XY                               42
[.XZ                 ].XZ                               43
[.YZ                 ].YZ                               44
[                    ]                                  45
[                    ]                                  46
[                    ]                                  47
[*-PREV-*            ]$S= $S=                           48
[*-EDIT-*            ]^C^C$S=X $S=EDIT                  49
[*-ZOOM-*            ]$S=X $S=ZOOM                      50
                                                        51
**ARC 3                                                 52
[ARC                 ]                                  53
[                    ]                                  54
[  -3P:              ]^C^CARC \\DRAG                    55
[  -SCE:             ]^C^CARC \C \DRAG                  56
[  -SCA:             ]^C^CARC \C \A DRAG                57
[  -CSE:             ]^C^CARC C \\DRAG                  58
[  -CSA:             ]^C^CARC C \\A DRAG                59
[  -CSL:             ]^C^CARC C \\L DRAG                60
[                    ]                                  61
[                    ]                                  62
[                    ]                                  63
[                    ]                                  64
[                    ]                                  65
[                    ]                                  66
[                    ]                                  67
[*-PREV-*            ]$S= $S=                           68
```

```
[*-EDIT-*       ]^C^C$S=X $S=EDIT                    69
[*-ZOOM-*       ]$S=X $S=ZOOM                        70
                                                     71
**CIRCLE     3                                       72
[CIRCLE:       ]                                     73
[              ]                                     74
[  -C,R:       ]^C^CCIRCLE                           75
[  -C,D:       ]^C^CCIRCLE \D                        76
[  -2P:        ]^C^CCIRCLE 2P                        77
[  -3P:        ]^C^CCIRCLE 3P                        78
[              ]                                     79
[              ]                                     80
[              ]                                     81
[              ]                                     82
[              ]                                     83
[              ]                                     84
[              ]                                     85
[              ]                                     86
[              ]                                     87
[*-PREV-*      ]$S= $S=                              88
[*-EDIT-*      ]^C^C$S=X $S=EDIT                     89
[*-ZOOM-*      ]$S=X $S=ZOOM                         90
                                                     91
                                                     92
**EDIT 3                                             93
[*-EDIT-*      ]                                     94
[              ]                                     95
[              ]                                     96
[ERASE:        ]$S=X $S=ERASE ^C^CERASE              97
[              ]                                     98
[              ]                                     99
[              ]                                    100
[              ]                                    101
[              ]                                    102
[              ]                                    103
[              ]                                    104
[              ]                                    105
[              ]                                    106
[              ]                                    107
[              ]                                    108
[*-PREV-*      ]$S= $S=                             109
[*-DRAW-*      ]^C^C$S=X $S=DRAW                    110
[*-ZOOM-*      ]$S=X $S=ZOOM                        111
                                                    112
**ERASE 3                                           113
[ERASE:        ]^C^CERASE                           114
```

9-28 Customizing AutoCAD

```
[                    ]                          115
Window                                          116
Last                                            117
Prev                                            118
Cross                                           119
Remove                                          120
Add                                             121
Undo                                            122
[                    ]                          123
[                    ]                          124
[                    ]                          125
[                    ]                          126
[                    ]                          127
[                    ]                          128
[*-PREV-*            ]$S= $S=                   129
[*-DRAW-*            ]^C^C$S=X $S=DRAW          130
[*-ZOOM-*            ]$S=ZOOM                   131
                                                132

**ZOOM 3                                        133
[*-ZOOM-*            ]                          134
[                    ]                          135
[                    ]                          136
[ZOOM:               ]'ZOOM                     137
[ -ALL               ]A                         138
[ -WIN               ]W                         139
[ -PREV              ]P                         140
[ -EXT               ]E                         141
[                    ]                          142
[                    ]                          143
[                    ]                          144
[                    ]                          145
[                    ]                          146
[                    ]                          147
[                    ]                          148
[*-PREV-*            ]$S= $S=                   149
[*-EDIT-*            ]^C^C$S=X $S=EDIT          150
[*-DRAW-*            ]$S=X $S=DRAW              151
                                                152

***LAYER  3                                     153
[*-LAYER*            ]                          154
[                    ]                          155
[                    ]                          156
[LAYER:              ]^C^CLAYER                 157
Make                                            158
New                                             159
Set                                             160
```

Linetype		161
Color		162
[List	]?;;	163
[	]	164
[	]	165
[	]	166
[	]	167
[	]	168
[*-PREV-*	]$S= $S=	169
[*-EDIT-*	]^C^C$S=X $S=EDIT	170
[*-DRAW-*	]$S=X $S=DRAW	171
		172
**X 3		173
[	]	174
[	]	175
[	]	176
[	]	177
[	]	178
[	]	179
[	]	180
[	]	181
[	]	182
[	]	183
[	]	184
[	]	185
[	]	186
[	]	187
[	]	188
[	]	189
[	]	190
[	]	191
		192
**OSNAP 2		193
[-OSNAPS-	]	194
[	]	195
[Center	]CEN $S=	196
[Endpoint	]END $S=	197
[Insert	]INS $S=	198
[Intersec	]INT $S=	199
[Midpoint	]MID $S=	200
[Nearest	]NEA $S=	201
[Node	]NOD $S=	202
[Perpend	]PER $S=	203
[Quadrant	]QUA $S=	204
[Tangent	]TAN $S=	205
[None	]NONE $S=	206

```
    [              ]                                    207
    [              ]                                    208
    [              ]                                    209
    [              ]                                    210
    [              ]                                    211
    [*-PREV-*      ]$S=                                  212
                                                         213

    ***TABLET1                                           214
    $S=X $S=LINE ^C^CLINE                                215
    $S=X $S=CIRCLE ^C^CCIRCLE                            216
    $S=X $S=ARC ^C^CARC                                  217
    $S=X $S=ZOOM ^C^CZOOM                                218
    $S=X $S=ERASE ^C^CERASE                              219
    $S=X $S=LAYER ^C^CLAYER                              220
```

Line 1
***SCREEN
This is the section label for the screen menu. The line numbers 1 through 213 are defined in this section of the menu file.

Line 3
[MENU-3]^C^C$S=X $S=S
In this menu item $S=X loads the submenu X and $S=S loads the submenu S.

Line 7
[DRAW   ]^C^C$S=X $S=DRAW
In this menu item $S=DRAW loads the submenu DRAW that is defined in the menu file.

Line 12
**DRAW 3
In this menu item DRAW is the name of the submenu. The 3 forces the submenu to be displayed on line 3. Since nothing is printed on the first two lines, [MENU-3] and [********] will be displayed on the screen all the time. If you select [MENU-3] from any menu, it will load the submenu S and display it on the screen. Similarly, if you pick [********] from any menu, it will load and display the OSNAP submenu.

Line 28
[*-PREV-*]$S= $S=
In this menu item $S= $S= loads the previous two submenus. One of them is the submenu X and the second submenu is the one that was displayed on the screen before the current menu.

Line 214
***TABLET1
TABLET1 is the section label for the tablet area 1. The line numbers 215 through 220 are defined in this section.

Line 215
$S=X $S=LINE ^C^CLINE

$S=X, defined in the screen menu section, loads the submenu X on the screen menu area. The purpose of loading the submenu X is to clear the screen so that there is no overlapping of the screen menu items. $S=LINE, defined in the screen menu section, loads the LINE submenu on the screen menu. ^C^CLINE cancels the existing command twice and then executes AutoCAD's LINE command.

When you select the LINE block from the digitizer, AutoCAD automatically clears the screen menu, loads the LINE submenu, and enters a LINE command. This makes it easier for the user to select the command options from the screen menu, since they are not on the digitizer template.

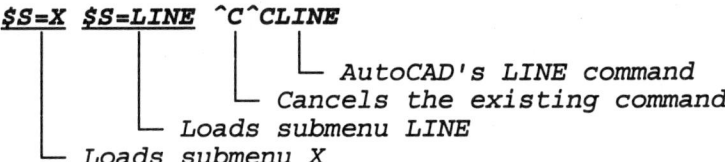

You can also use the automatic menu swapping feature of AutoCAD to load the corresponding screen menus. If you want to utilize this feature, the TABLET1 section must be written as:

```
***TABLET1                                                    214
$S=X ^C^CLINE                                                 215
$S=X ^C^CCIRCLE                                               216
$S=X ^C^CARC                                                  217
$S=X ^C^CZOOM                                                 218
$S=X ^C^CERASE                                                219
$S=X ^C^CLAYER                                                220
```

LONG MENU DEFINITIONS

You can put any number of commands in one screen menu line. There is no limit to the number of commands and the order in which they appear in the line, as long as they satisfy all the requirements of the command and the sequence of prompt entries. Again, you need to know the AutoCAD commands, the options in the command, the command prompts, and the entries for various prompts. If the statement cannot fit on one line, you can put a plus (+) sign at the end of the first line and continue with a second line. The command definition that involves several commands put together in one line is also called a macro. The following example illustrates a long menu item or a macro.

9-32 Customizing AutoCAD

Example 5

Write a screen menu command definition that performs the following functions (filename SM4.MNU):

Draw a border using polyline
Width =	0.01
Point-1	0,0
Point-2	12,0
Point-3	12,9
Point-4	0,9
Point-5	0,0

Initial drawing setup
Snap	0.25
Grid	0.5
Limits	12,9
Zoom	All

Before writing a menu you should know the commands, options, prompts, and the prompt entries required for the commands. Therefore, first study the commands involved in the above mentioned drawing setup.

POLYLINE
Command: **PLINE**
From point: **0,0**
Arc/Close/Halfwidth/Length/Undo/Width/<Endpoint of line>: **W**
Starting width <0.0000>: **0.01**
Ending width <0.01>: **Enter**
Arc/Close/Halfwidth/Length/Undo/Width/<Endpoint of line>: **12,0**
Arc/Close/Halfwidth/Length/Undo/Width/<Endpoint of line>: **12,9**
Arc/Close/Halfwidth/Length/Undo/Width/<Endpoint of line>: **0,9**
Arc/Close/Halfwidth/Length/Undo/Width/<Endpoint of line>: **C**

Screen menu command definition for PLINE
PLINE;0,0;W;0.01;;12,0;12,9;0,9;C

SNAP
Command: **SNAP**
Snap spacing or ON/OFF/Aspect/Rotate/Style <1.0000>: **0.25**

Screen menu command definition for SNAP
SNAP;0.25

GRID
Command: **GRID**

Grid spacing(X) or ON/OFF/Snap/Aspect <0.0000>: **0.5**

Screen menu command definition
GRID;0.5

> **LIMITS**
> Command: **LIMITS**
> ON/OFF/<Lower left corner>: **0,0**
> Upper right corner <12.00,9,00>: **12,9**

Screen menu command definition
LIMITS;0,0;12,9

> **ZOOM**
> Command: **ZOOM**
> All/Center/Dynamic/Extents/Left/Previous/Window/<Scale(X)>: **A**

Screen menu command definition
ZOOM;A

Now you can combine these individual command definitions to form a single screen menu command definition that will perform all the functions when you select NSETUP from the new screen menu.

> **Combined screen menu line**
>
> [-NSETUP-]PLINE;0,0;W;0.01;;12,0;12,9;0,9;C;+
> SNAP;0.25;GRID;0.5;LIMITS;+
> 0,0;12,9;ZOOM;A

Exercise 4

Write a screen menu item that will set up the following parameters for the UNITS command (filename SME4.MNU):

> System of units: Scientific
> Number of digits to right of decimal point: 2
> System of angle measure: Decimal
> Number of fractional places for display of angles: 2
> Direction for angle: 0
> Angles measured counterclockwise

MENU COMMAND REPETITION

AutoCAD has made provision for repeating a menu item until it is canceled by the user by pressing the CTRL and C keys from the keyboard or selecting another menu item. It is particularly

9-34 Customizing AutoCAD

useful when you are editing a drawing and using the same command several times. A command can be repeated if the menu command definition starts with an asterisk sign (*).

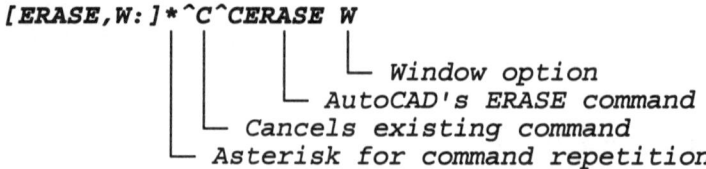

```
[ERASE,W:]*^C^CERASE W
```
- Window option
- AutoCAD's ERASE command
- Cancels existing command
- Asterisk for command repetition

If you select this menu item, AutoCAD will prompt you to enter two points to select the objects because the window option for the ERASE command requires two points. When you press Enter the objects you selected will be erased and the command will be repeated.

Example 6

Write a screen menu for the following AutoCAD commands. Provide for automatically repeating the commands (filename SM7.MNU).

LINE	LIST
ERASE	INSERT
TRIM	DIST

The following file is a listing of the screen menu for Example 6 that will repeat the selected command until canceled by pressing the CTRL and C keys.

```
[-REPEAT-    ]
[            ]
[            ]
[LINE:       ]*^C^CLINE
[ERASE:      ]*^C^CERASE
[TRIM:       ]*^C^CTRIM
[LIST:       ]*^C^CLIST
[INSERT:     ]*^C^CINSERT
[DIST:       ]*^C^CDIST
```

Note

If the command definition is incorrect and you happen to select that item, the screen prompt display will be repeated indefinitely. To get out of this infinite loop you must reboot the system. By pressing the Escape key or selecting another command does not stop display of the prompt. In Windows 95, press Ctrl+Alt+Del to close the program.

One of the major drawbacks of the menu command repetition is that it does not let you select a different command option. In the following menu item, the ERASE command uses the C (Crossing) option to select the objects. When the command is repeated, it does not allow you to use a different option for object selection.

[ERASE,C:]*^C^CERASE C

AUTOMATIC MENU SWAPPING

Screen menus can be automatically swapped by using the system variable **MENUCTL**. When the system variable **MENUCTL** is set to 1, AutoCAD automatically issues **the $S=CMDNAME** command, where CMDNAME is the command that loads the submenu. For example, if you select the LINE command from the digitizer or pull-down menu, or enter the LINE command at the keyboard, the CMDNAME command will load the LINE submenu and display it in the screen menu area. To use this feature, the command name and the submenu name must be the same. For example, if the name of the arc submenu is ARC and you select the ARC command, the arc submenu will be automatically loaded on the screen. However, if the submenu name is different (for example, MYARC) AutoCAD will not load the arc submenu on the screen. The default value of the MENUCTL system variable is 1. If you set the MENUCTL variable to 0, AutoCAD will not use the $S=CMDNAME command feature to load the submenus.

MENUECHO SYSTEM VARIABLE

If the system variable **MENUECHO** is set to 0, the commands that you select from the digitizer, screen menu, pull-down menu, or buttons menu will be displayed in the command prompt area. For example, if you select the CIRCLE command from the pull-down menu, AutoCAD will display _circle 3P/2P/TTR/<Center point>:. If you set the value of the MENUECHO variable to 1, AutoCAD will suppress the echo of the menu item and display **3P/2P/TTR/<Center point>:** only. You will notice that the _circle is not displayed when MENUECHO is set to 1. You can use ^P in the menu item to turn the echo on or off. The MENUECHO system variable can also be assigned a value of 2, 4, or 8, which controls the suppression of system prompts, disabling the ^P toggle, and debugging aid for DIESEL macros, respectively.

MENUS FOR FOREIGN LANGUAGES

Besides English, AutoCAD has several foreign language versions. If you want to write a menu that is compatible with other foreign language versions of AutoCAD, you must precede each command and keyword in the menu file with the underscore (_) character.

Examples
[New   ]^C^C_New
[Open   ]^C^C_Open
[Line   ]^C^C_Line
[Arc-SCA]^C^C_Arc;_C;_A

Any command or keyword that starts with the underscore character will be automatically translated. If you check the ACAD.MNU file, you will find that AutoCAD has made extensive use of this feature.

Example 7

Rewrite the screen menu of Example 1 so that the menu file is compatible with other foreign language versions of AutoCAD.

The following file is the listing of the menu file for Example 7:

```
***SCREEN
[ MENU-1    ]
[           ]
[           ]
[LINE       ]^C^C_LINE
[           ]
[CIR-C,R    ]^C^C_CIRCLE
[CIR-C,D    ]^C^C_CIRCLE;\_D
[CIR- 2P    ]^C^C_CIRCLE;_2P
[           ]
[ERASE      ]^C^C_ERASE
[MOVE       ]^C^C_MOVE
```

USE OF CONTROL CHARACTERS IN MENU ITEMS

You can use ASCII control characters in the command definition by using the caret sign (^) followed by the control character. For example, if you want to write a menu item that will toggle the SNAP off or on, you can use a caret sign followed by the control character B as shown in the following example.

```
[SNAP-TOG]^B
         │ └─ Control character for SNAP
         └─── Caret sign (^)
└─────────── Command label for SNAP toggle
```

Examples

^C	(Cancel)
^G	(Grid on/off)
^H	(Backspace)
^O	(Ortho on/off)
^T	(Tablet on/off)
^E	(Isoplane top/left/right)

^B is equivalent to pressing the CTRL and B keys from the keyboard that toggles the SNAP mode. SNAP-TOG is the menu item label that will be displayed on the screen menu. You can use any ASCII control character in the command definition. Following are some of the ASCII control characters:

^@	(ASCII code 0)
^[	(ASCII code 27)
^\	(ASCII code 28)
^]	(ASCII code 29)
^^	(ASCII code 30)
^-	(ASCII code 31)

SPECIAL CHARACTERS

The following is a list of the special characters that you can use in the AutoCAD menus:

Character	Description
***	Three asterisks indicate a section title
**	Two asterisks indicate a submenu
[ ]	Used to enclose a menu item label
;	Causes a Enter
Space	Space is equivalent to pressing the spacebar
\	AutoCAD pauses for user input
_	(Underscore character) translates the AutoCAD commands and keyword that follow it into english.
+	Used to continue the menu item definition to next line
=*	Forces the display of pull-down, cursor or image tile menus
*	Menu item repetition
$M=	Special character to load a DIESEL macro expression
$S=CMDNAME	Special command that loads a screen submenu
^B	Toggles snap (On/Off)
^C	Cancels the existing command
^D	Toggles coordinate dial (On/Off)
^E	Changes the isometric plane (Left/Right/Top)
^G	Toggles grid (On/Off)
^H	Causes a backspace
^O	Toggles ortho mode (On/Off)
^P	Toggles MENUECHO on or off
^Q	Echoes all prompts to the printer
^T	Toggles tablet mode (On/Off)
^V	Change current viewport
^Z	Suppresses the automatic addition of spacebar at the end of a menu item

Example 8

Write a screen menu for the following toggle functions (filename SM5.MNU):

```
ORTHO           SNAP
GRID            COORDINATE DIAL
TABLET          ISOPLANE
PRINTER
```

Before writing a screen menu for these toggle functions, it is important to know the control characters that AutoCAD uses to turn these functions on or off. The following is a list of the control characters for the functions in this example:

ORTHO	CTRL O
SNAP	CTRL B
GRID	CTRL G
COORDINATE DIAL	CTRL D
TABLET	CTRL T
ISOPLANE	CTRL E
PRINTER Echo	CTRL Q

The following file is a listing of the screen menu that toggle the given functions. You can use AutoCAD's EDIT command to write this file.

```
[-TOGGLE-   ]
[           ]
[           ]
[           ]
[ORTHO      ]^O        ----------   turns ORTHO on/off
[SNAP       ]^B        ----------   turns SNAP on/off
[GRID       ]^G        ----------   turns GRID on/off
[CO-ORDS    ]^D        ----------   turns COORDINATE dial on/off
[TABLET     ]^T        ----------   turns TABLET on/off
[           ]
[ISOPLANE   ]^E        ----------   turns ISOPLANE on/off
[PRINTER    ]^Q        ----------   turns PRINTER echo on/off
```

COMMAND DEFINITION WITHOUT ENTER OR SPACE

All command definitions that have been discussed so far use a semicolon (;) or a space for Enter. However, sometimes a user might want to define a menu item that is not followed by an Enter or a space. This is accomplished by using the backspace ASCII control character (^H). It is especially useful when you want to write a screen menu for a numeric keypad. ^H control character can be used with any character or a group of characters. The following is the format of menu item definition without an Enter or space:

```
[9]9X^H
 │ │ │└─ ^H for backspace
 │ │ └── X is erased because of ^H (backspace)
 │ │     (X could be substituted by any character)
 │ └──── Character returned by selecting this item
 └────── Menu item label
```

^H is the ASCII control character for backspace. When you select this item, ^H erases the previous character X and therefore returns 9 only. It is not necessary to have X preceding ^H. X is a dummy character; it could be any character. You can continue selecting more characters and when you are done, and you want to press Enter, use the keyboard or the digitizer.

Example 8

Write a screen menu for the following characters (filename SM6.MNU):

```
    0           5           .
    1           6           ,
    2           7           X
    3           8           Y
    4           9           Z
```

The following file is a listing of the screen menu of Example 9 that will enable the user to pick the characters without an Enter.

```
[-KEYPAD-    ]
[            ]
[            ]
[            ]
[0]0Y^H              ——————  returns 0
[1]1Y^H              ——————  returns 1
[2]2Y^H              ——————  returns 2
[3]3Y^H
[4]4Y^H
[5]5Y^H
[6]6Y^H
[7]7Y^H
[8]8Y^H
[9]9Y^H
[.].Y^H              ——————  returns period (.)
[,],Y^H              ——————  returns comma (,)
[]                   ——————  for space in the screen menu
[X]XX^H              ——————  returns X
[Y]YX^H              ——————  returns Y
[Z]ZZ^H              ——————  returns Z
```

Note

You can also use a backslash after the character to continue selecting more characters. For example, [2]2\ will return 2 and then pause for the user input, which could be another character. The following file uses a backslash (\) to let the user enter another character.

```
[-KEYPAD- ]
[           ]
[           ]
[           ]
[0]0\
[1]1\
[2]2\
```

9-40 Customizing AutoCAD

```
            [3]3\
            [4]4\
            [5]5\
            [6]6\
            [7]7\
            [8]8\
            [9]9\
            [.].\
            [,],\
            []
            [X]X\
            [Y]Y\
            [Z]Z\
```

MENU ITEMS WITH SINGLE OBJECT SELECTION MODE

The Single options for object selection, combined with the menu item repetition, can provide a powerful editing tool.

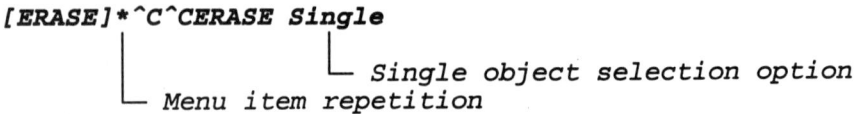

In this menu item the asterisk that follows the menu item label repeats the command. ^C^C cancels the existing command. ERASE is an AutoCAD command that erases the selected objects. The Single option enables the user to select the objects and then AutoCAD automatically terminates the object selection process. The ERASE command then erases the selected objects.

If the point you pick is in a blank area (if there is no object), AutoCAD automatically defaults to crossing or window option. If you pull the window to left, it is the crossing option and if you pull the box to right, it is the window option for selecting the objects. The asterisk in front of the menu item repeats the command.

USE OF AUTOLISP IN MENUS

You can combine AutoLISP variables and expressions with the menu items as a command definition. When you select that item from the menu, it will evaluate all the expressions and generate the necessary output. The following examples illustrate the use of AutoLISP variables and expressions. (For more information on AutoLISP, see Chapters 12 and 13.)

Screen Menus 9-41

Example 10

Write an AutoLISP program that draws a square and then write a screen menu that utilizes the AutoLISP variables and expressions to draw a square.

The following file is the listing of the AutoLISP program that generates a square. The program prompts the user to enter the starting point and the length of the side.

```
(DEFUN C:SQR()
(SETVAR "CMDECHO" 0)
(SETQ P1 (GETPOINT "\n ENTER STARTING POINT: "))
(SETQ S (GETDIST "\n ENTER LENGTH OF SIDE: "))
(SETQ P2 (LIST (+ (CAR P1) S) (CADR P1)))
(SETQ P3 (POLAR P2 (/ PI 2) S))
(SETQ P4 (POLAR P1 (/ PI 2) S))
(COMMAND "PLINE" P1 P2 P3 P4 "C")
(SETVAR "CMDECHO" 1)
(PRINC)
)
```

The following file is the listing of the screen menu file that utilizes the AutoLISP variables and expressions to draw a square:

```
[-SQUARE- ]
[            ]
[            ]
[SQUARE:  ](SETQ P1(GETPOINT "ENTER STARTING POINT: +
 "));\+
(SETQ S (GETDIST "ENTER LENGTH OF SIDE: "));\+
(SETQ P2 (LIST (+ (CAR P1) S) (CADR P1)))+
(SETQ P3 (POLAR P2 (/ PI 2) S))+
(SETQ P4 (POLAR P1 (/ PI 2) S));+
PLINE !P1 !P2 !P3 !P4 C
```

If you compare the AutoLISP program with the screen menu, you will notice that the statements that prompt the user to enter the information and the statements that do actual calculations are identical. Therefore, you can utilize AutoLISP variables and expressions in the menu file to customize AutoCAD.

DIESEL EXPRESSION IN MENUS

You can also define a DIESEL expression in the screen, tablet, pull-down, or button menu. When you select the menu item, it will automatically assign the value to the **MODEMACRO** system variable and then display the new status line. The following example illustrates the use of DIESEL expression in the screen menu.

Example 11

Write a screen menu that will load the DIESEL macro to display the following information in the status line:

Macro-1	Macro-2	Macro-3
Project name	Pline width	Dimtad
Drawing name	Fillet radius	Dimtix
Current layer	Offset distance	Dimscale

The following file is a listing of the screen menu that contains the definition of three DIESEL macros of Example 11. This menu can be loaded by using AutoCAD's MENU command and then entering the name of the menu file. If you select the first item, DIESEL1, it will automatically display the corresponding status line (Figure 9-10).

```
***screen
[*DIESEL*]
[DIESEL1:]^C^CMODEMACRO;$M=Cust-Acad,N:$(GETVAR,DWGNAME)+
,L:$(GETVAR,CLAYER);
[DIESEL2:]^C^CMODEMACRO;$M=PLWID:$(GETVAR,PLINEWID),+
FRAD:$(GETVAR,FILLETRAD),OFFSET:$(GETVAR,OFFSETDIST),+
LTSCALE:$(GETVAR,LTSCALE);
[DIESEL3:]^C^CMODEMACRO;$M=DTAD:$(GETVAR,DIMTAD),+
DTIX:$(GETVAR,DIMTIX),DSCALE:$(GETVAR,DIMSCALE);
```

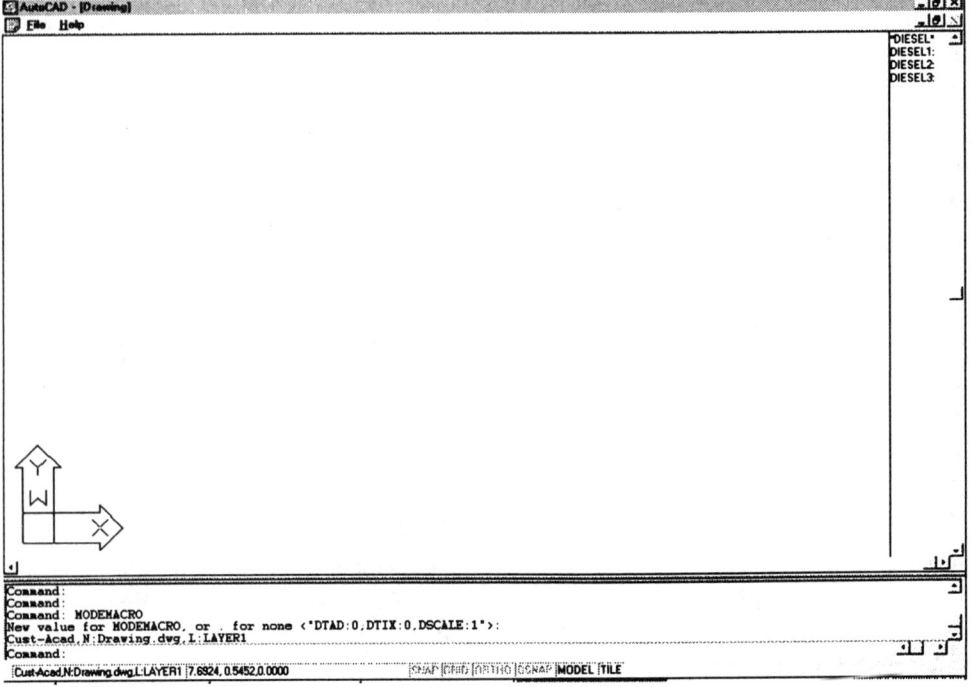

Figure 9-10 Screen menu and status line for Example 11

You can also use the $M= command to load a DIESEL macro. AutoCAD evaluates the DIESEL string expression that follows the equal sign (=) and the value returned by the DIESEL expression becomes a part of the menu item.

Example
[SCR-MODE]SCREENMODE $M=$(-,1,$(GETVAR,SCREENMODE))

In this example, AutoCAD evaluates the DIESEL string expression $(-,1,$(GETVAR, SCREENMODE)). The GETVAR command retrieves the value of the system variable SCREENMODE and subtracts the value from 1. If the value of the system variable SCREENMODE is 1, then the DIESEL expression returns 0, and if the value of SCREENMODE is 0, then the DIESEL expression returns 1. The value returned by the DIESEL expression is assigned to the system variable SCREENMODE. Therefore, this menu item can be used to change the screen display from the text to graphics mode or from the graphics to text mode.

REVIEW QUESTIONS

1. The name of the menu file that comes with the AutoCAD software package is _____.

2. The AutoCAD menu file can have up to _____ main sections.

3. The tablet menu can have up to _____ main sections.

4. The section label is designated by _____.

5. A submenu is designated by _____.

6. The part of the menu item that is inside the brackets is for _____ only.

7. Only the first _____ characters can be displayed on the screen menu.

8. If the number of characters inside the bracket exceeds eight, the screen menu _____ work.

9. In a menu file you can use _____ to cancel the existing command.

10. The AutoCAD _____ command is used to load a new menu file.

11. _____ is used to pause for input in a screen menu definition.

12. The menu item label _____ be a combination of uppercase and lowercase characters.

13. Submenu names can be _____ characters long.

14. The maximum number of accessible menu items in a screen menu depends on the _____.

15. Name two text editors that can be used to write menu files. _____, _____.

EXERCISES

Exercise 5
Design and write a screen menu for the following AutoCAD commands (file name **SME4.MNU**).

   POLYGON (Center)
   POLYGON (Edge)
   ELLIPSE (Center)
   ELLIPSE (Axis End point)
   CHAMFER
   EXPLODE
   COPY

Exercise 6
Write a screen menu file for the following AutoCAD commands. Use submenus if required (file name **SME5.MNU**).

   ARC   ROTATE
    -3P   ARRAY
    -SCE   DIVIDE
    -CSE   MEASURE
   BLOCK
   INSERT   LAYER
   WBLOCK   SET
   MINSERT   LIST

Exercise 7

Write a screen menu item that will set up the following layers, linetypes, and colors (file name **SME6.MNU**).

Layer Name	Color	Linetype
0	WHITE	CONTINUOUS
OBJECT	RED	CONTINUOUS
HIDDEN	YELLOW	HIDDEN
CENTER	BLUE	CENTER
DIM	GREEN	CONTINUOUS

Exercise 8

Write a screen menu for the commands in Figure 9-11. Use the menu item ******** to load Osnaps, and use the menu item MENU-7 to load the root menu (file name **SME7.MNU**).

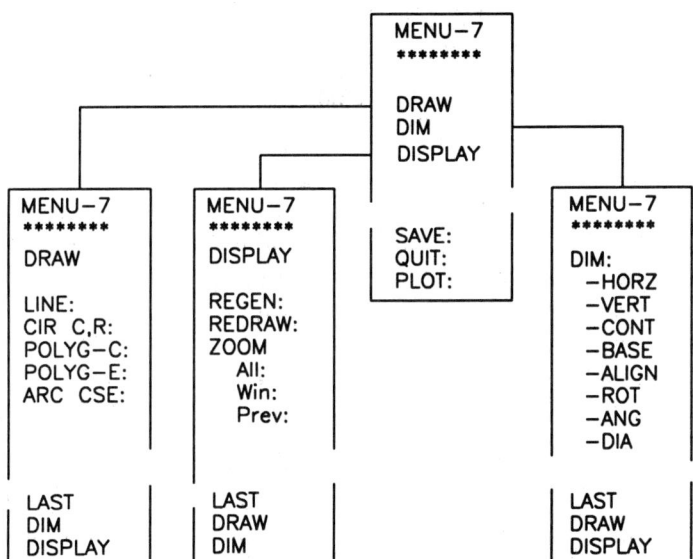

Figure 9-11 Screen menu displays

Exercise 9

Write a coordinated pull-down and screen menu for the following AutoCAD commands. Use a cascading menu for the LINE command options in the pull-down menu. When the user selects an item from the pull-down menu, the corresponding screen menu should be automatically loaded on the screen. (The layout of the screen menu and the pull-down menu are shown in Figures 9-12(a) and 9-12(b).)

LINE	ZOOM All	TIME
Continue	ZOOM Win	LIST

9-46 Customizing AutoCAD

Close   ZOOM Pre   DISTANCE
Undo   PAN   AREA
.X   DBLIST
.Y   STATUS
.Z
CIRCLE
ELLIPSE

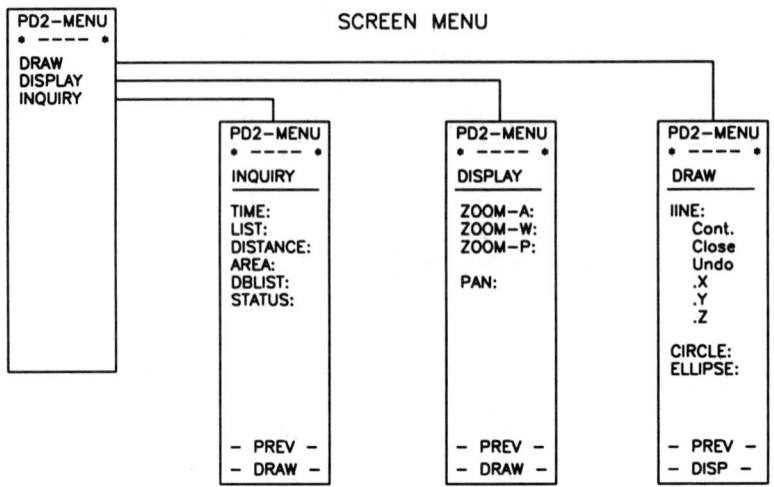

Figure 9-12(a) Design of screen menu for Exercise 9

Figure 9-12(b) Design of pull-down menu for Exercise 9

Exercise 10

Write a pull-down menu, a screen menu, and a tablet menu for the following AutoCAD commands. When you select a command from the template or the pull-down menu, the corresponding screen menu should be automatically loaded on the screen. (The layout of the screen is shown in Figure 9-13.)

LINE
PLINE
CIRCLE C,R
CIRCLE C,D
ELLIPSE AXIS ENDPOINT
ELLIPSE CENTER

BLOCK
WBLOCK
INSERT
BLOCK LIST
DDATTDEF
DDATTE

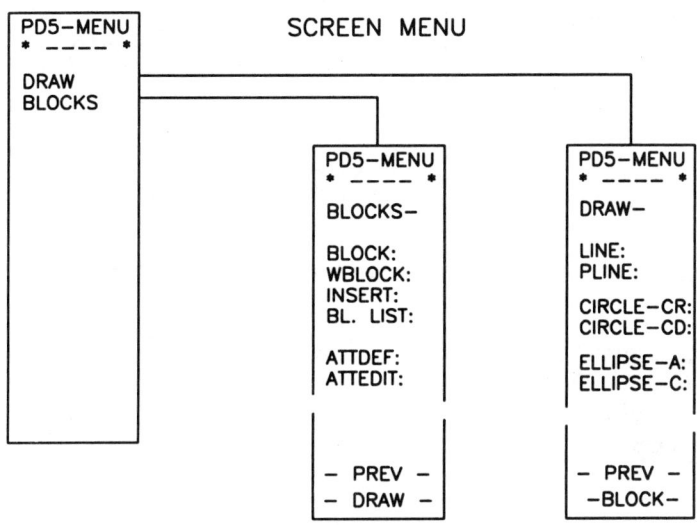

Figure 9-13 Design of screen menu for Exercise 10

Exercise 11

Write a screen, pull-down, and image tile menu for the commands shown in Figure 9-14. When the user selects a command from the pull-down menu, it should automatically load the corresponding menu on the screen.

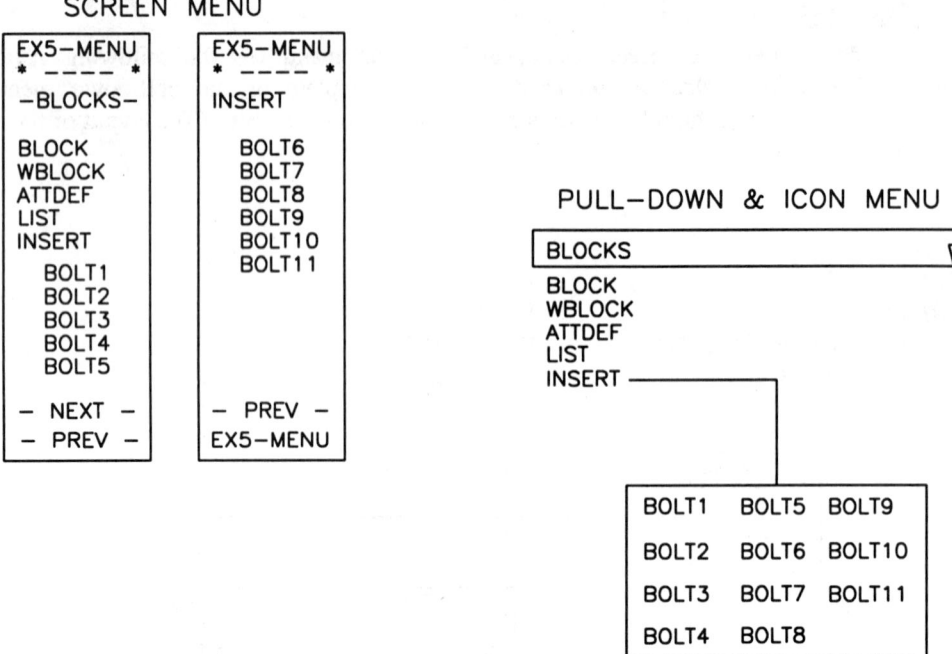

Figure 9-14 Design of screen, pull-down, and image tile menus for Exercise 11

Exercise 12

Write screen, tablet, pull-down, and image tile menu for inserting the following commands. B1 to B15 are the block names.

    BLOCK
    WBLOCK
    ATTDEF
    LIST
    INSERT

BL1	BL6	BL11
BL2	BL7	BL12
BL3	BL8	BL13
BL4	BL9	BL14
BL5	BL10	BL15

Exercise 13

Write a pull-down and screen menu for the following AutoCAD commands. Use submenus for the ARC and CIRCLE commands. When the user selects an item from the pull-down menu the corresponding screen menu should be automatically loaded on the screen.

Screen Menus 9-49

```
LINE                    BLOCK                QUIT
PLINE                   INSERT               SAVE
ARC                     WBLOCK               ----
    ARC 3P                                   PLOT
    ARC SCE
    ARC SCA
    ARC CSE
    ARC CSA
    ARC CSL
CIRCLE
    CIRCLE C,R
    CIRCLE C,D
    CIRCLE 2P
```

The layout shown in Figure 9-15A is one of the possible designs for this pull-down menu. The ARC and CIRCLE commands are in separate groups that will be defined as submenus in the menu file. The layout of the screen menu is shown in Figure 9-15B.

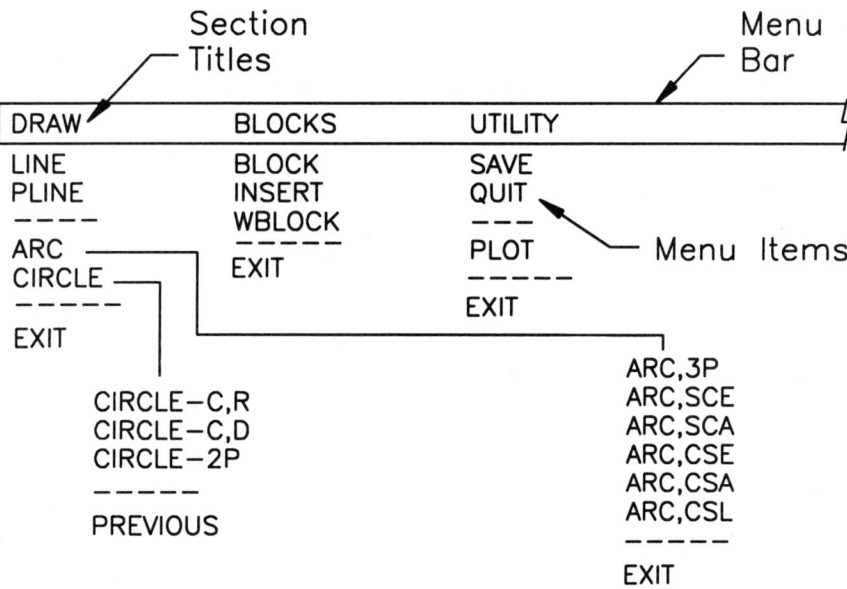

Figure 9-15A Design of pull-down menu for Exercise 13

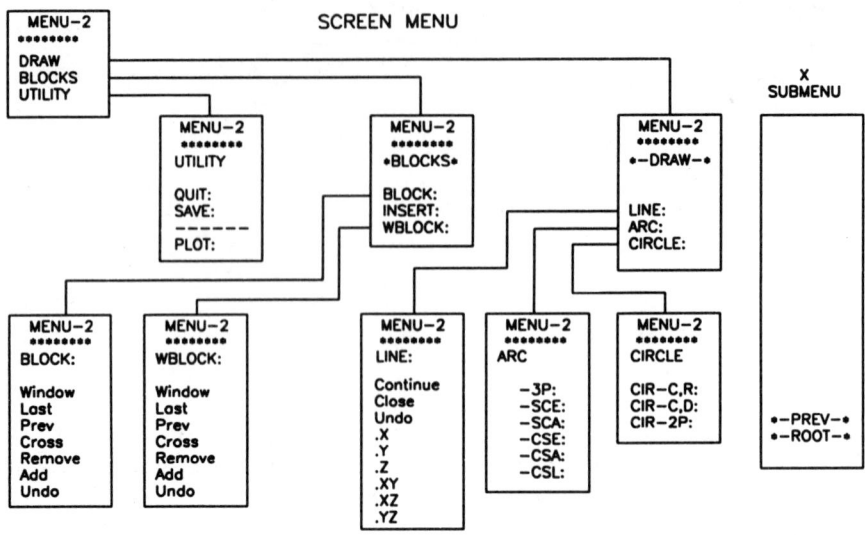

Figure 9-15B Design of screen menu

Chapter 10

Customizing the Standard AutoCAD Menu

Learning objectives

After completing this chapter, you will be able to:
- *Edit the standard AutoCAD menu file, ACAD.MNU.*
- *Load menus and submenus.*
- *Customize tablet areas.*
- *Customize buttons menus.*
- *Customize pull-down and cursor menus.*
- *Customize image tile menus and screen menus.*

THE STANDARD AUTOCAD MENU

The AutoCAD software package comes with a standard menu file: ACAD.MNU. AutoCAD stores the name of the menu that was used last in the system registry. When you start AutoCAD, the last used menu file is automatically loaded (Figure 10-1). The menu sections of a menu file are identified by section labels. The general format of the section label is *****section_name**. For example, if it is the screen menu section, the section label is ***SCREEN. The following is the list of the section labels.

***SCREEN		Screen menu
***TABLET(n)	n is from 1 to 4	Tablet menu
***IMAGE		Image tile menu
***POP(n)	n is from 0 to 16	Pull-down/Cursor menu
***BUTTONS(n)	n is from 1 to 4	Pointing device menu
***AUX(n)	n is from 1 to 4	System pointing device menu
***MENUGROUP		Menu file group name
***TOOLBARS		Toolbar definition
***HELPSTRING		Text displayed in the status bar
***ACCELERATORS		Accelerator key definitions

10-2 Customizing AutoCAD

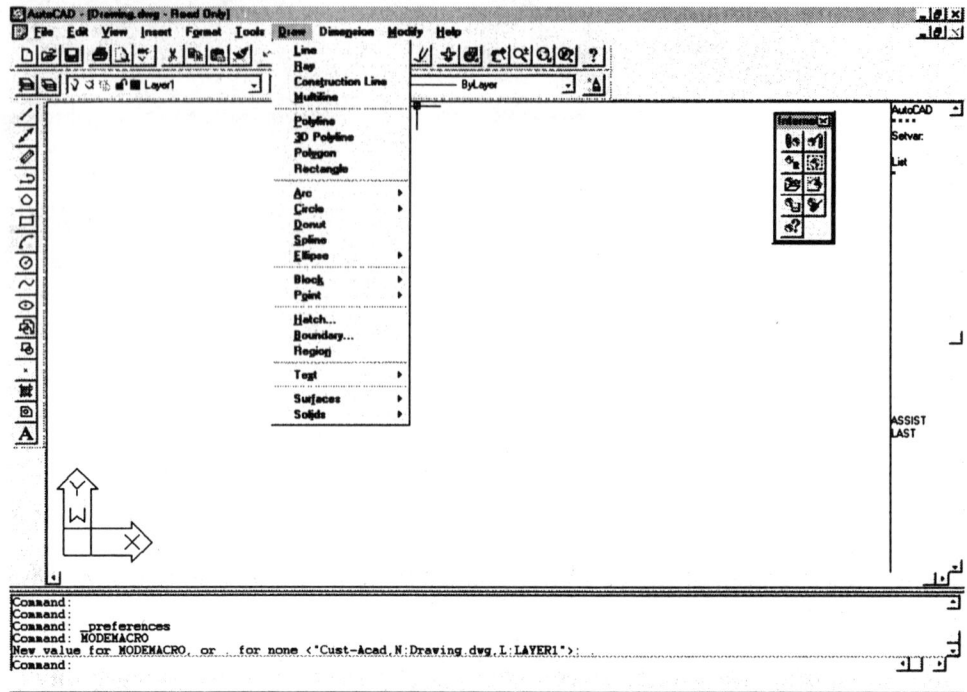

Figure 10-1 Screen display with standard AutoCAD menu loaded

The following file is a partial listing of ACAD.MNU file that shows some of the section labels:

> **Note**
> //AutoCAD Menu - Release 14.0, 28 April 1997
> //Copyright (C) 1986, 1987, 1988, 1989, 1990, 1991, 1992, 1994, 1996, 1997 by Autodesk, Inc.
>
> // NOTE: AutoCAD looks for an ".mnl" (Menu Lisp) file whose name is the same as that of the menu file, and loads it if found. If you modify this menu and change its name, you should copy acad.mnl to <yourname>.mnl, since the menu relies on AutoLISP routines found there.

***MENUGROUP=ACAD
//   Begin AutoCAD Digitizer Button Menus
***BUTTONS1
// Simple + button
// if a grip is hot bring up the Grips Cursor Menu (POP 17), else send a carriage return
$M=$(if,$(eq,$(substr,$(getvar,cmdnames),1,5),GRIP_),$P0=ACAD.GRIPS $P0=*);
$P0=SNAP $p0=*
^C^C
^B
^O
^G
^D

^E
^T

***BUTTONS2**
// Shift + button
$P0=SNAP $p0=*

***BUTTONS3**
// Control + button

***BUTTONS4**
// Control + shift + button
//   Begin System Pointing Device Menus
//
***AUX1**
// Simple button
// if a grip is hot bring up the Grips Cursor Menu (POP 17), else send a carriage return
$M=$(if,$(eq,$(substr,$(getvar,cmdnames),1,5),GRIP_),$P0=ACAD.GRIPS $P0=*);
$P0=SNAP $p0=*
^C^C
^B
^O
^G
^D
^E
^T

***AUX2**
// Shift + button
$P0=SNAP $p0=*
$P0=SNAP $p0=*

***AUX3**
// Control + button
$P0=SNAP $p0=*

***AUX4**
// Control + shift + button
$P0=SNAP $p0=*
//
//   Begin AutoCAD Pull-down Menus
***POP0**
**SNAP
// Shift-right-click
          [&Object Snap Cursor Menu]
ID_Tracking [Trac&king]_tracking

```
ID_From      [&From]_from
ID_MnPointFi [->Point Fi&lters]
ID_PointFilx [.X].X
ID_PointFily [.Y].Y
ID_PointFilz [.Z].Z
    |
    |
    |
***POP1
**FILE
ID_MnFile    [&File]
ID_New       [&New...\tCtrl+N]^C^C_new
ID_Open      [&Open...\tCtrl+O]^C^C_open
        [--]
ID_Save      [&Save\tCtrl+S]^C^C_qsave
ID_Saveas    [Save &As...]^C^C_saveas
ID_Export    [&Export...]^C^C_export
        [--]
ID_Config    [P&rinter Setup...]^C^C_config
ID_Preview   [Print Pre&view]^C^C_preview
ID_Print     [&Print...\tCtrl+P]^C^C_plot
        [--]
    |
    |
    |
***POP2
**EDIT
ID_MnEdit    [&Edit]
ID_U         [&Undo\tCtrl+Z]_u
ID_Redo      [&Redo\tCtrl+Y]^C^C_redo
        [--]
ID_Cutclip   [Cu&t\tCtrl+X]'_cutclip
ID_Copyclip  [&Copy\tCtrl+C]'_copyclip
ID_Copylink  [Copy &Link]^C^C_copylink
ID_Pasteclip [&Paste\tCtrl+V]'_pasteclip
ID_Pastesp   [Paste &Special...]^C^C_pastespec
ID_Erase     [Cle&ar\tDel]^C^C_erase
        [--]
ID_Links     [&OLE Links...]^C^C_olelinks

***POP3
**VIEW
ID_MnView    [&View]
ID_Redrawall [&Redraw]'_redrawall
ID_Regen     [Re&gen]^C^C_regen
ID_Regenall  [Regen &All]^C^C_regenall
```

```
            [--]
ID_MnZoom    [->&Zoom]
ID_ZoomRealt [&Realtime]'_zoom ;
            [--]
ID_ZoomPrevi [&Previous]'_zoom _p
ID_ZoomWindo [&Window]'_zoom _w
ID_ZoomDynam [&Dynamic]'_zoom _d
ID_ZoomScale [&Scale]'_zoom _s
ID_ZoomCente [&Center]'_zoom _c
            [--]
ID_ZoomIn    [&In]'_zoom 2x
ID_ZoomOut   [&Out]'_zoom .5x
            [--]
ID_ZoomAll   [&All]'_zoom _all
ID_ZoomExten [<-&Extents]'_zoom _e
|
|
|
|
***POP4
**INSERT
ID_MnInsert  [&Insert]
ID_Ddinsert  [&Block...]^C^C_ddinsert
        [--]
ID_Xref      [E&xternal Reference...]^C^C_xref
ID_Image     [Raster &Image...]^C^C_image
        [--]
ID_3dsin     [&3D Studio...]^C^C_3dsin
ID_Acisin    [&ACIS Solid...]^C^C_acisin
ID_Dxbin     [Drawing &Exchange Binary...]^C^C_dxbin
ID_Wmfin     [&Windows Metafile...]^C^C_wmfin
ID_Psin      [Encapsulated &PostScript...]^C^C_psin
        [--]
ID_Insertobj [&OLE Object...]^C^C_insertobj

***POP5
**FORMAT
ID_MnFormat  [F&ormat]
ID_Layer     [&Layer...]'_layer
ID_Ddcolor   [&Color...]'_ddcolor
ID_Linetype  [Li&netype...]'_linetype
        [--]
ID_Style     [Text &Style...]'_style
ID_Ddim      [&Dimension Style...]^C^C_ddim
ID_Ddptype   [&Point Style...]'_ddptype
ID_Mlstyle   [&Multiline Style...]^C^C_mlstyle
        [--]
```

```
ID_Ddunits    [&Units...]'_ddunits
ID_Thickness  [&Thickness]'_thickness
ID_Limits     [Dr&awing Limits]'_limits
              [--]
ID_Ddrename   [&Rename...]^C^C_ddrename

***POP6
**TOOLS
ID_MnTools    [&Tools]
ID_Spell      [&Spelling]^C^C_spell
ID_MnOrder    [->Display &Order]
ID_DrawordeF  [Bring to &Front]^C^C^P(ai_draworder "_f") ^P
ID_DrawordeB  [Send to &Back]^C^C^P(ai_draworder "_b") ^P
              [--]
ID_DrawordeA  [Bring &Above Object]^C^C^P(ai_draworder "_a") ^P
ID_DrawordeU  [<-Send &Under Object]^C^C^P(ai_draworder "_u") ^P
ID_Inquiry    [->In&quiry]
ID_Dist       [&Distance]^C^C_dist
ID_Area       [&Area]^C^C_area
  |
  |
  |
***POP7
**DRAW
ID_MnDraw     [&Draw]
ID_Line       [&Line]^C^C_line
ID_Ray        [&Ray]^C^C_ray
ID_Xline      [Cons&truction Line]^C^C_xline
ID_Mline      [&Multiline]^C^C_mline
              [--]
ID_Pline      [&Polyline]^C^C_pline
ID_3dpoly     [&3D Polyline]^C^C_3dpoly
ID_Polygon    [Pol&ygon]^C^C_polygon
ID_Rectang    [Rectan&gle]^C^C_rectang
              [--]
ID_MnArc      [->&Arc]
 _solprof

***POP8
**DIMENSION
ID_MnDimensi  [Dime&nsion]
ID_Dimlinear  [&Linear]^C^C_dimlinear
ID_Dimaligne  [Ali&gned]^C^C_dimaligned
ID_Dimordina  [&Ordinate]^C^C_dimordinate
              [--]
ID_Dimradius  [&Radius]^C^C_dimradius
```

ID_Dimdiamet [&Diameter]^C^C_dimdiameter
ID_Dimangula [&Angular]^C^C_dimangular
          [--]

***POP9
**MODIFY
ID_MnModify [&Modify]
ID_Ai_propch [&Properties...]^C^C_ai_propchk
ID_Matchprop [&Match Properties]^C^C_matchprop
ID_MnObject [->&Object]
ID_MnExterna [->&External Reference]
ID_Xbind   [&Bind...]^C^C_xbind

***POP10
**HELP
ID_MnHelp   [&Help]
ID_Help   [AutoCAD &Help Topics]'_help

***POP17
**GRIPS
// When a grip is hot, then display the following shortcut menu for grips. See also AUX1 menu.
// Note: This menu appears in the Menu Customization dialog,
// but it is intended for use as a right-click cursor menu.
          [&Grips Cursor Menu]
ID_Enter   [&Enter];
          [--]
ID_GripMove [&Move]_move
ID_GripMirro [M&irror]_mirror
ID_GripRotat [&Rotate]_rotate
ID_GripScale [Sca&le]_scale

// Begin AutoCAD ToolBars
//
***TOOLBARS
**TB_DIMENSION
ID_TbDimensi [_Toolbar("Dimension", _Floating, _Hide, 100, 130, 1)]
ID_Dimlinear [_Button("Linear Dimension", ICON_16_DIMLIN, ICON_24_DIMLIN)]^C^C_dimlinear

ID_Dimaligne [_Button("Aligned Dimension", ICON_16_DIMALI, ICON_24_DIMALI)]^C^C_dimaligned

**TB_DRAW
ID_TbDraw   [_Toolbar("Draw", _Left, _Show, 0, 0, 1)]
ID_Line   [_Button("Line", ICON_16_LINE, ICON_24_LINE)]^C^C_line
ID_Xline   [_Button("Construction Line", ICON_16_XLINE, ICON_24_XLINE)]^C^C_xline
ID_Mline   [_Button("Multiline", ICON_16_MLINE, ICON_24_MLINE)]^C^C_mline

**TB_EXTERNAL_DATABASE
ID_TbExtdb [_Toolbar("External Database", _Floating, _Hide, 100, 150, 1)]
ID_Aseadmin [_Button("Administration", ICON_16_ASEADM, ICON_24_ASEADM)]^C^C_aseadmin
ID_Aserows [_Button("Rows", ICON_16_ASEROW, ICON_24_ASEROW)]^C^C_aserows
ID_Aselinks [_Button("Links", ICON_16_ASELIN, ICON_24_ASELIN)]^C^C_aselinks
ID_Aseselect [_Button("Select Objects", ICON_16_ASESEL, ICON_24_ASESEL)]^C^C_aseselect

**TB_INSERT
ID_TbInsert [_Toolbar("Insert", _Floating, _Hide, 100, 190, 1)]
ID_Ddinsert [_Button("Insert Block", ICON_16_DINSER, ICON_24_DINSER)]^C^C_ddinsert

**TB_MODIFY
ID_TbModify [_Toolbar("Modify", _Left, _Show, 1, 0, 1)]
ID_Erase   [_Button("Erase", ICON_16_ERASE, ICON_24_ERASE)]^C^C_erase
ID_Copy   [_Button("Copy Object", ICON_16_COPYOB, ICON_24_COPYOB)]$M=$(if,$(eq,+
$(substr,$(getvar,cmdnames),1,4),grip),_copy,^C^C_copy)

**TB_MODIFY_II
ID_TbModifII [_Toolbar("Modify II", _Floating, _Hide, 100, 270, 1)]
ID_Draworder
[_Button("Draworder",ICON_16_DRWORD,ICON_24_DRWORD)]^C^C_draworder
         [--]
ID_Hatchedit [_Button("Edit Hatch", ICON_16_HATEDI, ICON_24_HATEDI)]^C^C_hatchedit
ID_Pedit   [_Button("Edit Polyline", ICON_16_PEDIT,

//   Begin AutoCAD Image Menus
//
***image
**image_3DObjects
[3D Objects]
[acad(Box3d,Box3d)]^C^Cai_box
[acad(Pyramid,Pyramid)]^C^Cai_pyramid
[acad(Wedge,Wedge)]^C^Cai_wedge
[acad(Dome,Dome)]^C^Cai_dome
[acad(Sphere,Sphere)]^C^Cai_sphere
[acad(Cone,Cone)]^C^Cai_cone
[acad(Torus,Torus)]^C^Cai_torus
[acad(Dish,Dish)]^C^Cai_dish
[acad(Mesh,Mesh)]^C^Cai_mesh

**image_poly
[Set Spline Fit Variables]
[acad(pm-quad,Quadric Fit Mesh)]'_surftype 5
[acad(pm-cubic,Cubic Fit Mesh)]'_surftype 6
[acad(pm-bezr,Bezier Fit Mesh)]'_surftype 8
[acad(pl-quad,Quadric Fit Pline)]'_splinetype 5
[acad(pl-cubic,Cubic Fit Pline)]'_splinetype 6

// AutoCAD Screen Menus
// There are two types of screen menus: command menus and options menus
// Command menus provide access to the lists of AutoCAD commands.
// Options menus provides access to the options available for individual commands.
// There are 22 lines between menu titles. This is one method for assuring that each time
// that a menu is called that it overwrites the previous menu.
//
// The organization of the command menus generally follows the organization of the pull-down menus.
// A command has a screen menu item only if it has a pull-down menu item.
// Command menus have, as much as possible, the same name as the equivalent pull-down menu.
// Command menus have, as much as possible, the same items in the same order as the pull-down menus.
//
// Command menus generally use the command name while pull-down menus offer a more descriptive title.
// Items in command menus that call other command menus are in upper case.

```
//
// Command menu names start with a number and are located after the special menus.
//
***SCREEN
**S
[AutoCAD ]^C^C^P(ai_rootmenus) ^P
[* * * * ]$S=ACAD.OSNAP
[FILE   ]$S=ACAD.01_FILE
[EDIT   ]$S=ACAD.02_EDIT
[VIEW 1 ]$S=ACAD.03_VIEW1
[VIEW 2 ]$S=ACAD.04_VIEW2
[INSERT ]$S=ACAD.05_INSERT
|
|
|
// The SNAP_TO menu can be called on high resolution displays to paint the snap
// options on the bottom of the screen menu.
// To add lines to the bottom of the Root menu delete the next line.
**SNAP_TO 28
[Endpoint]_endp
[Midpoint]_mid
[Intersec]_int
[App Int ]_appint
|
|
|
**ASSIST 3
[Last    ]_l
[Previous]_p
[All     ]_all
[Cpolygon]_cp
[Wpolygon]_wp
|
|
|
**09_DRAW1 3
[Line   ]^C^C_line
[Ray    ]^C^C_ray
[Xline  ]^C^C_xline
[Mline  ]^C^C_mline
|
|
|
// Here begins the command options menus
// Only commands that have command line options are included here.
// - Commands that immediately call dialog boxes do not have screen menus.
```

```
//   - Commands with no options (example: Ray) do not have screen menus.
//   Commands are listed in alphabetical order.
//   Commands and system variables are followed by a colon.
//   Command options and sub-options do not have colons.
//
**3D 3
[Solid:  ]^C^C_solid
[3Dface  ]^C^C_3dface
[3Dobjec:]$I=ACAD.image_3dobjects $I=ACAD.*

[Edge:   ]^C^C_edge
[3Dmesh: ]^C^C_3dmesh


[Revsurf:]^C^C_revsurf
[Tabsurf:]^C^C_tabsurf
[Rulsurf:]^C^C_rulesurf
[Edgsurf:]^C^C_edgesurf
|
|
|
**LINE 3
[Line:   ]^C^C_line

[1 Line: ]^C^C_line \\;

[Continue];
[Undo   ]_u
[Close  ]_c
|
|
|
//    Begin AutoCAD Tablet Menus
//    This is the TABLET1 menu.  You may put your own
//    macros and menu items here in these spaces.
//    All of the "blank" line items actually contain a
//    backslash so that no command is issued when you pick any
//    of them from the tablet.  Remove them if you want an Enter
//    to happen when they are selected, or place your own
//    macros in their place.

***TABLET1
**TABLET1STD
[A-1]\
[A-2]\
[A-3]\
```

[A-4]\
[A-5]\
|
|
|
|
***TABLET2
**TABLET2STD
// Row J View
^C^C_regen
'_zoom _e
'_zoom _a
'_zoom _w
'_zoom _p
[Draw]\
^C^C_box
^C^C_mtext
^C^C_circle
^C^C_line
^C^C
|
|
|
// TABLET3 menu.
// This tablet menu is 9 columns wide by 13 rows high.
// The equivalent area on the tablet drawing shows an empty column followed by three columns
// that are slightly narrower than the normal columns.
// The 9 tablet menu columns are used as follows:
// - 3 tablet menu columns for the blank column in the tablet drawing
// - 2 tablet menu columns for each of the 3 narrow tablet drawing columns
//
***TABLET3
**TABLET3STD
// Row 1
\
\
\
<<135
<<135
<<90
<<90
<<45
<<45
// Row 2
\
\

```
\
<<180
<<180
<\
<\
<<0
<<0
|
|
|
//    This is the TABLET4 menu.
//    It has been updated for this release.
//
```
***TABLET4**
**TABLET4STD
```
//  Row S
\
\
\
\
\
|
|
|
^C^C_align
^C^C_union
^C^C_subtract
^C^C_intersect
^C^C_xclip
^C^C_xbind
^C^C_imageadjust
^C^C_transparency
^C^C_imageclip
\
^C^C_preview
^C^C_purge
//  Row Y
^C^C_dimedit
^C^C_dimtedit
|
|
|
//  Where possible the ID name is the AutoCAD command name.
//  An ID is no more than 12 characters long.
//  Long command names are truncated (not abbreviated) to fit.
//  An ID names for a command opion appends the option to the command name:
```

10-14 Customizing AutoCAD

```
//  Example: the ID for Zoom Window is ID_ZoomWin
//  IDs are listed in alphabetical order.
//  Helps strings end with a colon, two spaces and the name of the command.
//
```
***HELPSTRINGS
ID_3darray   [Creates a three-dimensional array: 3darray]
ID_3dface   [Creates a three-dimensional face: 3dface]
ID_3dmesh   [Creates a free-form polygon mesh: 3dmesh]
ID_3dpoly   [Creates a polyline of straight line segments in three-dimensional space: 3dpoly]
ID_3dsin   [Imports a 3D Studio file: 3dsin]
ID_3dsurface [Creates three-dimensional surface objects using a dialog box]
ID_About   [Displays information about AutoCAD: about]
|
|
|
```
//  Keyboard Accelerators
//  If a keyboard accelerator is preceded by an ID string that references a menu item
//  in a pull-down menu, then the keyboard accelerator will run the command referenced
//  by that menu item.
//
```
***ACCELERATORS
// Toggle PICKADD
[CONTROL+"K"]$M=$(if,$(and,$(getvar,pickadd),1),'_pickadd 0,'_pickadd 1)
// Toggle Orthomode
[CONTROL+"L"]^O
// Next Viewport
[CONTROL+"R"]^V
// ID_Spell   ["\"F7\""]
// ID_PanRealti ["\"F11\""]
// ID_ZoomRealt ["\"F12\""]
ID_Copyclip [CONTROL+"C"]
ID_New   [CONTROL+"N"]
ID_Open   [CONTROL+"O"]
ID_Print   [CONTROL+"P"]
ID_Save   [CONTROL+"S"]
ID_Pasteclip [CONTROL+"V"]
ID_Cutclip   [CONTROL+"X"]
ID_Redo   [CONTROL+"Y"]

SUBMENUS

The number of items in a submenu section can be so large that they cannot be accommodated on one screen. For example, the maximum number of items that can be displayed on some of the display devices is 21. If the pull-down menu or the screen menu has more than 21 items, the menu items in excess of 21 are not displayed on the screen and cannot be accessed. Similarly, the maximum number of assignable blocks in tablet area-1 is 225. If the number of items in the TABLET1 or TABLET1ALT section is more than 225, the menu items in excess of 225 are not

Customizing the Standard AutoCAD Menu 10-15

assigned to any block on the template and are therefore inaccessible. The user can overcome this problem by using submenus that let the user define smaller groups of items within a menu section. When a submenu is selected, it loads the submenu items and displays them on the screen. In the case of a tablet menu the commands are assigned to different blocks on the template.

Submenu Definition

A submenu definition consists of two asterisk signs (**) followed by the name of the submenu. A menu can have any number of submenus and every submenu should have a unique name. The items that follow a submenu, up to the next section label or submenu label, belong to that submenu. The following is the format of a submenu label:

```
**Name
  │   └─ Name of the submenu
  └─ Two asterisk signs (**) designate a submenu
```

> **Note**
>
> The submenu name can be up to 31 characters long.
>
> The submenu name can consist of letters, digits, and the special characters like: $ (dollar), - (hyphen), and _ (underscore).
>
> The submenu name should not have any embedded blanks.
>
> The submenu names should be unique in a menu file.

Submenu Reference

The submenu reference is used to reference or load a submenu. It consists of a "$" sign followed by a letter that specifies the menu section. For example, the letter S specifies a screen menu, B specifies a button menu, I specifies an image tile menu, and Pn specifies a pull-down menu where n designates the number of the pull-down menu. Similarly, the letter that specifies a tablet menu section is Tn, where n designates the number of the tablet menu section. The menu section is followed by an "=" sign and the name of the submenu that the user wants to activate. The submenu name should be without "**". The following is the format of a submenu reference:

```
$Section=Submenu
 │   │      │   └─ Name of submenu
 │   │      └─ "=" sign
 │   └─ Menu section specifier
 └─ "$" sign
```

Example
```
$S=BLOCKS
   │   └─ Name of submenu
   └─ S specifies screen menu
```

Loading Screen Menus

You can load a menu that is defined in the screen menu section from any menu by using the following load command:

$S=(name1) $S=(name2)
        Where "name1" and "name2" are the names of the submenus

Example
```
$S=X  $S=INSERT
 |     |
 |     └─ Submenu name defined in screen menu section
 └─ Submenu name defined in screen menu section
 └─ S specifies screen menu
```

The first load command ($S=X) loads the submenu X that has been defined in the screen menu section of the menu file. The X submenu, defined in the screen menu section, contains 21 blank lines. When it is loaded, it clears the screen menu. The second load command ($S=INSERT) loads the submenu INSERT that has also been defined in the screen menu section of the menu file.

Loading Pull-down Menus

You can load a pull-down menu from any menu by using the following command:

$P(n)=(name) $P(n)=*
        Where "n" ranges from 1 to 10 (POP1 -- POP10)   and "name" is the name of the submenu defined in the pull-down menu.

Example

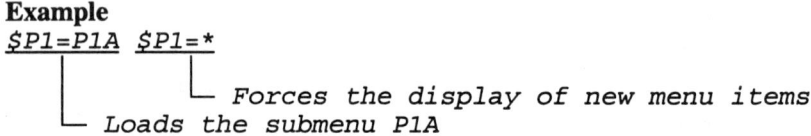
```
$P1=P1A  $P1=*
          └─ Forces the display of new menu items
 └─ Loads the submenu P1A
```

The first load command $P1=P1A loads the submenu P1A that has been defined in the POP1 section of the menu file. The second load command $P1=* is a special AutoCAD command that forces the new menu items to be displayed on the screen.

Loading Image Tile Menus

You can also load an image tile menu from any menu by using the following load command:

$I=(name) $I=*
        Where "name" is the name of the submenu defined in the image submenu

Example

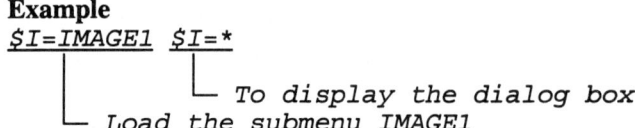
```
$I=IMAGE1  $I=*
            └─ To display the dialog box
 └─ Load the submenu IMAGE1
```

This menu item consists of two load commands. The first load command $I=IMAGE1 loads the image submenu IMAGE1 that has been defined in the image tile menu section of the file. The second load command $I=* displays the new dialog box on the screen.

CUSTOMIZING TABLET AREA-1

The tablet menu has four sections TABLET1, TABLET2, TABLET3, and TABLET4 (Figure 10-2).

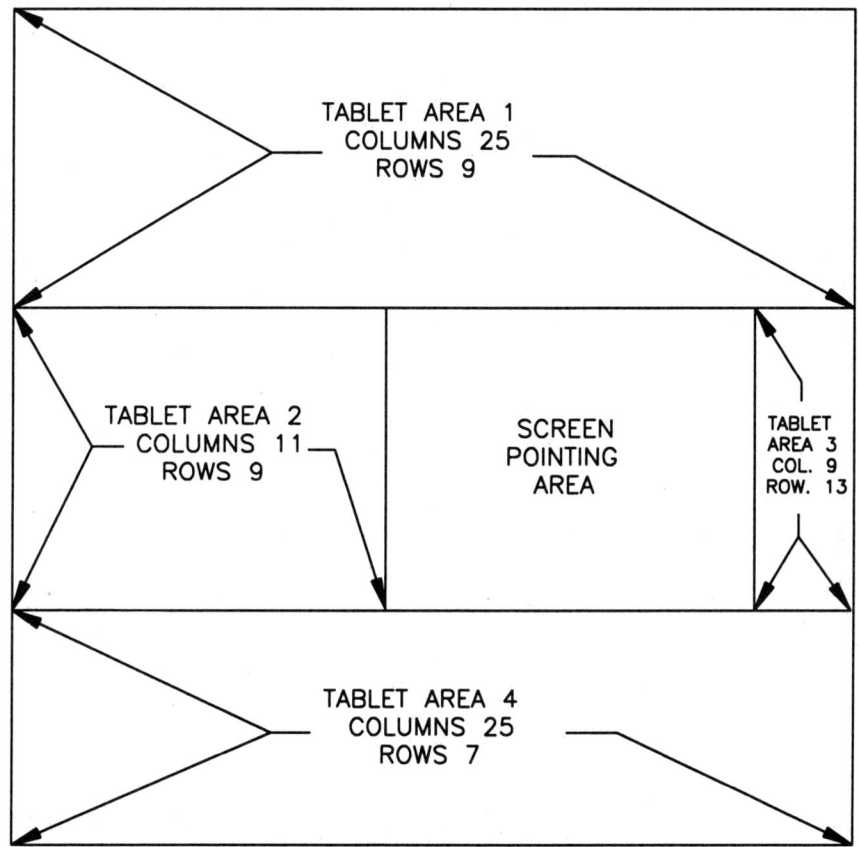

Figure 10-2 Four tablet areas of AutoCAD template

Tablet area-1 of the template has 25 columns numbered 1 to 25 and 9 rows designated by letters A through I. The total number of blocks for tablet area-1 is 225 (25 x 9 = 225). You can utilize this area to customize AutoCAD's tablet menu by assigning commands or macros to different blocks. Figure 10-3 shows tablet area-1 of the standard AutoCAD template. Before making any changes or additions to the tablet menu it is very important to figure out the commands that you want to add to the tablet menu and determine their location on the template. A well thought-out tablet design will save a lot of revision time in the long run. Figure 10-3 shows a drawing that has 25 columns and 9 rows. You may draw a similar drawing and make several copies of it for arranging the commands and designing the template.

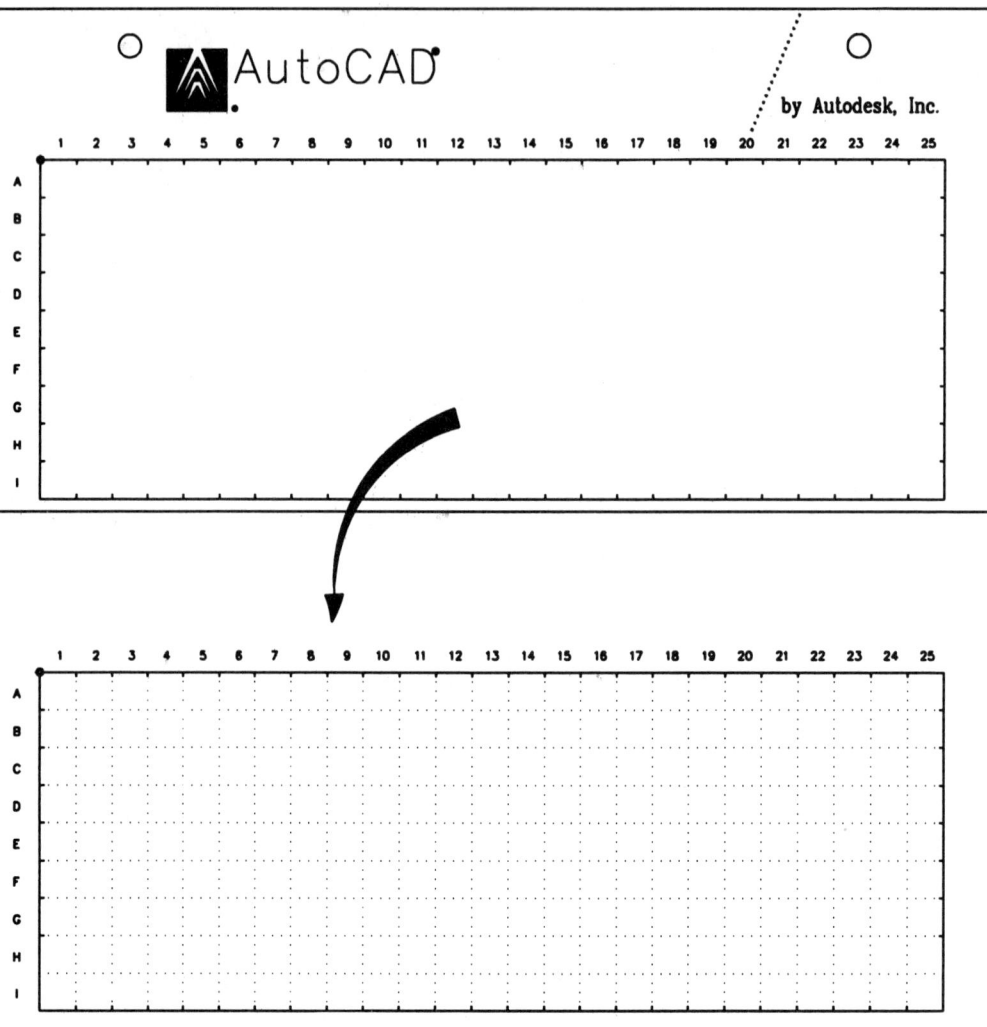

Figure 10-3 Tablet area-1 with 25 columns and 9 rows

Note

Before making any changes to the menu file make sure that the original file has been properly saved. You may also make a copy of the .i.ACAD.MNU; file and then make the changes in the new file. Use the following command to make a copy of the ACAD.MNU file. The name of the new menu file is CUSTOM.MNU. After you make the necessary changes you may load the new menu for testing by using AutoCAD's MENU command.

COPY ACAD.MNU CUSTOM.MNU

Example 1

Add the commands shown in Figure 10-4 to the TABLET1 section of CUSTOM.MNU file. B10 through B25 are the names of the blocks. The user should be able to insert the blocks with X-scale and Y-scale factors of 1.0 and a rotation angle of 0.

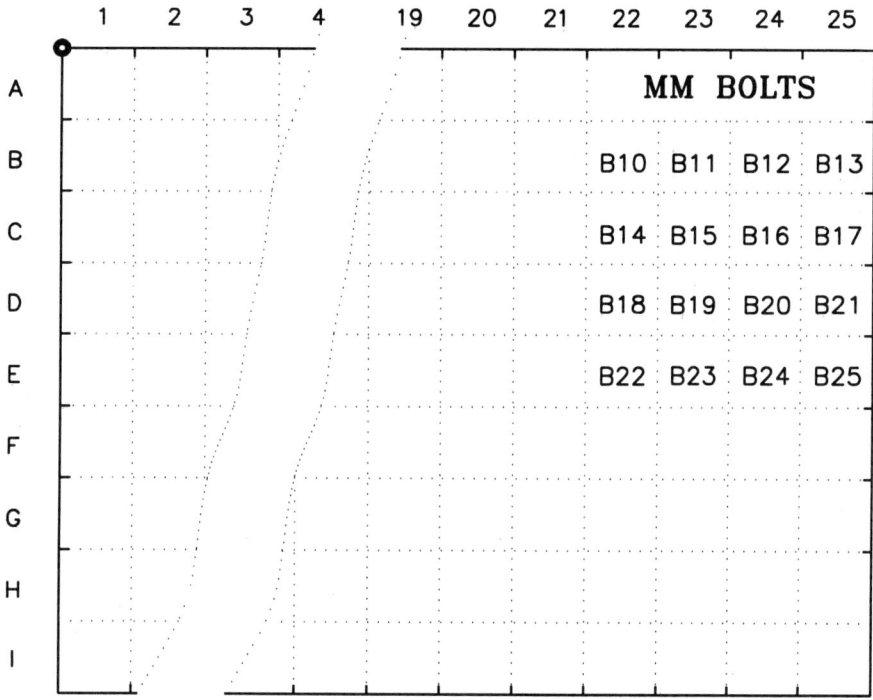

Figure 10-4 Commands assigned to tablet area-1

Before making any changes in the file, you need to determine the location of each block on the template. For example, the WBlock B10 is to be assigned to a block that is in row B and column number 22. Similarly, WBlock B17 is in row C and column 25. The following table shows the location of WBlocks in tablet area-1 of the template:

WBLOCK NAME	ROW	COLUMN	LOCATION IN TABLET1
B10	B	22	B-22
B11	B	23	B-23
B12	B	24	B-24
B13	B	25	B-25
B14	C	22	C-22
B15	C	23	C-23
B16	C	24	C-24
B17	C	25	C-25
B18	D	22	D-22

B19	D	23	D-23
B20	D	24	D-24
B21	D	25	D-25
B22	E	22	E-22
B23	E	23	E-23
B24	E	24	E-24
B25	E	25	E-25

You can use any word processor or text editor to load and edit the file. After CUSTOM.MNU file is loaded, search for ***TABLET1 section label. The letters and the numbers inside the brackets indicate the row and column. For example, in [A-1], A is for row number and 1 is for column number in the template area-1.

```
[A-1]
 | |
 | └─ Column number
 └─ Row
```

The location of the first WBlock B10 in the menu file is B-22. Locate the menu item B-22 in the menu file and add the INSERT command to the menu item command definition. The following file is a partial listing of the TABLET1 section of the menu file after editing:

***TABLET1
**TABLET1STD
[A-1]
[A-2]
[A-3]
[A-4]
[A-5]
[A-6]
[A-7]
[A-8]
[A-9]
[A-10]
[A-11]
[A-12]
[A-13]
[A-14]
[A-15]
[A-16]
[A-17]
[A-18]
[A-19]
[A-20]
[A-21]
[A-22]
[A-23]

[A-24]
[A-25]
[B-1]
[B-2]
[B-3]
[B-4]
[B-5]
[B-6]
[B-7]
[B-8]
[B-9]
[B-10
|
|
|
[B-22]^C^CINSERT;B10;\1.0;1.0;0
[B-23]^C^CINSERT;B11;\1.0;1.0;0
[B-24]^C^CINSERT;B12;\1.0;1.0;0
[B-25]^C^CINSERT;B13;\1.0;1.0;0
[C-1]
|
|
|
[C-21]
[C-22]^C^CINSERT;B14;\1.0;1.0;0
[C-23]^C^CINSERT;B15;\1.0;1.0;0
[C-24]^C^CINSERT;B16;\1.0;1.0;0
[C-25]^C^CINSERT;B17;\1.0;1.0;0
[D-1]
|
|
|
[D-20]
[D-21]
[D-22]^C^CINSERT;B18;\1.0;1.0;0
[D-23]^C^CINSERT;B19;\1.0;1.0;0
[D-24]^C^CINSERT;B20;\1.0;1.0;0
[D-25]^C^CINSERT;B21;\1.0;1.0;0
|
|
|
[E-21]
[E-22]^C^CINSERT;B22;\1.0;1.0;0
[E-23]^C^CINSERT;B23;\1.0;1.0;0
[E-24]^C^CINSERT;B24;\1.0;1.0;0
[E-25]^C^CINSERT;B25;\1.0;1.0;0

> **Note**
> To load the new menu CUSTOM.MNU, use AutoCAD's MENU command.
>
>   Command: **MENU**
>   Menu file name <ACAD>: **CUSTOM**
>
> When you select the block insert command from the template, AutoCAD will prompt you to enter an insertion point. The X-scale and Y-scale factors and the rotation angle are already defined in the command definition.

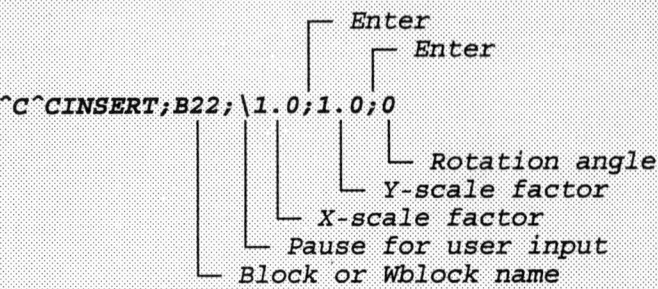

SUBMENUS

The number of items in the TABLET1STD or TABLET1ALT section of the menu file can be so large that may not be accommodated on one template. For example, the maximum number of assignable blocks in tablet area-1 is 225. If the number of items in the TABLET1STD or TABLET1ALT section is more than 225, the menu items in excess of 225 are not assigned to any block on the template and are therefore inaccessible. The user can overcome this problem by using submenus that let the user define any number of menu items in the tablet area-1 of the template. When the user selects the submenu, AutoCAD automatically loads the new submenu and assigns the commands to different blocks in template area-1. The format of the submenu reference is:

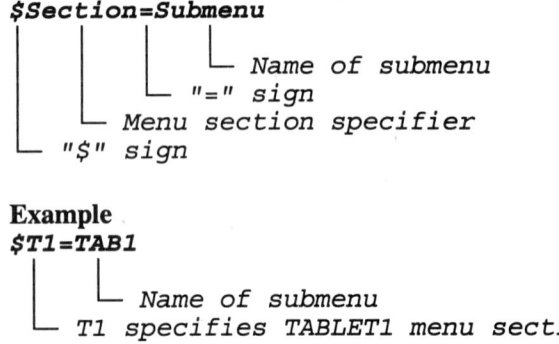

Example 2

Edit the CUSTOM.MNU file to add the commands shown in Figure 10-5. Use the submenus and make a provision for swapping the submenus. When the user selects a block insert command from the tablet menu, AutoCAD should automatically load the corresponding screen menu.

Customizing the Standard AutoCAD Menu 10-23

In this example it is assumed that there is no room for adding more commands to tablet area-1; therefore submenus have to be created to make room for the commands in excess of 225. Two submenus TABA and TABB have been defined in the TABLET1 section of the menu file CUSTOM.MNU. When you load the menu file CUSTOM.MNU, the first submenu TABA is automatically loaded and you can select the block insert commands from the template. You can load the submenu TABB by selecting the **Load TABB** block from the template. AutoCAD loads the submenu TABB and now you can select the commands from the new submenu. If you want to load the submenu TABA back, select the **Load TABA** from the template. The following file is a partial listing of the TABLET1 section of the menu file after inserting the new command definitions:

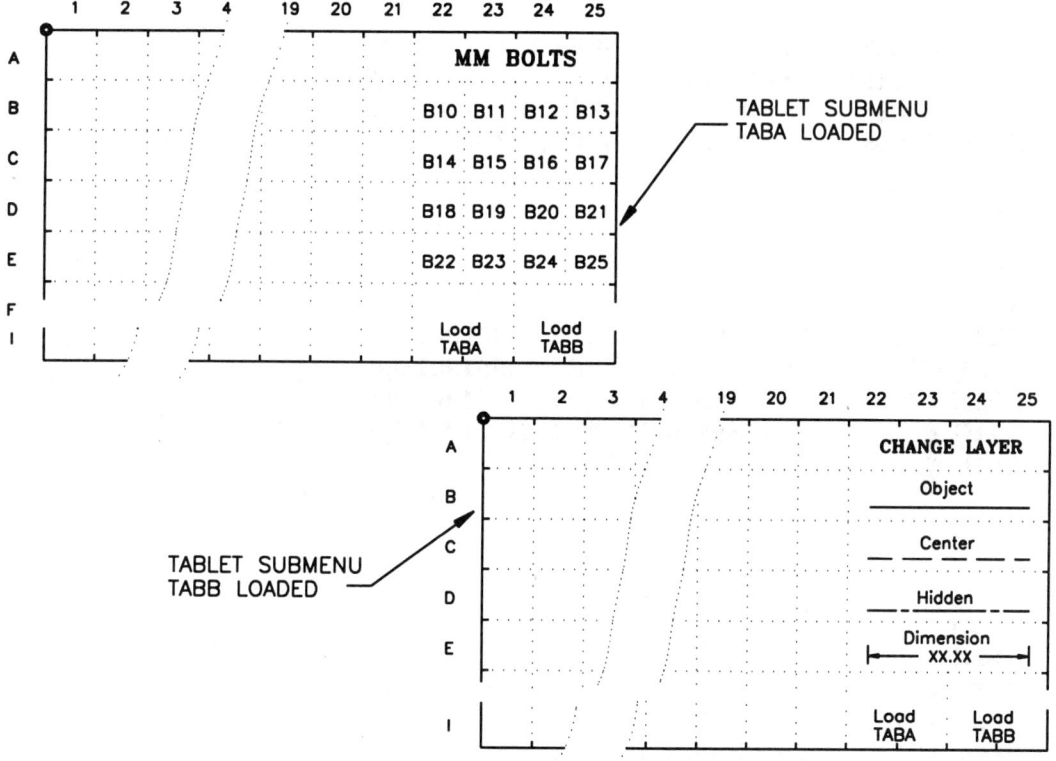

Figure 10-5 Commands assigned to tablet area-1

***TABLET1
**TABLET1STD
**TABA
[A-1]
[A-2]
[A-3]
[A-4]
[A-5]
[A-6]

[A-7]
[A-8]
[A-9]
[A-10]
 |
 |
 |
[B-18]
[B-19]

[B-20]
[B-21]
[B-22]^C^C$S=X $S=INSBLK INSERT;B10;\1.0;1.0;0
[B-23]^C^C$S=X $S=INSBLK INSERT;B11;\1.0;1.0;0
[B-24]^C^C$S=X $S=INSBLK INSERT;B12;\1.0;1.0;0
[B-25]^C^C$S=X $S=INSBLK INSERT;B13;\1.0;1.0;0
[C-1]
 |
 |
 |
[C-21]
[C-22]^C^C$S=X $S=INSBLK INSERT;B14;\1.0;1.0;0
[C-23]^C^C$S=X $S=INSBLK INSERT;B15;\1.0;1.0;0
[C-24]^C^C$S=X $S=INSBLK INSERT;B16;\1.0;1.0;0
[C-25]^C^C$S=X $S=INSBLK INSERT;B17;\1.0;1.0;0
[D-1]
 |
 |
 |
[D-20]
[D-21]
[D-22]^C^C$S=X $S=INSBLK INSERT;B18;\1.0;1.0;0
[D-23]^C^C$S=X $S=INSBLK INSERT;B19;\1.0;1.0;0
[D-24]^C^C$S=X $S=INSBLK INSERT;B20;\1.0;1.0;0
[D-25]^C^C$S=X $S=INSBLK INSERT;B21;\1.0;1.0;0
[E-1]
 |
 |
 |
[E-21]
[E-22]^C^C$S=X $S=INSBLK INSERT;B22;\1.0;1.0;0
[E-23]^C^C$S=X $S=INSBLK INSERT;B23;\1.0;1.0;0
[E-24]^C^C$S=X $S=INSBLK INSERT;B24;\1.0;1.0;0
[E-25]^C^C$S=X $S=INSBLK INSERT;B25;\1.0;1.0;0
[F-1]
 |

|
|
[H-19]
[H-20]
[H-21]
[H-22]^C^C$T1=TABA
[H-23]^C^C$T1=TABA
[H-24]^C^C$T1=TABB
[H-25]^C^C$T1=TABB

**TABB
[A-1]
[A-2]
[A-3]
[A-4]
[A-5]
[A-6]
[A-7]
[A-8]
|
|
|
[B-18]
[B-19]
[B-20]
[B-21]
[B-22]^C^CLAYER;SET;OBJECT;;
[B-23]^C^CLAYER;SET;OBJECT;;
[B-24]^C^CLAYER;SET;OBJECT;;
[B-25]^C^CLAYER;SET;OBJECT;;
[C-1]
|
|
|
[C-21]
[C-22]^C^CLAYER;SET;CENTER;;
[C-23]^C^CLAYER;SET;CENTER;;
[C-24]^C^CLAYER;SET;CENTER;;
[C-25]^C^CLAYER;SET;CENTER;;
[D-1]
|
|
|
[D-20]
[D-21]
[D-22]^C^CLAYER;SET;HIDDEN;;

[D-23]^C^CLAYER;SET;HIDDEN;;
[D-24]^C^CLAYER;SET;HIDDEN;;
[D-25]^C^CLAYER;SET;HIDDEN;;
[E-1]
|
|
|
[E-21]
[E-22]^C^CLAYER;SET;DIM;;
[E-23]^C^CLAYER;SET;DIM;;
[E-24]^C^CLAYER;SET;DIM;;
[E-25]^C^CLAYER;SET;DIM;;
[F-1]
|
|
[H-19]
[H-20]
[H-21]
[H-22]^C^C$T1=TABA
[H-23]^C^C$T1=TABA
[H-24]^C^C$T1=TABB
[H-25]^C^C$T1=TABB

Note
When you load the submenu TABB, the template overlay for tablet area-1 must be changed.

The menu items defined in H-22, H23 and H-24, H-25 load the submenus TABA and TABB respectively.

```
/^C^C$T1=TABA
       |      |
       |      └─ Loads tablet submenu TABA
       └─ Cancels the existing command twice
```

CUSTOMIZING TABLET AREA-2

Tablet area-2 has 11 columns and 9 rows. The columns are numbered from 1 to 11 and the rows are designated by the letters J through R as shown in Figure 10-6. The total number of blocks in tablet area-2 is 99 (11 x 9 = 99). There are no empty blocks in this area like tablet area-1 and the commands that have been assigned to the blocks are defined in the TABLET2 menu sections of the standard menu file. You can change or delete the command definition assigned to these blocks. You can even delete the entire TABLET2 menu sections and write your own menu that best fits your needs. However, you must be careful in developing a new menu and you may need quite some time to come up with a reliable menu.

Exercise 1

Assign the following commands to the blocks in the tenth column of tablet area-2 (Figure 10-6).

LINE
PLINE
ARC,CSE
ARC,SCE
ARC,CSA
CIRCLE-C,R
CIRCLE-C,D
CIRCLE-2P

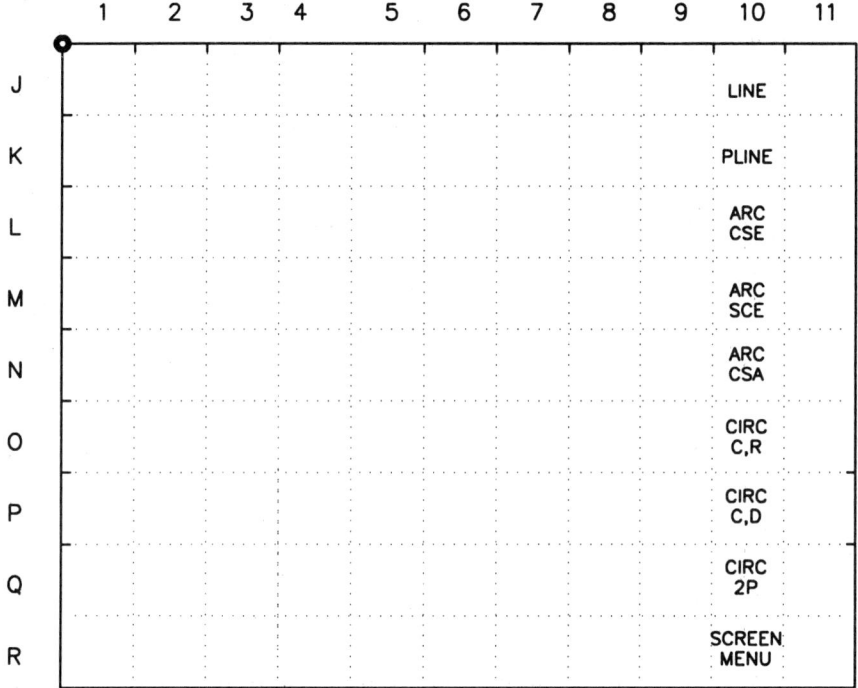

Figure 10-6 Commands assigned to tablet area-2

CUSTOMIZING TABLET AREA-3

Tablet area-3 has 9 columns and 13 rows. In Figure 10-7 the columns are numbered from 1 to 9, and the rows are numbered from 1 to 13. The total number of blocks in tablet area-3 are 117 (13 x 9 = 117). The size of the blocks in this area is smaller than the blocks in other sections of the tablet template. Also, the blocks are rectangular in shape, whereas in other tablet areas the blocks are square. You can modify or delete the command definitions assigned to these blocks. You can even delete the entire TABLET3 menu sections and write a menu that best fits your needs. However, you have to be careful in developing a new menu. The following example illustrates the editing process for adding commands to the TABLET3 section of the CUSTOM.MNU file.

Example 3

Add the following angles to the TABLET3 section of CUSTOM.MNU file. The layout of the tablet area-3 is shown in Figure 10-7.

<u>ANGLES</u>
30
120
210
330

Use your word processor or text editor to load CUSTOM.MNU file and search for **TABLET3 section of the menu file. Locate the lines that you want to edit and then assign the required command definitions to these lines. The following file is a partial listing of the TABLET3 section of the CUSTOM.MNU file after editing:

```
**TABLET3ALT
;
<<30
<<30
<<135
<<135
<<90
<<90
<<45
<<45
;
<<120
<<120
<<180
<<180
<\
<\
<<0
<<0
;
<<210
<<210
<<225
<<225
<<270
<<270
<<315
<<315
;
<<330
```

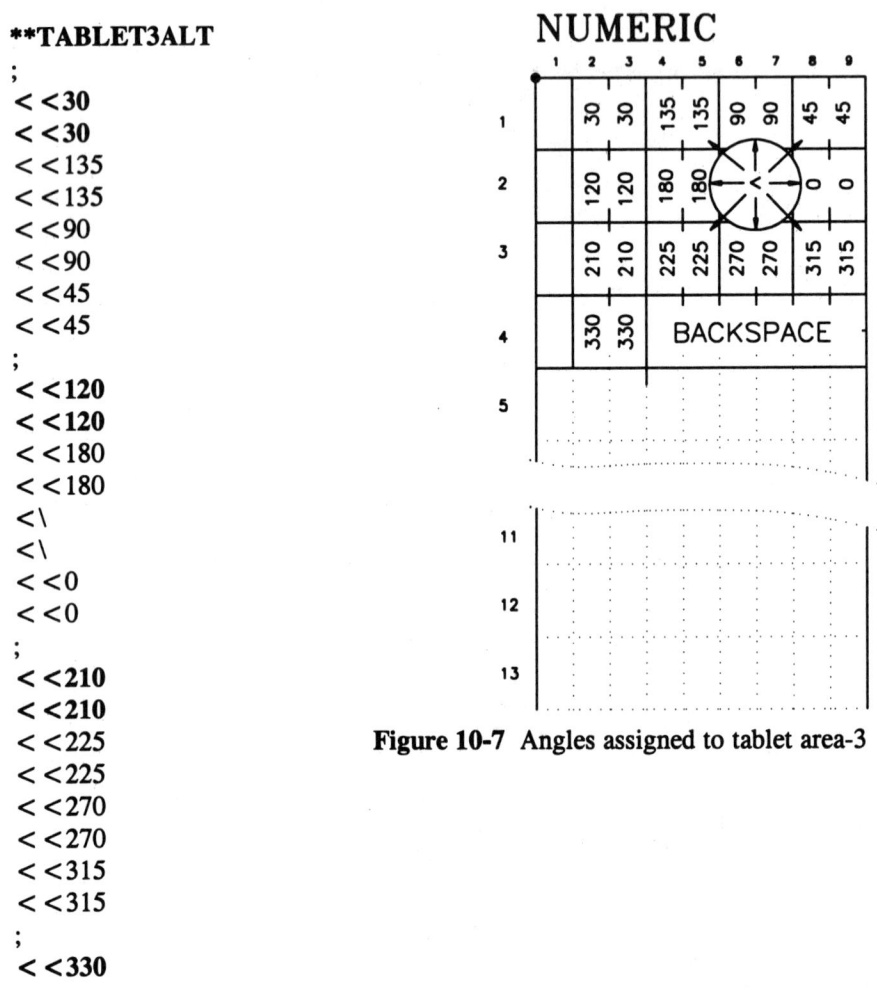

Figure 10-7 Angles assigned to tablet area-3

```
<<330
^H
^H
^H
^H
^H
^H
;
;
;
m\
m\
cm\
cm\
mm\
mm\
;
;
;
.\
.\
+\
+\
%%d\
%%d\
;
;
;
,\
,\
%%p\
%%p\
%%c\
%%c\
```

CUSTOMIZING TABLET AREA-4

Tablet area-4 has 25 columns and 7 rows. As shown in Figure 10-8, the columns are numbered from 1 to 25 and the rows are designated by the letters S through Y. The total number of blocks in tablet area-4 is 175 (7 * 25 = 175). You can modify or delete the command definition assigned to these blocks. You can even delete the entire TABLET4 menu sections and write a menu that best fits your needs. However, you have to be careful in developing a new menu.

10-30 Customizing AutoCAD

Exercise 2

Add the following commands to the TABLET4 section of CUSTOM.MNU file. The partial layout of tablet area-4 is shown in Figure 10-8.

1. Load and run the AutoLISP routines TRANA.LSP and TRANB.LSP. (It is assumed that the files TRANA.LSP and TRANB.LSP are predefined.)

2. Run the script files SCR1 and SCR2.

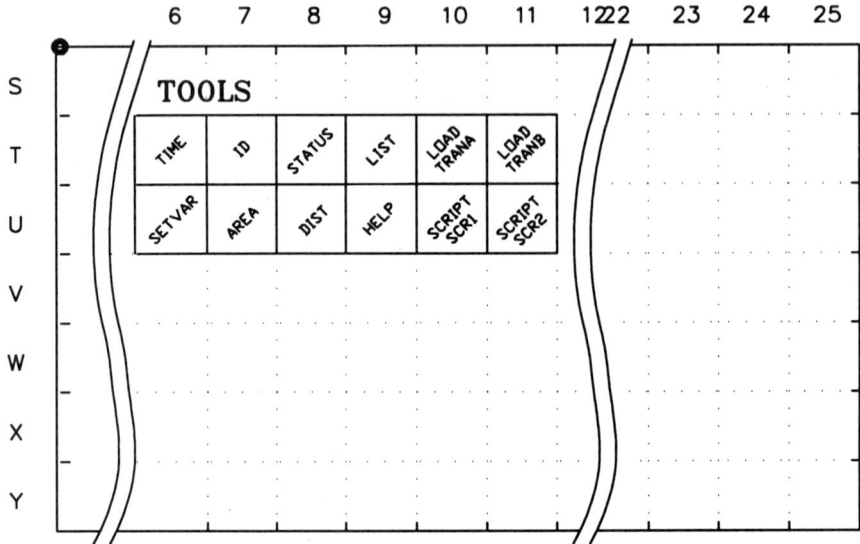

Figure 10-8 Commands assigned to tablet area-4

Note
^C^C(LOAD "TRANA");TRANA *In this menu item (LOAD "TRANA") loads the AutoLISP file TRANA. The TRANA that is outside the parenthesis, after the semicolon, executes the function TRANA.*

```
^C^C(LOAD "TRANA");TRANA
          |            |
          |            └─ Name of AutoLISP function
          └─ Loads AutoLISP program TRANA
```

^C^CSCRIPT;SCR1 *In this menu item SCR1 is the name of the script file, and SCRIPT is an AutoCAD command for running a script file.*

```
^C^CSCRIPT;SCR1
       |     |
       |     └─ Name of the script file
       └─ AutoCAD's SCRIPT command
```

CUSTOMIZING BUTTONS AND AUXILIARY MENUS

The standard AutoCAD menu file ACAD.MNU contains buttons and auxiliary menu sections. The buttons and auxiliary menu sections are identical and the first section has nine menu items. The following file is a listing of the buttons and auxiliary menu sections of the standard AutoCAD menu file, ACAD.MNU:

```
***BUTTONS1
;
$p0=*
^C^C
^B
^O
^G
^D
^E
^T

***BUTTONS2
$p0=*

***AUX1
;
$p0=*
^C^C
^B
^O
^G
^D
^E
^T

***AUX2
$p0=*
```

Note
The following table shows the function of the menu items that are defined in the BUTTONS section of the CUSTOM.MNU file.

MENU ITEM	FUNCTION
***BUTTONS	Section label
;	Enter
$p0=*	Displays the cursor menu
^C^C	Cancels the existing command twice
^B	Snap On/off (Ctrl B)

10-32 Customizing AutoCAD

^O   Ortho On/off (Ctrl O)
^G   Grid On/off (Ctrl G)
^D   Coordinate Dial On/off (Ctrl D)
^E   Isoplane (Ctrl E)
^T   Tablet On/off (Ctrl T)

Like any other menu you can make changes in the buttons menu to assign different commands to the buttons of the pointing device. The pointing devices generally come in 4-button or 10-button configuration. The first button is always the pick button and cannot be used for any other purpose. AutoCAD commands can be assigned to the remaining buttons. The following example describes the editing procedure for the BUTTONS section of CUSTOM.MNU file (CUSTOM.MNU file is a copy of ACAD.MNU file).

Example 4

Change the buttons sections of CUSTOM.MNU to assign the following commands to the four buttons of the pointing device (Figure 10-9):

         1. PICK   3. ZOOM Win
         2. Enter   4. ZOOM Prev

The ZOOM Window command assigned to button number 3 should automatically zoom in an area that is 2 units from the selected point.

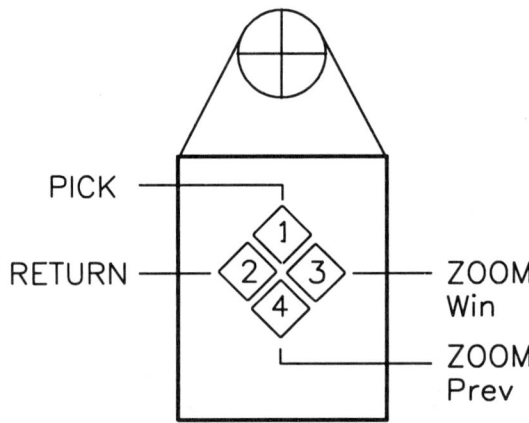

Figure 10-9 Commands assigned to 4 buttons of a pointing device

Before making any changes it is very important to know the commands and the prompt entries that are associated with the commands. Use your word processor to load the CUSTOM.MNU file and search for the ***BUTTONS1 section. The lines that follow the ***BUTTONS1 section are the lines that need to be edited to assign new commands to the buttons of the pointing device. The following file is a listing of the button menu section of the CUSTOM.MNU file after making the required changes:

```
***BUTTONS1                                              1
;                                                        2
'ZOOM;WIN;\@2,2                                          3
'ZOOM;PRE                                                4
^B                                                       5
^O                                                       6
^G                                                       7
^D                                                       8
^E                                                       9
^T                                                      10
```

Line 1
***BUTTONS1
This is the section label for the BUTTONS1 menu section.

Line 2
;
The semicolon (;) causes an Enter. It is like pressing the Enter key from the keyboard or template.

Line 3
'ZOOM;WIN;\@2,2
When you select this menu item, it will zoom as shown in Figure 10-10. The first point you select becomes the first corner of the zoom window and the other corner of the zoom window is 2.0,2.0 units from the selected point.

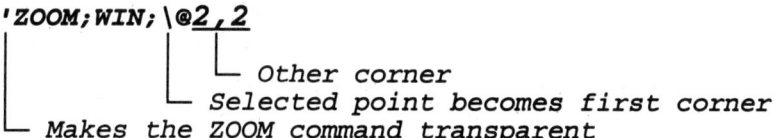

Before selecting the third button of the pointing device for this command, move the screen crosshairs to the point where you want to zoom and then click the button. AutoCAD will zoom in the area since the two corners of the window are defined in the menu item. The single quote (') in front of ZOOM makes the ZOOM Window command transparent.

Line 4
'ZOOM;PRE
This menu item defines the ZOOM command with Previous option. The single quote (') in front of the ZOOM command makes the ZOOM Previous command transparent.

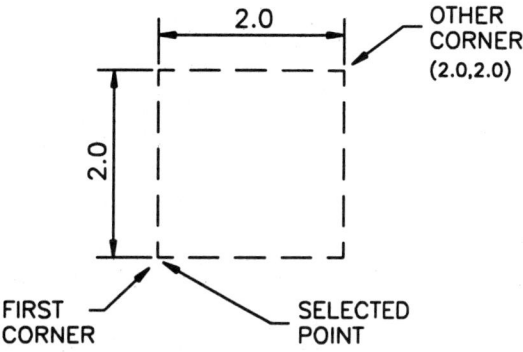

Figure 10-10 Zoom area

10-34 Customizing AutoCAD

> **Note**
> The commands on the first three lines (2 - 4), excluding the ***BUTTONS1 line, will be assigned to the second, third, and fourth button of the four-button pointing device. The remaining items will be ignored and do not affect the commands that are assigned to other buttons. Therefore, you can leave them in the file.

Example 5

Edit the buttons and the auxiliary menu sections of the menu file CUSTOM.MNU to add the following AutoCAD commands. The pointing device has 10 buttons (Figure 10-11) and button number 1 is used as a pick button. The blocks should be inserted with a scale factor of 1.00 and a rotation angle of 0 degrees. (The CUSTOM.MNU file is a copy of the ACAD.MNU menu file.)

1.	PICK BUTTON	2.	Enter	3.	CANCEL
4.	OSNAPS	5.	INSERT B1	6.	INSERT B2
7.	INSERT B3	8.	ZOOM Window	9.	ZOOM All
10.	ZOOM Previous				

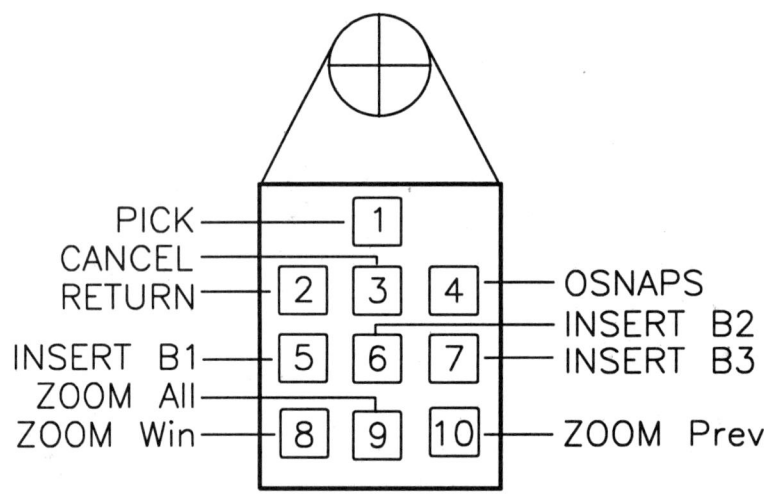

Figure 10-11 Commands assigned to different buttons of a 10-button pointing device

> **Note**
> B1, B2, B3 are the names of the blocks or wblocks that have already been created.
>
> It is assumed that the cursor menu POP0 has already been defined in the menu file.
>
> Use transparent ZOOM command for ZOOM Previous and ZOOM Window.

The following file is a listing of the buttons menu section of the CUSTOM.MNU file after making the required changes:

```
***BUTTONS                                         1
;                                                  2
^C^C                                               3
$P0=*                                              4
^C^CINSERT;B1;\1.0;1.0;0                           5
^C^CINSERT;B2;\1.0;1.0;0                           6
^C^CINSERT;B3;\1.0;1.0;0                           7
'ZOOM;Win                                          8
^C^CZOOM;All                                       9
'ZOOM;Prev                                        10
***AUX1                                           11
;                                                 12
^C^C                                              13
$P0=*                                             14
^C^CINSERT;B1;\1.0;1.0;0                          15
^C^CINSERT;B2;\1.0;1.0;01                         16
^C^CINSERT;B3;\1.0;1.0;0                          17
'ZOOM;Win                                         18
^C^CZOOM;All                                      19
'ZOOM;Prev                                        20
```

Line 3
^C^C
This command definition is assigned to button number 3. It cancels the existing command twice.

Line 4
$P0=*
In this menu item $P0=* is a special command that you can use to access the cursor menu. When you select this item, AutoCAD will display the cursor menu on the screen near the cursor location. The cursor menu that contains the object snap mode commands is defined in the POP0 menu section of ACAD.MNU file. This command definition is assigned to button number 4 of the pointing device.

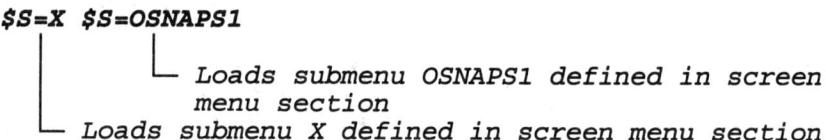

Line 5
^C^CINSERT;B1;\1.0;1.0;0
In this menu item ^C^C cancels the existing command twice. INSERT is an AutoCAD command that can be used to insert a Block or Wblock. B1 is the name of the block and the back slash (\) pauses for the user input. In this menu item, it is the insertion point of the block. The first 1.0 is the X-scale factor and the second 1.0 is the Y-scale factor of the block. The 0 at the end is for the rotation angle of the block.

10-36 Customizing AutoCAD

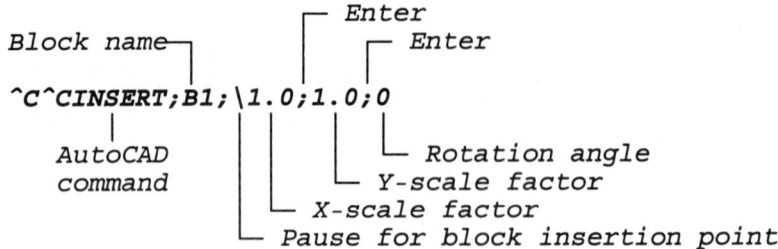

Line 9
^C^CZOOM;All

The command definition of this menu item is assigned to button number 9 of the pointing device. If you select this key, it will enter a ZOOM All command.

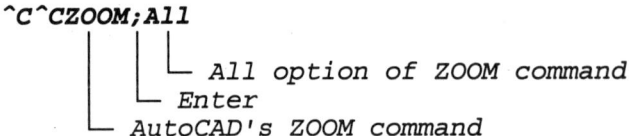

Line 10
'ZOOM;Prev

This menu item defines the transparent ZOOM Previous command. It is assigned to button number 10 of the pointing device.

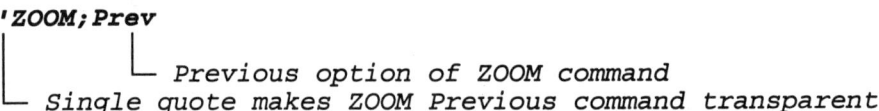

Line 11
***AUX1

In this line AUX1 is the section label for auxiliary menu section. Lines 12 through 20 are defined in this section.

CUSTOMIZING PULL-DOWN AND CURSOR MENUS

The pull-down and cursor menus are a part of the standard AutoCAD menu file (ACAD.MNU) that come with the AutoCAD software package. The ACAD.MNU file is automatically loaded when you get into the Drawing Editor, provided the standard configuration of AutoCAD has not been changed. In addition to POP0, the pull-down menu can have a maximum of 16 sections defined as POP1, POP2, POP3 -- --- --- --- POP16. The standard AutoCAD menu uses POP0 for the cursor menu and the first 9 sections, POP1 through POP9, for pull-down menus. The cursor menu can be loaded and displayed by using the command $P0=*.

CASCADING SUBMENUS IN PULL-DOWN MENUS

The cascading feature of AutoCAD allows the pull-down and cursor menus to be displayed in a hierarchical order that makes it easier to select submenus. To use the cascading feature in the pull-down and cursor menus, AutoCAD has provided some special characters. For example, the characters -> defines a cascaded submenu and <- designates the last item in the pull-down menu. The following table lists some the frequently used characters that can be used with the pull-down or cursor menus:

Character	Character Description
--	The item label consisting of two hyphens, automatically expand to fill the entire width of the pull down menu Example: [--]
+	Used to continue the menu item to the next line. This character has to be the last character of the menu item. Example: [Triang:]^C^CLine;1,1;+ 3,1;2,2;
->	This label character defines a cascaded submenu and it must precede the name of the submenu. Example: [->Draw]
<-	This label character designates the last item of the cascaded pull-down or cursor menu. The character must precede the label item. Example: [<-CIRCLE 3P]^C^CCIRCLE;3P
<-<-...	This label character designates the last item of the pull-down or cursor menu and also terminates the parent menu. The character must precede the label item. Example: [<-<-Center Mark]^C^C_dim;_center
$(	This label character can be used with the pull-down and cursor menus to evaluate a DIESEL expression. The character must precede the label item. Example: "$(if,$(getvar,orthomode),Ortho)"
~	This item grays-out the label item. The character must precede the item. Example: [~--]

The length of each menu bar title can be up to 14 characters long. Most display devices provide space for a maximum of 80 characters. Therefore, if there are 16 pull-down menus, the length of each menu title should be up to 5 characters long. If the combined length of all menu bar titles exceeds 80 characters, AutoCAD automatically truncates the characters from the longest menu title until it fits all menu titles in the menu bar. The following is a list of some additional features of the pull-down menu:

10-38 Customizing AutoCAD

1. The section labels of the pull-down menus are ***POP1 through ***POP16. The menu bar titles are displayed in the menu bar.

2. The pull-down menus can be accessed by selecting the menu title from the menu bar at the top of the screen.

3. A maximum of 999 menu items can be defined in the pull-down menu. This includes the items that are defined in the pull-down submenus. The menu items in excess of 999 are ignored.

4. The number of menu items that can be displayed on the screen depends on the display device that you are using. If the cursor or the pull-down menu contains more items than what can be accommodated on the screen, the excess items are truncated. For example, if your system can display 21 menu items, then the menu items in excess of 21 are automatically truncated.

5. If AutoCAD is not configured to show the status line, the pull-down menus, cursor menus, and the menu bar are automatically disabled.

Example 6

Edit the POP4 section of the pull-down menu to add a new insert command with the label NEW-INSERT (Figure 10-12). When the user selects NEW-INSERT from the POP4 pull-down menu, it should load and display a cascading submenu that contains the commands for inserting the following blocks:

```
            INSERT-BLOCKS      --------- Cascading submenu title
            DOOR1
            DOOR2
            ------
            WINDOW1
            WINDOW2
```

Note

The doors and windows are saved as WBLOCKS in the SYMBOLS subdirectory on the D drive.

User should be able to insert the block at the selected point with X-scale and Y-scale factors of 1.25 and rotation angle of 0.

Do not edit the ACAD.MNU file. Make a copy of the ACAD.MNU file and then edit the new file (CUSTOM.MNU).

Use your word processor to load the menu file CUSTOM.MNU and search for the **insert** command in the pull-down menu section ***POP4. Now, insert a line that contains the definition of the INSERT-BLOCKS cascading submenu as shown in Figure 10-12.

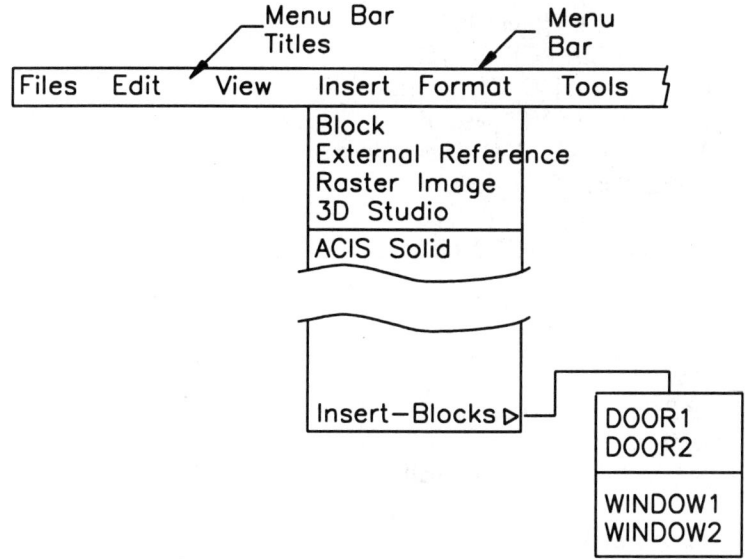

Figure 10-12 POP4 section of pull-down menu

The following file is a partial listing of CUSTOM.MNU file after editing the POP4 section of the pull-down menu and adding a new submenu:

```
***POP1
**FILE
ID_MnFile    [&File]
ID_New       [&New...\tCtrl+N]^C^C_new
ID_Open      [&Open...\tCtrl+O]^C^C_open
             [--]
ID_Save      [&Save\tCtrl+S]^C^C_qsave
ID_Saveas    [Save &As...]^C^C_saveas
ID_Export    [&Export...]^C^C_export
             [--]
ID_Config    [P&rinter Setup...]^C^C_config
ID_Preview   [Print Pre&view]^C^C_preview
ID_Print     [&Print...\tCtrl+P]^C^C_plot
             [--]
|
|
|
***POP2
**EDIT
ID_MnEdit    [&Edit]
ID_U         [&Undo\tCtrl+Z]_u
ID_Redo      [&Redo\tCtrl+Y]^C^C_redo
             [--]
```

```
ID_Cutclip   [Cu&t\tCtrl+X]'_cutclip
ID_Copyclip  [&Copy\tCtrl+C]'_copyclip
ID_Copylink  [Copy &Link]^C^C_copylink
ID_Pasteclip [&Paste\tCtrl+V]'_pasteclip
ID_Pastesp   [Paste &Special...]^C^C_pastespec
ID_Erase     [Cle&ar\tDel]^C^C_erase
         [--]
ID_Links     [&OLE Links...]^C^C_olelinks

***POP3
**VIEW
ID_MnView    [&View]
ID_Redrawall [&Redraw]'_redrawall
ID_Regen     [Re&gen]^C^C_regen
ID_Regenall  [Regen &All]^C^C_regenall
         [--]
ID_MnZoom    [->&Zoom]
ID_ZoomRealt [&Realtime]'_zoom ;
         [--]
ID_ZoomPrevi [&Previous]'_zoom _p
ID_ZoomWindo [&Window]'_zoom _w
ID_ZoomDynam [&Dynamic]'_zoom _d
ID_ZoomScale [&Scale]'_zoom _s
ID_ZoomCente [&Center]'_zoom _c
         [--]
ID_ZoomIn    [&In]'_zoom 2x
ID_ZoomOut   [&Out]'_zoom .5x
         [--]
ID_ZoomAll   [&All]'_zoom _all
ID_ZoomExten [<-&Extents]'_zoom _e
|
|
|
***POP4
**INSERT
ID_MnInsert  [&Insert]
ID_Ddinsert  [&Block...]^C^C_ddinsert
         [--]
ID_Xref      [E&xternal Reference...]^C^C_xref
ID_Image     [Raster &Image...]^C^C_image
         [--]
ID_3dsin     [&3D Studio...]^C^C_3dsin
ID_Acisin    [&ACIS Solid...]^C^C_acisin
ID_Dxbin     [Drawing &Exchange Binary...]^C^C_dxbin
ID_Wmfin     [&Windows Metafile...]^C^C_wmfin
ID_Psin      [Encapsulated &PostScript...]^C^C_psin
```

Customizing the Standard AutoCAD Menu 10-41

[--]
ID_Insertobj [&OLE Object...]^C^C_insertobj
[->Insert-Blocks]
  [Door1]^C^C_insert;d:/symbols/door1;\1.25;1.25;0
  [Door2]^C^C_insert;d:/symbols/door2;\1.25;1.25;0
  [--]
  [Window1]^C^C_inserd;d:/symbols/window1;\1.25;1.25;0
  [<-Window2]^C^C_insert;d:/symbols/window2;\1.25;1.25;0
  [--]

> **Note**
> [DOOR1]^C^CINSERT;D:/SYMBOLS/DOOR1;\1.25;1.25;
>
> *When you define the search path in the menu file, replace the back-slashes (\) by forward slashes (/). To load the file DOOR1 from the SYMBOLS subdirectory in the D drive the search path normally will be defined as D:\SYMBOLS\DOOR1. But the same statement in a menu file will be defined as D:/SYMBOLS/DOOR1. In the menu file back-slash is used for user input.*
>
> [->Insert-Blocks]
>
> *In this menu item the character -> defines a cascading submenu. When you select this item from the POP3 pull-down menu, AutoCAD will display the menu on the side of the pull-down menu.*
>
> [<-Window2]^C^C_insert;d:/symbols/window2;\1.25;1.25;0
>
> *In this menu item the character <- defines the last menu item in the cascading submenu.*

CURSOR MENUS

The cursor menus are similar to the pull-down menus, except that the cursor menu can contain only 499 item compared to 999 items in the pull-down menu. The section label of the cursor menu must be ***POP0. The cursor menus are displayed near or at the cursor location. Therefore, it can be used to provide a convenient and quick access to some of the frequently used commands. The following is a list of some of the features of the cursor menu:

1. The section label of the cursor menu is ***POP0. The menu bar title defined under this section label is not displayed in the menu bar.

2. On most systems, the menu bar title is not displayed at the top of the cursor menu. However, for compatibility reasons it is recommended to give a dummy menu bar title.

3. The cursor menu can be accessed through the $P0=* menu command only. This command can be issued by a menu item in another menu, such as the button menu, auxiliary menu, or the screen menu. The command can also be issued from an AutoLISP or ADS program.

4. A maximum of 499 menu items can be defined in the cursor menu. This includes the items that are defined in the cursor submenus. The menu items in excess of 499 are ignored.

5. The number of menu items that can be displayed on the screen depends on the system that you are using. If the cursor or the pull-down menu contains more items than what can be accommodated on the screen, the excess items are truncated. For example, if your system can display 21 menu items, then the menu items in excess of 21 are automatically truncated.

SUBMENUS

The number of items in a pull-down menu file can be so large that the items cannot be accommodated on one screen. For example, the maximum number of items that can be displayed on some of the display devices is 21. If the pull-down menu has more than 21 items, the menu items in excess of 21 are not displayed on the screen and cannot be accessed. The user can overcome this problem by using cascading menus or by swapping the pull-down submenus that let the user define smaller groups of items within a menu section. When a submenu is selected, it loads the submenu items and displays them on the screen.

Swapping Pull-down Menus

The pull-down menus that use AutoCAD's cascading feature are the most efficient and easy to write. The submenus follow a logical pattern that are easy to load and use without causing any confusion. It is strongly recommended to use the cascading menus whenever you need to write the pull-down or the cursor menus. However, AutoCAD provides the option to swap the submenus in the pull-down menus. These menus can sometimes cause distraction, because the original pull-down menu is completely replaced by the submenu when swapping the menus.

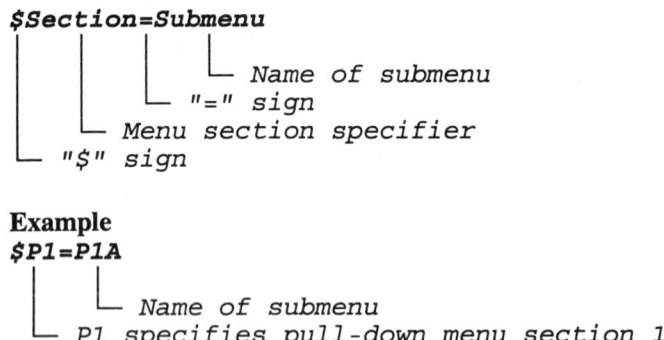

CUSTOMIZING IMAGE TILE MENUS

The image tile menus are extremely useful for inserting blocks and selecting a hatch pattern or a text font. You can also use the image tile menus to load an AutoLISP routine or a predefined macro. Therefore, the image tile menu is a powerful tool for customizing AutoCAD.

The image tile menus can be accessed from the pull-down, tablet, button, or screen menu. However, the image tile menus cannot be loaded by entering the command from the keyboard. When the user selects an image, a dialog box is displayed on the screen that contains **twenty images**. The names of the slide files associated with images appear on the left side of the dialog box with the scrolling bar that can be used to scroll the filenames. The title of the image tile menu is displayed at the top of the dialog box. When you activate the image tile menu an arrow appears on the screen that can be moved to select any image. You can select an image by selecting the slide filename from the dialog box and then selecting OK button from the dialog box or double-clicking on the slide file name.

When you select the slide file, AutoCAD highlights the corresponding image by drawing a rectangle around the image. You can also select an image by moving the arrow in the desired image and then pressing the pick button of the pointing device. The corresponding slide filename will be automatically highlighted and if you select OK button or double click on the image, the command associated with that menu item will be executed. You can cancel an image tile menu by pressing the CTRL and C keys or ESCAPE key on the keyboard, or selecting an image from the dialog box.

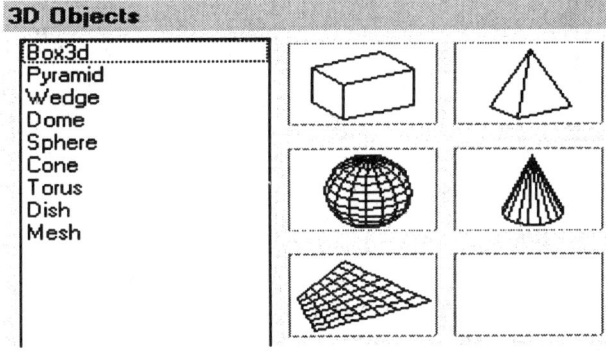

An image tile menu will work only if the system is configured so that the status line is not disabled. Otherwise, the image tile menus or the pull-down menus cannot be used. The image tile menu consists of a section label ***IMAGE. There is only one image tile menu section in a file and all the image tile menus are defined in this section.

```
***IMAGE
   │   └─ Section label for an image
   └─ Three asterisks (***) designate a section label
```

SUBMENUS

You can define an unlimited number of menu items in the image tile menu, but only 20 images will be displayed at a time. If the number of items exceeds 20, you can use the **Next** or **Previous buttons** of the dialog box to page through different pages of images. You can also define submenus that let the user define smaller groups of items within an image tile menu section. When a submenu is selected, it loads the submenu items and displays them on the screen.

10-44 Customizing AutoCAD

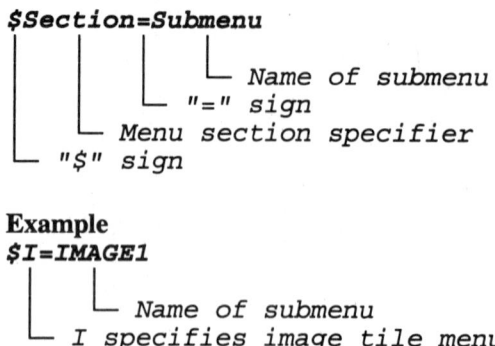

Example
```
$I=IMAGE1
     │    └─ Name of submenu
     └─ I specifies image tile menu
```

IMAGE TILE MENU ITEM LABELS

Like screen and pull-down menus, you can use the menu item labels in the pull-down menus. However, the menu item labels in the image tile menus use different formats and each format performs a particular function in the image tile menu. The menu item labels appear in the slide list box of the dialog box. The maximum number of characters that can be displayed in this box is 17. The characters in excess of 17 are not displayed in the list box. However, it does not affect the command that is defined with the menu item. The following are the different formats of the menu item labels:

[slidename]
In this menu item label format, **slidename** is the name of the slide that is displayed in the image. This name (slidename) is also displayed in the list box of the corresponding dialog box.

[slidename,label]
In this menu item label format, **slidename** is the name of the slide that is displayed in the image. However, unlike the previous format, the **slidename** is not displayed in the list box. It is the **label** text that is displayed in the list box. For example, if the menu item label is **[BOLT1,1/2-24UNC-3LG]**, **BOLT1** is the name of the slide and **1/2-24UNC-3LG** is the label that will be displayed in the list box.

[slidelib(slidename)]
In this menu item label format, **slidename** is the name of the slide in the slide library file **slidelib**. The slide (slidename) is displayed in the image and the slide filename (slidename) is also displayed in the list box of the corresponding dialog box.

[slidelib(slidename,label)]
In this menu item label format, **slidename** is the name of the slide in the slide library file **slidelib**. The slide (slidename) is displayed in the image and the **label** text is displayed in the list box of the corresponding dialog box.

[blank]
This menu item will draw a line that extends through the width of the list box. It also displays a blank image in the dialog box.

Customizing the Standard AutoCAD Menu 10-45

[label]
If the **label** text is preceded by a space, AutoCAD does not look for a slide. The **label** text is displayed in the list box only. For example, if the menu item label is [EXIT]^C, the label text (EXIT) will be displayed in the list box. If you select this item, the cancel command (^C) defined with the item will be executed. The **label** text is not displayed in the image of the dialog box.

Example 7

Write an image tile menu that can be accessed from POP11 for inserting the following blocks (see Figure 10-13):

BL1   BL4
BL2   BL5
BL3   BL6

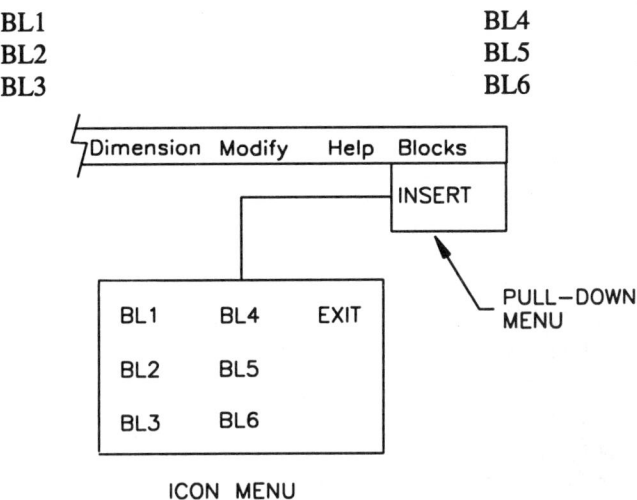

Figure 10-13 Commands defined in the image tile menu

Note
It is assumed that the slides have already been created and the names of the slides are the same as blocks.

Do not edit the ACAD.MNU file. Make a copy of the ACAD.MNU file and then edit the new file (CUSTOM.MNU).

Use your word processor to load the CUSTOM.MNU file and search for the ***IMAGE section label. The standard ACAD.MNU file does not have the POP10 section in the pull-down menu. Therefore, you can define POP10 just before ***IMAGE section label. Similarly, you can define the INSTBLK image tile menu after the image section label ***IMAGE. The following file is a partial listing of the CUSTOM.MNU file after editing the file.

    ***POP9
    **MODIFY
    ID_MnModify [&Modify]
    ID_Ai_propch [&Properties...]^C^C_ai_propchk

```
ID_Matchprop [&Match Properties]^C^C_matchprop
ID_MnObject  [->&Object]
ID_MnExterna [->&External Reference]
ID_Xbind     [&Bind...]^C^C_xbind
|
|
|
***POP10
**HELP
ID_MnHelp  [&Help]
ID_Help    [AutoCAD &Help Topics]'_help
|
|
|
***POP11
[BLOCKS]
[INSERT]^C^C$I=INSTBLK $I=*

***image
**INSTBLK
[INSERT CUSTOMIZED BLOCKS]
[BL1]^C^CINSERT;BL1;\1.0;1.0;0
[BL2]^C^CINSERT;BL2;\1.0;1.0;0
[BL3]^C^CINSERT;BL3;\1.0;1.0;0
[BL4]^C^CINSERT;BL4;\1.0;1.0;0
[BL5]^C^CINSERT;BL5;\1.0;1.0;0
[BL6]^C^CINSERT;BL6;\1.0;1.0;0
[ EXIT]^C^C
```

Note

[INSERT]^C^C$I=INSTBLK $I=*

In this menu item $I=INSTBLK loads the image submenu INSTBLK that has been defined in the image tile menu section. $I= forces the display of the dialog box with the images.*

```
[INSERT]^C^C$I=INSTBLK  $I=*
                |           |
                |           └── Displays dialog box with images
                └── Loads the submenu INSTBLK that
                    has been defined in the image tile menu
                    section
```

***IMAGE

*This is the section label of the Image tile menu. All image tile menus have to be defined in this section (***IMAGE) of the menu file.*

**INSTBLK

This is the submenu label. The name of the submenu is INSTBLK.

Customizing the Standard AutoCAD Menu 10-47

> **[INSERT CUSTOMIZED BLOCKS]**
> This menu item describes the contents of the image tile menu and it is displayed at the top of the dialog box.
>
> **[BL1]^C^CINSERT;BL1;\1.0;1.0;0**
> The BL1 that is inside the brackets is the name of the slide that is displayed in one of the images of the dialog box. ^C^C cancels the existing command twice and INSERT command inserts the block BL1 with the X,Y scale factor of 1.0 and the rotation angle of 0 degrees.

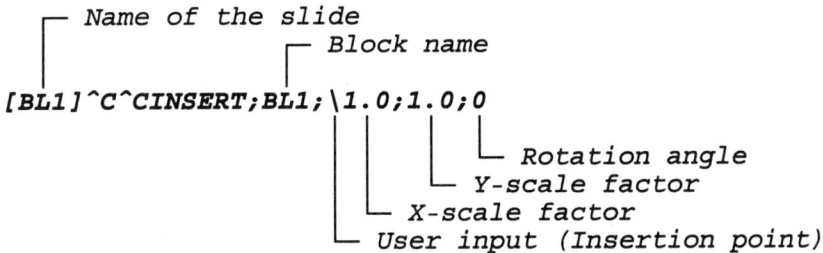

CUSTOMIZING THE SCREEN MENU

Like other menu sections, the screen menu section can be edited to modify or delete the existing commands or add new commands and submenus. Before writing a menu, it is very important to design a menu, know the exact sequence of AutoCAD commands, and know the prompts associated with the commands. You can edit the standard AutoCAD menu and arrange the commands in a way that provides the user an easy and quick access to most frequently used commands. A careful design will save quite some time in the long run. Therefore, it is strongly recommended to consider several possible alternatives and then select the one that is best suited for the job.

SUBMENUS

The number of items in the screen menu file can be so large that the items cannot be accommodated on one screen. For example, the maximum number of items that can be displayed on some of the display devices is 21. If the screen menu has more than 21 items, the menu items in excess of 21 are not displayed on the screen and therefore cannot be accessed. The user can overcome this problem by using submenus that let the user define smaller groups of items within a menu section. When a submenu is selected, it loads the submenu items and displays them on the screen.

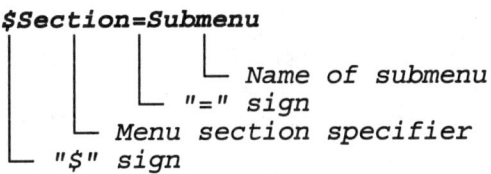

Example
```
$S=EDIT
  |   |
  |   └─ Name of submenu
  └─ S specifies screen menu section
```

Nested Submenus

When a submenu is activated, the current menu is copied to a stack. If you select another submenu, the submenu that was current will be copied or pushed to the top of the stack. The maximum number of menus that can be stacked is eight. If the stack size increases more than eight, the menu that is at the bottom of the stack is removed and forgotten. You can call the previous submenu by using the nested submenu call. The format of the call is:

```
$S=
 | | └─ "=" sign
 | └─ Screen menu specifier
 └─ "$" sign
```

The maximum number of nested submenu calls that AutoCAD can have is eight. Each time you call a submenu, this pops the last item off the stack and reactivates it.

> **Note**
> *To load the original menu (ACAD.MNU), load the menu file by using the MENU command.*
>
> **Command: Menu**
> Menu filename or . for none <SM1>: **ACAD**
>
> *If you need to use input from a keyboard or a pointing device use back slash "\". The system will pause for the user to enter data.*
>
> *There should be no space after the back-slash "\".*
>
> *The menu items, menu labels, and the command definition can be in upper case, lower case, or mixed.*
>
> *You can introduce spaces between the menu items to improve the readability of the menu file.*
>
> *If there are more items in the menu than the number of spaces available, the excess items are not displayed on the screen. For example, in the screen menu if the display device limits the number of items to 21, the items in excess of 21 will not be displayed on the screen, and are therefore inaccessible.*
>
> *If you configure AutoCAD and turn the screen prompt area off, you can increase the number of lines that can be displayed on the screen menu. On some devices it is 24 lines.*

Customizing the Standard AutoCAD Menu 10-49

Example 8

Edit the standard AutoCAD menu to add the commands as shown in Figure 10-14.

> **Note**
> *It is assumed that the image submenu **INSTBLK** has already been defined in the image tile menu section of the menu file.*
>
> *Do not edit the ACAD.MNU file. Make a copy of the ACAD.MNU file and then edit the new file (CUSTOM.MNU).*

You can use your word processor to load the menu file CUSTOM.MNU and search for ***SCREEN** section label. Add the new menu item at the end of the submenu **S** and then define the submenu **CUSTOM** and the menu items as shown in Figure 10-14. The following file is a partial listing of the **CUSTOM.MNU** file after editing:

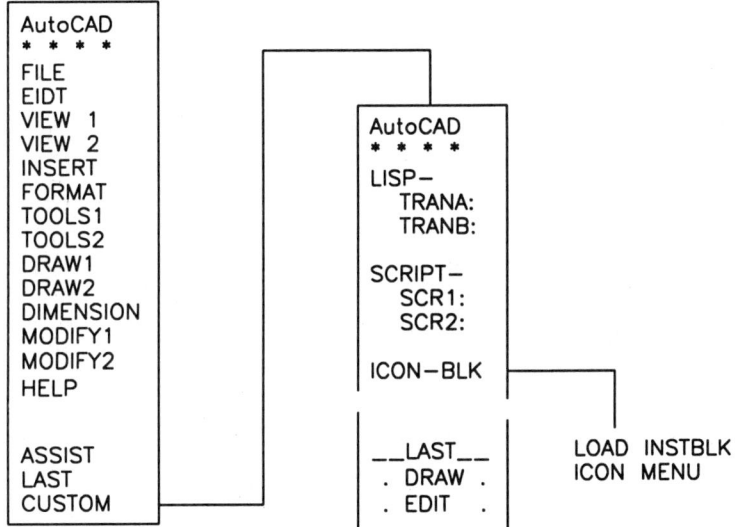

Figure 10-14 Modified screen menu

```
***IMAGE
**INSTBLK
[INSERT CUSTOMIZED BLOCKS]
[BL1]^C^CINSERT;BL1;\1.0;1.0;0
[BL2]^C^CINSERT;BL2;\1.0;1.0;0
[BL3]^C^CINSERT;BL3;\1.0;1.0;0
[BL4]^C^CINSERT;BL4;\1.0;1.0;0
[BL5]^C^CINSERT;BL5;\1.0;1.0;0
[BL6]^C^CINSERT;BL6;\1.0;1.0;0
```

[EXIT]^C^C

***SCREEN
**S
[AutoCAD]^C^C^P(ai_rootmenus) ^P
[* * * *]$S=ACAD.OSNAP
[FILE   ]$S=ACAD.01_FILE
[EDIT   ]$S=ACAD.02_EDIT
[VIEW 1]$S=ACAD.03_VIEW1
[VIEW 2]$S=ACAD.04_VIEW2
[INSERT]$S=ACAD.05_INSERT

**ASSIST 3
[Last   ]_l
[Previous]_p
[All   ]_all
[Cpolygon]_cp
[Wpolygon]_wp

[CUSTOM]^C^C$S=X $S=CUSTOM

**CUSTOM 3
[LISP-]
[TRANA:]^C^C(LOAD "TRANA");TRANA
[TRANB:]^C^C(LOAD "TRANB");TRANB
[ ]
[SCRIPT-]
[SCR1:]^C^CSCRIPT;SCR1
[SCR2:]^C^CSCRIPT;SCR2
[ ]
[IMAGE-BLK]^C^C$I=INSTBLK $I=*

Note
[CUSTOM]^C^C$S=X $S=CUSTOM *In this menu item $S=X loads the submenu X that has been defined in the screen menu section. $S=CUSTOM loads the submenu CUSTOM that has also been defined in the screen menu section of the menu file CUSTOM.MNU.*

[CUSTOM]^C^C$S=X $S=CUSTOM
                            └ *Loads submenu CUSTOM*
              └ *Loads submenu X*

**CUSTOM 3 *In this menu item CUSTOM is the name of the submenu. The 3 that follows the submenu name prints the menu items defined in the submenu CUSTOM from the line number 3. Nothing is printed on the first two lines, therefore the first two lines AutoCAD and * * * * stay on the screen.*

```
**CUSTOM 3
    │    └─ Start printing from line number 3
    └─ Submenu name
```

[IMAGE-BLK]^C^C$I=INSTBLK $I=* *In this menu item $I=INSTBLK loads the image submenu INSTBLK that has been defined in the image section of the menu file. $I=* forces the display of the new image tile menu on the screen.*

```
[IMAGE-BLK]^C^C$I=INSTBLK $I=*
              │            └─ Forces display of new
              │               image tile menu on screen
              └─ Loads the image submenu INSTBLK
```

REVIEW QUESTIONS

1. The AutoCAD menu file can have up to _____ sections.

2. Tablet menu can have up to _____ sections.

3. The section label is designated by _____.

4. The submenu label is designated by _____.

5. In a menu file you can use _____ to cancel the existing command.

6. Submenu names can be _____ characters long.

7. You _____ assign the same command to more than one block on the template.

8. AutoCAD's _____ command is used to configure the tablet menu template.

9. AutoCAD's _____ command is used to load a new menu.

10. You need to enter _____ points that are at _____ degrees to configure different tablet areas.

11. The commands are assigned to the buttons of the pointing device in the _____ order in which they appear in the buttons menu.

10-52 Customizing AutoCAD

12. The format of the command used for loading a submenu that has been defined in the screen menu section is _____.

13. The format of the command used for loading a submenu that has been defined in the pull-down menu section is _____.

14. The format of the command used for loading a submenu that has been defined in the image tile menu section is _____.

15. The command that is used to force the display of the current pull-down menu is _____.

EXERCISES

Exercise 3

Add the following commands to the TABLET1 section of the standard AutoCAD menu file ACAD.MNU. Figure 10-15 shows the layout of tablet area-1 of the template.

VIEW-POINTS

0,0,1	1,0,0	0,1,0
1,-1,1	1,1,1	-1,1,1

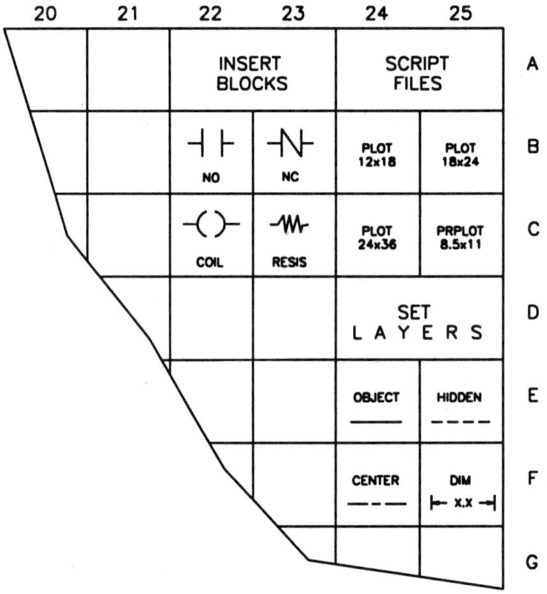

Figure 10-15 Commands assigned to tablet area-1

Exercise 4

Add the following commands to the TABLET1 section of the standard AutoCAD menu file ACAD.MNU. The layout of tablet area-1 is shown in Figure 10-16.

INSERT NO	PLOT 12x18	SETLAYER OBJ
INSERT NC	PLOT 18x24	SETLAYER HID
INSERT COIL	PLOT 24x36	SETLAYER CEN
INSERT RESIS	PRPLOT	SETLAYER DIM

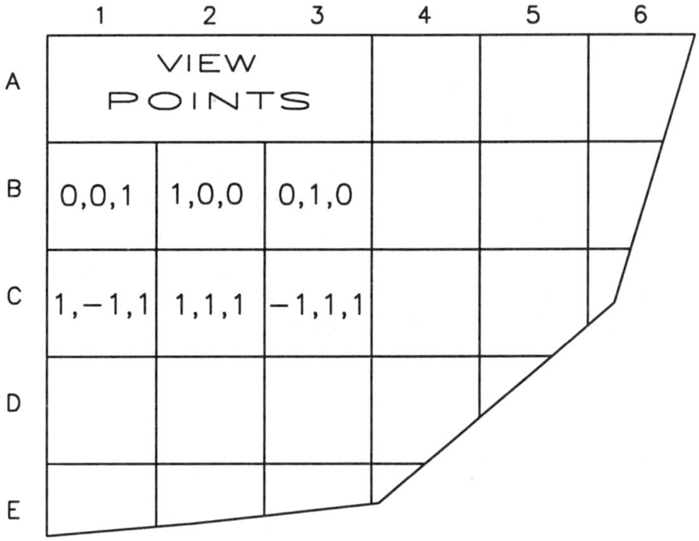

	1	2	3	4	5	6
A		VIEW POINTS				
B	0,0,1	1,0,0	0,1,0			
C	1,−1,1	1,1,1	−1,1,1			
D						
E						

Figure 10-16 Commands assigned to tablet area-1

Exercise 5

Write a button menu for the following AutoCAD commands (Figure 10-17). Add the commands to BUTTONS2 section of ACAD.MNU. The pointing device has 10 buttons; button number 1 is used for picking the points. The blocks are to be inserted with a scale factor of 1.00 and a rotation of 0 degrees (Filename BME1.MNU).

1. PICK BUTTON
2. Enter
3. CANCEL
4. OSNAPS
5. END PT
6. CENTER
7. NEAR
8. ZOOM Window
9. ZOOM Prev
10. PAN

10-54 Customizing AutoCAD

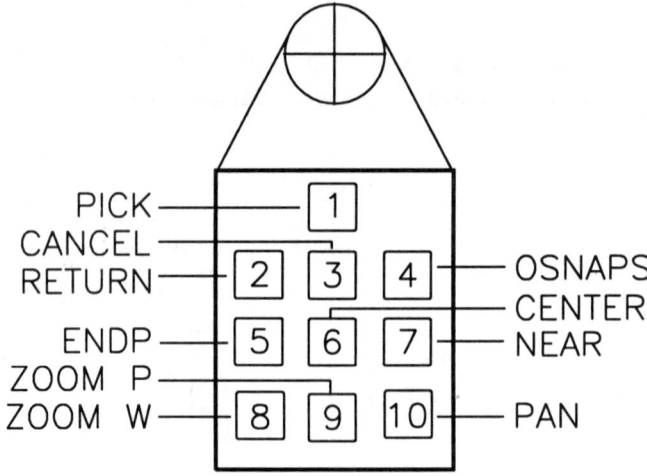

Figure 10-17 Commands assigned to different buttons of a pointing device

Exercise 6

Add the commands shown in Figure 10-18 to POP11 section of the standard AutoCAD menu ACAD.MNU

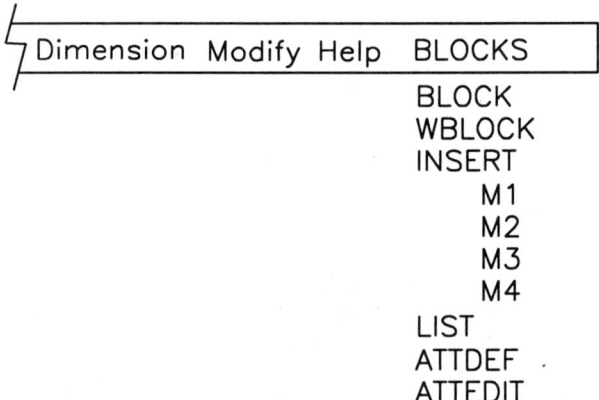

Figure 10-18 POP11 section of pull-down menu

Exercise 7

Write an image tile menu for inserting the following blocks that can be accessed through POP11 section. It is assumed that the blocks have already been created.

        SYMBOL-X   SYMBOL-Y   SYMBOL-Z
        LOGO-1   LOGO-2   LOGO-3
        TBLOCK-1   TBLOCK-2   TBLOCK-3

Chapter 11

Shapes and Text Fonts

Learning objectives

After completing this chapter, you will be able to:
♦ *Write shape files.*
♦ *Use vector length and direction encoding to write shape files.*
♦ *Compile and load shape/font files.*
♦ *Use special codes to define a shape.*
♦ *Write text font files.*

SHAPE FILES

AutoCAD provides a facility to define shapes and text fonts. These files are ASCII files with the extension **.SHP**. You can write these files using any text editor like Notepad.

Shape files contain information about the individual elements that constitute the shape of an object. The basic objects that are used in these files are lines and arcs. You can define any shape using these basic objects, and then insert them anywhere in a drawing. The shapes are easy to insert, and they take less disk space than blocks. However, there are some disadvantages to using shapes. For example, you cannot edit a shape or change it. Blocks, on the other hand, can be edited after exploding them with the AutoCAD EXPLODE command.

SHAPE DESCRIPTION

Shape description consists of the following two parts: **a header and a shape specification.**

Header

The header line has the following format:

    *SHAPE NUMBER, DEFBYTES, SHAPE NAME

```
*201,21,HEXBOLT
  |   |    |
  |   |    └─ Shape name
  |   └─ Number of data bytes in shape specification
  └─ Shape number
```

11-2 Customizing AutoCAD

Every header line starts with an asterisk (*), followed by the **SHAPE NUMBER**. The shape number is any number between 1 and 255 in a particular file, and these numbers cannot be repeated within the same file. However, the numbers can be repeated in another shape file with a different name. **DEFBYTES** is the number of data bytes used by the shape specification and includes the terminating zero. **SHAPE NAME** is the name of a shape, in uppercase letters. The name is ignored if the letters are lowercase. The file must not contain two shapes with the same name.

Shape Specification

The shape specification line contains the complete definition of the shape of an object. The shape is described with special codes, hexadecimal numbers, and decimal numbers. A hexadecimal number is designated by a leading zero (012), and a decimal number is a regular number without a leading zero (12). The data bytes are separated by a comma (,). The maximum number of data bytes is 2,000 bytes per shape, and in a particular shape file there can be more than one shape. The shape specification can have multiple lines. You should define the shape in some logical blocks and enter each block on a separate line. This makes it easier to edit and debug the files. The number of characters on any line must not exceed 80. The shape specification is terminated with a zero.

VECTOR LENGTH AND DIRECTION ENCODING

Figure 11-1 shows the vector direction codes. All vectors in this figure have the same length specification. The diagonal vectors have been extended to match the closest orthogonal vector. Let us assume that the endpoint of vector 0 is two grid units from the intersection point of vectors. The endpoint of vector 1 is one grid directly above the endpoint of vector 0. Therefore, the angle of vector 1 is 26.565 degrees ($Tan^{-1} 1/2 = Tan^{-1} 0.5 = 26.565$). Similarly, vector 2 is at 45 degrees ($Tan^{-1} 2/2 = Tan^{-1} 1 = 45$).

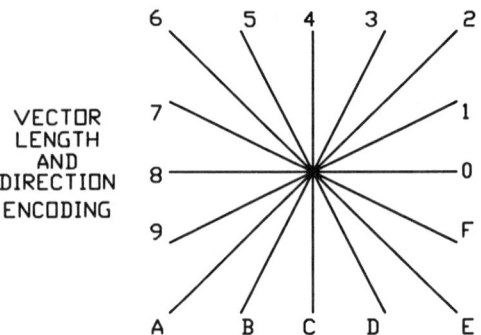

Figure 11-1 Vector length and direction encoding

All these vectors have the same magnitude or length specification. In other words, although their actual lengths vary, they are all considered one unit in definition. To define a vector you need its magnitude and direction. That means each shape specification byte contains vector length and a direction code. The maximum length of the vector is 15 units. Example 1 illustrates the use of vectors.

Example 1

Write a shape file for the resistor shown in Figure 11-2. The name of the file is **SH1.SHP**, and the shape name is RESIS.

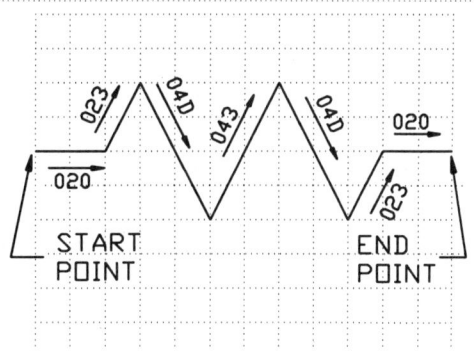

The following two lines define the shape file for the given resistor:

*201,8,RESIS
020,023,04D,043,04D,023,020,0

The first line is the **header line**; the second line is the **shape specification**.

Figure 11-2 Resistor

Header Line
*201,8,RESIS

*201 is the shape number, and 8 is the number of data bytes contained in the shape specification line. RESIS is the name of the shape.

Shape Specification
```
020,023,04D,043,04D,023,020,0
    │││
    ││└─ Direction code
    │└── Vector length
    └─── Hexadecimal notation
```

Each data byte in this line, except the terminating zero, has three elements. The first element (0) is the hexadecimal, the second is the length of the vector, and the third element is the direction code. For the first data byte, 020, the length of the vector is 2, and the direction is along the direction vector 0. Similarly, for the second data byte, 023, the first element, 0, is for hexadecimal; the second element, 2, is the length of the vector; and the third element, 3, is the direction code for the vector.

COMPILING AND LOADING SHAPE/FONT FILES

You can compile the shape file or the font file by using the COMPILE command.

    Command: **COMPILE**

When you enter this command, AutoCAD will display the **Select Shape or Font File** dialog box (Figure 11-3). From this dialog box, select the shape file that you want to compile.

11-4 Customizing AutoCAD

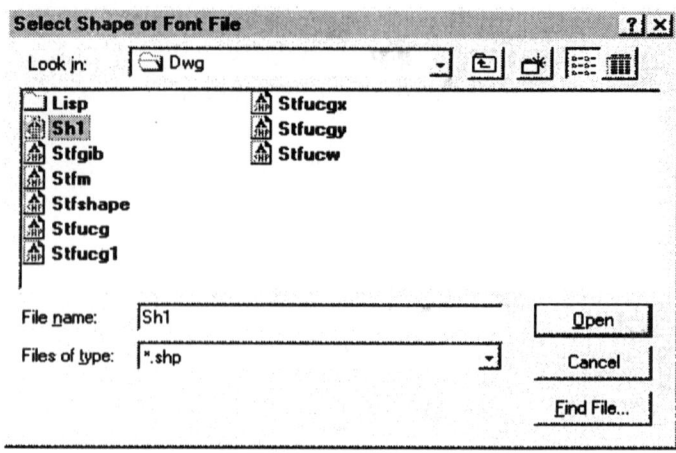

Figure 11-3 Select Shape or Font File dialog box

If the system variable **FILEDIA** is set to 0, you can enter the shape file name from the command line as follows:

Command: **COMPILE**
Enter NAME of shape file: **SH1**

AutoCAD will compile the file, and if the compilation process is successful, the following prompt will be displayed on the screen:

Compilation successful
Output file name.shx contains nn bytes

For Example 1, the name of the compiled output file is **SH1.SHX** and the number of bytes is 49. This is the file that is loaded when you use the AutoCAD LOAD command to load a shape. If AutoCAD encounters an error in compiling a shape file, an error message will be displayed, indicating the type of error and the line number where the error occurred.

To insert a shape in the drawing, you have to be in the drawing editor, and then use the LOAD command to load the shape file. You can select the file from the Select Shape File dialog box (Figure 11-4) or enter the name at the Command: prompt (FILEDIA=0).

Command: **LOAD**
Name of shape file to load (or ?): *Name of file.* **SH1**

SH1 is the name of the shape file for Example 1. Do **not** include the extension **.SHX** with the name because AutoCAD automatically assumes the extension. If the shape file is present, AutoCAD will display the shape names that are loaded. To insert the loaded shapes, use the AutoCAD SHAPE command:

Shapes and Text Fonts 11-5

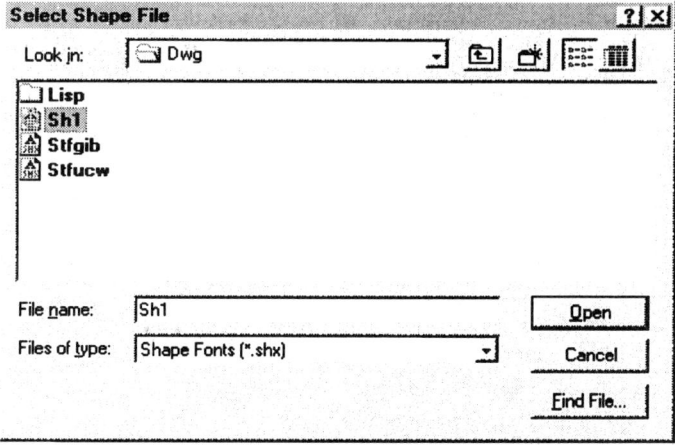

Figure 11-4 Select Shape File dialog box

Command: **SHAPE**
Shape name (or ?) <default>: *Shape name.*
Start point: *Shape origin.*
Height <1.0>: *Number or point.*
Rotation angle <0.0>: *Number or point.*

For Example 1, the shape name is RESIS. After you enter the information about the start point, height, and rotation, the shape will be displayed on the screen (Figure 11-5).

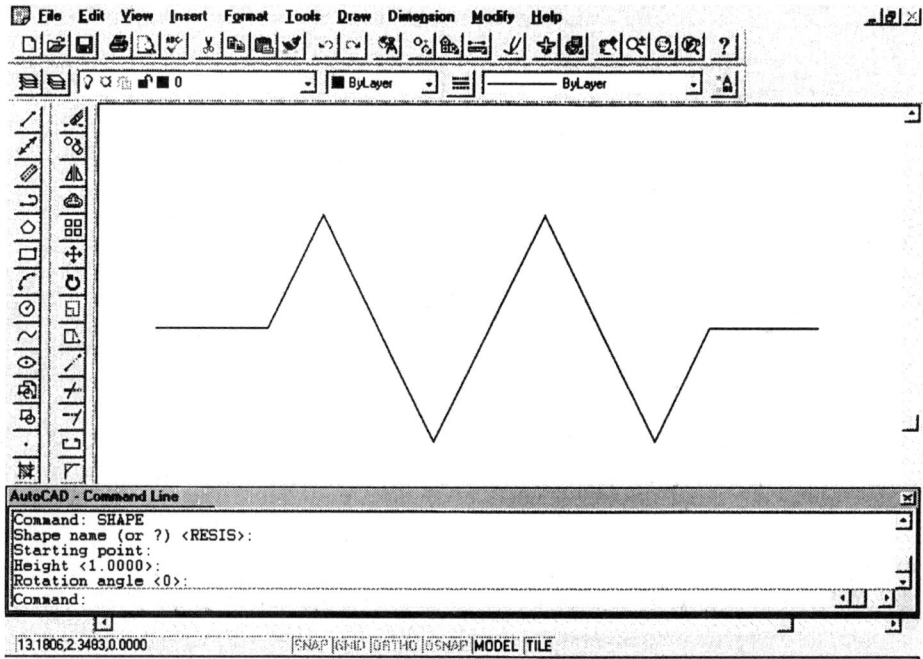

Figure 11-5 The shape (RESIS) inserted in the drawing

SPECIAL CODES

Generating shapes with the direction vectors has certain limitations. For example, you cannot draw an arc or a line that is not along the standard direction vectors. These limitations can be overcome by using special codes that add flexibility and give you better control over the shapes you want to create.

Standard Codes

000	End of shape definition
001	Activate draw mode (pen down)
002	Deactivate draw mode (pen up)
003	Divide vector lengths by next byte
004	Multiply vector lengths by next byte
005	Push current location from stack
006	Pop current location from stack
007	Draw subshape numbers given by next byte
008	X-Y displacement given by the next two bytes
009	Multiply X-Y displacement, terminated by (0,0)
00A or 10	Octant arc defined by next two bytes
00B or 11	Fractional arc defined by next five bytes
00C or 12	Arc defined by X-Y displacement and bulge
00D or 13	Multiple bulge-specified arcs
00E or 14	Process next command only if vertical text style

Code 000: End of Shape Definition

This code marks the end of a shape definition.

Code 001: Activate Draw Mode

This code turns the draw mode on. When you start a shape, the draw mode is on, so you do not need to use this code. However, if the draw mode has been turned off, you can use code 001 to turn it on.

Code 002: Deactivate Draw Mode

This code turns the draw mode off. It is used when you want to move the pen without drawing a line.

```
  1       2       3       4
```

Let's say the distance from point 1 to point 2, from point 2 to point 3, and from point 3 to point 4 is 2 units each. The shape specification for this line is:

020,002,020,001,020,0

The first data byte, 020, generates a line 2 units long along direction vector 0. The second data byte, 002, deactivates the draw mode; and the third byte, 020, generates a blank line 2 units long.

The fourth data byte, 001, activates the draw mode; and the next byte, 020, generates a line that is 2 units long along direction vector 0. The last byte, 0, terminates the shape description.

Example 2

Write a shape file to generate the character "G" as shown in Figure 11-6.

You can use any text editor to write a shape file. The name of the file is **CHRGEE** and the shape name is GEE. **In the following file, the line numbers are not a part of the file; they are for reference only.**

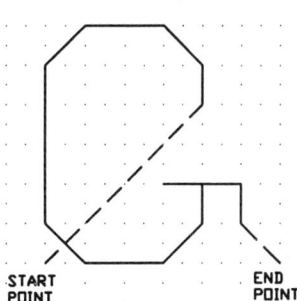

Figure 11-6 Shape of the character "G"

*215,20,GEE	1
002,042,	2
001,014,016,028,01A,	3
04C,01E,020,012,014,	4
002,018,	5
001,020,01C,	6
002,01E,0	7

Explanation

Line 1
*215,20,GEE
The first data byte contains an asterisk (*) and shape number 215. The second data byte is the number of data bytes contained in the shape specification, including the terminating 0. GEE is the name of shape.

Line 2
002,042,
The data byte 002 deactivates the draw mode (pen up), and the next data byte defines a vector 4 units long along direction vector 2.

Line 3
001,014,016,028,01A,
The data byte 001 activates the draw mode (pen down), and 014 defines a vector that is 1 unit long at 90 degrees (direction vector 4). The data byte 016 defines a vector that is 1 unit long along direction vector 6. The data byte 028 defines a vector that is 2 units long along direction vector 8 (180 degrees). The data byte 01A defines a unit vector along direction vector A.

Line 4
04C,01E,020,012,014,

The data byte 04C defines a vector that is 4 units long along direction vector C. The data byte 01E defines a direction vector that is 1 unit along direction vector E. The data byte 020 defines a direction vector that is 2 units long along direction vector 0 (0 degrees). The data byte 012 defines a direction vector that is 1 unit long along direction vector 2. Similarly, 014 defines a vector that is 1 unit long along direction vector 4.

Line 5
002,018,
The data byte 002 deactivates the pen (pen up), and 018 defines a vector that is 1 unit long along direction vector 8.

Line 6
001,020,01C,
The data byte 001 activates the pen (pen down), and 020 defines a vector that is 2 units long along direction vector 0. The data byte 01C defines a vector that is 1 unit long along direction vector C.

Line 7
002,01E,0
The data byte 002 deactivates the pen, and the next data byte, 01E, defines a vector that is 1 unit long along direction vector E. The data byte 0 terminates the shape specification.

Code 003: Divide Vector Lengths by Next Byte

This code is used if you want to divide a vector by a certain number. In Example 2, if you want to divide the vectors by 2, the shape description can be written as:

003,2,020,002,020,001,020,0

The first byte, 003, is the division code, and the next byte, 2, is the number by which all the remaining vectors are divided. The length of the lines and the gap between the lines will be equal to 1 unit now:

Also, the scale factors are cumulative within a shape. For example, if we insert another code, 003, in the preceding shape description, the length of the last vector, 020, will be divided by 4 (2 x 2):

```
003,2,020,002,020,001,003,2,020,0
    │                    │
    │                    └─ All the remaining
    │                       vectors are divided by 4 (2 x 2)
    └─ All the vectors
       are divided by 2
```

Here is the output of this shape file:

Code 004: Multiply Vector Lengths by Next Byte

This code is used if you want to multiply the vectors by a certain number. It can also be used to reverse the effect of code 003.

```
003,2,020,002,020,001,004,2,020,0
```

- `003,2` — Divides all the vectors on the right by 2
- `004,2` — Multiplies all the vectors on the right by 2

In this example, the code 003 divides all the vectors to the right by 2. Therefore, a vector that was 1 unit long will be 0.5 units long now. The second code, 004, multiplies the vectors to the right by 2. We know the scale factors are cumulative; therefore, the vectors that were divided by 2 earlier will be multiplied by 2 now. Because of this cumulative effect, the length of the last vector remains unchanged. This file will produce the following shape:

Codes 005 and 006: Location Save/Restore

Code 005 lets you save the current location of the pen, and code 006 restores the saved location. The following example illustrates the use of codes 005 and 006.

Example 3

Figure 11-7(a) shows three lines that are unit vectors and intersect at one point. After drawing the first line, the pen has to return to the origin to start a second vector. This is done using code 005, which saves the starting point (origin) of the first vector, and code 006, which restores the origin. Now, if you draw another vector, it will start from the origin. Since there are three lines, you need three code 005s and three code 006s. The following file shows the header line and the shape specification for generating three lines as shown in Figure 11-7(a):

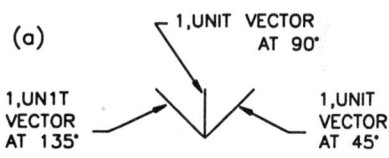

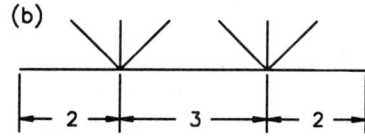

Figure 11-7 (a) Three unit vectors intersecting at a point; (b) repeating predefined subshapes

```
*210,10,POP1
005,005,005,012,006,014,006,016,006,0
```

- Saves origin three times
- Generates first vector
- Restores origin
- Generates second vector
- Restores origin

The number of saves (code 005) has to equal the number of restores (code 006). If the number of saves (code 005) is more than the number of restores (code 006), AutoCAD will display the following message when the shape is drawn:

Position stack overflow in shape (shape number)

Similarly, if the number of restores (code 006) is more than the number of saves (code 005), the following message will be displayed:

Position stack underflow in shape (shape number)

The maximum number of saves and restores you can use in a particular shape definition is four.

Code 007: Subshape

You can define a subshape like a subroutine in a program. To reference a subshape, the subshape code, 007, has to be followed by the shape number of the subshape. The subshape has to be defined in the same shape file, and the shape number has to be from 1 to 255.

```
              ┌─ Shape number
              │
*210,10,POP1
005,005,005,012,006,014,006,016,006,0
*211,8,SUB1
020,007,210,030,007,210,020,0
          │   └─ Shape number
          └─ Subshape reference
```

The shape that this example generates is shown in Figure 11-7(b).

Code 008: X-Y Displacement

In the previous examples, you might have noticed some limitations with the vectors. As mentioned earlier, you can draw vectors only in the 16 predefined directions, and the length of a vector cannot exceed 15 units. These restrictions make the shape files easier and more efficient, but at the same time, they are limiting. Therefore, codes 008 and 009 allow you to generate nonstandard vectors by entering the displacements along the X and Y directions. The general format for Code 008 is:

008, XDISPLACEMENT, YDISPLACEMENT
or
008, (XDISPLACEMENT, YDISPLACEMENT)

X and Y displacements can range from +127 to -128. Also, a positive displacement is designated by a positive (+) number, and a negative displacement is designated by a negative (-) number. The leading positive sign (+) is optional in a positive number. The parentheses are used to improve readability, but they have no effect on the shape specification.

Code 009: Multiple X-Y Displacements

Whereas code 008 allows you to generate nonstandard vectors by entering a single X and Y displacement, code 009 allows you to enter multiple X and Y displacements. It is terminated by a pair of 0 displacements (0,0). The general format is:

009,(XDISPL, YDISPL), (XDISP, YDISPL), . . ., (0,0)

Code 00A or 10: Octant Arc

If you divide 360 degrees into eight equal parts, each angle will be 45 degrees. Each 45-degree angle segment is called an **octant**, and the two lines that contain an octant are called an **octant boundary**. The octant boundaries are numbered from 0 to 7, as shown in Figure 11-8. The general format is:

```
                  ┌─ Hexadecimal notation
                  │
  10,(R,+/-0SN)
    │ │  │ ││└─ Number of octants
    │ │  │ │└── Starting octant boundary
    │ │  │ └─── Defines direction, + Counterclockwise, - Clockwise
    │ └──────── Radius of arc
```

10,(3,-043)

The first number, 10, is the code 00A for the octant arc. The second number, 3, is the radius of the octant arc. The negative sign indicates that the arc is to be generated in a clockwise direction. If it is positive (+) or if there is no sign, the arc will be generated in a counterclockwise direction. Zero is the hexadecimal notation, and the following number, 4, is the number of the octant boundary where the octant arc will start. The next element, 3, is the number of octants that this arc will extend. This example will generate the arc in Figure 11-9. The following is the listing of the shape file that will generate the shape shown in Figure 11-9:

*214,5,FOCT1
001,10,(3,-043),0

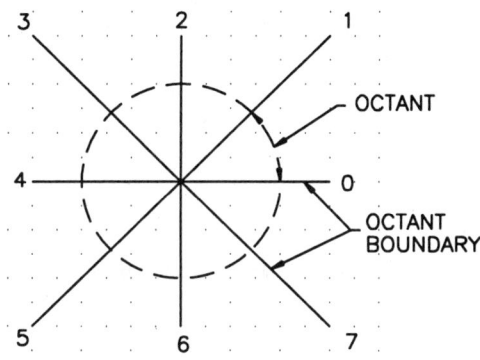

Figure 11-8 Octant boundaries

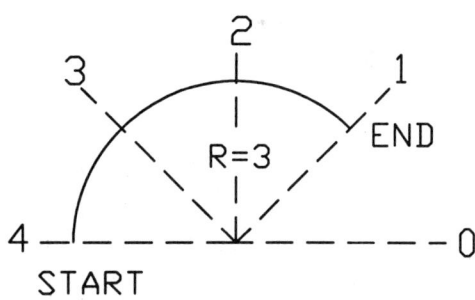

Figure 11-9 Octant arc

Code 00B or 11: Fractional Arc

You can generate a nonstandard fractional arc by using code 00B or 11. This code will allow you to start and end an arc at any angle. The definition uses five bytes, and the general format is:

11,(START OFFSET, END OFFSET, HIGHRADIUS, LOWRADIUS, +/-0SN)

The START OFFSET represents how far from an octant boundary the arc starts, and the END OFFSET represents how far from an octant boundary the arc ends. The HIGHRADIUS is zero if the radius is equal to or less than 255 units, and the LOWRADIUS is the radius of the arc. The positive (+) or negative (-) sign indicates, respectively, whether the arc is drawn counterclockwise or clockwise. The next element, S, is the number of the octant where the arc starts, and element N is the number of octants the arc goes through. The following example illustrates the fractional arc concept.

Example 4

Draw a fractional arc of radius 3 units that starts at a 20-degree angle and ends at a 140-degree angle (counterclockwise).

The solution involves the following steps:

1. Find the nearest octant boundary whose angle is less than 140 degrees. The nearest octant boundary is the number 4 octant boundary, whose angle is 135 degrees (3 * 45 = 135).

2. Calculate the end offset to the nearest whole number (integer):
   Start offset = (140 - 135) * 256/45 = 28.44 = 28

3. Find the nearest octant boundary whose angle is less than 20 degrees. The nearest octant boundary is 0 and its angle is 0 degrees.

4. Calculate the start offset to the nearest whole number:
   End offset = (20 - 0) * 256/45 = 113.7 = 114

5. Find the number of octants the arc passes through. In this example, the arc starts in the first octant and ends in the fourth octant. Therefore, the number of octants the arc passes through is four (counterclockwise).

6. Find the octant where the arc starts. In this example, it starts in the 0 octant.

7. Substitute the values in the general format of the fractional arc:

    11,(114,28,0,3,004)

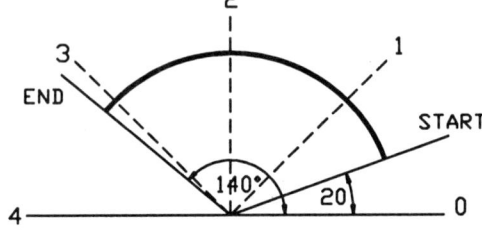

Figure 11-10 Fractional arc

The following shape file will generate the fractional arc shown in Figure 11-10:

*221,8,FOCT2
001,11,(114,28,0,3,004),0

Code 00C or 12: Arc Definition by Displacement and Bulge

Code C can be used to define an arc by specifying the displacement of the endpoint of an arc and the bulge factor. X and Y displacements may range from -127 to +127, and the bulge factor can also range from -127 to +127. A semicircle will have a bulge factor of 127, and a straight line will have a bulge factor of 0. If the bulge factor has a negative sign, the arc is drawn clockwise.

```
Bulge factor = ((2 * H)/D) * 127
                      │   └─ Displacement
                      └─ Height of arc
```

For a semicircle, 2H = D
Therefore, bulge = (D/D) * 127 = 127
For a straight line, H = 0
Therefore, bulge = (0/D) * 127 = 0

In Figure 11-11, the distance between the start point and the endpoint of an arc is 4 units, and the height is 1 unit. Therefore, the bulge can be calculated by substituting the values in the previously mentioned relation:

Bulge = (2 * 1/4) * 127 = 63.5 = 63
(integer)

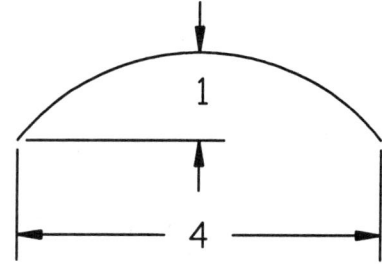

Figure 11-11 Calculating bulge

The following shape description will generate the arc shown in Figure 11-11.

```
*213,5,BULGE1
12,(4,0,-63),0
      │ │  └─ Bulge factor
      │ └─ Negative (-), generates clockwise arc
      │ └─ Y displacement
      └─ X displacement
```

Code 00D or 13: Multiple Bulge-Specified Arcs

Code 00D or 13 can be used to generate multiple arcs with different bulge factors. It is terminated by a (0,0). The following shape description defines the arc configuration of Figure 11-12:

```
*214,16,BULGE2
13,(4,0,-111),
(0,4,63),
(-4,0,-111),
(0,-4,63),(0,0),0
```

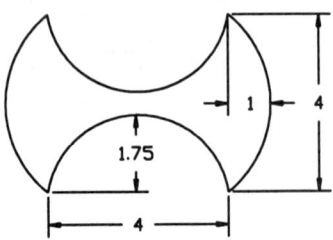

Figure 11-12 Different arc configuration

Code 00E or 14: Flag Vertical Text

Code 00E or 14 is used when the same text font description is to be used in both the horizontal and vertical orientations. If the text is drawn in a horizontal direction, the vector next to code 14 is ignored. If the text is drawn in a vertical position, the vector next to code 14 is not ignored. This lets you generate text in a vertical or horizontal direction with the same shape file.

For horizontal text, the start point is the lower left point, and the endpoint is on the lower right. In vertical text, the start point is at the top center, and the endpoint is at the bottom center of the text, as shown in Figure 11-13. At first, it appears that you need two separate shape files to define the shape of a horizontal and a vertical text. However, with code 14 you can avoid the dual shape definition.

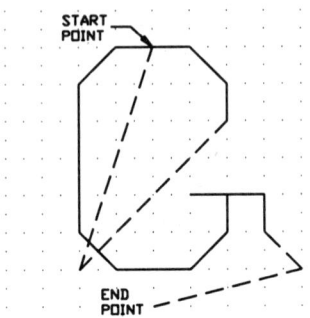

Figure 11-13 Pen movement for generating the character "G"

Figure 11-13 shows the pen movements for generating the text character "G". If the text is horizontal, the line that is next to code 14 is automatically ignored. However, if the text is vertical, the line is not ignored, resulting in resetting the start and endpoints appropriately for vertically aligned text.

```
1*15,28,FLAG
002,14,              ─── If text is horizontal, code 14
008,(-2,-6),             automatically ignores next line.
042,001,                 008, (-2,-6),
014,016,028,01A,
04C,01E,020,012,014,
002,018,
001,020,01C,
002,01E,
14,                  ─── If text is horizontal, code 14
008,(-4,-1),             automatically ignores next line.
0                        008, (-4,-1),
```

Example 5

Write a shape file for a hammer as shown in Figure 11-14a. (The name of the shape file is HMR.SHP and the shape name is HAMMER.) The following file defines the shape of a hammer. The line numbers are not a part of the file; they are for reference only.

*204,34,HAMMER	1
003,22	2
002,8,(2,-1),	3
001,024	4
8,(-1,4),	5
00A,(1,004),	6
8,(-1,-4),06C,	7
00C,(4,0,63),	8
044,8,(17,-1),	9
00C,(0,4,63),	10
8,(-17,-1),0	11

Explanation

Line 1

***204,34,HAMMER**

This is the header line that consists of shape number (204), number of data bytes in the shape specification (34), and name of the shape (HAMMER).

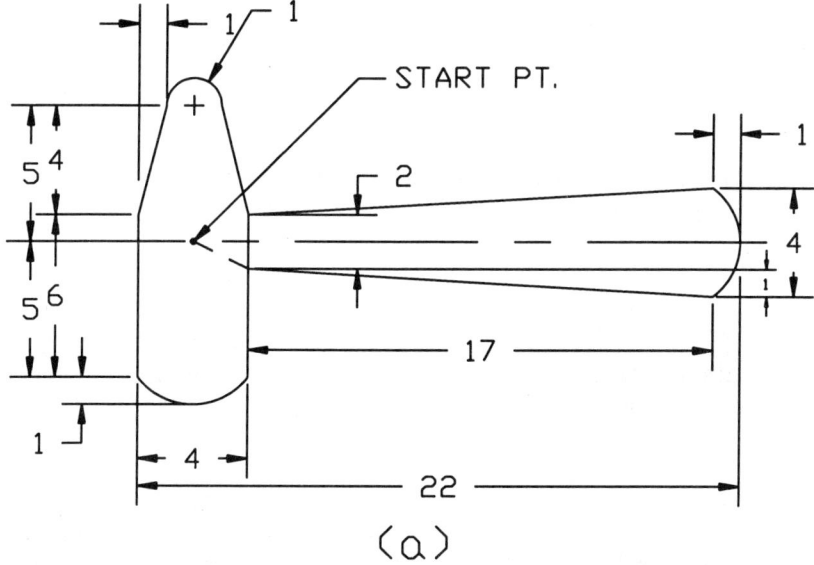

Figure 11-14a Dimensions of hammer in units

11-16 Customizing AutoCAD

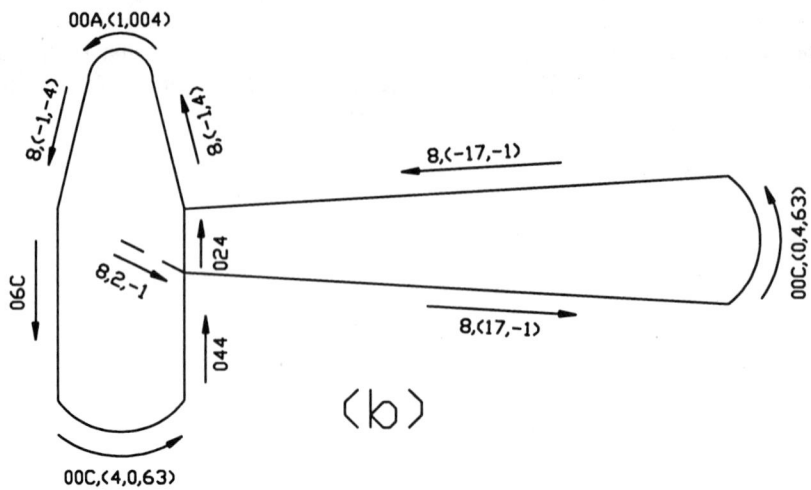

Figure 11-14b Pen movements for generating the shape of a hammer

Line 2
003,22
Data byte 003 has been used to divide the vectors by the next data byte, 22. This reduces the hammer shape to a unit size that facilitates the scaling operation when you insert the shape in a drawing.

Line 3
002,8,(2,-1),
Data byte 002 deactivates the pen (pen up), and the next data byte (code 008) defines a vector that has X-displacement of 2 units and Y-displacement of 1 unit. No line is drawn because the pen is deactivated.

Line 4
001,024
Data byte 001 activates the pen (pen down) and the next data byte defines a vector that is 2 units long along the direction vector 2.

Line 5
8,(-1,-4),
The first data byte, 8 (code 008), defines a vector whose X-displacement and Y-displacement is given by the next two data bytes. The X-displacement of this vector is -1 and the Y-displacement is -4 units.

Line 6
00A,(1,004),
Data byte 00A defines an octant arc that has a radius of 1 unit as defined by the next data byte. The first element (0) of data byte 004 is a hexadecimal notation. The second element (0) defines the starting octant of the arc, and the third element (4) defines the ending octant of the arc.

Line 7
8,(-1,-4),06C,
Data byte 8 (code 008) defines a vector that has X-displacement of -1 unit, and Y-displacement of -4 units. The next data byte defines a vector that is 6 units long along the direction vector C.

Line 8
00C,(4,0,63),
The first data byte (00C) defines an arc that has X-displacement of 4 units, Y-displacement of 0 units, and a bulge factor of 63.

$$\begin{aligned}\text{Bulge factor} &= (2 * H)/D * 127 \\ &= (2 * 1)/4 * 127 \\ &= 63.5 \\ &= 63 \text{ (integer)}\end{aligned}$$

Line 9
044,8,(17,-1),
The first data byte defines a vector that is 4 units long along the direction vector 4. The second data byte (8) defines a vector that has X-displacement of 17 units and Y-displacement of -1 unit.

Line 10
00C,(0,4,63),
The first data byte (00C) defines an arc that has X-displacement of 0 units, Y-displacement of 4 units, and a bulge factor of 63.

$$\begin{aligned}\text{Bulge factor} &= (2 * H)/D * 127 \\ &= (2 * 1)/2 * 127 \\ &= 63.5 \\ &= 63 \text{ (integer)}\end{aligned}$$

Line 11
8,(-17,-1),0
The first data byte (8) defines a vector that has X-displacement of -17 units and Y-displacement of -1 unit. The data byte 0 terminates the shape definition.

TEXT FONT FILES

In addition to shape files, AutoCAD provides a facility to create new text fonts. After you have created and compiled a text font file, text can be inserted in a drawing like regular text and using the new font. These text files are regular shape files with some additional information describing the text font and the line feed. The following is the general layout of the text font file:

Text font description
Line feed
Shape definition

Text Font Description

The text font description consists of two lines:

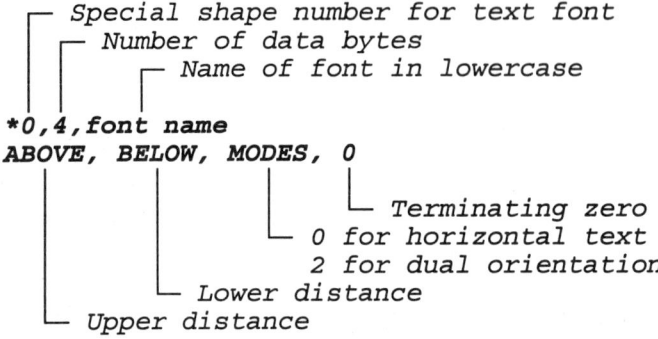

For example, if you are writing a shape definition for an uppercase M, the text font description would be:

    *0,4,ucm
    10,4,2,0

In the first line, the first data byte (0) is a special shape number for the text font, and every text font file will have this shape number. The next data byte (4) is the number of data bytes in the next line, and ucm is the shape name (name of font). The shape names in all text font files should be lowercase so that the computer does not have to save the names in memory. You can still reference the shape names for editing.

In the second line, the first data byte (10) specifies the height of an uppercase letter above the baseline. For example, in Figure 11-15 (page 11-20), the height of the letter M above the baseline is 10 units. The next data byte (4) specifies the distance of lowercase letters below the baseline. AutoCAD uses this information to scale the text automatically. For example, if you enter the height of text as 1 unit, the text will be 1 unit, although it was drawn 10 units high in the text font definition. The third data byte (2) defines the mode. It can have one of only two values, 0 or 2. If the text is horizontal, the mode is 0; if the text has dual orientation (horizontal and vertical), the mode is 2. The fourth data byte (0) is the zero that terminates the definition.

Line Feed

The line feed is used to space the lines so that characters do not overlap and so that a desired distance is maintained between the lines. AutoCAD has reserved the ASCII number 10 to define the line feed.

```
          ┌── (10) ASCII number reserved for line feed
          │   ┌── (5) Number of data bytes in shape specification
          │   │   ┌── (1f) Shape name
          │   │   │
         *10,5,1f
         2,8,(0,-14),0
         │   │     │
         │   │     └── Terminating zero
         │   └── Line feed of 14 units
         └── Deactivate pen (pen up)
```

In the first line, the first data byte (10) is the shape number reserved for line feed, and the next data byte (5) is the number of characters in the shape specification. The data byte 1f is the name of the shape.

In the second line, the first data byte (2) deactivates the pen. The next data byte (8) is a special code, 008, that defines a vector by X displacement and Y displacement. The third and fourth data bytes (0,-14) are the X displacement and Y displacement of the displacement vector; they produce a line feed that is 14 units below the baseline. The fifth data byte (0) is the zero that terminates the shape definition.

Shape Definition

The shape number in the shape definition of the text font corresponds to the ASCII code for that character. For example, if you are writing a shape definition for an uppercase M, the shape number is 77.

```
     *77,50,ucm
      │   │  │
      │   │  └── Shape name
      │   └── Number of data bytes
      └── Shape number - ASCII code of uppercase "M"
```

The ASCII codes can be obtained from the ASCII character table, which gives the ASCII codes for all characters, numbers, and punctuation marks.

32	space	56	8	80	P	104	h
33	!	57	9	81	Q	105	i
34	"	58	:	82	R	106	j
35	#	59	;	83	S	107	k
36	$	60	<	84	T	108	l
37	%	61	=	85	U	109	m
38	&	62	>	86	V	110	n
39	,	63	?	87	W	111	o

40	(	64	@	88	X	112	p
41	)	65	A	89	Y	113	q
42	*	66	B	90	Z	114	r
43	+	67	C	91	[	115	s
44	,	68	D	92	\	116	t
45	-	69	E	93	]	117	u
46	.	70	F	94	^	118	v
47	/	71	G	95	_	119	w
48	0	72	H	96	'	120	x
49	1	73	I	97	a	121	y
50	2	74	J	98	b	122	z
51	3	75	K	99	c	123	{
52	4	76	L	100	d	124		
53	5	77	M	101	e	125	}
54	6	78	N	102	f	126	~
55	7	79	O	103	g		

Example 6

Write a text font shape file (UCM) for an uppercase M as shown in Figure 11-15. The font file should be able to generate horizontal and vertical text. Each grid is 1 unit, and the directions of vectors are designated with leader lines. **In the following file, the line numbers at the right are not a part of the file; they are for reference only.**

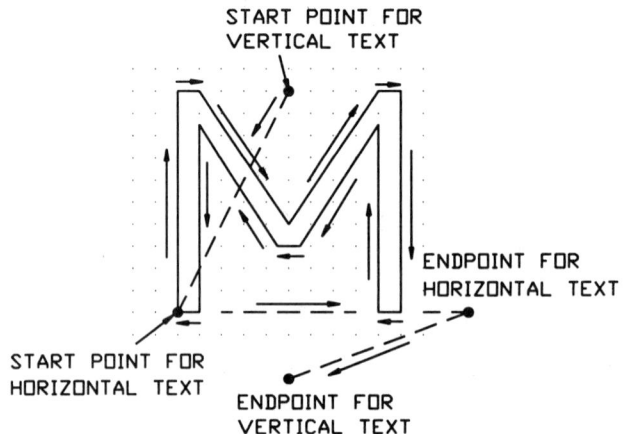

Figure 11-15 Shape and pen movement of uppercase "M"

```
*0,4,uppercase m                                        1
10,0,2,0                                                2
*10,13,lf                                               3
002,8,(0,-14),14,9,(0,14),(14,0),(0,0),0                4
*77,51,ucm                                              5
2,14,8,(-5,-10),                                        6
```

001,009,(0,10),(1,0),(4,-6),(4,6),(1,0),	7
(0,-10),(-1,0),(0,0),	8
003,2,	9
009,(0,17),(-7,-11),(-2,0),(-7,11),	10
(0,-17),(-2,0),(0,0),	11
002,8,(28,0),	12
004,2,	13
14,8,(-9,-4),0	14

Explanation

Line 1

*0,4,uppercase m

The first data byte (0) is the special shape number for the text font file. The next data byte (4) is the number of data bytes, and the third data byte is the name of the shape.

Line 2

10,0,2,0

The first data byte (10) represents the total height of the character M, and the second data byte (0) represents the length of the lowercase letters that extend below the base line. Data byte 2 is the text mode for dual orientation of the text (horizontal and vertical). If the text was required in the horizontal direction only, the mode would be 0. The fourth data byte (0) is the zero that terminates the definition of this particular shape.

Line 3

*10,13,lf

The first data byte (10) is the code reserved for line feed, and the second data byte (13) is the number of data bytes in the shape specification. The third data byte (lf) is the name of the shape.

Line 4

002,8,(0,-14),14,9,(0,14),(14,0),(0,0),0

The first data byte (002 or 2) is the code to deactivate the pen (pen up). The next three data bytes [8,(0,-14)] define a displacement vector whose X displacement and Y displacement are 0 and -14 units, respectively. This will cause a carriage return that is 14 units below the text insertion point of the first text line. This will work fine if the text is drawn in the horizontal direction only. However, if the text is vertical, the carriage return should produce a displacement to the right of the existing line. This is accomplished by the next seven data bytes. Data byte 14 ignores the next code if the text is horizontal. If the text is vertical, the next code is processed. The next set of data bytes (0,14) defines a displacement vector that is 14 units below the previous point, D1 in Figure 11-16.

Data bytes (14,0) define a displacement vector that is 14 units to the right, D2 in Figure 11-16. These four data bytes combined will result in a carriage return that is 4 units to the right of the existing line. The next set of data bytes (0,0) terminates the code 9, and the last data byte (0) terminates the shape specification.

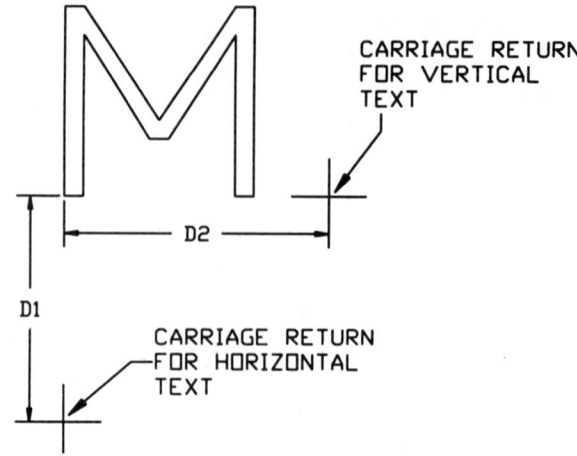

Figure 11-16 Carriage return for vertical and horizontal texts

Line 5
*77,51,ucm
The first data byte (77) is the ASCII code of the uppercase M. The second data byte (51) is the number of data bytes in the shape specification. The next data byte (ucm) is the name of the shape file in lowercase letters.

Line 6
2,14,8,(-5,-10),
The first data byte code (2) deactivates the pen (pen up), and the next data byte code (14) will cause the next code to be ignored if the text is horizontal. In the horizontal text, the insertion point of the text is the starting point of that text line (Figure 11-15). However, if the text is vertical, the starting point of the text is the upper middle point of the character M. This is accomplished by the next three data bytes [8,(-5,-10)], which displace the starting point of the text 5 units to the left (width of character M is 10) and 10 units down (height of character M is 10).

Lines 7,8
001,009,(0,10),(1,0),(4,-6),(4,6),(1,0),
(0,-10),(-1,0),(0,0),
The first byte (001) activates the draw mode (pen down), and the remaining bytes define the next seven vectors.

Lines 9,10,11
003,2,
009,(0,17),(-7,-11),(-2,0),(-7,11),
(0,-17),(-2,0),(0,0),
The inner vertical line of the right leg of the character M is 8.5 units long, and you cannot define a vector that is not an integer. However, you can define a vector that is 2 x 8.5 = 17 units long and then divide that vector by 2 to get a vector 8.5 units long. This is accomplished by code 003 and the next data byte, 2. All the vectors defined in the next two lines will be divided by 2.

Line 12
002,8,(28,0),
The first data byte (002) deactivates the draw mode, and the next three data bytes define a vector that is 28/2 = 14 units to the right. This means that the next character will start 14 - 10 = 4 units to the right of the existing character so that it will produce a horizontal text.

Line 13
004,2,
The code 004 multiplies the vectors that follow it by 2; therefore, it nullifies the effect of code 003,2.

Line 14
14,8,(-9,-4),0
If the text is vertical, the next letter should start below the previous letter. This is accomplished by data bytes 8,(-9,-4), which define a vector that is -9 units along the X axis and -4 units along the Y axis. The data byte 0 terminates the definition of the shape.

To load the shape file, use the LOAD command as discussed in Example 1. Use the STYLE command to define a text style file that uses the font (UCM) created in this example. Now, you can use the TEXT or DTEXT command and enter the text as shown in Figure 11-17.

Figure 11-17 Using the defined text font

Example 7

Write a text font shape file for the lowercase m as shown in Figure 11-18. The font file should be able to generate horizontal and vertical text. Each grid is 1 unit, and the direction of the vectors is designated with the leader lines.

11-24 Customizing AutoCAD

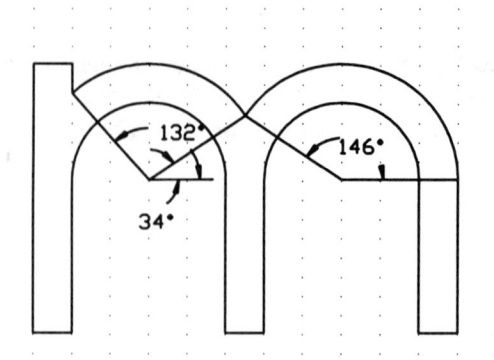

Figure 11-18 Shape of lowercase character, m

The following file is a listing of the text font shape file for Example 7. The line numbers are not a part of the file; they are shown here for reference only.

*0,4,lower-case m	1
14,3,2,0	2
*10,13,lf	3
002,8,(0,-18),14,9,(0,18),(27,0),(0,0),0	4
*109,57,lcm	5
2,14,8,(-11,-14),	6
005,005,001,020,084,	7
00A,(4,-044),	8
08C,020,084,	9
00A,(4,-044),	10
08C,020,084,	11
00B,(0,62,0,6,004),	12
00B,(193,239,0,6,003),	13
006,9,(0,14),(2,0),(0,0)	14
003,5,07C,004,5	15
006,2,8,(27,0),	16
14,8,(-16,-5),0	17

The format of most of the lines is the same as in the previous example, except for the following lines, which use save/restore origin, octant, and fractional arcs.

Line 7
005,005,001,020,084,
The first and second data bytes (005) are used to save the location of the point twice. The remaining data bytes activate the draw mode and define the vectors.

Line 8
00A,(4,-044),

The first data byte code (00A) is the code for octant arc and the second data byte (4) defines the radius of the arc. The negative sign (-) in the third data byte generates an arc in a clockwise direction. The first element (0) is a hexadecimal notation. The second element defines the starting octant and the third element (4) defines the number of octants that the arc passes through.

Line 12
00B,(0,62,0,6,004),

The first data byte (00B) is the code for the fractional arc that is defined by the next five data bytes. The second data byte (0) is the starting offset of the first arc, as shown in the following calculations:

1st Arc
Starting angle	= 0	Ending angle = 146
Starting octant	= 0	Ending octant = 4
Starting offset	= (0-0)*256/45	
	= 0	
Ending offset	= (146-135)*256/45	
	= 62.57	
	= 62 (integer)	

The third data byte (62) is the ending offset of the arc and the fourth data byte (0) is the high radius. The fifth data byte (6) defines the radius of the arc. The second element (0) of the next data byte is the starting octant and the third element (4) is the number of octants the arc goes through.

Line 13
00B,(193,239,0,6,003),

The first data byte (00B) is the code for the fractional arc. The remaining data bytes define various parameters of the fractional arc as explained earlier. The offset angles have been obtained from the following calculations:

2nd Arc
Starting angle	= 34	Ending angle = 132
Starting octant	= 0	Ending octant = 3
Starting offset	= (34-0)*256/45	
	= 193.4	
	= 193 (integer)	
Ending offset	= (132-90)*256/45	
	= 238.9	
	= 239 (integer)	

> **Note**
> *Since the offset values have been rounded, it is not possible to describe an arc that is very accurate. Therefore, in this example the origin has been restored after two arcs were drawn. This origin was then used to draw the remaining lines.*

Line 14
006,9,(0,14),(2,0),(0,0)
The first data byte (006) restores the previously saved point, and the remaining data bytes define the vectors using code 009.

REVIEW QUESTIONS

1. The basic objects used in shape files are _____ and _____.

2. Shapes are easy to insert and take less disk space than _____. However, there are certain disadvantages to using shapes. For example, you cannot _____ a shape.

3. The shape number could be any number between 1 and _____ in a particular file, and these numbers cannot be repeated within the same file.

4. The shape file may not contain two _____ with the same name.

5. A hexadecimal number is designated by a leading _____.

6. The maximum number of data bytes is _____ bytes per shape.

7. To define a vector, you need its magnitude and _____.

8. Do not include the extension with the _____.

9. To load the shape file, use the AutoCAD _____ command.

10. Generating shapes with direction vectors has some limitations. For example, you cannot draw an arc or a line that is not along the _____ vectors. These limitations can be overcome by using the _____, which add a lot of flexibility and give you better control over the shapes you want to create.

11. Code 001 activates the _____ mode, and code _____ deactivates the draw mode.

12. The byte that follows the division code divides the _____ vectors.

13. Code 004 is used if you want to multiply the vectors by a certain number. It can also be used to _____ the effect of code 003.

14. Scale factors are _____.

15. The number of saves (code 005) must be equal the number of _____ code _____ .

16. The maximum number of saves and restores you can use in a particular shape definition is _____ .

17. You can define a subshape like a subroutine in a program. To reference the subshape, use code _____ .

18. Vectors can be drawn in the 16 predefined directions only, and the length of the vector cannot exceed _____ units.

19. A nonstandard fractional arc can be generated by using code 00B or _____ .

20. Code _____ can be used to define an arc by specifying the displacement of the endpoint of an arc and the bulge factor.

21. Bulge factor can range from -127 to _____ .

22. Code 00E or _____ is used when the same text font description is to be used in both horizontal and vertical orientation.

23. The text files are regular _____ files with some additional information about the text font description and the line feed.

24. The shape names in all text font files should be lowercase so that the computer does not have to save the names in its _____ .

25. The line feed is used to space the lines so that the characters do not _____ .

26. The shape number in the shape definition of the text font corresponds to the _____ code for that character.

EXERCISES

Exercise 1

Write a shape file for the uppercase M shown in Figure 11-19.

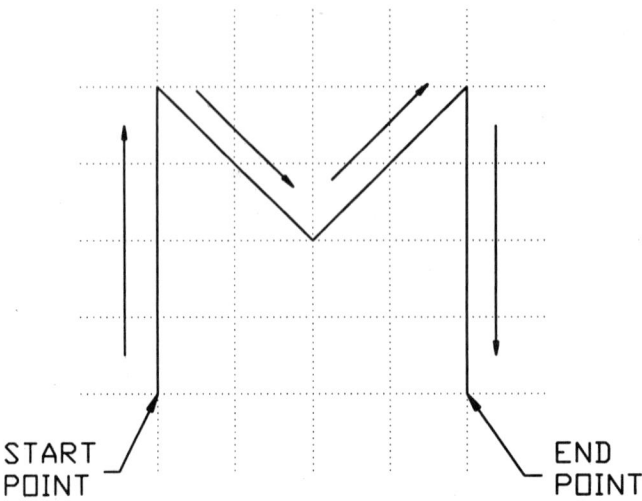

Figure 11-19 Uppercase letter M

Exercise 2

Write a shape file for generating the tapered gib-head key shown in Figure 11-20.

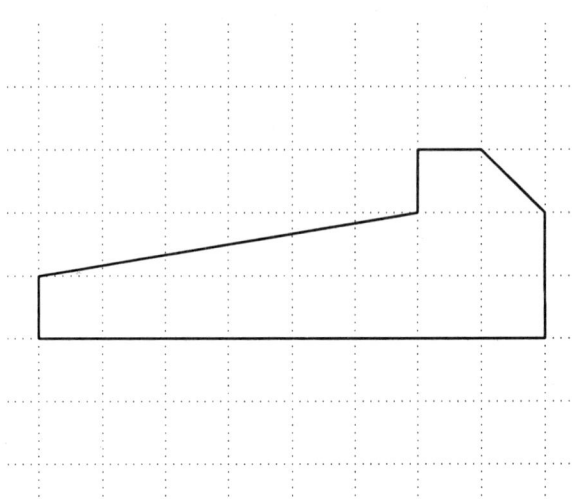

Figure 11-20 Tapered gib-head key

Exercise 3

Write a text font shape file for the uppercase G shown in Figure 11-21.

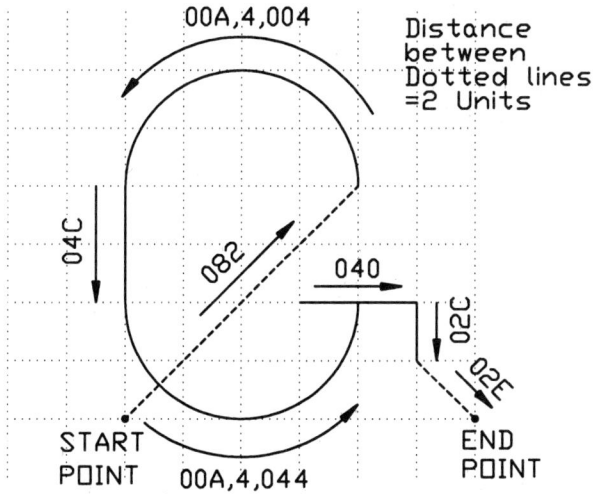

Figure 11-21 Uppercase letter G

Exercise 4

Write a text font shape file for the uppercase W shown in Figure 11-22. The font file should be able to generate horizontal and vertical text.

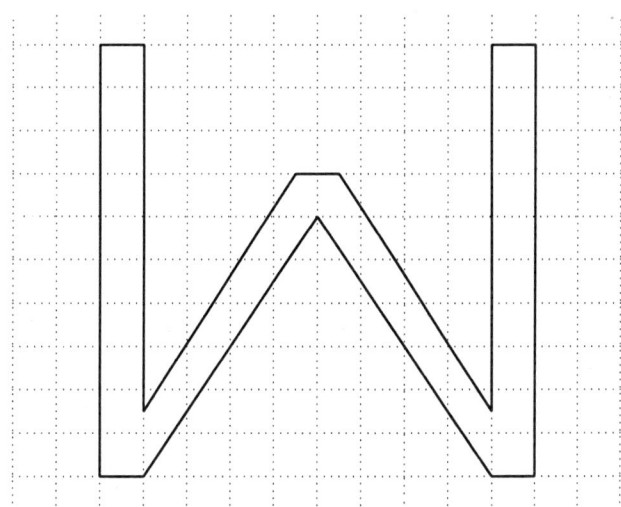

Figure 11-22 Uppercase letter W

Chapter 12

AutoLISP

Learning objectives

After completing this chapter, you will be able to:
♦ *Perform mathematical operations using AutoLISP.*
♦ *Use trigonometrical functions in AutoLISP.*
♦ *Understand the basic AutoLISP functions and their applications.*
♦ *Load and run AutoLISP programs.*
♦ *Use flowcharts to analyze problems.*
♦ *Test a condition using conditional functions.*
♦ *Use the Persistent AutoLISP feature of AutoCAD Release 14.*

ABOUT AutoLISP

Developed by Autodesk, Inc., **AutoLISP** is an implementation of the **LISP** programming language (LISP is an acronym for **LISt Processor**.) The first reference to LISP was made by John McCarthy in the April 1960 issue of *The Communications of the ACM*.

Except for **FORTRAN** and **COBOL**, most of the languages developed in the early 1960s have become obsolete. But LISP has survived and has become a leading programming language for artificial intelligence (AI). Some of the dialects of the LISP programming language are Common LISP, BYSCO LISP, ExperLISP, GCLISP, IQLISP, LISP/80, LISP/88, MuLISP, TLCLISP, UO-LISP, Waltz LISP, and XLISP. XLISP is a public-domain LISP interpreter. The LISP dialect that resembles AutoLISP is Common LISP. The AutoLISP interpreter is embedded within the AutoCAD software package. However, AutoCAD LT and AutoCAD versions 2.17 and lower lack the AutoLISP interpreter; therefore, you can use the AutoLISP programming language only with AutoCAD Release 2.18 and up.

The AutoCAD software package contains most of the commands used to generate a drawing. However, some commands are not provided in AutoCAD. For example, AutoCAD has no command to draw a rectangle or make global changes in the drawing text objects. With AutoLISP you can write a program in the AutoLISP programming language that will draw a rectangle or make global or selective changes in the drawing text objects. As a matter of fact, you can use AutoLISP to write any program or embed it in the menu and thus customize your system to make it more efficient.

The AutoLISP programming language has been used by hundreds of third-party software developers to write software packages for various applications. For example, the author of this text has developed a software package, **SMLayout**, that generates flat layouts of various geometrical shapes like transitions, intersection of pipes and cylinders, elbows, cones, and tank heads. There is demand for AutoLISP programmers as consultants for developing application software and custom menus.

This chapter assumes that you are familiar with AutoCAD commands and AutoCAD system variables. However, you need not be an AutoCAD or programming expert to begin learning AutoLISP. This chapter also assumes that you have no prior programming knowledge. If you are familiar with any other programming language, learning AutoLISP may be easy. A thorough discussion of various functions and a step-by-step explanation of the examples should make it fun to learn. This chapter discusses the most frequently used AutoLISP functions and their application in writing a program. For those functions not discussed in this chapter, refer to the **AutoLISP Programmers Reference Manual** from Autodesk. AutoLISP does not require any special hardware. If your system runs AutoCAD, it will also run AutoLISP. To write AutoLISP programs you can use any text editor.

MATHEMATICAL OPERATIONS

A mathematical function constitutes an important feature of any programming language. Most of the mathematical functions commonly used in programming and mathematical calculations are available in AutoLISP. You can use AutoLISP to add, subtract, multiply, and divide numbers. You can also use it to find the sine, cosine, and arctangent of angles expressed in radians. There is a host of other calculations you can do with AutoLISP. This section discusses the most frequently used mathematical functions supported by the AutoLISP programming language.

Addition

Format (+ **num1 num2 num3 - - -**)

This function (+) calculates the sum of all the numbers to the right of the plus (+) sign (num1 + num2 + num3 + . . .). The numbers can be integers or real. If the numbers are integers, then the sum is an integer. If the numbers are real, the sum is real. However, if some numbers are real and some are integers, the sum is real. In the following display, all numbers in the first two examples are integers, so the result is an integer. In the third example, one number is a real number (50.0), so the sum is a real number.

Examples
Command: (+ 2 5)   returns 7
Command: (+ 2 30 4 50)   returns 86
Command: (+ 2 30 4 50.0)   returns 86.0

Subtraction

Format **(- num1 num2 num3 - - -)**

This function (-) subtracts the second number from the first number (num1 - num2). If there are more than two numbers, the second and subsequent numbers are added and the sum is subtracted from the first number [num1 - (num2 + num3 + . . .)]. In the first of the following examples, 14 is subtracted from 28 and returns 14. Since both numbers are integers, the result is an integer. In the third example, 20 and 10.0 are added, and the sum of these two numbers (30.0) is subtracted from 50, returning a real number, 20.0.

Examples
Command: (- 28 14)   returns 14
Command: (- 25 7 11)   returns 7
Command: (- 50 20 10.0)   returns 20.0
Command: (- 20 30)   returns -10
Command: (- 20.0 30.0)   returns -10.0

Multiplication

Format **(* num1 num2 num3 - - -)**

This function (*) calculates the product of the numbers to the right of the asterisk (num1 x num2 x num3 x . . .). If the numbers are integers, the product of these numbers is an integer. If one of the numbers is a real number, the product is a real number.

Examples
Command: (* 2 5)   returns 10
Command: (* 2 5 3)   returns 30
Command: (* 2 5 3 2.0)   returns 60.0
Command: (* 2 -5.5)   returns -11.0
Command: (* 2.0 -5.5 -2)   returns 22.0

Division

Format **(/ num1 num2 num3 - - -)**

This function (/) divides the first number by the second number (num1/num2). If there are more than two numbers, the first number is divided by the product of the second and subsequent numbers [num1 / (num2 x num3 x . . .)]. In the fourth of the following examples, 200 is divided by the product of 5 and 4 [200 / (5 * 4)].

Examples
Command: (/ 30)   returns 30
Command: (/ 3 2)   returns 1
Command: (/ 3.0 2)   returns 1.5
Command: (/ 200.0 5.5)   returns 36.363636
Command: (/ 200 -5)   returns -40
Command: (/ -200 -5.0)   returns 40.0

INCREMENTED, DECREMENTED, AND ABSOLUTE NUMBERS

Incremented Number

Format **(1+ number)**

This function **(1+)** adds 1 (integer) to **number** and returns a number that is incremented by 1. In the second example below, 1 is added to -10.5 and returns -9.5.

Examples
(1+ 20)	returns 21
(1+ -10.5)	returns -9.5

Decremented Number

Format **(1- number)**

This function **(1-)** subtracts 1 (integer) from the **number** and returns a number that is decremented by 1. In the second example below, 1 is subtracted from -10.5 and returns -11.5

Examples
(1- 10)	returns 9
(1- -10.5)	returns -11.5

Absolute Number

Format **(abs num)**

The **abs** function returns the absolute value of a number. The number may be an integer number or a real number. In the second example below, the function returns 20 because the absolute value of -20 is 20.

Examples
(abs 20)	returns 20
(abs -20)	returns 20
(abs -20.5)	returns 20.5

TRIGONOMETRIC FUNCTIONS
sin
Format **(sin angle)**

The **sin** function calculates the sine of an angle, where the angle is expressed in radians. In the second of the following examples, the **sin** function calculates the sine of pi (180 degrees) and returns 0.

Examples
Command: (sin 0)   returns 0.0
Command: (sin pi)   returns 0.0
Command: (sin 1.0472)   returns 0.866027

cos
Format **(cos angle)**

The **cos** function calculates the cosine of an angle, where the angle is expressed in radians. In the third of the following examples, the **cos** function calculates the cosine of pi (180 degrees) and returns -1.0.

Examples
Command: (cos 0)   returns 1.0
Command: (cos 0.0)   returns 1.0
Command: (cos pi)   returns -1.0
Command: (cos 1.0)   returns 0.540302

atan
Format **(atan num1)**

The **atan** function calculates the arctangent of **num1**, and the calculated angle is expressed in radians. In the second of the following examples, the **atan** function calculates the arctangent of 1.0 and returns 0.785398 (radians).

Examples
Command: (atan 0.5)   returns 0.463648
Command: (atan 1.0)   returns 0.785398
Command: (atan -1.0)   returns -0.785398

You can also specify a second number in the atan function:

Format **(atan num1 num2)**

If the second number is specified, the function returns the arctangent of (num1/num2) in radians. In the first of the following examples the first number (0.5) is divided by the second number (1.0), and the **atan** function calculates the arctangent of the dividend (0.5/1.0 = 0.5).

Customizing AutoCAD

Examples

Command: (atan 0.5 1.0)	returns 0.453648 radians
Command: (atan 2.0 3.0)	returns 0.588003 radians
Command: (atan 2.0 -3.0)	returns 2.55359 radians
Command: (atan -2.0 3.00)	returns -0.588003 radians
Command: (atan -2.0 -3.0)	returns -2.55359 radians
Command: (atan 1.0 0.0)	returns 1.5708 radians
Command: (atan -0.5 0.0)	returns -1.5708 radians

angtos

Format **(angtos angle [mode [precision]])**

The **angtos** function returns the angle expressed in radians in a string format. The format of the string is controlled by the **mode** and **precision** settings.

Examples

(angtos 0.588003 0 4)	returns "33.6901"
(angtos 2.55359 0 4)	returns "146.3099"
(angtos 1.5708 0 4)	returns "90.0000"
(angtos -1.5708 0 2)	returns "270.00"

> **Note**
>
> In *(angtos angle [mode [precision]])*
>
> **angle** *is angle in radians*
>
> **mode** *is the angtos mode that corresponds to the AutoCAD system variable AUNITS*
>
> *The following modes are available in AutoCAD:*
>
> | ANGTOS MODE | EDITING FORMAT |
> |---|---|
> | 0 | Decimal degrees |
> | 1 | Degrees/minutes/seconds |
> | 2 | Grads |
> | 3 | Radians |
> | 4 | Surveyor's units |
>
> **Precision** *is an integer number that controls the number of decimal places. Precision corresponds to the AutoCAD system variable **AUPREC**. The minimum value of **precision** is zero and the maximum is four.*
>
> *In the first example above, **angle** is 0.588003 radians, **mode** is 0 (angle in degrees), and **precision** is 4 (four places after decimal). The function will return 33.6901.*

RELATIONAL STATEMENTS

Programs generally involve features that test a particular condition. If the condition is true, the program performs certain functions, and if the condition is not true, then the program performs other functions. For example, the relational statement (if (< x 5)) tests true if the value of the variable x is less than 5. This type of test condition is frequently used in programming. The following section discusses various relational statements used in AutoLISP programming.

Equal to

Format (= atom1 atom2 - - - -)

This function (=) checks whether the two atoms are equal. If they are equal, the condition is true and the function will return T. Similarly, if the specified atoms are not equal, the condition is false and the function will return nil.

Examples
(= 5 5)	returns T
(= 5 4.9)	returns nil
(= 5.5 5.5 5.5)	returns T
(= "yes" "yes")	returns T
(= "yes" "yes" "no")	returns nil

Not equal to

Format (/= atom1 atom2 - - - -)

This function (/=) checks whether the two atoms are not equal. If they are not equal, the condition is true and the function will return T. Similarly, if the specified atoms are equal, the condition is false and the function will return **nil**.

Examples
(/= 50 4)	returns T
(/= 50 50)	returns nil
(/= 50 -50)	returns T
(/= "yes" "no")	returns T

Less than

Format (< atom1 atom2 - - - -)

This function (<) checks whether the first atom **(atom1)** is less than the second atom **(atom2)**. If it is true, then the function will return **T**. If it is not, the function will return **nil**.

Examples
(< 3 5)	returns T
(< 5 3 4 2)	returns nil
(< "x" "y")	returns T

Less than or equal to

Format (<= atom1 atom2 - - - -)

This function (<=) checks whether the first atom (**atom1**) is less than or equal to the second atom (**atom2**). If it is, the function will return **T**. If it is not, the function will return **nil**.

Examples

(<= 10 15)	returns T
(<= "c" "b")	returns nil
(<= -2.0 0)	returns T

Greater than

Format (> atom1 atom2 - - - -)

This function (>) checks whether the first atom (**atom1**) is greater than the second atom (**atom2**). If it is, the function will return **T**. If it is not, then the function will return **nil**. In the first example below, 15 is greater than 10. Therefore, this relational function is true and the function will return **T**. In the second example, 10 is greater than 9, but this number is not greater than the second 9; therefore, this function will return **nil**.

Examples

(> 15 10)	returns T
(> 10 9 9)	returns nil
(> "c" "b")	returns T

Greater than or equal to

Format (>= atom1 atom2 - - - -)

This function (>=) checks whether the first atom (**atom1**) is greater than or equal to the second atom (**atom2**). If it is, the function returns **T**; otherwise, it will return **nil**. In the first example below, 78 is greater than 50, but 78 is not equal to 50; therefore, it will return **nil**.

Examples

(>= 78 50)	returns T
(>= "x" "y")	returns T

defun, setq, getpoint, AND Command FUNCTIONS

defun

The **defun** function is used to define a function in an AutoLISP program. The format of the **defun** function is:

```
(defun name [argument])
         │         └─ Argument list
         └─ Name of the function
```

Examples
(defun ADNUM ( )

Defines a function ADNUM with no arguments or local symbols. This means that all variables used in the program are global variables. A global variable does not lose its value after the programs ends.

(defun ADNUM (a b c)
Defines a function ADNUM that has three arguments: **a**, **b**, and **c**. The variables **a**, **b**, and **c** receive their value from outside the program.

(defun ADNUM (/ a b)
Defines a function ADNUM that has two local variables: **a** and **b**. A local variable is one that retains its value during program execution and can be used within that program only.

(defun C:ADNUM ( )
With **C:** in front of the function name, the function can be executed by entering the name of the function at the AutoCAD Command: prompt. If **C:** is not used, the function name has to be enclosed in parentheses.

Note
AutoLISP contains some built-in functions. Do not use any of those names for function or variable names. The following is a list of some of the names reserved for AutoLISP built-in functions. (Refer to the AutoLISP Programmer's Reference manual for a complete list of AutoLISP built-in functions.)

abs	ads	alloc
and	angle	angtos
append	apply	atom
ascii	assoc	atan
atof	atoi	distance
equal	fix	float
if	length	list
load	member	nil
not	nth	null
open	or	pi
read	repeat	reverse
set	type	while

setq

The **setq** function is used to assign a value to a variable. The format of the **setq** function is:

(setq Name Value [Name Value].......)

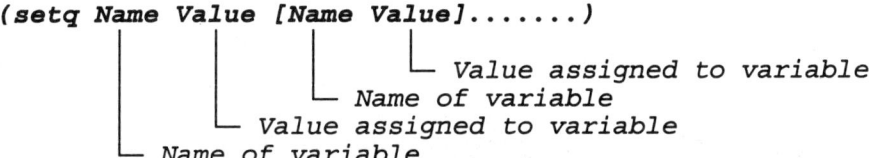

The value assigned to a variable can be any expression (numeric, string, or alphanumeric). If the value is a string, the string length cannot be exceed 100 characters.

Command: (setq X 12)
Command: (setq X 6.5)
Command: (setq X 8.5 Y 12)

In this last expression, the number 8.5 is assigned to the variable **X**, and the number 12 is assigned to the variable **Y**.

Command: (setq answer "YES")

In this expression the string value "YES" is assigned to the variable **answer**.

The **setq** function can also be used in conjunction with other expressions to assign a value to a variable. In the following examples the **setq** function has been used to assign values to different variables.

(setq pt1 (getpoint "Enter start point: "))
(setq ang1 (getangle "Enter included angle: "))
(setq answer (getstring "Enter YES or NO: "))

> **Note**
> *AutoLISP uses some built-in function names and symbols. Do not assign values to any of those functions. The following functions are valid ones, but must never be used because the pi and angle functions that are reserved functions will be redefined.*
>
> *(setq pi 3.0)*
> *(setq angle (. . .))*

getpoint

The **getpoint** function pauses to enable you to enter the X, Y coordinates or X, Y, Z coordinates of a point. The coordinates of the point can be entered from the keyboard or by using the screen cursor. The format of the **getpoint** function is:

```
(getpoint [point] [prompt])
                │         └─ Prompt to be displayed on screen
                └─ Enter a point, or select a point
```

Example
(setq pt1 (getpoint))
(setq pt1 (getpoint "Enter starting point"))

> **Note**
> *You cannot enter the name of another AutoLISP routine in response to the getpoint function.*
>
> *A 2D or a 3D point is always defined with respect to the current user coordinate system (UCS).*

Command

The **Command** function is used to execute standard AutoCAD commands from within an AutoLISP program. The AutoCAD command name and the command options have to be enclosed in double quotation marks. The format of the **Command** function is:

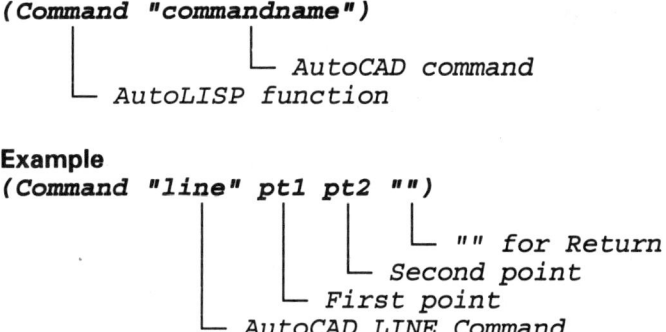

Example
(Command "line" pt1 pt2 "")
— "" for Return
— Second point
— First point
— AutoCAD LINE Command

> **Note**
> *Prior to AutoCAD Release 12, the **Command** function **could not be used** to execute the AutoCAD PLOT command. For example: (Command "plot" . . .) was not a valid statement. In AutoCAD Release 14 and Release 13, you can use plot with the Command function (Command "plot" . . .).*
>
> *The **Command** function cannot be used to enter data with the AutoCAD DTEXT or TEXT command. (You can issue the DTEXT and TEXT command with the **Command** function. You can also enter text height and text rotation, but you cannot enter the text when DTEXT or TEXT prompts for text entry.)*
>
> *You cannot use the input functions of AutoLISP with the **Command** function. The input functions are **getpoint, getangle, getstring,** and **getint**. For example, (Command "getpoint" . . .) or (Command "getangle" . . .) are not valid functions. If the program contains such a function, it will display an error message when the program is loaded.*

Example 1

Write a program that will prompt you to select three points of a triangle and then draw lines through those points to generate the triangle shown in Figure 12-1.

Most programs consist of essentially three parts: **input**, **output**, and **process**. **Process** includes what is involved in generating the desired output from the given input (Figure 12-2). Before writing a program, you must identify these three parts. In this example, the **input** to the program is the coordinates of the three points. The desired **output** is a triangle. The **process** needed to

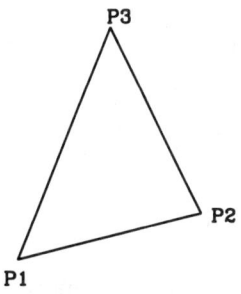

Figure 12-1 Triangle P1, P2, P3

generate a triangle is to draw three lines from P1 to P2, P2 to P3, and P3 to P1. Identifying these three sections makes the programming process less confusing.

The process section of the program is vital to the success of the program. Sometimes it is simple, but sometimes it involves complicated calculations. If the program involves many calculations, divide them into sections (and perhaps subsections) that are laid out in a logical and systematic order. Also, remember that programs need to be edited from time to time, perhaps by other programmers. Therefore, it is wise to document the programs as clearly and unambiguously as possible so that other programmers can understand what the program is doing at different stages of its execution. Give sketches, and identify points where possible.

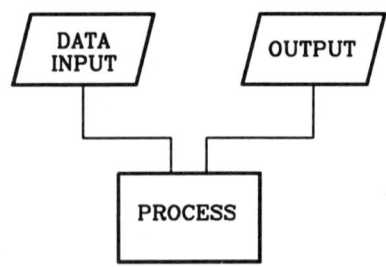

Figure 12-2 Three elements of a program

Input
Location of point P1
Location of point P2
Location of point P3

Output

Triangle P1, P2, P3

Process
Line from P1 to P2
Line from P2 to P3
Line from P3 to P1

The following file is a listing of the AutoLISP program for Example 1. **The line numbers at the right are not a part of the program; they are shown here for reference only.**

```
;This program will prompt you to enter three points                    1
;of a triangle from the keyboard, or select three points               2
;by using the screen cursor. P1, P2, P3 are triangle corners.          3
                                                                       4
(defun c:TRIANG1()                                                     5
   (setq P1 (getpoint "\n Enter first point of Triangle: "))           6
   (setq P2 (getpoint "\n Enter second point of Triangle: "))          7
   (setq P3 (getpoint "\n Enter third point of Triangle: "))           8
   (Command "LINE" P1 P2 P3 "C")                                       9
)                                                                     10
```

Explanation
Lines 1-3
The first three lines are comment lines describing the function of the program. These lines are important because they make it easier to edit a program. Comments should be used when needed.

All comment lines must start with a semicolon (;). These lines are ignored when the program is loaded.

Line 4
This is a blank line that separates the comment section from the program. Blank lines can be used to separate different modules of a program. This makes it easier to identify different sections that constitute a program. The blank lines have no effect on the program.

Line 5
(defun c:TRIANG1( )
In this line, **defun** is an AutoLISP function that defines the function **TRIANG1**. **TRIANG1** is the name of the function. The **c:** in front of the function name, **TRIANG1**, enables it to be executed like an AutoCAD command. If the **c:** is missing, the **TRIANG1** command can be executed only by enclosing it in parentheses (TRIANG1). The TRIANG1 function has three global variables (P1, P2, and P3). When you first write an AutoLISP program, it is a good practice to keep the variables global, because after you load and run the program, you can check the values of these variables by entering exclamation point (!) followed by the variable name at AutoCAD Command prompt (Command: !P1). Once the program is tested and it works, you should make the variable local, (defun c:TRIANG1(/ P1 P2 P3)).

Line 6
(setq P1 (getpoint "\n Enter first point of Triangle: "))
In this line, the **getpoint** function pauses for you to enter the first point of the triangle. The prompt, **Enter first point of Triangle**, is displayed in the prompt area of the screen. You can enter the coordinates of this point at the keyboard or select a point by using the screen cursor. The **setq** function then assigns these coordinates to the variable **P1**. **\n** is used for the carriage return so that the statement that follows \n is printed on the next line ("n" stands for "newline").

Lines 7 and 8
(setq P2 (getpoint "\n Enter second point of Triangle: "))
(setq P3 (getpoint "\n Enter third point of Triangle: "))
These two lines prompt you to enter the second and third corners of the triangle. These coordinates are then assigned to the variables P2 and P3. \n causes a carriage return so that the input prompts are displayed on the next line.

Line 9
(Command "LINE" P1 P2 P3 "C")
In this line, the **Command** function is used to enter the AutoCAD LINE command and then draw a line from P1 to P2 and from P2 to P3. "C" (for "close" option) joins the last point, P3, with the first point, P1. All AutoCAD commands and options, when used in an AutoLISP program, have to be enclosed in double quotation marks. The variables P1, P2, P3 are separated by a blank space.

Line 10
This line consists of a closing parenthesis that completes the definition of the function, TRIANG1. This parenthesis could have been combined with the previous line. It is good practice to keep it on

a separate line so that any programmer can easily identify the end of a definition. In this program there is only one function defined, so it is easy to locate the end of a definition. But in some programs a number of definitions or modules within the same program might need to be clearly identified. The parentheses and blank lines help to identify the start and end of a definition or a section in the program.

LOADING AN AUTOLISP PROGRAM

There are generally two names associated with an AutoLISP program: the program file name and the function name. For example, **TRIANG.LSP** is the name of the file, not a function name. All AutoLISP file names have the extension **.LSP**. An AutoLISP file can have one or several functions defined within the same file. For example, TRIANG1 in Example 1 is the name of a function. To execute a function, the AutoLISP program file that defines that function must be loaded. Use the following command to load an AutoLISP file when you are in the drawing editor:

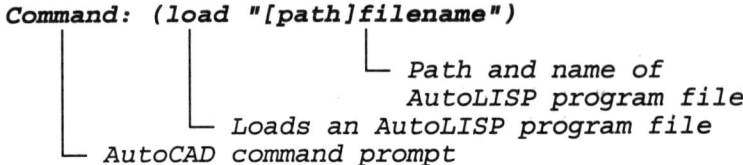

The AutoLISP file name and the optional path name must be enclosed in double quotes. The **load** and **filename** must be enclosed in parentheses. If the parentheses are missing, AutoCAD will try to load a shape or a text font file, not an AutoLISP file. The space between **load** and **filename** is not required. If AutoCAD is successful in loading the file, it will display the name of the function in the Command: prompt area of the screen.

C:TRIANG1

To run the program, type the name of the function at the AutoCAD Command: prompt, and press ENTER **(Command: TRIANG1)**. If the function name does not contain **C:** in the program, you can run the program by enclosing the function name in parentheses:

Command: TRIANG1 or **Command: (TRIANG1)**

Note
Use a forward slash when defining the path for loading an AutoLISP program. For example, if the AutoLISP file TRIANG is in the LISP subdirectory on the C drive, use the following command to load the file. You can also use a double backslash (\\) in place of the forward slash.

Command (load "c:/lisp/triang") or **Command (load "c:\\lisp\\triang")**

You can also load an AutoLISP program using the Load **AutoLISP, ADS,** and **ARX Files** dialog box (Figure 12-3). This dialog box can be invoked by selecting Load Applications in the Tools pull-down menu or by entering **APPLOAD** at AutoCAD Command prompt.

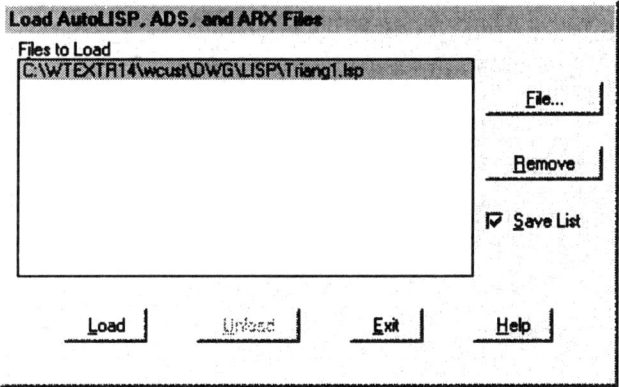

Figure 12-3 Loading AutoLISP files using the Load AutoLISP, ADS, and ARX Files dialog box

Exercise 1

Write an AutoLISP program that will draw a line between two points (Figure 12-4). The program must prompt the user to enter the X and Y coordinates of the points.

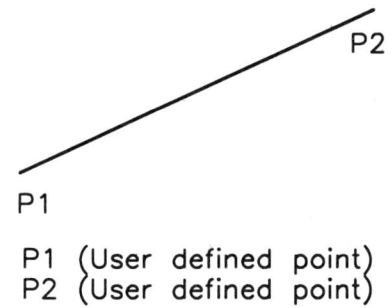

P1 (User defined point)
P2 (User defined point)

Figure 12-4 Draw line from point P1 to P2

getcorner, getdist, AND setvar FUNCTIONS
getcorner
The **getcorner** function pauses for you to enter the coordinates of a point. The coordinates of the point can be entered at the keyboard or by using the screen crosshairs. This function requires a base point, and it displays a rectangle with respect to the base point as you move the screen crosshairs around the screen. The format of the **getcorner** function is:

```
(getcorner point [prompt])
            |       |
            |       L Prompt displayed on screen
            L Base point
```

Examples
(getcorner pt1)
(setq pt2 (getcorner pt1))
(setq pt2 (getcorner pt1 "Enter second point: "))

> **Note**
>
> *The base point and the point that you select in response to the getcorner function are located with respect to the current UCS.*
>
> *If the point you select is a 3D point with X, Y, and Z coordinates, the Z coordinate is ignored. The point assumes current elevation as its Z coordinate.*

getdist

The **getdist** function pauses for you to enter distance, and it then returns the distance as a real number. The format of the **getdist** function is:

```
(getdist [point] [prompt])
                    │
                    └── Any prompt that needs to be
                        displayed on screen
          └── First point for distance
```

Examples
(getdist)
(setq dist (getdist))
(setq dist (getdist pt1))
(setq dist (getdist "Enter distance"))
(setq dist (getdist pt1 "Enter second point for distance"))

The distance can be entered by selecting two points on the screen. For example, if the assignment is **(setq dist (getdist))**, you can enter a number or select two points. If the assignment is **(setq dist (getdist pt1))**, where the first point (pt1) is already defined, you need to select the second point only. The getdist function will always return the distance as a real number. For example, if the current setting is architectural and the distance is entered in architectural units, the getdist function will return the distance as a real number.

setvar

The **setvar** function assigns a value to an AutoCAD system variable. The name of the system variable must be enclosed in double quotes. The format of the **setvar** function is:

```
(setvar "variable-name" value)
                          │
                          └── Value to be assigned to
                              the system variable
          └── AutoCAD system variable
```

Examples
(setvar "cmdecho" 0)
(setvar "dimscale" 1.5)
(setvar "ltscale" 0.5)
(setvar "dimcen" -0.25)

Example 2

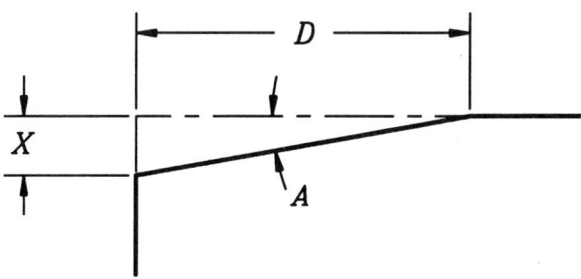

Figure 12-5 Chamfer with angle A and distance D

Write an AutoLISP program that will generate a chamfer between two given lines by entering the chamfer angle and the chamfer distance. To generate a chamfer, AutoCAD uses the values assigned to system variables **CHAMFERA** and **CHAMFERB**. When you select the AutoCAD CHAMFER command, the first and second chamfer distances are automatically assigned to the system variables **CHAMFERA** and **CHAMFERB**. The CHAMFER command then uses these assigned values to generate a chamfer. However, in most engineering drawings, the preferred way to generate the chamfer is to enter the chamfer length and the chamfer angle, as shown in Figure 12-5.

Input
First chamfer distance (D)
Chamfer angle (A)

Output
Chamfer between any two
selected lines

Process
1. Calculate second chamfer distance
2. Assign these values to the system variables **CHAMFERA** and **CHAMFERB**
3. Use AutoCAD CHAMFER command to generate chamfer

Calculations
x/d   = tan a
x =   d * (tan a)
   =   d * [(sin a) / (cos a)]

The following file is a listing of the program for Example 3. **The line numbers on the right are not a part of the file; they are for reference only.**

```
;This program generates a chamfer by entering              1
;the chamfer angle and the chamfer distance                2
;                                                          3
(defun c:chamf (/ d a)                                     4
(setvar "cmdecho" 0)                                       5
   (graphscr)                                              6
   (setq d (getdist "\n Enter chamfer distance: "))        7
   (setq a (getangle "\n Enter chamfer angle: "))          8
   (setvar "chamfera" d)                                   9
   (setvar "chamferb" (* d (/ (sin a) (cos a))))          10
   (command "chamfer")                                    11
   (setvar "cmdecho" 1)                                   12
   (princ)                                                13
)                                                         14
```

Explanation

Line 7
(setq d (getdist "\n Enter chamfer distance: "))
The **getdist** function pauses for you to enter the chamfer distance, then the **setq** function assigns that value to variable d.

Line 8
(setq a (getangle "\n Enter chamfer angle: "))
The **getangle** pauses for you to enter the chamfer angle, then the **setq** function assigns that value to variable a.

Line 9
(setvar "chamfera" d)
The **setvar** function assigns the value of variable d to AutoCAD system variable **chamfera**.

Line 10
(setvar "chamferb" (* d (/ (sin a) (cos a))))
The **setvar** function assigns the value obtained from the expression **(* d (/ (sin a) (cos a)))** to the AutoCAD system variable **chamferb**.

Line 11
(command "chamfer")
The **Command** function uses the AutoCAD **CHAMFER** command to generate a chamfer.

Exercise 2
Write an AutoLISP program that will generate the drawing shown in Figure 12-6. The program should prompt the user to enter points P1 and P2 and diameters D1 and D2.

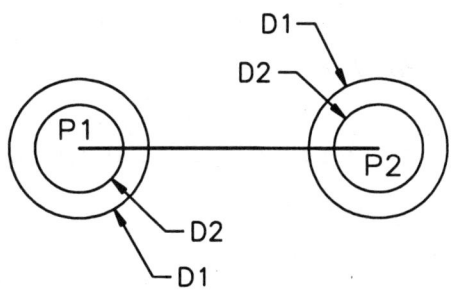

Figure 12-6 Concentric circles with connecting line

list FUNCTION

In AutoLISP the **list** function is used to define a 2D or 3D point. The list function can also be designated by using the single quote character ('), if the expression does not contain any variables or undefined items.

Examples

(Setq x (list 2.5 3 56))	returns 2.5, 3.56
(Setq x '(2.5 3.56))	returns 2.5, 3.56

car, cdr, AND cadr FUNCTIONS

car

The **car** function returns the first element of a list. If the list does not contain any elements, the function will return **nil**. The format of the **car** function is:

```
(car list)
      │   └─ List of elements
      └─ Returns the first element
```

Examples

(car '(2.5 3 56))	returns 2.5
(car '(x y z))	returns X
(car '((15 20) 56))	returns (15 20)
(car '( ))	returns nil

The single quote mark signifies a list.

cdr

The **cdr** function returns a list with the first element removed from the list. The format of the **cdr** function is:

(cdr list)
  └── List of elements
 └── Returns a list with the first element removed

Examples

(cdr '(2.5 3 56)	returns (3 56)
(cdr '(x y z))	returns (Y Z)
(cdr '((15 20) 56))	returns (56)
(cdr '( ))	returns nil

cadr

The **cadr** function performs two operations, **cdr** and **car**, to return the second element of the list. The **cdr** function removes the first element, and the **car** function returns the first element of the new list. The format of the **cadr** function is:

(cadr list)
  └── List of elements
 └── Performs two operations (car (cdr '(x y z))

Examples

(cadr '(2 3))	returns 3
(cadr '(2 3 56))	returns 3
(cadr '(x y z))	returns y
(cadr '((15 20) 56 24))	returns 56

In these examples, **cadr** performs two functions:

(cadr '(x y z))   = (car (cdr '(x y z)))
                   = (car (y z)) returns y

> **Note**
>
> *In addition to the functions car, cdr, and cadr, several other functions can be used to extract different elements of a list. Following is a list of these functions, where the function f consists of a list '((x y) z w)).*
>
> **(setq f '((x y) z w))**
>
> (caar f) = (car (car f))   returns x
> (cdar f) = (cdr (car f))   returns y
> (cadar) = (car (cdr (car f)))   returns y
> (cddr f) = (cdr (cdr f))   returns w
> (caddr f) = (car (cdr (cdr f)))   returns w

graphscr, textscr, princ, AND terpri FUNCTIONS

graphscr
The **graphscr** function switches from the text window to the graphics window, provided the system has only one screen. If the system has two screens, this function is ignored.

textscr
The **textscr** function switches from the graphics window to the text window, provided the system has only one screen. If the system has two screens, this function is ignored.

princ
The **princ** function prints (or displays) the value of the variable. If the variable is enclosed in double quotes, the function prints (or displays) the expression that is enclosed in the quotes. The format of the **princ** function is:

(princ [variable or expression])

Examples
(princ)	prints a blank on screen
(princ a)	prints the value of variable **a** on screen
(princ "Welcome")	prints **Welcome** on screen

terpri
The **terpri** function prints a new line on the screen, just as \n does. This function is used to print the line that follows the **terpri** function.

Example
(setq p1 (getpoint "Enter first point: "))(terpri)
(setq p2 (getpoint "Enter second point: "))

The first line (Enter first point:) will be displayed in the screen's command prompt area. The **terpri** function causes a carriage return; therefore, the second line (Enter second point:) will be displayed on a new line, just below the first line. If the terpri function is missing, the two lines will be displayed on the same line (Enter first point: Enter second point:).

Example 3

Write a program that will prompt you to enter two opposite corners of a rectangle and then draw the rectangle on the screen as shown in Figure 12-7.

Input	Output
Coordinates of point P1	Rectangle
Coordinates of point P3	

Process
1. Calculate the coordinates of the points P2 and P4.

2. Draw the following lines.
   Line from P1 to P2
   Line from P2 to P3
   Line from P3 to P4
   Line from P4 to P1

The X and Y coordinates of points P2 and P4 can be calculated using the **car** and **cadr** functions. The **car** function extracts the X coordinate of a given list, and the **cadr** function extracts the Y coordinate.

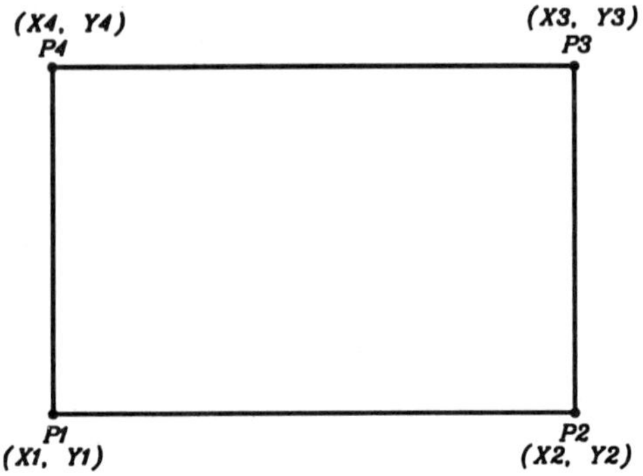

X coordinate of point p2
x2 = x3
x2 = car (x3 y3)
x2 = car p3

Y coordinate of point p2
y2 = y1
y2 = CADR (x1 y1)
y2 = CADR p1

X coordinate of point p4
x4 = x1
x4 = car(x1 y1)
x4 = car p1

Y coordinate of point p4
y4 = y3
y4 = cadr (x3 y3)
y4 = cadr p3

Figure 12-7 Rectangle P1 P2 P3 P4

Therefore, points p2 and p4 are:
p2 = (list (car p3) (cadr p1))
p4 = (list (car p1) (cadr p3))

The following file is a listing of the program for Example 3. **The line numbers at the right are for reference only; they are not a part of the program.**

```
;This program will draw a rectangle. User will            1
;be prompted to enter the two opposite corners            2
;                                                         3
(defun c:RECT1(/ p1 p2 p3 p4)                             4
    (graphscr)                                            5
    (setvar "cmdecho" 0)                                  6
    (prompt "RECT1  command  draws  a rectangle")(terpri) 7
    (setq p1 (getpoint "Enter first corner"))(terpri)     8
    (setq p3 (getpoint "Enter opposite corner"))(terpri)  9
```

```
        (setq p2 (list (car p3) (cadr p1)))           10
        (setq p4 (list (car p1) (cadr p3)))           11
     (command "line" p1 p2 p3 p4 "c")                 12
     (setvar "cmdecho" 1)                             13
     (princ)                                          14
)                                                     15
```

Explanation

Lines 1-3

The first three lines are comment lines that describe the function of the program. All comment lines that start with a semicolon are ignored when the program is loaded.

Line 4

(defun c:RECT1(/ p1 p2 p3 p4)

The **defun** function defines the function **RECT1**.

Line 5

(graphscr)

This function switches the text screen to the graphics screen, if the current screen happens to be a text screen. Otherwise, this function has no effect on the display screen.

Line 6

(setvar "cmdecho" 0)

The **setvar** function assigns the value 0 to the AutoCAD system variable **cmdecho**, which turns the echo off. When **cmdecho** is off, AutoCAD command prompts are not displayed in the command prompt area of the screen.

Line 7

(prompt "RECT1 command draws a rectangle")(terpri)

The **prompt** function will display the information in double quotes ("RECT1 command draws a rectangle"). The function **terpri** causes a carriage return so that the next text is printed on a separate line.

Line 8

(setq p1 (getpoint "Enter first corner"))(terpri)

The **getpoint** function pauses for you to enter a point (the first corner of the rectangle), and the **setq** function assigns that value to variable p1.

Line 9

(setq p3 (getpoint "Enter opposite corner"))(terpri)

The **getpoint** function pauses for you to enter a point (the opposite corner of the rectangle), and the **setq** function assigns that value to variable p3.

Line 10

(setq p2 (list (car p3) (cadr p1)))

The **cadr** function extracts the Y coordinate of point p1, and the **car** function extracts the X coordinate of point p3. These two values form a list, and the **setq** function assigns that value to variable p2.

Line 11
(setq p4 (list (car p1) (cadr p3)))
The **cadr** function extracts the Y coordinate of point p3, and the **car** function extracts the X coordinate of point p1. These two values form a list, and the **setq** function assigns that value to variable p4.

Line 12
(command "line" p1 p2 p3 p4 "c")
The command function uses the AutoCAD **LINE** command to draw lines between points p1, p2, p3, and p4. The c (close) joins the last point, p4, with the first point, p1.

Line 13
(setvar "cmdecho" 1)
The **setvar** function assigns a value of 1 to the AutoCAD system variable **cmdecho**, which turns the echo on.

Line 14
(princ)
The **princ** function prints a blank on the screen. If this line is missing, AutoCAD will print the value of the last expression. This value does not affect the program in any way. However, it might be confusing at times. The **princ** function is used to prevent display of the last expression in the command prompt area.

Line 15
The closing parenthesis completes the definition of the function **RECT1** and ends the program.

> **Note**
> *In this program the rectangle is generated after you define the two corners of the rectangle. The rectangle is not dragged as you move the screen crosshairs to enter the second corner. However, the rectangle can be dragged by using the getcorner function, as shown in the following program listing:*

```
;This program will draw a rectangle with the
;drag mode on and using getcorner function
;
(defun c:RECT2(/ p1 p2 p3 p4)
   (graphscr)
   (setvar "cmdecho" 0)
   (prompt "RECT2 command draws a rectangle")(terpri)
   (setq p1 (getpoint "enter first corner"))(terpri)
   (setq p3 (getcorner p1 "Enter opposite corner" ))(terpri)
   (setq p2 (list (car p3) (cadr p1)))
```

```
    (setq p4 (list (car p1) (cadr p3)))
(command "line" p1 p2 p3 p4 "c")
(setvar "cmdecho" 1)
(princ)
)
```

getangle AND getorient FUNCTIONS

getangle

The **getangle** function pauses for you to enter the angle; then it returns the value of that angle in radians. The format of the **getangle** function is:

(getangle [point] [prompt])

       │   └── Any prompt that needs to be
       │   displayed on screen
       └── First point of the angle

Examples
(getangle)
(setq ang (getangle))
(setq ang (getangle pt1));-------------- pt1 is a predefined point
(setq ang (getangle "Enter taper angle"))
(setq ang (getangle pt1 "Enter second point of angle"))

The angle you enter is affected by the angle setting. The angle settings can be changed using the AutoCAD **UNITS** command or by changing the value of the AutoCAD system variables **ANGBASE** and **ANGDIR**. Following are the default settings for measuring an angle:

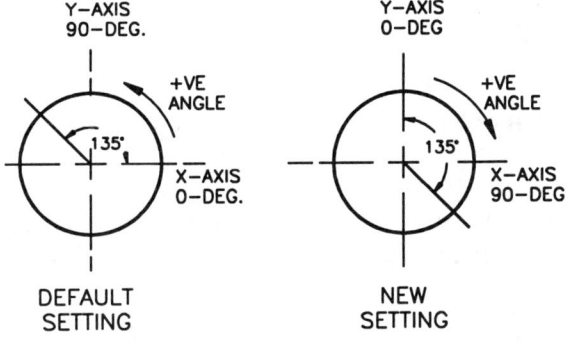

Figure 12-8(a)   Figure 12-8(b)

The angle is measured with respect to the positive X-axis (3 o'clock position). The value of this setting is saved in the AutoCAD system variable **ANGBASE**.

The angle is positive if it is measured in the counterclockwise direction and is negative if it is measured in the clockwise direction. The value of this setting is saved in the AutoCAD system variable **ANGDIR**.

If the angle has a default setting [Figure 12-8(a)], the **getangle** function will return 2.35619 radians for an angle of 135.

Example
(setq ang (getangle "Enter angle"))   returns 2.35619 for an angle of 135 degrees

Figure 12-8(b) shows the new settings of the angle, where the Y-axis is 0 degrees and the angles measured clockwise are positive. The **getangle** function will return 3.92699 for an angle of 135 degrees. The getangle function calculates the angle in the counterclockwise direction, **ignoring the direction set in the system variable ANGDIR**, with respect to the angle base as set in the system variable **ANGBASE** [Figure 12-9(b)].

Example
(setq ang (getangle "Enter angle"))   returns 3.92699

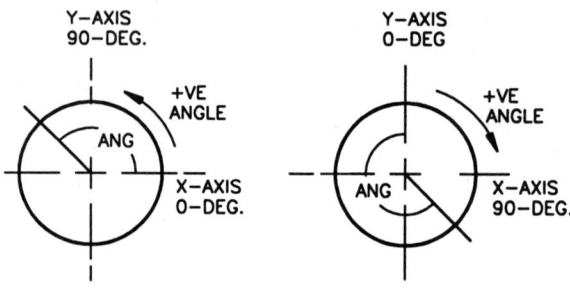

Figure 12-9(a)   Figure 12-9(b)

getorient

The **getorient** function pauses for you to enter the angle, then it returns the value of that angle in radians. The format of the **getorient** function is:

```
(getorient [point] [prompt])
                    │
                    └─ Any prompt that needs to be
                       displayed on the screen
             └─ First point of the angle
```

Examples

(getorient)
(setq ang (getorient))
(setq ang (getorient pt1))
(setq ang (getorient "Enter taper angle"))
(setq ang (getorient pt1 "Enter second point of angle"))

The **getorient** function is just like the **getangle** function. Both return the value of the angle in radians. However, the **getorient** function always measures the angle with a positive X-axis (3 o'clock position) and in a counterclockwise direction. **It ignores the ANGBASE and ANGDIR settings.** If the settings have not been changed, as shown in Figure 12-10A (default settings for ANGDIR and ANGBASE), for an angle of 135 degrees the **getorient** function will return 2.35619 radians. If the settings are changed, as shown in Figure 12-10B, for an angle of 135 degrees the **getorient** function will return 5.49778 radians. Although the settings have been changed where the angle is measured with the positive Y-axis and in a clockwise direction, the **getorient** function ignores the new settings and measures the angle from positive X-axis and in a counterclockwise direction.

> **Note**
> *For the getangle and getorient functions you can enter the angle by typing the angle at the keyboard or by selecting two points on the screen. If the assignment is (setq ang (getorient pt1)), where the first point pt1 is already defined, you will be prompted to enter the second point. You can enter this point by selecting a point on the screen or by entering the coordinates of the second point.*
>
> *180 degrees is equal to pi (3.14159) radians. To calculate an angle in radians use the following relation:*
>
> **Angle in radians; = (pi x angle)/180**

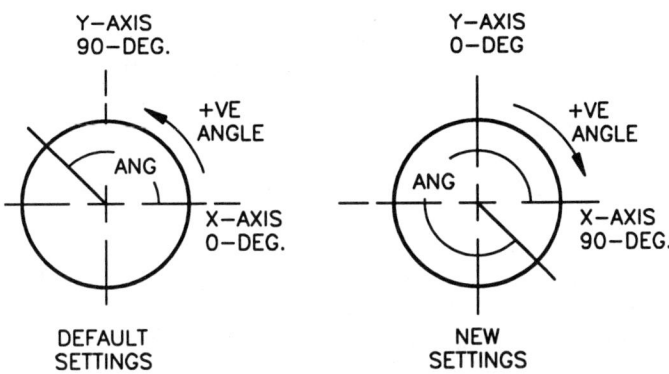

Figure 12-10A   Figure 12-10B

getint, getreal, getstring, AND getvar FUNCTIONS

getint

The **getint** function pauses for you to enter an integer. The function always returns an integer, even if the number you enter is a real number. The format of the **getint** function is:

(getint [prompt])

                └── *Optional prompt that you want to display on screen*

Examples
(getint)
(setq numx (getint))
(setq numx (getint "Enter number of rows: "))
(setq numx (getint "\n Enter number of rows: "))

getreal

The **getreal** function pauses for you to enter a real number and it always returns a real number, even if the number you enter is an integer. The format of the **getreal** function is:

(getreal [prompt])

                └── *Optional prompt that is displayed on screen*

Examples
(getreal)
(setq realnumx (getreal))
(setq realnumx (getreal "Enter num1: "))
(setq realnumx (getreal "\n Enter num2: "))

getstring

The **getstring** function pauses for you to enter a string value, and it always returns a string, even if the string you enter contains numbers only. The format of the **getstring** function is:

(getstring [prompt])

                └── *Optional prompt that is displayed on screen*

Examples
(getstring)
(setq answer (getstring))
(setq answer (getstring "Enter Y for yes, N for no:))
(setq answer (getstring "\n Enter Y for yes, N for no:))

> **Note**
> *The maximum length of the string is 132 characters. If the string exceeds 132 characters, the excess is ignored.*

getvar

The **getvar** function lets you retrieve the value of an AutoCAD system variable. The format of the **getvar** function is:

```
(getvar "variable")
           └─ AutoCAD system variable name
```

Examples

(gatvar)

(getvar "dimcen")	returns 0.09
(getvar "ltscale")	returns 1.0
(getvar "limmax")	returns 12.00,9.00
(getvar "limmin")	returns 0.00,0.00

> **Note**
> *The system variable name should always be enclosed in double quotes.*
>
> *You can retrieve only one variable value in one assignment. To retrieve the values of several system variables, use a separate assignment for each variable.*

polar AND sqrt FUNCTIONS

polar

The **polar** function defines a point at a given angle and distance from the given point (Figure 12-11). The angle is expressed in radians, measured positive in the counterclockwise direction (assuming default settings for **ANGBASE** and **ANGDIR**). The format of the **polar** function is:

```
(polar point angle distance)
         |     |        └─ Distance of the point from
         |     |           the referenced point
         |     └─ Angle the point makes with the
         |        referenced point
         └─ Reference point
```

Example

(polar pt1 ang dis)
(setq pt2 (polar pt1 ang dis))
(setq pt2 (polar '(2.0 3.25) ang dis))

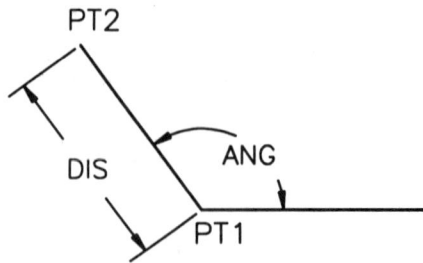

Figure 12-11 Using the **polar** function to define a point

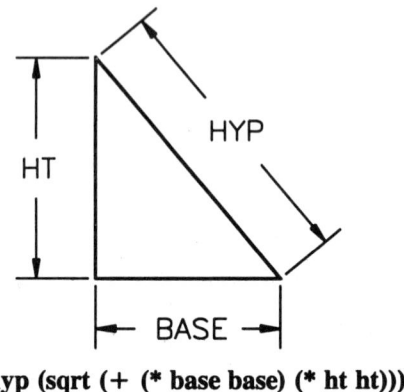

(setq hyp (sqrt (+ (* base base) (* ht ht))))

Figure 12-12 Application of the **sqrt** function

sqrt

The **sqrt** function calculates the square root of a number, and the value this function returns is always a real number. The format of the **sqrt** function is:

```
(sqrt number)
       └─ Number you want to find the
          square root of (real or integer)
```

Examples
(sqrt 144)   returns 12.0
(sqrt 144.0)   returns 12.0
(setq x (sqrt 57.25))   returns 7.566373
(setq x (sqrt (* 25 36.5)))   returns 30.207615
(setq x (sqrt (/ 7.5 (cos 0.75)))) returns 3.2016035
(setq hyp (sqrt (+ (* base base) (* ht ht))))

Example 4

Write an AutoLISP program that will draw an equilateral triangle outside a circle (Figure 12-13). The sides of the triangle are tangent to the circle. The program should prompt you to enter the radius and the center point of the circle.

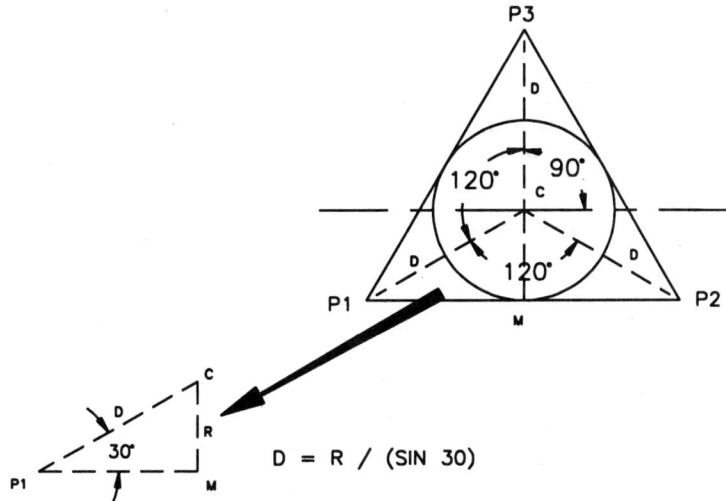

Figure 12-13 Equilateral triangle outside a circle

The following file is the listing of the AutoLISP program for Example 4.

```
;This program will draw a triangle outside
;the circle with the lines tangent to circle
;
(defun dtr (a)
   (* a (/ pi 180.0))
)
(defun c:trgcir(/ r c d p1 p2 p3)
(setvar "cmdecho" 0)
(graphscr)
   (setq r(getdist "\n Enter circle radius: "))
   (setq c(getpoint "\n Enter center of circle: "))
   (setq d(/ r (sin(dtr 30))))
   (setq p1(polar c (dtr 210) d))
   (setq p2(polar c (dtr 330) d))
   (setq p3(polar c (dtr 90) d))
(command "circle" c r)
(command "line" p1 p2 p3 "c")
(setvar "cmdecho" 1)
(princ)
)
```

Exercise 3

Write an AutoLISP program that will draw an isosceles triangle P1,P2,P3. The base of the triangle (P1,P2) makes an angle B with the positive X-axis (Figure 12-14). The program should prompt you to enter the starting point, P1, length L1, and angles A and B.

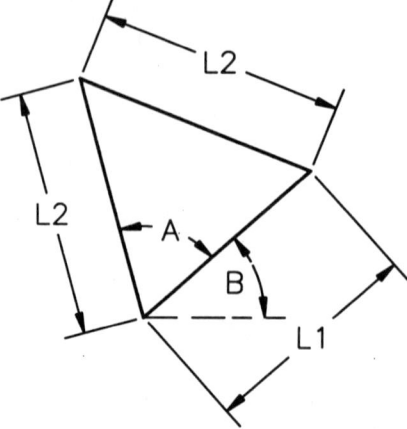

Figure 12-14 Isosceles triangle at an angle

Exercise 4

Write a program that will draw a slot with centerlines. The program should prompt you to enter slot length, slot width, and the layer name for the centerlines (Figure 12-15).

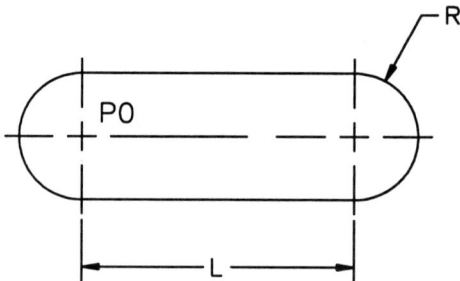

Figure 12-15 Slot of length L and radius R

itoa, rtos, strcase, AND prompt FUNCTIONS

itoa

The **itoa** function changes an integer into a string and returns the integer as a string. The format of the **itoa** function is:

```
(itoa number)
       └─ The integer number that you want to convert
          into a string
```

Examples
(itoa 89)   returns "89"
(itoa -356)   returns "-356"

(setq intnum 7)
(itoa intnum)   returns "7"

(setq intnum 345)
(setq intstrg (itoa intnum))   returns "345"

rtos

The **rtos** function changes a real number into a string and the function returns the real number as a string. The format of the **rtos** function is:

```
(rtos realnum)
        └─ The real number that you want to
           convert into a string
```

Examples
(rtos 50.6)   returns "50.6"
(rtos -30.0)   returns "-30.0"
(setq realstrg (rtos 5.25))   returns "5.25"

(setq realnum 75.25)
(setq realstrg (rtos realnum)) returns "75.25"

The **rtos** function can also include mode and precision. The format of the **rtos** function with mode and precision is:

```
(rtos realnum [mode] [precision])
        │       │         └─ Number of decimal places
        │       │            or denominator of fractional units
        │       └─ Unit mode, like decimal, scientific
        └─ Real number
```

strcase

The **strcase** function converts the characters of a string into uppercase or lowercase. The format of the **strcase** function is:

```
(strcase string [true])
                  │
                  └─ If it is not nil, all characters
                     are converted to lowercase
         └─ String that needs to be converted to
            uppercase or lowercase
```

The **true** is optional. If it is missing or if the value of **true** is nil, the string is converted to uppercase. If the value of true is not nil the string is converted to lowercase.

Examples
(strcase "Welcome Home")   returns "WELCOME HOME"

(setq t 0)
(strcase "Welcome Home" t)   returns "welcome home"

(setq answer (strcase (getstring "Enter Yes or No: ")))

prompt

The **prompt** function is used to display a message on the screen in the command prompt area. The contents of the message must be enclosed in double quotes. The format of the **prompt** function is:

```
(prompt message)
         │
         └─ Message that you want to display on the screen
```

Examples
(prompt "Enter circle diameter: ")
(setq d (getdist (prompt "Enter circle diameter: ")))

Note
On a two-screen system the prompt function displays the message on both screens.

Example 5

Write a program that will draw two circles of radii r1 and r2, representing two pulleys that are separated by a distance d. The line joining the centers of the two circles makes an angle a with the X-axis, as shown in Figure 12-16.

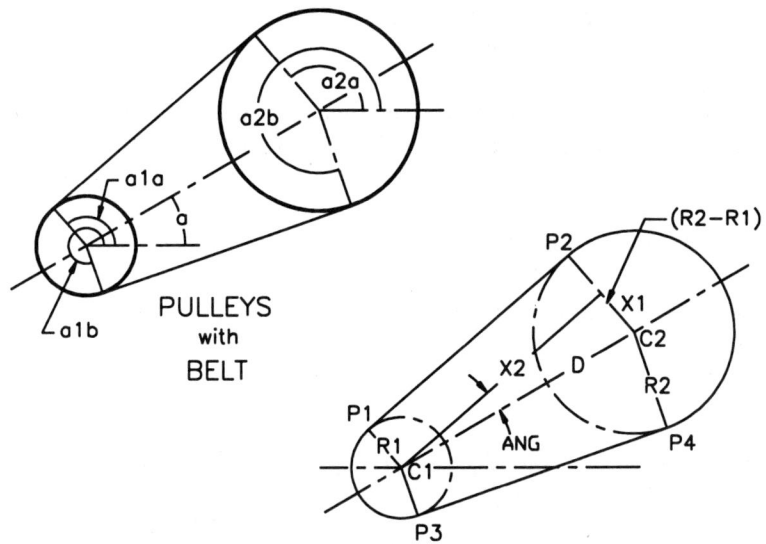

Figure 12-16 Two circles with tangent lines

Input
Radius of small circle - r1
Radius of large circle - r2
Distance between circles - d
Angle of center line - a
Center of small circle - c1

Output
Small circle of radius - r1
Large circle of radius - r2
Lines tangent to circles

Process
1. Calculate distance x1, x2
2. Calculate angle ang
3. Locate point c2 with respect to point c1
4. Locate points p1, p2, p3, p4
5. Draw small circle with radius r1 and center c1
6. Draw large circle with radius r2 and center c2
7. Draw lines p1 to p2 and p3 to p4

Calculations
x1 = r2 - r1
x2 = SQRT [d**2 - (r2 - r1)**2]
tan ang = x1 / x2
ang = atan (x1 / x2)
a1a = 90 + a + ang
a1b = 270 + a - ang

$$a2a = 90 + a + ang$$
$$a2b = 270 + a - ang$$

The following file is a listing of the AutoLISP program for Example 5. **The line numbers on the right are not a part of the file. These numbers are for reference only.**

```
;This program draws a tangent (belt) over two                              1
;pulleys that are separated by a given distance.                           2
                                                                           3
;This function changes degrees into radians                                4
(defun dtr (a)                                                             5
  (* a (/ pi 180.0))                                                       6
)                                                                          7
                                                                           8
;The belt function draws lines that are tangent to circles                 9
(defun c:belt(/ r1 r2 d a c1 x1 x2 c2 p1 p2 p3 p4)                         10
  (setvar "cmdecho" 0)                                                     11
  (graphscr)                                                               12
  (setq r1(getdist "\n Enter radius of small pulley: "))                   13
  (setq r2(getdist "\n Enter radius of larger pulley: "))                  14
  (setq d(getdist "\n Enter distance between pulleys: "))                  15
  (setq a(getangle "\n Enter angle of pulleys: "))                         16
  (setq c1(getpoint "\n Enter center of small pulley: "))                  17
  (setq x1 (- r2 r1))                                                      18
  (setq x2 (sqrt (- (* d d) (* (- r2 r1) (- r2 r1)))))                     19
  (setq ang (atan (/ x1 x2)))                                              20
  (setq c2 (polar c1 a d))                                                 21
  (setq p1 (polar c1 (+ ang a (dtr 90)) r1))                               22
  (setq p3 (polar c1 (- (+ a (dtr 270)) ang) r1))                          23
  (setq p2 (polar c2 (+ ang a (dtr 90)) r2))                               24
  (setq p4 (polar c2 (- (+ a (dtr 270)) ang) r2))                          25
                                                                           26
;The following line draw cirles and lines                                  27
  (command "circle" c1 p3)                                                 28
  (command "circle" c2 p2)                                                 29
  (command "line" p1 p2 "")                                                30
  (command "line" p3 p4 "")                                                31
  (setvar "cmdecho" 1)                                                     32
  (princ))                                                                 33
```

Explanation
Line 5
(defun dtr (a)
In this line, the **defun** function defines a function, **dtr (a)**, that converts degrees into radians.

Line 6
(* a (/ pi 180.0))
(/ pi 180) divides the value of **pi** by 180, and the product is then multiplied by angle a (180 degrees is equal to **pi** radians).

Line 10
(defun c:belt(/ r1 r2 d a c1 x1 x2 c2 p1 p2 p3 p4)
In this line, the function **defun** defines a function, c:belt, that generates two circles with tangent lines.

Line 18
(setq x1 (- r2 r1))
In this line, the function **setq** assigns a value of r2 - r1 to variable x1.

Line 19
(setq x2 (sqrt (- (* d d) (* (- r2 r1) (- r2 r1)))))
In this line, **(- r2 r1)** subtracts the value of r1 from r2 and **(* (- r2 r1) (- r2 r1))** calculates the square of (- r2 r1). **(sqrt (- (* d d) (* (- r2 r1) (- r2 r1))))** calculates the square root of the difference, and **setq x2** assigns the product of this expression to variable x2.

Line 20
(setq ang (atan (/ x1 x2)))
In this line, **(atan (/ x1 x2))** calculates the arctangent of the product of **(/ x1 x2)**. The function **setq ang** assigns the value of the angle in radians to variable **ang**.

Line 21
(setq c2 (polar c1 a d))
In this line, **(polar c1 a d)** uses the **polar** function to locate point c2 with respect to c1 at a distance d and making an angle **a** with the positive X-axis.

Line 22
(setq p1 (polar c1 (+ ang a (dtr 90)) r1))
In this line, **(polar c1 (+ ang a (dtr 90)) r1))** locates point p1 with respect to c1 at a distance r1 and making an angle **(+ ang a (dtr 90))** with the positive X-axis.

Line 28
(command "circle" c1 p3)
In this line, the **Command** function uses the AutoCAD **CIRCLE** command to draw a circle with center c1 and a radius defined by the point p3.

Line 30
(command "line" p1 p2 "")
In this line, the **Command** function uses the AutoCAD **LINE** command to draw a line from p1 to p2. The pair of double quotes ("") at the end introduces a Return, which terminates the LINE command.

Exercise 5

Write an AutoLISP program that will draw two lines tangent to two circles, as shown in Figure 12-17. The program should prompt you to enter the circle diameters and the center distance between the circles.

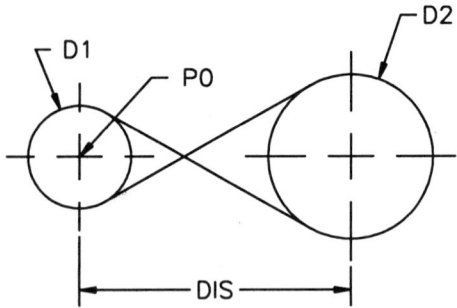

Figure 12-17 Circles with two tangent lines

FLOWCHARTS

A **flowchart** is a graphical representation of an algorithm. It can be used to analyze a problem systematically. It gives a better understanding of the problem, especially if the problem involves some conditional statements. It consists of standard symbols that represent a certain function in the program. For example, a rectangle is used to represent a process that takes place when the program is executed. The blocks are connected by lines indicating the sequence of operations. Figure 12-18 gives the standard symbols that can be used in a flowchart.

CONDITIONAL FUNCTIONS

The relational functions discussed earlier in the chapter establish a relationship between two atoms. For example, (< x y) describes a test condition for an operation. To use such functions in a meaningful way a conditional function is required. For example, (if (< x y) (setq z (- y x)) (setq z (- x y))) describes the action to be taken when the condition is true (T) and when it is false (nil). If the condition is true, then z = y - x. If the condition is not true, then z = x - y. Therefore, conditional functions are very important for any programming language, including AutoLISP.

if

The **if** function (Figure 12-19) evaluates the first expression (then) if the specified condition returns T, and it evaluates the second expression (else) if the specified condition returns nil. The format of the **if** function is:

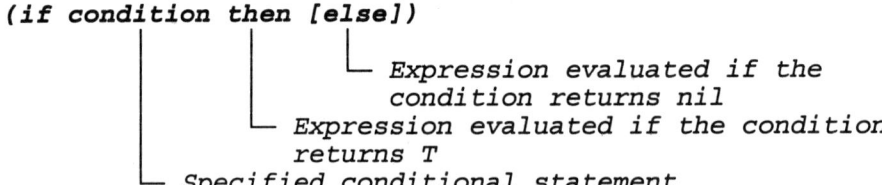

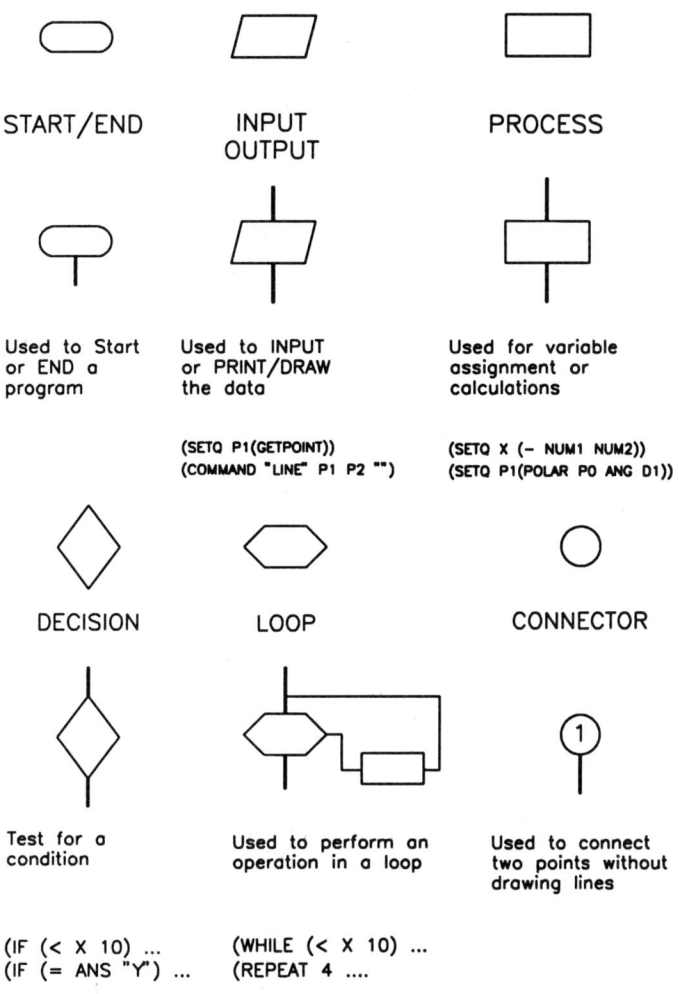

Figure 12-18 Flowchart symbols

Examples
(if (= 7 7) ("true"))   returns "true"
(if (= 5 7) ("true") ("false"))   returns "false"

(setq ans "yes")
(if (= ans "yes") ("Yes") ("No"))   returns "Yes"

(setq num1 8)
(setq num2 10)
(if (> num1 num2)
   (setq x (- num1 num2))
   (setq x (- num2 num1))
)   returns 2

12-40 Customizing AutoCAD

Figure 12-19 if function

Example 6

Write an AutoLISP program that will subtract a smaller number from a larger number. The program should also prompt you to enter two numbers.

Input
Number (num1)
Number (num2)

Output
x = num1 - num2
or
x = num2 - num1

Process
If num1 > num2 then x = num1 - num2
If num1 < num2 then x = num2 - num1

The flowchart in Figure 12-20 describes the process involved in writing the program using standard flowchart symbols.

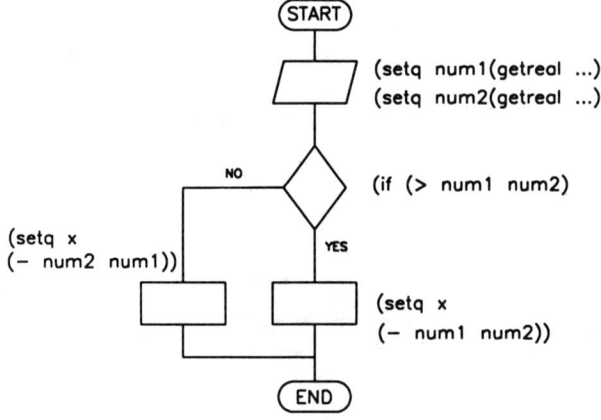

Figure 12-20 Flowchart for Example 6

The following file is a listing of the program for Example 6. **The line numbers are not a part of the file; they are for reference only.**

```
;This program subtracts smaller number                    1
;from larger number                                       2
;                                                         3
(defun c:subnum( )                                        4
   (setvar "cmdecho" 0)                                   5
   (setq num1 (getreal "\n Enter first number: "))        6
   (setq num2 (getreal "\n Enter second number: "))       7
   (if (> num1 num2)                                      8
      (setq x (- num1 num2))                              9
      (setq x (- num2 num1))                             10
   )                                                     11
   (setvar "cmdecho" 1)                                  12
   (princ)                                               13
)                                                        14
```

Explanation
Line 8
(if (> num1 num2)
In this line, the **if** function evaluates the test expression **(> num1 num2)**. If the condition is true, it returns T; if the condition is not true, it returns nil.

Line 9
(setq x (- num1 num2))
This expression is evaluated if the test expression **(if (> num1 num2)** returns T. The value of variable num2 is subtracted from num1, and the resulting value is assigned to variable x.

Line 10
(setq x (- num2 num1))
This expression is evaluated if the test expression **(if (> num1 num2)** returns nil. The value of variable num1 is subtracted from num2, and the resulting value is assigned to variable x.

Line 11
)
The closing parenthesis completes the definition of the **if** function.

Example 7

Write an AutoLISP program that will enable you to multiply or divide two numbers (Figure 12-21). The program should prompt you to enter the choice of multiplication or division. The program should also display an appropriate message if you do not enter the right choice. The following file is a listing of the AutoLISP program for Example 7:

```
;This program multiplies or divides two given numbers
(defun c:mdnum()
    (setvar "cmdecho" 0)
    (setq num1 (getreal "\n Enter first number: "))
    (setq num2 (getreal "\n Enter second number: "))
    (prompt "Do you want to multiply or divide. Enter M or D: ")
    (setq ans (strcase (getstring)))
    (.i.if ;(= ans "M")
        (setq x (* num1 num2))
    )
    (if (= ans "D")
        (setq x (/ num1 num2))
    )
    (if (and (/= ans "D")(/= ans "M"))
        (prompt "Sorry! Wrong entry, Try again")
    )
    (setvar "cmdecho" 1)
    (princ))
```

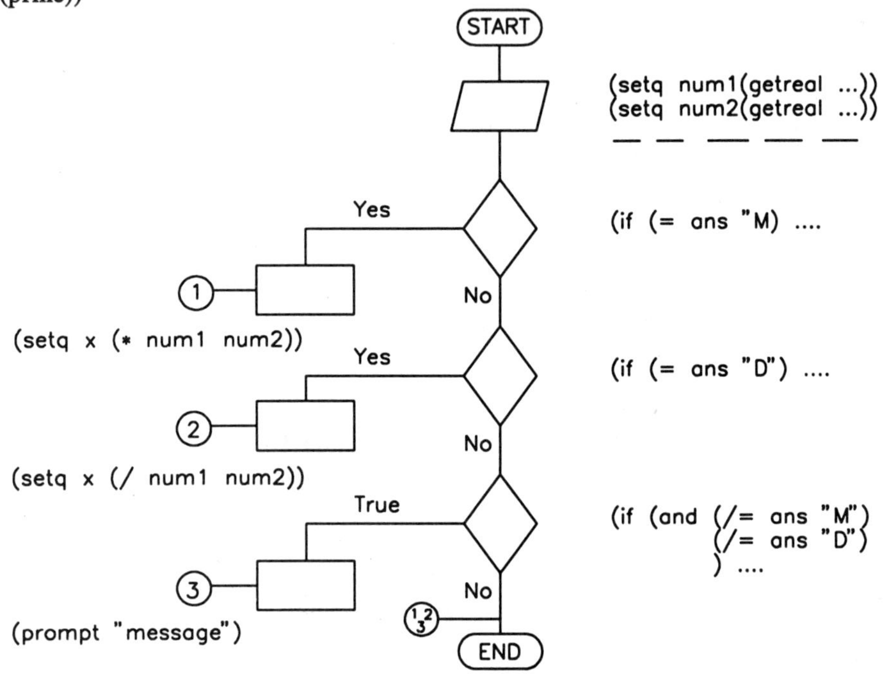

Figure 12-21 Flow diagram for Example 7

progn

The **progn** function can be used with the **if** function to evaluate several expressions. The format of **progn** function is:

(progn expression expression . . .)

The **if** function evaluates only one expression if the test condition returns T. The **progn** function can be used in conjunction with the **if** function to evaluate several expressions.

Example
```
(if (= ans "yes")
   (progn
     (setq x (sin ang))
     (setq y (cos ang))
     (setq tanang (/ x y))
))
```

while

The **while** function (Figure 12-22) evaluates a test condition. If the condition is true (expression does not return nil), the operations that follow the while statement are repeated until the test expression returns nil. The format of the **while** function is:

(while testexpression operations)

— Operations to be performed until the test expression returns nil
— Expression that tests a condition

Examples
```
(while (= ans "yes")
    (setq x (+ x 1))
    (setq ans (getstring "Enter yes or no: "))
)
(while (< n 3)
    (setq x (+ x 10))
    (setq n (1+ n))
)
```

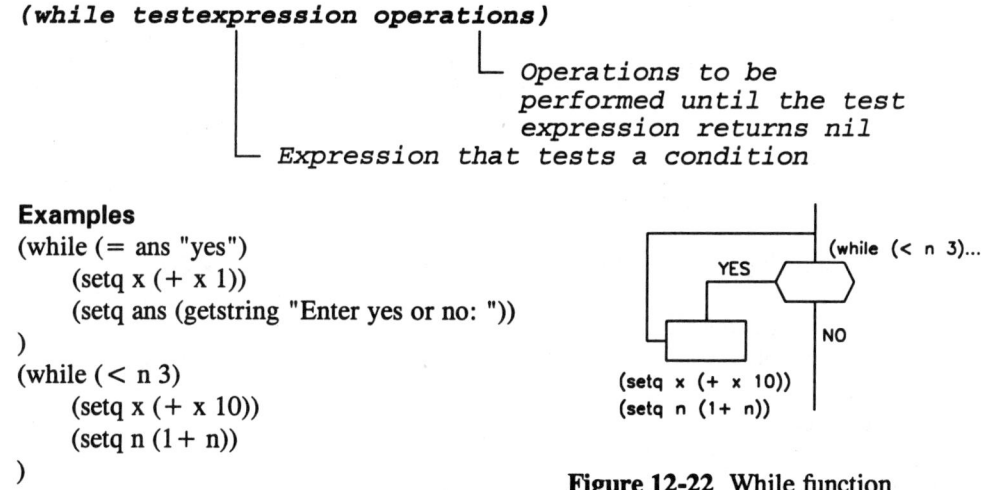

Figure 12-22 While function

Example 8

Write an AutoLISP program that will find the nth power of a given number. The power is an integer. The program should prompt you to enter the number and the nth power (Figure 12-23).

Input
Number x
nth power n

Output
product x^n

Process
1. Set the value of t = 1 and c = 1
2. Multiply t * x and assign that value to the variable t

3. Repeat the process until the counter c is less than or equal to n

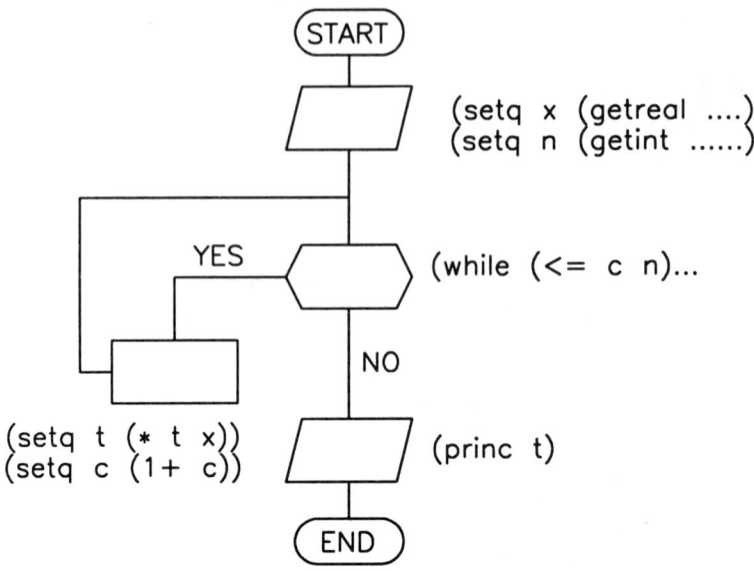

Figure 12-23 Flowchart for Example 8

The following file is a listing of the AutoLISP program for Example 8.

```
;This program calculates the nth
;power of a given number
(defun c:npower()
    (setvar "cmdecho" 0)
    (setq x(getreal "\n Enter a number: "))
    (setq n(getint "\n Enter Nth power-integer number: "))
    (setq t 1) (setq c 1)
    (while (<= c n)
       (setq t (* t x))
       (setq c (1+ c))
    )
    (setvar "cmdecho" 1)
    (princ t)
)
```

Example 9

Write an AutoLISP program that will generate the holes of a bolt circle (Figure 12-24). The program should prompt you to enter the center point of the bolt circle, the bolt circle diameter, the bolt circle hole diameter, the number of holes, and the start angle of the bolt circles.

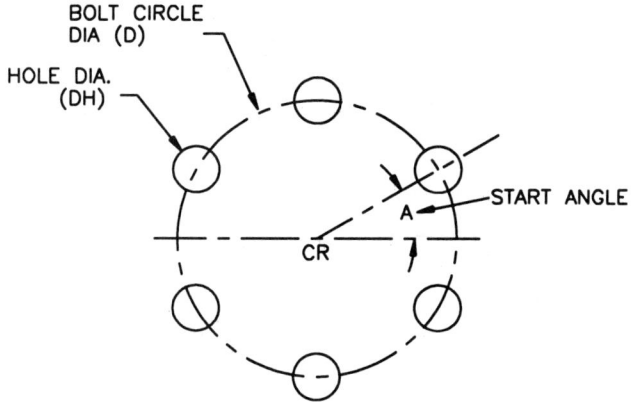

Figure 12-24 Bolt circle with six holes

```
;This program generates the bolt circles
;
(defun c:bc1( )
(graphscr)
(setvar "cmdecho" 0)
   (setq cr(getpoint "\n Enter center of Bolt-Circle: "))
   (setq d(getdist "\n Dia of Bolt-Circle: "))
   (setq n(getint "\n Number of holes in Bolt-Circle: "))
   (setq a(getangle "\n Enter start angle: "))
   (setq dh(getdist "\n Enter diameter of hole: "))
   (setq inc(/ (* 2 pi) n))
   (setq ang 0)
   (setq r (/ dh 2))
(while (< ang (* 2 pi))
   (setq p1 (polar cr (+ a inc) (/ d 2)))
   (command "circle" p1 r)
   (setq a (+ a inc))
   (setq ang (+ ang inc))
   )
(setvar "cmdecho" 1)
(princ)
)
```

repeat

The **repeat** function evaluates the expressions **n** number of times as specified in the **repeat** function (Figure 12-25). The variable **n** must be an integer. The format of the **repeat** function is:

```
repeat n
       └─ n is an integer
that defines the number of
times the expressions are
to be evaluated
```

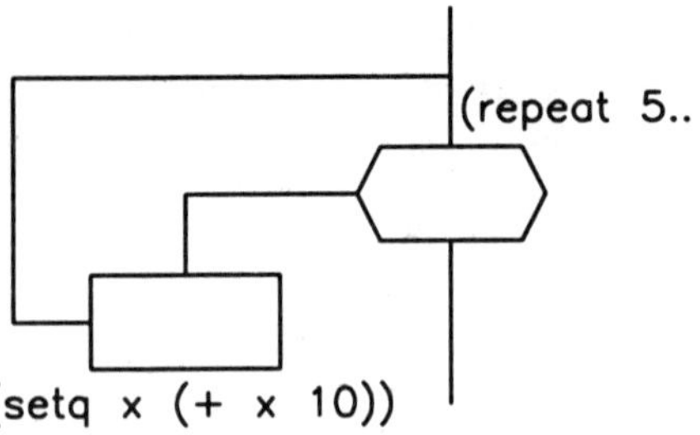

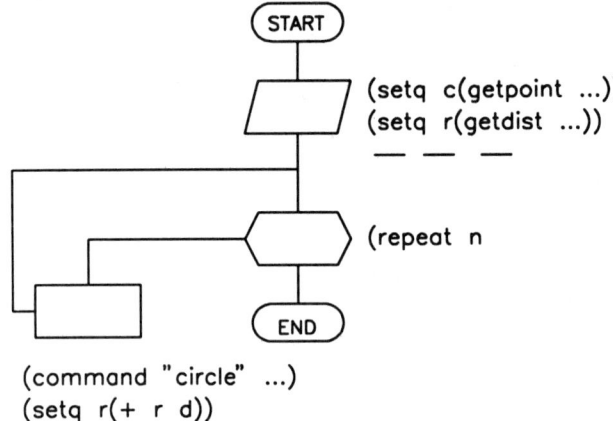

Figure 12-25 Repeat function

Example
(repeat 5
  (setq x (+ x 10))
)

Example 10

Write an AutoLISP program that will generate a given number of concentric circles. The program should prompt you to enter the center point of the circles, the start radius, and the radius increment (Figure 12-26).

Figure 12-26 Flowchart for Example 10

The following file is a listing of the AutoLISP program for Example 10.

```
;This program uses the repeat function to draw
;a given number of concentric circles.
(defun c:concir( )
(graphscr)
(setvar "cmdecho" 0)
(setq c (getpoint "\n Enter center point of circles: "))
(setq n (getint "\n Enter number of circles: "))
```

```
(setq r (getdist "\n Enter radius of first circle: "))
(setq d (getdist "\n Enter radius increment: "))
(repeat n
  (command "circle" c r)
  (setq r (+ r d))
)
(setvar "cmdecho" 1)
(princ)
)
```

PERSISTENT AUTOLISP*

The term Persistent AutoLISP means that the AutoLISP programs remain loaded from previous drawings. For example, if you are working on a project and you have loaded an AutoLISP program, the program will autoload if you start another drawing. You can enable this feature through the compatibility tab of the Preferences dialog box by selecting the check box (no check mark in the box) for **Reload AutoLISP Between Drawings option**.

Example 11

Write an AutoLISP program that will generate a flat layout drawing of a transition and then dimension the layout. The transition and the layout without dimensions are shown in Figure 12-27.

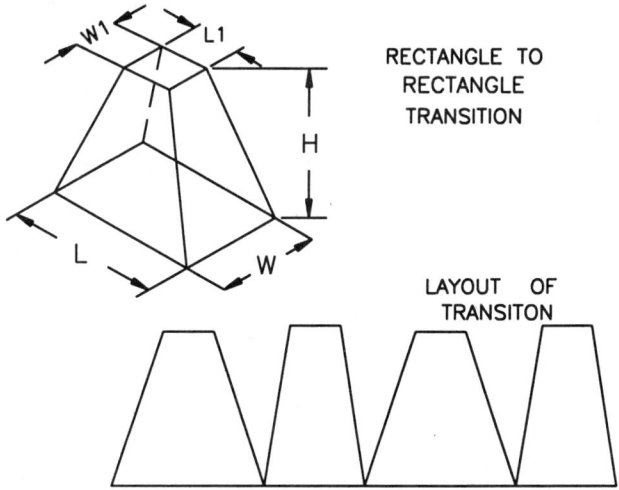

Figure 12-27 Flat layout of a transition

The following file is a listing of the AutoLISP program for Example 11. Programs do not need to be in lowercase letters. They can be uppercase or a combination of uppercase and lowercase.

```
;This program generates flat layout of
;a rectangle to rectangle transition
```

```
;
(defun c:TRANA(/)
(graphscr)
(setvar "cmdecho" 0)

(setq L (getdist "\n Enter length of bottom rectangle: "))
(setq W (getdist "\n Enter width of bottom rectangle: "))
(setq H (getdist "\n Enter height of transition: "))
(setq L1 (getdist "\n Enter length of top rectangle: "))
(setq W1 (getdist "\n Enter width of top rectangle: "))

(setq x1 (/ (- w w1) 2))
(setq y1 (/ (- l l1) 2))
(setq d1 (sqrt (+ (* h h) (* x1 x1))))
(setq d2 (sqrt (+ (* d1 d1) (* y1 y1))))
(setq s1 (/ (- l l1) 2))
(setq p1 (sqrt (- (* d2 d2) (* s1 s1))))
(setq s2 (/ (- w w1) 2))
(setq p2 (sqrt (- (* d2 d2) (* s2 s2))))

(setq t1 (+ l1 s1))
(setq t2 (+ l w))
(setq t3 (+ l s2 w1))
(setq t4 (+ l s2))
(setq pt1 (list 0 0))
(setq pt2 (list s1 p1))
(setq pt3 (list t1 p1))
(setq pt4 (list l 0))
(setq pt5 (list t4 p2))
(setq pt6 (list t3 p2))
(setq pt7 (list t2 0))
(command "layer" "make" "ccto" "c" "1" "ccto" "")
(command "line" pt1 pt2 pt3 pt4 pt5 pt6 pt7 "c")

(setq sf (/ (+ l w) 12))
(setvar "dimscale" sf)
(setq c1 (list 0 (- 0 (* 0.75 sf))))
(setq c7 (list (- 0 (* 0.75 sf)) 0))
(setq c8 (list (- l (* 0.75 sf)) 0))

(command "layer" "make" "cctd" "c" "2" "cctd" "")
(command "dim" "hor" pt1 pt2 c1 "" "base" pt3 "" "base" pt4 "" "exit")
(command "dim" "hor" pt4 pt5 c1 "" "base" pt6 "" "base" pt7 "" "exit")
(command "dim" "vert" pt1 pt2 pt2 "" "exit")
(command "dim" "vert" pt4 pt5 pt5 "" "exit")
(command "dim" "aligned" pt1 pt2 c7 "" "exit")
```

```
(command "dim" "aligned" pt4 pt5 c8 "" "exit")
(setvar "cmdecho" 1)
(princ))
```

Example 12

Write an AutoLISP program that can generate a flat layout of a cone as shown in Figure 12-28. The program should also dimension the layout.

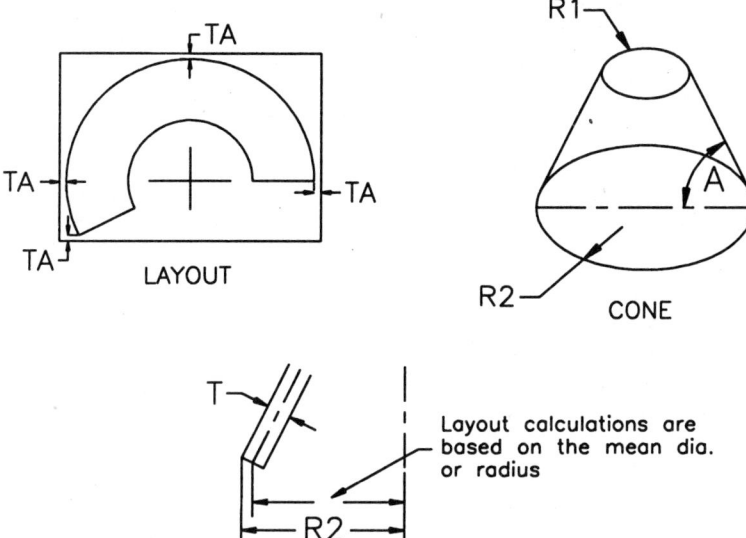

Figure 12-28 Flat layout of a cone

The following file is a listing of the AutoLISP program for Example 12:

```
;This program generates layout of a cone
;
;DTR function changes degrees to radians
(defun DTR (a)
 (* PI (/ A 180.0))
)

;RTD Function changes radians to degrees
(defun rtd (a)
(* a (/ 180.0 pi))
)

(defun tan (a)
(/ (sin a) (cos a))
)
(defun c:cone-1p(/)
```

```
(graphscr)
(setvar "cmdecho" 0)
(setq r2 (getdist "\n enter outer radius at larger end: "))
(setq r1 (getdist "\n enter inner radius at smaller end: "))
(setq t (getdist "\n enter sheet thickness:-"))
(setq a (getangle "\n enter cone angle:-"))

;this part of the program calculates various parameters
;needed in calculating the strip layout
(setq x0 0)
(setq y0 0)
(setq sf (/ r2 3))
(setvar "dimscale" sf)
(setq ar a)
(setq tx (/ (* t (sin ar)) 2))
(setq rx2 (- r2 tx))
(setq rx1 (+ r1 tx))
(setq w (* (* 2 pi) (cos ar)))
(setq rl1 (/ rx1 (cos ar)))
(setq rl2 (/ rx2 (cos ar)))

;this part of the program calculates the x-coordinate
;of the points
  (setq x1 (+ x0 rl1)
        x3 (+ x0 rl2)
        x2 (- x0 (* rl1 (cos (- pi w))))
        x4 (- x0 (* rl2 (cos (- pi w))))
  )

;this part of the program calculates the y-coordinate
;of the points
  (setq y1 y0
        y3 y0
        y2 (+ y0 (* rl1 (sin (- pi w))))
        y4 (+ y0 (* rl2 (sin (- pi w))))
        )

  (setq p0 (list x0 y0)
        p1 (list x1 y1)
        p2 (list x2 y2)
        p3 (list x3 y3)
        p4 (list x4 y4)
        )
(command "layer" "make" "ccto" "c" "1" "ccto" "")
(command  "arc" p1 "c" p0 p2)
(command  "arc" p3 "c" p0 p4)
```

```
        (command "line" p1 p3 "")
        (command "line" p2 p4 "")

        (setq f1 (/ r2 24))
        (setq f2 (/ r2 2))
        (setq d1 (list (+ x3 f2) y3))
        (setq d2 (list x0 (- y0 f2)))

    (command "layer" "make" "cctd" "c" "2" "cctd" "")
    (setvar "dimtih" 0)
    (command "dim" "hor" p0 p1 d2 "" "baseline" p3 "" "baseline" p2 "" "baseline" p4 "" "exit")
    (command "dim" "vert" p0 p2 d1 "" "baseline" p4 "" "exit")
    (setvar "dimscale" 1)
    (setvar "cmdecho" 1)
    (princ)
    )
```

REVIEW QUESTIONS

1. Evaluate the following AutoLISP functions:

    Command: (+ 2 30 5 50)   returns _____
    Command: (+ 2 30 4 55.0)   returns _____

    (- 20 40)   returns _____
    (- 30.0 40.0)   returns _____

    (* 72 5 3 2.0)   returns _____
    (* 7 -5.5)   returns _____

    (/ 299 -5)   returns _____
    (/ -200 -9.0)   returns _____

    (1- 99)   returns _____
    (1- -18.5)   returns _____

    (abs -90)   returns _____
    (abs -27.5)   returns _____

    (sin pi)   returns _____
    (sin 1.5)   returns _____
    (cos pi)   returns _____
    (cos 1.2)   returns _____
    (atan 1.1 0.0)   returns _____ radians

12-52 Customizing AutoCAD

    (atan -0.4 0.0)   returns _____ radians
    (angtos 1.5708 0 5)   returns _____
    (angtos -1.5708 0 3)   returns _____

    (< "x" "y")   returns _____
    (>= 80 90 79)   returns _____

2. The **setq** function is used to assign a value to _____.

3. The _____ function pauses to enable you to enter the X, Y coordinates or X, Y, Z coordinates of a point.

4. The _____ function is used to execute standard AutoCAD commands from within an AutoLISP program.

5. In an AutoLISP expression the AutoCAD command name and the command options have to be enclosed in double quotation marks. (T/F).

6. The **getdist** function pauses for you to enter a _____ and it then returns the distance as a real number.

7. The _____ function assigns a value to an AutoCAD system variable. The name of the system variable must be enclosed in _____.

8. The **cadr** function performs two operations, _____ and _____, to return the second element of the list.

9. The _____ function prints a new line on the screen just as \n.

10. The _____ function pauses for you to enter the angle, then it returns the value of that angle in radians.

11. The _____ function always measures the angle with a positive X-axis (3 o'clock position) and in a counterclockwise direction.

12. The _____ function pauses for you to enter an integer. The function always returns an integer, even if the number that you enter is a real number.

13. The _____ function lets you retrieve the value of an AutoCAD system variable.

14. The _____ function defines a point at a given angle and distance from the given point (Figure 12-10).

15. The _____ function calculates the square root of a number and the value this function returns is always a real number.

16. The _____ function changes a real number into a string and the function returns the real number as a string.

17. The **if** function evaluates the test expression (**> num1 num2**). If the condition is true it returns _____ ; if the condition is not true it returns.

18. The _____ function can be used with the **if** function to evaluate several expressions.

19. The **while** function evaluates the test condition. If the condition is true (expression does not return nil) the operations that follow the while statement are _____ until the test expression returns _____ .

20. The **repeat** function evaluates the expressions n number of times as specified in the **repeat** function. The variable n must be a real number. (T/F)

EXERCISES

Exercise 6

Write an AutoLISP program that will draw three concentric circles with center C1 and diameters D1, D2, D3 (Figure 12-29). The program should prompt you to enter the coordinates of center point C1 and the circle diameters D1, D2, D3.

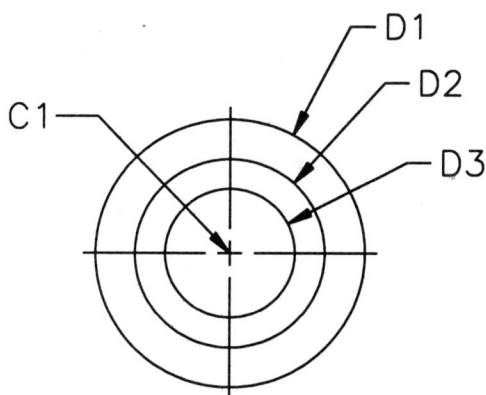

Figure 12-29 Three concentric circles with diameters D1, D2, D3

Exercise 7

Write an AutoLISP program that will draw a line from point P1 to point P2 (Figure 12-30). Line P1, P2 makes an angle A with the positive X-axis. Distance between the points P1 and P2 is L. The diameter of the circles is D1 (D1 = L/4).

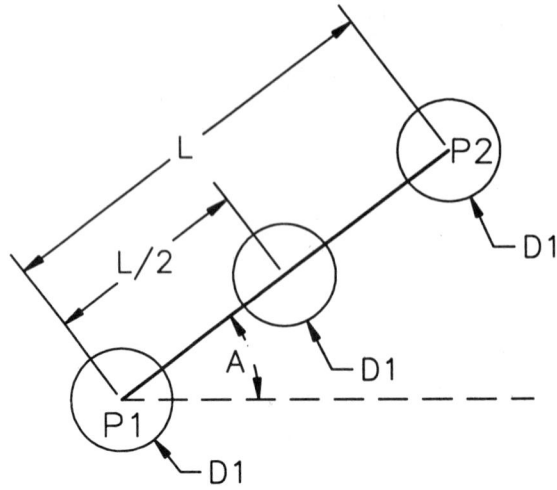

Figure 12-30 Circles and line making an angle A with X-axis

Exercise 8

Write an AutoLISP program that will draw an isosceles triangle P1, P2, P3 (Figure 12-31). The program should prompt you to enter the starting point P1, length L1, and the included angle A.

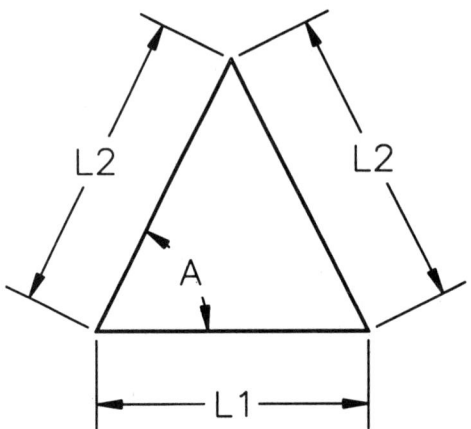

Figure 12-31 Isosceles triangle

Exercise 9

Write an AutoLISP program that will draw a parallelogram with sides S1, S2 and angle W as shown in Figure 12-32. The program should prompt you to enter the starting point PT1, lengths S1, S2, and the included angle W.

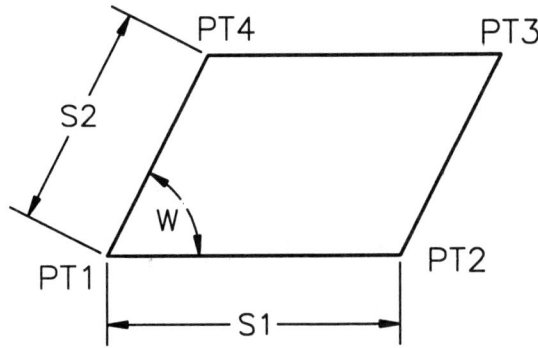

Figure 12-32 Parallelogram with sides S1, S2, and angle W

Exercise 10

Write an AutoLISP program that will draw a square of sides S and a circle tangent to the four sides of the square as shown in Figure 12-33. The base of the square makes an angle, ANG, with the positive X-axis. The program should prompt you to enter the starting point P1, length S, and angle ANG.

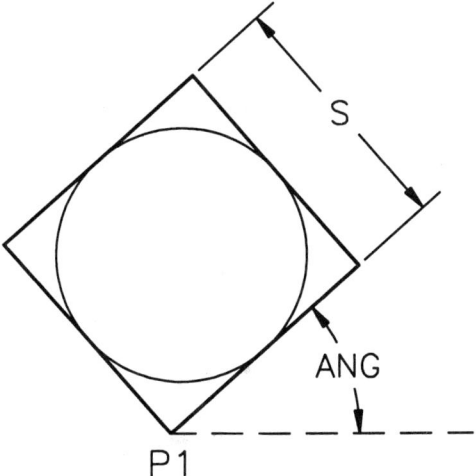

Figure 12-33 Square of side S at an angle ANG

Exercise 11

Write an AutoLISP program that will draw an equilateral triangle inside the circle (Figure 12-34). The program should prompt you to enter the radius and the center point of the circle.

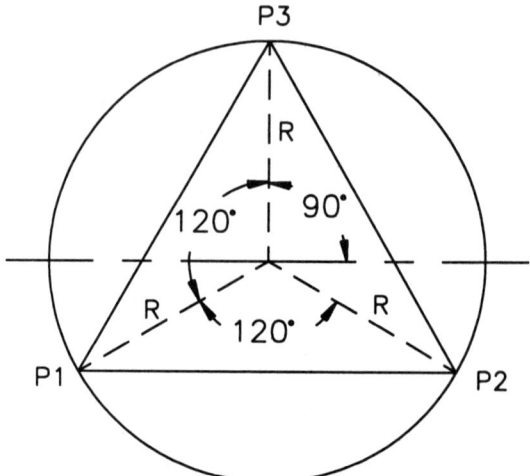

Figure 12-34 Equilateral triangle inside a circle

Exercise 12

Write an AutoLISP program that will delete all objects contained within the upper (limmax) and lower (limmin) limits. Use AutoCAD's SETVAR and ERASE commands to delete the objects.

Exercise 13

Write an AutoLISP program that will draw two lines tangent to two circles as shown in Figure 12-35. The program should prompt you to enter the circle diameters and the center distance between the circles.

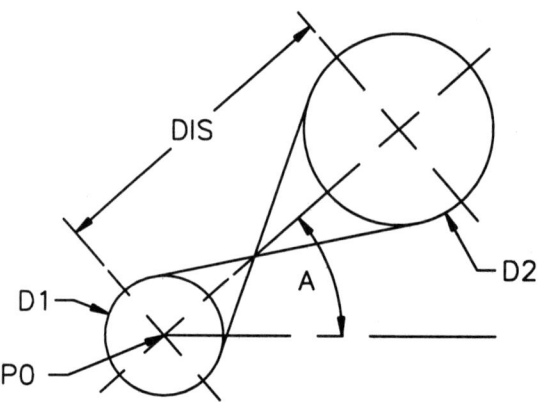

Figure 12-35 Circle with tangent lines at an angle A

Exercise 14

Write a program that will draw a slot with center lines. The program should prompt you to enter slot length, slot width, and the layer name for center lines (Figure 12-36).

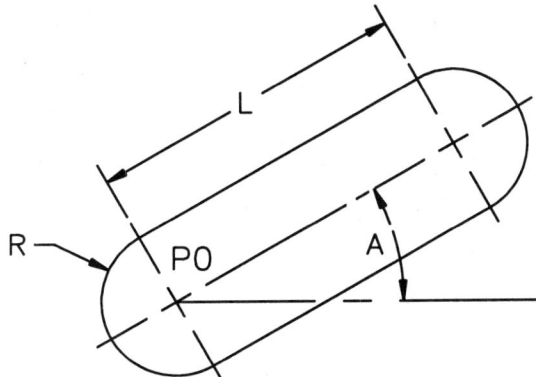

Figure 12-36 Slot of length L and radius R

Exercise 15

Write an AutoLISP program that will draw a line and then generate a given number of lines (N), parallel to the first line (Figure 12-37).

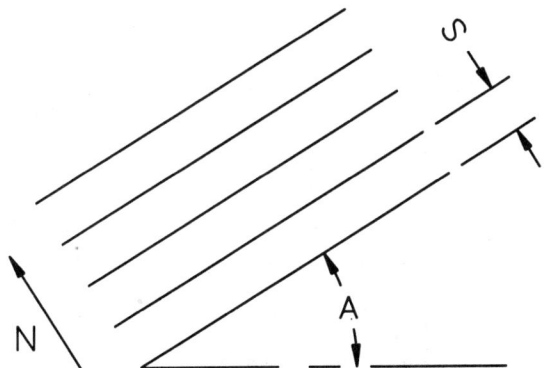

Figure 12-37 N number of lines offset at a distance S

Exercise 16

Write an AutoLISP program to draw a circle with center lines. The program should prompt for the diameter of circle, center of circle, and the angle of center lines as shown in Figure 12-38.

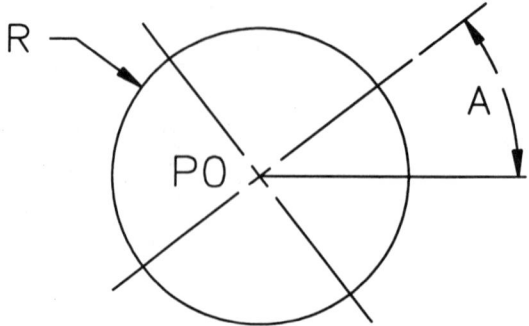

Figure 12-38 Circle with center lines at an angle A

Exercise 17

Write a program to draw a keyway slot. The program should prompt you to enter the width of slot, depth of slot, angle of slot, and starting point (Figure 12-39).

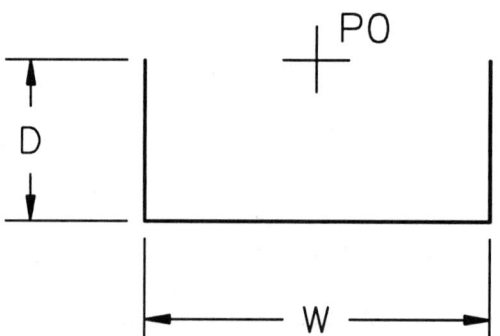

Figure 12-39 Keyway slot of width W and depth D

Exercise 18

Write an AutoLISP program that will draw the figure as shown in Figure 12-40 with center lines and dimensions. Assume, L5=D1, L3=1.5*D1, L6=10*D1, L1= L6-D1, L4=L3+D1.

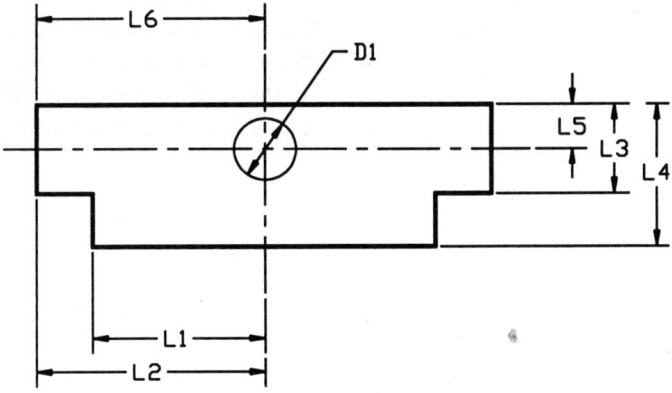

Figure 12-40 Drawing for Exercise 18

Exercise 19

Write an AutoLISP program that will draw a hub with the key slot as shown in Figure 12-41. The program should prompt the user to enter the values for P0 (Center of hub/shaft, D1 (Diameter of shaft), D2 (Outer diameter of hub), W (Width of key), and H (Height of key). The program should also draw the center lines in Center layer (Green color) and draw dimension T and W in Dim layer (Magenta color).

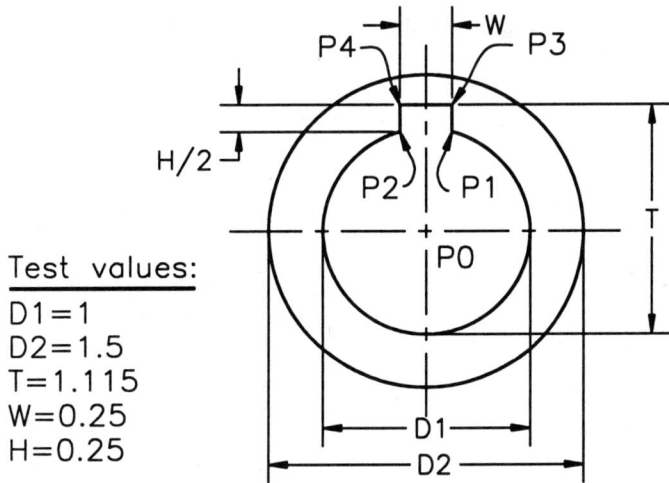

Figure 12-41 Drawing for Exercise 19

Exercise 20

Write an AutoLISP program that will draw the two views of a bushing as shown in Figure 12-42. The program should prompt you to enter the starting point P0, lengths L1, L2, and the bushing diameters ID, OD, HD. The distance between the front view and the side view of bushing is DIS (DIS = 1.25 * HD). The program should also draw the hidden lines in the HID layer and center lines in the CEN layer. The center lines should extend 0.75 units beyond the object line.

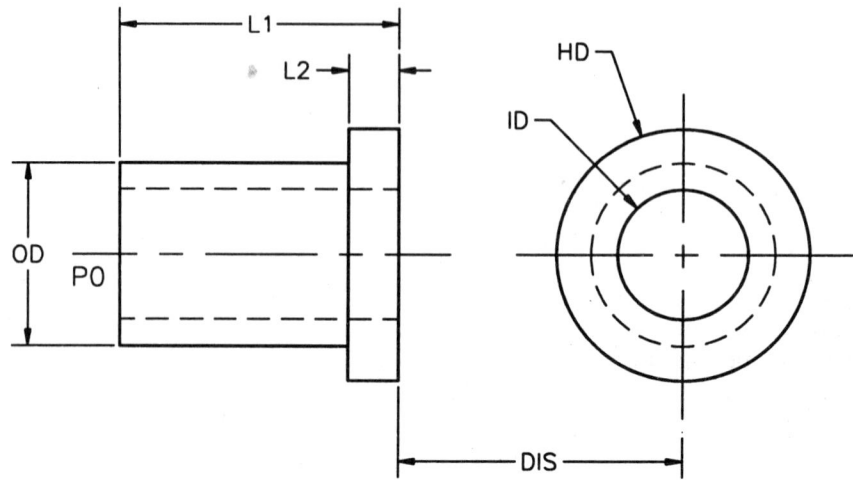

Figure 12-42 Two views of bushing

Chapter 13

AutoLISP: Editing the Drawing Database

Learning objectives

After completing this chapter, you will be able to:
- *Edit the drawing database using AutoLISP.*
- *Use the* **ssget**, **sslength**, **ssname**, **entget**, **assoc**, **cons**, **subst**, *and* **entmod** *functions.*
- *Retrieve information from the drawing database.*
- *Edit the database and substitute the values back into the drawing database.*

EDITING THE DRAWING DATABASE

In addition to writing programs to create new commands, you can use the AutoLISP programming language to edit the drawing database. This is a powerful tool to make changes in a drawing. For example, you can write a program that will delete all text objects in the drawing or change the layer and color of all circles just by entering one command. Once you understand how AutoCAD stores the information of the drawing objects and how it can be retrieved and edited, you can manipulate the database any way you want to, limited only by your imagination.

This chapter discusses some of the commands that are frequently used to edit the drawing database. For other commands not discussed in this section, please refer to the online reference manual, *AutoLISP Programmers Reference*, published by Autodesk.

ssget

The **ssget** function enables you to select any number of objects in a drawing. The object selection modes (window, crossing, previous, last, etc.) and the points that define the corners of the window can be included in the **ssget** assignment. The format of the **ssget** function is:

```
(ssget [selection-mode] [point1 point2])
                                    │
                                    └─ Second point of
                                       window (optional)
                         └─ First point of window (optional)
            └─ Object selection mode (w,c,l,p,etc.)
```

Examples

(ssget)	For general object selection
(ssget "L")	For selecting last object
(ssget "p")	For selecting previous selection set
(ssget "w" (list 0 0) (list 12.0 9.0))	Object selection using window object selection mode, where the window is defined by points 0,0 and 12.0,9.0
(ssget "c" pt1 pt2)	Object selection using crossing object selection mode, where the window is defined by predefined points pt1 and pt2

Example 1

Write an AutoLISP program that will erase all objects within the drawing limits, limmax, and limmin. Use the **ssget** function to select the objects.

The following file is a listing of the AutoLISP program for Example 1. The line numbers are not a part of the program; they are shown here for reference only.

```
;This program will delete all objects                    1
;that are within the drawing limits                      2
;                                                        3
(defun c:delall()                                        4
   (setvar "cmdecho" 0)                                  5
   (setq pt1 (getvar "limmin"))                          6
   (setq pt2 (getvar "limmax"))                          7
   (setq ss1 (ssget "c" pt1 pt2))                        8
   (command "erase" ss1 "")                              9
   (command "redraw")                                   10
   (setvar "cmdecho" 1)                                 11
   (princ)                                              12
)                                                       13
```

Lines 1-3
The first three lines are comment lines that describe the function of the program. Notice that all comment lines start with a semicolon (;).

Line 4
(defun c:delall()
In this line the **defun** function defines function delall.

Line 6
(setq pt1 (getvar "limmin"))
The **getvar** function secures the value of the lower left corner of the drawing limits (limmin) and the **setq** function assigns that value to variable pt1.

Line 7
(setq pt2 (getvar "limmax"))
The **getvar** function secures the value of the upper right corner of the drawing limits (limmax) and the **setq** function assigns that value to variable pt2.

Line 8
(setq ss1 (ssget "c" pt1 pt2))
The **ssget** function uses the "crossing" objection selection mode to select the objects that are within or touching the window defined by points pt1 and pt2. The **setq** function then assigns this object selection set to variable ss1.

Line 9
(command "erase" ss1 "")
The **command** function uses AutoCAD's **ERASE** command to erase the predefined object selection set ss1.

Line 10
(command "redraw")
In this line the **command** function uses AutoCAD's **REDRAW** command to redraw the screen and get rid of the blip marks left after erasing the objects.

ssget "X"

The **ssget "X"** function enables you to select specified types of objects in the entire drawing database, even if the layers are frozen or turned off. The format of the **ssget "X"** function is:

```
(ssget "X" specified-criteria)
           │         │
           │         └─ List of the specified criteria
           │            for selecting the objects
           └─ Filter mode of the ssget function
```

Examples
(ssget "X" (list (cons 0 "TEXT")))   returns a selection set that consists of all TEXT objects
                                         in the drawing

(ssget "X" (list (cons 7 "ROMANC"))) returns a selection set that consists of all TEXT objects
                                         in the drawing with the text style name ROMANC

(ssget "X" (list (cons 0 "LINE")))   returns a selection set that consists of all LINE objects
                                         in the drawing

(ssget "X" (list (cons 8 "OBJECT")))   returns a selection set that consists of all objects in the OBJECT layer

The **ssget "X"** function can contain more than one selection criteria. This option can be used to select a specific set of objects in a drawing. For example, if you want to select the LINE objects in the OBJECT layer, there are two selection criteria. The first is that the object has to be a LINE; the second is that the LINE object has to be in the OBJECT layer. As shown in Example 1, these two selection criteria can be combined to filter out the objects that satisfy these two conditions.

(ssget "X" (list (cons 0 "LINE")(cons 8 "OBJECT")))

Group Codes for ssget "X"

The following table is a list of AutoCAD group codes that can be used with the function **ssget "X"**:

Group Code	Code Function
0	Object type
2	Block name for block reference
3	Dimension object DIMSTYLE name
6	Linetype name
7	Text style name
8	Layer name
38	Elevation
39	Thickness
62	Color number
66	Attributes
210	3D extrusion direction

Example 2

Write an AutoLISP program that will erase all text objects in a drawing on a specified layer. Use the filter option of the **ssget** function (ssget "X") to select text objects in the specified layer.

The following file is a listing of the AutoLISP program for Example 2:

```
;This program will delete all text
;in the user-specified layer
;
(defun c:deltext()
   (setvar "cmdecho" 0)
   (setq layer (getstring "\n Enter layer name: "))
   (setq ss1 (ssget "x" (list (cons 8 layer) (cons 0 "text"))))
   (command "erase" ss1 "")
   (command "redraw")
   (setvar "cmdecho" 1)
   (princ))
```

sslength

The **sslength** function determines the number of objects in a selection set and returns an integer corresponding to the number of objects found. The format of the **sslength** function is:

```
(sslength selection-set)
                └── Name of the selection set
```

Examples
(setq ss1 (ssget))
(setq num (sslength ss1))   returns the number of objects in the predefined selection set ss1

(setq ss2 (ssget "l"))
(setq num (sslength ss2))   returns the number of objects (1) in selection set ss2, where selection set ss2 has been defined as the last object in the drawing

ssname

The **ssname** function returns the name of the object, from a predefined selection set, as referenced by the index that designates the object number. The name of the object returned by this function is in the hexadecimal format (such as 60000014). The format of the **ssname** function is:

```
(ssname selection-set index)
            │              └── Index designates the object
            │                   number in a selection set
            └── A predefined selection set
```

Examples
(setq ss1 (ssget))
(setq index 0)
(setq entname (ssname ss1 index))   returns the name of the first object contained in the predefined selection set ss1

Note
If the index is 0, the ssname function returns the name of the first object in the selection set. Similarly, if the index is 1, it returns the name of the second object.

entget

The **entget** function retrieves the object list from the object name. The name of the object can be obtained by using the function ssname. The format of the **entget** function is:

```
(entget object-name)
              └── Name of the object obtained by ssname
```

Examples
(setq ss1 (ssget))
(setq index 0)
(setq entname (ssname ss1 index))
(setq entlist (entget entname))   returns the list of the first object from the variable, entname, and assigns the list to the variable, entlist

assoc

The **assoc** function searches for a specified code in the object list and returns the element that contains that code. The format of the **assoc** function is:

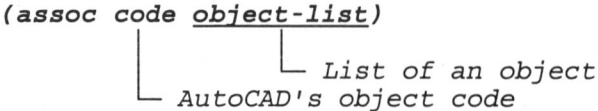

Examples
(setq ss1 (ssget))
(setq index 0)
(setq entname (ssname ss1 index))
(setq entlist (entget entname))
(setq entasso (assoc 0 entlist))   returns the element associated with AutoCAD's object code 0, from the list defined by the variable entlist

cons

The **cons** function constructs a new list from the given elements or lists. The format of the **cons** function is:

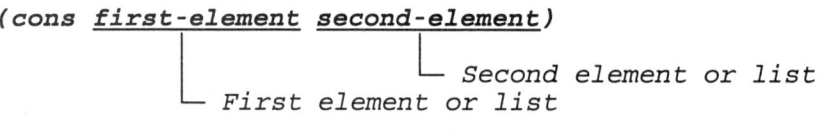

Examples
(cons 'x 'y)   returns (X . Y)
(cons '(x y) 'z)   returns ((X Y) . Z)
(cons '(x y z) '(0.5 5.0))   returns ((X Y Z) 0.5 5.0)

subst

The **subst** function substitutes the new item in place of old items. The old items can be a single item or multiple items, provided they are in the same list. The format of the **subst** function is:

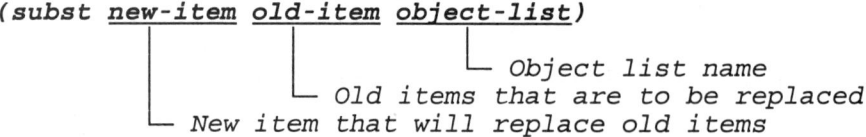

Examples
(setq entlist '(x y x))
(setq newlist (subst '(z) '(x) entlist)   returns (z y z); the **subst** function replaces x in the
                                           object list (entlist) by z

entmod
The **entmod** function updates the drawing by writing the modified list back to the drawing database. The format of the **entmod** function is:

(entmod *object-list*)
         └─ *Name of the modified object list*

Example 3
Write an AutoLISP program that will enable you to change the height of a text object. The program should prompt you to enter the new height of the text.

Input	Output
New text height	Text with new text height
Text object	

Process
1. Select the text object; obtain the name of the object using the function **ssname**.
2. Extract the list of the object using the function **entget**.
3. Separate the element associated with AutoCAD object code 0 from the list using the function **assoc**.
4. Construct a new element where the height of text is changed to a new height using the function **cons**.
5. Substitute the new element back into the original list using the **subst** function.
6. Update the drawing database using the **entmod** function.

The following file is a listing of the AutoLISP program for Example 3. The line numbers are not a part of the program; they are shown here for reference only.

```
;This program changes the height of the                        1
;selected text, only one text at a time.                        2
;                                                               3
(defun c:chgtext1()                                             4
(setvar "cmdecho" 0)                                            5
(setq newht (getreal "\n Enter new text height: "))             6
(setq ss1 (ssget))                                              7
(setq name (ssname ss1 0))                                      8
(setq ent (entget name))                                        9
```

```
    (setq oldlist (assoc 40 ent))                              10
    (setq conlist (cons (car oldlist) newht))                  11
    (setq newlist (subst conlist oldlist ent))                 12
    (entmod newlist)                                           13
    (setvar "cmdecho" 1)                                       14
    (princ)                                                    15
)                                                              16
```

HOW THE DATABASE IS RETRIEVED AND EDITED

To change the objects in a drawing you need to understand the structure of the drawing database and how it can be manipulated. Once you understand this concept it is easy, and sometimes fun, to edit the drawing database and the drawing. The following step-by-step explanation describes the process involved in changing the height of a selected text object in a drawing. Assume that the text that needs to be edited is "CHANGE TEXT" and that this text is already drawn on the screen. The height of the text is 0.3 units. Before going through the following steps, load the AutoLISP program from Example 3, and run it so that the variables are assigned a value.

Step 1
Select the text using the function **ssget** or **ssget "X"** and assign it to variable ss1. AutoCAD creates a selection set that could have one or more objects. In line 7 **(setq ss1 (ssget))** of the program for Example 3, the selection set is assigned to variable ss1. Use the following command to check the variable ss1:

    Command: !ss1
    <Selection set: 2>

Step 2
There could be several objects in a selection set and these objects need to be separated, one at a time, before any change is made to an object. This is made possible using the function **ssname** which extracts the name of an object. The index number used in the function **ssname** determines the object whose name is being extracted. For example, if the index is 0 the **ssname** function will extract the name of the first object, if the index is 1 the **ssname** function will extract the name of second object, and so on. In line 8 **(setq name (ssname ss1 0))** of the program, the **ssname** extracts the name of the first object and assigns it to the variable name. Use the following command to check the variable, name:

    Command: !name
    <Object name: 60000018>

Step 3
Extract the object list using the function **entget**. In line 9 **(setq ent (entget name))** of the program, the value of the list has been assigned to the variable ent. Use the following command to check the value of the variable ent.

Command: !ent
((-1.<Object name: 600000018> (0 . "TEXT") (8 . "0") (10 4.91227 5.36301 0.0) **(40 . 0.3)** (1 . "CHANGE TEXT") (50 .0.0) (41 . 1.0) (51 .0.0) (7 . "standard") (71 .0)) (72 . 1) (11 6.51227 5.36302 0.0) (210 0.0 0.0 1.0))

This list contains all the information about the selected text object (CHANGE TEXT), but you are only interested in changing the height of the text. Therefore, you need to identify the element that contains the information about the text height (40 . 0.3) and separate that from the list.

Step 4
Use the function **assoc** to separate the element that is associated with code 40 (text height). The statement in line 10 **(setq oldlist (assoc 40 ent))** of the program uses the **assoc** function to separate the value and assign it to the variable oldlist. Use the following command to check the value of this variable:

Command: **!oldlist**
(40 . 0.3)

Step 5
The **(40 . 0.3)** element consists of the code for text (40), and the text height (0.3). To change the height of the old text, the text height value (0.3) needs to be replaced by the new value. This is accomplished by constructing a new list as described in line 11 **(setq conlist (cons (car oldlist) newht))** of the program. This line also assigns the new element to variable conlist. For example, if the value assigned to variable newht is 0.5, the new element will be (40 . 0.5). Use the following command to check the value of conlist.

Command: **!conlist**
(40 . 0.5)

Step 6
After constructing the new element, use the **subst** function to substitute the new element back into the original list, ent. This is accomplished by line 12 **(setq newlist (subst conlist oldlist ent))** of the program. Use the following command to check the value of the variable newlist:

Command: **!newlist**
((-1.<Object name: 600000018> (0 . "TEXT") (8 . "0") (10 4.91227 5.36301 0.0) **(40 . 0.5)** (1 . "CHANGE TEXT") (50 .0.0) (41 . 1.0) (51 .0.0) (7 . "standard") (71 .0)) (72 . 1) (11 6.51227 5.36302 0.0) (210 0.0 0.0 1.0))

Step 7
The last step is to update the drawing database. Do this by using the function **entmod** as shown in line 13 **(entmod newlist)** of the program.

Example 4

Write an AutoLISP program that will enable you to change the height of all text objects in a drawing (Figure 13-1). The program should prompt you to enter the new text height.

The following file is a listing of the AutoLISP program for this example. The line numbers are not a part of the program; they are for reference only.

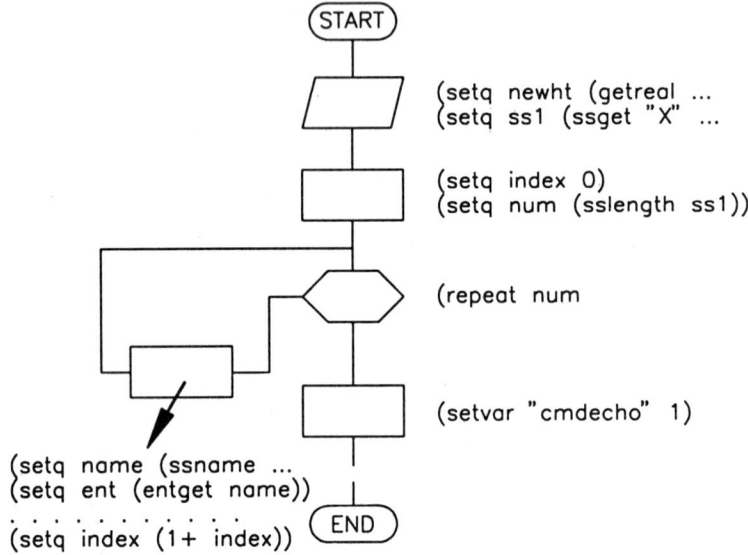

Figure 13-1 Flowchart for Example 4

```
;This program changes the height of                                    1
;all text objects in a drawing.                                        2
;                                                                      3
(defun c:chgtext2()                                                    4
   (setvar "cmdecho" 0)                                                5
   (setq newht (getreal "\n Enter new text height: "))                 6
   (setq ss1 (ssget "x" (list (cons 0 "text"))))                       7
   (setq index 0)                                                      8
   (setq num (sslength ss1))                                           9
   (repeat num                                                         10
      (setq name (ssname ss1 index))                                   11
      (setq ent (entget name))                                         12
      (setq oldlist (assoc 40 ent))                                    13
      (setq conlist (cons (car oldlist) newht))                        14
      (setq newlist (subst conlist oldlist ent))                       15
      (entmod newlist)                                                 16
      (setq index (1+ index))                                          17
   )                                                                   18
   (setvar "cmdecho" 1)                                                19
```

```
      (princ)                                                          20
)                                                                      21
```

Line 7
(setq ss1 (ssget "x" (list (cons 0 "text"))))
The **ssget** "X" function filters the text objects from the drawing database. The **setq** function assigns that selected set of text objects to variable **ss1**.

Line 8
(setq index 0)
The **setq** function sets the value of the **index** variable to 0. This variable is used later to select different objects.

Line 9
(setq num (sslength ss1))
The function **sslength** determines the number of objects in the selection set ss1 and the **setq** function assigns that number to the **num** variable.

Line 10
(repeat num
The **repeat** function will repeat the processes defined within the repeat function **num** number of times.

Example 5

Write an AutoLISP program that will enable you to change the height of the selected text objects in a drawing (Figure 13-2). The program should prompt you to enter the new text height.

```
;This program changes the height of the
;selected text objects.
;
(defun c:chgtext3()
(setvar "cmdecho" 0)
(setq newht (getreal "\n Enter new text height: "))
(setq ss1 (ssget))
(setq index 0)
(setq num (sslength ss1))
(repeat num
   (setq name (ssname ss1 index))
   (setq ent (entget name))
   (setq ass (assoc 0 ent))
   (setq index (1+ index))
   (If (= "TEXT" (cdr ass))
       (progn
       (setq oldlist (assoc 40 ent))
       (setq conlist (cons (car oldlist) newht))
```

13-12 Customizing AutoCAD

```
            (setq newlist (subst conlist oldlist ent))
            (entmod newlist)
            )
         )
      )
(setvar "cmdecho" 1)
(princ)
)
```

Flowchart

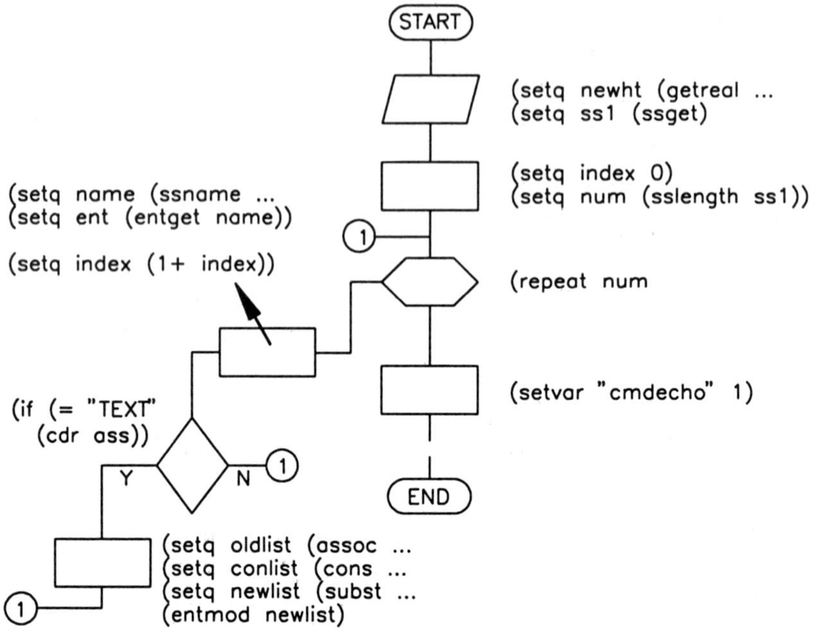

```
                                    ( START )
                                        |
                                       /  /  (setq newht (getreal ...
                                      /__/   (setq ss1 (ssget)
                                        |
                                     [      ]  (setq index 0)
                                     [      ]  (setq num (sslength ss1))
                                        |
  (setq name (ssname ...              /‾‾‾\
  (setq ent (entget name))     ──(1)─<     >   (repeat num
                                         \___/
  (setq index (1+ index))                  |
          ↑                            [      ]  (setvar "cmdecho" 1)
        [      ]                          |
           |                            ( END )
  (if (= "TEXT"      /\
      (cdr ass))    /  \
              Y ───<    >─── N ──(1)
                    \  /
                     \/
                     |
                [      ]  (setq oldlist (assoc ...
          (1)── [      ]  (setq conlist (cons ...
                [      ]  (setq newlist (subst ...
                         (entmod newlist)
```

Figure 13-2 Flowchart of Example 5

REVIEW QUESTIONS

1. In addition to writing programs to create new commands, you can use the AutoLISP programming language to edit the drawing database. (T/F)

2. The _____ function enables you to select any number of objects in a drawing.

3. The _____ function enables you to select specified types of objects in the entire drawing database, even if the layers are frozen or turned off.

4. The _____ function determines the number of objects in a selection set and returns an integer corresponding to the number of objects found.

5. The _____ function returns the name of the object, from a predefined selection set, as referenced by the index that designates the object number.

6. The _____ function retrieves the object list from the object name.

7. The _____ function searches for a specified code in the object list and returns the element that contains that code.

8. The _____ function constructs a new list from the given elements or lists.

9. The _____ function substitutes the new item in place of old items.

EXERCISES

Exercise 1
Write an AutoLISP program that will enable you to change the layer of the selected objects in a drawing. The program should prompt you to enter the new layer name.

Exercise 2
Write an AutoLISP program that will change the text style name of the selected text objects in a drawing. The program should prompt you to enter the new text style.

Exercise 3
Write an AutoLISP program that will change the layer of the selected objects in a drawing to a new layer. You should be able to enter the new layer by selecting an object in that layer.

Chapter 14

Programmable Dialog Boxes Using Dialog Control Language

Learning objectives

After completing this chapter, you will be able to:
- *Write programs using dialog control language.*
- *Use predefined attributes.*
- *Load a dialog control language (DCL) file.*
- *Display new dialog boxes.*
- *Use standard button subassemblies.*
- *Use AutoLISP functions to control dialog boxes.*
- *Manage dialog boxes with AutoLISP.*
- *Use tiles, buttons, and attributes in DCL programs.*

DIALOG CONTROL LANGUAGE

Dialog control language (**DCL**) files are ASCII files that contain the descriptions of dialog boxes. A DCL file can contain the description of a single or multiple dialog boxes. There is no limit to the number of dialog box descriptions that can be defined in a DCL file. The suffix of a DCL file is **.DCL** (such as **DDOSNAP.DCL**.)

This chapter assumes that you are familiar with AutoCAD commands, AutoCAD system variables, and AutoLISP programming. You need not be a programming expert to learn writing programs for dialog boxes in DCL or to control the dialog boxes through AutoLISP programing. However, knowledge of any programming language should help you to understand and learn DCL. This chapter introduces you to the basic concepts of developing a dialog box, frequently used attributes, and tiles. A thorough discussion of DCL functions and a step-by-step explanation of examples should make it easy for you to learn DCL. For those functions not discussed in this chapter, you can refer to the *AutoCAD Customization Guide from Autodesk*. To write programs in DCL, you need no special software or hardware. If AutoCAD is installed on your computer, you can write DCL files. To write DCL files, you can use any text editor.

DIALOG BOX

A dialog control language (DCL) file contains the description of how the dialog boxes will appear on the screen. These boxes can contain buttons, text, lists, edit boxes, rows, columns, sliders, and images. A sample dialog box is shown in Figure 14-1.

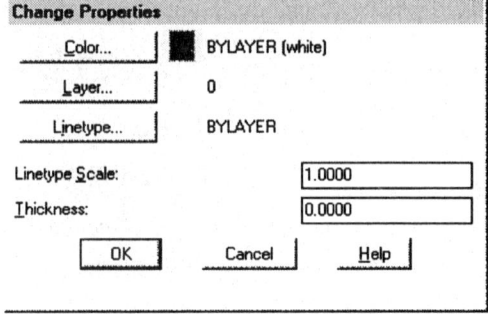

You do not need to specify the size and layout of a dialog box or its component parts. Sizing is done automatically when the dialog box is loaded on the screen. By itself, the dialog box cannot perform the functions it is designed for. The functions of a dialog box are controlled by a program written in the AutoLISP programming language or with the AutoCAD development system (ADS), or ARX. For example, if you load a dialog box and select the Cancel button, it will not perform the cancel operation. The instructions associated with a button or any part of the dialog box are handled through the

Figure 14-1 Change Properties dialog box (**DDCHPROP**)

functions provided in AutoLISP, ADS, or ARX. Therefore, AutoLISP and ADS are needed to control the dialog boxes, and you should have, in addition to an understanding of DCL, a good knowledge of AutoLISP or ADS to develop new dialog boxes or edit existing ones.

Dialog boxes are not dependent on the platform; therefore, they can run on any system that supports AutoCAD. However, depending on the graphical user interface (GUI) of the platform, the appearance of the dialog boxes might change from one system to another. The functions defined in the dialog box will still work without making any changes in the dialog box or the application program (AutoLISP or ADS) that uses these dialog boxes.

DIALOG BOX COMPONENTS

The two major components of a dialog box are the tiles and the box itself. The tiles can be arranged in rows and columns in any desired configuration. They can also be enclosed in boxes or borders to form subassemblies, giving them a tree structure Figure 14-2(b). The basic tiles, such as buttons, lists, edit boxes, and images, are predefined by the programmable dialog box (PDB) facility of AutoCAD. These buttons are described in the file **base.dcl**. The layout and function of a tile is determined by the attribute assigned to it. For example, the height attribute controls the height of the tile. Similarly, the label attribute specifies the text that is associated with the tile. Some of the components of a dialog box are shown in Figure 14-2(a). Following that figure is a list of the predefined tiles and their format in DCL.

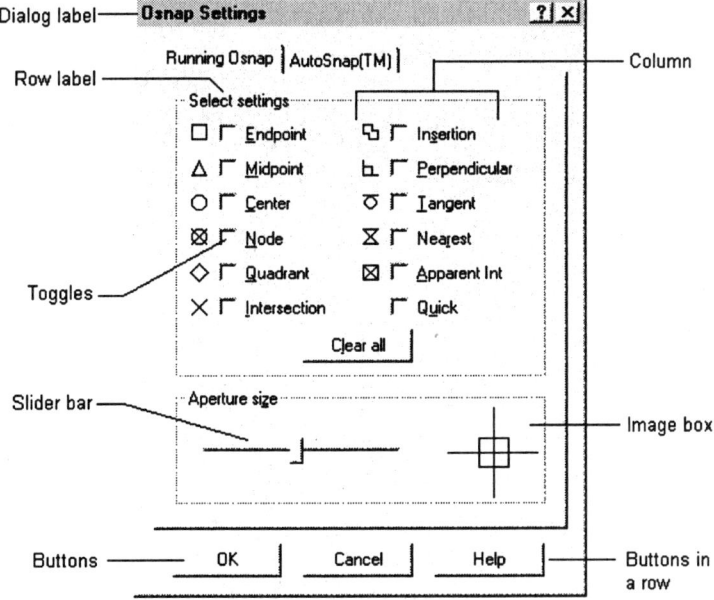

Figure 14-2(a) Components of a dialog box

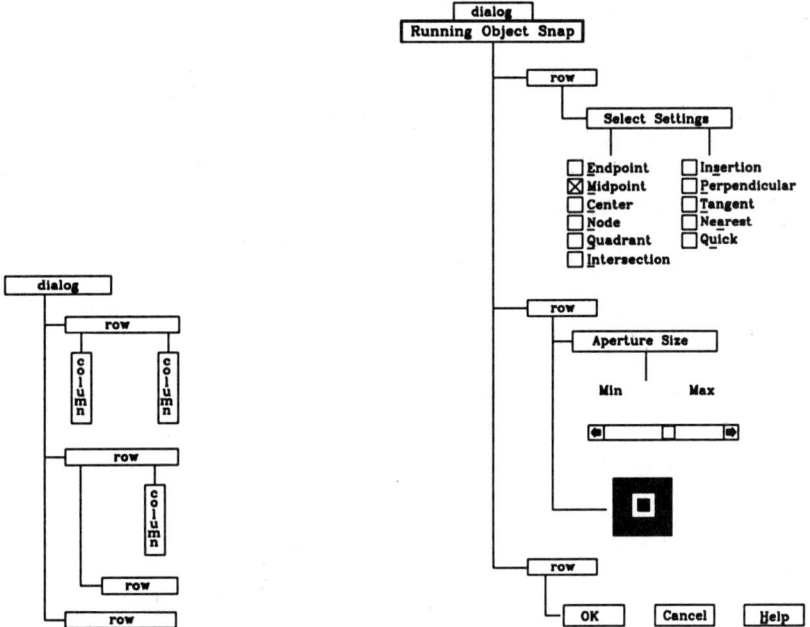

Figure 14-2(b) Tree structure of a dialog box

Predefined Tile	DCL Format
Button	button
Edit box	edit_box
Image button	image_button
List box	list_box
Pop-up list	popup_list
Radio button	radio_button
Slider	slider
Toggle	toggle
Column	column
Boxed column	boxed_column
Row	row
Boxed row	boxed_row
Radio column	radio_column
Boxed radio column	boxed_radio_column
Radio row	radio_row
Boxed radio row	boxed_radio_row
Image	image
Text	text
Spacer	spacer

BUTTON AND TEXT TILES

Button Tile

Format in DCL: **button**

The button tile consists of a rectangular box that resembles a push button. The button's label appears inside the button. For example, in the OK button of a dialog box, the label OK appears inside the button. If you select the OK button in a dialog box, it performs the functions defined in the dialog box and clears the dialog box from the screen. Similarly, if you select the cancel button, it cancels the dialog box without taking any action.

> **Note**
> *A dialog box should contain at least one OK button or a button that is equivalent to it. This allows you to exit the dialog box when you are done using it.*

Text Tile

Format in DCL: **text**

The text tile is used to display information or a title in a dialog box. It has limited application in the dialog boxes because most of the tiles have their own label attributes for tiling. However, if you need to display any text string in the dialog box, you can do so with the text tile. The text tile is used extensively in AutoCAD alert boxes to display warnings or error messages.

> **Note**
> *An alert box must contain an OK button or a Cancel button to end the dialog box.*

TILE ATTRIBUTES

The appearance, size, and function performed by a dialog box tile depend on the attributes that have been assigned to the tile. For example, if a button has been assigned the **fixed_width** attribute, the width of the box surrounding the button will not stretch through the entire length of the dialog box. Similarly, the **height** attribute determines the height of the tile, and the **key** attribute assigns a name to the tile that is then used by the application program. The tile attribute consists of two parts: the name of the attribute and the value assigned to the attribute. For example, in the expression **fixed_width = true**, **fixed_width** is the name of the attribute and **true** is the value assigned to the attribute. Attribute names are like variable names in programming, and the values assigned to these variables must be of a specific type.

Types of Attribute Values

Integer. Unlike integer values in programming (1, 15, 22), the numeric values assigned to attributes can be both integers and real numbers.

> **Examples**
> width = 15
> height = 10

Real Number. The real values assigned to attributes should always be fractional real numbers with a leading digit.

> **Example**
> aspect_ratio = 0.75

Quoted String. The string values assigned to an attribute consist of a text string that is enclosed in double quotes.

> **Examples**
> key = "accept"
> label = "OK"

Reserved Word

Dialog control language uses some reserved words as identifiers. These identifiers are alphanumeric characters that start with a letter.

> **Examples**
> is_default = true
> fixed_width = true

> **Note**
> *Reserved names are case-sensitive. For example, "is_default = true" is not the same as "is_default = True". The reserved word is "true", not "True" or "TRUE".*

> *Like reserved words, attribute names are also case-sensitive. For example, the attribute is "key", not "Key" or "KEY".*

PREDEFINED ATTRIBUTES

To facilitate writing programs in dialog control language, AutoCAD has provided some predefined attributes that are defined in the programmable dialog box (PDB) package that comes with the AutoCAD software. Some of these attributes can be used with any tile and some only with a particular type of tile. The values assigned to various attributes in a DCL file are used by the application program to handle the tiles or the dialog box. Therefore, you must use the correct attributes and assign an appropriate value to these attributes. The following is a list of some of the frequently used predefined attributes defined in the PDB facility of AutoCAD:

action	key
alignment	label
allow_accept	layout
aspect_ratio	list
color	max_value
edit_limit	min_value
edit_width	mnemonic
fixed_height	multiple_select
fixed_width	small_increment
height	tabs
is_cancel	value
is_default	width

key, label, AND is_default ATTRIBUTES

key Attribute

Format in DCL
key

Examples
key = "accept"
key = "XLimit"

The **key** attribute assigns a name to a tile. The name must be enclosed in double quotes. This name can then be used by the application program to handle the tile. A dialog box can have any number of key values, but within a particular dialog box the values used for the key attributes must be unique. In the example key = "accept", a string value "accept" is assigned to the **key** attribute. If there is another key attribute in the dialog box, you must assign it a different value.

label Attribute

Format in DCL
label

Examples
label = "OK"
label = "Hallo DCL Users"

Sometimes it is necessary to display a label in a dialog box. The label attribute can be used in a boxed column, boxed radio column, boxed radio row, boxed row, button tile, dialog box, or edit box. Some of the frequently used label attributes are described next.

Use of the label Attribute in a Dialog Box. When the label attribute is used in a dialog box, it is displayed in the top border or the title bar of the dialog box. The label must be a string enclosed in double quotes. Use of the label in a dialog box is optional; in case of default, no title is displayed in the dialog box.

Example
welcome : dialog {
    label = "Sample Dialog Box";

In this example, the label "Sample Dialog Box" is displayed at the top of the dialog box.

Use of the label Attribute in a Boxed Column. When the label attribute is used in a boxed column, the label is displayed within a box in the upper left corner of the column; the box consists of a single line at the top of the column. The label must be a quoted string, and the default is a set of quoted string (" "). If the default is used, only the box is displayed, without any label.

Example
: text {
   label = "Welcome to the world of DCL";
}

In this example, the label "Welcome to the world of DCL" is displayed within a box in the upper left corner of the column.

Use of the label Attribute in a Button. When the label attribute is used in a button, the label is displayed inside the button. The label must be a quoted string and has no default.

Example
: button {
  key = "accept";
  label = "OK";
}

In this example, the label "OK" is displayed inside and in the center of the button tile.

is_default Attribute

Format in DCL   **Example**
is_default   is_default = true

The **is_default** attribute is used for the button of a dialog box. In the example, the value assigned to the **is_default** attribute is **true**. Therefore, this button will be automatically selected when you press Enter. For example, if you load a dialog box on the screen, one way to exit and accept the values of the dialog box is to select the OK button. You can accomplish the same thing by pressing Enter (accept key). This action is made possible by assigning the value, **true**, to the **is_default** attribute. In a dialog box the default button can be recognized by the thick border drawn around the text string.

> **Note**
> *In a dialog box, only one button can be assigned the true value for the is_default attribute.*

fixed_width AND alignment ATTRIBUTES

fixed_width Attribute

Format in DCL   Example
fixed_width   fixed_width = true

This attribute controls the width of the tile. If the value of this attribute is set to true, the width of the tile does not extend across the complete width of the dialog box. The width of the tile is automatically adjusted to the length of the text string that is displayed in the tile.

alignment Attribute

Format in DCL   Example
alignment   alignment = centered
                                 alignment = right

The value assigned to the **alignment** attribute determines the horizontal or vertical position of the tile in a row or column. For a row, the values that can be assigned to this attribute are left, right, and centered. The default value of the alignment attribute is left; that forces the tile to be displayed left-justified. For a column, the possible values of the **alignment** attribute are top, bottom, and centered. The default value is centered.

Example 1

Using dialog control language (DCL), write a program for the following dialog box (Figure 14-3). The dialog box has two text labels and an OK button to end the dialog box.

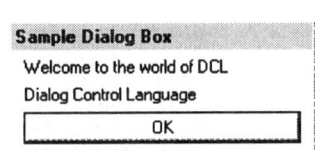

Figure 14-3 Dialog box for Example 1

The following file is a listing of the DCL file for the dialog box of Example 1. The name of this DCL file is **dclwel1.dcl**. The line numbers are not a part of the file; they are for reference only.

```
welcome1 : dialog {                                    1
          label = "Sample Dialog Box";                 2
          : text {                                     3
              label = "Welcome to the world of DCL";   4
```

```
                    }                                               5
                    : text {                                        6
                         label = "Dialog Control Language";         7
                    }                                               8
                    : button {                                      9
                         key = "accept";                           10
                         label = "OK";                             11
                         is_default = true;                        12
                    }                                              13
              }                                                    14
```

Explanation

Line 1
welcome1 : dialog {
In this line, **welcome1** is the name of the dialog and the definition of the dialog box is contained within the braces. The open brace in this line starts the definition of this dialog box.

Line 2
label = "Sample Dialog Box";
In this line, **label** is the label attribute, and **"Sample Dialog Box"** is the string value assigned to the label attribute. This string will be displayed in the title bar of the dialog box. The label description must be enclosed in quotes. If this line is missing, no title is displayed in the title bar of the dialog box.

Lines 3-5
: text {
label = "Welcome to the world of DCL";
}
These three lines define a text tile with the label description. In the first line, **text** refers to the text tile. The line that follows it, **label = "Welcome to the world of DCL";** defines a label for this tile that will be displayed left-justified in the dialog box. The closing brace completes the definition of this tile.

Lines 9-13
: button {
key = "accept";
label = "OK";
is_default = true;
}
These five lines define the attributes of the button tile. In the first line, **button** refers to the button tile. In the second line, **key = "accept";** specifies an ASCII name, **"accept"**, that will be used by the application program to refer to this tile. The next line, **is_default = true;**, specifies that this button is the default button. It is automatically selected if you press Enter at the keyboard.

The closing brace in line 14 completes the definition of the dialog box.

LOADING A DCL FILE

As with an AutoLISP file, you can load a DCL file from the AutoCAD drawing editor. A DCL file can contain the definition of one or several dialog boxes. There is no limit to the number of dialog boxes you can define in a DCL file. The format of the command for loading a dialog box is:

> **Note**
> *When you use the load_dialog and new_dialog functions to load and display a DCL program, the AutoCAD screen will freeze. To avoid this, it is recommended to use the AutoLISP program to load and display a DCL program. See "Using AutoLISP Function to Load a DCL File" at the end of this section.)*

```
(load_dialog filename)
         │            └─ Name of the DCL file, with or without
         │                              the file extension (.dcl)
         └─ Load command for loading a dialog file
```

Example
(load_dialog "dclwel1.dcl") or (load_dialog "dclwel1")

In this example, **dclwel1** is the name of the DCL file and **.dcl** is the file extension for DCL files.

The file name can be with or without the DCL file extension (**welcome1** or **welcome1.dcl**). The load function returns an integer value that is used as a handle in the **new_dialog** function and the **unload_dialog** function. This integer will be referred to as **dcl_id** in subsequent sections. You don't need to use the name dcl_id for this integer; it could be any name (dclid, or just id).

DISPLAYING A NEW DIALOG BOX

The load_dialog function loads the DCL file, but it does not display it on the screen. The format of the command used to display a new dialog box is:

```
(new_dialog dlgname dcl_id)
     │           │        └─ Integer returned in the
     │           │                   load_dialog function
     │           └─ Name of the dialog box
     └─ Command to display a dialog box
```

Example
(new_dialog "welcome1" 1)

In this example, assume that dcl_id is 1, an integer returned by the load_dialog function.

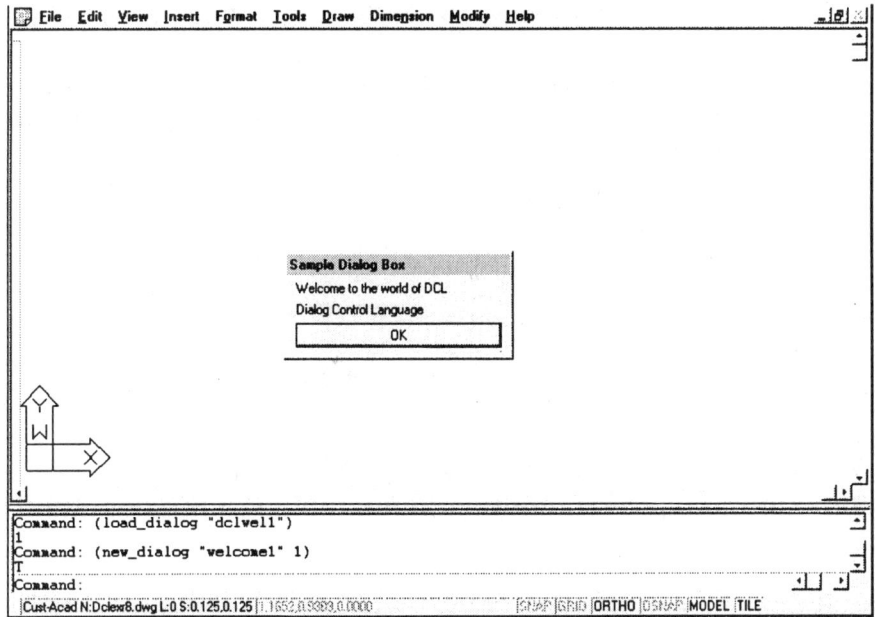

Figure 14-4 Dialog box as displayed on the screen

You can use the (load_dialog) and (new_dialog) commands to load the DCL file in Example 1. In the following command sequence, assume that the integer (dcl_id) returned by the (load_dialog) function is 3. Figure 14-4 shows the dialog box after entering the following two command lines:

Command: (load_dialog "dclwel1.dcl")
3
Command: (new_dialog "welcome1" 3)

In this example, notice that the OK button stretches across the complete width of the dialog box. This button would look better if its width were limited to the width of the string "OK". This can be accomplished by using the **fixed_width = true;** attribute in the definition of this button. You can also use the **alignment = centered** attribute to display the OK button label center-justified. The following file is a listing of the DCL file where the OK label for the OK button is center-justified and the width of the box does not stretch across the width of the dialog box (Figure 14-5).

```
welcome2 : dialog {
        label = "Sample Dialog Box";
        : text {
            label = "Welcome to the world of DCL";
        }
        : text {
            label = "Hello - DCL";
        }
        : button {
```

```
            key = "accept";
            label = "OK";
              is_default = true;
            fixed_width = true;         ◄————— (Controls width)
            alignment = centered;       ◄————— (Controls
        }                                              Justification)
}
```

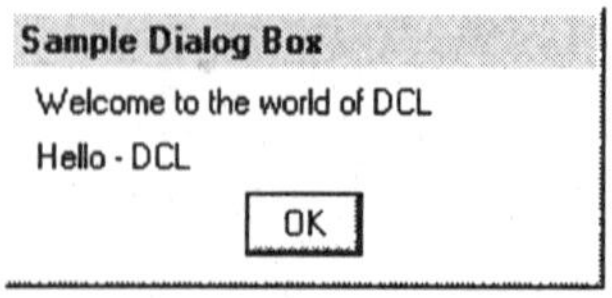

Figure 14-5 Dialog box with fixed width for OK button

Using AutoLISP Function to Load a DCL File

You can use the following AutoLISP program to load and display the dialog box without freezing the screen.

```
(defun c:load_dcl( / dcl_id )
    (setq dcl_id (load_dialog "dclwel1.dcl"))    (Loads the DCL file)
    (new_dialog "welcome1" dcl_id)               (Initializes the dialog box)
    (start_dialog)                               (Displays the dialog box)
    (princ)
)
```

Where: **load_dcl** is the name of the AutoLISP function.
         dclwel1 is the name of the DCL file that you want to load.
         welcome1 is the name of the dialog as defined in the DCL file.

USE OF STANDARD BUTTON SUBASSEMBLIES

Some standard button subassemblies are predefined in the **base.dcl** file. You can use these standard buttons in your DCL file to maintain consistency among various dialog boxes. One such predefined button is **ok_cancel**, which displays the OK and Cancel buttons in the dialog box (Figure 14-6). The following file is the listing of the DCL file of Example 1, using the **ok_cancel** predefined standard button subassembly:

```
welcome3 : dialog {
        label = "Sample Dialog Box";
        : text {
            label = "Welcome to the world of DCL";
        }
        : text {
            label = "Dialog Control Language";
            alignment = right;
        }
        ok_cancel;
}
```

Programmable Dialog Boxes Using Dialog Control Language   14-13

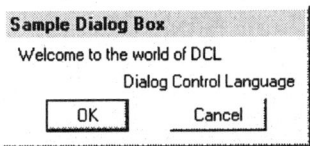

Figure 14-6 Dialog box with OK and Cancel buttons

The following is a list of the standard button subassemblies that are predefined in the **base.dcl** file. These buttons are also referred to as **dialog exit buttons** because they are used to exit a dialog box.

OK Button
Format in DCL: **ok_only**

OK and Cancel
Format in DCL: **ok_cancel**

OK, Cancel, and Help Buttons
Format in DCL: **ok_cancel_help**

AUTOLISP FUNCTIONS

load_dialog
The AutoLISP function **load_dialog** is used to **load a DCL file** that is specified in the load_dialog function. In the following examples, the name of the file that AutoCAD loads is "dclwel1". The extension (**.dcl**) of the file is optional. When the DCL file is loaded successfully, AutoCAD returns an integer that identifies the dialog box.

Format in DCL
(load_dialog filename)

Examples
(load_dialog "dclwel1.dcl")
(load_dialog "dclwel1")

unload_dialog
The AutoLISP function **unload_dialog** is used to **unload a DCL file** that is specified in the unload_dialog function. The file is specified by the variable (dcl_id) that identifies a DCL file.

Format in DCL
(unload_dialog dcl_id)

Example
(unload_dialog dcl_id)

new_dialog
The AutoLISP function **new_dialog** is used to initialize a dialog box and then display it on the screen. In the following file, **"welcome1"** is the name of the **dialog box**. (Note: welcome1 is not the name of the DCL file.) The variable **dcl_id** contains an integer value that is returned when the DCL file is loaded.

Format in DCL
(new_dialog "dialogname" dcl_id)

Example
(new_dialog "welcome1" dcl_id)

start_dialog

The AutoLISP function **start_dialog** is used in an AutoLISP program to accept user input from the dialog box. For example, if you select OK from the dialog box, the start_dialog function retrieves the value of that tile and uses it to take action and to end the dialog box.

Format in DCL
(start_dialog)

done_dialog

The AutoLISP function **done_dialog** is used to terminate display of the dialog box from the screen. This function must be defined within the action expression, as shown in the following example.

Format in DCL
(done_dialog)

Example
(action_tile "accept" "(done_dialog)")

action_tile

The AutoLISP function **action_tile** is used to associate an action expression with a tile in the dialog box. In the following example, the **action_tile** function associates the tile "accept" with the action expression (done_dialog) that terminates the dialog box. The "accept" is the name of the tile assigned to the OK button in the DCL file.

Format in DCL
(action_tile tile-name action-expression)

Example
(action_tile "accept" "(done_dialog)")

MANAGING DIALOG BOXES WITH AUTOLISP

When you load the DCL file of Example 1 and select the OK button, it does not perform the desired function (exit from the dialog box). This is because a dialog box cannot, by itself, execute the AutoCAD commands or the functions assigned to a tile. An application program is required to handle a dialog box. These application programs can be written in AutoLISP or ADS. The functions defined in AutoLISP and ADS can be used to load a DCL file, display the dialog box on the screen, prompt user input, set values in tiles, perform action associated with user input, and execute AutoCAD commands. Example 2 describes the use of an AutoLISP program to handle a dialog box.

Example 2

Write an AutoLISP program that will handle the dialog box and perform the functions shown in the dialog box of Example 1.

The following file is a listing of the DCL file of Example 1. This DCL file defines the dialog box **welcome1**, which contains only one action tile: "OK". If you select the **OK** button, you must be able to exit the dialog box. As just mentioned, the dialog box will not perform by itself the function assigned to the **OK** button unless you write an application that will execute the functions defined in the dialog box.

```
welcome1 : dialog {
          label = "Sample Dialog Box";
          : text {
              label = "Welcome to the world of DCL";
          }
          : text {
              label = "Dialog Control Language";
          }
          : button {
              key = "accept";
              label = "OK";
              is_default = true;
          }
}
```

The following file is a listing of the AutoLISP program that loads the DCL file (**dclwel1**) of Example 1, displays the dialog box **welcome1** on, and defines the action for the **OK** button. **The line numbers are not a part of the program; they are for reference only.**

```
(defun C:welcome ( / dcl_id)                        1
(setq dcl_id (load_dialog "dclwel1.dcl"))           2
(new_dialog "welcome1" dcl_id)                      3
(action_tile                                        4
    "accept"                                        5
    "(done_dialog)")                                6
(start_dialog)                                      7
(unload_dialog dcl_id)                              8
(princ)                                             9
)                                                   10
```

Explanation
Line 1
(defun C:welcome (/ dcl_id)
In this line, **defun** is an AutoLISP function that defines the function **welcome**. With the C: in front of the function name, the **welcome** function can be executed like an AutoCAD command. The **welcome** function has one local variable, **dcl_id**.

Line 2
(setq dcl_id (load_dialog "dclwel1.dcl"))
In this line, **(load_dialog "dclwel1.dcl")** loads the DCL file **dclwel1.dcl** and returns a positive integer. The **setq** function assigns this integer value to the local variable **dcl_id**.

Line 3
(new_dialog "welcome1" dcl_id)
In this line, the AutoLISP function **new_dialog** loads the dialog box **welcome1** that is defined in the DCL file (line 1 of DCL file). The variable **dcl_id** is an integer that identifies the DCL file.

Note
The dialog name (welcome1) in the AutoLISP program must be the same as the dialog name in the DCL file (welcome1).

Lines 4-6
(action_tile
  "accept"
  "(done_dialog)")

In the DCL file of Example 1, the **OK** button has been assigned an ASCII name, **accept** (key = "accept"). The first two lines associate the key (OK button) with the action expression. The action_tile initializes the association between the OK button and the action expression (done_dialog). If you select the **OK** button from the dialog box, the AutoLISP program reads that value and performs the function defined in the statement (done_dialog). The done_dialog function ends the dialog box.

Line 7
(start_dialog)

The **start_dialog** function enables the AutoLISP program to accept your input from the dialog box.

Line 8
(unload_dialog dcl_id)

This statement unloads the DCL file identified by the integer value of dcl_id.

Lines 9 and 10
(princ)
)

The **princ** function displays a blank on the screen. You use this to prevent display of the last expression in the command prompt area of the screen. If the **princ** function is not used, AutoCAD will display the value of the last expression. The closing parenthesis in the last line completes the definition of the welcome function.

ROW AND BOXED ROW TILES

Row Tile
Format in DCL: **row**

In a DCL file, several tiles can be grouped together to form a composite row or a composite column that is treated as a single tile. A row tile consists of several tiles grouped together in a horizontal row.

Boxed Row Tile
Format in DCL: **boxed_row**

In a boxed row, the tiles are grouped together in a row and a border is drawn around them, forming a box shape. If the boxed row has a label, it will be displayed left-justified at the top of the box, above the border line. If no label attribute is defined, only the box is displayed around the tile. On some systems, depending on the graphical user interface (GUI), the label may be displayed inside the border.

COLUMN, BOXED COLUMN, AND TOGGLE TILES

Column Tile

Format in DCL: **column**

In a column tile, the tiles are grouped together in a vertical column to form a composite tile.

Boxed Column Tile

Format in DCL: **boxed_column**

In a boxed column, the tiles are grouped together in a column and a border is drawn around the tiles. If the boxed column has a label, it will be displayed left-justified at the top of the box, above the border line. If there is no label attribute defined, only the box is displayed around the tile. On some systems, depending on the graphical user interface (GUI), the label may be displayed inside the border.

Toggle Tile

Format in DCL: **toggle**

A toggle tile in a dialog box displays a small box on the screen with an optional label on the right of the box. Although the label is optional, you should label the toggle box so that users know what function is assigned to the toggle box. A toggle box has two states: on and off. When the function is turned on, a check mark is displayed in the box. Similarly, when a function is turned off, no check mark is displayed. The on or off state of the toggle box represents a Boolean value. When the toggle box is on, the Boolean value is 1; when the toggle box is off, the Boolean value is 0.

MNEMONIC ATTRIBUTE

Format in DCL
mnemonic

Example
mnemonic = "U"

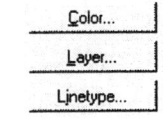

A dialog box can have several tiles with labels. One of the ways you can select a tile is by using the arrow to highlight the tile and then pressing the accept key. This is possible if you have a pointing device like a digitizer or a mouse. If you do not have a pointing device, it may not be possible to select a tile. However, you can select a tile by using the mnemonic key assigned to the tile. For example, the mnemonic character for the Color tile is C. If you press the C key at the keyboard, it will highlight the Color tile. The mnemonic character is designated by underlining one of the characters in the label. The underlining is done automatically once you define the mnemonic attribute for the tile. You can select only one character **in the label** as a mnemonic character and you must use different mnemonic characters for different tiles. If a dialog box has two tiles with the same mnemonic character, only one tile will be selected when you press the mnemonic key. The mnemonic characters are not case-sensitive; therefore, they can be uppercase (C) or lowercase (c). However, the character in the label you select as a mnemonic character should be capitalized for easy identification.

> **Note**
> *Some graphical user interfaces (GUIs) do not support mnemonic attributes. On such systems, you cannot select a tile by pressing the mnemonic key. However, you can still use pointing devices to select the tile.*

Example 3

Write a DCL program for the object snap dialog box shown in Figure 14-7(a). The object snap tiles are arranged in two columns in a boxed row.

Before writing a program, especially a DCL program for a dialog box, you should determine the organization of the dialog box. It is given in Example 3. But when you develop a dialog box yourself, you must be careful when organizing the tiles in the dialog box. The structure of the DCL program depends on the desired output. In Example 3, the desired output is shown in Figure 14-7(a). This dialog box has two rows and two columns. The first row has two columns and the second row has no columns, as shown in Figure 14-7(b).

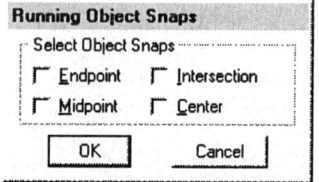

Figure 14-7(a) Dialog box for object snaps

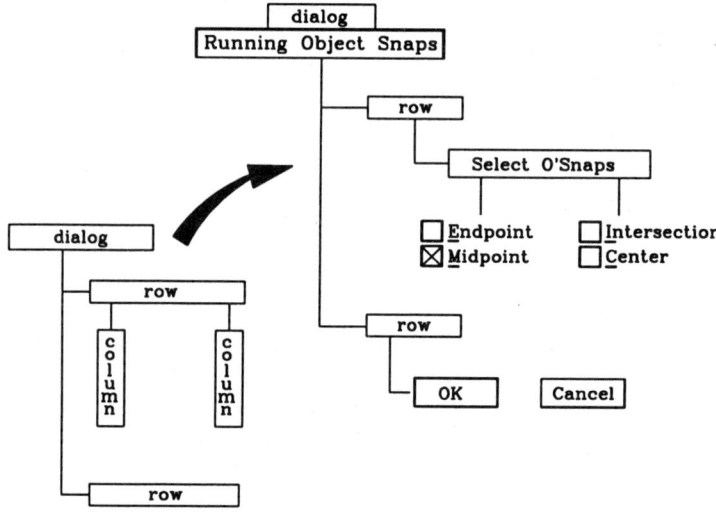

Figure 14-7(b) Dialog box for object snaps

The following file is a listing of the DCL file for Example 3. **The line numbers are not a part of the file; they are shown for reference only.**

```
osnapsh : dialog {                              1
        label = "Running Object Snaps";         2
      : boxed_row {                             3
        label = "Select Osnaps";                4
        : column {                              5
                : toggle {                      6
                 label = "Endpoint";            7
                 key = "Endpoint";              8
                 mnemonic = "E";                9
                 fixed_width = true;           10
                }                              11
                : toggle {                     12
                 label = "Midpoint";           13
                 key = "Midpoint";             14
                 mnemonic = "M";               15
                 fixed_width = true;           16
                }                              17
        }                                      18
      : column {                               19
                : toggle {                     20
                 label = "Intersection";       21
                 key = "Intersection";         22
                 mnemonic = "I";               23
                 fixed_width = true;           24
                }                              25
                : toggle {                     26
                 label = "Center";             27
                 key = "Center";               28
                 mnemonic = "C";               29
                 fixed_width = true;           30
                }                              31
            }                                  32
        }                                      33
        ok_cancel;                             34
}                                              35
```

Explanation
Lines 3 and 4
: boxed_row {
 label = "Select Osnaps";
The **boxed_row** is a predefined cluster tile that draws a border around the tiles. The second line, **label** = "Select Osnaps";, will display the label (**Select Osnaps**) left-justified at the top of the box. The **label** is a predefined DCL attribute.

Lines 5-7
: column {
 : toggle {

            label = "Endpoint";

The **column** is a predefined tile that will arrange the tiles within it into a vertical column. The **toggle** is another predefined tile that displays a small box in the dialog box with an optional text on the right side of the box. The **label** attribute will display the label (**Endpoint**) to the right of the toggle box.

Lines 8-11
**key = "Endpoint";
mnemonic = "E";
fixed_width = true;
}**
The **key** attribute assigns a name (**Endpoint**) to the tile. This name is then used by the application program to handle the tile. The second line, **mnemonic = "E";**, defines the keyboard mnemonic for **Endpoint**. This causes the character **E** of **Endpoint** to be displayed underlined in the dialog box. The attribute **fixed_width** controls the width of the tile. If the value of this attribute is **true**, the tile does not stretch across the width of the dialog box. The closing brace (**}**) on the next line completes the definition of the toggle tile.

Lines 31-35
            }
        }
    }
    ok_cancel;
}
The closing brace on line 32 completes the definition of the toggle tile, and the closing brace on line 33 completes the definition of the column of line 20. The closing brace on line 34 completes the definition of the boxed_row of line 3. The predefined tile, **ok_cancel**, displays the **ok** and **cancel** tiles in the dialog box. The closing brace on the last line completes the definition of the dialog box.

After writing the program, use the following commands to load the DCL file and display the dialog box on the screen. The name of the file is assumed to be **osnapsh.dcl**. If AutoCAD is successful in loading the file, it will return an integer. In the following example, the integer AutoCAD returns is assumed to be 1. The screen display after loading the dialog box is shown in Figure 14-8.

    Command: (load_dialog "osnapsh.dcl")
    1
    Command: (new_dialog "osnapsh" 1)

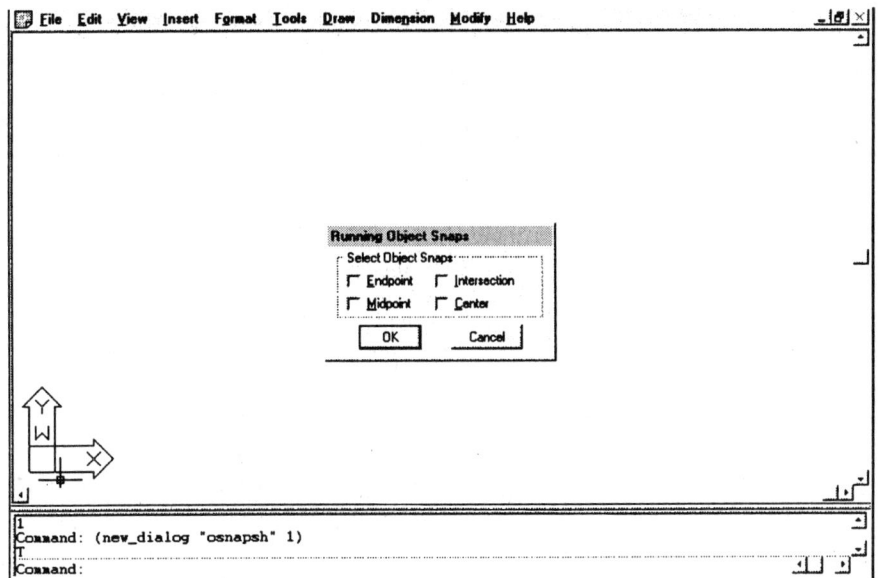

Figure 14-8 Screen display with dialog box for Example 3

AUTOLISP FUNCTIONS

logand and logior

These AutoLISP functions, **logand** and **logior**, are used to obtain the result of **logical bitwise AND** and **logical bitwise inclusive OR** of a list of numbers.

Examples of logand function
(logand 2 7)   will return   2
(logand 8 15)   will return   8
(logand 6 15 7)   will return   6
(logand 6 15 1)   will return   0
(logand 1 4)   will return   0
(logand 1 5)   will return   1

Examples of logior function
(logior 2 7)   will return   7
(logior 8 15)   will return   15
(logior 6 15 7)   will return   15
(logior 6 15 1)   will return   15
(logior 1 4)   will return   5
(logior 1 5)   will return   5

One application of bit-codes is found in the object snap modes. The following is a list of bit-codes assigned to different object snap modes:

Snap Mode	Bit Code	Snap Mode	Bit Code
None	0	Intersection	32
Endpoint	1	Insertion	64
Midpoint	2	Perpendicular	128
Center	4	Tangent	256
Node	8	Nearest	512
Quadrant	16	Quick	1024

The system variable **OSMODE** can be used to specify a snap mode. For example, if the value assigned to **OSMODE** is 4, the object snap mode is center. The bit-codes can be combined to produce multiple object snap modes. For example, you can get endpoint, midpoint, and center object snaps by adding the bit-codes of endpoint, midpoint, and center (1 + 2 + 4 = 7) and assigning that value (7) to the **OSMODE** system variable. In AutoLISP, the existing snap modes information can be extracted by using the logand function. For example, if **OSMODE** is set to 7, the logand function can be used to check different bit-codes that represent object snaps.

```
(logand 1 7)      will return    1    (Endpoint)
(logand 2 7)      will return    2    (Midpoint)
(logand 4 7)      will return    4    (Center)
```

atof and rtos Functions

The **atof** function converts a string into a real and the **rtos** function converts a number into a string.

Examples
```
(atof "3.5")      will return    3.5
(rtos 8 15)       will return    "8.1500"
```

get_tile and set_tile Functions

The **get_tile** function retrieves the current value of a dialog box tile and the **set_tile** function sets the value of a dialog box tile.

Examples
(get_tile "midpoint")
(set_tile "xsnap" value)

Example 4

Write an AutoLISP program that will handle the dialog box and perform the functions as described in the dialog box of Example 3.

The following file is a listing of the AutoLISP program for Example 4. **The line numbers are not a part of the program; they are for reference only.**

```
;;Lisp program for setting Osnaps                                    1
;;Dialog file name is osnapsh.dcl                                    2
(defun c:osnapsh ( / dcl_id)                                         3
```

```
    (setq dcl_id (load_dialog "osnapsh.dcl"))                4
    (new_dialog "osnapsh" dcl_id)                            5
    ;;Get the existing value of object snaps and             6
    ;;then write the values to the dialog box                7
    (setq osmode (getvar "osmode"))                          8
    (if (= 1 (logand 1 osmode))                              9
        (set_tile "Endpoint" "1")                           10
    )                                                       11
    (if (= 2 (logand 2 osmode))                             12
        (set_tile "Midpoint" "1")                           13
    )                                                       14
    (if (= 32 (logand 32 osmode))                           15
        (set_tile "Intersection" "1")                       16
    )                                                       17
    (if (= 4 (logand 4 osmode))                             18
        (set_tile "Center" "1")                             19
    )                                                       20
                                                            21
    ;;Read the values as set in the dialog box and          22
    ;;assign those values to AutoCAD variable osmode        23
    (defun setvars ()                                       24
    (setq osmode 0)                                         25
    (if (= "1" (get_tile "Endpoint"))                       26
        (setq osmode (logior osmode 1))                     27
    )                                                       28
    (if (= "1" (get_tile "Midpoint"))                       29
        (setq osmode (logior osmode 2))                     30
    )                                                       31
    (if (= "1" (get_tile "Intersection"))                   32
        (setq osmode (logior osmode 32))                    33
    )                                                       34
    (if (= "1" (get_tile "Center"))                         35
        (setq osmode (logior osmode 4))                     36
    )                                                       37
      (setvar "osmode" osmode)                              38
    )                                                       39
                                                            40
    (action_tile "accept" "(setvars) (done_dialog)")        41
    (start_dialog)                                          42
    (princ)                                                 43
    )                                                       44
```

Explanation

Lines 1 and 2
;;Lisp program for setting Osnaps
;;Dialog file name is osnapsh.dcl
These lines are comment lines, and all comment lines start with a semicolon. AutoCAD ignores the lines that start with a semicolon.

Lines 3-5
(defun c:osnapsh (/ dcl_id)

```
(setq dcl_id (load_dialog "osnapsh.dcl"))
(new_dialog "osnapsh" dcl_id)
```
In line 3, **defun** is an AutoLISP function that defines the **osnapsh** function. Because of the **c:** in front of the function name, the **osnapsh** function can be executed like an AutoCAD command. The **osnapsh** function has one local variable, **dcl_id**. In line 4, the **(load_dialog "osnapsh.dcl")** loads the DCL file **osnapsh.dcl** and returns a positive integer. The **setq** function assigns this integer to the local variable, dcl_id. In line 5, the AutoLISP **new_dialog** function loads the dialog **osnapsh that is defined in the DCL file** (line 1 of the DCL file). The variable **dcl_id** has an integer value that identifies the DCL file.

Line 8
```
(setq osmode (getvar "osmode"))
```
In this line, **getvar "osmode"** has the value of the AutoCAD system variable **osmode**, and the **setq** function sets the **osmode** variable equal to that value. Note that the first **osmode** is just a variable, whereas the second **osmode**, in quotes (**"osmode"**), is the system variable.

Lines 9-11
```
(if (= 1 (logand 1 osmode))
   (set_tile "Endpoint" "1")
)
```

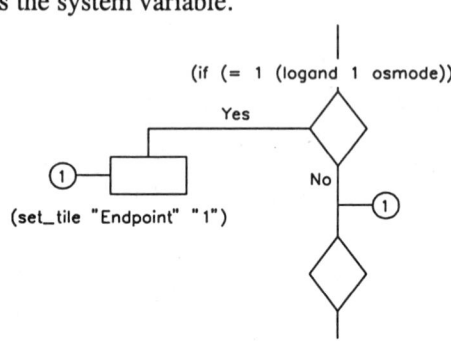

In line 9, the **(logand 1 osmode)** will return 1 if the bit-code of endpoint (1) is a part of the **OSMODE** value. For example, if the value assigned to **OSMODE** is 7 (1 + 2 + 4 = 7), (logand 1 7) will return 1. If **OSMODE** is 6 (2 + 4 = 6), then (logand 1 6) will return 0. The AutoLISP **if** function checks whether the value returned by **(logand 1 osmode)** is 1. If the function returns **T** (true), the instructions described in the second line are carried out. Line 10 sets the value of the **Endpoint** tile to 1; that displays a check mark in the toggle box. The closing parenthesis in line 11 completes the definition of the **if** function. If the expression **(if (= 1 (logand 1 osmode))** returns **nil**, the program skips to line 12 of the program.

Lines 24 and 25
```
(defun setvars ( )
(setq osmode 0)
```
Line 24 defines a **setvars** function, and line 25 sets the value of the **osmode** variable to zero.

Lines 26-28
```
(if (= "1" (get_tile "Endpoint"))
   (setq osmode (logior osmode 1))
)
```
In line 26, the **(get_tile "Endpoint")** obtains the value of the toggle tile named **Endpoint**. If the **Endpoint** toggle tile is on, the value it returns is 1; if the toggle tile is off, the value it returns is 0. In line 27, the **setq** function sets the value of **osmode** to the value returned by **(logior osmode**

1). For example, if the initial value of **osmode** is 0, then **(logior osmode 1)** will return 1. Similarly, if the initial value of **osmode** is 1, then **(logior osmode 2)** will return 3.

Lines 41 and 42
(action_tile "accept" "(setvars) (done_dialog)")
(start_dialog)
Line 42, **(start_dialog)**, starts the dialog box. In the dialog file the name assigned to the **OK** button is **"accept"**. When you select the **OK** button in the dialog box, the program executes the **setvars** function that updates the value of the **osmode** system variable and sets the selected object snaps.

PREDEFINED RADIO BUTTON, RADIO COLUMN, BOXED RADIO COLUMN, AND RADIO ROW TILES

Predefined Radio Button Tile
Format in DCL: **radio_button**

The **radio button** is a predefined active tile. Radio buttons can be arranged in a row or in a column. The unique characteristic of radio buttons is that only one button can be selected at a time. For example, if there are buttons for scientific, decimal, and engineering units, only one can be selected. If you select the decimal button, the other two buttons will be turned off automatically. Because of this special characteristic, radio buttons must be used only in a radio row or in a radio column. The label for the radio button is optional; in most systems the label appears to the right of the button.

Predefined Radio Column Tile
Format in DCL: **radio_column**

The radio column is an active predefined tile where the radio button tiles are arranged in a column and only one button can be selected at a time. When the radio buttons are arranged in a column, the buttons are next to each other vertically and are easy to select. Therefore, you should arrange radio buttons in a column to make it easy to select a radio button.

Predefined Boxed Radio Column Tile
Format in DCL: **boxed_radio_column**

The **boxed radio column** is an active predefined tile where a border is drawn around the radio column.

Predefined Radio Row Tile

Format in DCL: **radio_row**

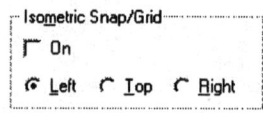

The **radio row** consists of radio button tiles arranged in a row. Only one button can be selected at a time. The radio row can become quite long if there are several radio button tiles and labels. In selecting a radio button, the cursor travel will increase because the buttons are not immediately next to each other. Therefore, you should avoid using radio button tiles in a row, especially if there are more than two.

Example 5

Write a DCL program for a dialog box that will enable you to select different units and unit precision as shown in Figure 14-9. Also, write an AutoLISP program that will handle the dialog box.

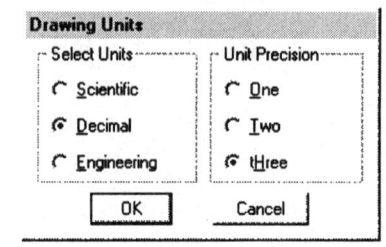

The following file is a listing of the DCL program for the dialog box shown in Figure 14-9. The name of the dialog box is **dwgunits**. **The line numbers are not a part of the file; they are shown here for reference only.**

Figure 14-9 Dialog box for Example 5

```
dwgunits : dialog {                              1
  label = "Drawing Units";                       2
  : row {                                        3
    : boxed_column {                             4
      label = "Select Units";                    5
      : radio_column {                           6
        : radio_button {                         7
          key = "scientific";                    8
          label = "Scientific";                  9
          mnemonic = "S";                       10
        }                                       11
        : radio_button {                        12
          key = "decimal";                      13
          label = "Decimal";                    14
          mnemonic = "D";                       15
        }                                       16
        : radio_button {                        17
          key = "engineering";                  18
          label = "Engineering";                19
          mnemonic = "E";                       20
        }                                       21
      }                                         22
    }                                           23
    : boxed_column {                            24
      label = "Unit Precision";                 25
      : radio_column {                          26
```

```
                : radio_button {                                27
                    key = "one";                                28
                    label = "One";                              29
                    mnemonic = "O";                             30
                }                                               31
                : radio_button {                                32
                    key = "two";                                33
                    label = "Two";                              34
                    mnemonic = "T";                             35
                }                                               36
                : radio_button {                                37
                    key = "three";                              38
                    label = "Three";                            39
                    mnemonic = "h";                             40
                }                                               41
            }                                                   42
        }                                                       43
    }                                                           44
    ok_cancel;                                                  45
}                                                               46
```

Explanation

Lines 4 and 5
: boxed_column {
  label = "Select Units";

The **boxed_column** is an active predefined tile that draws a border around the column. Line 5, **label** = "Select Units";, displays the label (**Select Units**) at the top of the column. The **label** is a predefined DCL attribute.

Lines 6-8
: radio_column {
  : radio_button {
  key = "scientific";

The **radio_column** is a predefined active tile that will arrange the tiles within it in a vertical column and draw a border around the column. The **radio_button** is another active predefined tile that displays a radio button in the dialog box, with an optional text to the right of the button. The **key** attribute assigns a name (**scientific**) to the tile. This name is then used by the application program to handle the tile.

Lines 9-11
  label = "Scientific";
  mnemonic = "S";
}

The **label** attribute will display the label (**Scientific**) on the right of the toggle box. The second line, **mnemonic** = "S";, defines the keyboard mnemonic for **Scientific**. This causes the letter **S** of **Scientific** to be displayed underlined in the dialog box. The closing brace (}) on the next line completes the definition of the toggle tile.

14-28 Customizing AutoCAD

Lines 41-46
```
            }
          }
        }
      }
  ok_cancel;
}
```
The closing brace on line 41 completes the definition of the radio button and the closing brace on the second line completes the definition of the radio column. The closing brace on the next line completes the definition of the boxed_column, and the closing brace on the next line completes the definition of the row on line 3 of the DCL file. The predefined tile, **ok_cancel**, displays the **OK** and **Cancel** tiles in the dialog box. The last closing brace completes the definition of the dialog box.

Use the following commands to load and display the dialog file on the screen. The name of the file is assumed to be **dwgunits.dcl**, and the name of the dialog is **dwgunits**. If AutoCAD is successful in loading the file, it will return an integer. In the following examples, the integer that AutoCAD returns is assumed to be 5. The screen display after loading the dialog box is shown in Figure 14-10.

Command: (load_dialog "osnapsh.dcl")
5
Command: (new_dialog "osnapsh" 5)

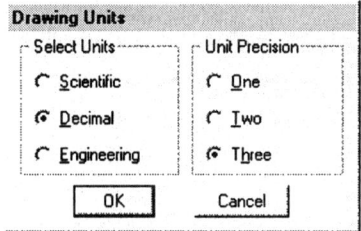

Figure 14-10 Dialog box for Example 5

The following file is a listing of the AutoLISP program that loads, displays, and handles the dialog box for Example 5. **The line numbers are not a part of the file; they are shown here for reference only.**

```
;;Lisp program dwgunits.lsp for setting units              1
;;and precision. Dialog file name dwgunits.dcl             2
;                                                          3
(defun c:dwgunits ( / dcl_id)                              4
(setq dcl_id (load_dialog "dwgunits.dcl"))                 5
(new_dialog "dwgunits" dcl_id)                             6
;                                                          7
;;Get the existing values of lunits and luprec             8
;;and turn the corresponding radio_button on               9
;                                                         10
```

```
    (setq lunits (getvar "lunits"))                         11
    (if (= 1 lunits)                                        12
        (set_tile "scientific" "1")                         13
    )                                                       14
    (if (= 2 lunits)                                        15
        (set_tile "decimal" "1")                            16
    )                                                       17
    (if (= 3 lunits)                                        18
        (set_tile "engineering" "1")                        19
    )                                                       20
    ;                                                       21
    (setq luprec (getvar "luprec"))                         22
    (if (= 1 luprec)                                        23
        (set_tile "one" "1")                                24
    )                                                       25
    (if (= 2 luprec)                                        26
        (set_tile "two" "1")                                27
    )                                                       28
    (if (= 3 luprec)                                        29
        (set_tile "three" "1")                              30
    )                                                       31
                                                            32
    ;;Read the value of the radio_buttons and               33
    ;;assign it to AutoCAD lunit and luprec variables       34
    ;                                                       35
    (action_tile "scientific" "(setq lunits 1)")            36
    (action_tile "decimal" "(setq lunits 2)")               37
    (action_tile "engineering" "(setq lunits 3)")           38
    ;                                                       39
    (action_tile "one" "(setq luprec 1)")                   40
    (action_tile "two" "(setq luprec 2)")                   41
    (action_tile "three" "(setq luprec 3)")                 42
    (action_tile "accept" "(done_dialog)")                  43
    ;                                                       44
    (start_dialog)                                          45
    (setvar "lunits" lunits)                                46
    (setvar "luprec" luprec)                                47
    (princ)                                                 48
)                                                           49
```

Explanation
Lines 1-3
;;Lisp program dwgunits.lsp for setting units
;;and precision. Dialog file name dwgunits.dcl
;
The first three lines of this program are comment lines, and all comment lines start with a semicolon. AutoCAD ignores them.

Lines 4-6
(defun c:dwgunits (/ dcl_id)
(setq dcl_id (load_dialog "dwgunits.dcl"))

14-30 Customizing AutoCAD

(new_dialog "dwgunits" dcl_id)
In line 4, **defun** is an AutoLISP function that defines the **dwgunits** function, which has one local variable, **dcl_id**. The **c:** in front of the function name, **dwgunits**, makes the **dwgunits** function act like an AutoCAD command. In the next line, **(load_dialog "dwgunits.dcl")** loads the DCL file **dwgunits.dcl** and returns a positive integer. The **setq** function assigns this integer to the local variable, dcl_id. In the next line, the AutoLISP **new_dialog** function displays the **dwgunits** dialog box that is defined in the DCL file (line 1 of DCL file). The **dcl_id** variable is an integer that identifies the DCL file.

Line 11
(setq lunits (getvar "lunits"))
In this line, **getvar "lunits"** has the value of the AutoCAD system variable, **lunits**, and the **setq** function sets the **lunits** variable equal to that value. The first **lunits** is a variable, whereas the second **lunits**, in quotes (**"lunits"**), is a system variable.

Lines 12-14
(if (= 1 lunits)
   (set_tile "scientific" "1")
)
The **if** function (AutoLISP function) checks whether the value of the variable **lunits** is 1. If the function returns **T** (true), the instructions described in the next line are carried out. Line 13 sets the value of the tile named **"scientific"** equal to 1; that turns the corresponding radio button on. The closing parenthesis in line 14 completes the **if** function. If the **if** function, **(if (= 1 lunits)**, returns **nil**, the program skips to line number 14.

Line 22
(setq luprec (getvar "luprec"))
In this line, **getvar "luprec"** has the value of the AutoCAD system variable **luprec**, and the **setq** function sets the **luprec** variable equal to that value. The first **luprec** is a variable, whereas the second **luprec**, in quotes (**"luprec"**), is a system variable.

Lines 23-25
(if (= 1 luprec)
   (set_tile "one" "1")
)
In line 23, the **if** function checks whether the value of the **luprec** variable is 1. If the function returns **T** (true), the instructions described in the next line are carried out. This line sets the value of the tile named **"one"** to 1; that turns the corresponding radio button on. The closing parenthesis in the next line completes the definition of the **if** function. If the **if** function returns **nil**, the program skips to line 25.

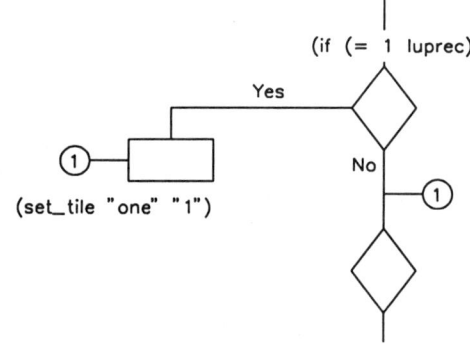

Line 36
(action_tile "scientific" "(setq lunits 1)")
If the radio button tile named **"scientific"** is turned on, the **setq** function sets the value of the AutoCAD system variable **lunits** to 1. The **lunits** system variable controls the drawing units. The following is a list of the integer values that can be assigned to the **lunits** system variable:

1	Scientific	4	Architectural
2	Decimal	5	Fractional
3	Engineering		

Line 40
(action_tile "one" "(setq luprec 1)")
If the radio button tile named **"one"** is turned on, the **setq** function sets the value of the AutoCAD system variable **luprec** to 1. The **lunits** system variable controls the number of decimal places in a decimal number or the denominator of a fractional or architectural unit.

Lines 43-47
(action_tile "accept" "(done_dialog)")
;
(start_dialog)
(setvar "lunits" lunits)
(setvar "luprec" luprec)
In the dialog file, the name assigned to the **OK** button is **"accept"**. When you select the **OK** button in the dialog box, the program executes the function defined in the **dwgunits** dialog box. The **(start_dialog)** function starts the dialog box. The **(setvar "lunits" lunits)** and **(setvar "luprec" luprec)** set the values of the **lunits** and **luprec** system variables to lunits and luprec, respectively.

EDIT BOX TILE
Format in DCL: **edit_box**

The **edit box** is a predefined active tile that enables you to enter or edit a single line of text. If the text is longer than the length of the edit box, the text will automatically scroll to the right or left horizontally. The label for the edit box is optional, and it is displayed to the left of the edit box.

width AND edit_width ATTRIBUTES

width Attribute
Format in DCL: **width**   **Example:** width = 22

The **width** attribute is used to keep the width of the tile to a desired size. The value assigned to the **width** attribute can be a real number or an integer that represents the distance in character width. The value assigned to the **width** attribute defines the minimum width of the tile. For example, in

14-32 Customizing AutoCAD

the example the minimum width of the tile is 22. However, the tile will automatically stretch if more space is available. It will retain the size of 22 only if the **fixed_width** attribute is assigned to the tile. The **width** attribute can be used with any tile.

edit_width Attribute

Format in DCL: **edit_width**   **Example:** edit_width = 10

The **edit_width** attribute is used with the predefined edit box tiles, and it determines the size of the edit box in character width units. If the width of the edit box is 0, or if it is not specified and the fixed_width attribute is not assigned to the tile, the tile will automatically stretch to fill the available space. When the edit box is stretched, the PDB facility inserts spaces between the edit box and the label so that the box is right-justified and the label is left justified.

Example 6

Write a DCL program for a dialog box (Figure 14-11) that will enable you to turn the snap and grid on and off. You should also be able to edit the X and Y values of snap and grid. Also, write an AutoLISP program that will load, display, and handle the dialog box.

The following file is a listing of the DCL file for Example 6.

```
dwgaids : dialog {
  label = "Drawing Aids";
  : row {
    : boxed_column {
      label = "SNAP";
      fixed_width = true;
      width = 22;
      : toggle {
        label = "On";
        mnemonic = "O";
        key = "snapon";
      }
      : edit_box {
        label = "X-Spacing";
        mnemonic = "X";
        key = "xsnap";
        edit_width = 10;
      }
      : edit_box {
        label = "Y-Spacing";
        mnemonic = "Y";
        key = "ysnap";
        edit_width = 10;
      }
    }
    : boxed_column {
      label = "GRID";
      fixed_width = true;
      width = 22;
```

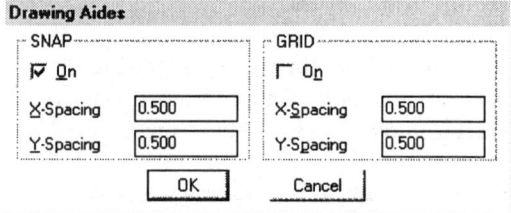

Figure 14-11 Dialog box for Example 6

```
                : toggle {
                  label = "On";
                  mnemonic = "n";
                  key = "gridon";
                  }
                : edit_box {
                  label = "X-Spacing";
                  mnemonic = "S";
                  key = "xgrid";
                  edit_width = 10;
                  }
                : edit_box {
                  label = "Y-Spacing";
                  mnemonic = "p";
                  key = "ygrid";
                  edit_width = 10;
                  }
                }
          }
    ok_cancel;
}
```

The following file is a listing of the AutoLISP program for Example 6. When the program is loaded and run, it will load, display, and control the dialog box (Figure 14-12).

```
;;Lisp program for Drawing Aids dialog box
;;Dialog file name is dwgaids.dcl
;
(defun c:dwgaids( / dcl_id snapmode xsnap ysnap
 orgsnapunit gridmode gridsnap xgrid ygrid orggridunit)
(setq dcl_id (load_dialog "dwgaids.dcl"))
(new_dialog "dwgaids" dcl_id)

;;Get the existing value of snapmode and snapunit
;;and write those values to the dialog box
(setq snapmode (getvar "snapmode"))
(if (= 1 snapmode)
    (set_tile "snapon" "1")
    (set_tile "snapon" "0")
    )
(setq orgsnapunit (getvar "snapunit"))
(setq xsnap (car orgsnapunit))
(setq ysnap (cadr orgsnapunit))
(set_tile "xsnap" (rtos xsnap))
(set_tile "ysnap" (rtos ysnap))
;
;;Get the existing value of gridmode and gridunit
;;and write those values to the dialog box
(setq gridmode (getvar "gridmode"))
(if (= 1 gridmode)
    (set_tile "gridon" "1")
    (set_tile "gridon" "0"))
```

14-34 Customizing AutoCAD

```
    (setq orggridunit (getvar "gridunit"))
    (setq xgrid (car orggridunit))
    (setq ygrid (cadr orggridunit))
    (set_tile "xgrid" (rtos xgrid))
    (set_tile "ygrid" (rtos ygrid))
    ;;Read the values set in the dialog box and
    ;;then change the associated AutoCAD variables
    (defun setvars ()
    (setq xsnap (atof (get_tile "xsnap")))
    (setq ysnap (atof (get_tile "ysnap")))
    (setvar "snapunit" (list xsnap ysnap))
    (if (= "1" (get_tile "snapon"))
        (setvar "snapmode" 1)
        (setvar "snapmode" 0))
    (setq xgrid (atof (get_tile "xgrid")))
    (setq ygrid (atof (get_tile "ygrid")))
    (setvar "gridunit" (list xgrid ygrid))
    (if (= "1" (get_tile "gridon"))
        (progn
           (setvar "gridmode" 0)
           (setvar "gridmode" 1))
        (setvar "gridmode" 0)
    ))
    (action_tile "accept" "(setvars) (done_dialog)")
    (start_dialog)
    (princ))
```

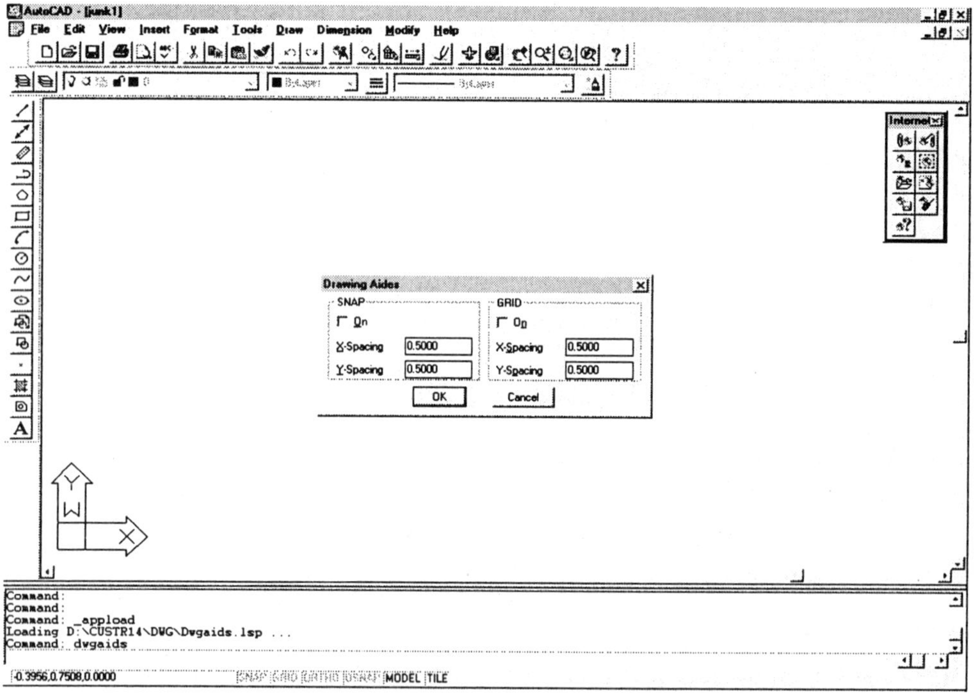

Figure 14-12 Screen display with the dialog box for Example 6

SLIDER AND IMAGE TILES

Slider Tile

Format in DCL
slider

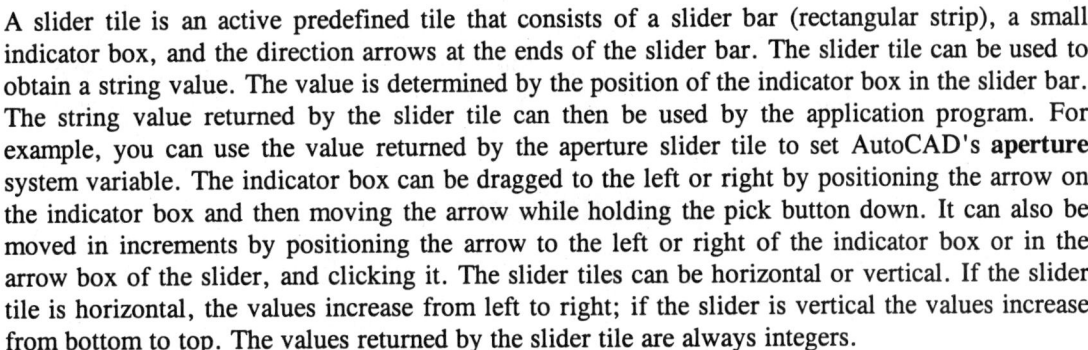

A slider tile is an active predefined tile that consists of a slider bar (rectangular strip), a small indicator box, and the direction arrows at the ends of the slider bar. The slider tile can be used to obtain a string value. The value is determined by the position of the indicator box in the slider bar. The string value returned by the slider tile can then be used by the application program. For example, you can use the value returned by the aperture slider tile to set AutoCAD's **aperture** system variable. The indicator box can be dragged to the left or right by positioning the arrow on the indicator box and then moving the arrow while holding the pick button down. It can also be moved in increments by positioning the arrow to the left or right of the indicator box or in the arrow box of the slider, and clicking it. The slider tiles can be horizontal or vertical. If the slider tile is horizontal, the values increase from left to right; if the slider is vertical the values increase from bottom to top. The values returned by the slider tile are always integers.

Image Tile

Format in DCL
image

The image tile is a predefined tile that is used to display graphical information. For example, it can be used to display the aperture box, linetypes, icons, and text fonts in dialog boxes. It consists of a rectangular box and the vector graphics that are displayed within the box.

min_value, max_value, small_increment, and big_increment Attributes

min_value and max_value Attributes

Format in DCL
min_value
max_value

Examples
min_value = 2
max_value = 15

The **min_value** attribute and the **max_value** attribute are predefined attributes that specify the minimum and maximum value that the **slider** tile will return. In the previous example, the minimum value is 2 and the maximum value is 15. If you do not assign these attributes to a slider tile, the slider automatically assumes the default values. For the **min_value** attribute the default value is 0, for the **max_value** attribute the default value is 10,000.

small_increment and big_increment Attributes

Format in DCL
small_increment
big_increment

Examples
small_increment = 1
big_increment = 1

The value assigned to the **small_increment** attribute and the **big_increment** attribute determines the increment value of the slider incremental control. For example, if the increment is 1, the values returned by the slider will be in the increment of 1. If these attributes are not assigned to a slider tile, the slider automatically assumes the default values. The default value of **small_increment** is one one-hundredth (1/100) of the slider range and the default value of **big_increment** is one-tenth (1/10) of the slider range.

aspect_ratio and color Attributes

aspect_ratio Attribute

Format in DCL
aspect_ratio

Example
aspect_ratio = 1

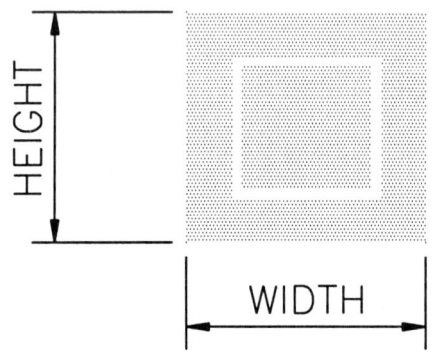

The aspect_ratio is a predefined attribute that can be used with an image. The aspect ratio is the ratio of the width of the image box to the height of the image box (width/height). You can control the size of the image box by assigning an integer value to the width attribute and the height attribute. You can also control the size of the image box by assigning an integer value to one of the attributes (width or height) and assigning a real or integer value to the aspect_ratio attribute. For example, if the value assigned to the height attribute is five and the value assigned to the aspect_ratio is 0.5, the width of the image box will be 2.5 (width = height x aspect_ratio). In this case, you do not need to specify the width attribute.

color Attribute

Format in DCL
color

Example
color = 2

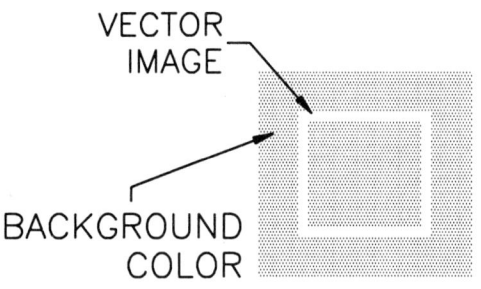

The color is a predefined attribute that can be used with the image box to control the background color. The integer number assigned

Programmable Dialog Boxes Using Dialog Control Language 14-37

to the color attribute specifies an AutoCAD color index. In the above example, the background color is yellow (color number 2). The color of the vector image that is drawn inside the image box is determined by the color specified in the vector_image attribute.

Example 7

Write a DCL program for the dialog box in Figure 14-13A that will enable you to change the size of the aperture box.

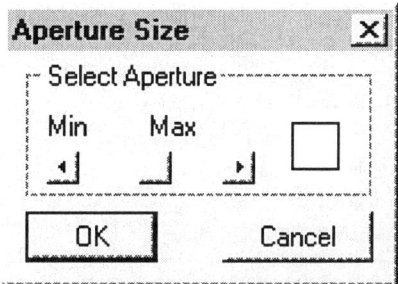

Figure 14-13A Dialog box for Example 7

The dialog box shown in Figure 14-13A has two rows and two columns. The first column has two items, label (**Min .... Max**) and the **slider bar**. The second column has only one item (the image box). The second row has the **OK** and **Cancel** buttons. As shown in Figure 14-13B, the dialog box also has a dialog label (Aperture Size) and a row label (Select Aperture). The following file is a listing of the DCL file for Example 7. The line numbers are not a part of the file; they are shown here for reference only.

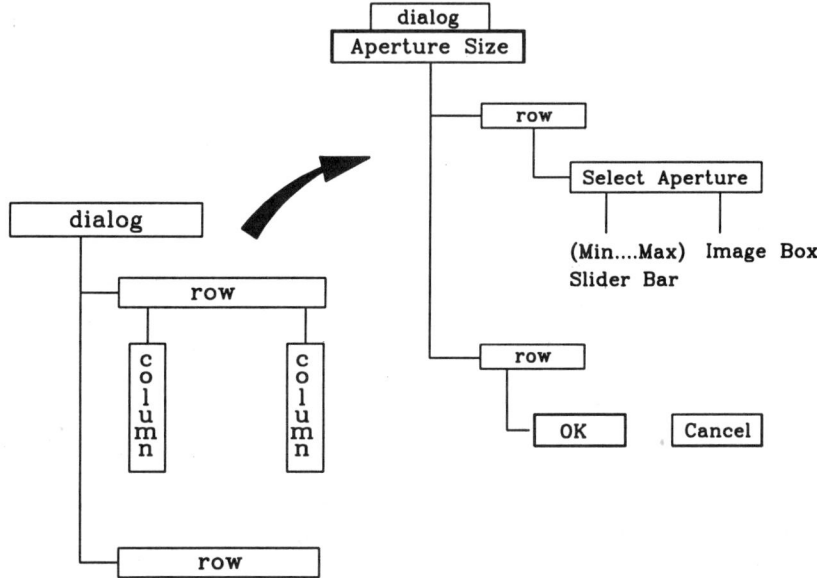

Figure 14-13B Layout of dialog box for Example 7

```
aprtsize : dialog {                                    1
  label = "Aperture Size";                             2
  : boxed_row {                                        3
    label = "Select Aperture";                         4
  : column {                                           5
      fixed_width = true;                              6
      : text {                                         7
        label = "Min            Max";                  8
        alignment = centered;                          9
      }                                               10
      : slider {                                      11
        key = "aperture_slider";                      12
        min_value = 2;                                13
        max_value = 17;                               14
        width = 15;                                   15
        height = 1;                                   16
        big_increment = 1;                            17
        fixed_width = true;                           18
        fixed_height = true;                          19
      }                                               20
    }                                                 21
    : image {                                         22
      key = "aperture_image";                         23
      aspect_ratio = 1;                               24
      width = 5;                                      25
      color = 2;                                      26
    }                                                 27
  }                                                   28
  ok_cancel;                                          29
}                                                     30
```

Lines 11, 12
: slider {
key = "aperture_slider";
In the first line, : **slider** starts the definition of the slider tile. The **slider** is an active predefined tile that consists of a slider bar, a small indicator box, and the direction arrows at the ends of the slider bar. The second line, **key = "aperture_slider"**, assigns the name, **aperture_slider**, to this slider tile.

Lines 13, 14
min_value = 2;
max_value = 17;
The **min_value** is a predefined attribute that specifies the minimum value that the **slider** tile will return. Similarly, the **max_value** attribute specifies the maximum value that the slider tile will return. In the above two lines, the minimum value is 2 and the maximum value is 17.

Lines 15-17
width = 15;
height = 1;
big_increment = 1;
The first line specifies the width of the slider tile and the second line specifies the height of the slider tile. The third line, **big_increment = 1;**, defines the increment of the indicator box in the slider bar.

Lines 22, 23
: image {
 key = "aperture_image";
The first line, **: image {**, starts the definition of the image box and the second line assigns a name, **aperture_image**, to the image tile. The image is a predefined tile that is used to display graphical information.

Lines 24-26
aspect_ratio = 1;
width = 5;
color = 2;
The first line, **aspect_ratio = 1;**, defines the ratio of the width of the image box to the height of the image box. The second line, **width = 5;**, specifies the width of the image box. Since the aspect ratio is 1, the height of the image box is 5 (width/height = aspect_ratio). The third line, **color = 2;**, assigns the AutoCAD color number 2 (yellow) to the background of the image box.

AUTOLISP FUNCTIONS

dimx_tile and dimy_tile

The AutoLISP function, **dimx_tile**, obtains the width dimension (x_aperture) of the specified image box along the X-axis and the **dimy_tile** function obtains the height dimension (y_aperture) along the Y-axis. In the following examples, "aperture_image" is the name of the image tile that is assigned by using the **key** attribute in the DCL file (key = "aperture_image").

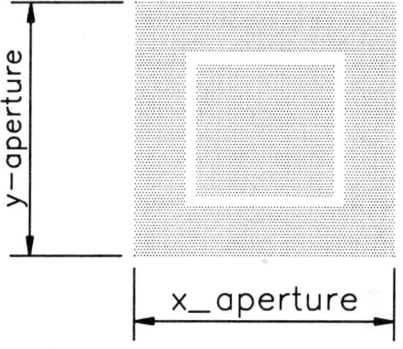

Format
(dimx_tile tilename)
(dimy_tile tilename)

Examples
(dimx_tile "aperture_image")
(dimy_tile "aperture_image")

vector_image

The AutoLISP function, **vector_image**, draws a vector (line) in the active image box, between the points that are defined in the **vector_image** function. In the following example, AutoCAD will draw a line from point 1,1 to point 3,3 of the image box and the color of this vector will be AutoCAD's color number 1 (red).

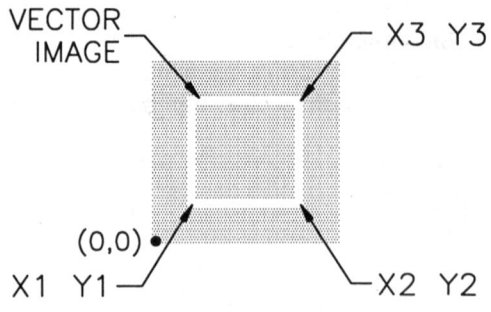

Format
(vector_image x1 y1 x2 y2 color)

Example

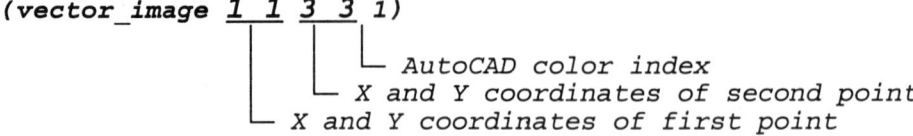

fill_image

The AutoLISP function, **fill_image**, fills the rectangle in the active image box with the color specified in the **fill_image** function. For example, in the following example the active image box will be filled with yellow color (AutoCAD's color number 2). The filled rectangle is determined by the coordinates of the first point (x1 y1) and the second point (x2 y2), the two opposite corners of the rectangle.

Format
(fill_image x1 y1 x2 y2 color)

Example
(fill_image 0 0 x_aperture y_aperture 2)

start_image

The AutoLISP function, **start_image**, starts the image whose name is specified in the **start_image** function. In the following example, the name of the image tile is **"aperture_image"** which is assigned by using **key** attribute in the DCL file (key = "aperture_image").

Format   **Example**
(start_image)   (start_image "aperture_image")

end_image

The AutoLISP function **end_image** ends the active image whose name is specified in the **end_image** function. In the following example, the name of the active image tile is "aperture_image".

Format
(end_image)

Example
(end_image "aperture_image")

$value

The **$value** function is an AutoLISP action expression that retrieves the string value from the tile of a dialog box. The tile could be an edit box or a toggle box. In Example 8, the **$value** function retrieves the current value of the active tile and the setq function assigns that value to the variable, aprt_size.

Format
$value

Example
(setq aprt_size $value)

Example 8

Write an AutoLISP program that will load, display, and control the dialog box of Example 7 (Figure 14-13A). The flowchart and screen display for this example are shown in Figures 14-14 and 14-15.

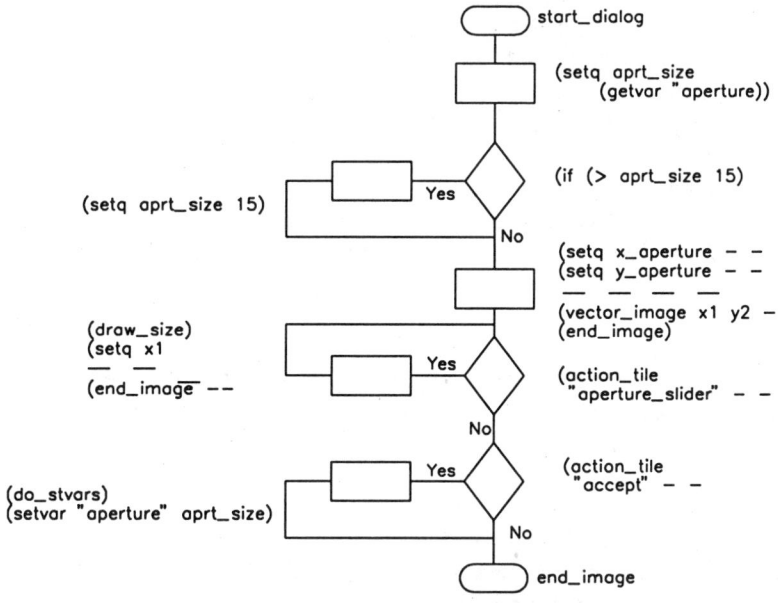

Figure 14-14 Flowchart for Example 8

The following file is a listing of the AutoLISP file for Example 8. The line numbers are not a part of the file; they are shown here for reference only.

```
;;APRTSIZE.LSP, AutoLISP program for Aperture
;; Dialog Box. DCL file name -- APRTSIZE.DCL
;
(defun c:aprtsize ( )
  (setq dcl_id (load_dialog "aprtsize"))
  (new_dialog "aprtsize" dcl_id)
;
;Obtain value of the system variable "aperture",
;calculate X and Y values of vector image, and draw
;vector image in the image box.
  (setq aprt_size (getvar "aperture"))
  (if (> aprt_size 15)
    (setq aprt_size 15)
    )
  (setq x_aperture (dimx_tile "aperture_image"))
  (setq y_aperture (dimy_tile "aperture_image"))
  (set_tile "aperture_slider" (itoa aprt_size))
  (setq x1 (- (/ x_aperture 2) aprt_size))
  (setq x2 (+ (/ x_aperture 2) aprt_size))
  (setq y1 (- (/ y_aperture 2) aprt_size))
  (setq y2 (+ (/ y_aperture 2) aprt_size))
  (start_image "aperture_image")
  (fill_image 0 0 x_aperture y_aperture 2)
  (vector_image x1 y1 x2 y1 1)
  (vector_image x2 y1 x2 y2 1)
  (vector_image x2 y2 x1 y2 1)
  (vector_image x1 y2 x1 y1 1)
  (end_image)
  (action_tile "aperture_slider"
     "(draw_size (setq aprt_size (atoi $value)))")
  (action_tile "accept" "(do_setvars)(done_dialog)")
  (start_dialog)
(princ)
)
;Set aperture variable "aperture" equal to aprt_size
(defun do_setvars ( )
  (setvar "aperture" aprt_size)
  )
;
;Calculate the X and Y coordinates of vector image
;and draw the aperture image in the image box.
(defun draw_size (aprt_size)
  (setq x1 (- (/ x_aperture 2) aprt_size))
  (setq x2 (+ (/ x_aperture 2) aprt_size))
  (setq y1 (- (/ y_aperture 2) aprt_size))
  (setq y2 (+ (/ y_aperture 2) aprt_size))
```

```
      (start_image "aperture_image")                              47
      (fill_image 0 0 x_aperture y_aperture -2)                   48
      (vector_image x1 y1 x2 y1 1)                                49
      (vector_image x2 y1 x2 y2 1)                                50
      (vector_image x2 y2 x1 y2 1)                                51
      (vector_image x1 y2 x1 y1 1)                                52
      (end_image)                                                 53
    )                                                             54
```

Figure 14-15 Screen display with the dialog box for Example 8

Lines 11-13
(setq aprt_size (getvar "aperture"))
(if (> aprt_size 15)
   (setq aprt_size 15)
In the first line, the AutoLISP function, **getvar**, obtains the value of the **"aperture"** system variable and the **setq** function assigns that value to the **aprt_size** variable. The second line,**(if (> aprt_size 15)**, checks whether the value of the variable aprt_size is greater than 15. If it is, the third line, **(setq aprt_size 15)**, sets the value of the **aprt_size** variable equal to 15. The value of the "aperture" system variable must be equal to

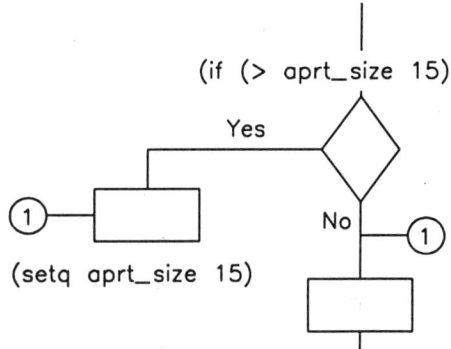

or less than 15 to display the vector image in the image box because the maximum size of the image box in this example is 15.

Lines 15, 16
(setq x_aperture (dimx_tile "aperture_image"))
(setq y_aperture (dimy_tile "aperture_image"))
In the first line, the AutoLISP function, **dimx_tile**, retrieves the X-dimension of the image tile and the setq function assigns that value to the **x_aperture** variable. Similarly, the **dimy_tile** function retrieves the Y-dimension of the image tile.

Line 17
(set_tile "aperture_slider" (itoa aprt_size))
The AutoLISP function, **itoa**, changes the integer value of the **aprt_size** variable into a string and the **set_tile** function assigns that value to "aperture_slider". The **aperture_slider** is the name of the slider tile in the DCL file (DCL file of Example 7).

Lines 18-21
(setq x1 (- (/ x_aperture 2) aprt_size))
(setq x2 (+ (/ x_aperture 2) aprt_size))
(setq y1 (- (/ y_aperture 2) aprt_size))
(setq y2 (+ (/ y_aperture 2) aprt_size))

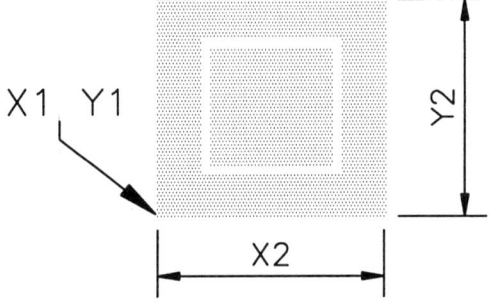

The first line calculates the X-coordinate of the lower left corner of the aperture box. The value is calculated by dividing the **x_aperture** distance by 2 and then subtracting the distance **aprt_size** from it. The aprt_size has the same value as that of the system variable, "aperture". Aperture is the distance of the sides of aperture box from the cross-hair lines, measured in pixels. Similarly, the second line calculates the X-coordinate of the lower_right corner. The next two lines calculate the Y-coordinates of these two points.

Lines 22, 23
(start_image "aperture_image")
(fill_image 0 0 x_aperture y_aperture 2)
The AutoLISP function, **start_image**, starts the image box and **aperture_image** is the name of the image box as defined in the DCL file. The **fill_image** function fills the image within the specified coordinates with AutoCAD color index 2 (yellow).

Lines 27, 28
(vector_image x1 y2 x1 y1 1)
(end_image)
The AutoLISP function, **vector_image**, draws a vector (line) from the point with coordinates x1 y2 to the point with coordinates x1 y1. The **end_image** function ends the image.

Lines 29-31
(action_tile "aperture_slider"
  "(draw_size (setq aprt_size (atoi $value)))")
(action_tile "accept" "(do_setvars)(done_dialog)")

The AutoLISP action expression **$value** retrieves the string value of the active slider tile and the **atoi** function returns the integer value of the string. This value is then used by the **draw_size** function to draw the vector image of the aperture box. The **aperture_slider** is the name of the slider tile defined in the DCL file. The third line executes the **do_setvars** function and ends the dialog box when you select OK from the dialog box. The **draw_size** and **do_setvars** functions are defined in the program.

Lines 36, 37
(defun do_setvars ( )
  (setvar "aperture" aprt_size))

The **defun** function defines the **do_setvars** function. The **setvar** function sets the value of the AutoCAD system variable, "aperture", equal to **aprt_size**.

Lines 42, 43
(defun draw_size (aprt_size)
  (setq x1 (- (/ x_aperture 2) aprt_size))

The first line defines the **draw_size** function with one argument, **aprt_size**, and the second line calculates the value of X-coordinate of the first vector.

REVIEW QUESTIONS

Indicate whether the following statements are true or false.

1. A dialog control language (DCL) file can contain descriptions of multiple files. (T/F)

2. In a DCL file, you do not need to specify the size of a dialog box. (T/F)

3. Dialog boxes are not dependent on the platform. (T/F)

4. Dialog boxes can contain several OK buttons. (T/F)

5. The numeric values assigned to the attributes in a DCL file can be both integers and real numbers. (T/F)

6. The reserved names in DCL are case-sensitive. (T/F)

7. The label attribute used in a button tile has no default value. (T/F)

8. In a dialog box, only one button can be assigned the true value for the **is_default** attribute. (T/F)

9. The **load_dialog** function displays the dialog box on the screen. (T/F)

10. You cannot select a tile by using the mnemonic key assigned to the tile. (T/F)

11. Mnemonic characters are case-sensitive. (T/F)

12. Bit-codes cannot be combined to produce multiple object snaps. (T/F)

13. The radio button is a predefined attribute. (T/F)

14. You should arrange the radio buttons in a column setting. (T/F)

15. The edit box active tile allows you to enter or edit multiple lines of text. (T/F)

16. The width attribute will make the tile stretch automatically to fill the entire width of the dialog box. (T/F)

17. The slider tile returns a string value. (T/F)

18. The aspect ratio is the ratio of the width of the image box to the height of the image box. (T/F)

19. The color attribute can be used with an image box to control the background color. (T/F)

20. The vector_image function can be used to draw vectors in the Drawing Editor. (T/F)

21. The AutoLISP function, $value, retrieves the real value from the tile of a dialog box. (T/F)

Fill in the blanks.

22. The basic tiles, such as buttons, edit boxes, and images, are predefined by the _____ facility of AutoCAD.

23. The label of a button appears _____ the button.

24. The _____ attribute assigns a name to the tile.

25. The label attribute in a boxed column must be a _____ string.

26. The _____ attribute controls the width of a tile.

27. The format of the command for loading a dialog box is _____.

28. The AutoLISP function _____ is used to initialize a dialog box and then display it on the screen.

29. The AutoLISP function _____ is used to accept user input from the dialog box.

30. The AutoLISP function _____ is used to associate an action expression with a tile in the dialog box.

31. The AutoLISP function _____ can be used to obtain the result of logical bitwise AND.

32. The edit_width attribute determines the size of the edit box in _____ units.

33. The image tile is used to display _____ information.

34. The default value of the max_value attribute is _____.

35. The default value of the big_increment attribute is _____ of the slider range.

36. The AutoLISP function _____ can be used to fill the image box with a color.

EXERCISES

Exercise 1

Using dialog control language (DCL), write a program for the dialog box in Figure 14-16. Also, write an AutoLISP program that will load, display, and control the dialog box and perform the functions shown in the dialog box.

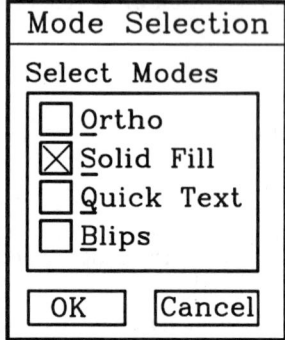

Figure 14-16 Dialog box for mode selection

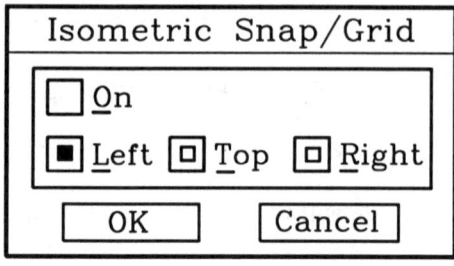

Figure 14-17 Dialog box for isometric snap/grid

Exercise 2

Write a DCL program for the isometric snap/grid dialog box shown in Figure 14-17. Also, write an AutoLISP program that will load, display, and control the dialog box and perform the functions shown in the dialog box.

Exercise 3

Write a DCL program for the dialog box in Figure 14-18 that will enable you to insert a block. Also, write an AutoLISP program that will load, display, and control the dialog box. The values shown in the edit boxes for insertion point, scale, and rotation are the default values.

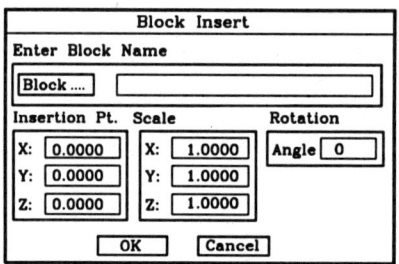

Figure 14-18 Dialog box for inserting blocks

Exercise 4

Write a DCL program for the dialog box in Figure 14-19 that will enable you to select the angle and angle precision. Also, write an AutoLISP program that will load, display, and control the dialog box.

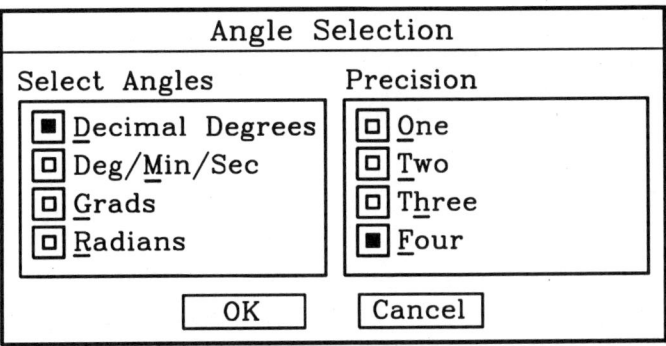

Figure 14-19 Dialog box for angle selection

Exercise 5

Write a DCL program for the dialog box in Figure 14-20 that will enable you to insert a block. Also, write an AutoLISP program that will load, display, and control the dialog box. The values shown in the edit boxes for insertion point, scale, and rotation are the default values.

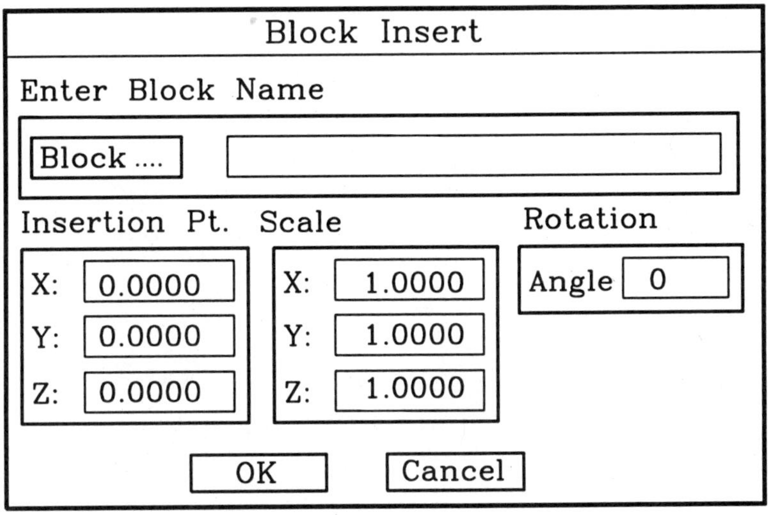

Figure 14-20 Dialog box for inserting blocks

Chapter 15

DIESEL:
A String Expression Language

Learning objectives

After completing this chapter, you will be able to:
♦ *Use DIESEL to customize a status line.*
♦ *Use the* **MODEMACRO** *system variable.*
♦ *Write macro expressions using DIESEL.*
♦ *Use AutoLISP with MODEMACRO.*

DIESEL

DIESEL (Direct Interpretively Evaluated String Expression Language) is a string expression language. It can be used to display a user-defined text string (macro expression) in the status line by altering the value of the AutoCAD system variable **MODEMACRO**. The value assigned to **MODEMACRO** must be a string, and the output it generates will be a string. It is fairly easy to write a macro expression in DIESEL, and it is an important tool for customizing AutoCAD. However, DIESEL is slow, and it is not intended to function like AutoLISP or DCL. You can use AutoLISP to write and assign a value to the **MODEMACRO** variable, or you can write the definition of the **MODEMACRO** expression in the menu files. A detailed explanation of the DIESEL functions and the use of DIESEL in writing a macro expression is given later in this chapter.

STATUS LINE*

When you are in AutoCAD, displayed is a status line at the bottom of the graphics screen (Figure 15-1). This line contains some useful information and tools that will make it easy to change the status of some AutoCAD functions. To change the status, you must double-click on the buttons. For example, if you want to display grid lines on the screen, double-click on the GRID button. Similarly, if you want to switch to paper space, double-click on MODEL. The status line contains the following information:

15-2 Customizing AutoCAD

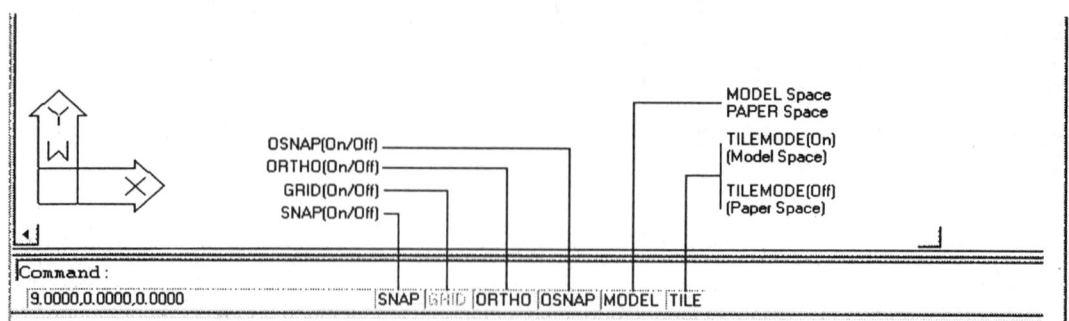

Figure 15-1 Default status line display

Coordinate Display. The coordinate information displayed in the status line can be static or dynamic. If the coordinate display is static, the coordinate values displayed in the status line change only when you specify a point. However, if the coordinate display is dynamic (default setting), AutoCAD constantly displays the absolute coordinates of the graphics cursor with respect to the UCS origin. AutoCAD can also display the polar coordinates (length<angle) if you are in an AutoCAD command.

SNAP. If SNAP is on, SNAP is displayed in the status line; otherwise, it is not displayed.

GRID. If GRID is on, grid lines are displayed on the screen.

ORTHO. If ORTHO is on, ORTHO is displayed in the status line; otherwise, it is not displayed.

OSNAP. If OSNAP is on, OSNAP is displayed in the status line. When OSNAP is on, you can use the running object snaps. If OSNAP is off, not displayed in the status line, the running object snaps are temporarily disabled. The status of OSNAP (Off or On) does not prevent you from using regular object snaps.

MODEL and PAPER Space. AutoCAD displays MODEL in the status line when you are working in the model space. If you are working in the paper space, AutoCAD will display PAPER in place of MODEL.

TILE. AutoCAD displays TILE, if TILEMODE is on (1). When TILEMODE is on, you are in the model space. If TILE is off, you are in the paper space (TILEMODE is 0).

MODEMACRO SYSTEM VARIABLE

The AutoCAD system variable **MODEMACRO** can be used to display a new text string in the status line. You can also display the value returned by a macro expression using the DIESEL language, which is discussed in a later section of this chapter. MODEMACRO is a system variable and you can assign a value to this variable by entering MODEMACRO at the Command prompt or by using the SETVAR command. For example, if you want to display **Customizing AutoCAD** in the status line, enter **SETVAR** at the Command: prompt and then press Enter. AutoCAD will prompt you to enter the name of the system variable. Enter **MODEMACRO** and then press Enter again. Now you can enter the text you want to display in the status line. After you enter **Customizing AutoCAD** and press Enter, the status line will display the new text.

Command: **MODEMACRO**
or
Command: **SETVAR**
Variable name or ?: **MODEMACRO**
New value for MODEMACRO, or . for none <"">: **Customizing AutoCAD**

You can also enter MODEMACRO at the Command: prompt and then enter the text that you want to display in the status line.

Command: **MODEMACRO**
New value for MODEMACRO, or . for none <"">: **Customizing AutoCAD**

Once the value of the **MODEMACRO** variable is changed, it retains that value until you enter a new value, start a new drawing, or open an existing drawing file. If you want to display the standard text in the status line, enter a period (.) at the prompt **New value for MODEMACRO, or . for none <"">:**. The value assigned to the **MODEMACRO** system variable is not saved with the drawing, in any configuration file, or anywhere in the system.

Command: MODEMACRO
New value for MODEMACRO, or . for none <"">:

CUSTOMIZING THE STATUS LINE

The information contained in the status line can be divided into two parts: toggle functions and coordinate display. The toggle functions part consists of the status of Snap, Grid, Ortho, Osnap, Model Space, and Tile (Figure 15-1). The coordinate display displays the X, Y, and Z coordinates of the cursor. The status line can be customized to your requirements by assigning a value to the AutoCAD system variable **MODEMACRO**. The value assigned to this variable is displayed left-justified in the staus bar at the bottom of the AutoCAD window. The number of characters that can be displayed in the status line depends on the system display and the size of the AutoCAD window. The coordinate display field cannot be changed or edited.

The information displayed in the status line is a valuable resource. Therefore, you must be careful when selecting the information to be displayed in the status line. For example, when working on a project, you may like to display the name of the project in the status line. If you are using several dimensioning styles, you could display the name of the current dimensioning style (DIMSTYLE) in the status line. Similarly, if you have several text files with different fonts, the name of the current text file (TEXTSTYLE) and the text height (TEXTSIZE) can be displayed in the status line. Sometimes, in 3D drawings, if you need to monitor the viewing direction (VIEWDIR), the camera coordinate information can be displayed in the status line. Therefore, the information that should be displayed in the status line depends on you and the drawing requirements. AutoCAD lets you customize this line and have any information displayed in the status line that you think is appropriate for your application.

MACRO EXPRESSIONS USING DIESEL

You can also write a macro expression using DIESEL to assign a value to the **MODEMACRO** system variable. The macro expressions are similar to AutoLISP functions, with some differences. For example, the drawing name can be obtained by using the AutoLISP statement **(getvar dwgname)**. In DIESEL, the same information can be obtained by using the macro expression **$(getvar,dwgname)**. However, unlike the case with AutoLISP, the DIESEL macro expressions return only string values. The format of a macro expression is:

$(function-name,argument1,argument2, ...)

Example
$(getvar,dwgname)

Here, **getvar** is the name of the DIESEL string function and **dwgname** is the argument of the function. There must not be any spaces between different elements of a macro expression. For example, spaces between the $ sign and the open parentheses are not permitted. Similarly, there must not be any spaces between the comma and the argument, **dwgname**. All macro expressions must start with a $ sign.

The following example illustrates the use of a macro expression using DIESEL to define and then assign a value to the **MODEMACRO** system variable.

Example 1

Using the AutoCAD MODEMACRO command, redefine the status line to display the following information in the status line:

Project name (Cust-Acad)
Name of the drawing (DEMO)
Name of the current layer (OBJ)

Note that in this example the project name is Cust-Acad, the drawing name is DEMO, and the current layer name is OBJ.

Before entering the MODEMACRO command, you need to determine how to retrieve the required information from the drawing database. For example, here the project name **(Cust-Acad)** is a user-defined name that lets you know the name of the current project. This project name is not saved in the drawing database. The name of the drawing can be obtained using the DIESEL string function GETVAR **$(getvar,dwgname)**. Similarly, the GETVAR function can also be used to obtain the name of the current layer, **$(getvar,clayer)**. Once you determine how to retrieve the information from the system, you can use the **MODEMACRO** system variable to obtain the new status line. For Example 1, the following DIESEL expression will define the required status line.

Command: MODEMACRO
New value for MODEMACRO, or . for none<"">: Cust-Acad N:$(GETVAR,dwgname)
L:$(GETVAR,clayer)

Explanation

Cust-Acad
Cust-Acad is assumed to be the project name you want to display in the status line.

N:$(GETVAR,dwgname)
Here, N: is used as an abbreviation for the drawing name. The GETVAR function retrieves the name of the drawing from the system variable **dwgname** and displays it in the status line, next to N:.

L:$(GETVAR,clayer)
Here L: is used as an abbreviation for the layer name. The GETVAR function retrieves the name of the current layer from the system variable **clayer** and displays it in the status line.

The new status line is shown in Figure 15-2.

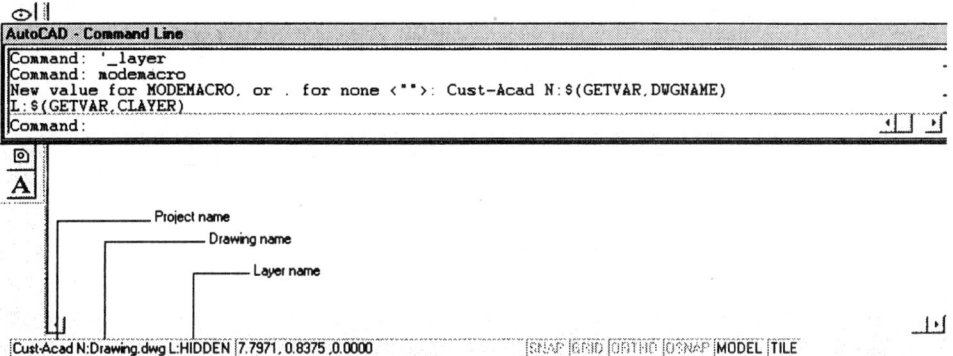

Figure 15-2 Status line for Example 1

Example 2

Using the AutoCAD MODEMACRO command, redefine the status line to display the following information in the status line:

    Name of the current textstyle
    Size of text
    User-elapsed time in minutes

In this example the abbreviations for text style, text size, and user-elapsed time in minutes are TSTYLE:, TSIZE:, and ETM:, respectively.

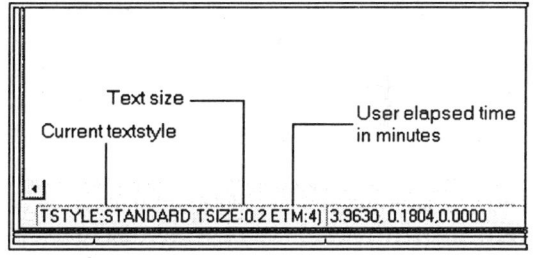

Command: **MODEMACRO**
New value for MODEMACRO, or . for none<"">: **TSTYLE:$(GETVAR,TEXTSTYLE)**
                      TSIZE:$(GETVAR,TEXTSIZE)
                      ETM:$(FIX,$(*,60,$(*,24,$(GETVAR,TDUSRTIMER))))

Explanation

TSTYLE:$(GETVAR,TEXTSTYLE)
The **GETVAR** function obtains the name of the current textstyle from the system variable **TEXTSTYLE** and displays it next to TSTYLE: in the status line.

TSIZE:$(GETVAR,TEXTSIZE)
The **GETVAR** function obtains the current size of the text from the system variable **TEXTSIZE** and then displays it next to TSIZE: in the status line.

ETM:$(FIX,$(*,60,$(*,24,$(GETVAR,TDUSRTIMER))))
The **GETVAR** function obtains the user-elapsed time from the system variable **TDUSRTIMER** in the following format:

<Number of days>.<Fraction>

Example
0.03206400 (time in days)
To change this time into minutes, multiply the value obtained from the system variable **TDUSRTIMER** by 24 to change it into hours, and then multiply the product by 60 to change the time into minutes. To express the minutes value without a decimal, determine the integer value using the DIESEL string function FIX.

Example
Assume that the value returned by the system variable **TDUSRTIMER** is 0.03206400. This time is in days. Use the following calculations to change the time into minutes, and then express the time as an integer:

0.03206400 days x 24 = 0.769536 hr
0.769536 hr x 60 = 46.17216 min
integer of 46.17216 min = 46 min

USING AUTOLISP WITH MODEMACRO

Sometimes the DIESEL expressions can be as long as those shown in Example 1 and Example 2. It takes time to type the DIESEL expression, and if you make a mistake in entering the expression, you have to retype it. Also, if you need several different status line displays, you need to type them every time you want a new status line display. This can be time-consuming and sometimes confusing.

To make it convenient to change the status line display, you can use AutoLISP to write a DIESEL expression. It is easier to load an AutoLISP program, and it also eliminates any errors that might be caused by typing a DIESEL expression. The following example illustrates the use of AutoLISP to write a DIESEL expression to assign a new value to the **MODEMACRO** system variable.

Example 3

Using AutoLISP, redefine the value assigned to the **MODEMACRO** system variable to display the following information in the status line:

Name of the current text style
Size of text
User-elapsed time in minutes

In this example the abbreviations for text style, text size, and user-elapsed time in minutes are TSTYLE:, TSIZE:, and ETM:, respectively.

The following file is a listing of the AutoLISP program for Example 3. The name of the file is **ETM.LSP. The line numbers are not a part of the file; they are shown here for reference only.**

```
(defun c:etm ( )                                 1
(setvar "MODEMACRO"                              2
(strcat                                          3
  "TSTYLE:$(getvar,textstyle)"                   4
  " TSIZE:$(getvar,textsize)"                    5
  " ETM:$(fix,$(*,60,$(*,24,                     6
  $(getvar,tdusrtimer))))"                       7
  )                                              8
 )                                               9
)                                                10
```

Explanation

Line 3
(strcat
The AutoLISP function **strcat** links the string value of lines 4 - 7 and returns a single string that becomes a DIESEL expression for the MODEMACRO command.

Line 4
"TSTYLE:$(getvar,textstyle)"
This line is a DIESEL expression in which getvar, a DIESEL string function, retrieves the value of the system variable **textstyle** and **$(getvar,textstyle)** is replaced by the name of the textstyle. For example, if the textstyle is STANDARD, the line will return "TSTYLE:STANDARD". This is a string because it is enclosed in quotes.

Lines 6 and 7
" ETM:$(fix,$(*,60,$(*,24,
$(getvar,tdusrtimer))))"
These two lines return **ETM:** and the time in minutes as a string. The **fix** is a DIESEL string function that changes a real number to an integer.

To load this AutoLISP file (**ETM.LSP**), use the following commands. In this example, the file name and the function name are the same (ETM).

Command: **(load "ETM")**
ETM
Command: **ETM**

DIESEL EXPRESSIONS IN MENUS

You can also define a DIESEL expression in the screen, tablet, pull-down, or button menu. When you select the menu item, it will automatically assign the value to the MODEMACRO system variable and then display the new status line. The following example illustrates the use of the DIESEL expression in the screen menu.

Example 4

Write a DIESEL macro for the screen menu that displays the following information in the status line (Figure 15-3):

Macro-1	**Macro-2**	**Macro-3**
Project name	Pline width	Dimtad
Drawing name	Fillet radius	Dimtix
Current layer	Offset distance	Dimscale

The following file is a listing of the screen menu that contains the definition of three DIESEL macros for Example 4. This menu can be loaded using AutoCAD's MENU command and then entering the name of the file. If you select the first item, DIESEL1, it will automatically display the new status line.

```
***screen
[*DIESEL*]
[DIESEL1:]^C^CMODEMACRO;$M=Cust-Acad,N:$(GETVAR,DWGNAME)+
,L:$(GETVAR,CLAYER);
[DIESEL2:]^C^CMODEMACRO;$M=PLWID:$(GETVAR,PLINEWID),+
FRAD:$(GETVAR,FILLETRAD),OFFSET:$(GETVAR,OFFSETDIST),+
LTSCALE:$(GETVAR,LTSCALE);
[DIESEL3:]^C^CMODEMACRO;$M=DTAD:$(GETVAR,DIMTAD),+
DTIX:$(GETVAR,DIMTIX),DSCALE:$(GETVAR,DIMSCALE);
```

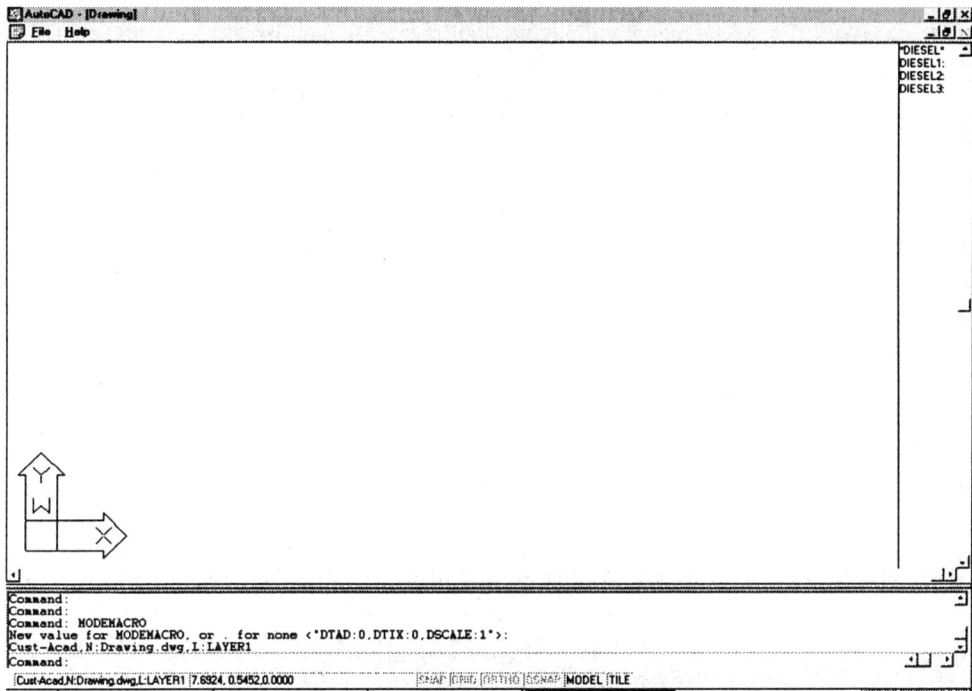

Figure 15-3 Status line for Example 4

MACROTRACE SYSTEM VARIABLE

The MACROTRACE variable is an AutoCAD system variable that can be used to debug a DIESEL expression. The default value of this variable is 0 (off). It is turned on by assigning a value of 1 (on). When on, the MACROTRACE system variable will evaluate all DIESEL expressions and display the results in the command prompt area. For example, if you have defined several DIESEL expressions in a drawing session, all of them will be evaluated at the same time and the messages, if any, will be displayed in the command prompt area.

Example
Command: **MODEMACRO**
New value for MODEMACRO, or . for none < " " > : **$(getvar,dwgname),$(getvar clayer)**

Note that in this DIESEL expression a comma is missing between getvar and clayer. If the MACROTRACE system variable is on, the following message will be displayed in the command prompt area:

Eval: $(GETVAR,DWGNAME)
= = = = >UNNAMED
Eval: $(GETVAR CLAYER)
Err: $(GETVAR CLAYER)??

This error message gives you an idea about the location of the error in a DIESEL expression. In the previous example, the first part of the expression successfully returns the name of the drawing (unnamed) and the second part of the expression results in the error message. This confirms that there is an error in the second part of the DIESEL expression. You can further determine the cause of the error by comparing it with the error messages in the following table:

Error Message	Description
$?	Syntax error
$?(func,??)	Incorrect argument to function
$(func)??	Unknown function
$(++)	Output string too long

DIESEL STRING FUNCTIONS

Like AutoLISP, you can use DIESEL functions to do some mathematical operations, retrieve the values from the drawing database, and display the values in the status line in a predetermined order. For example, you can add, subtract, multiply, and divide the numbers. You can also obtain the values of some system variables and display them in the status line. The maximum number of parameters that the DIESEL expression can contain is 10. This number includes the name of the function. The following section discusses some frequently used DIESEL string functions.

Addition

Format **$(+,num1,num2,num3 - - -)**

This function (+) calculates the sum of the numbers that are to the right of the plus (+) sign. The numbers can be integers or real.

Examples
$(+,2,5)   returns 7
$(+,2,5,50.75)   returns 57.75

> **Note**
> *You can test the calculations by assigning the DIESEL expression to the MODEMACRO system variable.*

Command: **MODEMACRO**
New value for MODEMACRO, or . for none<"">: $(+,2,5)

AutoCAD will return a value of 7, which will be displayed in the status line. It is important to note that the values returned by the DIESEL string expressions are string values.

Subtraction

Format **$(-,num1,num2,num3,- - -)**

This function (-) subtracts the second number from the first number. If there are more than two numbers, the second and the subsequent numbers are added and their sum is subtracted from the first number.

Examples
$(-,28,14)   returns 14
$(-,25,7,11.5)   returns 6.5

Multiplication

Format **$(*,num1,num2,num3,- - -)**

This function (*) calculates the product of the numbers that are to the right of asterisk.

Examples
$(*,2,5)   returns 10
$(*,2,5,3.0) returns 30.0
$(*,2,5,3.25)   returns 32.5

Division

Format **$(/,num1,num2,num3 - - -)**

This function (/) divides the first number by the second number. If there are more than two numbers, the first number is divided by the product of the second and subsequent numbers.

Examples
(/ 3 2)   returns 1.5
(/ 3.0 2)   returns 1.5
(/ 200 5 4)   returns 10
(/ 200.0 5.5)   returns 36.363636
(/ 200 -5)   returns -40
(/ -200 -5.0)   returns 40.0

Relational Statements

Some DIESEL expressions involve features that test a particular condition. If the condition is true the expression performs a certain function; if the condition is not true, the expression performs another function. The following section discusses various relational statements used in DIESEL expressions.

Equal to

Format **$(=,num1,num2)**

This function (=) checks whether the two numbers are equal. If they are, the condition is true and the function will return **1**. Similarly, if the specified numbers are not equal, the condition is false and the function will return **0**.

Examples

$(=,5,5)	returns 1
$(=,5,4.9)	returns 0
$(=,5,-5)	returns 0

Not equal to

Format $(!=,num1,num2)

This function (!=) checks whether the two numbers are **not equal**. If they are not equal, the condition is true and the function will return **1**; otherwise the function will return **0**.

Examples

$(!=,50,4)	returns 1
$(!=,50,50)	returns 0
$(!=,50,-50)	returns 1

Less than

Format $(<,num1,num2)

This function (<) checks whether the first number **(num1)** is less than the second number **(num2)**. If it is true, the function will return **1**; otherwise the function will return **0**.

Examples

$(<,3,5)	returns 1
$(<,5,3)	returns 0
$(<,3.0,5)	returns 1

Less than or equal to

Format $(<=,num1,num2)

This function (<=) checks whether the first number **(num1)** is less than or equal to the second number **(num2)**. If it is true, the function will return **T**; otherwise the function will return **nil**.

Examples

$(<=,10,15)	returns 1
$(<=,19,10)	returns 0
$(<=,-2.0,0)	returns 1

Greater than

Format **$(>,num1,num2)**

This function (**>**) checks whether the first number (**num1**) is greater than the second number (**num2**). If it is true, the function will return **1**; otherwise the function will return **0**.

Examples
$(>,15,10)   returns 1
$(>,20,30)   returns 0

Greater than or equal to

Format **$(>=,num1,num2)**

This function (**>=**) checks whether the first number (**num1**) is greater than or equal to the second number (**num2**). If it is true, the function will return **1**; otherwise the function will return **0**.

Examples
$(>=,78,50)   returns 1
$(>=,78,88)   returns 0

eq function

Format **$(eq,value1,value2)**

This function (**eq**) checks whether the two string values are equal. If they are, the condition is true and the function will return **1**; otherwise the function will return **0**.

Examples
$(eq,5,5)   returns 1
$(eq,yes,yes)   returns 1
$(eq,yes,no)   returns 0

angtos

Format **$(angtos,angle[,mode,precision])**

The **angtos** function returns the angle expressed in radians in a string format. The format of the string is controlled by the **mode** and **precision** settings.

Examples
$(angtos,0.588003,0,4)   returns 33.6901
$(angtos,-1.5708,1,2)   returns 270d0'

> **Note**
> The following modes are available in AutoCAD:
>
> | ANGTOS MODE | EDITING FORMAT |
> |---|---|
> | 0 | Decimal degrees |
> | 1 | Degrees/minutes/seconds |
> | 2 | Grads |
> | 3 | Radian |
> | 4 | Surveyor's units |
>
> **Precision** is an integer number that controls the number of decimal places. Precision corresponds to the AutoCAD system variable, **AUPREC**. The minimum value of **precision** is 0, the maximum is 4.

eval function

Format **$(eval,string)**

The **eval** function passes the string to the DIESEL evaluator and the result obtained after evaluating the string is returned and displayed in the status line.

Examples

$(eval,welcome)	returns welcome
$(eval,$(getvar,dimscale))	returns value of dimscale

fix function

Format **$(fix,num)**

The **fix** function converts a real number into an integer by truncating the digits after the decimal.

Examples

$(fix,42.573)	returns 42
$(fix,-23.50)	returns -23

getvar function

Format **$(getvar,varname)**

The **getvar** function retrieves the value of an AutoCAD system variable.

Examples

$(gatvar,dimtad)	returns value of dimtad
$(getvar,clayer)	returns current layer name

rtos function

    Format $(rtos,number)
    or $(rtos,number,mode,precision)

The **rtos** function changes a given number into a real number and the format of the real number is determined by the mode and precision values.

Examples
$(rtos,50)   returns 50
$(rtos,1.5,5,4)   returns 1 1/2

Note

The following linear unit modes are available in AutoCAD:

MODE VALUE	STRING FORMAT
1	Scientific
2	Decimal
3	Engineering
4	Architectural
5	Fractional

Precision is an integer number that controls the number of decimal places. Precision corresponds to AutoCAD system variable, LUPREC, and the mode corresponds to LUNITS. If the mode and precision values are omitted, AutoCAD uses the current values of LUNITS and LUPREC as set with the units command.

if function

    Format **$(if,condition,then,else)**

The **if** function evaluates the first expression (then), if the value returned by the specified condition is non zero. The second expression (else) is evaluated if the specified condition returns 0.

```
(if condition then [else])
 │      │         │     └─ Expression evaluated if the
 │      │         │          condition returns 0
 │      │         └─ Expression evaluated if
 │      │              the condition returns non zero
 │      └─ Specified conditional statement
```

Examples
$(if,$(=,7,7),true)   returns true
$(if,$(=,5,7),true,false)   returns false
$(if,1.5,true,false)   returns true

strlen function

Format **$(strlen,string)**

The **strlen** function returns an integer number that designates the number of characters contained in the specified string.

Example
$(strlen,Customizing AutoCAD)   returns 19

linelen function

Format **$(linelen)**

The **linelen** function returns an integer number that lets you know the number of characters that can be accommodated in the status line. This number varies and depends on the system. However, on most systems the number of characters that can be displayed in the layer-mode section of the status line is 38.

Example
$(linelen)   returns 38 (**depends on the system**)

upper function

Format **$(upper,string)**

The **upper** function returns the specified string in uppercase.

Example
$(upper,Customizing)   returns CUSTOMIZING

edtime function

Format **$(edtime,date,display-format)**

The **edtime** function can be used to edit the date and display it in a format specified by display-format. For example, the Julian date obtained from AutoCAD's system variable, DATE, is obtained in the form 2449013.85156759. The **edtime** function can be used to display this date in an understandable form. The following table gives the format of the phrases that can be used with the **edtime** function:

FORMAT	OUTPUT	FORMAT	OUTPUT
D	5	H	2
DD	05	HH	02
DDD	Tue	MM	23

DDDD	Tuesday	SS	12
M	1	MSEC	325
MO	11	AM/PM	PM
MON	Nov	am/pm	am
MONTH	November	A/P	P
YY	92	a/p	p
YYYY	1992		

Example
$(edtime,$(getvar,date),DDDD","DD MONTH YY - HH:MMAM/PM)
                               returns Monday,25 January 93 - 08:52PM

In the above example, the **getvar** function retrieves the date in Julian format. The **edtime** function displays the date as specified by the display-format. The following illustration shows the corresponding fields of the display-format and the value returned by the **edtime** function:

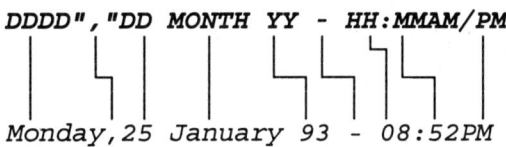

15-18 Customizing AutoCAD

REVIEW QUESTIONS

Indicate whether the following statements are true or false.

1. DIESEL (direct interpretively evaluated string expression language) is a string expression language. (T/F)

2. The value assigned to the **MODEMACRO** variable is a string, and the output it generates is not a string. (T/F)

3. You cannot define a DIESEL expression in the screen menu. (T/F)

4. The coordinate information displayed in the status line can be dynamic only. (T/F)

5. Once the value of the **MODEMACRO** variable is changed, it retains that value until you enter a new value, start a new drawing, or open an existing drawing file. (T/F)

6. The number of characters that can be displayed in the layer mode field on any system is 38. (T/F)

7. The coordinate display field cannot be changed or edited. (T/F)

8. You can write a macro expression using DIESEL to assign a value to the **MODEMACRO** system variable. (T/F)

9. In DIESEL, the drawing name can be obtained by using the macro expression **$(getvar,dwgname)**. (T/F)

10. You cannot use AutoLISP to write a DIESEL expression. (T/F)

Fill in the blanks.

11. All macro expressions must start with a _____ sign.

12. The display-format DDD will return Monday as _____.

13. The DIESEL expression, $(upper,AutoCAD), will return _____.

14. The DIESEL expression, $(strlen,AutoCAD), will return _____.

15. The DIESEL expression, $(if,$(=,3,2),yes,no), will return _____.

16. The DIESEL expression, $(fix,-17.75), will return _____.

17. The DIESEL expression, $(eq,Customizing,customizing), will return _____.

18. The DIESEL expression, $(/,81,9,9), will return _____.

19. The MACROTRACE system variable can be used to _____ a DIESEL expression.

20. The status line can be turned off by _____ AutoCAD.

EXERCISES

Exercise 1
Using the AutoCAD MODEMACRO command, redefine the status line to display the following information in the status line:

   Your name
   Name of drawing

Exercise 2
Using the AutoCAD MODEMACRO command, redefine the status line to display the following information in the status line:

   Name of the current dimension style
   Dimension scale factor (dimscale)
   User-elapsed time in hours

The abbreviations for dimstyle, dimscale, and user-elapsed time in hours are DIMS:, DIMFAC:, and ETH:, respectively.

Exercise 3
Using **AutoLISP**, redefine the status line to display the following information in the status line:

   Name of the current dimension style
   Dimension scale factor (dimscale)
   User-elapsed time in hours

Chapter 16

VISUAL BASIC

Learning objectives

After completing this chapter, you will be able to:
- *Install the AutoCAD Preview, Visual Basic for Applications (VBA) Edition.*
- *Load and run sample VBA projects..*
- *Utilize the Visual Basic Editor.*
- *Understand and use AutoCAD objects.*
- *Use object properties.*
- *Apply and use AutoCAD methods.*

ABOUT VISUAL BASIC

The original BASIC was developed in 1963. BASIC (Beginners All-purpose Symbolic Instruction Code) was intended to be an accessible, user friendly programming language. BASIC was the language supplied with many of the early microcomputers starting in the late 1970's. It continued development until the advent of Windows. In 1991 the first version of **Visual Basic (VB)** appeared followed by new versions spaced about a year apart. The current version 5, like its predecessors lets even a beginning programmer create a user friendly graphical user interface (GUI) with a small fraction of the effort required previously. It used to take a team of programmers working with a language like C that came with a stack of reference material nearly a foot thick to do the same thing.

Another advantage inherent to Visual Basic is that engineers may have old BASIC programs used in sizing components. This code could be used with minor input/output modifications in new Visual Basic modules to parametrically design and draft the component.

Autodesk has licensed **Visual Basic for Applications (VBA)** from Microsoft as have dozens of other large software companies. The VBA language tools are either identical or very similar to those in the stand-alone Visual Basic 5.0. VBA runs right inside AutoCAD, and at a significantly higher speed than a VB program running externally or an internal AutoLISP program. This is because of the way a VBA program communicates with the AutoCAD core. Programmers who learn to write macros for Microsoft Office97 only have to learn AutoCAD-specific functions to program VBA in AutoCAD.

A new capability going along with VBA is the ability to communicate with other applications such as Microsoft Excel, Microsoft Word, Microsoft Access and Visual Basic 5.0 using **ActiveX Automation**. Automation lets you access and manipulate the objects and functionality of

AutoCAD from Excel or another application supporting ActiveX. Alternately, the objects and functionality of Excel for example can be used inside AutoCAD. This cross-application macro programming capability does not exist in AutoLISP. Of course, before you can use AutoCAD's objects in some other software supporting ActiveX you must make the application aware that the **AutoCAD Object Library** is available on the computer.

Release 14 is the first version of AutoCAD that supports Visual Basic for Applications. Autodesk calls the VBA implementation a Preview edition. The Preview designation is an acknowledgement that Autodesk is still working on integrating VBA into AutoCAD. One of the limitations of this implementation of ActiveX is that you cannot call AutoLISP routines from a VBA routine nor get or set the values of AutoLISP variables. Another limitation is that you cannot record macros. Recording keystrokes to create macros is common in other applications. Only one VBA project may be open at a time in release 14. There is no equivalent in VBA to the "COMMAND" statement of AutoLISP which is true for a number of other statements as well. According to AutoCAD's Visual Basic help system "VBA is not viewed as a replacement for AutoLISP and there are no current plans to provide equivalent functionality for all of AutoLISP's API's." API stands for application programming interface.

INSTALLING VBA

You must find and install VBA yourself. It is not installed automatically when you install AutoCAD. VBA is only included with the International English language versions of Release 14. The published purpose is to "set the foundation for AutoCAD customers to explore Visual Basic as a method to programmatically customize their CAD environment through ActiveX Automation." Autodesk offers no official support but is "extremely interested" in feedback, particularly bug reports from customers. Their email address is: *vba.support@autodesk.com*. All comments will be carefully considered, although they caution you may not receive an explicit reply.

To install the Preview addition for Windows 95 and Windows NT 4.0:

1. Insert the AutoCAD Release 14 Product CD into the CD-ROM drive.
2. Click the Start button.
3. Select Run.
4. Enter e:\vbainst\setup.exe. Click OK.
5. Follow the screen prompts

Note: Substitute your own CD-ROM drive letter for e:.

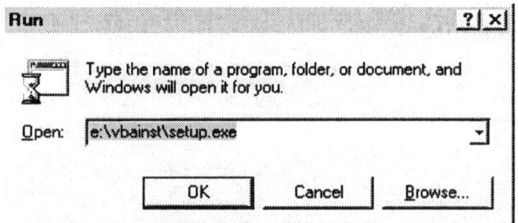

Figure 16-1 Run Item from Start Menu

There are over three million programmers using Visual Basic. A great many more have programmed in BASIC but not VB. This chapter assumes you have some familiarity with BASIC or programming concepts such as those covered in the AutoLISP chapter of this book and are comfortable looking up commands and syntax in the help system.

OBJECTS

Visual Basic uses **objects** to facilitate what it can accomplish. Examples of objects include drawings or documents, geometric elements such as lines or circles, and user interface controls to handle input and output to programs or macros. The "visual" part of Visual Basic refers to the familiar user interface controls such as check boxes, radio controls, and dialog boxes to open or save files seen in windows applications. These controls are screen objects which may be dragged from a "toolbox" onto the background object called a form. An example of a simple form is the background window of a dialog box. In the course of dragging a control from the Toolbox onto a form, the coding which provides the functionality of the control is simultaneously added to the form object. Several aspects of true object-oriented programming (OOP) languages such as C++ are missing from Visual Basic. So VB is considered an object-based, rather than an object-oriented programming language. The missing aspects are more than made up for by the development tools and environments available to the user.

Functions known as **methods** have been defined in the AutoCAD object library to perform an action on an object, for example draw a line in a drawing. The method AddLine adds a line object to a drawing. **Properties** are functions that set or return information about the state of an object. In the example of a line drawn in a drawing object the Color, Layer, StartPoint, and EndPoint would be some of the properties of the Line object.

ADD METHOD

The way to draw in paper space, model space, or in a block is to use the **Add method** such as **AddCircle, AddLine, AddArc,** and **AddText**. Instead of thinking of drawing in the model space of the current drawing, think in terms of adding a geometric object. **Dot notation** is required to clarify on which object the method is acting. The dot notation reference starts with the most global object first narrowing to the right.

AddCircle

The **AddCircle** method requires a predefined center point and radius. Both of these arguments are required. Other input options used in the AutoCAD circle command such as three-points, two-points, or tangent-tangent-radius require additional programming in VBA to calculate the center point and radius for the AddCircle method. The format of the **AddCircle** method is:

```
ThisDrawing.ModelSpace.AddCircle centerpoint, radius
                                           |          |
    Center point, double precision vector──┘          |
        Radius, double precision─────────────────────┘
```

Notice that the arguments of the AddCircle method follow a required space. For the examples below, point1 and point2 are predefined points consisting of a vector of double precision coordinate values. These may be defined with assignment statements as shown in Example 1 on page 16-5. The pound sign (#) signifies a double precision number in Basic.

Example
ThisDrawing.ModelSpace.AddCircle point1, 3#

AddLine

The **AddLine method** requires two predefined endpoints. The **AddLine** method has a syntax similar to AddCircle:

ThisDrawing.ModelSpace.AddLine firstpoint, secondpoint

　　First point, double precision vector ┘
　　　　Second point, double precision vector ┘

Example
ThisDrawing.ModelSpace.AddLine point1, point2

AddArc
The **AddArc** method format is:

ThisDrawing.ModelSpace.AddArc ctrpt, radius, StartAng, EndAng

　Center point ┘
　　Radius (distance) ┘
　　　Arc starting angle in radians ┘
　　　　Arc ending angle in radians ┘

Example
ThisDrawing.ModelSpace.AddArc point1, 4#, 0#, 1.570796327

AddText
The **AddText** method requires a predefined string, insertion point and text height. The **AddText** method syntax is:

ThisDrawing.ModelSpace.AddText textString$, point, textHeight

　Actual text to be displayed ┘
　　Position point double precision vector ┘
　　　Text Height. Must be positive number ┘

Example
ThisDrawing.ModelSpace.AddText ".063 TYP, 4 PLACES", point1, 0.25#

FINDING HELP ON METHODS AND PROPERTIES

You will be disappointed when you try the help menu in the VBA Integrated Development Environment for help with AddLine, AddCircle, or any of the particular methods or properties needed to write a parametric program. You can easily find general information such as on Methods, help on Visual Basic key words and many non-AutoCAD VBA programming examples taken from Excel, Word, or PowerPoint. AutoCAD Release 14 contains excellent **VBA specific help** with copious examples, however. It's just not well integrated with the VBA IDE help system or accessible from the Index or Find tabs there. The Help menu item is has the title "Microsoft Visual Basic Help" and contains the Word, Excel, and PowerPoint Visual Basic References but not one for AutoCAD, probably because of the add-in status of the preview. The **AutoCAD Visual Basic Reference** is called the **Automation User's Guide** and **Automation Reference** and is available in the AutoCAD drawing editor's help system under **ActiveX Automation**.

The Index and Find tabs of this help menu are very useful. To get help in the VBA IDE, you can use the **Object Browser** which can be accessed from the **View** pull-down menu. Select **Project** in the drop down list box which shows <All Libraries> by default. Select object **ThisDrawing** in the left window and **ModelSpace** in the right one as shown in Figure 16-2. The question mark help button on the toolbar or **F1** takes you to a screen where you can access the ModelSpace Collection of help screens on methods, properties and examples. Alternately, the function key **F1** will get help on any word typed in the code window.

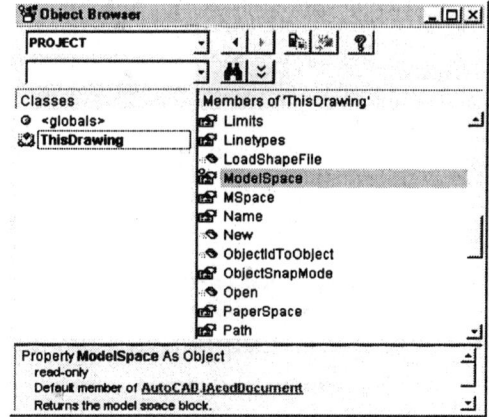

Figure 16-2 Object Browser

LOADING AND SAVING VBA PROJECTS

In order to run Visual Basic for Applications you need to be sure VBA has been separately installed. You can tell by the addition of a pull down menu titled V<u>B</u>A as shown in Figure 16-3.

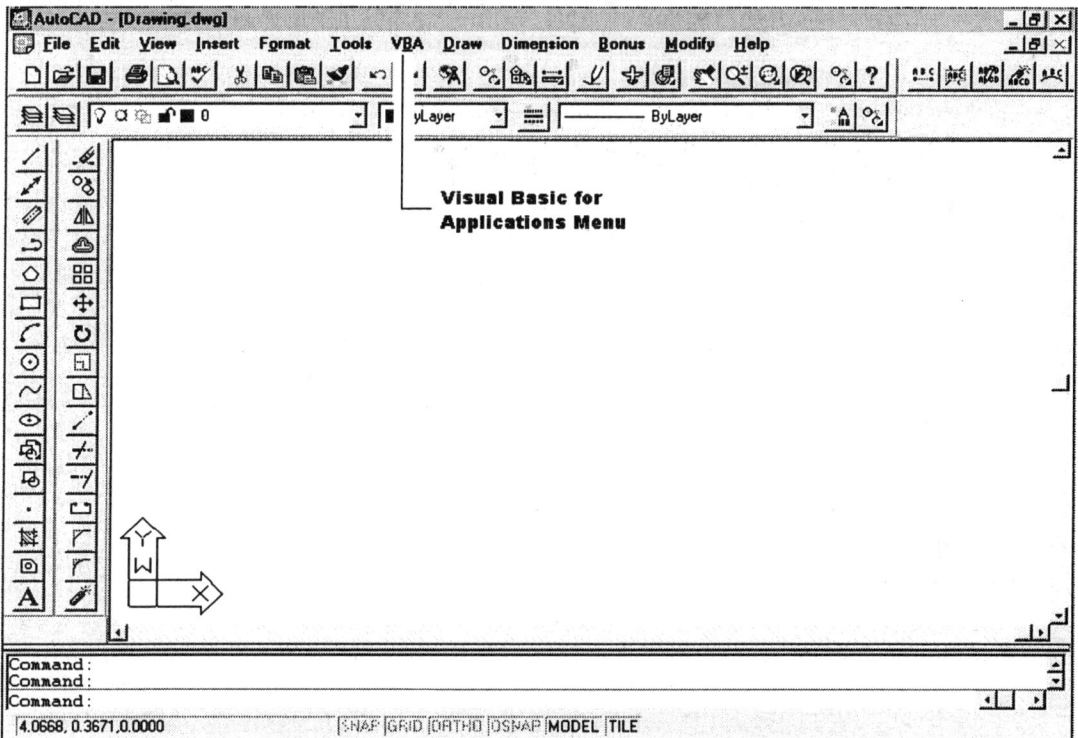

Figure 16-3 AutoCAD Drawing Editor VBA Menu

The **VBA** menu item **Show VBA IDE** is one way to access the **Integrated Development Environment (IDE)** which consists of a number of useful windows which can be sized, shown,

16-6 Customizing AutoCAD

hidden, or otherwise customized to suit your needs. Figure 16-4 shows the IDE with the VBA project for Example 1 loaded. There are six windows shown. The top left most window is the Project Explorer which allows navigation among windows. Note the View Code, View Object, and Toggle Folders Icons at the top of the window. The bottom left menu is the Properties Window which shows object properties and is a convenient way to change them.

The top right window is the UserForm Window where the "Visual" user interface is constructed using the Toolbox. The next window down on the right is the UserForm code window where Basic language program code resides which runs when triggered by events such as a mouse click on a control button. The next window down is the Module code window where mathematical procedures with the .bas extension are generally kept. The bottom right Immediate and Watch Windows are used for debugging.

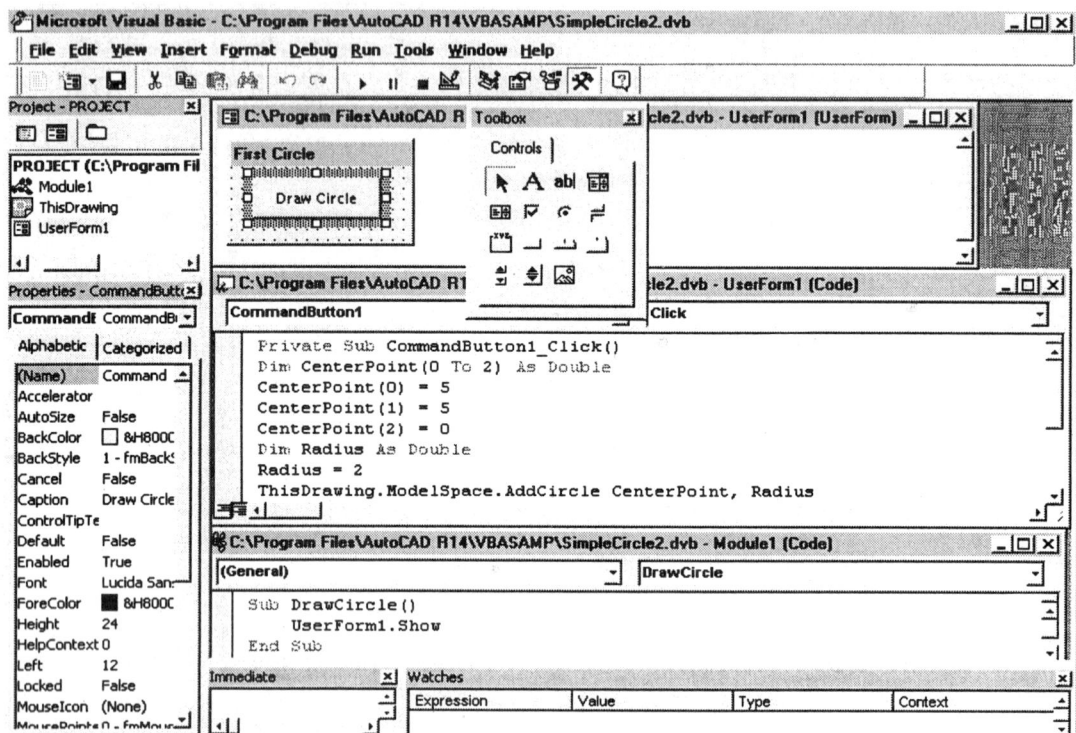

Figure 16-4 VBA Integrated Development Environment

You are advised to save your VBA program before testing it. The preview seems more likely to crash from programming errors than a similar Visual Basic 5 program would be. To save a project from the VBA Integrated Design Environment use the **File** pull down menu item **Save PROJECT** or select the button on the toolbar that looks like a floppy disk. Menu items may be more quickly accessed with Alt and the underlined letter.

Example 1

Write a program that will draw a circle centered at 5,5,0 with radius 2 as shown in Figure 16-6.

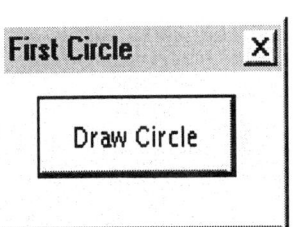

Figure 16-5 Screen Interface Form for Example 1

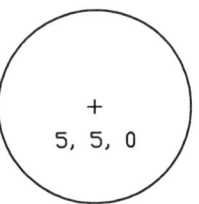

Figure 16-6 Circle, Center at 5,5,0; Radius 2

Once you are in the IDE you should insert a Module and a UserForm using the **Insert** menu item **Module** and **UserForm** respectively. The default names will be Module1 and UserForm1. Click the form shown in Figure 16-5 to make it the active window. To insert a control such as a command button onto the form, you click the corresponding button on the toolbox. The second button from the left in the third row of the toolbox inserts the command button control as shown in Figure 16-7.

Next click the form at the desired position. Resize the UserForm and CommandButton to your taste and edit the caption on the button or in the properties window at the bottom right of Figure 16-6 to say something like "Draw Circle". Click the form background to get out of editing the caption when you are finished.

A **project** is the name given to the forms, controls, modules, and the programming making up a Visual Basic program. To finish a project it is necessary to write the code underlying the visual control(s). One way to get to the code window for the CommandButton control is to double click on the button and enter the code for CommandButton1_Click(). Visual Basic is **event driven**. Lines 1 through 16 from Example 1 correspond to the code that runs when the command button is clicked with the mouse.

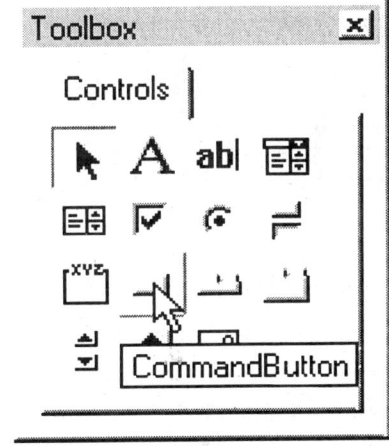

Figure 16-7 Toolbox

The Sub DrawCircle behind Module1 can be defined for the purpose of allowing the module to be run from the **VBA** menu item **Run Macro** as shown in Figure 16-8.

The following file is a listing of the Visual Basic program for Example 1. **The line numbers at the right are not a part of the programming; they are shown here for reference only.**

Command Button Code

```
'UserForm1 Code to Draw Circle, Radius 2                           1
'Centered at 5,5,0. Trigger is mouse click on                      2
'command button marked Draw Circle                                 3
Private Sub CommandButton1_Click()                                 4
Dim CenterPoint(0 To 2) As Double                                  5
Dim Radius As Double                                               6
'Data                                                              7
CenterPoint(0) = 5                                                 8
CenterPoint(1) = 5                                                 9
CenterPoint(2) = 0                                                10
Radius = 2                                                        11
                                                                  12
'OLE Automation Object Call                                       13
ThisDrawing.ModelSpace.AddCircle CenterPoint, Radius              14
Unload Me                                                         15
End Sub                                                           16
```

Explanation

Lines 1 to 3
The first lines are comments or remarks describing the function of the program. Comments make understanding and modifying a program easier and should be used liberally. Comments start with a Rem or use an apostrophe ('). These lines are ignored when the program is run.

Line 4
Private Sub CommandButton1_Click()
This line defines where Sub CommandButton1_Click() starts. The subroutine is executed when CommandButton1 is clicked with the mouse. It is the code that draws the circle when the command button is clicked. There is no need to type this line as VBA generates it automatically as soon as the command button control is added to the form.

Lines 5 and 6
Dim CenterPoint(0 To 2) As Double
Dim Radius As Double
These two lines are necessary to establish the double precision variable type for the arguments needed by the AddCircle method. The default type is Variant which will not work with AddCircle.

Lines 8 through 11
CenterPoint(0) = 5
CenterPoint(1) = 5
CenterPoint(2) = 0
Radius = 2
Here the circle center and radius are assigned values.

Line 14
ThisDrawing.ModelSpace.AddCircle CenterPoint, Radius
This line applies the AddCircle method to the ModelSpace object which is a part of the ThisDrawing object. Note the required space between the key word AddCircle and the first argument, CenterPoint.

Line 15
Unload Me
This is a method which removes UserForm1 from memory and returns the focus back to AutoCAD.

Line 16
End Sub
Like line 4, this line is generated automatically.

Module1 Code
The following code is entered in the Module1 code window.

```
'Module 1 General Declarations                      1
Sub DrawCircle()                                    2
UserForm1.Show                                      3
End Sub                                             4
```

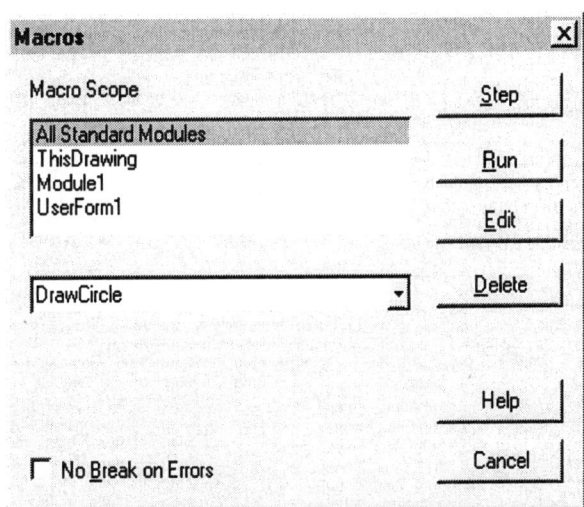

Figure 16-8 AutoCAD Drawing Editor **VBA** Pull-down Menu **Run Macro** Item

Explanation
Line 1 through 4
Sub DrawCircle()
UserForm1.Show
End Sub

This subroutine's function is to create a Macro name, DrawCircle, which shows up in the lower list box of the Run Macro dialog box run from AutoCAD's V<u>B</u>A pull-down menu item Run <u>M</u>acro shown in Figure 16-8. These four lines of code belong to the Module1 object. Another way to run a project is to select the pull down menu **Run** item **Run Sub/UserForm** in the VBA IDE or the Run Button with the triangle from the toolbar of the VBA IDE.

GetPoint, GetDistance, and GetAngle METHODS

GetPoint Method
The **GetPoint method** allows you to enter the X, Y coordinates or X, Y, Z coordinates of a point. The coordinates of the point can be entered from the keyboard or by using the screen cursor. The format of the **GetPoint** method is:

```
P = ThisDrawing.Utility.GetPoint([Point], [Prompt])
```
Enter a point, or select a point ⟶
Reference Point, Rubber-Band Origin ⟶
Prompt to be Displayed on Screen ⟶

Example
pnt1 = ThisDrawing.Utility.GetPoint("Enter 1st Point")
Pt2 = ThisDrawing.Utility.GetPoint(Pnt1,"Enter 2nd Point")

GetDistance Method
The **GetDistance method** lets you enter a distance on the command line, a distance from a given point, or two points and it then returns the distance as a double precision number. The format of the **GetDistance** method is:

```
d = ThisDrawing.Utility.GetDistance([point], [prompt])
```
Type distance or select 1 or 2 points ⟶
Reference point, rubber band origin ⟶
prompt to be displayed on screen ⟶

GetAngle Method
The **GetAngle method** allows you to enter an angle; either from the keyboard in degrees or by selecting two points. In the case of selecting points, the positive horizontal direction is taken as one leg of the angle, the first point selected as the vertex and the second point defines the second leg. If the point argument is specified, AutoCAD uses this point as the first point or angle vertex. The GetAngle method returns the value of the angle in radians as a double precision value. The format of the **GetAngle** Method is:

```
ang = ThisDrawing.Utility.GetAngle([point], [prompt])
                                      |         |          |
                  Angle in radians ────┘         |          |
              Vertex point with horizontal ──────┘          |
           Screen prompt to clarify angle selection ────────┘
```

Examples
a1 = ThisDrawing.Utility.GetAngle(,"Enter taper angle in degrees")
ang = ThisDrawing.Utility.GetAngle(pt1)' pt1 is a predefined point
ThisDrawing.Utility.GetAngle pt1, "Enter second point of angle"

The angle you enter is affected by the angle setting. The angle settings can be changed by changing the value of the AutoCAD system variables **ANGBASE** and **ANGDIR**. The default settings for measuring an angle are as follows:

The angle is measured with respect to the positive X-Axis (3 o'clock position). The value for 3 o'clock corresponds to the current value of **ANGBASE**, the AutoCAD system variable being 0. **ANGBASE** could be set in any of 4, 90 degree quadrant directions.

The angle is positive if it is measured in the counterclockwise direction and is negative if it is measured in the clockwise direction. The value of this setting is saved in the AutoCAD system variable **ANGDIR**. The **GetOrientation** method has the same syntax as the GetAngle method but ignores **ANGBASE** and **ANGDIR** system variables. The 0 angle is always at 3 o'clock and angles are always positive counterclockwise.

Example 2

Write a program that will draw a triangle with user supplied vertices P1, P2, and P3 as in Figure 16-10. This program is to use the GetPoint and AddLine methods.

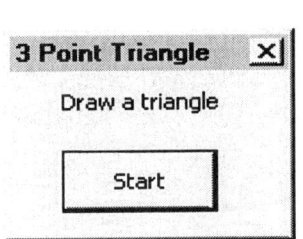

Figure 16-9 Screen Interface Form for Example 2

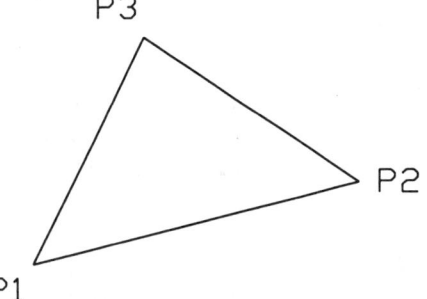

Figure 16-10 Triangle with user defined points

Once again from the IDE a Module and UserForm are inserted using the **Insert** menu item **Module** and **UserForm** respectively. The command button is inserted from the toolbox onto the form with the mouse to produce an interface form similar to Figure 16-9. The following file is a listing of the project for Example 2. **The line numbers at the right are not a part of the programming; they are shown here for reference only.**

Command Button Code

```
'The function of this routine is to draw                                    1
'a triangle form 3 user specified points                                    2
'The trigger is a mouse click on the command                                3
'button labeled Start.                                                      4
Private Sub CommandButton1_Click()                                          5
Dim pnt1 As Variant                                                         6
Dim pnt2 As Variant                                                         7
Dim pnt3 As Variant                                                         8
UserForm1.Hide                                                              9
'Returns a point in WCS                                                    10
pnt1 = ThisDrawing.Utility.GetPoint(,"Provide the First Point: ")          11
'Returns a point in WCS and draws a rubber-band line from point pnt1       12
pnt2 = ThisDrawing.Utility.GetPoint(pnt1, "Second Point? ")                13
Dim point1(0 To 2) As Double                                               14
Dim point2(0 To 2) As Double                                               15
Dim point3(0 To 2) As Double                                               16
'Changes from Variant to Double                                            17
For I = 0 To 2: point1(I) = pnt1(I): point2(I) = pnt2(I): Next I           18
ThisDrawing.ModelSpace.AddLine point1, point2                              19
pnt3 = ThisDrawing.Utility.GetPoint(pnt2, "3rd Point? ")                   20
For I = 0 To 2: point3(I) = pnt3(I): Next I 'from Variant to Double        21
ThisDrawing.ModelSpace.AddLine point2, point3                              22
ThisDrawing.ModelSpace.AddLine point3, point1                              23
Unload Me                                                                  24
End Sub                                                                    25
```

Explanation

Lines 1-4, 10, 12, and 17
These lines are comments or remarks describing the function of the following line or lines. Comments make understanding and modifying a program easier and should be used liberally. Comments start with a Rem or use an apostrophe ('). These lines are ignored when the program is run.

Line 5
Private Sub CommandButton1_Click()
This line defines where Sub CommandButton1_Click() starts. The subroutine code is executed when CommandButton1 is clicked with the mouse.

Lines 6 through 8
Dim pnt1 As Variant
Dim pnt2 As Variant
Dim pnt3 As Variant
Variant is the variable type returned by the GetPoint method. These three points are obtained this way.

Line 9
UserForm1.Hide
This statement hides the screen interface form and returns focus to AutoCAD.

Line 11, 13 and 20
pnt1 = ThisDrawing.Utility.GetPoint(,"Provide the First Point: ")
pnt2 = ThisDrawing.Utility.GetPoint(pnt1, "Second Point? ")
pnt3 = ThisDrawing.Utility.GetPoint(pnt2, "3rd Point? ")
The GetPoint method is used without a reference point in line 15. The point may be input either with the keyboard or mouse. When used with a reference point we see a rubber-band line from the reference point to the mouse cursor position.

Lines 14 through 16
Dim point1(0 To 2) As Double
Dim point2(0 To 2) As Double
Dim point3(0 To 2) As Double
These are the numeric double precision vector definitions for the user specified points.

Lines 18 and 21
For I = 0 To 2: point1(I) = pnt1(I): point2(I) = pnt2(I): Next I
For I = 0 To 2: point3(I) = pnt3(I): Next I
The x, y and z components of the numeric user specified points pick up their values from a conversion from the corresponding variant component which is a string or character form. The colon (:) allows the programmer to put more than one statement on a line. It is only recommended where the multiple statements on the line can be thought of as carrying out a single purpose. The For...Next loop repeats each assignment statement exactly three times.

Lines 19, 22, and 23
ThisDrawing.ModelSpace.AddLine point1, point2
ThisDrawing.ModelSpace.AddLine point2, point3
ThisDrawing.ModelSpace.AddLine point3, point1
The **AddLine method** requires the two point arguments to be double precision vectors with 3 components. Note the required space before the first point. If the AddLine method were used to create a (named) line object whose properties were of interest, parentheses would be used and the syntax would be instead:

16-14 Customizing AutoCAD

```
line1=ThisDrawing.ModelSpace.AddLine(firstpoint, secondpoint)
                    First point, double precision vector
                        Second point, double precision vector
```

Line 24 and 25
Unload Me
End Sub
These lines remove UserForm1 from memory, return the focus back to AutoCAD and end the subroutine.

Module1 Code
The following code is entered in the Module1 code window.

Option Explicit	1
Sub Triangle() 'Module1	2
UserForm1.Show	3
End Sub	4

Explanation
Line 1
Option Explicit
The inclusion of this line forces explicit declaration of all variable types. This minimizes common inconsistent variable usage errors and typographic errors as they are quickly caught at run time. Otherwise, undeclared variable types would be Variant by default.

Lines 2 through 4
Sub Triangle() 'Module1
UserForm1.Show
End Sub
This subroutine's function is to create a Macro name, Triangle, that can be run from AutoCAD's **VBA** pull-down menu item **Run Macro**. Lines 1-4 are a part of Module1.

PolarPoint and AngleFromXAxis METHODS

PolarPoint Method
The **PolarPoint method** defines a point at a given angle and distance from a given point. It has the syntax:

```
P = ThisDrawing.Utility.PolarPoint(Point, Angle, Distance)
              Reference point, rubber-band origin
                    Angle in radians, double precision
                         Distance from point, double precision
```

Visual Basic 16-15

AngleFromXAxis Method

The **AngleFromXAxis method** calculates the angle of a line defined by two points from the horizontal axis in radians. The format of the **AngleFromXAxis method** is:

```
ang = ThisDrawing.Utility.AngleFromXAxis(point1, point2)
                      Start point of the line
                    End point of the line
```

Exercise 1

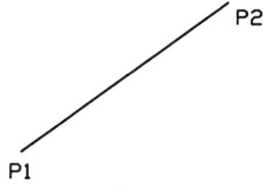

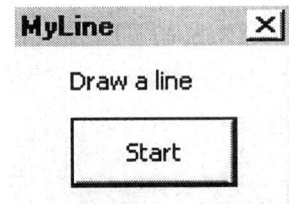

Figure 16-11 Line with user defined end points

Figure 16-12 Screen Interface Form for Exercise 1

Write a Visual Basic program that will draw a line between two points as in Figure 16-11. The user may either enter coordinates or choose points with the mouse. The graphical screen should draw a rubber-band on the screen to the current position of the mouse cursor if the second point is entered with the mouse. Use a graphical user interface similar to the one shown in Figure 16-12.

Example 3

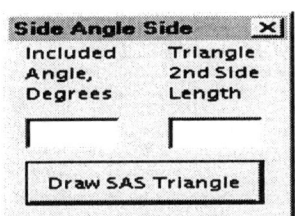

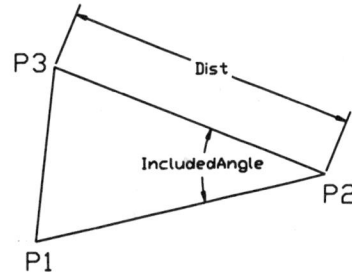

Figure 16-13 Screen Interface Form for Example 3

Figure 16-14 Side Angle Side Triangle

Write a program in the AutoCAD VBA IDE that will draw a triangle based on a given line produced from two points P1 and P2, on an included angle and on a length of the second side shown in Figure 16-14. A UserForm1 is created by selecting the **Insert** pull down menu item **UserForm**. To create TextBoxes from the toolbox, click on the TextBox button then click on the

16-16 Customizing AutoCAD

form. The labels and control button on the form are created the same way to give the results in Figure 16-13. Insert a Module1 also.

The following file is a listing of the Visual Basic program for Example 3. **The line numbers at the right are not a part of the programming; they are shown here for reference only.**

Command Button Code

```
Const PI = 3.141592654                                                       1
Public IncludedAngle As Double  'pi-converted input angle                    2
Public Angle As Double   'included angle                                     3
Public Dist As Double    'length of 2nd side                                 4
'This module draws a triangle from 2 sides and an included angle SAS         5
'Program trigger is the command button labeled "Draw SAS Triangle"           6
'Additional feature is use of text box for included angle input              7
Private Sub CommandButton1_Click()                                           8
UserForm1.Hide                                                               9
Dim p1 As Variant '1st point of base                                        10
Dim p2 As Variant '2nd point of base                                        11
Dim FirstPoint(0 To 2) As Double    '1st point base                         12
Dim SecondPoint(0 To 2) As Double   '2nd point base                         13
Dim ThirdPoint(0 To 2) As Double    'opposite vertex point                  14
p1 = ThisDrawing.Utility.GetPoint(, "Enter or select 1st base point:")      15
For i = 0 To 2: FirstPoint(i) = p1(i): Next i                               16
p2 = ThisDrawing.Utility.GetPoint(p1, "Enter  select 2nd base point:")      17
 For i = 0 To 2: SecondPoint(i) = p2(i): Next i                             18
ThisDrawing.ModelSpace.AddLine FirstPoint, SecondPoint 'draw base line      19
If textBox1.Text = "" Then                                                  20
 Angle = ThisDrawing.Utility.GetAngle(p2, "Enter  select angle from Horiz.:") 21
 Else                                                                       22
 Angle=ThisDrawing.Utility.AngleFromXAxis(FirstPoint, SecondPoint)+IncludedAngle 23
End If                                                                      24
If textBox2.Text = "" Then                                                  25
 Dist = ThisDrawing.Utility.GetDistance(p2, "Enter  select dist. from base point:") 26
End If                                                                      27
p3 = ThisDrawing.Utility.PolarPoint(SecondPoint, Angle, Dist)               28
For i = 0 To 2: ThirdPoint(i) = p3(i): Next i                               29
ThisDrawing.ModelSpace.AddLine SecondPoint, ThirdPoint                      30
ThisDrawing.ModelSpace.AddLine FirstPoint, ThirdPoint                       31
Unload Me                                                                   32
End Sub                                                                     33
                                                                            34
Private Sub textBox1_Change()                                               35
IncludedAngle = PI - Val(textBox1.Text) * PI / 180                          36
End Sub                                                                     37
                                                                            38
Private Sub textBox2_Change()                                               39
```

```
        Dist = Val(textBox2.Text)                                                    40
    End Sub                                                                          41
```

Explanation
Line 1
Const PI = 3.141592654
The constant pi is used in this routine so is declared to demonstrate this type of statement.

Line 2 through 4
Public IncludedAngle As Double
Public Angle As Double
Public Dist As Double
These variables are declared to be public which means any procedure can access them from any module in the project (file) without passing the argument in a parameter list. Also three variables are declared to be double precision as required by the AutoCAD methods in which they will be employed.

Lines 5-8
'This module draws a triangle from 2 sides and an included angle SAS
'Program trigger is the command button labeled "Draw SAS Triangle"
'Additional feature is use of text box for included angle input
Private Sub CommandButton1_Click()
These lines give the purpose and trigger of the procedure CommandButton1_Click which starts with line 8. Purpose and trigger remarks are useful for all event driven subroutines.

Line 9
UserForm1.Hide
This statement hides the screen interface form and returns focus to AutoCAD. Otherwise the user would have to close the form with the button at the top right corner of the window to proceed with entering input in AutoCAD.

Lines 10 through 14
Dim p1 As Variant '1st point of base
Dim p2 As Variant '2nd point of base
Dim FirstPoint(0 To 2) As Double '1st point of base
Dim SecondPoint(0 To 2) As Double '2nd point of base
Dim ThirdPoint(0 To 2) As Double 'opposite vertex point
Five variables are declared to be variant or double precision as required by the AutoCAD methods in which they will be employed. Note use of same line remarks.

Lines 15 and 17
p1 = ThisDrawing.Utility.GetPoint(, "Enter or select 1st base point:")
p2 = ThisDrawing.Utility.GetPoint(p1, "Enter select 2nd base point:")
These lines get input from the AutoCAD graphic screen or command line for the two points defining the base side of the triangle.

16-18 Customizing AutoCAD

Lines 16, 18, and 29
For i = 0 To 2: FirstPoint(i) = p1(i): Next i
For i = 0 To 2: SecondPoint(i) = p2(i): Next i
For i = 0 To 2: ThirdPoint(i) = p3(i): Next i
Here the variant data type is converted to double precision.

Lines 20-24 and 35-37
If textBox1.Text = "" Then
Angle = ThisDrawing.Utility.GetAngle(p2, "Enter or select angle from Horiz.:")
Else Angle=ThisDrawing.Utility.AngleFromXAxis(FirstPoint,SecondPoint)+IncludedAngle
End If
Private Sub textBox1_Change()
IncludedAngle = PI - Val(textBox1.Text) * PI / 180 'in radians
End Sub
The If ... Then ... Else statement checks TextBox1 for an entry. If no angle has been entered on the UserForm the GetAngle method is used. If a numeric angle was entered on the form it is converted by the subroutine textBox1_Change() to a radian included angle. Adding the radian angle returned by the AngleFromXAxis method gives the AutoCAD polar angle for the vertex (third) point.

Lines 25-27
If textBox2.Text = "" Then
Dist = ThisDrawing.Utility.GetDistance(p2, "Enter or select dist. from base point:")
End If
If no numeric entry for the second side length was entered on the user form these lines use the GetDistance method.

Lines 39-41
Private Sub textBox2_Change()
Dist = Val(textBox2.Text)
End Sub
If a numeric entry for the second side length was entered on the user form these lines convert from text to numeric form.

Lines 19, 30 and 31
ThisDrawing.ModelSpace.AddLine FirstPoint, SecondPoint
ThisDrawing.ModelSpace.AddLine SecondPoint, ThirdPoint
ThisDrawing.ModelSpace.AddLine FirstPoint, ThirdPoint
These methods draw the first, second, and third sides of the triangle.

Module1 Code

The following code is entered in the Module1 code window.

Option Explicit	1
Sub SAS() 'Module1	2
UserForm1.Show	3
End Sub	4

Exercise 2

Write a program in the AutoCAD VBA IDE that will draw a triangle based on a given line produced from two points P1 and P2, on an adjacent angle at one end and another angle at the other end as shown in Figure 16-16. Employ a UserForm similar to the one shown in Figure 16-15.

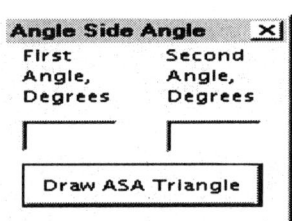

Figure 16-15 Screen Interface Form for Exercise 2

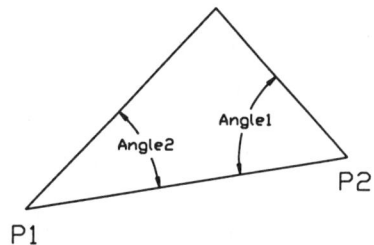

Figure 16-16 Angle Side Angle Triangle with user defined points defining the base

Exercise 3

Write a program in the AutoCAD VBA IDE that will draw a triangle based on a given line produced from two points P1 and P2, and two distances which are the lengths of the sides adjoining each end as in Figure 16-18. Employ a UserForm similar to the one shown in Figure 16-17.

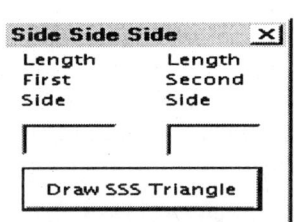

Figure 16-17 Screen Interface Form for Exercise 3

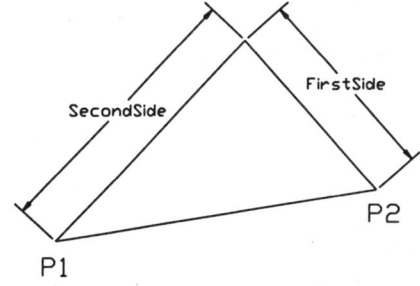

Figure 16-18 Side Side Side Triangle with user defined points defining the base

ADDITIONAL VBA EXAMPLES

More VBA examples are included with the AutoCAD Preview, VBA Edition. These are considered sample files and are located in the Vbasamp subdirectory of the AutoCAD R14 directory or from the AutoCAD R14 CD ROM directory e:\Vbainst\Vbasamp. A text file called Vbasamp.txt contains a description of a half dozen files using VBA. If you have Microsoft Excel and Word there is an additional cross-application example in the directory.

There are an additional four cross-application samples in the Sample\Activex subdirectory of the AutoCAD R14 directory or from the AutoCAD R14 CD ROM directory e:\Acad\Sample\Activex. Substitute the drive letter for your CD ROM for e:. One of these applications uses Excel and the other three are intended to be run from Visual Basic 4 or from a compiled exe version.

REVIEW QUESTIONS

1. _____ lets you access and manipulate the objects and functionality of AutoCAD from Excel or another application supporting ActiveX.

2. Before you can use AutoCAD's objects in some other software supporting ActiveX you must make the application aware that the AutoCAD _____ Library is available on the computer.

3. There are over _____ programmers using Visual Basic.

4. Visual Basic is considered an object-_____, rather than an object-oriented programming language.

5. Functions known as _____ have been defined in the AutoCAD object library to perform an action on an object, for example draw a line in a drawing.

6. _____ are functions that set or return information about the state of an object.

7. The VBA menu item Show VBA IDE is one way to access the _____ .

8. A _____ is the name given to the forms, controls, modules, and the programming making up a Visual Basic program.

9. _____ forces explicit declaration of all variable types which minimizes common inconsistent variable usage errors.

10. _____ is the variable type returned by the GetPoint method.

11. The _____ method always measures the angle with a positive X-axis (3 o'clock position) and in a counterclockwise direction.

12. The _____ method allows you to enter the X, Y coordinates or X, Y, Z coordinates of a point.

13. The _____ method lets you retrieve the value of an AutoCAD system variable.

14. The _____ method defines a point at a given angle and distance from the given point.

15. The _____ method lets you enter a distance on the command line, a distance from a given point, or two points.

16. The _____ method allows you to enter an angle; either from the keyboard in degrees or by selecting two points.

17. The _____ method calculates the angle of a line defined by two points from the horizontal axis in radians.

EXERCISES

Exercise 4

Write an Visual Basic program that will draw a square of sides S and a circle tangent to the four sides of the square as shown in Figure 16-19. The base of the square makes an angle, ANG, with the positive X-axis. The program should allow you to enter the starting point P1, length S, and angle ANG on a user interface form or in the AutoCAD graphical interface.

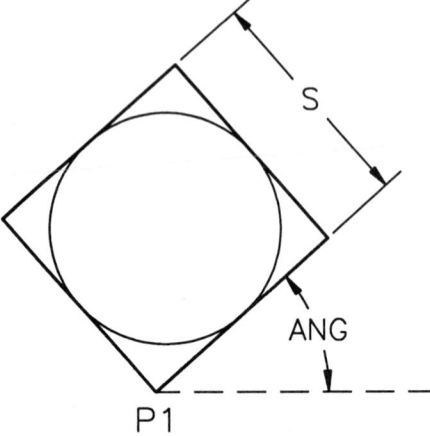

Figure 16-19 Square of side S at an angle ANG

Exercise 5

Write a Visual Basic program that will draw an equilateral triangle inside the circle (Figure 16-20).

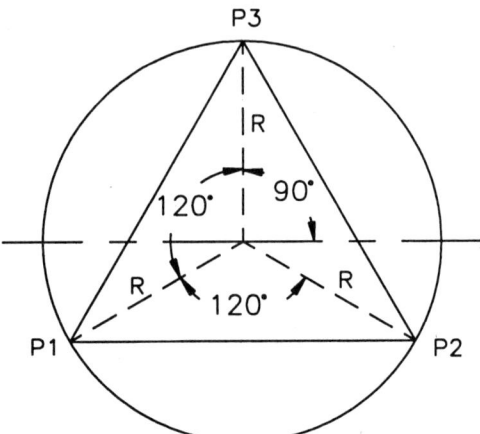

Figure 16-20 Equilateral triangle inside a circle

Exercise 6

Write a program that will draw a slot (Figure 16-21) with center lines. The program should allow you to enter slot length, slot width, and the layer name for center lines on a user interface form or in the AutoCAD graphical interface.

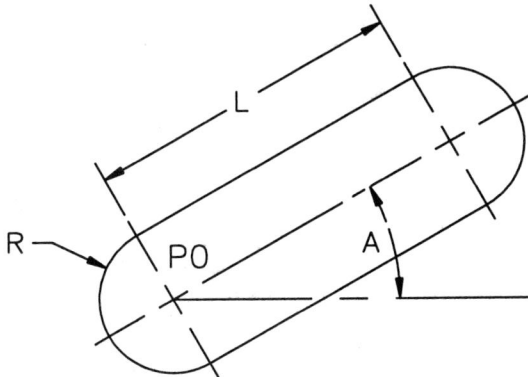

Figure 16-21 Slot of length L and radius R

Exercise 7

Write an Visual Basic program that will draw the two views of a bushing as shown in Figure 16-22. The program should allow you to enter the starting point P0, lengths L1, L2, and the bushing diameters ID, OD, HD on a user interface form or in the AutoCAD graphical interface. The distance between the front view and the side view of bushing is DIS (DIS = 1.25 * HD). The program should also draw the hidden lines in the HID layer and center lines in the CEN layer. The center lines should extend 0.75 units beyond the object line.

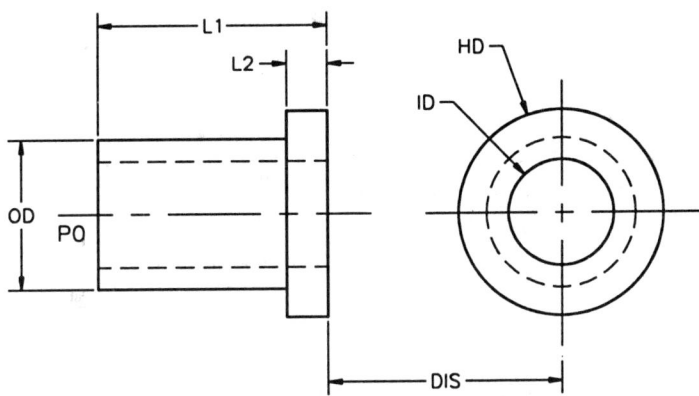

Figure 16-22 Two views of bushing

Chapter 17

Accessing External Databases, AutoCAD SQL Extension (ASE)

Learning objectives

After completing this chapter, you will be able to:
- *Understand ASE and SQL and their uses.*
- *Understand databases, DBMS, and relational databases.*
- *Set up the database environment.*
- *Define key columns and set link paths and isolation levels.*
- *Access and edit a database.*
- *Link a database with drawing entities.*
- *Display attributes in a drawing.*
- *Create selection sets.*
- *Use SQL statements to access an external database.*
- *Generate reports from exported data.*

AUTOCAD SQL2 ENVIRONMENT (ASE)

SQL is an acronym for **structured query language**. It is often referred to as **sequel**. SQL is a format in computer programming that lets the user ask questions about a database according to specific rules. **The AutoCAD SQL environment** (ASE) lets you access and manipulate the data that is stored in the external database and link data from the database to objects in a drawing. For example, the data in the table of Figure 17-1 can be accessed from within AutoCAD. Once you access the table, you can manipulate the data. The connection is made through a database management system (DBMS) like dBASE, PARADOX, FoxPro, or some other DBMS program. The DBMS programs have their own methodology for working with databases. However, the ASE commands work the same way regardless of the database being used. This is made possible by the ASE drivers that come with AutoCAD software.

In AutoCAD Release 14, **SQL2** has been incorporated. The enhanced features of SQL2 make it superior to SQL. The fundamental change is that the SQL model is based on **DBMS, databases,** and **tables**, while the SQL2 model is based on **environments, catalogs,** and **schemas**.

Environment
The **environment** is composed of the DBMS, the accessible databases, and the users and programs that are granted access to the databases. An environment can have zero or more catalogs.

Catalog
A **catalog** is simply a database. It defines the collection of base tables. A catalog can have one or more schemas. The name of the catalog relates to the directory path name that is the location of the database.

Schema
A **schema** is a collection of database components for a user. A schema can have one or more tables. The name of the schema relates to the catalog subdirectory that contains the database tables.

Session
Session is a new concept introduced in SQL2. A session is required for the management of temporary data. Session management statements are provided by SQL2 in order to establish the environment for an SQL session, catalog name, schema session, authorization identifier, and the local time zone. When the client is connected to the server by an SQL application, the session begins. The SQL environment is established by the connect statement.

Transaction
When a sequence of SQL statements are executed, a **transaction** takes place. A transaction is terminated with the SQL COMMIT or SQL ROLLBACK operation.

UNDERSTANDING DATABASES

Database
A **database** is a collection of data that is arranged in a logical order. For example, there are six computers in an office, and we want to keep a record of these computers on a sheet of paper. One of the ways of recording the computer information is to make a table with rows and columns as shown in Figure 17-1. Each column will have a heading that specifies a certain feature of the computer, such as COMP_CFG, CPU, HDRIVE, or RAM. Once the columns are labeled, the computer data can be placed in the columns for each computer. By doing this, we have created a database on a sheet of paper that contains information on our computers. The same information can be stored in a computer; this is generally known as a computerized database.

COMPUTER

COMP_CFG	CPU	HDRIVE	RAM	GRAPHICS	INPT_DEV
1	486/33	300MB	8MB	SUPER VGA	DIGITIZER
2	286/12	60MB	640K	VGA	MOUSE
3	MACIIC	40MB	2MB	STANDARD	MOUSE
4	386SX/16	80MB	4MB	VGA	MOUSE
5	386/33	300MB	6MB	VGA	MOUSE
6	SPARC2	600MB	16MB	STANDARD	MOUSE

Figure 17-1 A table containing computer information

Database Management System

The **database management system** (DBMS) is a program or a collection of programs (software) used to manage the data in the database. For example, PARADOX, dBASE, INFORMIX, and ORACLE are database management systems.

Relational Database

INVENTRY

INV_ID	TYPE	DESCRIPT	MFR	MODEL	COMP_CFG	PRICE	ROOM	EMP_ID
1	Furniture	6x3 Couch	Couches	Lounge	1	800	101	1000
2	Furniture	Adj. Chair	Chairs	Comp	2	200	101	1000
3								
4								
5								
6								

EMPLOYEE

COMPUTER

COMP_CFG	CPU	HDRIVE	RAM	GRAPHICS	INPUT
1	486/33	300MB	8MB	SUPER VGA	DIGITIZER
2	286/12	60MB	640K	VGA	MOUSE
3	MACIIC	40MB	2MB	STANDARD	MOUSE
4	386SX/16	80MB	4MB	VGA	MOUSE
5	386/33	300MB	6MB	VGA	MOUSE
6	SPARC2	600MB	16MB	STANDARD	MOUSE

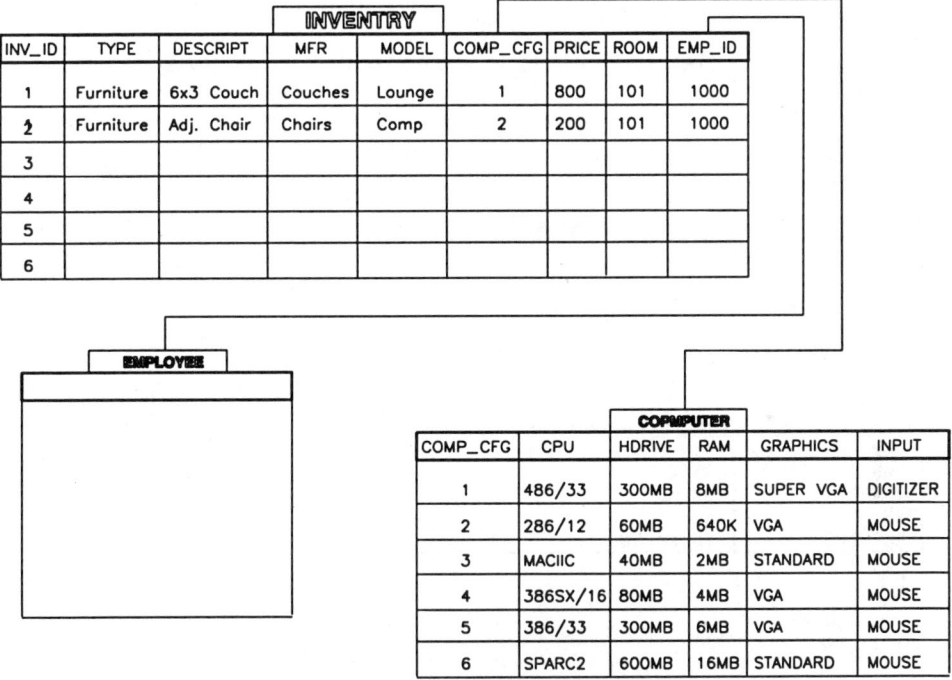

Figure 17-2 Relational database model

A **database** may consist of several tables, each table containing a set of data, and there may exist a relation between the tables. For example, in Figure 17-2, the INVENTORY table contains a column for COMP_CFG and its values (1, 2, . . .). The data for these computers is defined in another table (COMPUTER). A relation exists between the INVENTORY and COMPUTER tables. This model of a database where a relation exists between tables is called a **relational database**.

Components of a Table

A database **table** is a two-dimensional data structure that consists of rows and columns.

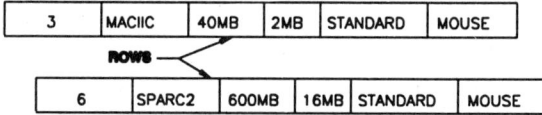

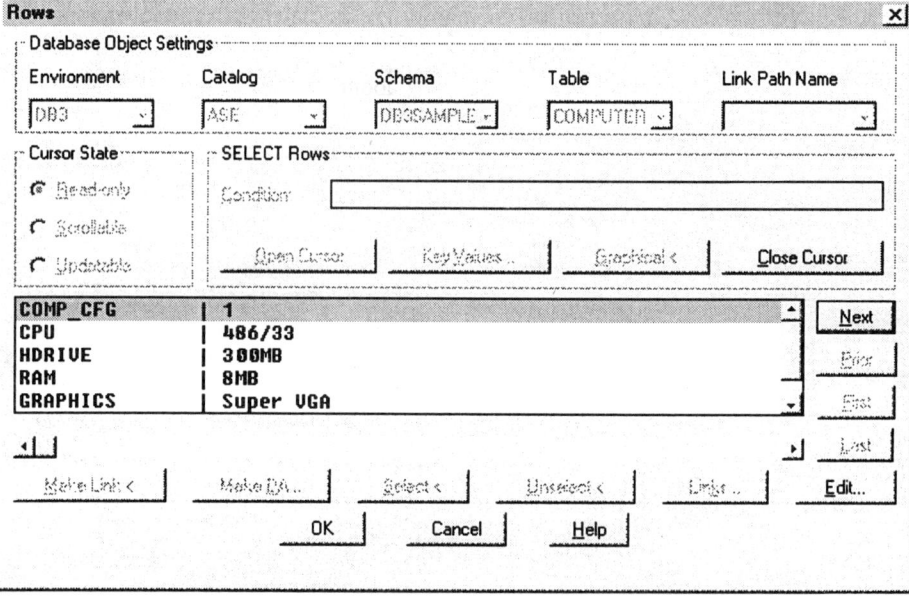

Figure 17-3 Rows in a table (the horizontal group of data)

Figure 17-4 Rows dialog box

Row. The horizontal group of data is called a **row**. For example, Figure 17-3 shows a table, and just below the full table are two rows of the table. Each value in a row defines an attribute of the item. For example, in Figure 17-4, the attributes assigned to COMP_CFG (1) include 486/33, 300MB, 8MB, etc. These attributes are arranged in the first row of the table (see Figure 17-3). A row is also referred to as a **record**.

Column. A vertical group of data (attribute) is called a **column**. Three of the columns in Figure 17-3 are shown in Figure 17-5. HDRIVE is the column heading that represents a feature of the computer, and the HDRIVE attributes of each computer are placed vertically in this column. A column is sometimes called a **field**.

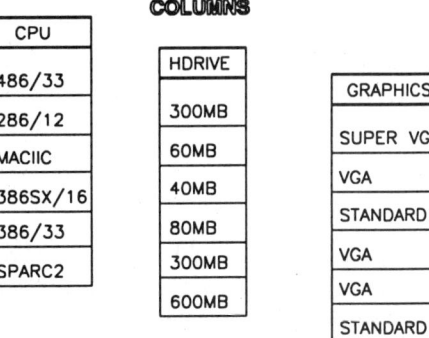

Figure 17-5 Columns in a table (the vertical group of data)

DEFINING KEYS

A **key** is used to link a specified row in the table with an entity in the drawing. The key acts like an identification tag for locating and linking a row. For example, if you want to link the second row of the table in Figure 17-3 to a drawing entity, you must first identify the row. You can do it by specifying the column name (COMP_CFG) and the value in that column (2). The name of the column (COMP_CFG) and the value in the column (2) become the **key** for the second row. Sometimes it may not be enough to identify a row by specifying one key. For example, if two computers have the same value in the COMP_CFG column (2), one key will not be enough to identify the second row. In this case, you can identify the second row by defining a second key. The second key could consist of another column label (CPU) and a value under that column (286/12). Selecting two or more columns and their values to define a key creates a **compound key**. Keys can be defined in the **Link Path Names** dialog box. (This process is discussed in detail later in this chapter. At present, we are concerned only about the idea behind defining keys.)

ISOLATION LEVELS

In a database, you can relate or isolate the data in one transaction with the data from some other transaction. As the name suggests, the concept behind setting isolation levels is to establish isolation. Isolation of data returned by a query transaction when many users try to address the same data simultaneously depends on the isolation levels. Isolation level influences the data returned by sequential transactions on the basis of transaction type. There are three types of transactions: **dirty read transaction, nonrepeatable read transaction**, and **phantom read transaction**.

Dirty Read Transaction

In the case of a **dirty read transaction**, transaction A makes changes to a row, but does not save the changes to the database. In other words, the SQL command COMMIT is not executed. Now another transaction (transaction B) performs a read on the row. Transaction A then executes a ROLLBACK command, which results in the undoing of the changes made previously to the row. This results in transaction B trying to read a row that did not exist.

Nonrepeatable Read Transaction

In the case of a **nonrepeatable read transaction**, transaction A performs a read on a row. Transaction B makes changes to the row and saves the changes to the database with the COMMIT command. Transaction A tries to read this row again. This results in transaction A trying to read a row that has been modified or deleted.

Phantom Read Transaction

In the case of a **phantom read transaction**, transaction A performs a read operation on a set of rows. Transaction B executes an SQL statement that results in the insertion of a new row meeting the search condition specified in transaction A. Next, transaction A again tries to read the rows selected with the search condition specified in transaction A. This results in the selection of different rows by transaction A in the first and second reads.

If you specify a serializable isolation level, the transactions are performed in a serial manner. Therefore, the problems faced with dirty read transactions, nonrepeatable read transactions, and phantom read transactions can be avoided.

ESTABLISHING THE DATABASE ENVIRONMENT

You can set the database environment according to the information contained in the **asi.ini** file. This file is present in the same directory in which the **acad.exe** file exists. For more information on this file, refer to Chapter 7 of the Installation Guide book. This file can be customized to your requirements by editing it using an editor. If you want to use the dialog boxes with ASE commands, you must set the **CMDDIA** variable to 1 (on). If it is off, the values will be displayed on the text screen. It is recommended that you use the dialog boxes because they provide an efficient way for handling ASE commands and data. Use the following command to check or assign a new value to the **CMDDIA** system variable:

Command: **CMDDIA**
New value for CMDDIA <1>: ↵   *(0 = off, 1 = on)*

Toolbar:	External Database, Administration
Pull-down:	Tools, External Database, Administration
Command:	ASEADMIN

 The following instructions need to be carried out to set up an environment:

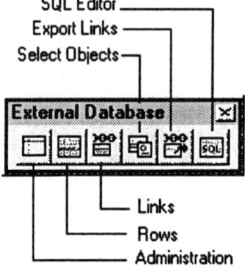

Figure 17-6 Options in the External Database toolbar

1. Invoke the **ASEADMIN** command. When this command is invoked, the **Administration** dialog box (Figure 17-7) is displayed on the screen.

Accessing External Database, AutoCAD SQL Extension (ASE) 17-7

2. All the environments supported by the system are listed in the Database Objects: list. Select the environment from this list. For example, select the DBMS **DB3** by clicking on it (Figure 17-7).

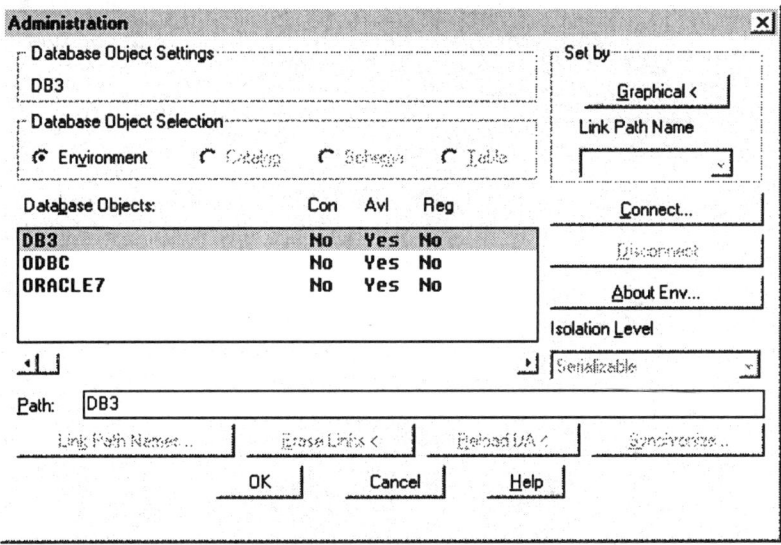

Figure 17-7 DB3 selected in the Administration dialog box

3. Now, you need to set up a connection. To do so, select the Connect button. The **Connect to Environment** dialog box (Figure 17-8) is displayed. Specify your name in the User Name: edit box and your password in the Password: edit box (if required by the selected DBMS). Select the OK button.

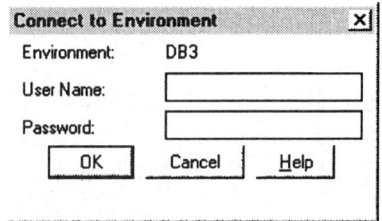

Figure 17-8 Connect to Environment dialog box

4. Next, select the Catalog radio button in the Database Object Selection area of the Administration dialog box (Figure 17-7). All the catalogs are listed in the Database Objects: list. Select the catalog name **ASE** by double-clicking on it. In the same manner, select the schema **DB3SAMPLE** and the table **EMPLOYEE**.

You have connected AutoCAD to the database (DB3) table using the default schema. You can now view the data in the external database from within AutoCAD.

Note

If the environment is connected, it means that the DBMS driver is loaded in memory.

If the environment, catalog, or schema has an associated table with defined key columns, it is registered.

17-8 Customizing AutoCAD

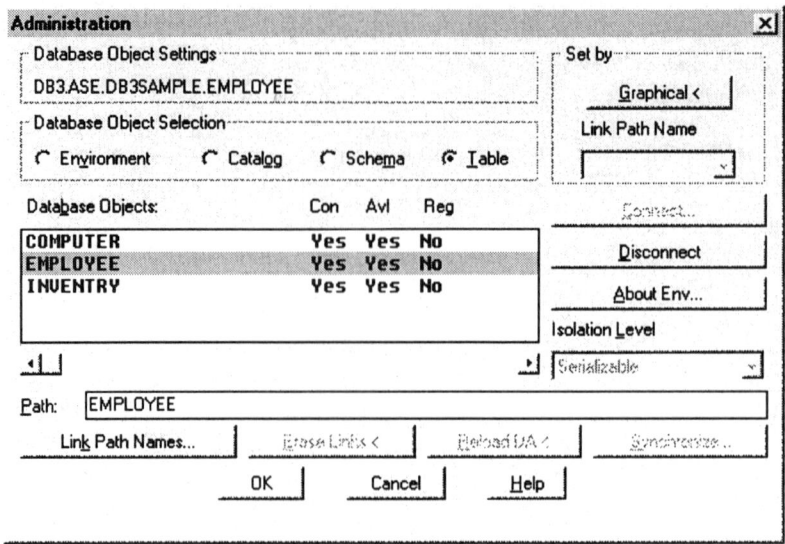

Figure 17-9 Administration dialog box

The database object has now been defined. At this stage, the settings in the Administration dialog box are as shown in Figure 17-9. Now, define a link path name.

1. Select the Link Path Names... button. The **Link Path Names** dialog box (Figure 17-10) is displayed.

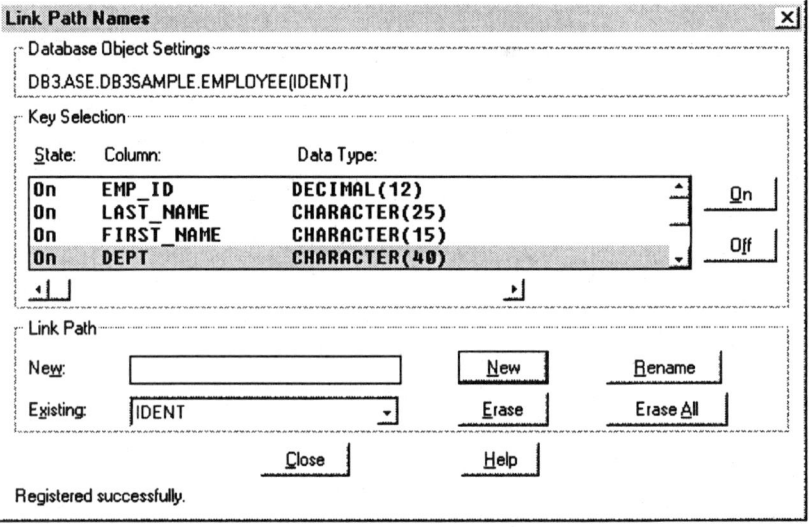

Figure 17-10 Creating the IDENT link in the Link Path Names dialog box

2. Select the column that defines the link path name for the table from the Key Selection area of the Link Path Name box. In our case, we have selected the **EMP_ID** column (Figure 17-10). Select the On button.

Accessing External Database, AutoCAD SQL Extension (ASE) 17-9

3. Perform the second step for every column you desire to have in the link path name.

4. Specify the link path name for the selected column(s) in the **New:** edit box, and then select the **New** button. The link path name specified is **IDENT**. You will notice that the link path name (IDENT) is displayed in the **Existing:** box. Select the **Close** button. The Link Path Names dialog box is removed from the screen. Select the OK button in the Administration dialog box.

The links in the present drawing session can be unloaded with the **ASEUNLOAD** command.

ACCESSING DATA IN EXTERNAL DATABASES

After setting up the environment, it is possible to access data in an external database as defined by the current database object (i.e., environment, catalog, schema, table). You can access the database with SQL statements or with the interface provided by AutoCAD. The prevailing database object selection can be altered with the use of any external database command. With the **ASESQLED** command, you can change the environment, catalog, and schema through the user interface; however, you can perform SQL operations on all the tables.

Selection of Rows

Toolbar:	External Database, Rows
Pull-down:	Tools, External Database, Rows
Command:	ASEROWS

Selection of a row can be achieved with the **ASEROWS** command. When this command is invoked, the **Rows** dialog box (Figure 17-4) is displayed.

Note
Some operations modify only the database and do not influence the drawing; for example, inserting a row, viewing a row, changing the values in the data fields, and creating selection sets.

Some operations influence the drawing directly by modifying the graphic, information or by adding links to the database of the drawing.

First set the cursor state. This can be done by selecting the desired radio button in the Cursor State area of the Rows dialog box. It is the cursor state that governs the accessibility of rows belonging to the selection set. There are three options.

Read-only. In this case, you cannot update. In other words, the current row cannot be deleted or modified.

Scrollable. With this option, any row can be accessed in the selection set of rows. However, you cannot update.

Updatable. If this option is selected, you can modify the next rows of the selection set. Scrolling through the selected rows is not possible.

Only one row, the **current row**, can be manipulated at a time. You can navigate between the rows of a table with the use of the Next, Prior, First, and Last buttons. If you have specified key values and you want to make a row current, use a key to search for a particular row or rows. In this case, the selection of a row is carried out on the basis of its key value. This can be accomplished by selecting the Key Values... button in the Rows dialog box. Make sure the Cursor State is set to Updatable. The **Select Row by Key Values** dialog box (Figure 17-11) is displayed. Select the key column. Then, in the Values: edit box enter the value for the key column, and press Enter. Observe that the key value entered in the Values: edit box is displayed in the list box.

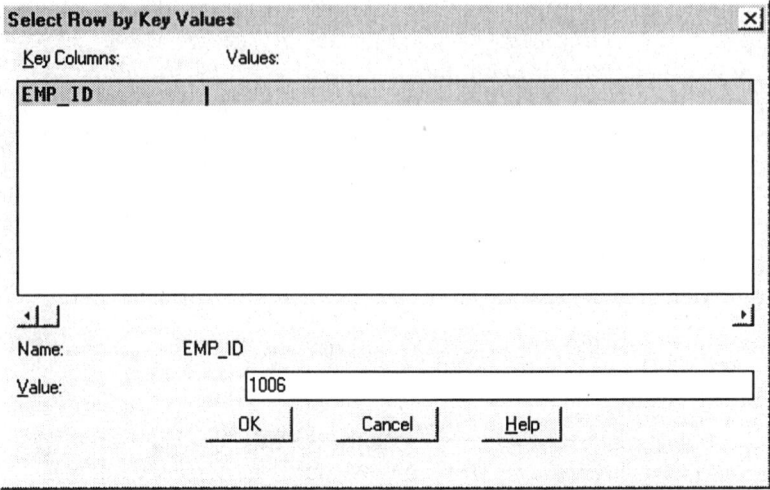

Figure 17-11 Select Row by Key Values dialog box

Next select the OK button. The Select Row by Key Values dialog box is removed from the screen. In the list box of the Rows dialog box, the selected row is displayed. Also a row can be selected graphically by selecting the object to which the row is linked.

Editing Data in a Table

Periodically there is a need to change (update) the data in a database. For example, in the EMP1 table, there might be an employee who has been transferred to some other department, or an employee whose room number has changed. These changes need to be incorporated in the database so that the database contains accurate and up-to-date information. Editing information in the table can be carried out by selecting (making current) the row whose data needs to be updated and then making the changes. In AutoCAD, the Updating data operation can be performed with the use of the **ASEROWS** command. The process involved is as follows:

1. Invoke the ASEROWS command, The **Rows** dialog box is displayed. Depending on the specification of database objects in the **Administration** dialog box, the current database objects are listed in the Database Object Settings area of the Rows dialog box.

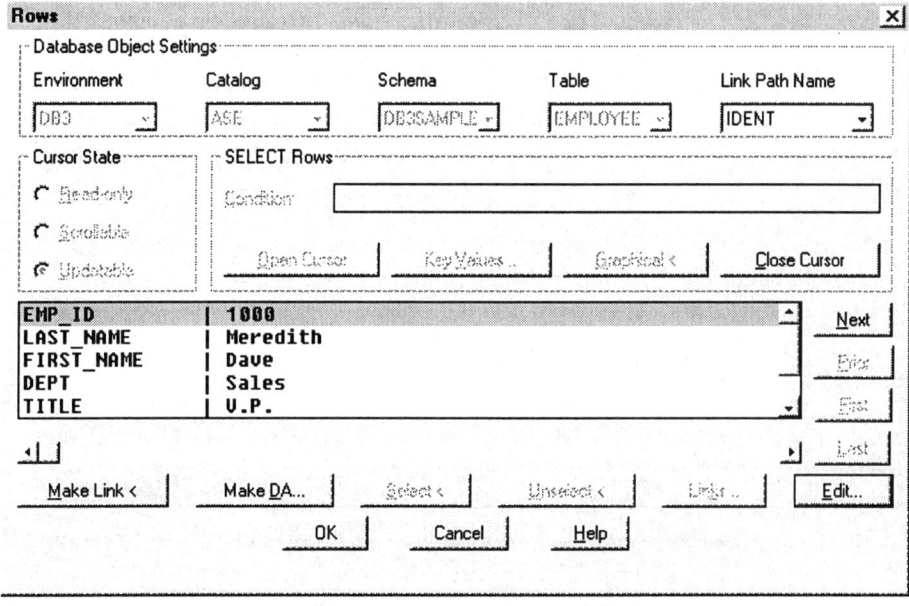

Figure 17-12 Rows dialog box after opening the cursor

2. Select the database objects of your choice from the various drop-down lists.

3. Next, select the Updatable radio button in the Cursor State area in order to make all the rows accessible and editable.

4. Select the Open Cursor button so that all the rows are selected. Now navigate to the row that is to be modified. You can use the Next navigation option placed on the right side of the list box to reach the desired row. The settings are shown in Figure 17-12.

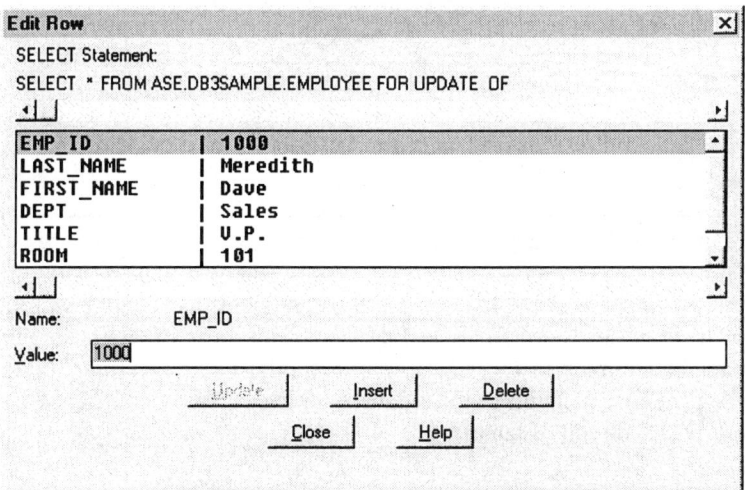

Figure 17-13 Edit Row dialog box

17-12   Customizing AutoCAD

5. Next, select the Edit... button. The **Edit Row** dialog box (Figure 17-13) appears on screen.

6. Select the column that is to be updated from the list box. Notice that the column name is displayed in the Name: field just below the list box, and its value is displayed in the Value: edit box.

7. Enter the new value in the Value: edit box and press Enter. This updates the value of the current column. The value of the next column is now listed in the Value: edit box, and you can change its value. In the same manner, all the columns of the selected row can be edited. After making the necessary changes, select the Update button and then select the Close button.

Rows can also be deleted from the table. To delete a row, open the Rows dialog box (Figure 17-12), set the Cursor State to Updatable, and make the row to be deleted current. Next, select the Edit... button. The Edit Row dialog box (Figure 17-14) is displayed. Select the Delete button. The Confirm dialog box is displayed. Select the OK button to delete the row from the table. After this, select the Close button in the Edit Row dialog box. The row is deleted from the table.

Just as rows can be deleted, rows can also be inserted (added to a table). Open the Rows dialog box, set the Cursor State to Updatable and then select the Edit... button. The Edit Row dialog box is displayed (Figure 17-14).

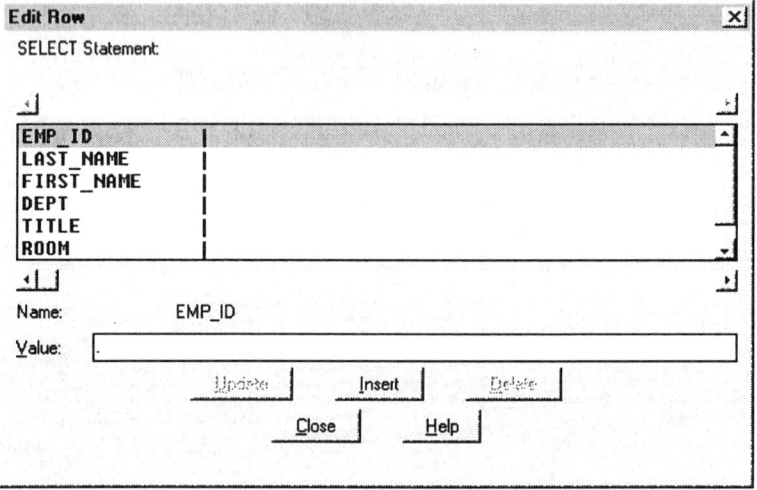

Figure 17-14 Using the Edit Row dialog box to insert a new row

Since a cursor was not opened, in the list box there are no values next to the column names. Now, select the column from the list box. Enter its new value and press Enter. In this manner, you can enter values for all the columns of the new row. After this, select the Insert button and then the Close button so that the modification to the database is saved. The newly inserted row is added to the table at the last position (end of table).

LINKING A DATABASE WITH A DRAWING

You can use ASE commands to link the information in a database with the drawing entities. Once you define the link, AutoCAD stores that information with the drawing. You can also edit, delete, or view the link and the information associated with the link in the database. To understand the process involved in linking a drawing entity with the database, consider the following example.

Example 1

Given an office with six computers (Figure 17-15) and a database that contains information about these computers, you are required to link the database information with the six computers in the office.

When you install AutoCAD, it automatically creates a **.DBF** subdirectory and copies three database files: **COMPUTER.DBF**, **EMPLOYEE.DBF**, and **INVENTRY.DBF**. You can use the **COMPUTER** database file for this example, or you can create your own file if you have access to any database program. We also need a drawing that has six computers; therefore, make a drawing similar to the drawing in Figure 17-15. The following steps will take you through the process involved in linking the computers with the database:

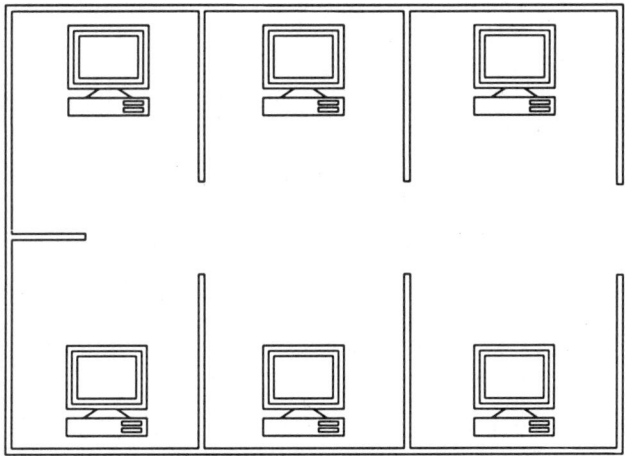

Figure 17-15 Floor plan of an office with six computers

1. Establish the database environment as described earlier (in the section "Establishing the Database Environment"). Select COMPUTER instead of EMP1 (Figure 17-9). In the Link Path Names dialog box, select CPU from the Key Selection area and set it On (Figure 17-10); enter CONFIG as link path name. Select the Close button in the Link Path Names dialog box, and then select OK in the Administration dialog box.

2. Invoke the **Rows** dialog box using any of the methods discussed previously.

3. Select a row from the current table using any row selection option listed in the Rows dialog box (select Open Cursor button). The information about the first row in displayed in the dialog box. Now you are ready to link a drawing entity with the database.

17-14 Customizing AutoCAD

4. Select the Make Link < button. The Rows dialog box is removed from the screen and AutoCAD prompts you to select the object you want to link to the specified row. In this manner, you can create a link by assigning the current row of the current table to a graphics entity in the drawing.

5. Use the Next button in the Rows dialog box to make the second row current, and then use the Make Link button to create a link between the second row and the second computer. Use the same procedure to link the subsequent rows with the remaining computers.

Editing Links (ASELINKS)

Toolbar:	External Database, Links
Pull-down:	Tools, External Database, Links
Command:	ASELINKS

You can use the **ASELINKS** command to edit and delete links in the drawing and also to create blocks in the drawing. The process involved in editing a link is as follows:

1. Invoke the ASELINKS command.

2. Next, AutoCAD prompts you to select the object whose link you want to edit or delete. After you select the object, the **Links** dialog box (Figure 17-16) will be displayed. The dialog box displays information about the link.

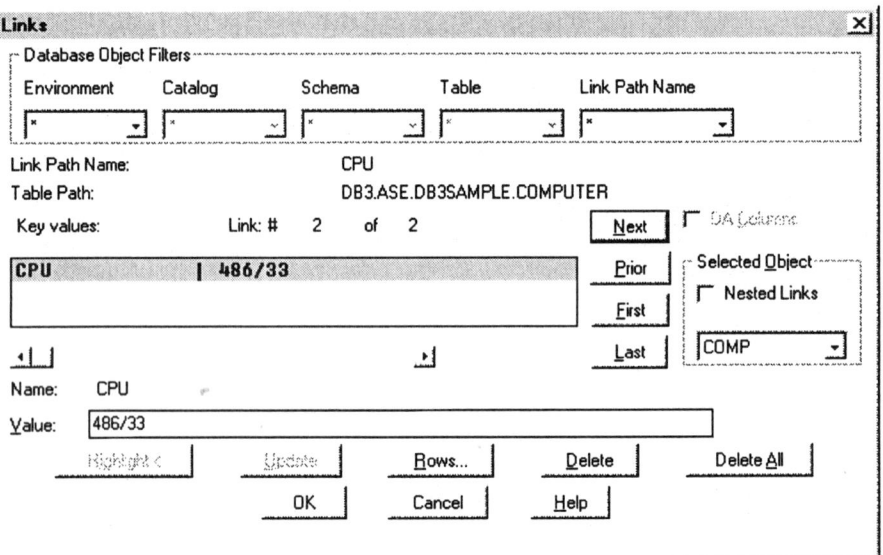

Figure 17-16 Using the Links dialog box to delete a link

The key value of the row linked to the object is displayed in the Key values: list. In case there is more than one link to the object, you can select the link you want to edit by using the navigation buttons.

3. Select the database object that defines the extent of the change. Links can be edited for all environments or for some particular environment, catalog, schema, table, or link path name. By default, all the lists contain an asterisk (*); in this case, all the links to the selected object are accessible for editing and viewing.

4. In the Value: edit box, change the value of the key column to the new value and press Enter. For example, you could enter the value 486/33 (the value must be defined in **COMPUTER.DBF**) and press Enter.

5. You can view the row associated to the new key column value by selecting the Rows... button in the Links dialog box. The Rows dialog box is displayed. Next, select the OK button; the Rows dialog box is removed from the screen.

6. Select the Update button in the Links dialog box to update the link.

Deleting Links

Just as you can edit a link, it is also possible to delete a link in the drawing and also to delete blocks in the drawing. The procedure for deletion of a link is as follows:

1. Invoke the ASELINKS command using any one of the methods discussed previously.

2. Select the object whose link is to be deleted. The selected object may be linked to more than one row; in such a case, you can use the navigation buttons on the right side of the list box to navigate to the link information you want to delete.

3. Next, select the Delete button. With this, the link information displayed in the Key values: list box is deleted. In case more than one link exists for the selected object, select Delete All to delete all link information.

Instead of using the ASELINKS command, which prompts you to select the object first and then select the link you want to delete, you could delete a link by selecting the row to which an object is linked. This can be accomplished in the following manner:

1. Invoke the ASEROWS command.

2. The Rows dialog box is displayed. Choose the row whose link is to be deleted.

3. Next, select the Links button. The Links dialog box is displayed.

4. Now, edit or delete the link from the object associated with the current row as required.

CREATING DISPLAYABLE ATTRIBUTES

A displayable attribute is nothing but an AutoCAD block having text fields linked to a row in the database. The text fields contain the column value of the current row. The text appears at the location specified as the starting point on the drawing. The text style, text justification, text height, and rotation angle can also be controlled for the displayable attributes. The attributes that are displayed do not have any link with the drawing entity. They are independent text entities that can be edited without affecting the database information or the link between the drawing entity and the table. The instructions involved in creating displaying attributes are as follows:

1. Invoke the Rows dialog box (Figure 17-9). The Rows dialog box is displayed.

2. Select the Graphical button in the Rows dialog box and select the AutoCAD object whose attributes you want to display.

3. Select the MakeDA button in the Row dialog box; the **Make Displayable Attribute** dialog box (Figure 17-17) is displayed. This dialog box has two list boxes, **Table Columns:** and **DA Columns:**. The Table Columns: list box displays the titles of all columns in the table.

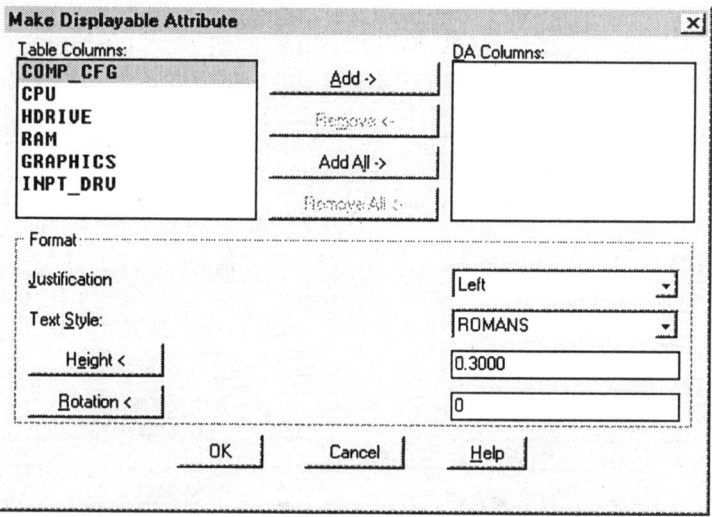

Figure 17-17 Make Displayable Attribute dialog box

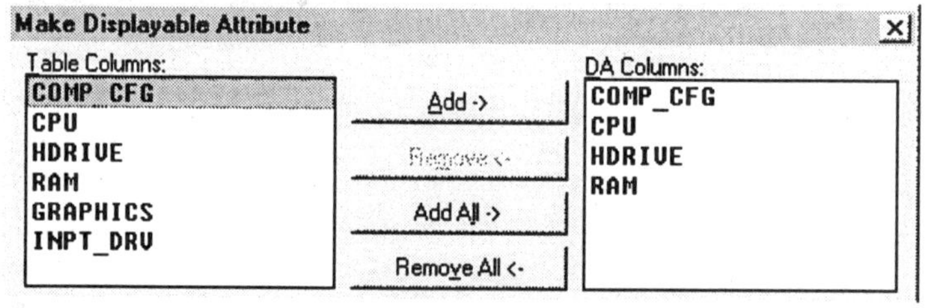

Figure 17-18 Make Displayable Attribute dialog box after selecting the table column

4. Place the attributes that you want to display on the graphics screen in the DA Columns: list. You can do this by highlighting the title item in the Table Columns: list box, and then selecting the Add -< button, or you can double-click on the item. If you want to display all the attributes on the graphics screen, select the Add All -< button. You can also remove the items in the DA Columns: list box by highlighting the item, and then selecting the Remove -< button. Select the Remove All button to remove the items in the DA Columns: list box.

5. Set the text style, text justification, text height, and rotation angle for the displayable attributes.

6. Select the OK button.

7. Select a point on the screen where you want the displayable attribute to be displayed, and then select OK in the Rows dialog box. With this, the displayable attributes are displayed in the drawing at the specified position.

For example, in the Rows dialog box of Figure 17-4 we have selected the DB3 environment, catalog ASE, schema SAMPLE, and table COMPUTER as the database objects. After this, the first row of the table has been selected. In the Make Displayable Attribute dialog box (Figure 17-17), we have selected COMP_CFG, CPU, HDRIVE, and RAM as displayable attribute columns. Next, we have placed the displayable attributes (specified column values of the first row) for the first row along with the figure of the first computer on the drawing. Similarly, displayable attributes for all the computers have been placed along with the remaining computers. The figure thus obtained is Figure 17-19.

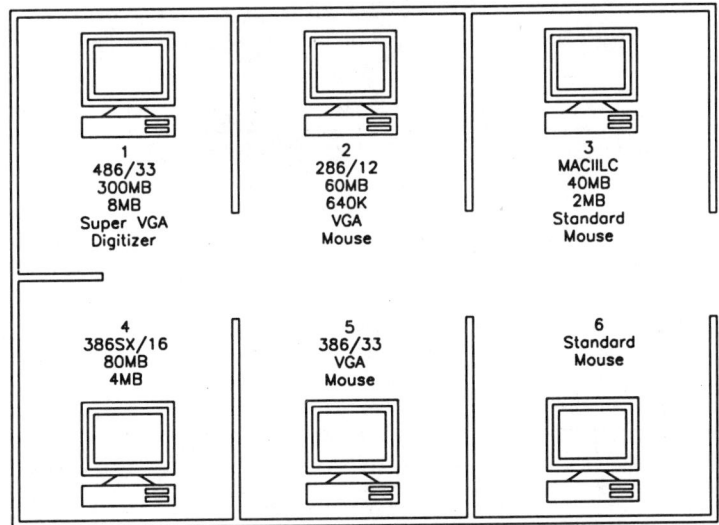

Figure 17-19 Displaying row attributes on the graphics screen

EDITING ROWS

As mentioned earlier, you can access and modify the database from within the AutoCAD drawing editor regardless of the database you are using. You can modify the database through the dialog boxes or from the command line. The following example illustrates the editing of rows.

Example 2

Given an office with six computers and a database that contains information about these computers, you are required to modify the data in the row that is linked to the second computer in the office. (Use the drawing and data from Example 1.) The following is the new data for this computer:

COMP_CFG	2
CPU	P5-60
HDRIVE	500MB
RAM	16MB
GRAPHICS	ATI
INPT_DEV	MOUSE

The first step is to find the key that links the second computer in the office with the database. This can be accomplished by invoking the **ASELINKS** command, which was discussed earlier in the chapter. When you enter this command, AutoCAD will prompt you to select the object. Select the second computer. After selecting the object, the **Links** dialog box is displayed (Figure 17-20). The key is COMP_CFG, and the attribute value is 2.

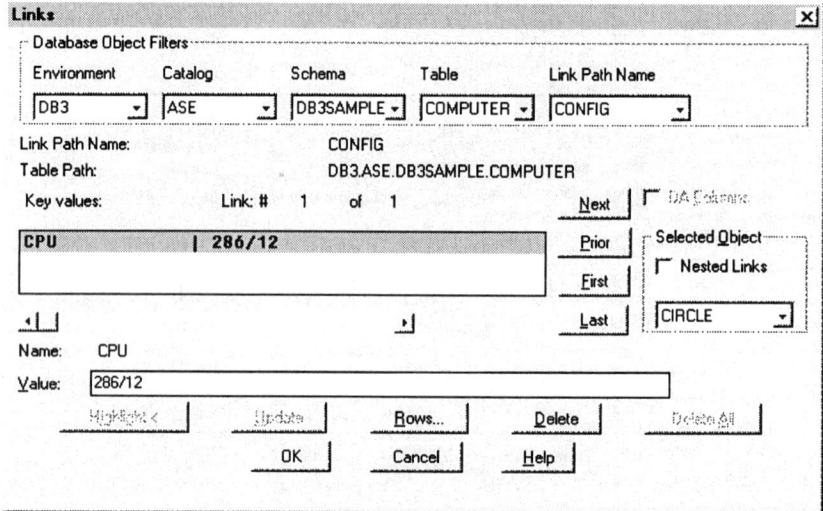

Figure 17-20 Links dialog box

The second step is to make the row current, which can be done with the **ASEROWS** command. When you enter this command, the **Rows** dialog box will be displayed. Use the Key Values...

button or the Open Cursor button to set, as the current row, the row whose COMP_CFG attribute value is 2. **Make sure you select the Updatable radio button in the Cursor State box.** You can also make a row current by selecting the Graphical in the Row dialog box, and then selecting the computer (second computer) in the office.

The last step is to edit the row and enter new values as given in Example 2. You can edit the row of a table by selecting the Edit... button. This invokes the **Edit Row** dialog box (Figure 17-21). Make the changes in the values as required, select Update button, and then close the Edit Row dialog box. The data associated with the selected computer is automatically updated. Now, select the Make DA button in the Rows dialog box and insert the new attributes on the screen.

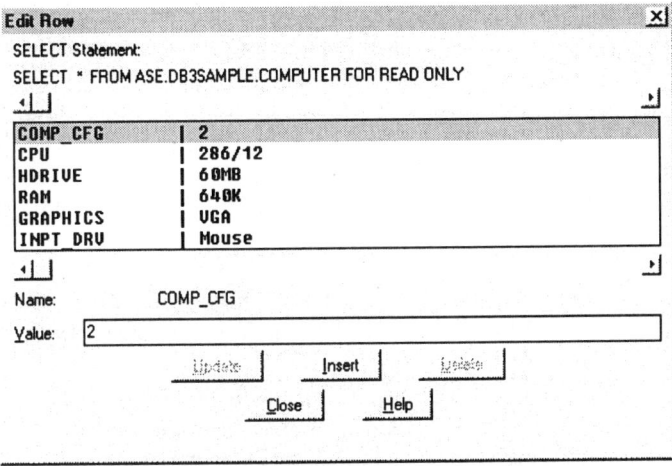

Figure 17-21 Edit Row dialog box

Note
The objects linked to the row being edited manifest the changes.

When a row linked to a displayable attribute is edited, invoke the Reload DA option of the ASEADMIN command to update the displayable attributes with the new values assigned to the row.

If data has to be changed for just one linked object, use the ASEROWS command to form a new row for that particular linked object and assign the new attribute values to this row. Then, use the ASELINKS command to link the object to the new row.

FORMING SELECTION SETS

It is possible to locate objects on the drawing on the basis of the linked nongraphic information. For example, you can locate the object that is linked to the first row of the COMPUTER table or to the first and second rows of the COMPUTER table. You can highlight specified objects or form a selection set of the selected objects.

1. Invoke the ASEROWS command.

17-20　Customizing AutoCAD

2. Select a row that is linked to an object in the drawing. For example, select the second row of the COMPUTER table. One of the ways you can accomplish this is by selecting the Open Cursor button, and then selecting the Next button until you find the required row in the table.

3. Now, select the Select button. The object linked to the second row of the COMPUTER table is highlighted in the drawing. You can add more objects to the selection set of objects linked to the current row.

Toolbar:	External Database, Select Objects
Pull-down:	Tools, External Database, Select Objects
Command:	ASESELECT

　You can also form a selection set on the basis of a combination of nongraphic and graphic data.

1. Invoke the ASESELECT command. The **Select Objects** dialog box (Figure 17-22) is displayed.

2. Set Environment as DB3, Catalog as ASE (in Figure 17-22 it is XYZ), Schema as DB3SAMPLE (in Figure 17-22 it is SAMPLE), and Table as COMPUTER.

3. In the Condition: edit box, enter the SELECT statement **GRAPHICS='VGA'** as shown in Figure 17-22, and then select the SELECT button. This creates selection set A, which has all the objects in the drawing linked to rows having VGA as the graphics.

4. Now, select the Intersect button and then the Graphical < button. The Select Objects dialog box is removed from the screen, and AutoCAD issues Select Objects prompts.

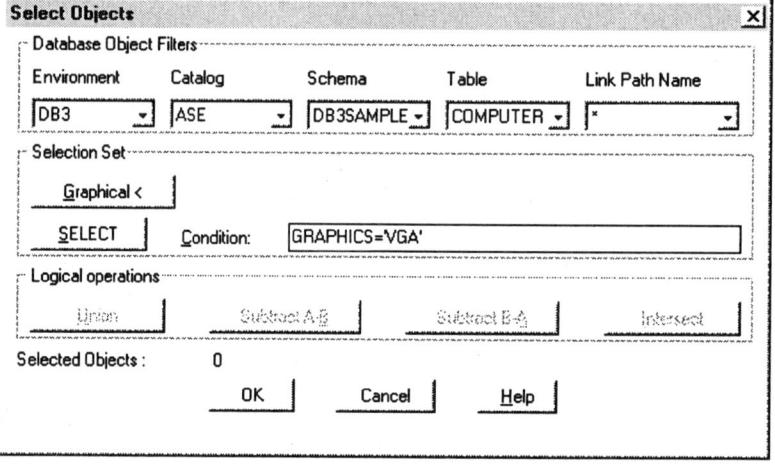

Figure 17-22 Select Objects dialog box

5. Select all computers, and then press Enter. This process forms selection set B. The Select Objects dialog box is displayed. Since we selected Intersect for the logical operation, AutoCAD forms a selection set based on the intersection of selection sets A and B. Hence, the selection set formed by the intersection of selection sets A and B contains all objects from set B linked to rows in which VGA is the graphics.

ASESELECT Command Options

A Union B. With this option, you can create a selection set of objects containing all the objects in the (A) and (B) selection sets.

Subtract A-B. With this option you can create a selection set of objects that contains objects obtained upon subtracting the second selection set (B) from the first selection set (A); in other words, the objects in (A) which are not in (B).

Subtract B-A. With this option you can create a selection set of objects that contains objects obtained upon subtracting the first selection set (A) from the second selection set (B); in other words, the objects in (B) which are not in (A).

A Intersect B. With this option, you can create a selection set of objects belonging to both (A) and (B) selection sets.

If you want to view the objects in the selection set, invoke the **SELECT** command, and then choose the Previous option. This command can be invoked from the pull-down menu (select Edit, Select Objects, Previous) or by entering **Previous (P)** at the Select Objects: prompt of the **SELECT** command. The objects in the selection set formed by the intersection of selection sets (A) and (B) (previously formed) are highlighted.

USING SQL STATEMENTS (ASESQLED)

Toolbar:	External Database, SQL Editor
Pull-down:	Tools, External Database, SQL Editor
Command:	ASESQLED

The **ASESQLED** command can be used to communicate with the external database using SQL statements. This is an important application that lets you search through the database and retrieve the information as specified in the SQL statements. Additionally, the SQL statements can be used to insert values into the table; delete rows from the table; change a single row in a table; create, drop, or alter a table; create or drop an index; create a view; and drop view functions. Example 3 explains some of the applications of ASESQLED command.

Example 3

Select all the rows in the EMPLOYEE table where EMP_ID is less than 1005.

1. Invoke the ASESQLED command. The **SQL Editor** dialog box (Figure 17-23) is displayed.

2. Specify the database object setting. Next, in the SQL: edit box enter the following SQL statement:

> Select * from employee where emp_id < 1005

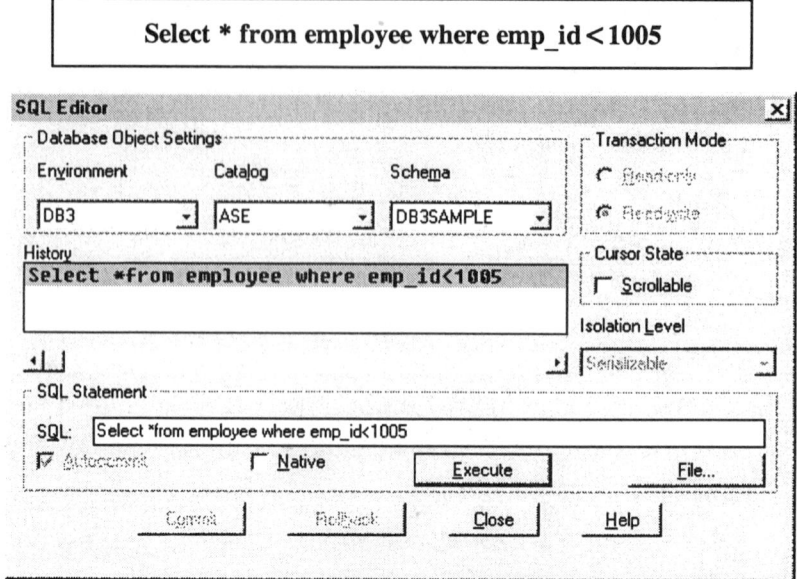

Figure 17-23 SQL Editor dialog box

3. Select the Execute button. The result obtained upon execution of the SQL statement is displayed in the **SQL Cursor** dialog box (Figure 17-24). In our case, all the rows whose EMP_ID is less than 1005 are selected and can be viewed with the help of navigation buttons.

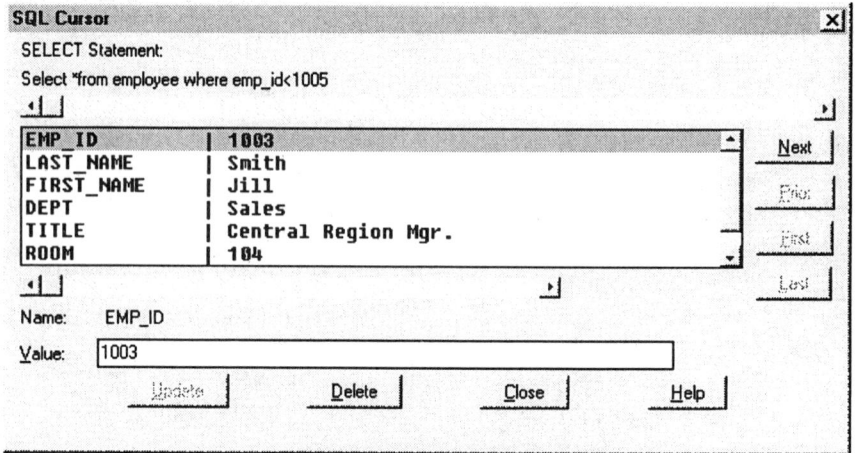

Figure 17-24 SQL Cursor dialog box

The value for a column in the selected row can be updated by selecting the column and then changing the value in the Value: edit box. You have the option of deleting rows from the SQL Cursor dialog box with the Delete button. For update and deletion operations, the cursor must be set to Updatable, and the Close button must be selected to make the changes permanent to the database. Now, let us say that we want to change the departments of all the employees whose emp_id is less than 1005 to accounting. This task can be performed in the following manner:

1. Enter the following SQL statement in the SQL: edit box of the SQL Editor dialog box (Figure 17-23):

> Update employee set dept = 'Accounting' where emp_id < 1005

2. Select the Execute button. The result obtained upon execution of the SQL statement is displayed in the SQL Cursor dialog box. In our case, all the rows whose EMP_ID is less than 1005 are selected and can be viewed with the help of navigation buttons.

3. Select the Execute button in the SQL Editor dialog box. The number of rows that are updated will be displayed at the lower left corner of the dialog box.

To check if the data has been updated, enter the following SQL statement in the SQL: edit box, and then select Execute:

> select * from employee where emp_id < 1005

The SQL Cursor dialog box is displayed. Notice that in the **DEPT** column of each selected row the **Accounting** value is displayed.

Example 4

Given an office with six computers and a database that contains information about these computers, you are required to do the following tasks (use the drawing and database from Example 2):

1. Locate the computers that have 2MB of RAM.
2. Locate the computers that have 16MB of RAM and 500MB hard drive.
3. Update the computers that have 16MB of RAM to 32MB.

Locating Computers with 2MB of RAM
Use the following instructions to locate the computers that have 2MB of RAM.

1. Enter the **ASEROW** command and press Enter. The Rows dialog box (Figure 17-25) is displayed.

17-24 Customizing AutoCAD

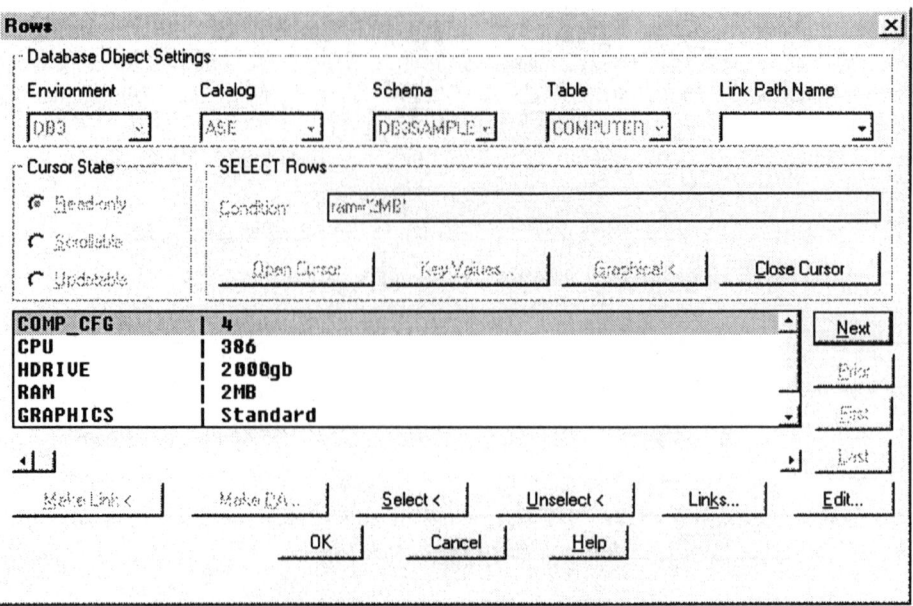

Figure 17-25 Rows dialog box

2. Enter the following statement in the Condition edit box. The attribute values enclosed in single quotes ('2MB') are case-sensitive and must be typed exactly as they appear in the table. The rest of the statement can be uppercase or lowercase.

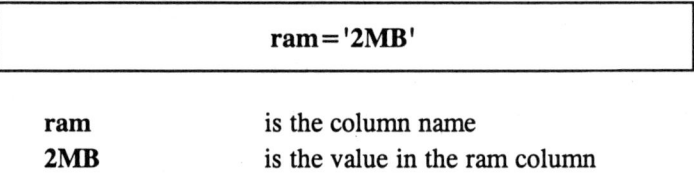

    ram   is the column name
    2MB   is the value in the ram column

3. Select the Open Cursor button. The rows that have 16MB value for the RAM column are selected, and the first of the selected rows is displayed in the list box. With the navigation buttons you can check the other selected rows. In our case, only one row must be checked.

4. Select the Links... button. The Links dialog box (Figure 17-26) is displayed.

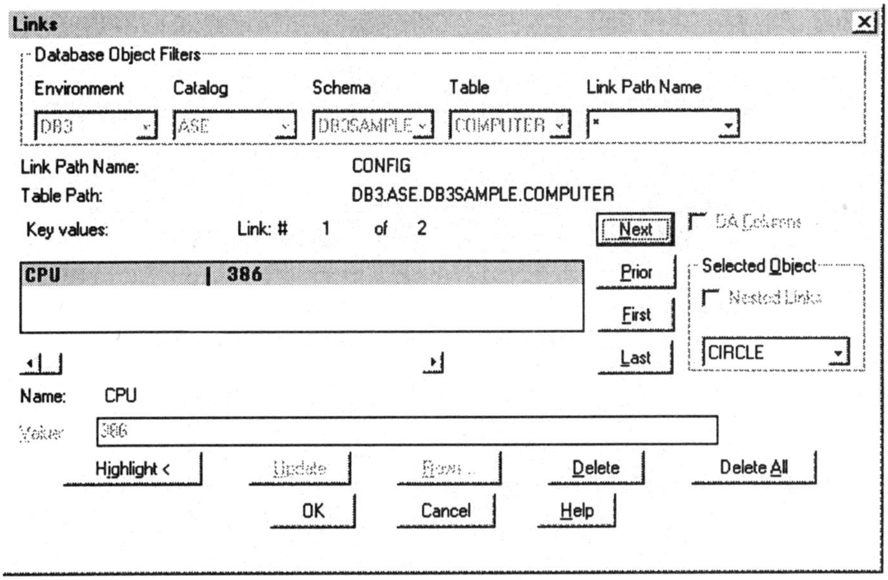

Figure 17-26 Links dialog box

5. Select the Highlight < button. The computer with 2MB value for the RAM column is highlighted.

Locating Computers with 16MB of RAM and 500MB of hard drive. For the second part of the example, we have to locate the computers that have 16MB of RAM and 500MB hard drive. The condition statement in this case contains multiple search criteria. For example, if there are six computers in the office and we want to locate the computers that have 16MB of RAM and 500MB hard drive, there are two conditions that the database search should satisfy. We can write a condition statement that will search through the database and locate the items that satisfy the specified criteria. The procedure involved is similar to the first part of the example. The only difference is that the statement to be entered in the Condition edit box of the Rows dialog box is:

> ram='16MB' and hdrive='500MB'

Updating the Computers that have 16MB of RAM to 32MB. You can use the SQL statements to update the values in the table. For example, we can write an SQL statement that will search the "**COMPUTER**" table for "**16MB**" values and then replace those values with "**32MB**". Use the following instructions to update the "**COMPUTER**" table:

1. Enter **ASESQLED** command and press Enter. The SQL Editor dialog box is displayed.

2. Enter the following SQL statement in the SQL: edit box. The statement must be entered exactly as it appears here.

> update computer set RAM='32MB' where
> RAM='16MB'

Where:
    computer   is the table name
    RAM   in the column name

3. Select the Execute button from the SQL Editor dialog box.

4. To check if the data has been updated, enter the following SQL statement in the SQL: edit box. The statement must be entered exactly as it appears here.

> select comp_cfg,ram from computer where ram='32MB'

The values in the **comp_cfg** and **ram** columns will be displayed in the SQL Cursor dialog box.

GENERATING REPORTS FROM EXPORTED DATA

On the basis of the information linked to the drawing, you can generate reports using the external database commands. AutoCAD does not support a direct reporting feature. Hence, reports are created with DBMS and report-writer softwares. The external database commands provided by AutoCAD make the procedure involved in the creation of reports easy.

Toolbar:	External Database, Export Links
Pull-down:	Tools, External Database, Export Links
Command:	ASEEXPORT

With external database commands, it is possible to link more than one object to a row. This lessens the size of the database in case objects do not have distinct attributes. In this case, it is not possible to determine how many objects are linked to one row from the DBMS. The **ASEEXPORT** command can be used to determine the number of objects linked to a single row by exporting link information in the drawing and xref and blocks in the drawing, in different formats. This information can be fused with the data in the tables to generate a report.

1. Invoke the **ASEEXPORT** command.

2. AutoCAD PROMPTS you to select objects whose link information you want to export. After completing the selection of objects, the **Export Links** dialog box (Figure 17-27) is displayed.

3. Select the scope of the linked data to be produced in the report by making adequate selections in the Database Object Filters area.

Accessing External Database, AutoCAD SQL Extension (ASE) 17-27

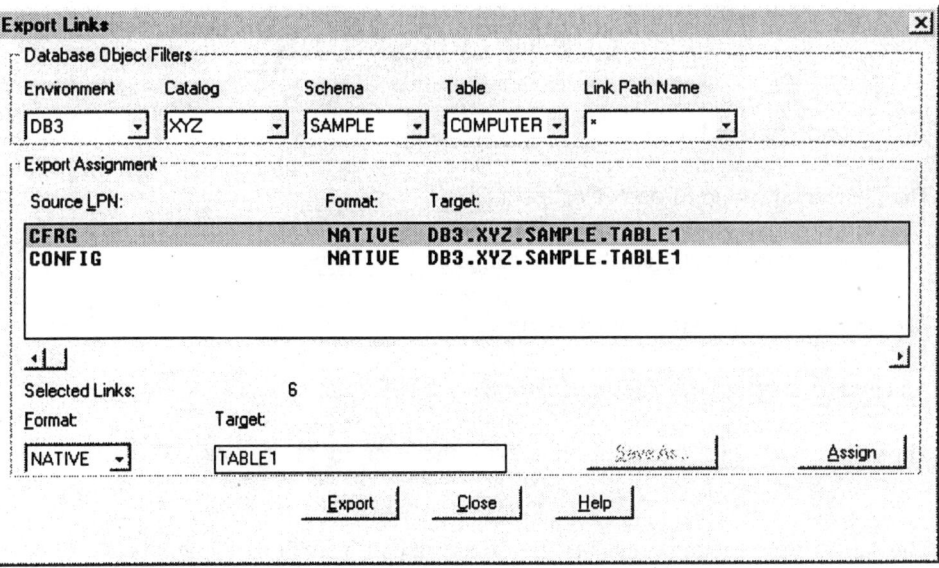

Figure 17-27 Export Links dialog box

4. Select a Link Path Name in the list box.

5. In the Format: drop-down list, select NATIVE as the file format of the exported information.

6. In the Target: edit box, specify the name of the table you want to form.

7. Select the Assign button to assign the defined file format and target path name to the selected link path name.

8. If you want to export links of some other link path name, repeat steps 3 through 7.

9. Select the Export button. It will create a DBF file with the path and file name you specified.

The export operation is repeated for each link path name assignment.

REVIEW QUESTIONS

1. SQL is often referred to as _____.

2. The horizontal group of data is called a _____.

3. A vertical group of data (attribute) is called a _____.

4. The **ASE** commands are defined in the _____ file.

5. An environment can have zero or more catalogs. (T/F)

6. The _____ command allows the user to set a row current in the predefined (current) table.

7. The name of the catalog relates to the directory path name that is the location of the database. (T/F)

8. Once you define a link, AutoCAD does not store that information with the drawing. (T/F)

9. The location of database files can be defined by setting an environment variable that tells AutoCAD where to find the files. (T/F)

10. If you want to use the dialog boxes with ASE commands, you must set the **CMDDIA** variable to 0 (off). (T/F)

11. The **Make Link** option in the Rows dialog box enables you to create a link by assigning the current row of the current table to a graphics entity in the drawing. (T/F)

12. The **ASESQLED** command can be used to communicate with an external database using SQL statements. (T/F)

13. The SQL statements let you search through the database and retrieve the information as specified in the SQL statements. (T/F)

14. What does **ASE** stand for? _____

15. What is a **database**? _____

16. What is a **database management system** (DBMS)? _____

17. A schema cannot have one or more tables. (T/F)

18. The name of the schema relates to the catalog subdirectory that contains the database tables. (T/F)

19. When a sequence of SQL statements are executed, a _____ takes place.

20. A transaction is terminated with the SQL _____ or SQL _____ operation.

21. A row is also referred as a _____

22. A column is sometimes called a _____

23. The _____ acts like an identification tag for locating and linking a row.

24. Selecting two or more columns and their values to define a key is called _____ .

25. Keys can be defined in the _____ dialog box.

26. Isolation of data returned by a query transaction when many users try to address the same data simultaneously does not depend on the isolation levels. (T/F)

27. Only one row can be manipulated at a time. This row is called the _____ .

28. It is possible to make a row current by making use of a key to search for a particular row or rows if you have specified key values. (T/F)

29. You can reestablish data integrity with the use of the _____ option of the ASEADMIN command.

30. The _____ command can be used to determine the number of objects linked to a single row by exporting link information in the drawing and xref and blocks in the drawing, in different formats.

EXERCISES

Exercise 1

In this exercise, you will set the environment to DB3, the catalog to XYZ, the schema to SAMPLE database, and the EMPLOYEE table current; then, edit the second row of the table (EMP_ID=1001, Williams). You will also add a new row to the table, set the new row current, and then view it.

The row to be added has the following values:

EMP_ID	1100
LAST_NAME	McLees
FIRST_NAME	David
Dept	Technology
TITLE	Programmer
ROOM	108
EXT	8001

Exercise 2

Load the drawing ASETUT from the **TUTORIAL** subdirectory and use the SAVEAS command to save the drawing as SQLEX2. (You can also make a copy of the ASETUT drawing and name it SQLEX2; then, load the new drawing, SQLEX2.) After loading the drawing, perform the following operations.

1. Use the **ASEADMIN** command to set the environment to DB3, the catalog to XYZ, the schema to SAMPLE, and the table to COMPUTER.
2. Make a copy of the computer in room 106 and place it in room 108.
3. Use the **ASELINKS** command to view the existing link of the copied computer (computer in room 108).
4. Delete any existing link.
5. Use the **ASEROW** command to set COMP_CFG number 2 current.
6. Use the **Make Link** option in the Rows dialog box to link the current computer to the computer in room 108.
7. Use the **Make DA** option in the Rows dialog box to display the attributes of the computer on the screen.
8. Using the **ASESQLED** command, enter the following SQL statement and highlight the computer in room 108.

> select emp_id,last_name from employee where room='108'

Chapter 18

Defining Block Attributes

Learning objectives

After completing this chapter, you will be able to:
- Understand what attributes are and how to define attributes with a block.
- Edit attribute tag names.
- Insert blocks with attributes and assign values to attributes.
- Extract attribute values from the inserted blocks.
- Control attribute visibility.
- Perform global and individual editing of attributes.
- Insert a text file in a drawing to create bill of material.

ATTRIBUTES

AutoCAD has provided a facility that allows the user to attach information to blocks. This information can then be retrieved and processed by other programs for various purposes. For example, you can use this information to create a bill of material, find the total number of computers in a building, or determine the location of each block in a drawing. Attributes can also be used to create blocks (such as title blocks) with prompted or preformatted text, to control text placement. The information associated with a block is known as **attribute value** or simply **attribute.** AutoCAD references the attributes with a block through tag names.

Before you can assign attributes to a block, you must create an attribute definition by using the **DDATTDEF** or **ATTDEF** command. The attribute definition describes the characteristics of the attribute. You can define several attribute definitions (tags) and include them in the block definition. Each time you insert the block, AutoCAD will prompt you to enter the value of the attribute. The attribute value automatically replaces the attribute tag name. The information (attribute values) assigned to a block can be extracted and written to a file by using AutoCAD's **DDATTEXT** or **ATTEXT** command. This file can then be inserted in the drawing as a table or processed by other programs to analyze the data. The attribute values can be edited by using the **DDATTE** or **ATTEDIT** command. The display of attributes can be controlled with **ATTDISP** command.

DEFINING ATTRIBUTES
DDATTDEF Command

Pull-down:	Draw, Block, Define Attribute
Command:	DDATTDEF

When you invoke the DDATTDEF command, the **Attribute Definition** dialog box (Figure 18-1) is displayed. The block attributes can be defined through this dialog box. When you create an attribute definition, you must define the mode, attributes, insertion point, and text information for each attribute. All this information can be entered in the dialog box. Following is a description of each area of the Attribute Definition dialog box.

Figure 18-1 Attribute Definition dialog box

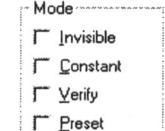

Mode. The **Mode** area of the Attribute Definition dialog box has four options: Invisible, Constant, Verify, and Preset. These options determine the display and edit features of the block attributes. For example, if an attribute is invisible, the attribute is not displayed on the screen. Similarly, if an attribute is constant, its value is predefined and cannot be changed. These options are described below:

Invisible. This option lets you create an attribute that is not visible on the screen. This mode is useful when you do not want the attribute values to be displayed on the screen to avoid cluttering the drawing. Also, if the attributes are invisible, it takes less time to regenerate the drawing. If you want to make the invisible attribute visible, use the **ATTDISP** command discussed later in this chapter [section "Controlling Attribute Visibility (ATTDISP) COMMAND)"].

Constant. This option lets you create an attribute that has a fixed value and cannot be changed after block insertion. When you select this mode, the **Prompt:** edit box and the **Verify** and **Preset** check boxes are disabled.

Verify. This option allows you to verify the attribute value you have entered when inserting a block. If the value is incorrect, you can correct it by entering the new value.

Preset. This option allows you to create an attribute that is automatically set to default value. The prompt is not displayed and an attribute value is not requested when you insert a block with attributes using this option to define a block attribute. Unlike a constant attribute, the preset attribute value can later be edited.

Attribute. The **Attribute** area of the Attribute Definition dialog box (Figure 18-2) has three edit boxes: Tag, Prompt, and Value. To enter a value, you must first select the corresponding edit box and then enter the value. You can enter up to 256 characters in these edit boxes.

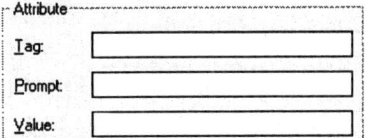

Figure 18-2 Attribute area of the Attribute Definition dialog box

Tag. This is like a label that is used to identify an attribute. For example, the tag name COMPUTER can be used to identify an item. The tag names can be uppercase, lowercase, or both. Any lowercase letters are automatically converted into uppercase. The tag name cannot be null. Also, the tag name must not contain any blank spaces. You should select a tag name that reflects the contents of the item being tagged. For example, the tag name COMP or COMPUTER is an appropriate name for labeling computers.

Prompt. The text that you enter in the **Prompt:** edit box is used as a prompt when you insert a block that contains the defined attribute. If you have selected the Constant option in the Mode area, the **Prompt:** edit box is disabled because no prompt is required if the attribute is constant. If you do enter nothing in the **Prompt:** edit box, the entry made in the Tag: edit box is used as the prompt.

Value. The entry in the **Value:** edit box defines the default value of the specified attribute; that is, if you do not enter a value, it is used as the value for the attribute. The entry of a value is optional.

Insertion Point. The **Insertion Point** area of the Attribute Definition dialog box (Figure 18-3) lets you define the insertion point of block attribute text. You can define the insertion point by entering the values in the **X:**, **Y:**, and **Z:** edit boxes or by selecting **Pick Point <** button. If you select this button, the dialog box clears, and you can enter the X, Y, and Z values of the insertion point at the command line or specify the point by selecting a point on the screen. When you are done specifying the insertion point, the **Attribute Definition** dialog box reappears.

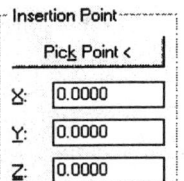

Figure 18-3 Insertion Point area of the Attribute Definition dialog box

Just under the Insertion Point area of the dialog box is a check box labeled **Align below previous attribute**. You can use this box to place the subsequent attribute text just below the previously defined attribute automatically. This check box is disabled if no attribute has been defined. When you select this check box, the Insertion Point area and the Text Options areas are disabled because AutoCAD assumes previously defined values for text such as text height, text style, text justification, and text rotation. Also, the text is automatically placed on the following line. After insertion, the attribute text is responsive to the setting of the **MIRRTEXT** system variable.

Text Options. The **Text Options** area of the Attribute Definition dialog box lets you define the justification, text style, height, and rotation of the attribute text. To set the text justification, select justification type in the **Justification:** drop-down list. Similarly, you can use the **Text Style:** drop-down list to define the text style. You can specify the text height and text rotation in the **Height <** and **Rotation <** edit boxes. You can also define the text height by selecting the Height < button. If you select this button, AutoCAD temporarily exits the dialog box and lets you enter the value from the command line. Similarly, you can define the text rotation by selecting the Rotation < button and then entering the rotation angle at the command line.

> **Note**
>
> *The text style must be defined before it can be used to specify the text style.*
>
> *If you select a style that has the height predefined, AutoCAD automatically disables the Height < edit box.*
>
> *If you have selected the Align option for the text justification, the Height < and Rotation < edit box are disabled.*
>
> *If you have selected the Fit option for the text justification, the Rotation < edit box is disabled.*

After you complete the settings in the **Attribute Definition** dialog box and choose **OK**, the attribute tag text is inserted in the drawing at the specified insertion point.

ATTDEF Command

You can define block attributes through the command line by entering the **ATTDEF** command.

    Command: **ATTDEF**
    Attribute modes - Invisible:N   Constant:N   Verify:N   Preset:N
    Enter (ICVP) to change, or press ENTER when done:

The default value of all attribute modes is N (No). To reverse the default mode, enter I, C, V, or P. For example, if you enter I, AutoCAD will change the Invisible mode from N to Y (Yes). This will make the attribute visible. After setting the modes, press Enter to go to the next prompts where you can enter the attribute tag, attribute prompt, and the attribute values.

    Attribute tag:
    Attribute prompt:
    Default attribute value:

The entry at this prompt defines the default value of the specified attribute; that is, if you do not enter a value, it is used as the value for the attribute. The entry of a value is optional. If you have selected the Constant mode, AutoCAD displays the following prompt:

Attribute value:

Next, AutoCAD displays the following text prompts:

Justify/Style/<Start point>:
Height<0.200>:
Rotation angle<0>:

After you respond to these prompts, the attribute tag text will be placed at the specified location. If you press Enter at Justify/Style/<Start point>:, AutoCAD will automatically place the subsequent attribute text just below the previously defined attribute, and it assumes previously defined text values such as text height, text style, text justification, and text rotation. Also, the text is automatically placed on the following line.

Example 1

In this example, you will define the following attributes for a computer and then create a block using the BLOCK command. The name of the block is COMP.

Mode	Tag name	Prompt	Default value
Constant	ITEM		Computer
Preset, Verify	MAKE	Enter make:	CAD-CIM
Verify	PROCESSOR	Enter processor type:	Unknown
Verify	HD	Enter Hard-Drive size:	100MB
Invisible, Verify	RAM	Enter RAM:	4MB

1. Draw the computer as shown in Figure 18-4. Assume the dimensions, or measure the dimensions of the computer you are using for AutoCAD.

2. Invoke the **DDATTDEF** command. The **Attribute Definition** dialog box is displayed (Figure 18-5).

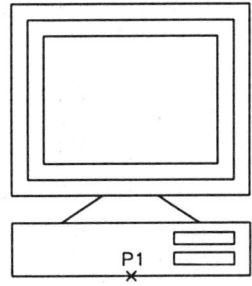

Figure 18-4 Drawing for Example 1

3. Define the first attribute as shown in the preceding table. Select **Constant** in the **Mode area** because the mode of the first attribute is constant. In the **Tag:** edit box, enter the tag name, ITEM. Similarly, enter Computer in the **Value:** edit box. The **Prompt:** edit box is disabled because the variable is constant.

18-6 Customizing AutoCAD

4. In the **Insertion Point** area, select the **Pick Point** < button to define the text insertion point. Select a point below the drawing of the computer.

5. In the **Text Options** area, select the justification, style, height, and rotation of the text.

6. Select the **OK** button when you are done entering information in the Attribute Definition dialog box.

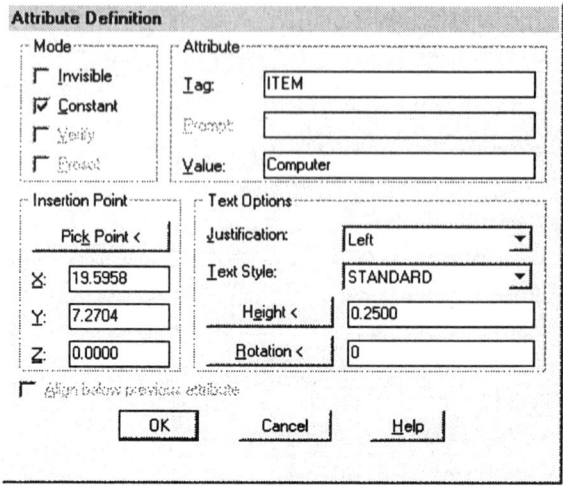

Figure 18-5 Enter information in the Attribute Definition dialog box

7. Enter **DDATTDEF** at the Command: prompt to invoke the Attribute Definition dialog box. Enter the Mode and Attribute information for the second attribute shown in the table at the beginning of Example 1. You need not define the Insertion Point and Text Options. Check the **Align below previous attribute** box that is located just below the Pick Point < area. When you check this box, the Insertion Point and Text Justification areas are disabled. AutoCAD places the attribute text just below the previous attribute text (Figure 18-6).

Figure 18-6 Define attributes below the computer drawing

8. Define the remaining attributes.

9. Use the **BLOCK** command to create a block. The name of the block is COMP, and the insertion point of the block is P1, midpoint of the base. When you select the objects for the block, make sure you also select the attributes.

EDITING ATTRIBUTE TAGS
Using the DDEDIT Command

Toolbar:	Modify II, Edit Text
Pull-down:	Modify, Object, Text
Command:	DDEDIT

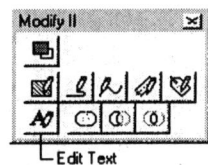

Figure 18-7 Edit Text icon in the Modify toolbar

The **DDEDIT** command lets you edit text and attribute definitions. After invoking this command, AutoCAD will prompt you to select an annotation object. If you select an attribute definition, the **Edit Attribute Definition** dialog box, listing the tag name, prompt, and default value of the attribute, is displayed (Figure 18-8). You can select the edit boxes and enter the changes. Once you are done making the required changes, select the **OK** button in the dialog box. After you exit the dialog box, AutoCAD will prompt you to select text or attribute object (Attribute tag). If you are done editing, press Enter to return to the command line.

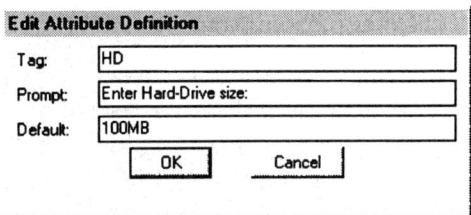

Figure 18-8 Edit Attribute Definition dialog box

> Command: **DDEDIT**
> <Select an annotation object>/Undo: *Select the attribute tag.*

Using the CHANGE Command

You can also use the **CHANGE** command to edit text or attribute objects. The following is the command prompt sequence for the CHANGE command:

> Command: **CHANGE**
> Select objects: *Select attribute objects.*
> Select objects: ↵
> Properties/<Change point>: ↵
> Enter text insertion point: ↵
> Text style: Current
> New style or press ENTER for no change: ↵
> New height <current>: ↵
> New rotation angle <0>: ↵
> New tag <current>: *Enter new tag name or* ↵.
> New prompt <current>: *Enter new prompt or* ↵.
> New default value <current>: *Enter new default or* ↵.

INSERTING BLOCKS WITH ATTRIBUTES
Using the Dialog Box

The value of the attributes can be specified during block insertion, either at the command line or in the **Edit Attribute Definition** dialog box. The Edit Attribute Definition dialog box is invoked by setting the system variable **ATTDIA** to 1 and then using the INSERT command. The default value for ATTDIA is 0 which disables the dialog box.

 Command: **ATTDIA**
 New value for ATTDIA<0>: **1**

 Command: **INSERT**
 Block name (or ?): *Enter block name which has attributes in it.*
 Insertion point: *Specify the insertion point.*
 X scale factor<1>/Corner/XYZ: *Enter X scale factor.*
 Y scale factor (default)=X): *Enter Y scale factor.*
 Rotation angle<0>: *Enter rotation angle.*

After you respond to these prompts, AutoCAD will display the **Enter Attributes** dialog box (Figure 18-9), displaying the prompts and their default values which have been entered at the time of attribute definition. If there are more attributes, they can be accessed by using the **Next** or **Previous** buttons. You can enter the attribute values in the edit box located next to the attribute prompt. The block name is displayed at the top of the dialog box. After entering the new attribute values, select the **OK** button; AutoCAD will place these attribute values at the specified location.

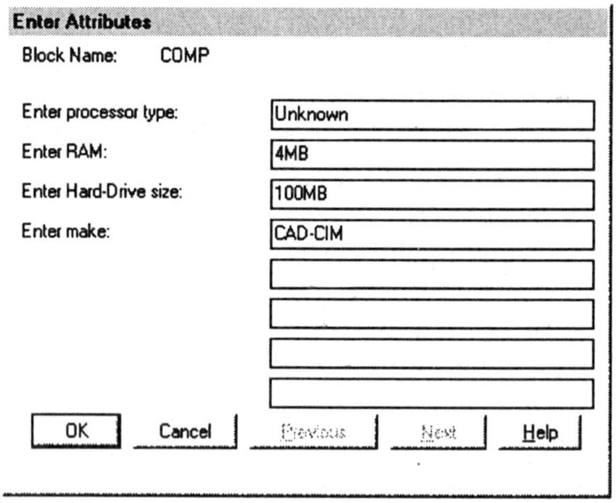

Figure 18-9 Enter attribute values in the Enter Attributes dialog box

Note
If you use the dialog box to define the attribute values, the Verify mode is ignored because the Enter Attributes dialog box allows the user to examine and edit the attribute values.

Using the Command Line

You can also define attributes from the command line by setting the system variable **ATTDIA** to 0 (default value). When you use the INSERT command with **ATTDIA** set to 0, AutoCAD does not display the Enter Attributes dialog box. Instead, AutoCAD will prompt you to enter the attribute values for various attributes that have been defined in the block. To define the attributes from the command line, enter the INSERT command at the Command: prompt. After you define the insertion point, scale, and rotation, AutoCAD will display the following prompt:

Enter attribute values

It will be followed by the prompts that have been defined with the block using the ATTDEF command. For example:

Enter processor type <Unknown>:
Enter RAM <4MB>:
Enter Hard-Drive size <100MB>:

Example 2

In this example, you will use the INSERT command to insert the block (COMP) that was defined in Example 1. The following is the list of the attribute values for computers.

ITEM	MAKE	PROCESSOR	HD	RAM
Computer	Gateway	486-60	150MB	16MB
Computer	Zenith	486-30	100MB	32MB
Computer	IBM	386-30	80MB	8MB
Computer	Del	586-60	450MB	64MB
Computer	CAD-CIM	Pentium-90	100 Min	32MB
Computer	CAD-CIM	Unknown	600MB	Standard

1. Make the floor plan drawing as shown in Figure 18-10 (assume the dimensions).

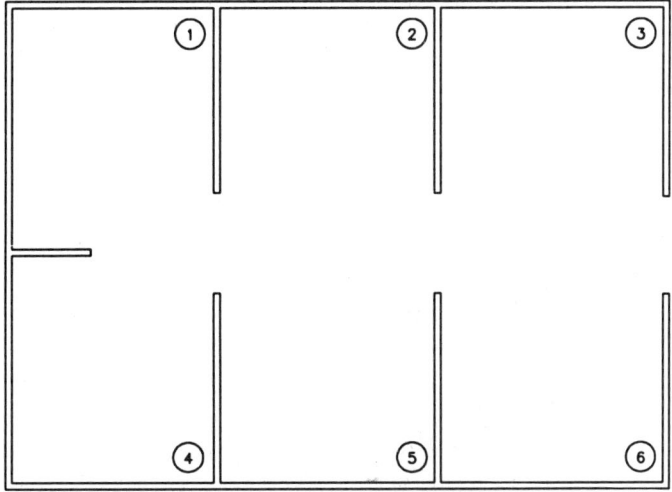

Figure 18-10 Floor plan drawing for Example 2

2. Set the system variable **ATTDIA** to 1. Use the **INSERT** command to insert the blocks, and define the attribute values in the Enter Attributes dialog box (Figure 18-11).

18-10 Customizing AutoCAD

Command: **INSERT**
Block name (or ?): **COMP**
Insertion point: *Specify the insertion point.*
X scale factor<1>/Corner/XYZ: ↵
Y scale factor (default)=X): ↵
Rotation angle<0>: ↵

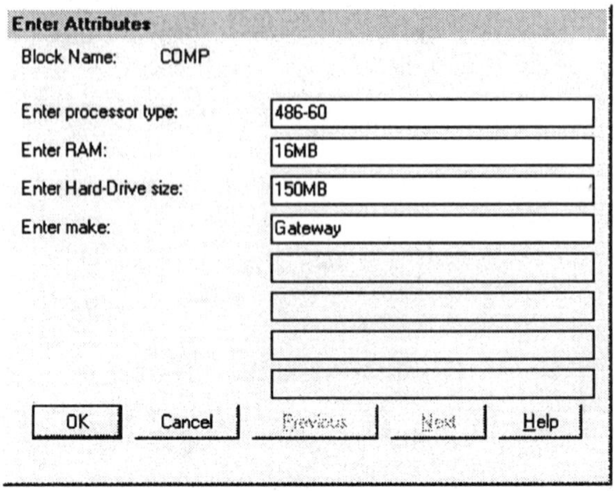

Enter Attributes dialog box showing:
Block Name: COMP
Enter processor type: 486-60
Enter RAM: 16MB
Enter Hard-Drive size: 150MB
Enter make: Gateway

Figure 18-11 Enter attribute values in the Enter Attributes dialog box

3. Repeat the **INSERT** command to insert other blocks, and define their attribute values as shown in Figure 18-12.

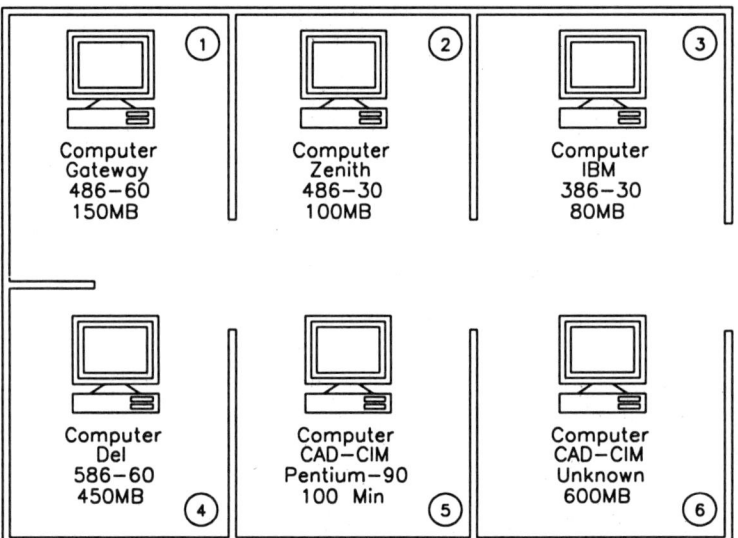

Figure 18-12 The floor plan after inserting blocks and defining their attributes

4. Save the drawing for further use.

EXTRACTING ATTRIBUTES
Using the Dialog Box (DDATTEXT)

| Command: DDATTEXT |

To use the **Attribute Extraction** dialog box (Figure 18-13) for extracting the attributes, enter **DDATTEXT** at the Command: prompt. The information about the file format, template file, and output file must be entered in the dialog box to extract the defined attribute. Also, you must select the blocks whose attribute values you want to extract.

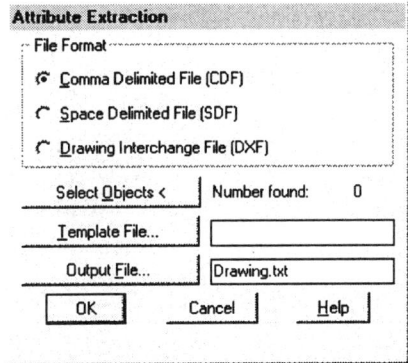

Figure 18-13 Attribute Extraction dialog box

File Format. This area of the dialog box lets you select the file format of the extracted data. You can select Comma Delimited File, Space Delimited File, or Drawing Interchange File. The format selection is determined by the application that you plan to use to process the data.

Comma Delimited File (CDF). In CDF format, each character field is enclosed in single quotes, and the records are separated by a delimiter (comma by default). CDF file is a text file with the extension **.TXT**.

Space Delimited File (SDF). In SDF format, the records are of fixed width as specified in the template file. The records are not separated by a comma, and the character fields are not enclosed in single quotes. The SDF file is a text file with the extension **.TXT**.

Drawing Interchange File (DXF). If you select the Drawing Interchange File format, the template file name and the **Template File** edit box in the **Attribute Extraction** dialog box are automatically disabled. The file created by this option contains only block references, attribute values, and end-of-sequence objects. The extension of these files is **.DXX**.

Template File. The **template file** allows you to specify the attribute values you want to extract and the information you want to retrieve about the block. It also lets you format the display of the extracted data. The file can be created by using any text editor, such as Notepad, Windows Write, or WordPad. You can also use a word processor or a database program to write the file. The template file must be saved as an **ASCII** file and the extension of the file must be **.TXT**. The following are the fields that you can specify in a template file (the comments given on the right are for explanation only; they must not be entered with the field description):

```
BL:LEVEL      Nwww000     (Block nesting level)
BL:NAME       Cwww000     (Block name)
BL:X          Nwwwddd     (X coordinate of block insertion point)
BL:Y          Nwwwddd     (Y coordinate of block insertion point)
BL:Z          Nwwwddd     (Z coordinate of block insertion point)
```

BL:NUMBER	Nwww000	*(Block counter)*
BL:HANDLE	Cwww000	*(Block handle)*
BL:LAYER	Cwww000	*(Block insertion layer name)*
BL:ORIENT	Nwwwddd	*(Block rotation angle)*
BL:XSCALE	Nwwwddd	*(X scale factor of block)*
BL:YSCALE	Nwwwddd	*(Y scale factor of block)*
BL:ZSCALE	Nwwwddd	*(Z scale factor of block)*
BL:XEXTRUDE	Nwwwddd	*(X component of block's extrusion direction)*
BL:YEXTRUDE	Nwwwddd	*(Y component of block's extrusion direction)*
BL:ZEXTRUDE	Nwwwddd	*(Z component of block's extrusion direction)*
Attribute tag		*(The tag name of the block attribute)*

The extract file may contain several fields. For example, the first field might be the item name and the second field might be the price of the item. Each line in the template file specifies one field in the extract file. Any line in a template file consists of the name of the field, the width of the field in characters, and its numerical precision (if applicable). For example:

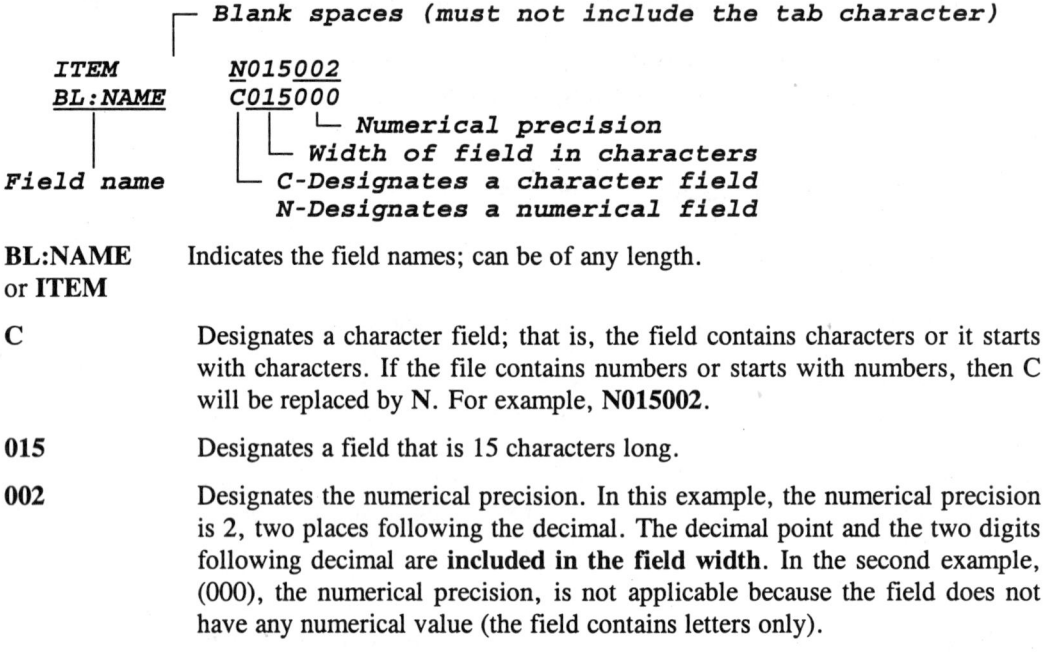

BL:NAME or ITEM Indicates the field names; can be of any length.

C Designates a character field; that is, the field contains characters or it starts with characters. If the file contains numbers or starts with numbers, then C will be replaced by N. For example, **N015002**.

015 Designates a field that is 15 characters long.

002 Designates the numerical precision. In this example, the numerical precision is 2, two places following the decimal. The decimal point and the two digits following decimal are **included in the field width**. In the second example, (000), the numerical precision, is not applicable because the field does not have any numerical value (the field contains letters only).

Note
You can put any number of spaces between the field name and the character C or N (ITEM N015002). However, you must not use the tab characters. Any alignment in the fields must be done by inserting spaces after the field name.

In the template file, a field name must not appear more than once. The template file name and the output file name must be different.

The template file must contain at least one field with an attribute tag name because the tag names determine which attribute values are to be extracted and from which blocks. If several blocks have different block names but the same attribute tag, AutoCAD will extract attribute values from all selected blocks. For example, if there are two blocks in the drawing with the attribute tag PRICE, then when you extract the attribute values, AutoCAD will extract the value from both blocks (if both blocks were selected). To extract the value of an attribute, the tag name must match the field name specified in the template file. AutoCAD automatically converts the tag names and the field names to uppercase letters before making a comparison.

Example 3

In this example, you will write a template file for extracting the attribute values as defined in Example 2. These attribute values must be written to a file **COMPLST1.TXT** and the values arranged as shown in the following table:

Field width in characters

< 10 >	< 12 >	< 10 >	< 12 >	< 10 >	< 10 >
COMP	Computer	Gateway	486-60	150MB	**16MB**
COMP	Computer	Zenith	486-30	100MB	**32MB**
COMP	Computer	IBM	386-30	80MB	**8MB**
COMP	Computer	Del	586-60	450MB	**64MB**
COMP	Computer	**CAD-CIM**	Pentium-90	100 Min	**32MB**
COMP	Computer	**CAD-CIM**	**Unknown**	600MB	Standard

1. Load the drawing you saved in Example 2.

2. Use the Windows **Notepad** to write the following template file. You can use any text editor or word processor to write the file. After writing the file, save it as an ASCII file under the file name **TEMP1.TXT**. Exit the Notepad and access AutoCAD.

BL:NAME	C010000	*(Block name, 10 spaces)*
Item	C012000	*(Item, 12 spaces)*
Make	C010000	*(Computer make, 10 spaces)*
Processor	C012000	*(Processor type, 12 spaces)*
HD	C010000	*(Hard drive size, 10 spaces)*
RAM	C010000	*(RAM size, 10 spaces)*

3. Use the **DDATTEXT** command to invoke the Attribute Extraction dialog box (Figure 18-14), and select the Space Delimited File (SDF) radio button.

4. Choose the **Select Objects <** button to select the objects (blocks) present on the screen. You can select the objects by using the Window or Crossing option. After selection is complete, right-click your pointing device to display the dialog box again.

5. In the Template File... edit box, enter the name of the template file, **TEMP1.TXT**.

6. In the Output File... edit box, enter the name of the output file, **COMPLST1.TXT**.

7. Choose the **OK** button in the Attribute Extraction dialog box.

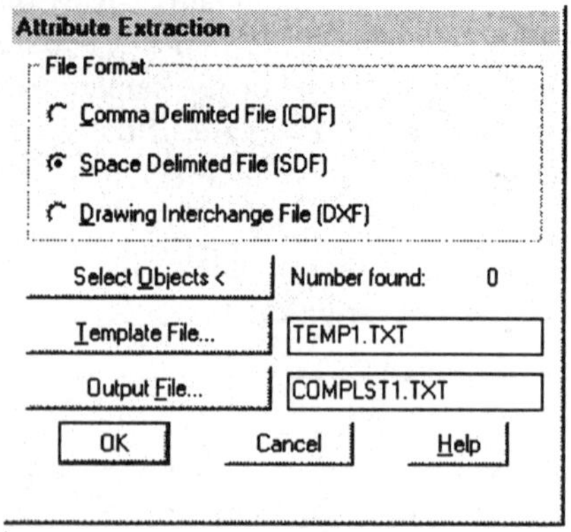

Figure 18-14 Enter information in the Attribute Extraction dialog box

8. Use the Notepad again to list the output file, **COMPLST1.TXT**. The output file will be similar to the file shown at the beginning of Example 3.

Using the Command Line (ATTEXT)

You can also extract the attributes from the command line by entering the **ATTEXT** command at the Command: prompt. When you enter this command, AutoCAD will first prompt you to enter the type of output file. You could select CDF, SDF, DXF, or Objects. If you select the Objects option, AutoCAD will prompt you to select the objects whose attributes you want to extract. After you select the objects, AutoCAD will return the prompt, this time without the Objects option. The following is the command prompt sequence for this command.

Command: **ATTEXT**
CDF, SDF, or DXF Attribute extract (or Objects)? <C>: **E**
Select objects: *Select objects.*
Select objects: ↵
CDF, SDF, or DXF Attribute extract? <C>: **C or S**

If you enter C or S (CDF or SDF), AutoCAD will prompt you to enter the name of the template file. (Template files were discussed earlier in this chapter in the subsection "Template File.")

Template file <default>: *Enter the name of the template file.*

After you enter the template file name, AutoCAD will prompt you to enter the name of the extract file. The extract file is the output file where you want to write the attribute values. This prompt will be displayed for any of the attribute extract formats (CDF, SDF, or DXF).

Extract file name <drawing name>: *Enter the name of the output file.*

If you do not enter a file name, AutoCAD assumes that the name of the extract file is the same as the drawing file name with the file extension **.TXT** or **.DXX**, depending on the attribute extract format. If you enter the file name and have selected the CDF or SDF file extraction format, the file extension must be **.TXT**; if you have selected the **.DXF** format, the file extension must be **.DXX**. After you enter the file name, AutoCAD will extract the attribute values from the blocks and write the data to the output file.

CONTROLLING ATTRIBUTE VISIBILITY (ATTDISP COMMAND)

Pull-down:	View, Display, Attribute Display
Command:	ATTDISP

The **ATTDISP** command allows you to change the visibility of all attribute values. Normally, the attributes are visible unless they are defined invisible by using the Invisible mode. The invisible attributes are not displayed, but they are a part of the block definition. The prompt sequence is:

Command: **ATTDISP**
Normal/ON/OFF <current>:

If you enter ON and press Enter, all attribute values will be displayed, including the attributes that are defined with the Invisible mode.

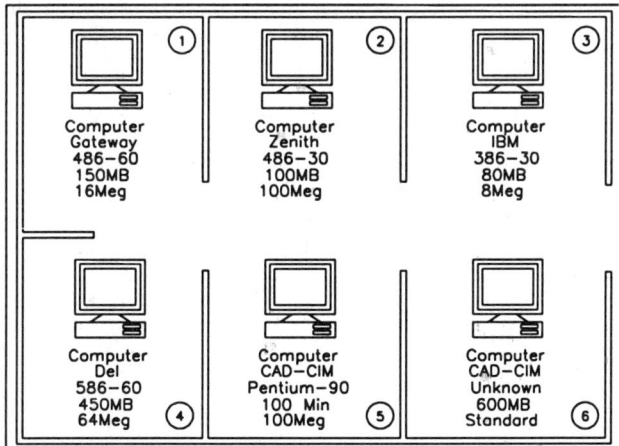

Figure 18-15 Using the ATTDISP command to make RAM attribute values visible

If you select OFF, all attribute values will become invisible. Similarly, if you enter N (Normal), AutoCAD will display the attribute values the way they are defined; that is, the attributes that were defined invisible will stay invisible and the attributes that were defined visible will become visible. In Example 2, the RAM attribute was defined with the Invisible mode; therefore, the

RAM values are not displayed with the block. If you want to make the RAM attribute values visible (Figure 18-15), enter the ATTDISP command and then turn it ON.

EDITING ATTRIBUTES (DDATTE COMMAND)

Toolbar:	Modify II, Edit Attribute
Pull-down:	Modify, Object, Attribute, Single...
Command:	DDATTE

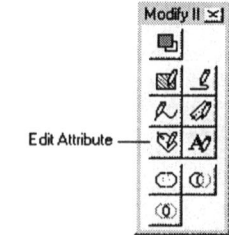

Figure 18-16 Attribute toolbar

The **DDATTE** command allows you to edit the block attribute values through the **Edit Attributes** dialog box. When you enter this command, AutoCAD prompts you to select the block whose values you want to edit. After selecting the block, the Edit Attribute dialog box is displayed. The dialog box shows the prompts and the attribute values of the selected block. If an attribute has been defined with Constant mode, it is not displayed in the dialog box because a constant attribute value cannot be edited. To make any changes, select the edit box and enter the new value. After you select the **OK** button, the attribute values are updated in the selected block.

   Command: **DDATTE**
   Select block: *Select a block with attributes*.

If the selected block has no attributes, AutoCAD will display the alert message **Block has no attributes**. Similarly, if the selected object is not a block, AutoCAD again displays the alert message **Object is not a block**.

> **Note**
> *You cannot use the DDATTE command to do global editing of attribute values.*
>
> *You cannot use the DDATTE command to modify position, height, or style of the attribute value.*

Example 4

In this example you will use the DDATTE command to change the attribute of the first computer (16MB to 16 Meg), which is located in Room-1.

1. Load the drawing that was created in Example 2. The drawing has six blocks with attributes. The name of the block is COMP, and it has six defined attributes, one of them invisible. Zoom in so that the first computer is displayed on the screen (Figure 18-17).

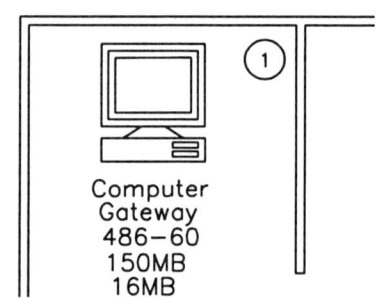

Figure 18-17 Zoomed view of the first computer

2. At the AutoCAD Command: prompt, enter the DDATTE command. AutoCAD will prompt you to select a block. Select the block, first computer located in Room-1.

   Command: **DDATTE**
   Select block: *Select a block.*

   AutoCAD will display the **Edit Attributes** dialog box (Figure 18-18), which shows the attribute prompts and the attribute values.

3. Edit the values, and select the OK button in the dialog box. When you exit the dialog box, the attribute values are updated.

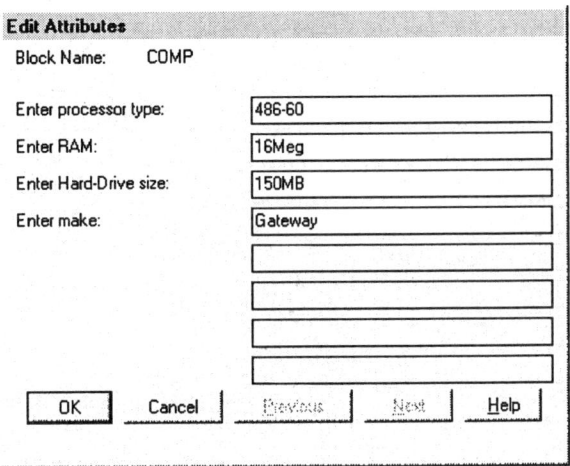

Figure 18-18 Editing attribute values using the Edit Attributes dialog box

EDITING ATTRIBUTES (ATTEDIT COMMAND)

Pull-down:	Modify, Object, Attribute, Global
Command:	ATTEDIT

The **ATTEDIT** command allows you to edit the attribute values independently of the blocks that contain the attribute reference. For example, if there are two blocks, COMPUTER and TABLE, with the attribute value PRICE, you can globally edit this value (PRICE) independently of the block that references these values. You can also edit the attribute values one at a time. For example, you can edit the attribute value (PRICE) of the block TABLE without affecting the value of the other block, COMPUTER.

Global Editing of Attributes

When you enter the ATTEDIT command, AutoCAD displays the following prompt:

   Command: **ATTEDIT**
   Edit attributes one at a time? <Y>: **N**

Global edit of attribute values

If you enter N at this prompt, it means that you want to do the global editing of the attributes. However, you can restrict the editing of attributes by block names, tag names, attribute values, and visibility of attributes on the screen.

Editing Visible Attributes Only

After you select global editing, AutoCAD will display the following prompt:

Edit only attributes visible on screen? <Y>: **Y**

If you enter Y at this prompt, AutoCAD will edit only those attributes that are visible and displayed on the screen. The attributes might have been defined with the Visible mode, but if they are not displayed on the screen they are not visible for editing. For example, if you zoom in, some of the attributes may not be displayed on the screen. Since the attributes are not displayed on the screen, they are invisible and cannot be selected for editing.

Editing All Attribute

If you enter N at the earlier-mentioned prompt, AutoCAD flips from graphics to text screen and displays the following message on the screen:

Drawing must be regenerated afterwards

Now, AutoCAD will edit all attributes even if they are not visible or displayed on the screen. Also, changes that you make in the attribute values are not reflected immediately. Instead, the attribute values are updated and the drawing regenerated after you are done with the command.

Editing Specific Blocks

Although you have selected global editing, you can confine the editing of attributes to specific blocks by entering the block name at the prompt. For example:

Block name specification<*>: **COMP**

When you enter the name of the block, AutoCAD will edit the attributes that have the given block (COMP) reference. You can also use the wild-card characters to specify the block names. If you want to edit attributes in all blocks that have attributes defined, press Enter.

Editing Attributes with Specific Attribute Tag Names

Like blocks, you can confine attribute editing to those attribute values that have the specified tag name. For example, if you want to edit the attribute values that have the tag name MAKE, enter the tag name at the following AutoCAD prompt:

Attribute tag specification<*>: **MAKE**

When you specify the tag name, AutoCAD will not edit attributes that have a different tag name, even if the values being edited are the same. You can also use the wild-card characters to specify the tag names. If you want to edit attributes with any tag name, press Enter.

Editing Attributes with a Specific Attribute Value

Like blocks and attribute tag names, you can confine attribute editing to a specified attribute value. For example, if you want to edit the attribute values that have the value 100MB, enter the value at the following AutoCAD prompt:

Attribute value specification<*>: **100MB**

When you specify the attribute value, AutoCAD will not edit attributes that have a different value, even if the tag name and block specification are the same. You can also use the wild-card characters to specify the attribute value. If you want to edit attributes with any value, press Enter.

Sometimes the value of an attribute is null, and these values are not visible. If you want to select the null values for editing, make sure you have not restricted the global editing to visible attributes. To edit the null attributes, enter \ at the following prompt:

Attribute value specification<*>: \

After you enter this information, AutoCAD will prompt you to select the attributes. You can select the attributes by selecting individual attributes or by using one of the object selection options (Window, Crossing, etc).

Select Attributes: *Select the attribute values.*

After you select the attributes, AutoCAD will prompt you to enter the string you want to change and the new string. AutoCAD will retrieve the attribute information, edit it, and then update the attribute values.

String to change:
New string:

The following is the complete command prompt sequence of the **ATTEDIT** command. It is assumed that the editing is global and for visible attributes only.

Command: **ATTEDIT**
Edit attributes one at a time?<Y>: **N**
Global edit of attribute values
Edit only attributes visible on screen?<Y>: **Y**
Block name specification<*>:
Attribute tag specification<*>:
Attribute value specification<*>:
String to change:
New string:

18-20 Customizing AutoCAD

Example 5

In this example, you will use the drawing from Example 2 to edit the attribute values that are **highlighted** in the following table. The tag names are given at the top of the table (ITEM, MAKE, PROCESSOR, HD, RAM). The RAM values are invisible in the drawing.

ITEM	MAKE	PROCESSOR	HD	RAM
COMP Computer	Gateway	486-60	150MB	**16MB**
COMP Computer	Zenith	486-30	100MB	**32MB**
COMP Computer	IBM	386-30	80MB	**8MB**
COMP Computer	Del	586-60	450MB	**64MB**
COMP Computer	**CAD-CIM**	Pentium-90	100 Min	**32MB**
COMP Computer	**CAD-CIM**	**Unknown**	600MB	Standard

Make the following changes in the **highlighted** attribute values (Figure 18-19).

1. Change Unknown to Pentium.
2. Change CAD-CIM to Compaq.
3. Change MB to Meg for all attribute values that have the tag name RAM. (No changes should be made to the values that have the tag name HD.)

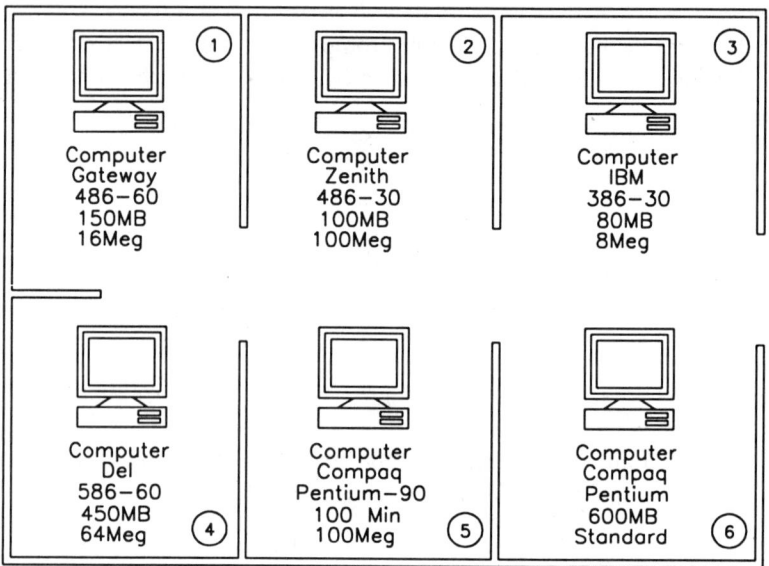

Figure 18-19 Using ATTEDIT to change the attribute values

The following is the prompt sequence to change the attribute value **Unknown** to **Pentium**.

1. Enter the **ATTEDIT** command at the Command: prompt. At the next prompt, enter N (No).

   Command: **ATTEDIT**
   Edit attributes one at a time? <Y>: **N**

Global edit of attribute values

2. We want to edit only those attributes that are visible on the screen, so press Enter at the following prompt:

   Edit only attributes visible on the screen? <Y>: ↵

3. As shown in the table, the attributes belong to a single block, COMP. In a drawing, there could be more blocks. To confine the attribute editing to the COMP block only, enter the name of the block (COMP) at the next prompt.

   Block name specification <*>: COMP

4. At the next two prompts, enter the attribute tag name and the attribute value specification. When you enter these two values, only those attributes that have the specified tag name and attribute value will be edited.

   Attribute tag specification <*>: **Processor**
   Attribute value specification <*>: **Unknown**

5. Next, AutoCAD will prompt you to select attributes. Use the Crossing option to select all blocks. AutoCAD will search for the attributes that satisfy the given criteria (attributes belong to the block COMP, the attributes have the tag name Processor, and the attribute value is Unknown). Once AutoCAD locates such attributes, they will be highlighted.

   Select Attributes:

6. At the next two prompts, enter the string you want to change, and then enter the new string.

   String to change: **Unknown**
   New string: **Pentium**

7. The following is the command prompt sequence to change the make of the computers from **CAD-CIM** to **Compaq**.

   Command: **ATTEDIT**
   Edit attributes one at a time? <Y>: **N**
   Global edit of attribute values
   Edit only attributes visible on the screen? <Y>: ↵
   Block name specification <*>: **COMP**
   Attribute tag specification <*>: **MAKE**
   Attribute value specification <*>:
   Select Attributes:
   String to change: **CAD-CIM**
   New string: **Compaq**

8. The following is the command prompt sequence to change **MB** to **Meg**.

   Command: **ATTEDIT**
   Edit attributes one at a time?<Y>: **N**
   Global edit of attribute values

   At the next prompt, you must enter N because the attributes you want to edit (tag name, RAM) are not visible on the screen.

   Edit only attributes visible on the screen?<Y>: **N**
   Drawing must be regenerated afterwards
   Block name specification<*>: **COMP**

   At the next prompt, about the tag specification, you must specify the tag name because the text string MB also appears in the hard drive size (tag name, HD). If you do not enter the tag name, AutoCAD will change all MB attribute values to Meg.

   Attribute tag specification<*>: **RAM**
   Attribute value specification<*>:
   Select Attributes:
   String to change: **MB**
   New string: **Meg**

9. Use the **ATTDISP** command to display the invisible attributes on the screen.

   Command: **ATTDISP**
   Normal/ON/OFF/<current>: **ON**

Individual Editing of Attributes

The **ATTEDIT** command can also be used to edit the attribute values individually. When you enter this command, AutoCAD will prompt **Edit attributes one at a time? <Y>:**. At this prompt, press Enter to accept the default or enter Y. The next three prompts are about block specification, attribute tag specification, and attribute value specification, which were discussed in the previous section. These options let you limit the attributes for editing. For example, if you specify a block name, AutoCAD will limit the editing to those attributes that belong to the specified block. Similarly, if you also specify the tag name, AutoCAD will limit the editing to the attributes in the specified block and with the specified tag name.

   Command: **ATTEDIT**
   Edit attributes one at a time?<Y>: ↵
   Block name specification<*>: ↵
   Attribute tag specification<*>: ↵
   Attribute value specification<*>: ↵
   Select Attributes:

At the **Select Attributes:** prompt, select the objects by choosing the objects or by using an object selection option such as Window, Crossing, WPolygon, CPolygon, or Box. By using these options you can further limit the attribute values selected for editing. After you select the objects, AutoCAD will mark the first attribute it can find with an X. The next prompt is:

Value/Position/Height/Angle/Style/Layer/Color/Next<N>:

Value. The Value option lets you change the value of an attribute. To change the value, enter V at this prompt. AutoCAD will display the following prompt:

Change or Replace?<R>

The Change option allows you to change a few characters in the attribute value. To select the Change option, enter Change or C at the prompt. AutoCAD will display the next prompt:

String to change:
New string:

At the **String to change:** prompt, enter the characters you want to change and press Enter. At the next prompt, **New string:**, enter the new string.

> **Note**
> *You can use ? and * in the string value. When these characters are used in string values, AutoCAD does not interpret them as wild-card characters.*

To use the Replace option, enter R or press Enter at the **Change to Replace? <R>:** prompt. AutoCAD will display the following prompt:

New Attribute value:

At this prompt, enter the new attribute value. AutoCAD will replace the string bearing the X mark with the new string. If the new attribute is null, the attribute will be assigned a null value.

Position, Height, Angle. You can change the position, height, or angle of an attribute value by entering, respectively, P, H, or A at the following prompt:

Value/Position/Height/Angle/Style/Layer/Color/Next<N>:

The Position option lets you define the new position of the attribute value. AutoCAD will prompt you to enter the new starting point, center point, or endpoint of the string. If the string is aligned, AutoCAD will prompt for two points. You can also define the new height or angle of the text string by entering, respectively, H or A at the prompt.

Layer and Color. The Layer and Color options allow you to change the layer and color of the attribute. For a color change, you can enter the new color by entering a color number (1 through 255), a color name (red, green, etc.), BYLAYER, or BYBLOCK.

18-24 Customizing AutoCAD

Example 6

In this example, you will use the drawing in Example 2 to edit the attributes individually (Figure 18-20). Make the following changes in the attribute values.

a. Change the attribute value 100 Min to 100MB.
b. Change the height of all attributes with the tag name RAM to 0.075 units.

1. Load the drawing which you had saved in Example 2.

2. At the AutoCAD Command: prompt, enter the **ATTEDIT** command. The following is the command prompt sequence to change the value of 100 Min to 100MB.

   Command: **ATTEDIT**
   Edit attributes one at a time? <Y>: ↵
   Block name specification <*>: **COMP**
   Attribute tag specification <*>: ↵
   Attribute value specification <*>: ↵
   Select Attributes: *Select the attribute*.
   Value/Position/Height/Angle/Style/Layer/Color/Next <N>: **V**
   Change or Replace? <R>: **C**
   String to change: \ **Min**
   New string: **100MB**

When AutoCAD prompts **String to change:**, enter the characters you want to change. In this example, the characters **Min** are preceded by a space. If you enter a space, AutoCAD displays the next prompt, **New string:**. If you need a leading blank space, the character string must start with a backslash (\), followed by the desired number of blank spaces.

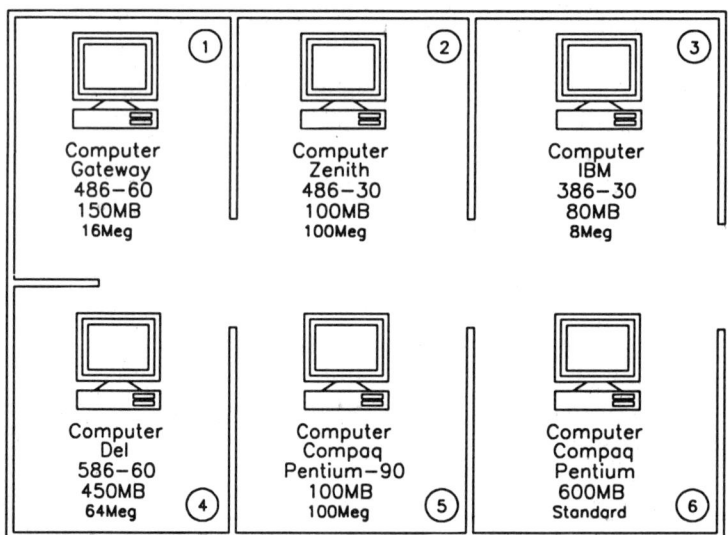

Figure 18-20 Using ATTEDIT to change the attribute values individually

3. To change the height of the attribute text, enter the ATTEDIT command as just shown. When AutoCAD displays the following prompt, enter H for height.

   Value/Position/Height/Angle/Style/Layer/Color/Next <N>: **H**
   New Height: 0.075

After you enter the new height and press Enter, AutoCAD will change the height of the text string that has the X mark. AutoCAD will then repeat the last prompt. Use the **Next** option to move the X mark to the next attribute. To change the height of other attribute values, repeat these steps.

INSERTING TEXT FILES IN THE DRAWING
Using MTEXT Command

Toolbar:	Draw, Multiline Text
Pull-down:	Draw, Text, Multiline Text
Command:	MTEXT

You can insert a text file by selecting the **Import Text** option in the **Multiline Text Editor** dialog box. You can invoke this dialog box by using the MTEXT command.

   Command: MTEXT

Next, AutoCAD prompts you to enter the insertion point and other corner of the paragraph text box. After you enter these points, the **Multiline Text Editor** dialog box appears on screen. To insert the text file COMPLST1.txt (created in Example 3), choose the **Import Text** button. AutoCAD displays the **Open** dialog box (Figure 18-21).

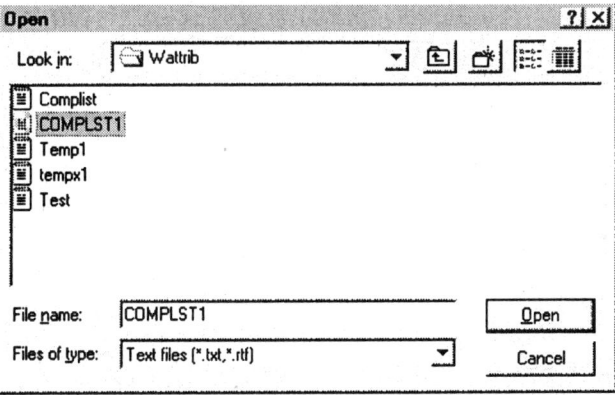

Figure 18-21 Open dialog box

In this dialog box, you can select the text file COMPLST1 and then choose the **Open** button. The imported text is displayed in the text area of the Multiline Text Editor dialog box (Figure 18-22). Note that only ASCII files are properly interpreted.

18-26 Customizing AutoCAD

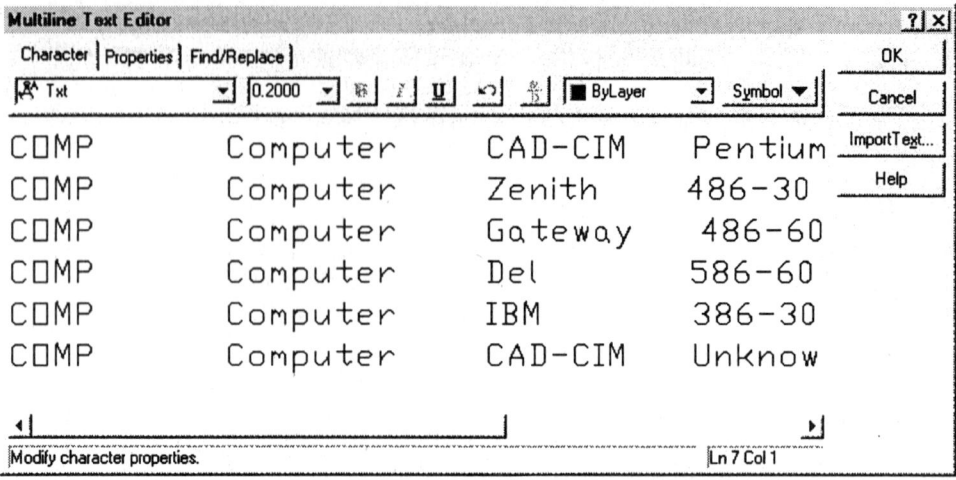

Figure 18-22 Multiline Text Editor dialog box displaying the imported text

Now choose the OK button to get the imported text in the selected area on the screen (Figure 18-23).

```
COMP    Computer    CAD-CIM    Pentium-90    100Min    32MB
COMP    Computer    Zenith     486-30        100MB     32MB
COMP    Computer    Gateway    486-60        150MB     16MB
COMP    Computer    Del        586-60        450MB     64MB
COMP    Computer    IBM        386-30        80MB      8MB
COMP    Computer    CAD-CIM    Unknown       600MB     Standard
```

Figure 18-23 Imported text file on the screen

You can also use the Multiline Text Editor dialog box to change the text style, height, direction, width, rotation, and attachment.

REVIEW QUESTIONS

1. Give two major uses of defining block attributes._____

2. The information associated with a block is known as _____ or _____.

3. You can define the block attributes by entering _____ at the **Command:** prompt.

4. The most convenient way to define the block attributes is by using the **Attribute Definition** dialog box, which can be invoked by entering _____.

5. What are the options in the **Mode area** of the **Attribute Definition** dialog box?_____

6. The **Constant** option lets you create an attribute that has a fixed value and cannot be changed later. (T/F)

7. What is the function of the **Preset** option? _____

8. The attribute value is requested when you use the Preset option to define a block attribute. (T/F)

9. Name the three edit boxes in the **Attribute area** of the Attribute Definition dialog box. ____

10. The **tag** is like a label that is used to identify an attribute. (T/F)

11. The tag names can only be uppercase. (T/F)

12. The tag name cannot be null. (T/F)

13. The tag name can contain a blank space. (T/F)

14. If you select the Constant option in the Mode area of the Attribute Definition dialog box, the Prompt: edit box is _____ because no prompt is required if the attribute is _____.

15. If you do not enter anything in the Prompt: edit box, the entry made in the Tag: edit box is used as the prompt. (T/F)

16. What option, button, or check box should you select in the **Attribute Definition** dialog box to automatically place the subsequent attribute text just below the previously defined attribute?_____

17. The text style must be defined before it can be used to specify the text style. (T/F)

18. If you select a style that has the height predefined, AutoCAD automatically disables the Height < edit box in the Attribute Definition dialog box. (T/F)

19. If you have selected the Fit option for the text justification, the _____ edit box is disabled.

20. What is the difference between the **ATTDEF** and the **DDATTDEF** commands?_____

21. The _____ command lets you edit both text and attribute definitions.

18-28 Customizing AutoCAD

22. The value of the block attributes can be specified in the Edit Attribute Definition dialog box. (T/F)

23. The Edit Attribute Definition dialog box is invoked by using the _____ or _____ command with the system variable _____ set to 1.

24. If you use the Enter Attributes dialog box to define the attribute values, the Verify mode is _____ because the Enter Attributes dialog box allows the user to examine and edit the attribute values.

25. You can also define attributes from the command line by setting the system variable _____ to 0.

26. To use the Attribute Extraction dialog box for extracting the attributes, enter _____ at the Command: prompt.

27. You can select Comma Delimited File, Space Delimited File, or Drawing Interchange File. The format selection is determined by the text editor you use. (T/F)

28. In the Comma Delimited File, each character field is enclosed in _____ and each record is separated by a _____.

29. The **Verify** option allows you to verify the attribute value you have entered when inserting a block. (T/F)

30. Unlike a constant attribute, the **Preset** attribute cannot be edited. (T/F)

31. For tag names, any lowercase letters are automatically converted to uppercase. (T/F)

32. The entry in the **Value:** edit box of the Attribute Definition dialog box defines the _____ of the specified attribute.

33. If you have selected the **Align** option for the text justification, the Height < and Rotation < edit boxes are _____.

34. You can use the _____ command to edit text or attribute definitions.

35. The default value of the **ATTDIA** variable is _____, which disables the dialog box.

36. In the **Space Delimited File**, the records are of fixed width as specified in the _____ file.

37. In the _____ File, the records are not separated by a comma and the character fields are not enclosed in single quotes.

38. You must not use the _____ character in template files. Any alignment in the fields must be done by inserting spaces after the field name. (T/F)

39. You cannot use the **DDATTE** command to modify the position, height, or style of the attribute value. (T/F)

40. The _____ command allows you to edit the attribute values independently of the blocks that contain the attribute reference.

41. The **ATTEDIT** command can also be used to edit the attribute values individually. (T/F)

42. You can use ? and * in the string value. When these characters are used in string values, AutoCAD does not interpret them as wild-card characters. (T/F)

43. You can insert the text file in the drawing by entering the _____ command at the AutoCAD Command: prompt.

EXERCISES

Exercise 1

In this exercise, you will define the following attributes for a resistor and then create a block using the BLOCK command. The name of the block is RESIS.

Mode	Tag name	Prompt	Default value
Verify	RNAME	Enter name	RX
Verify	RVALUE	Enter resistance	XX
Verify, Invisible	RPRICE	Enter price	00

1. Draw the resistor as shown in Figure 18-24.
2. Enter **DDATTDEF** at the AutoCAD Command: prompt to invoke the Attribute Definition dialog box.
3. Define the attributes as shown in the preceding table, and position the attribute text as shown in Figure 18-24.
4. Use the **BLOCK** command to create a block. The name of the block is RESIS, and the insertion point of the block is at the left end of the resistor. When you select the objects for the block, make sure you also select the attributes.

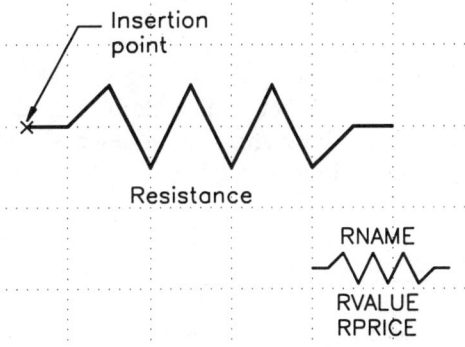

Figure 18-24 Drawing of resistor for Exercise 1

Exercise 2

In this exercise, you will use the INSERT command to insert the block that was defined in Exercise 1 (RESIS). Following is the list of the attribute values for the resistances in the electric circuit.

RNAME	RVALUE	RPRICE
R1	35	0.32
R2	27	0.25
R3	52	0.40
R4	8	0.21
RX	10	0.21

1. Draw the electric circuit diagram as shown in Figure 18-25 (assume the dimensions).

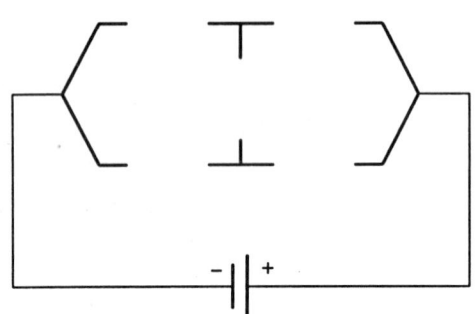

Figure 18-25 Drawing of the electric circuit diagram without resistors for Exercise 2

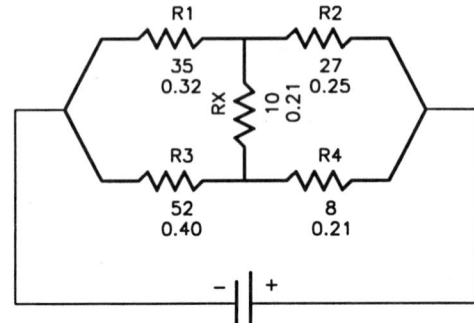

Figure 18-26 Drawing of the electric circuit diagram with resistors for Exercise 2

2. Set the system variable **ATTDIA** to 1. Use the **INSERT** command to insert the blocks, and define the attribute values in the **Enter Attributes** dialog box.
3. Repeat the **INSERT** command to insert other blocks, and define their attribute values as given in the table. Save the drawing as **ATTEXR2.DWG**.

Exercise 3

In this exercise, you will write a template file for extracting the attribute values as defined in Exercise 2. These attribute values must be written to a file **RESISLST.TXT** and arranged as shown in the following table.

	Field width in characters		
< 10 >	< 10 >	< 10 >	< 10 >
RESIS	R1	35	0.32
RESIS	R2	27	0.25
RESIS	R3	52	0.40
RESIS	R4	8	0.21
RESIS	RX	10	0.21

1. Load the drawing **ATTEXR2** that you saved in Exercise 2.
2. Use the Windows Notepad to write the template file. After writing the file, save it as an ASCII file under the file name **TEMP2.TXT**.
3. In the AutoCAD screen, use the **DDATTDEF** command to invoke the Attribute Extraction dialog box, and select the Space Delimited File (SDF) radio button.
4. Select the objects (blocks). You can also select the objects by using the Window or Crossing option.
5. In the Template File edit box, enter the name of the template file, **TEMP2.TXT**.
6. In the Output File edit box, enter the name of the output file, **RESISLST.TXT**.
7. Select the OK button in the Attribute Extraction dialog box.
8. Use the Windows Notepad to list the output file, **RESISLST.TXT**. The output file should be similar to the file shown in the beginning of Exercise 3.

Exercise 4

In this exercise, you will use the DDATTE or ATTEDIT command to change the attributes of the resistances that are highlighted in the following table. You will also extract the attribute values and insert the text file in the drawing.

1. Load the drawing **ATTEXR2** that was created in Exercise 2. The drawing has five resistances with attributes. The name of the block is RESIS, and it has three defined attributes, one of them invisible.
2. Use the AutoCAD DDATTE or ATTEDIT command to edit the values that are **highlighted** in the following table.

RESIS	R1	**40**	0.32
RESIS	R2	**29**	0.25
RESIS	R3	52	**0.45**
RESIS	R4	8	**0.25**
RESIS	**R5**	10	0.21

3. Extract the attribute values, and write the values to a text file.
4. Use the MTEXT command to insert the text file in the drawing.

Exercise 5

Use the information given in Exercise 3 to extract the attribute values, and write the data to the output file. The data in the output file should be Comma Delimited CDF. Use the DDATTEXT and ATTEXT commands to extract the attribute values.

Exercise 6

In this exercise, you will draw the circuit diagram as shown in Figure 18-27, define the attributes, and then extract the attributes to create a bill of materials.

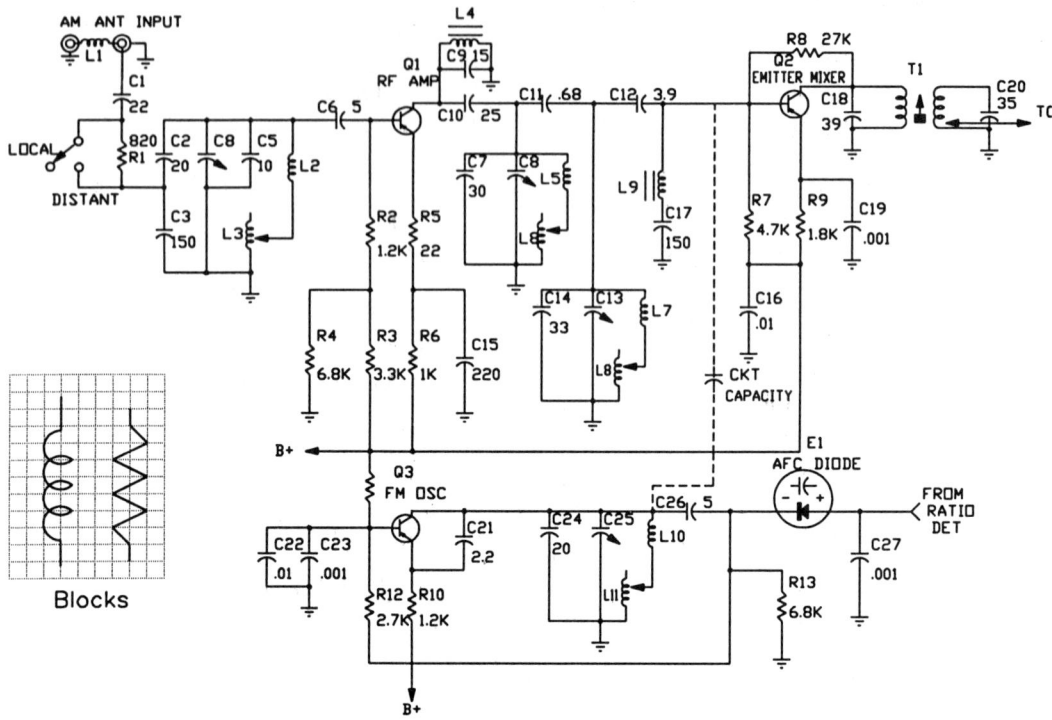

Figure 18-27 Drawing of the circuit diagram for Exercise 6

Chapter 19

Rendering

Learning objectives

After completing this chapter, you will be able to:
♦ *Understand rendering and why to render the objects.*
♦ *Configure, load, and unload AutoCAD Render.*
♦ *Insert and modify light sources.*
♦ *Select rendering type and render objects.*
♦ *Define and render scenes.*
♦ *Modify scenes.*
♦ *Use AutoCAD Render light sources.*
♦ *Attach materials to objects.*
♦ *Define materials and modify existing materials.*
♦ *Replay and print renderings.*

RENDERING

A rendered image makes it easier to visualize the shape and size of a three-dimensional (3D) object, compared to a wireframe image or a shaded image. A rendered object also makes it easier to express your design ideas to other people. For example, if you want to make a presentation of your project or a design, you do not need to build a prototype. You can use the rendered image to explain your design much more clearly because you have complete control over the shape, size, color, and surface material of the rendered image. Additionally, any required changes can be incorporated into the object, and the object can be rendered to check or demonstrate the effect of those changes. Thus, rendering is a very effective tool for communicating ideas and demonstrating the shape of an object. You can create a rendered image of a 3D object by using **AutoCAD's RENDER Command**. It allows you to control the appearance of the object by defining the surface material and reflective quality of the surface and by adding lights to get the desired effects.

Determining which Sides are to be Rendered in a Model
Some of the faces of a 3D model, such as the back faces and the hidden faces, need not be rendered since it would amount to an unnecessary waste of time. In the process of rendering, AutoCAD determines the front faces and the back faces of the 3D model with the use of **normals** on each face. A **vector** perpendicular to each face on a 3D model and whose direction is outward toward space is known as **normal**. If a face has been drawn in the clockwise direction, then the normal points inward; if a face has been drawn counterclockwise, the normal points outward.

Now, depending on the location of your viewpoint, if the normal of a face points away from the viewpoint, the face is a back face. As mentioned before, rendering of such faces is not advisable (since those faces are not visible from the viewpoint) and can be avoided by invoking the Discard Back Faces option (explained later in this chapter). The faces that conceal other faces are discarded. In this manner, the time required to render objects can be decreased by discarding faces that need not be rendered.

Points to be Remembered while Defining a Model

1. To make the rendering process as time-efficient as possible, you must use the fewest possible faces to define a plane.

2. There should be consistency in your drafting technique. Models formed of a complex mixture of faces, extruded lines, and wireframe meshes should be avoided.

3. If you are rendering circles, ellipses, or arcs, set the **VIEWRES** command to a high number. This way, the circles, arcs, or ellipses will appear smoother and the rendering of such objects will be better. But increasing the value of **VIEWRES** increases the time taken to render the objects. The smoothness of rendered curved solids depends on **FACETRES** variable.

4. If you are using the Smooth Shading option (available in the **Render** and **Rendering Preferences** dialog box, explained later), then you should specify the mesh density in such a manner that the angle described by normals of any two adjoining faces is less than 45 degrees. This is because if the angle is greater than 45 degrees, then after rendering, an edge is displayed between the faces even when the Smooth Shading option is active.

LOADING AND UNLOADING AUTOCAD RENDER

When you select any AutoCAD **RENDER** command, AutoCAD Render is loaded automatically. AutoCAD will display the following message in the command prompt area:

Command: **RENDER**
Loading Landscape Object module.
Initializing Render...
Initializing preferences... done

If you do not need AutoCAD Render, you can unload it by entering the **ARX** command at the Command: prompt. AutoCAD acknowledges the unloading of Render by issuing the message **"Render successfully unloaded."** After unloading Render, you can reload AutoCAD Render by invoking the **RENDER** command or any other command associated with rendering (such as SCENE or LIGHT).

Command: **ARX**
Enter an option (?/Load/Unload/Options): **U**
Unload ARX file: **RENDER**
render successfully unloaded.

ELEMENTARY RENDERING

Toolbar:	Render, Render
Pull-down:	View, Render, Render
Command:	RENDER

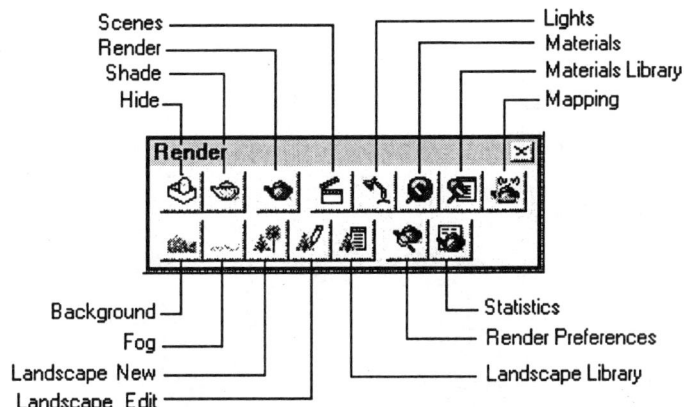

Figure 19-1 Render toolbar

In this section you are going to perform an ordinary rendering. You will encounter many terms of which you will not be aware, but you need not bother about them at this stage. All the terms are explained later in this chapter. In this rendering, lights, materials, and other advanced features of rendering will not be applied. Only a default distant light is used. This will give you a feel of rendering in the simplest possible manner. Perform the following steps:

1. Using the SPHERE command, create four spheres on the screen as shown in Figure 19-2.
2. Invoke the **Render** dialog box (Figure 19-3).
3. In the Render dialog box, select the **Render** button. AutoCAD will render all objects that are on the screen.

After some time, all the spheres are rendered and displayed on the screen (Figure 19-4). Press any key to exit the rendered image screen and return to the AutoCAD drawing screen.

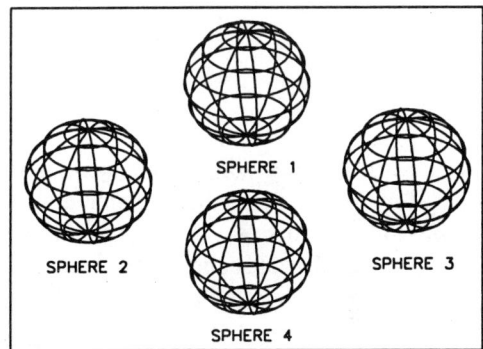

Figure 19-2 Drawing to be rendered

19-4 Customizing AutoCAD

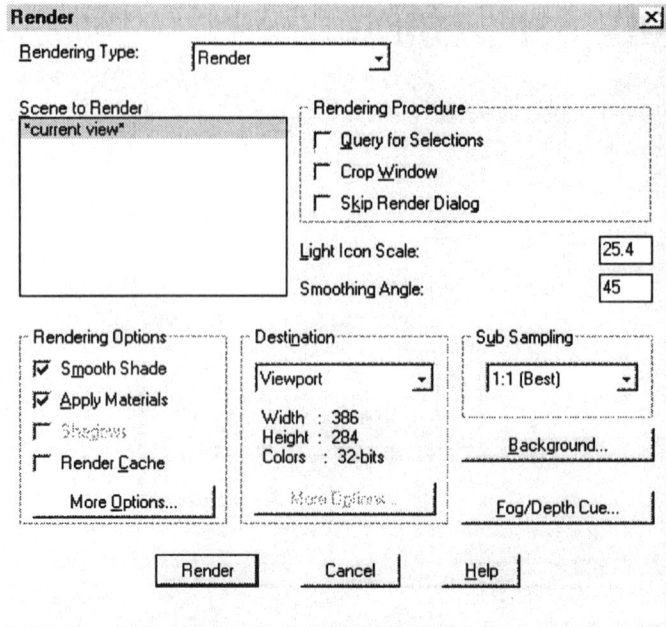

Figure 19-3 Render dialog box

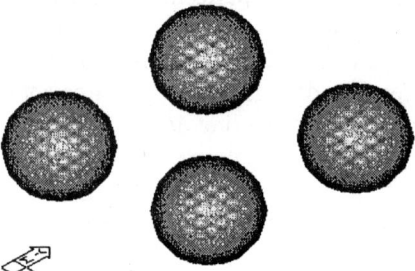

Figure 19-4 Rendering all spheres

If you want to render only some spheres (for example, SPHERE 2 and SPHERE 3), select the **Query for Selections** in the **Render** dialog box. When you select the Render button, AutoCAD will prompt you to select the objects that you want to render. Select the objects and press Enter; you will notice that only the selected spheres are rendered and displayed on the screen.

Exercise 1

In this exercise you will render the drawing that was created in Example 2 of Chapter 23, Solid Modeling. Load and render the drawing. Figure 19-5 shows the object before rendering; Figure 19-6 shows the object after rendering.

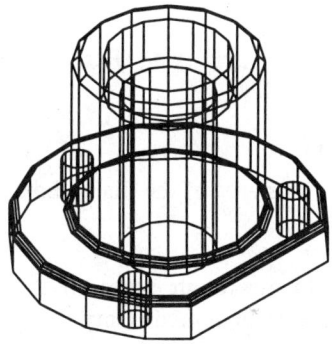

Figure 19-5 Drawing for Exercise 1

Figure 19-6 After rendering

SELECTING DIFFERENT PROPERTIES FOR RENDERING

Toolbar:	Render, Preferences
Pull-down:	View, Render, Preferences
Command:	RPREF

AutoCAD Render allows you to select various properties for the rendering. This can be achieved through the **Rendering Preferences** dialog box (Figure 19-7). The various sections of the dialog box are described next.

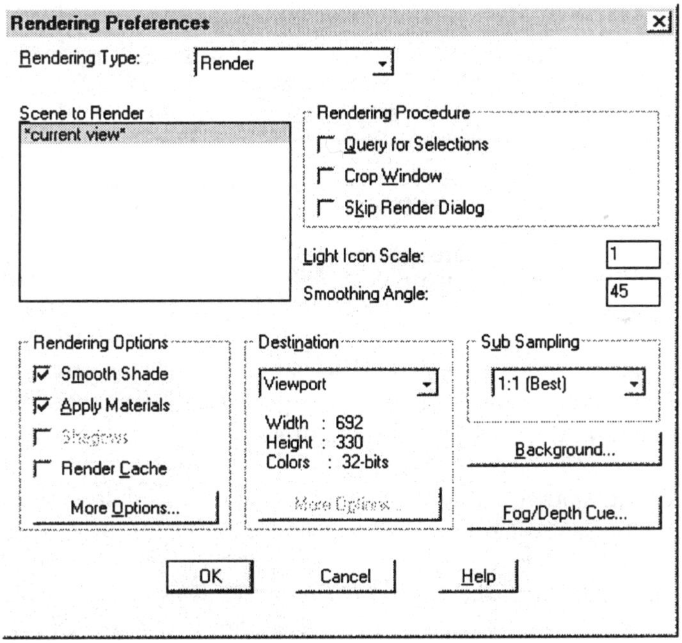

Figure 19-7 Rendering Preferences dialog box

Rendering Type

In the Rendering Preferences dialog box you can select the type of rendering (Render, Photo Real, Photo Raytrace) you want from the **Rendering Type** drop-down list.

Rendering Options

Various rendering options are provided in the Rendering Options area of the dialog box. These are described next.

Smooth Shade. The **Smooth Shade** allows you to smooth the rough edges. If this option is enabled, the rough-edged appearance of a multifaceted surface is smoothed. Only polygon meshes are affected by Smooth Shading. The surface normals are determined, and colors across two or more adjoining faces are blended.

Apply Materials. The Apply Materials option allows you to assign surface materials to objects. If this option is disabled, then the objects in the drawing are assigned the *GLOBAL* material.

Shadows. When you select this option, AutoCAD generates shadows. This option applies only to Photo Real and Photo Raytrace rendering.

Render Cache. Selecting this option results in writing rendering information to a cache file on the hard disk that can be used for subsequent renderings. This eliminates the need for AutoCAD to recalculate (tessellate) the objects for rendering. This saves time, especially when rendering solids.

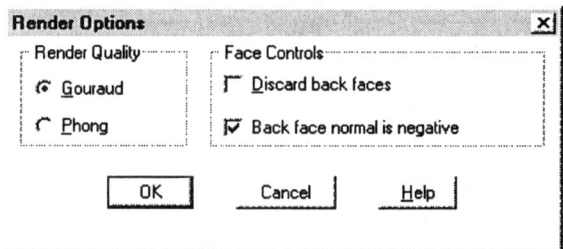

Figure 19-8 Render Options dialog box

More Options. If you select the More Options button, the **Render Options** dialog box (Figure 19-8) is displayed. The following is a brief description of these additional options.

    Render Quality Area. In the **Render Quality** area of this dialog box, you can choose either the Gouraud or the Phong option. These options determine the quality of shading that will be used if you have activated the Smooth Shading option.

    Phong. The **Phong** option creates a high-quality rendered image. The light intensity is calculated for all the pixels, and hence, more accurate highlights are generated.

    Gouraud. On the other hand, the **Gouraud** option produces a lower-quality rendered image. Light intensity at each vertex is determined, and intermediate intensities are interpolated. The advantage of the Gouraud render option is that the rendering occurs faster as compared with the Phong render option.

Face Controls Area. The **Face Controls** area governs the definition of faces of a 3D solid.

Discard Back Faces. If you select the **Discard Back Faces** option, the back faces of a 3D solid object are not taken into consideration (i.e., they are made invisible), and hence, they are omitted from the calculations for the rendering. Rendering time can be reduced by enabling this option.

Back Face Normal is Negative. The **Back Face Normal is Negative** option can be used to define the back faces in a drawing. If this option is enabled, the faces with negative normal vectors are treated as back faces and discarded. When this option is disabled, the selection faces treated as back faces by AutoCAD are just reversed.

Rendering Procedures

The **Rendering Procedure** of the Rendering Preferences dialog box has the following options:

Query for Selections. If you select the **Query for Selections** option, AutoCAD prompts you to select objects to render.

Crop Window. When you select this option, AutoCAD will prompt you to select a window. Only the objects inside the specified window are rendered.

Skip Render Dialog. In the Rendering Procedure area, if the **Skip Render Dialog** option is invoked, the current view is rendered without displaying the Render dialog box.

Light Icon Scale. The **Icon Scale** edit box can be used to set the size of the light blocks in the drawing.

Smoothing Angle. The **Smoothing Angle** option lets you specify the angle defined by two edges. The default value for smoothing an angle is 45 degrees. Angles of less than 45 degrees are smoothed. Angles greater than 45 degrees are taken as edges.

Destination

The **Destination** area of the Rendering Preferences dialog box allows you to specify the destination for the rendered image output.

Viewport Option. If you select the **Viewport** option, AutoCAD renders to the current viewport.

Render Window Option. This option renders to the AutoCAD for Windows Render window.

File Option. The **File** option lets you output the rendered image to a file.

More Options. The **More Options** button can be used to set the configuration for the output file through the **File Output Configuration** dialog box (Figure 19-9).

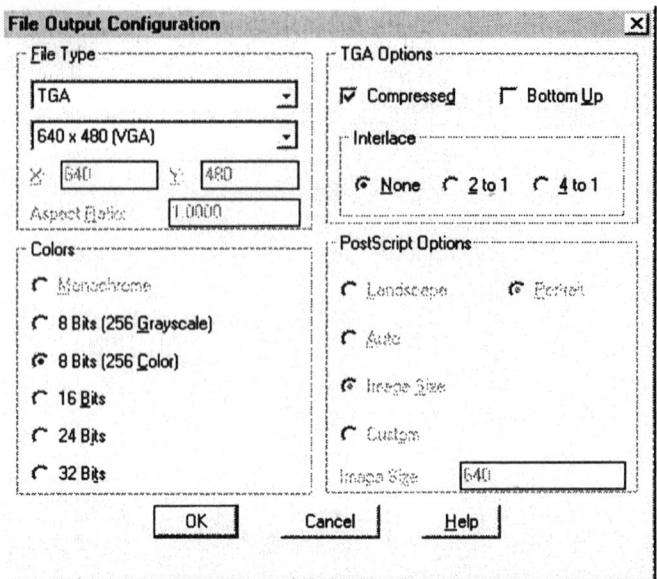

Figure 19-9 File Output Configuration dialog box

File Type Area. In this area you can specify the output file type and rendering resolution. The file formats allowed are TGA, PCX, SUN, FITS, PostScript, TIFF, FAX G III, and IFF. The screen resolution can also be specified in this area. The aspect ratio of the output file can be specified in the **Aspect Ratio** edit box.

Colors Area. The colors in the output file can be specified in this area.

TGA Options Area. The **Compressed** option lets you specify compression for those file types that allow compression. The **Bottom Up** option lets you specify the scan line start point as bottom left instead of top left.

Interlace Area. Selecting the **None** option turns off line interlacing. The **2 to 1** and **4 to 1** options turn on interlacing.

PostScript Options Area. The **Landscape** and **Portrait** options in this area specify the orientation of the file. The **Auto** option automatically scales the image. The **Custom** option sets the image size in pixels. The **Image Size** edit box uses the explicit image size.

Sub Sampling

The Sub Sampling edit box controls the quality and the rendering time by reducing the number of pixels to be rendered. A 1:1 ratio (default) produces the best quality rendering. If you are just testing a rendering, you can change the value to reduce rendering time.

Background

The background option allows you to add a background to a rendering. You can use solid colors, gradient colors, or an image file as a background.

Fog/Depth Cue

You can use the Fog option to add a misty effect to a rendering. You can also assign a color to the fog.

AUTOCAD RENDER LIGHT SOURCE

Lights are vital to rendering a realistic image of an object. Without proper lighting, the rendered image may not show the features the way you expect. Colors and surface reflection can be set for the lights with RGB or HLS color systems. AutoCAD Render supports the following four light sources: ambient light, point light, distant light, and spotlight.

Ambient Light

You can visualize **ambient light** as the natural light in a room that equally illuminates all surfaces of the objects. Ambient light does not have a source and hence has no location or direction. However, you can increase or decrease the intensity of ambient light or completely turn it off. Normally, you should set the ambient light to a low value because high values give a washed-out look to the image. If you want to create a dark room or a night scene, turn off the ambient light. With ambient light alone you cannot render a realistic image. Figure 19-10 shows an object illuminated by ambient light.

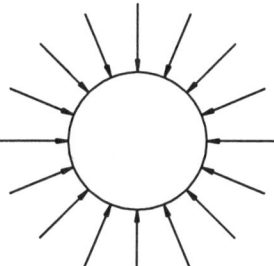

Figure 19-10 Ambient light provides constant illumination

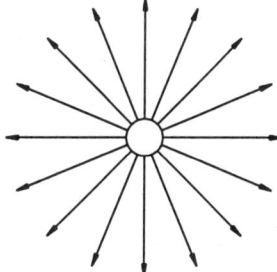

Figure 19-11 Point light emits light in all directions

Point Light

A **point light** source emits light in all directions, and the intensity of the emitted light is uniform. You can visualize an electric bulb as a point light source. In AutoCAD Render, a point source does not cast a shadow because the light is assumed to be passing **through** the object. The intensity of light radiated by a point source decreases over distance. This phenomenon is called **attenuation**. Figure 19-11 shows a light source that radiates light in all directions.

Spotlight

A spotlight emits light in the defined direction with a cone-shaped light beam (Figure 19-12). The direction of the light and the size of the cone can be specified. The phenomenon of **attenuation** (falloff) also applies to spotlights. This light is used mostly to highlight particular features and portions of the model. If you want to simulate a soft lighting effect, set the falloff cone angle a few degrees larger than the hot-spot cone angle.

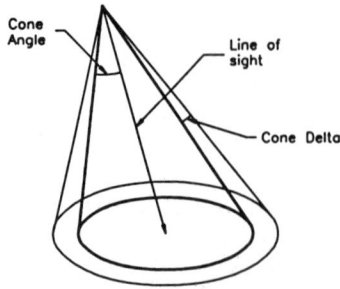

Figure 19-12 Spotlight

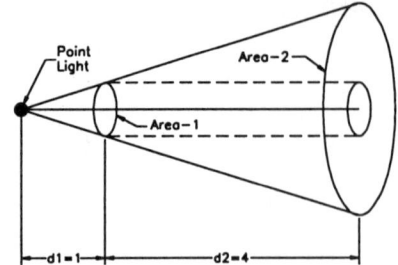

Figure 19-13 Light intensity decreases with distance

Attenuation

Light intensity is the amount of light falling per unit area. The intensity of light is directly proportional to the degree of brightness of the object. The intensity of light decreases as the distance increases. This phenomenon, called **attenuation**, occurs only with spotlights and point light. In Figure 19-13, the light is emitted by a point source. Assume that the amount of light incident on Area-1 is I. Therefore, the intensity of light on Area-1 = I/Area. As the light travels farther from the source, it covers a larger area. The amount of light falling on Area-2 is same as on Area-1, but the area is larger. Therefore, the intensity of light for Area-2 is smaller (Intensity of light for Area-2 = I/Area). Area-1 will be brighter than Area-2 because of higher light intensity. AutoCAD Render has three options for controlling the light falloff: None, Inverse Linear, and Inverse Square.

None
If you select the **None** option for light falloff, the brightness of objects is independent of distance. This means that objects that are far away from the point light source will be as bright as objects close to the light source.

Inverse Linear
In this option, the light falling on the object (brightness) is inversely proportional to the distance of the object from the light source (Brightness = 1/Distance). As distance increases, brightness decreases. For example, let us assume the intensity of the light source is **I** and the object is located at a distance of 2 units from the light source. Brightness or intensity = **I**/2. If the distance is 8 units, the intensity (light falling on the object per unit area) = **I**/8. The brightness is a linear function of the distance of the object from the light source.

Inverse Square
In this option, the light falling on the object (brightness) is inversely proportional to the square of the distance of the object from the light source (Brightness = $1/Distance^2$). For example, let us assume the intensity of the light source is **I** and the object is located at a distance of 2 units from the light source. Brightness or intensity = $I/(2)^2$ = **I**/4. If the distance is 8 units, the intensity (light falling on the object per unit area) = $I/(8)^2$ = **I**/64.

Distant Light

A **distant light** source emits a uniform parallel beam of light in a single direction only (Figure 19-14). The intensity of the light beam does not decrease with distance; it remains constant. For example, the sun's rays can be assumed to be a distant light source because the light rays are parallel. When you use a distant light source in a drawing, the location of the light source does not matter; only the **direction is critical**. Distant light is used mostly to light objects or a backdrop uniformly and for getting the effect of sunlight.

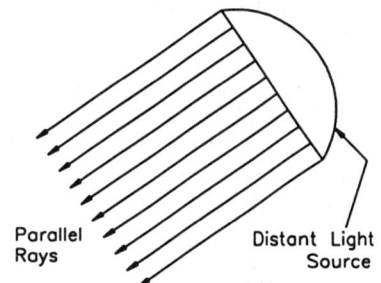

Figure 19-14 Distant light source

INSERTING AND MODIFYING LIGHTS

Toolbar:	Render, Lights
Pull-down:	View, Render, Light
Command:	LIGHT

In a rendering, the lights and lighting effects are important to create a realistic representation of an object. The sides of an object that face the light must appear brighter and the sides that are on the other side of the object must be darker. This smooth gradation of light produces a realistic image of the object. If the light intensity is uniform over the entire surface, the rendered object probably will not look realistic. For example, if you use the SHADE command to shade an object, that object does not look realistic because the displayed model lacks any gradation of light. Any number of lights can be installed in a drawing. The color, location, and direction of all the lights can be specified individually. As mentioned before, you can specify attenuation for point lights and spotlights. AutoCAD also allows you to change the color, position, and intensity of any light source. The only limitation is that light types cannot be changed. For example, you cannot change a distant light into a point light. The following sections describe how to insert, position, and modify lights.

Inserting Distant Light

As mentioned before, the lights play an important role in producing a realistic representation of a 3D model. The gradation effect produced by the lights makes the object look realistic. In this example, you will insert a distant light source and then render the object.

1. Use the **VPOINT** command to change the viewpoint to (1,-1,1).

   Command: **VPOINT** ↵
   Rotate/<viewpoint> <current>: **1,-1,1** ↵

2. Select Render from the View pull-down menu.
3. Select the **Lights...** option from the cascading menu. The **Lights** dialog box is displayed (Figure 19-15). You can use any one of the methods just described to invoke the Lights dialog box.

4. Select the **Distant Light** option from the drop-down list, and then select **New** to insert a new distant light. The **New Distant Light** dialog box is displayed [Figure 19-16(a)].
5. Enter the name of the distant light (**D1**) in the New Distant Light dialog box. Leave the intensity at its default value (1.00).
6. Select the OK button to exit this dialog box, then select OK button again to exit the Lights dialog box.
7. Now render the image.

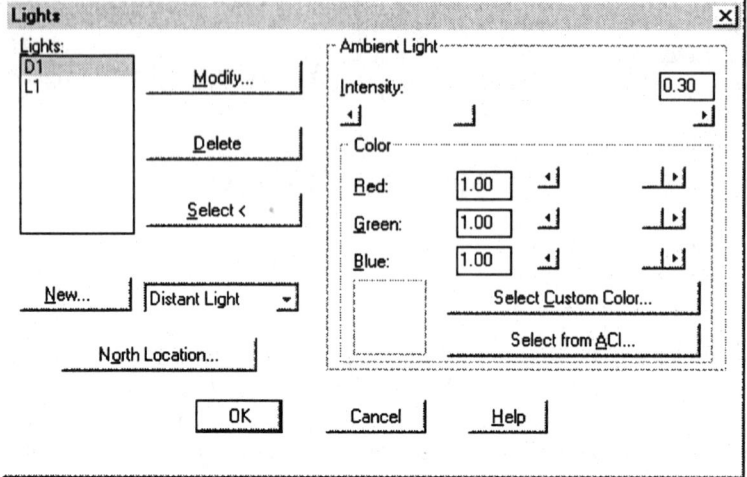

Figure 19-15 Lights dialog box

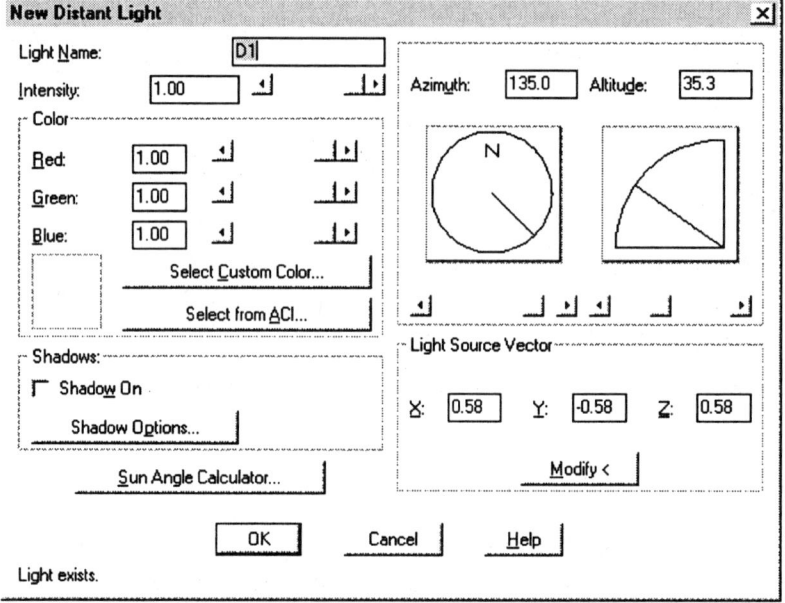

Figure 19-16(a) New Distant Light dialog box

Modifying Distant Light

In this rendering, notice that the light falls on the object from the viewpoint direction and the right front corner is brighter than the rest of the object. We would like to modify the light source so that the light falls on the object from the right at an angle. When the light source is at an angle, the top surface also receives some light. Notice that as the oblique angle increases, the amount of light reflected by the oblique surface decreases.

1. Select the **Lights** icon from the Render toolbar.
2. Select the light (D1) from the **Lights** dialog box. If there is only one light, the light is automatically highlighted.
3. Select the **Modify...** button. The **Modify Distant Light** dialog box is displayed [Figure 19-16(b)].

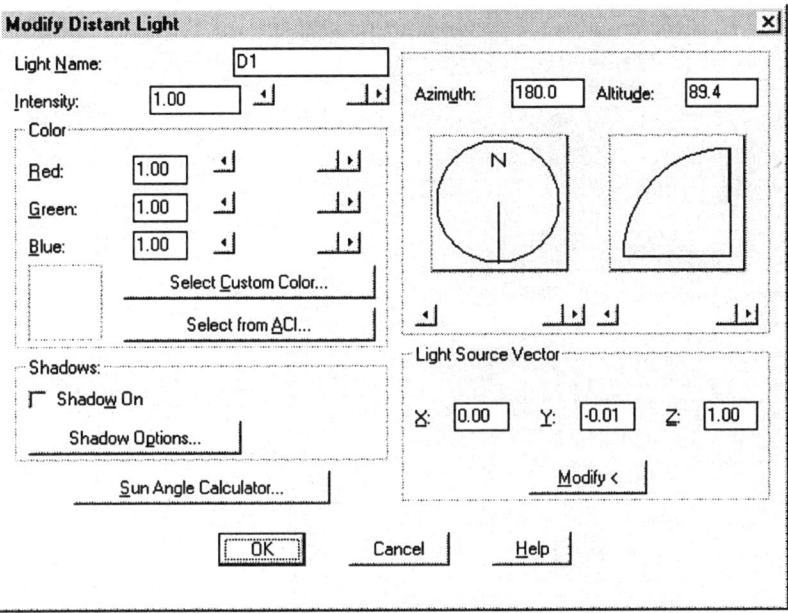

Figure 19-16(b) Modify Distant Light dialog box

Note
Double-clicking on the light name is equivalent to selecting the light name and the Modify... option.

4. Select the Sun Angle Calculator button to display the **Sun Angle Calculator** dialog box (Figure 19-17). Change the date and clock time, if needed.

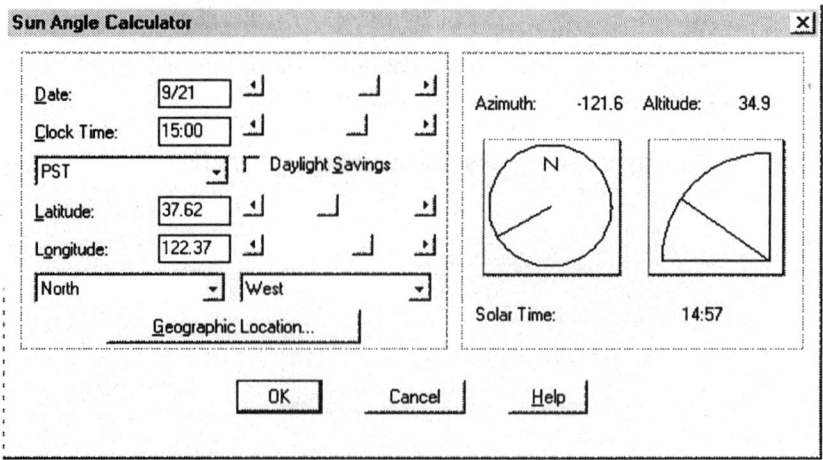

Figure 19-17 Sun Angle Calculator dialog box

5. Select the Geographic Location button to display the Geographic Location dialog box (Figure 19-18). You can specify the location by choosing the name of the city in the City list box or selecting a point in the map.

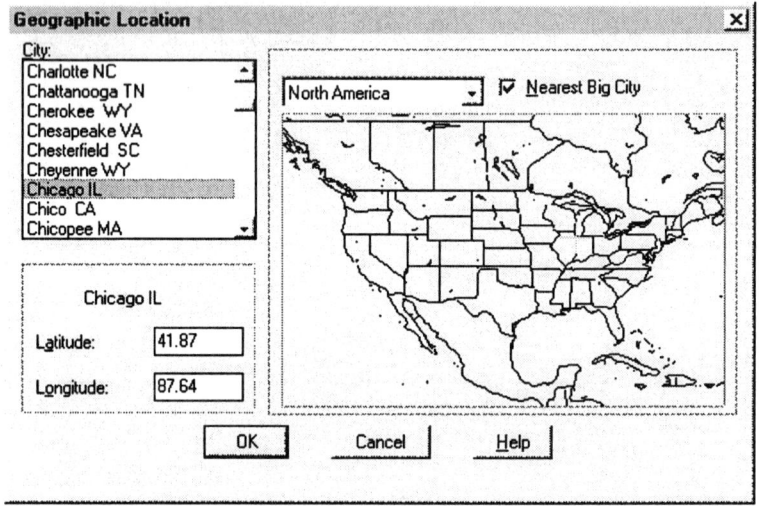

Figure 19-18 Geographic Location dialog box

6. Select **OK** to exit the Geographic Location dialog box and select OK again to exit the Sun Angle Calculator dialog box.
7. Select **OK** to exit the Modify Distant Lights dialog box and select OK again to exit the Lights dialog box.
8. Render the object.

Notice that the sides (vertical faces) of the object and the horizontal surfaces have different degrees of brightness. This is because the oblique angle that the vertical surfaces make with the direction of the light rays is smaller than the oblique angle that the horizontal surfaces make. Maximum brightness occurs when the object is perpendicular to the direction of the light rays.

Inserting Point Light

In Example 1 you will insert a point light source. Then you will render the object.

Example 1

Insert a point light source and render the object of Exercise 1 with Inverse Linear and Inverse Square options.

1. Load the drawing of Exercise 1.
2. Select the **Lights** icon from the Render toolbar.
3. Select the **Point Light** option from the Lights dialog box.
4. Select the **New...** button. The **New Point Light** dialog box is displayed.
5. Enter the name of the point light (**P1**) and set the intensity (**6.0**).
6. Select the **Modify <** option from the New Point Light dialog box.
7. Enter the light location (**-2,0,5**).
8. In the New Point Light dialog box, select **Inverse Linear**, and then select the **OK** button to exit the dialog box. Select **OK** to exit the Lights dialog box.
9. Change the viewpoint to (**1,-1,1**), and then render the object.

Notice that the top surface of the cylindrical part and the top surface of the flange are equally bright. Next, change the falloff to Inverse Square:

10. Select the **Lights** icon from the Render toolbar.
11. Select the point light (**P1**) and then select the **Modify** button.
12. The **Modify Point Light** dialog box will be displayed. Select **Inverse Square**, and then select **OK** to exit the dialog box. Select **OK** to exit the Lights dialog box.
13. Render the object again. Now, notice that the top surface of the cylindrical part is brighter than the top surface of the flange.

DEFINING AND RENDERING A SCENE

Toolbar:	Render, Scenes
Pull-down:	View, Render, Scene
Command:	SCENE

The rendering depends on the view that is current and the lights that are defined in the drawing. Sometimes the current view or the lighting setup may not be enough to show all features of an object. You might need different views of the object with a certain light configuration to show different features of the object. When you change the view or define the lights for a rendering, the previous setup is lost. You can save the rendering information by defining a **scene**. For each scene, you can assign a view and the lights. When you render a particular view, AutoCAD Render uses the view information and the lights that were assigned to that scene. It ignores the lights that were not defined in the scene. Defining scenes makes it convenient to render different views with the required lighting arrangement. The following example describes the process of defining scenes and assigning views and lights to the scenes.

Example 2

In this example, you will draw a rectangular box and a sphere that is positioned at the top of the box. Next, you will insert lights, define views and scenes, and then render the scenes. You may start a new drawing or continue with the current drawing. Before defining a scene, perform the following steps in a new drawing:

1. Draw a rectangular box with length = 3, width = 3, and height = 1.5.
2. Use the **UCS** command to define the new origin at the center of the top face.
3. Draw a sphere of 1.5 radius.
4. Move the sphere so that the bottom of the sphere is resting at the top face of the box, as shown in Figure 19-19.

Next, insert the lights. In this example, you will insert a point light and a distant light. Let us first insert a distant light source.

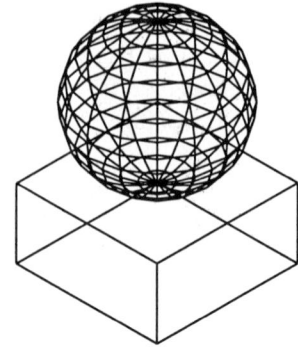

Figure 19-19 Drawing for Example 2

1. Select the **Render** option from the **View** pull-down menu and then select the **Lights...** option from the cascading menu. The **Lights** dialog box is displayed.
2. Select the **Distant** option, and then select **New** to insert a new distant light. The **New Distant Light** dialog box is displayed.
3. Enter the name of the distant light (**D1**) in the **Light Name:** edit box, and press Enter. Now select Modify from the New Distant Light dialog box.
4. Enter the light target point (direction TO) **(0,0,0)** and the light location (direction FROM) **(3,-3,4)**. Leave the intensity at 1.0.
5. Select the **OK** button to exit the dialog box. Select **OK** to exit the Lights dialog box.
6. Change the viewpoint to **1,-1,1**, and then render the object.

After inserting a distant light source, you need to insert a point light.

1. Select the **Lights** icon from the Render toolbar.
2. Select the **Point Light** option from the Lights dialog box and select the **New...** button. The **New Point Light** dialog box is displayed.
3. Enter the name of the point light (**P1**) and set the intensity **(6.0)**.
4. Select the **Modify** option from the dialog box and enter the light location **(0,0,4)**.
5. In the New Point Light dialog box, select **Inverse Linear**, and then select **OK** to exit the dialog box. Select **OK** to exit the Lights dialog box.
6. Change the viewpoint to **1,-1,1**, and then render the object.

Now you will create three scenes. The first scene, SCENE1, contains distant light D1; the second scene, SCENE2, contains point light P1. The third scene, SCENE3, will contain the view

VIEW1, distant light D1, and point light P1. Create the first scene by performing the following steps:

1. Select the **Scenes** icon from the Render toolbar.

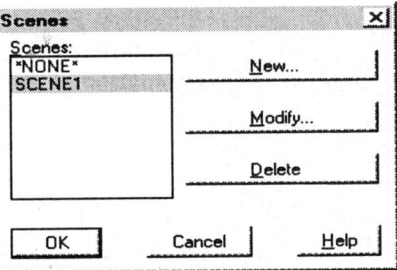

Figure 19-20 Scenes dialog box

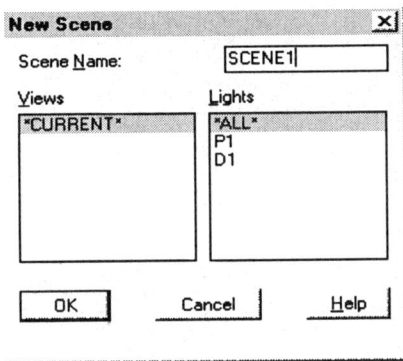

Figure 19-21 New Scene dialog box

2. Select **New...** from the Scenes dialog box (Figure 19-20). The **New Scene** dialog box (Figure 19-21) will be displayed.
3. Enter the name of the scene (SCENE1) in the **Scene Name:** edit box. Select Distant Light **D1** in the lights area, and then select the **OK** button to exit the dialog box.
4. Select the **OK** button in the Scenes dialog box.

Similarly, create the second scene, **SCENE2**, and assign the point light **P1** to the scene. Select the OK button to close the Scenes dialog box.

Next, create the third scene, **SCENE3**:

1. Use the **VPOINT** command to change the viewpoint to 1,-0.5,1 and use the **VIEW** command to save the view as **VIEW1**. Select OK.
2. Invoke the **Scenes** dialog box and select **New** from the Scenes dialog box to display the **New Scene** dialog box.
3. Enter the name of the scene (SCENE3) in the **Scene Name:** edit box. Select *ALL* to assign Distant Light D1 and Point Light P1 to the scene. Now select **VIEW1**.
4. Select the **OK** button from the Scenes dialog box.

19-18 Customizing AutoCAD

You have created three scenes, and each scene has been assigned some lights. SCENE3 has been assigned a view in addition to the lights. You can render any scene, after making the scene current, as follows:

1. Invoke the **Scenes** dialog box (Figure 19-22).
2. Select on the scene name you want to render, and then select **OK** to exit the dialog box.
3. Render the object.

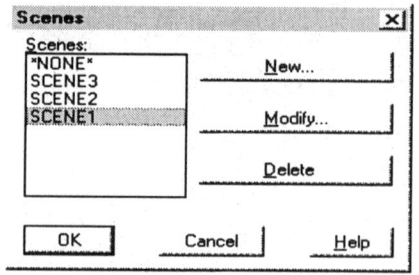

Figure 19-22 Scenes dialog box displaying the scene names

Modifying a Scene

You can modify a scene by changing the view and the lights that are assigned to that scene. When you render the object, AutoCAD Render will use the newly assigned view and lights to render the object.

1. Select the **Scenes** icon from the Render toolbar; the Scenes dialog box with the scene names is displayed on the screen.
2. Select the scene, **SCENE3**, and then select **Modify....** The **Modify Scene** dialog box (Figure 19-23) is displayed.
3. Select on **CURRENT**, and then select **OK** to exit the Modify Scene dialog box. This will assign the current view to SCENE3.
4. Select the **OK** button to exit the Scenes dialog box.
5. Render the object.

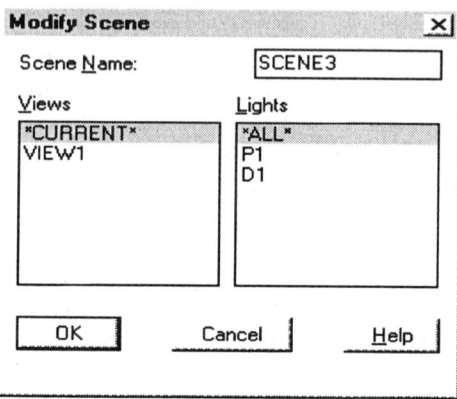

Figure 19-23 Modify Scene dialog box displaying the views and lights

OBTAINING RENDERING INFORMATION

Toolbar:	Render, Statistics
Pull-down:	View, Render, Statistics
Command:	STATS

 You can obtain information about the last rendering. When you enter the STATS command, AutoCAD will display the **Statistics** dialog box (Figure 19-24). Information is provided about the name of the current scene, the last rendering type used, the time taken to produce the last rendering, the number of faces processed by the last rendering, and the number of triangles processed by the last rendering. The information contained in the dialog box cannot be edited. However, the information can be saved to a file by checking the Save Statistics to File box and then entering the name of the file in the edit box. The file is saved as an ASCII file. In case a file by the specified name already exists, AutoCAD adds the present information to that file. You can use the EDIT function of any text editor to read the file.

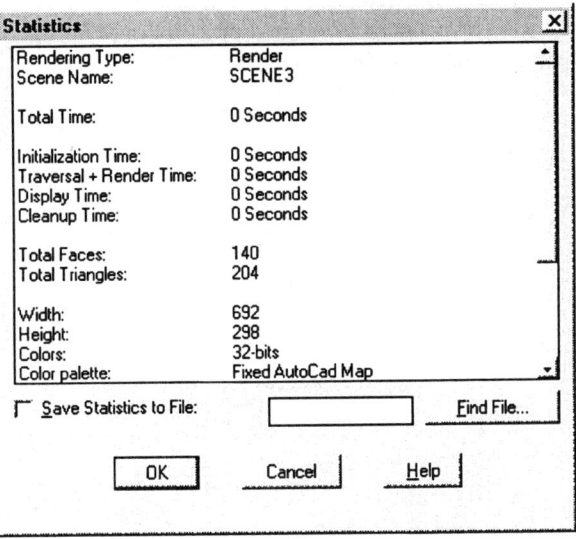

Figure 19-24 Statistics dialog box

ATTACHING MATERIALS

You can assign materials to objects, to blocks, to layers, and to an AutoCAD color index (ACI). Descriptions of each of them follows.

Attaching Material to an Object

Toolbar:	Render, Materials Library
Pull-down:	View, Render, Materials Library
Command:	MATLIB

Any object is made up of some material; hence, to get the actual (realistic) rendered image of the object, it is necessary to assign materials to the surface of the object. AutoCAD supports different materials such as bronze, copper, brass, steel, and plastic. These materials are located in material libraries such as **RENDER.MLI**. You can assign materials to objects, to blocks, to layers, and to an AutoCAD color index (ACI). By default, only ***GLOBAL*** material is assigned to a new drawing. However, if you have an object on the screen to which you want to attach a material that is already defined in the library of materials provided by AutoCAD (**RENDER.MLI**), you need to import the desired material to the drawing and then attach it to the object. For example, if you have a sphere in the current drawing and you want to attach Aqua Glaze to it, the following steps need to be executed.

1. Invoke the **Materials Library** dialog box (Figure 19-25).
2. Once the Materials Library dialog box is displayed, select the material you want from the **Library List** box. In our case, we will select AQUA GLAZE from the list. (In case the

material you want to import is in a different library (other than **RENDER.MLI**), select the Open... button. The **Library File** dialog box is displayed. Now you can specify the library file from which you want to select the material.) If you want to check how this material will appear after rendering, select the **Preview** button. A rendered sphere with an Aqua Glaze material surface appears in the preview image tile in the dialog box.

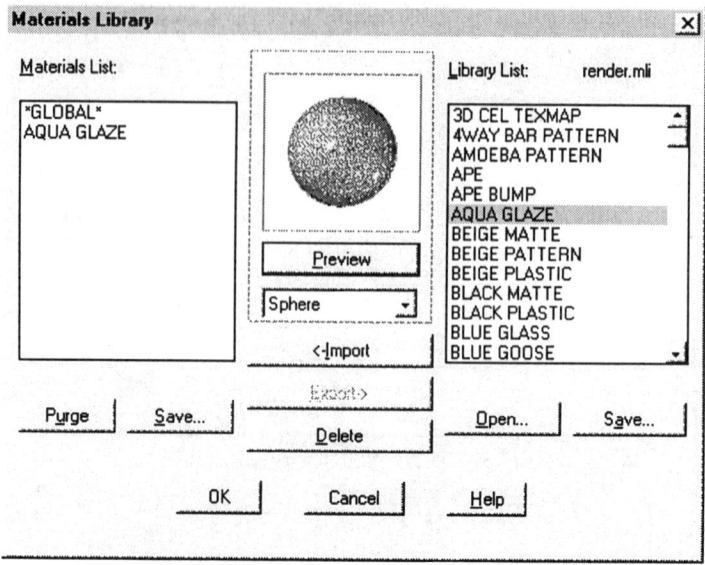

Figure 19-25 Materials Library dialog box

3. Next, choose the **<-Import** button to import the Aqua Glaze material into the drawing. Once you do so, the entry Aqua Glaze is automatically displayed in the **Materials List** column. After this, you can save the materials list in a file by selecting the **Save...** button and then specifying the **.mli** file in the **Library File** dialog box. Next, choose the **OK** button.

Note
When you open a new drawing, the Materials List column contains only the *GLOBAL* entry, which is the default material assigned to a new drawing.

When you import a material from the library to the drawing, the material and its properties are copied to the list of materials in the drawing, and in no case is the material deleted from the library list. Unattached materials can be deleted from the Materials List by selecting the **Purge** button in the **Materials Library** dialog box.

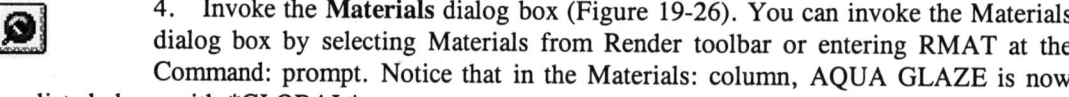

4. Invoke the **Materials** dialog box (Figure 19-26). You can invoke the Materials dialog box by selecting Materials from Render toolbar or entering RMAT at the Command: prompt. Notice that in the Materials: column, AQUA GLAZE is now listed along with *GLOBAL*.

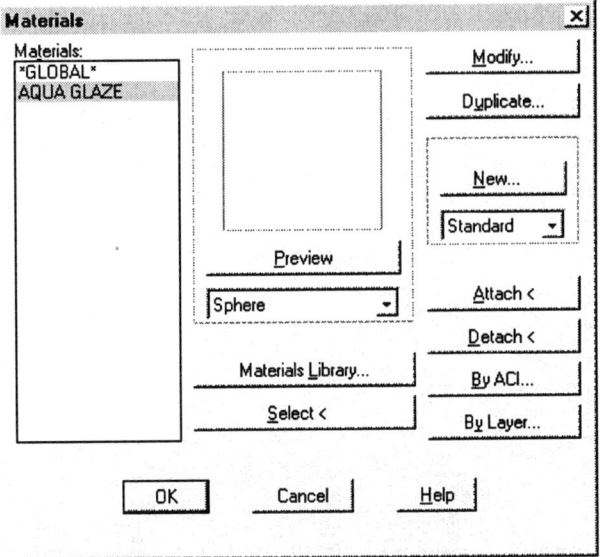

Figure 19-26 Materials dialog box

5. Select AQUA GLAZE from the list of materials, and then choose the **Attach <** button. AutoCAD clears the dialog box and issues the following prompt:

   Select objects to attach "AQUA GLAZE" to:

6 Select the sphere. The **Materials** dialog box is again displayed. Select the OK button. Now if you render the drawing, you will see that the Aqua Glaze material has been attached to the sphere.

Assigning Materials to the AutoCAD Color Index (ACI)

It is possible to attach materials to the AutoCAD color index (ACI). This can be realized in the following manner:

1. Invoke the **Materials** dialog box (Figure 19-26).
2. Select the **By ACI...** button. The **Attach by AutoCAD Color Index** dialog box (Figure 19-27) is displayed.
3. Select the material to be attached to an ACI from the **Materials:** column of the Materials dialog box.
4. Select the desired ACI from the Select ACI: list and select the **Attach** button.
5. Select the **OK** button to complete the process of attaching the specified material to objects with the specified ACI.

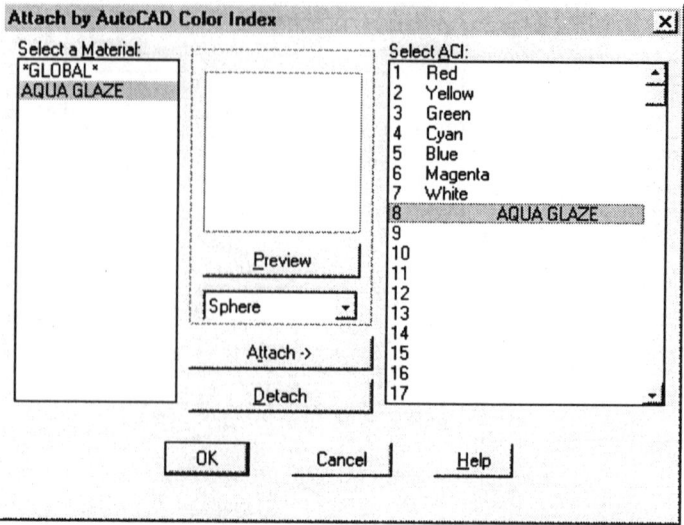

Figure 19-27 Attach by AutoCAD Color Index dialog box

Assigning Materials to Layers

Materials can be associated with layers. This option is useful when you want all the objects on a specified layer to have the assigned material. This can be realized in the following manner:

1. Invoke the **Materials** dialog box.
2. Select the By Layer... button. The **Attach by Layer** dialog box (Figure 19-28) is displayed.
3. Select the material to be attached to all objects on a particular layer.

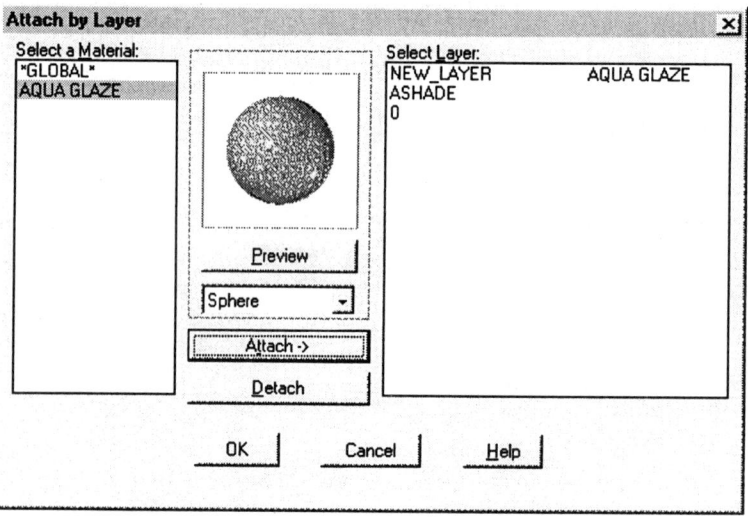

Figure 19-28 Attach by Layer dialog box

4. Select the desired layer and choose the **Attach** button.
5. Choose the **OK** button.

With this, the specified material is attached to all objects on the specified layer. AutoCAD attaches a material to an object while rendering according to the following order:

1. Materials explicitly attached to objects have highest priority.
2. Next, materials attached by ACI are considered.
3. Finally, materials attached by layer are considered.
4. GLOBAL material is used if no material is attached to an object.

If you assign different materials to different objects and then combine those objects into a single block, then, upon rendering the block, the different components of the block are rendered according to the materials individually assigned to them. Now, if different components of a block have been created on different layers to which different materials have been attached, and you attach a material to the layer on which the block exists, then the block still will be rendered according to the materials attached to different layers on which the individual parts of the block were created. However, if there is some component in the block to which no material has been attached, then the material attached to the block will be attached to this component. In such a case, if no material has been attached to the block, then the component in the block to which no material has been attached will be assigned the *GLOBAL* material.

DETACHING MATERIALS

If you want to assign a different material to an object, it is important first to dissociate the material previously attached to it; only then can you attach another material to it. Also, sometimes it might be required that no material be attached to an object. To dissociate the material attached to an object, select the **Detach <** button in the **Materials** dialog box. If you have attached a material by ACI (AutoCAD Color Index), then to detach this material, select the **By ACI...** button in the **Materials** dialog box. The **Attach by ACI** dialog box is displayed. Select the **Detach** button. Similarly, if you want to detach a material attached by layer, select the **By Layer...** button in the **Materials** dialog box. The **Attach by Layer** dialog box is displayed. Select the **Detach** button.

CHANGING THE PARAMETERS OF A MATERIAL

It is possible to change the parameters of a material, such as color and reflection properties, to meet your requirements. The following steps need to be performed:

1. Invoke the **Materials** dialog box.
2. Select the name of the material you want to modify from the list of material names.
3. Select the **Modify...** button. The **Modify Standard Material** dialog box appears (Figure 19-29). If you have specified the wrong material name, you can rectify this problem by entering the name of the material you want to modify in the **Material Name:** edit box.

19-24 Customizing AutoCAD

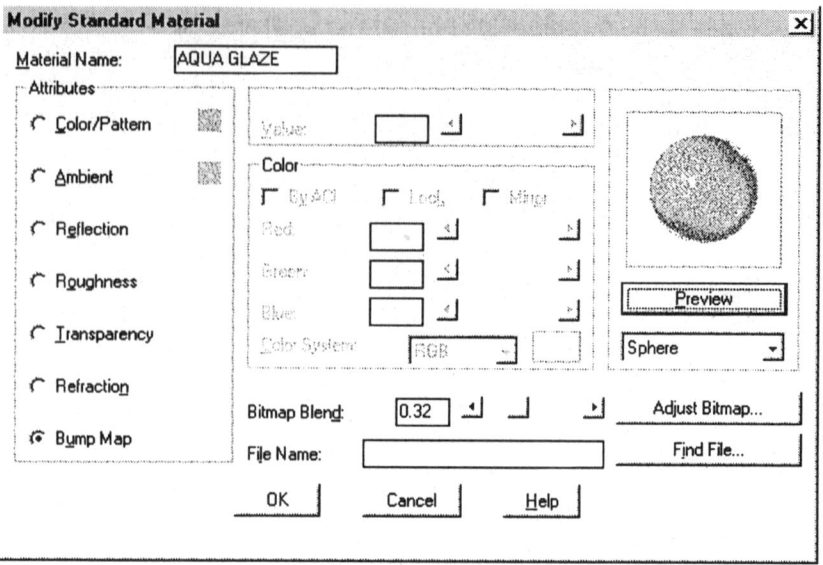

Figure 19-29 Modify Standard Material dialog box

The Modify Standard Material dialog box is comprised of the following options.

Attributes Area

The Attributes area specifies the attributes of a material. You can modify an attribute of a material by selecting it in the Attributes area and then making the alterations to it. The following attributes are listed in this area.

Color Pattern. You can change the diffuse (base) color of the material with this attribute. The intensity of the color can be changed by changing the value in the Value control with the help of the slider bar. The color can be changed by using the RGB or HLS slider bar in the Color area in this dialog box. You can also change the color with the help of the color wheel or the AutoCAD Color Index (ACI) options. The change made to the color is reflected in the color swatch next to the Color option in the Attributes area.

Ambient. By selecting this attribute, you can change the ambient color (shadow) of the material. Changes to the intensity can be made by changing the value in the Value control box with the help of the slider bar. The color can be changed by using the RGB or HLS slider bar in the Color area in this dialog box. You can also change the color with the help of the color wheel or the AutoCAD Color Index (ACI) options. The change made to the color is reflected in the color swatch next to the Ambient option in the Attributes area.

Reflection. With this attribute, you can modify the reflective (highlight or specular) color of the material. Changes to intensity can be made by changing the value in the Value control box by using the scroll bar to increase or decrease the value. Changes to the reflective color can be made by changing the RGB or HLS slider bar in the Color area in this dialog box. You can also change the color with the help of the color wheel or the AutoCAD Color Index (ACI) options. The change made to the color is reflected in the color swatch next to the Reflection option in the Attributes area.

Roughness. With this attribute you can modify the shininess or roughness level of the material. Changes can be made by changing the value in the Value control. The size of the material's reflective highlight alters by changing the Roughness level. As you increase the level of roughness, the highlight also increases.

Transparency. With this attribute, you can change the transparency of an object. The degree of transparency can be controlled by changing the value in the **Value:** edit box.

Refraction. With this attribute, you can change the refractive characteristics of a material. The degree of refraction can be controlled by changing the value in the **Value:** edit box.

Bump Map. When you select this option, you can use the **File Name:** edit box to enter the name of a bump map that you want to assign to the object.

> **Note**
> *If you want to see the effect the changes will have on the properties of an object after rendering, select the Preview button. A rendered sphere is displayed in the Preview image tile that reflects the effect of the present properties of the material.*

DEFINING NEW MATERIALS

Sometimes you may want to define a new material. To define a new material in AutoCAD, you have to perform the following steps:

1. Invoke the **Materials** dialog box.
2. Select the **New...** button. The **New Standard Material** dialog box (Figure 19-30) appears.
3. Enter the name you want to assign to the new material in the **Material Name:** edit box. The material name cannot be more than 16 characters long, and there should be no other material by this name.
4. Just as you modified the attributes of a previously defined material in the Modify Standard Material dialog box, set the color and values for the Color, Ambient, Reflection, and Roughness attributes of the new material to your requirement.
5. To examine the effect of rendering with the newly defined material, select the **Preview** button. A rendered sphere depicting the effect of rendering with the new material is displayed in the image tile.
6. After defining the attributes, select the **OK** button. In this manner, a new material has been defined.

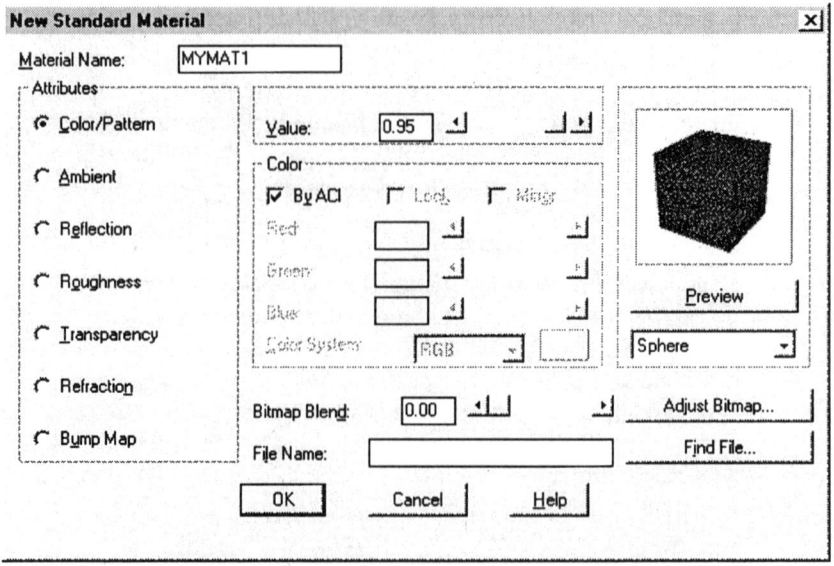

Figure 19-30 New Standard Material dialog box

EXPORTING A MATERIAL FROM A DRAWING TO THE LIBRARY OF MATERIALS

Sometimes you may have a material attached to a drawing, but the material may not be present in the library of materials. One way this situation can arise is when you have transferred a drawing that has materials assigned to it from one system to another system that does not have the materials in its library of materials. It is possible to export a material from a drawing to the library of materials. The following steps need to be performed:

1. Open the drawing that has the material you want to export.
2. Invoke the **Materials Library** dialog box.
3. Select the material you want to export from the **Materials List:** column.
4. Select the **Export->** button. You will notice that the material name is appended to the list of materials in the Library List: column.
5. With the **Save...** button just below the Library List: column, save the material in the present drawing to a library.

SAVING A RENDERING

A rendered image can be saved by rendering to a file or by rendering to the screen and then saving the image. Redisplaying a saved rendered image requires very little time compared to the time involved in rendering.

Saving a Rendering to a File

You can save a rendering directly to a file. One advantage of doing so is that you can replay the rendering in a short time. Another advantage is that when you render to the screen, the resolution of the rendering is limited by the resolution of your current display. Now if you render to a file, you can render to a higher resolution than that of your current display. Later on you can play this rendered image on a computer having a higher-resolution display. The rendered images can be saved in different formats, such as TGA, TIFF, BMP, PostScript, X11, PBM, PGM, PPM, BMP, PCX, SUN, FITS, FAX G III, and IFF. To perform this operation, carry out the following steps:

1. Invoke the **Render** dialog box from the pull-down menu (select Tools, Render, Render) or by entering **RENDER** at the Command: prompt.
2. In the **Destination** area, select **File** in the pop-up window.
3. Select the **More Options...** button. The **File Output Configuration** dialog box is displayed.
4. Specify the file type, rendering resolution, colors, and other options. Then select the **OK** button. The File Output Configuration dialog box is cleared from the screen.
5. Select the **Render** button in the Render dialog box. The **Rendering File** dialog box is displayed. Specify the name of the file to which you want to save the rendering, then select the **OK** button.

Saving a Viewport Rendering

| **Pull-down:** | View, Display Image, Save |
| **Command:** | SAVEIMG |

A rendered image in the viewport can be saved with the **SAVEIMG** command. In this case, the file formats that can be used are TGA, TIFF, and BMP. To perform this operation, carry out the following steps:

1. Invoke the **Render** dialog box and select the **Render** button to render the object.
2. Invoke the **Save Image** dialog box.
3. Specify the file format (TGA, TIFF, BMP), size, and offsets for the image, then select the **OK** button.
4. The **Image File** dialog box is displayed.
5. Specify the name of the file to which you want to save the rendering, and then select the **OK** button. In this way, a rendered image in the viewport can be saved in the specified file format.

Saving a Render-Window Rendered Image

A rendered image in the Render window can be saved with the **SAVE** command, which is available in the Render window. In this case, the file format in which the Render-window rendered image can be saved is bitmap (**.BMP**). To perform this operation carry out the following steps:

1. Invoke the **Render** dialog box, and select the **Render Window** option in the Destination area. Select the Render button to render the object. In our case, we have a single sphere to be rendered. Shortly after you select the object (sphere), the **Render** window is displayed. The rendered object is displayed in this window.
2. From the File pull-down menu, select **Options**. The **Window Render Options** dialog box is displayed. In this dialog box, you can set the size and color depth of the rendered image. In our case, the size is 640 x 480 and the color depth is 8 bit.
3. Next, from select the **Save** icon in Render window to save the rendered image. The **Save BMP** dialog box is displayed. Enter the file name in the **File Name:** edit box, and then select the OK button.

The problem with bitmap images is that when you scale them up, the images become nonuniform (blocky) and cannot be printed properly. On the other hand, if you scale down these images, they lose information.

REPLAYING A RENDERED IMAGE

In the previous section, saving a rendered image was explained. In this section, we will explain how to replay the saved rendered image.

Replaying a Rendered Image to a Viewport

| Pull-down: | View, Display Image, View |
| Command: | REPLAY |

If you have saved the rendered image in a TGA, TIFF, or BMP format, you can use the **REPLAY** command to display the image on the screen. This operation can be performed in the following manner.

1. Invoke the **Replay** command to display the **Replay** dialog box.
2. Select the file you want to replay and select the **OK** button.
3. Next, the **Image Specification** dialog box (Figure 19-31) is displayed. You can define the size and offsets, or take the default full screen size for displaying the rendered image.
4. After specifying the size and offsets, select the **OK** button. The selected image is displayed on the screen per the specified size and offsets.

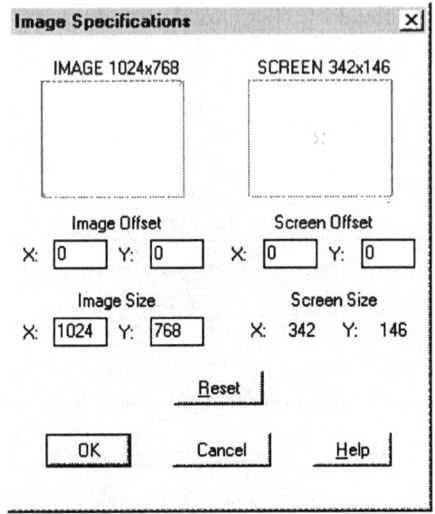

Figure 19-31 Image Specification dialog box

Replaying a Rendered Image to the Windows Render Window

1. Select the Open option in the Render Window File pull-down menu.
2. **The Render Open** dialog box (Figure 19-32) is displayed. Select the file you want AutoCAD to replay, and then select the OK button. The selected image is displayed.

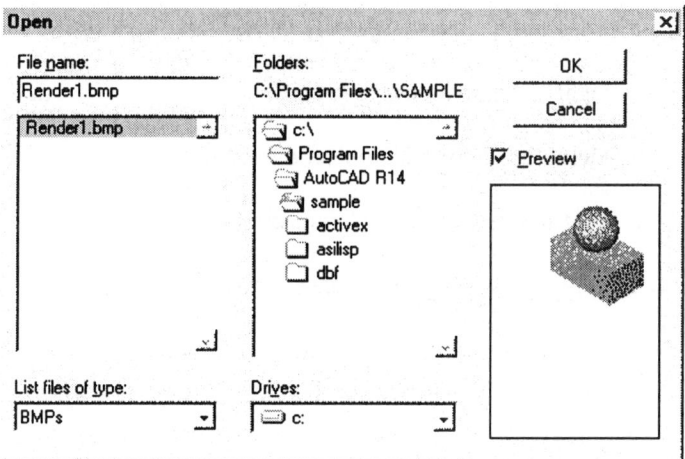

Figure 19-32 Render Open dialog box

REVIEW QUESTIONS

1. Rendering is not an effective tool for communicating design ideas. (T/F)

2. When you select any AutoCAD Render command, AutoCAD Render is loaded automatically. (T/F)

3. You cannot render a drawing without inserting any lights or defining reflective characteristics of the surface. (T/F)

4. The Gouraud render option saves computation time, but does not generate a good quality image. (T/F)

5. You can increase or decrease the intensity of ambient light, but you cannot turn it completely off. (T/F)

6. AutoCAD Render allows you to control the appearance of an object by defining the reflective quality of the surface and by adding lights to get the desired effects. (T/F)

7. The gradation effect produced by the lights does not make any difference in the appearance of the object. (T/F)

8. By default, AutoCAD Render uses smooth shading to render the image. (T/F)

9. In the _____ option, the light falling on the object (brightness) is inversely proportional to the distance of the object from the light source (Brightness = 1/Distance).

19-30 Customizing AutoCAD

10. _____ light does not have a source, and hence, has no location or direction.

11. The intensity of the light radiated by a point source _____ with the distance.

12. You can create a rendered image of a 3D object by using the _____ command.

13. You can load AutoCAD Render from the Command: prompt by entering _____.

14. The intensity of light _____ as distance increases.

15. In the _____ option, the light falling on the object (brightness) is inversely proportional to the square of the distance of the object from the light source (Brightness = 1/Distance^2).

16. A distant light source emits _____ beam of light in one direction only.

17. The intensity of a light beam from a(n) _____ light source does not change with the distance.

18. What are the names of the two rendering shading types available in AutoCAD Render?
    1._____, 2._____

19. As the oblique angle _____, the amount of light reflected by the oblique surface decreases.

20. By defining _____, it is convenient to render different views with the required lighting arrangement.

21. Materials can only be assigned explicitly to objects. (T/F)

22. When you import a material from the library of materials to a drawing, the material and its properties are copied to the list of materials in the drawing. (T/F)

23. When you import a material from the library of materials to a drawing, it gets deleted from the library of materials. (T/F)

24. New materials can be defined. (T/F)

25. A material existing in a drawing can be exported to the library of materials (**.mli** file). (T/F)

26. A rendered image can be saved by _____ or by _____.

27. Redisplaying a saved rendered image takes _____ time as compared to the time involved in rendering.

28. You can output a _____ directly to a hard-copy device if the hard-copy device driver software is loaded and the AutoCAD renderer is configured to output to hard copy.

29. Name the light sources that AutoCAD Render supports. _____

30. Define intensity. _____

31. What is attenuation? _____

32. A rendered image makes it easier to visualize the shape and size of a 3D object. (T/F)

33. You can unload AutoCAD Render by invoking the **ARX** command. (T/F)

34. Falloff occurs only with a distant source of light. (T/F)

35. You cannot assign a view to a scene. (T/F)

36. You can modify a scene by changing the lights that are assigned to that scene. (T/F)

37. You can use a spotlight to create a light source that emits light in a conical fashion in the defined direction. (T/F)

38. The Phong render option produces a _____ quality image, but it is computationally _____ and takes more time to display the image.

39. Name the three methods that AutoCAD Render has provided for controlling light falloff:
1._____, 2._____, 3._____

40. A _____ light source emits light in all directions, and the intensity of the emitted light is uniform.

41. A rendered image in the viewport can be saved with the _____ command.

42. Materials can be imported from a(n) _____ file.

43. A predefined material cannot be modified. (T/F)

44. By default, only _____ material is assigned to a new drawing.

45. Materials once attached to an object cannot be detached. (T/F)

46. In case you have saved the rendered image in a TGA, TIFF, or BMP format, you can use the _____ command to display the image.

EXERCISES

Exercise 2

Make the drawing shown in Figure 19-33. This drawing will then be used for rendering. You can also load this drawing from the accompanying disk. The name of the drawing is REND. The following is the description of one of the methods of making this drawing.

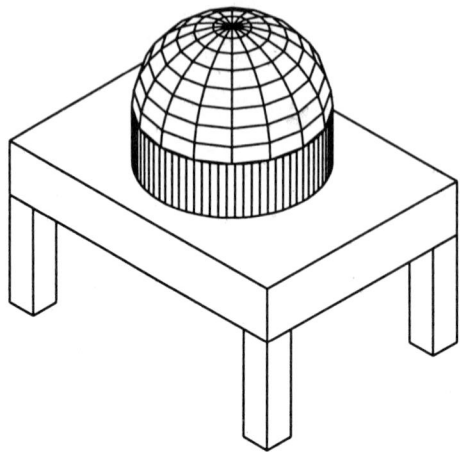

Figure 19-33 Drawing for Exercise 2

1. Draw a rectangle 2.75 x 2.0.
2. Use the CHANGE command to change the thickness to 0.25.
3. Use the 3DFACE command to draw a 3dface at the top of the rectangular box.
4. Draw another rectangle 0.25 x 0.25.
5. Use the CHANGE command to change the thickness to 1.0.
6. Copy the leg of the table to four corners.
7. Define the center point of the table as the new origin.
8. Draw a circle at the top surface of the table. The center of the circle must be at the center of the table.
9. Use the CHANGE command to change the thickness to 0.5.
10. Draw a dome that has a diameter of 1.5. The bottom edge of the dome must coincide with the top edge of the cylinder, as shown in Figure 19-33.
11. To render the drawing, first position a distant light source at (3,-2,5) and the target at (0,0,0). Now render the drawing.
12. Position a point light source at (-1,0,5) with the falloff set to Inverse Square. Render the object again.
13. Modify the distant and the point light sources to study the effect of lighting on the rendering.

Exercise 3

Generate the 3D drawings as shown in Figures 19-34 and 19-35. Next, render the 3D view of the drawing after inserting the lights at appropriate locations so as to get a realistic 3D image of the object.

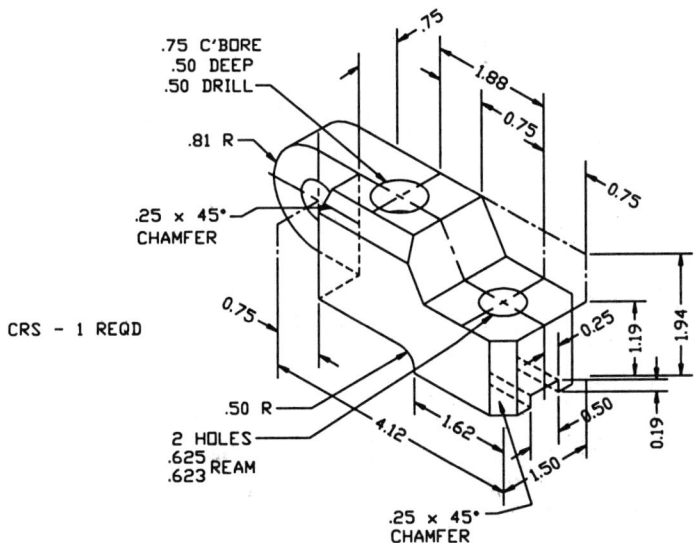

Figure 19-34 Drawing for Exercise 3

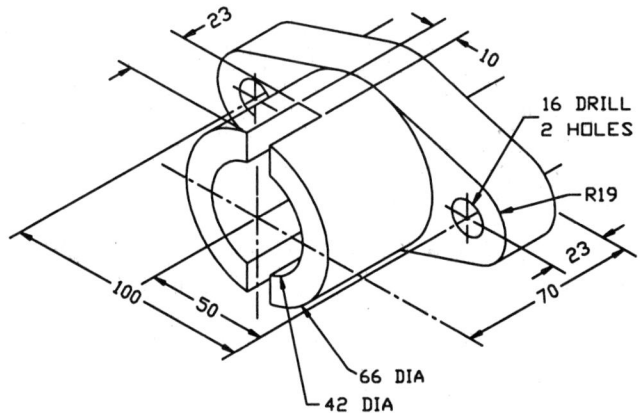

Figure 19-35 Drawing for Exercise 3

Exercise 4

a. Draw different objects on the screen, and assign different materials explicitly to them. Then render the drawing.
b. Detach the materials previously assigned to the objects and assign new materials to them.

Exercise 5

Attach copper material to some layer on your system. Draw objects on this layer, and then render the drawing. Observe the results carefully.

Exercise 6

a. Define a new material named NEWMAT. Assign desired attribute values to it in the New Standard Material dialog box, and then attach this material to an object and render it to a viewport.
b. Save the rendered image in the desired size and offset. Give it any name.
c. Redisplay the saved image on the screen.

Exercise 7

Generate the 3D drawing shown in Figure 19-36. (Assume a value for the missing dimensions.) Next, render the 3D view of the drawing after inserting the lights at appropriate locations so as to get a realistic 3D image of the object.

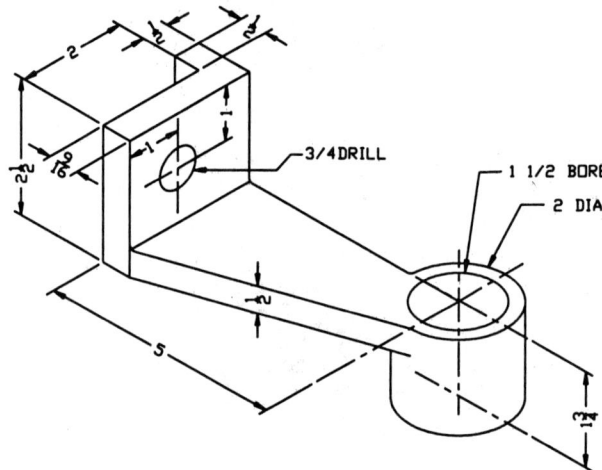

Figure 19-36 Drawing for Exercise 7

Chapter 20

AutoCAD on the Internet

Learning objectives

After completing this chapter, you will be able to:
- *Launch the Web Browser.*
- *Attach URLs to objects.*
- *Select, list, and remove URLs.*
- *Create DWF files.*
- *View DWF files.*
- *Access a drawing on the Internet.*
- *Insert a block from the Internet.*
- *Save a drawing to the Internet.*

AUTOCAD ON THE INTERNET

The Internet is the most important way to exchange digital information around the world. The best-known uses for the Internet are email (electronic mail) and the Web (short for "WORLDWIDE Web"). Email lets users exchange messages and data at very low cost. The Web brings together text, graphics, audio, and movies in an easy to use format. Other uses of the Internet include FTP (file transfer protocol for effortless binary file transfer), Gopher (presents data in a structured, subdirectory-like format), and USENET, a collection of more than 10,000 news groups.

AutoCAD allows the user to interact with the Internet in several ways. Release 14 is able to launch a Web browser from within AutoCAD. Release 14 can create DWF (short for "drawing Web format) files for viewing drawings in 2D format on Web pages. Release 14 can open, insert, and save drawings to and from the Internet.

Launching the Web Browser (BROWSER Command)

The BROWSER command lets you start a Web browser from within AutoCAD. By default, the BROWSER command uses whatever brand of Web browser program registered in your computer's Windows operating system. AutoCAD lists the name of the browser before prompting you for the URL (short for "uniform resource locator"). The URL is the Web site address, such as http://www.autodesk.com. (More about URLs in the next section.) The BROWSER command can be used in scripts, toolbar or menu macros, and AutoLISP routines to automatically access the Internet.

20-2 Customizing AutoCAD

Type the BROWSER command or select Help and then Connect to Internet from the menu bar (Figure 20-1).

   Command: **BROWSER**
   Default Browser:C:\INTERNET\NETSCAPE\PROGRAM\NETSCAPE.EXE
   Browse <www.autodesk.com>: *Enter the URL*

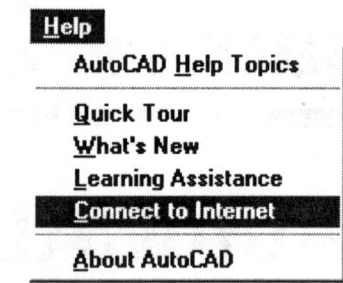

Figure 20-1 Connecting to Internet using the Help pull-down menu

The default URL is Autodesk's own Web site. After you type the URL and press Enter, AutoCAD launches the Web browser and contacts the Web site, such as Netscape Communicator and the Autodesk Web site shown in Figure 20-2.

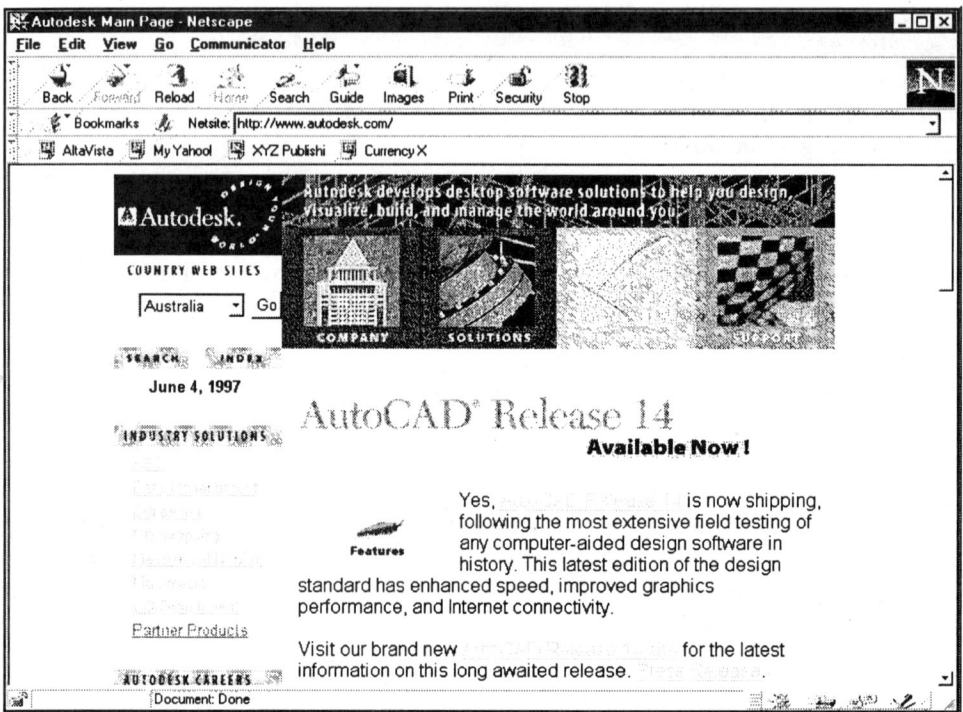

Figure 20-2 Netscape Communicator displaying the Autodesk Web site

Note
If the BROWSER command (or any other command mentioned in this chapter) does not work, you need to load it into AutoCAD. The BROWSER command is located in the program file called Browser.Arx. Use the APPLOAD command to locate and load the program, as shown in Figure 20-3.

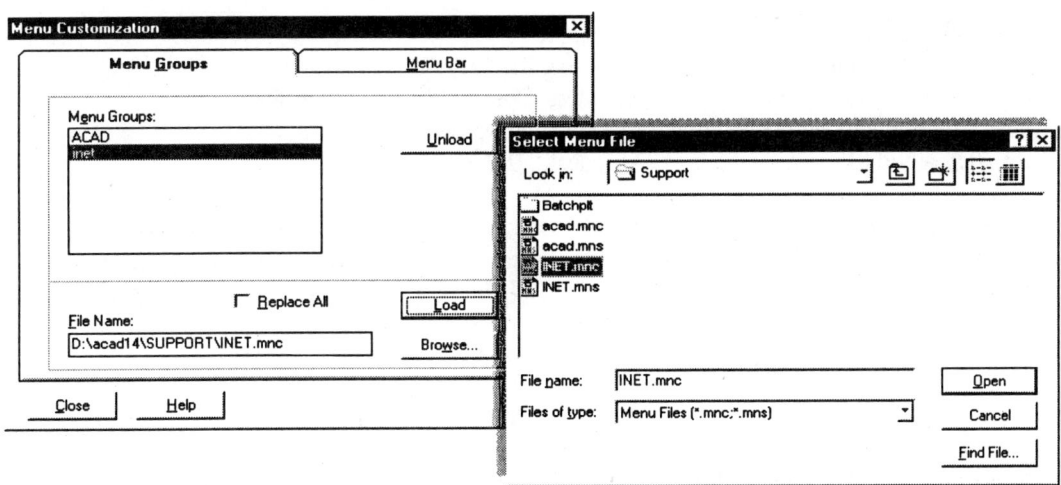

Figure 20-3 Grip location of various objects

Four programs contain Internet-related commands for AutoCAD Release 14.

ObjectARx Program	Command Names
Browser.Arx	Browser
Dwfiu.Arx	Attachurl
	Detachurl
	Listurl
	Selecturl
Dwfout.Arx	Dwfout
	Dwfoutd
Internet.Arx	Inetcfg
	Inethelp
	Inserturl
	Openurl
	Saveurl

The Uniform Resource Locator

As mentioned earlier, the file naming system of the Internet is known as URL, short for "uniform resource locator." The URL system allows you to find any resource (a file) on the Internet. Example resources include a text file, a Web page, a program file, an audio or movie clip -- in short, anything you might also find on your own computer. The primary difference is that these resources are located on somebody else's computer. A typical URL looks like the following examples:

20-4 Customizing AutoCAD

Example URL	Meaning
http://www.autodesk.com	Autodesk Primary Web Site
http://data.autodesk.com	Autodesk Data Publishing Web Site
news://adesknews.autodesk.com	Autodesk News Server
ftp://ftp.autodesk.com	Autodesk FTP Server

Note that the "http://" prefix is not required. Most of today's Web browser automatically add in the routing prefix, which saves you a few keystrokes. URLs can access several different kinds of resources -- such as Web sites, email, news groups -- but always take on the same general format:

   scheme://netloc

The "scheme" accesses the specific resource on the Internet, including these:

Scheme	Meaning
file://	Files on your computer's hard drive or local network
ftp://	File Transfer Protocol (downloading files)
http://	Hyper Text Transfer Protocol (Web sites)
mailto://	Electronic mail (email)
news://	Usenet news (news groups)
telnet://	Telnet protocol
gopher://	Gopher protocol

The "://" characters indicate a network address. Autodesk recommends that these formats for specifying URL-style filenames with the BROWSER command:

Web Site	http://servername/pathname/filename
FTP Site	ftp://servername/pathname/filename
Local File	file:///drive:/pathname/filename
or	file:///drive|/pathname/filename
or	file://\\localPC\pathname\filename
or	file:////localPC/pathname/filename
Network File	file://localhost/drive:/pathname/filename
or	file://localhost/drive|/pathname/filename

The terminology can be confusing and is explained by the following examples. "Servername" is like "www.autodesk.com". The "pathname" is the same as a subdirectory or folder name. The "drive:" is the driver letter, such as C: or D:. A "local file" is a file located on your computer. The "localhost" is the name of the network host computer. If you are not sure of the network name, use Windows Explorer to check the Network Neighborhood for the network names of computers.

How URLs are Used in AutoCAD

URLs are used indirectly by the Web browser in several different ways. The first method places URLs in the drawing for use by the browser when the drawing is exported in DWF format. To help make the process clearer, here are the steps that you need to go through to make use of URLs:

Step 1: Open a drawing in AutoCAD.
Step 2. Place URLs in the drawing with the ATTACHURL command.
Step 3: Export the drawing with the DWFOUT command.
Step 4: Copy the DWF file to your Web site.
Step 5: Start your Web browser with the BROWSER command.
Step 6: View the DWF file and select a hyperlink spot.

The second method uses URLs to access drawings over the Internet. For example, the INSERTURL command uses the URL the user types to locate and insert a drawing as a block. If the Web site contains both the DWF files (that you are viewing) and the original DWG file, then you can drag the DWG file into the current AutoCAD drawing.

A third purpose of URLs is to let you create links between files. By simply clicking on a link, you automatically access additional information. For example, clicking on the parts list in the drawing might bring up the original Excel file used to create the part list. Clicking on a standard detail might bring up the local building code. Clicking on a side view might bring up the 3D perspective view.

However, there is one significant drawback to using URLs in AutoCAD. You cannot use a URL directly within a drawing to create hyperlinks inside of AutoCAD. The examples given previously only work when viewing the drawing in DWF format with a Web browser. When working with URLs in AutoCAD, you may come across these other limitations:

* AutoCAD does not check that the URL you type is valid.

* If you attach a URL to a block, be aware that the URL data is lost when you scale the block unevenly, stretch the block, or explode it.

* You cannot attach a URL to rays and xlines since the URL would be infinitely long, something the DWF format cannot handle.

* Note that wide polylines have a one-pixel wide URL and not the full width of the polyline.

Note
When you need on-line help with AutoCAD's Internet commands, use the INETHELP command or click the help icon on the Internet Utilities toolbar.

Command: **INETHELP**

AutoCAD displays the AutoCAD Internet Utilities help window (Figure 20-4). Select one of the gray buttons for help on one of the topics.

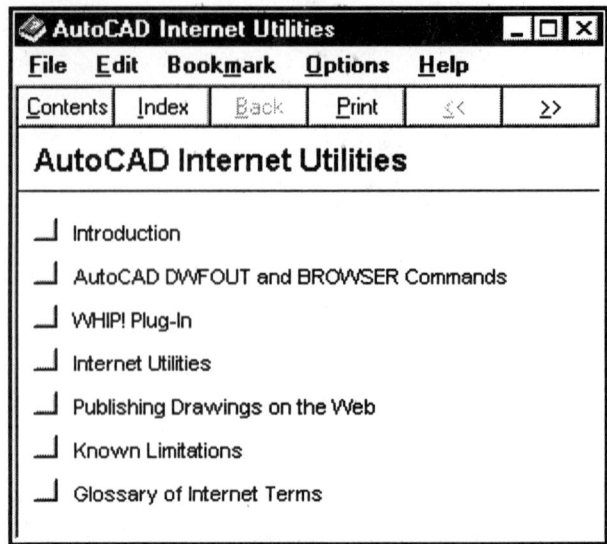

Figure 20-4 AutoCAD Internet Utilities help window

Attaching URLs to Objects (ATTACHURL Command)

The ATTACHURL command allows you to attach one or more URLs to objects and rectangular areas in a drawing. Note that you cannot use URLs in a drawing. Instead, after one or more URLs are inserted in the drawing, you must export the drawing in DWF format using the DWFOUT command. Then, when the DWF file is displayed by a Web browser, the URL locations take on special meaning. The location of the URL is sometimes called a "hyperlink." When you select the hyperlink, the Web browser automatically accesses the related URL file location.

The ATTACHURL command lets you attach URLs to objects or rectangular areas. The "object" URL is best for attaching hyperlinks to one or more objects. The "area" URL is best for placing a hyperlink around a group of objects or in an area where there are no objects.

To attach a URL to an object in the current drawing, start the ATTACHURL command:

    Command: ATTACHURL
    URL by (Area/<Objects>): **O**
    Select objects: *Select an object*
    1 found Select object: *Press Enter*
    Enter URL: *Type a URL and press Enter*

AutoCAD gives no indication that an object has a URL attached. Note that AutoCAD's ATTACHURL command lets you select more than one object for the URL. Attaching a URL to an object is sometimes called a "1D link." It is a one-dimensional link because you select a single object to activate it within the Web browser. (Technically, the URL is stored in that object's extended entity data or "xdata," for short.)

Since AutoCAD gives no visible indication which objects contain a URL, you might prefer to place the URL in a rectangular area, which shows up in the drawing as a red rectangle. To create a rectangular URL that covers an area, use the A option of the ATTACHURL command:

> Command: **ATTACHURL**
> URL by (Area/<Objects>): **A**
> First corner: *Select*
> Other corner: *Select*
> Enter URL: *Type a URL and press Enter*

Attaching the URL to an area is sometimes called a "2D link." It is a two-dimensional link because you can select anywhere in that area to activate the hyperlink. When the URL is an area, AutoCAD creates a layer named "URLLAYER" (with the color red), places the rectangle, and stores the URL as xdata of that rectangle.

Be careful that you do not overlap URLs (either objects or areas) since you could hyperlink to the wrong URL.

If you find the URL rectangles distracting, turn off layer URLLAYER. Do not erase or freeze that layer, since it will not be exported by the DWFOUT command.

Selecting a URL (SELECTURL Command)

Although you can see the rectangle of area URLs, object URLs and the URLs themselves are invisible. For this reason, AutoCAD has the SELECTURL command, which highlights all objects and areas that have URLs attached.

> Command: **SELECTURL**

AutoCAD highlights all objects that have a URL, including URL area rectangles. Depending on your computer's display system, the highlighting shows up as dashed lines or another color.

Listing a URL (LISTURL Command)

Now that you know where the URLed objects are, you use the LISTURL command to find out what the URLs are.

> Command: **LISTURL**
> Select objects: *Select*
> 1 found Select Objects: *Press Enter*
> URL for selected object is: *http://www.autodesk.com*

You cannot use the LIST command since it does not display xdata. If you really want to see the full details, use the XDLIST command found in the Bonus | Tools | List Entity Xdata menu. This program lists all extended entity data found in the selected objects.

    Command: **XDLIST**
    Select object: *Select object*
    Application name <*>: *Press Enter*
    * Registered Application Name: PE_URL
    * Code 1000, ASCII string: *http://www.autodesk.com*
    Object has 16355 bytes of Xdata space available.

Removing a URL (DETACHURL Command)

To remove a URL from an object, use the DETACHURL command.

    Command: **DETACHURL**
    Select objects: *Select*
    1 found Select Objects: *Press Enter*

When you select the rectangle of an area URL, AutoCAD erases the rectangle and reports, "DetachURL, deleting the area." If there are no more area URLs remaining, AutoCAD also purges the URLLAYER layer name.

Example 1

In this example, AutoCAD's URL tools will be employed to place several URLs in a drawing (Figure 20-5). The URLs are listed, one is deleted, then the drawing is exported in DWF format.

1. Start AutoCAD and open any drawing file.

2. Place an area URL anywhere in the drawing.

    Command: **ATTACHURL**
    URL by (Area/<Objects>): **A**
    First corner: *Select*
    Other corner: *Select*
    Enter URL: *http://www.autodesk.com*

AutoCAD places a red rectangle in the drawing.

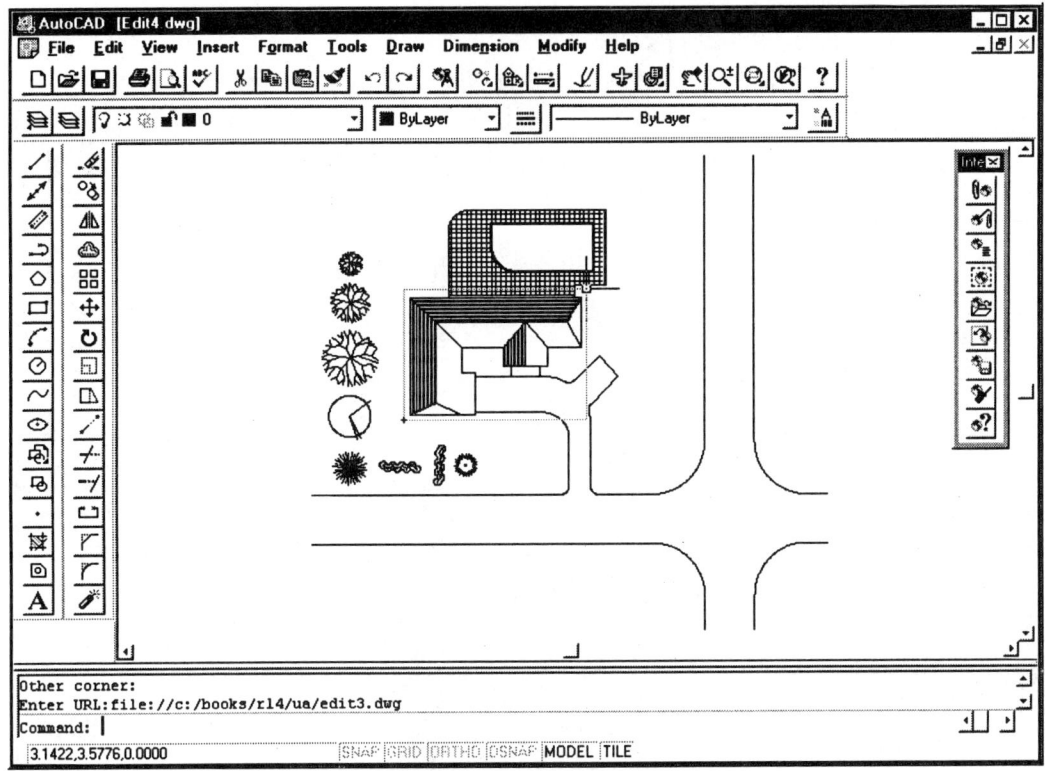

Figure 20-5 Area URL

3. Now attach a URL to an object in the drawing.

   Command: **ATTACHURL**
   URL by (Area/<Objects>): **O**
   Select objects: *Select any object*
   1 found Select object: *Press Enter*
   Enter URL: *http://data.autodesk.com*

4. You cannot see which objects have a URL attached. Use the SELECTURL command to make them visible.

   Command: **SELECTURL**

AutoCAD highlights the area URL and the object URL, as shown in figure 20-6.

20-10 Customizing AutoCAD

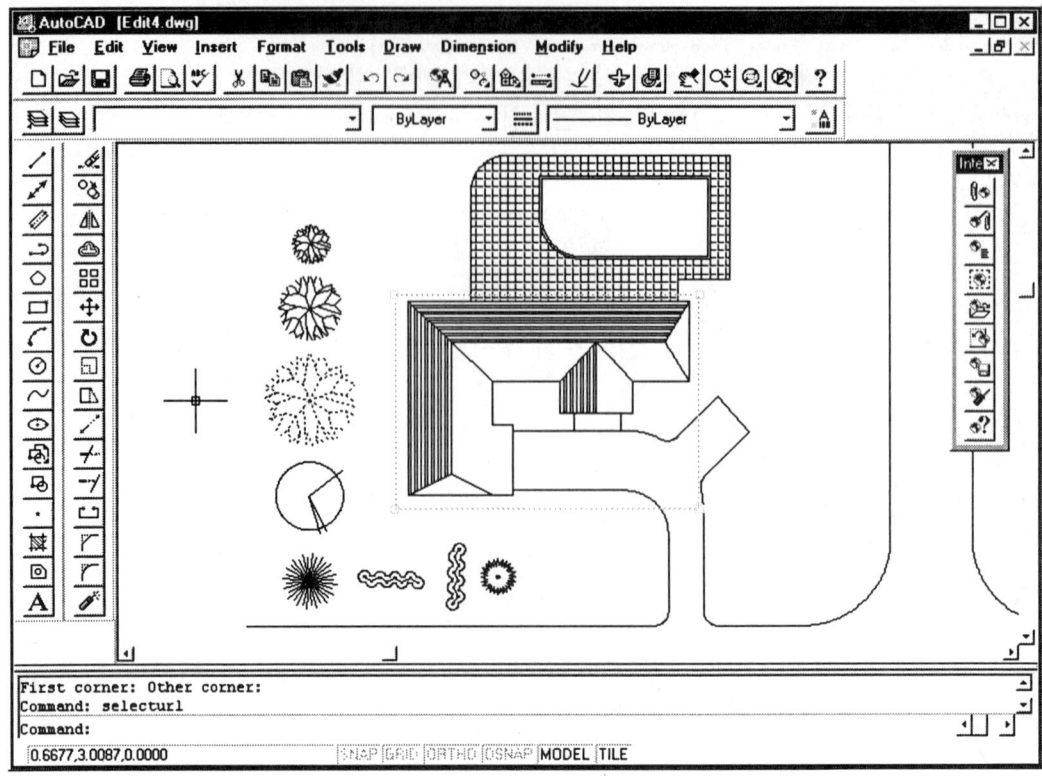

Figure 20-6 Highlighted URLs

6. If you don't remember the names of the URLs, use the LISTURL command.

   Command: **LISTURL**
   URL for selected object is: *http://www.autodesk.com*
   URL for selected object is: *http://data.autodesk.com*

   Since the URL objects were already selected (highlighted), the LISTURL command did not prompt you to select objects.

7. Press [Esc] twice to remove the highlighting.

8. If you don't need a URL, remove it.

   Command: **DETACHURL**
   Select objects: *Select an object*
   1 found Select Objects: *Press Enter*

9. Keep the drawing open for Exercise 2.

THE DRAWING WEB FORMAT

To display AutoCAD drawings on the Internet, Autodesk invented a new file format called "drawing Web format," or DWF for short. The DWF file has several benefits and some drawbacks. The DWF file is compressed as much as eight times smaller than the original DWG drawing file so that it takes less time to transmit over the Internet, particularly with the relatively slow telephone modem connections. The DWF format is more secure, since the original drawing is not being displayed. Another user cannot tamper with the original DWG file.

However, the DWF format has some drawbacks. You must go through the extra step of translating from DWG to DWF. DWF files cannot display rendered or shaded drawings. DWF is a flat, 2D file format; therefore, it does not preserve 3D data, although you can export a 3D view. The early versions of DWF (version 2.x and earlier) did not handle paper space objects.

To view a DWF file on the Internet, your Web browser needs a "plug-in." A plug-in is a software extension that lets a Web browser handle a variety of file formats. Autodesk makes the DWF plug-in freely available from its Web site. It's a good idea to regularly check the following URL for updates to the DWF plug-in, which is updated about twice a year at http://www.autodesk.com. (The exact location of the DWF plug-in is not given since Autodesk changes its Web site around every year. Simply use the Web site's Index to search for the plug-in.)

To view AutoCAD DWG and DXF files on the Internet, your Web browser needs a DWG-DXF plug-in from a third-party developer since Autodesk does not make one available. The plug-ins are available free for non-commercial use from the following vendors:

    SoftSource -- http://www.softsource.com/
    California Software Labs -- http://www.cswl.com

Creating a DWF File (DWFOUT Command)

To create a DWF file, type the DWFOUT command or select File, Export, DWF from the menu bar. AutoCAD displays the File dialog box (Figure 20-7). The Options button lets you choose the following:

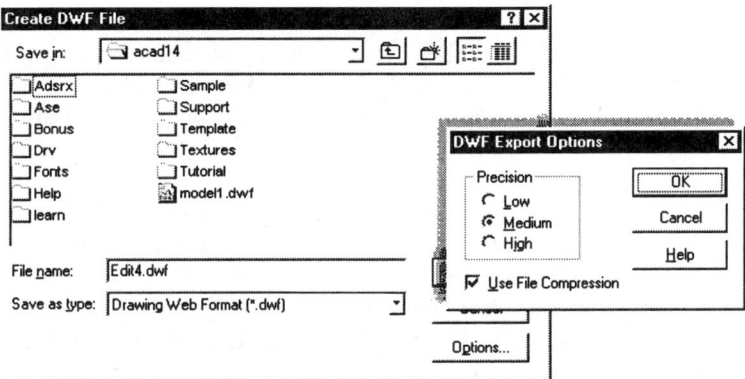

Figure 20-7 DWF Export Options Dialog Box

Precision. Unlike AutoCAD DWG files, which are based on real numbers, DWF files are saved using integer numbers. The Low precision setting saves the drawing using 16-bit integers, which is adequate for all but the most complex drawings. The file is about 40% smaller than High precision, which means a 40% faster transmission time over the Internet. Medium precision saves the DWF file using 20-bit integers. High precision saves using 32-bit integers.

Use File Compression. Compression further reduces the size of the DWF file. You should always use compression, unless you know that another application cannot decompress the DWF file.

As an alternative, you can use the DWFOUTD command, which does not display the file dialog box. DWFOUTD is meant for use in scripts, toolbar or menu macros, and AutoLISP routines. It does not give you the opportunity to select any output options.

AutoCAD itself cannot display DWF files, nor can DWF files be converted back to DWG format without using file translation software from a third-party vendor.

Example 2

1. If AutoCAD is not running, start it now and open a drawing.

2. Now export the drawing as a DWF file.

   Command: **DWFOUT**

3. When AutoCAD displays the Create DWF File dialog box, click the Options button.

4. In the DWF Export Options dialog box, select Low precision.

5. Click OK twice. AutoCAD saves the drawing as a DWF file.

Viewing DWF Files

In order to view a DWF file, you need to use a Web browser with a special "plug-in" that allows the browser to correctly interpret the file. (Remember: you cannot view a DWF file with AutoCAD.) Autodesk has named their DWF plug-in "Whip!," short for "Windows HIgh Performance." The plug-in should not be confused with AutoCAD's display driver, even though both are named "Whip!" The following table summarizes the difference between the two products:

Feature	**Whip Plug-in**	**Whip Display Driver**
Meant for:	Viewing DWF files.	Displaying AutoCAD drawings.
Works in:	Web browsers.	AutoCAD Release 13 and 14.
DWF file:	Read, view, and print.	Export only.
Available from:	Autodesk Web site.	Included with AutoCAD.

To help reduce confusion, we'll call the Whip plug-in the "DWF plug-in."

Autodesk updates the DWF plug-in approximately twice a year. Each update includes some new features. In summary, all versions of the Whip plug-in perform the following functions:

* Views DWF files created by AutoCAD within a browser.
* Right-clicking the DWF image displays a cursor menu with commands.
* Real-time pan and zoom lets you change the view of the DWF file as quickly as a drawing file in AutoCAD R14.
* Embedded hyperlinks let you display other documents and files.
* File compression means that a DWF file appears in your Web browser faster than the equivalent DWG drawing file would.
* Print the DWF file alone or along with the entire Web page.
* Works with Netscape Navigator v3.x or Microsoft Internet Explorer v3.x. A separate plug-in is required, depending upon which of the two browsers you use.

At the time of writing this book, DWF plug-in Release 2 was current and adds these features:

* Views DWF files created by AutoCAD R14 and R13.
* Allows you to "drag and drop" a DWG file from a Web site into AutoCAD R14 as a new drawing or as a block.
* Views percentage- or pixel-specified DWF files; views DWF file in a specified browser frame; views a named view stored in the DWF file; and can specify a view using x,y-coordinates.
* Turn the user interface on and off.
* Send DWF file information to CGI scripts.
* Supports Netscape Communicator v4.0 (previously named Navigator) and Microsoft Internet Explorer v4.0 Web browsers. A separate plug-in is required for either browser.
* Raster images are supported in DWF files; images are rendered as a purple outline during pans and zooms to help speed up the display.
* TrueType fonts are supported.

To use the Whip plug-in with a Web browser, your computer should have a fast CPU, such as an 80486, Pentium, or Pentium Pro. Your computer must be running Windows 95, Windows NT v3.51 (with service pack 5), or Windows NT v4.0. The computer should display at least 256 colors. The Web browser must be at least Netscape v3.0x or Explorer v3.0x.

If you don't know whether the DWF plug-in is installed in your Web browser, select Help | About Plug-ins from the browser's menu bar. You may need to scroll through the list of plug-ins to find something like this:

> *WHIP!«*
> *File name: C:\INTERNET\NETSCAPE\PROGRAM\plugins\npdwf.dll*
> *Autodesk Drawing Web Format File*
> *Mime Type   Description   Suffixes Enabled*
> *Drawing/x-dwf Drawing Web Format file dwf   Yes*

20-14 Customizing AutoCAD

If the plug-in is not installed, or is an older version, then you need to download it from Autodesk's Web site at http://www.autodesk.com/products/. For Netscape users, the file is quite large at 3.5MB and takes about a half-hour to download using a typical 28.8Kbaud modem. The file you download from the Autodesk Web site is a self-extracting installation file with a name such as Whip2.Exe. After the download is complete, start the program and follow the instructions on the screen.

> **Note**
> *If your computer has an older version of the DWF plug-in for Netscape, you must uninstall it before installing the newer version.*
>
> *If the Netscape Web browser is running, close it before installing the DWF plug-in.*

For Internet Explorer users, the DWF plug-in is an ActiveX control. Explorer's auto-download feature automatically installs the control the first time your browser accesses a Web page that includes a DWF file. The download time is about ten minutes with a 28.8Kbps modem.

DWF Plug-in Commands

To display the DXF plug-in's commands, move the cursor over the DXF image and press the mouse's right button. This displays a cursor menu with commands to PAN, ZOOM, NAMED VIEWS. To select a command, move the cursor over the command name and press the left mouse button.

Pan is the default command. Press the left mouse button and move the mouse. The cursor changes to an open hand to signal that you can pan the view around the drawing. This is exactly the same as real-time panning in AutoCAD. Naturally, panning only works when you are zoomed in; it does not work in Full View mode.

Zoom is like the ZOOM Realtime command in AutoCAD. The cursor changes to a magnifying glass. Hold down the left mouse button and move the cursor up (to zoom in) and down to zoom out.

Zoom to Rectangle is the same as ZOOM Window in AutoCAD. The cursor changes to a plus-sign. Click the left mouse button at one corner, then drag the rectangle to specify the size of the new view.

Fit to Window is the same as AutoCAD's Zoom Extents command. You see the entire drawing.

Full View causes the Web browser to display the DWF image as large as possible all by itself. This is useful for increasing the physical size of a small DWF image. When you are done viewing the large image, right-click and select the BACK command or click the browser's BACK button to return to the previous screen.

NAMED VIEWS works only when the original DWG drawing file contained named views created with the VIEW command. Selecting NAMED VIEWS displays a "non-modal" dialog box that allows you to select a named view. (A "non-modal" dialog box remains on the screen; unlike

AutoCAD's "modal" dialog boxes, you do not need to dismiss a non-modal dialog box to continue working.) Double-click a named view to see it; click OK to dismiss the dialog box.

Highlight URLs displays a highlight box around all objects and areas with URLs in the image. This helps you see where the hyperlinks are. To read the associated URL, pass the cursor over a highlight area and look at the URL on the browser's status line. (A shortcut is to hold down the [Shift] key, which causes the URLs to highlight until you release the [Shift] key).

PRINT prints the DWF image alone. To print the entire Web page (including the DWF image), use the browser's PRINT button.

SaveAs saves the DWF file in three formats to your computer's hard drive: DWF, BMP (Windows bitmap), or DWG. Saving in DWG (AutoCAD drawing file) format only works when a copy of the DWG file is available at the same subdirectory as the DWF file. You cannot use the SaveAs command until the entire DWF file has finished being transmitted to your computer.

WHIP displays information about the DWF file, including DWF file revision number, description, author, creator, source filename, creation time, modification time, source creation time, source modification time, current view left, current view right, current view bottom, and current view top.

Back is almost the same as clicking the browser's Back button. It works differently when the DWF image is displayed in a frame.

Forward is almost the same as clicking the browser's Forward button. When the DWF image is in a frame, only that frame goes forward.

Drag and Drop

The DWF plug-in allows you to perform several "drag and drop" functions. Drag and drop is when you use the mouse to drag an object from one application to another.

Hold down the [Ctrl] key to drag a DWF file from the browser into AutoCAD. Recall that AutoCAD cannot translate a DWF file into DWG drawing format. For this reason, this form of drag and drop only works when the originating DWG file exists in the same subdirectory as the DWF file. (This may change in a future release of the DWF plug-in when DWGdirectory and DWFdirectory options are implemented.) Note that this drag and drop function works only for AutoCAD Release 14; it does not work with AutoCAD Release 13 or earlier.

Another drag and drop function is to drag a DWF file from the Windows Explorer (or File Manager) into the Web browser. This causes the Web browser to load the DWF plug-in, then display the DWF file. Once displayed, you can execute all of the commands listed in the previous section.

Finally, you can also drag and drop a DWF file from Windows Explorer (or File Manager) into AutoCAD. This causes AutoCAD to launch another program that is able to view the DWF file.

20-16 Customizing AutoCAD

This does not work if you have no other software on your computer system capable of viewing DWF files.

Example 3

Once the plug-in is installed in your Web browser, you can view the DWF file by dragging it from Windows Explorer into the browser.

1. If your Web browser has the DWF plug-in installed, start the Web browser now.

2. Switch to Windows Explorer (or File Manager) and look for the DWF file you created in Example 2. If necessary, use the Search function to find *.DWF.

3. Dragging the DWF file from Windows Explorer (or the File Manager) into the browser. The browser will take several seconds to first load the plug-in, then load and display the DWF file.

4. Place the cursor in the image and right-click the mouse button to bring up the cursor menu.

5. Practice selecting several commands, such as Zoom, Pan, and Print.

6. To remove the image, click the browser's Back button.

Embedding a DWF File

To let others view your DWF file over the Internet, you need to embed the DWF file in a Web page. There are several approaches to embedding a DWF file in a Web page.

The quickest way is to use the <embed> tag in a very simple manner. <Embed> is an HTML (short for "hyper text markup language") code for embedding an object in a Web page.

   <embed src="filename.dwf">

Replace "filename.dwf" with the name of the DWF file. Remember to keep the quotation marks in place. To specify the size of the DWF image on the Web page, you include the Width and Height options.

   <embed width=800 height=600 name=description src="filename.dwf">

Replace 800 and 600 with any appropriate numbers, such as 100 and 75 for a "thumbnail" image, 300 and 200 for a small image, or 640 and 480 for a medium-size image. These width and height values are measured in pixels.

The Name option displays a textual description of the image when the browser does not load images. For example, you might replace "description" with the DWF filename.

If you use Netscape Navigator, you can include a description of where to get the Whip plug-in if the Web browser is lacking it.

*<embed   width=800   height=600   name=description   src="filename.dwf"
pluginspage=http://www.autodesk.com/products/autocad/whip/whip.htm>*

If you use Microsoft Internet Explorer, you must add the <object> and <param> tags.

*<object width=600 height=400 classid= "clsid:b2be75f3-9197-11cf-abf4-08000996e931"
codebase= "ftp://ftp.autodesk.com/pub/autocad/plugin/whip.cab#Version=2,0,0,0">
<param name= "description" value= "filename.dwf">
<embed   width=600   height=400   name=description   src= "filename.dwf"
pluginspage=http://www.autodesk.com/products/autocad/whip/whip.htm>
</object>*

Accessing a Drawing on the Internet (OPENURL Command)

When a drawing is stored on the Internet, you access it from within AutoCAD using the OPENURL command. Instead of specifying the file's location with the usual drive-subdirectory-file name format, such as C:\ACAD14\FILENAME.DWG, you use the URL format. Recall that the URL is the universal file naming system used by the Internet to access any file located on any computer hooked up to the Internet.

Command: **OPENURL**

Figure 20-8 Open DWG from URL dialog box

The dialog box prompts you to type the URL (Figure 20-8). For your convenience, AutoCAD fills in the preliminary "http://", which is used for accessing Web pages. If you plan to access a different resource on the Internet, erase the http:// and replace it. For example, you might want to replace it with ftp:// to access a drawing at an FTP site.

Use the following as templates for typing the URL for opening a drawing file:

Drawing Location	**Template URL**
Web or HTTP Site	http://servername/pathname/filename.dwg
FTP Site	ftp://servername/pathname/filename.dwg

Local File   file:///drive:/pathname/filename.dwg
Network File   file://localhost/drive:/pathname/filename.dwg

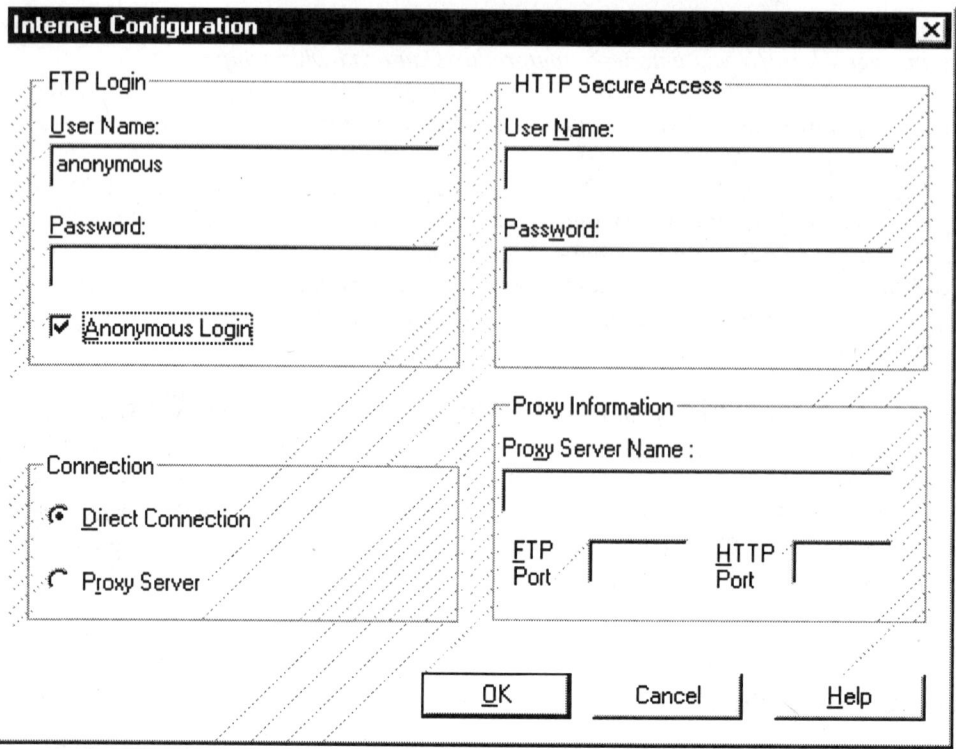

Figure 20-9 Internet Configuration dialog box

You need only click the Options button when you want AutoCAD to deal with "secure" sites that ask you for a username and password. (This Internet Configuration dialog box (Figure 20-9) is the same one that appears when you type the INETCFG command.) Very often, you do not need them. For example, most FTP sites allows "anonymous" logins where you need no password and the username is "anonymous." Most Web sites require no login at all. For security reasons, the passwords are not retained once you exit AutoCAD.

To access a secure FTP site, turn off Anonymous Login and type your username and password.

To access a secure Web site (also called a HTTP site), type your username and password. If you prefer, you can leave them blank and AutoCAD will display a User Authentication dialog box later, where you can enter the username and password.

Select Direct Connection when your computer connects with the Internet via an Internet service provider. When your computer connects to the Internet through a proxy server, select Proxy Server. (Most typically, larger firms have a "proxy server" that acts as a gateway between the firm's internal network and the external Internet.) Ask your system administrator for the details of how to configure the proxy server, including the proxy server name, default FTP port, and default HTTP port.

Click OK to dismiss the Internet Configuration dialog box.

To open the drawing, click Open. If necessary, AutoCAD prompts you to save the current drawing before opening the drawing from the Internet. During the file transfer, AutoCAD displays a dialog box (Figure 20-10) to report the progress. If your computer uses a 28.8Kbps modem, you should allow about ten minutes per megabyte of drawing file size. If your computer has access to a faster T1 connection to the Internet, you should expect a transfer speed of about one minute per megabyte.

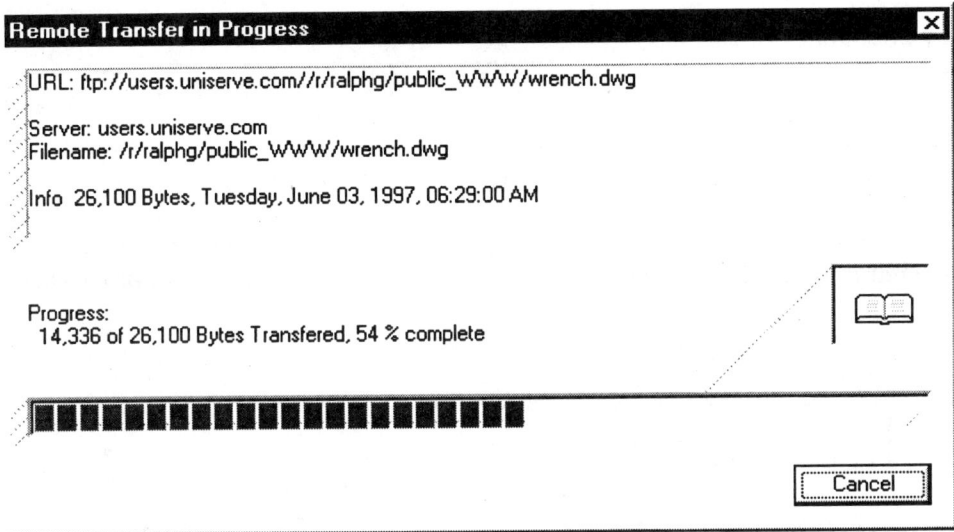

Figure 20-10 Remote Transfer in Progress dialog box

It may be helpful to understand that OPENURL command does not copy the file from the Internet location directly into AutoCAD. Instead, it copies the file from the Internet to your computer's designated Temporary subdirectory, such as C:\WIN95\TEMP (and then loads the drawing from the hard drive into AutoCAD). This is known as "caching." It helps to speed up the processing of the drawing, since the drawing file is now located on your computer's fast hard drive, instead of the relatively slow Internet.

Inserting a Block from the Internet (INSERTURL Command)

When a block (symbol) is stored on the Internet, you can access it from within AutoCAD using the INSERTURL command. The INSERTURL command works just like the OPENURL command, except that the external block is inserted into the current drawing.

Command: **INSERTURL**

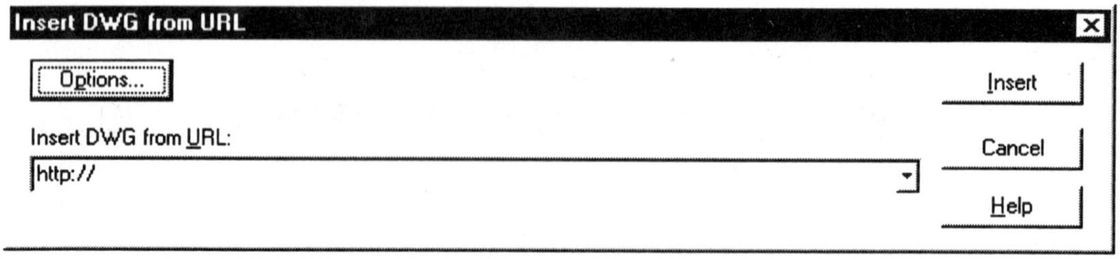

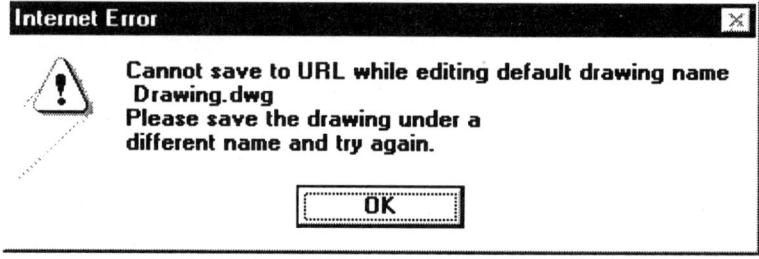

Figure 20-11 Insert DWG from URL dialog box

The dialog box prompts you to type the URL (Figure 20-11). After you click the Insert button, AutoCAD retrieves the file and continues with the INSERT command's familiar prompts.

Insert Block name (or ?): *c:\win95\temp\filename.dwg*
Insertion point: *Specify insertion point*
X scale factor <1> / Corner / XYZ: *Press Enter*
Y scale factor (default=X): *Press Enter*
Rotation angle <0>: *Press Enter*

Saving a Drawing to the Internet (SAVEURL Command)

When you are finished editing a drawing in AutoCAD, you can save it to a file server on the Internet with the SAVEURL command. If you inserted the drawing from the Internet (using INSERTURL) into the default DRAWING.DWG drawing, AutoCAD insists you first save the drawing to your computer's hard drive (Figure 20-12).

Figure 20-12 Internet Error dialog box

Command: **SAVEURL**

Figure 20-13 Save DWG to URL dialog box

The dialog box prompts you to type the URL. For your convenience, AutoCAD fills in the preliminary "ftp://", which is used for writing a file to an FTP site (Figure 20-13). The reason that "http://" is not displayed is because you have to write a file to Web site using FTP (file transfer protocol). Nor will AutoCAD allow you to use "file://" or other URL prefixes.

When a drawing of the same name already exists at that URL, AutoCAD warns you, just like when you use the SAVEAS command (Figure 20-14). Recall from the OPENURL command that AutoCAD uses your computer system's Temporary subdirectory, hence the reference to it in the dialog box.

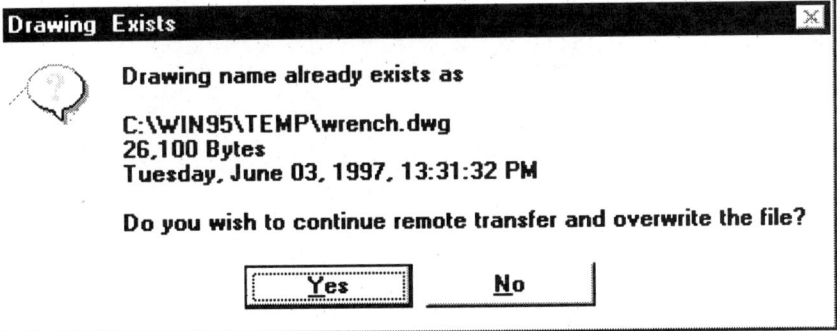

Figure 20-14 Drawing Exists dialog box

Example 4

In this example, AutoCAD's Internet tools will be employed to open a drawing called Wrench.Dwg from the Autodesk Press Web site on the Internet. (The example assumes your computer has access to the Internet.)

1. Start AutoCAD. When the Start New Drawing dialog box is displayed, click Start From Scratch and OK.

2. If your computer is not logged on to the Internet, do so now. For example, if you access the Internet through an Internet Service Provider, use the Dial-up Networking feature of Windows now. Click the Connect button.
3. Switch back to AutoCAD and open a drawing using a URL.
   Command: **OPENURL**

4. Once you invoke the command, the Open DWG from URL dialog box appears. If your computer has access to the Internet, type the following URL:

   http://www.autodeskpress.com/   /wrench.dwg

5. Click Open. AutoCAD displays the Remote Transfer in Progress dialog box. The file is 25KB in size and should take a few seconds to download.
6. Once the Wrench drawing appears, it looks just like any drawing you open from your computer's hard drive (Figure 20-15).

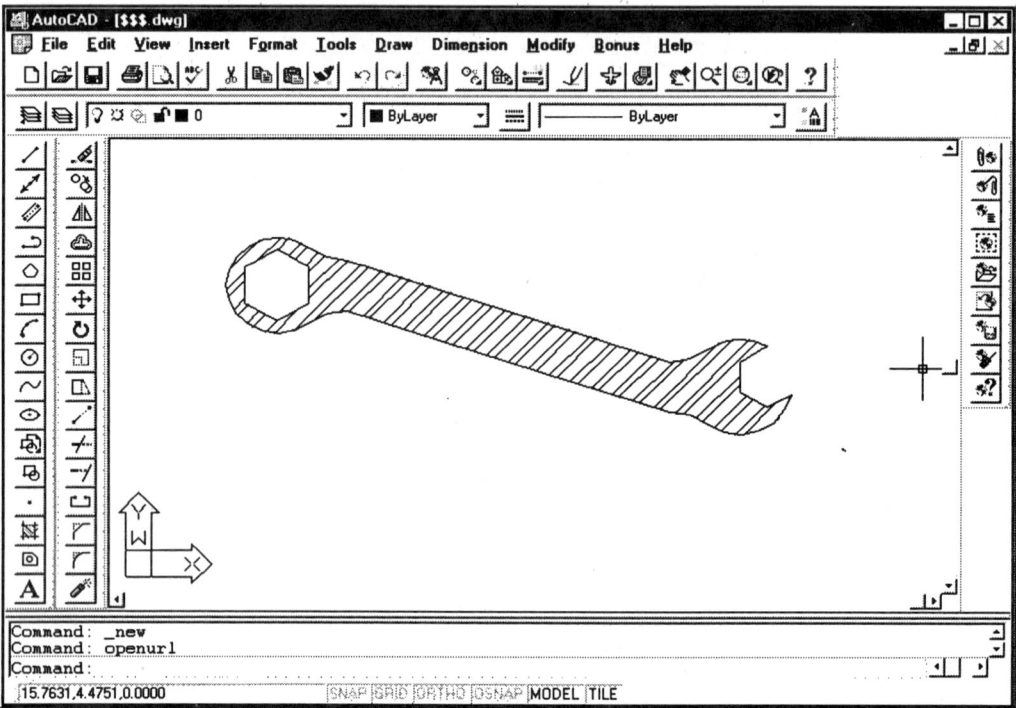

Figure 20-15 Wrench.Dwg opened from the Internet

REVIEW QUESTIONS

1. AutoCAD is able to launch a Web browser from within AutoCAD. (T/F)_____

2. DWF is short for _____

3. The purpose of DWF files is to view 2D drawings where? _____

4. URL is short for _____

5. Give one example of a URL of a Web page. _____

6. What is the purpose of a URL? _____

7. Give three examples of URL schemes: _____

8. FTP is short for _____

9. What is a "local host"? _____

10. URLs can be used in an AutoCAD drawing. (T/F)

11. The purpose of URLs is to let you create _____ between files.

12. When you attach a URL to a block, the URL data is _____ when you scale the block unevenly, stretch the block, or explode it.

13. You can/cannot attach a URL to rays and xlines.

14. The ATTACHURL command allows you to attach a URL to _____ and _____.

15. The location of the URL is also called a _____.

16. To see the location of URLs in a drawing, use the _____ command.

17. Rectangular URLs are stored on layer _____.

18. The LISTURL command tells you _____.

19. The DETACHURL command removes a _____ from an object.

20. Compression in the DWF file causes it to take _____ time to transmit over the Internet

21. DWF is created from a _____ file using the _____ command.

22. A "plug-in" lets a Web browser _____.

23. A Web browser can view DWG drawing files over the Internet. (T/F)

24. _____ is an HTML tag for embedding graphics in a Web page.

25. "Whip" is short for _____

26. The Whip plug-in and the Whip display driver are the same. (T/F)

27. A file being transmitted over the Internet via a 28.8Kpbs modem takes about _____ minutes per megabyte.

28. To open a drawing located on the Internet, use the _____ command.

29. An "anonymous" FTP login requires _____ for the password and _____ for the username.

30. Passwords are retained when you exit AutoCAD. (T/F)

31. The SAVEURL command saves the _____ to the Internet.

32. Give two examples of a URL _____
    _____

33. What is the purpose of the ATTACHURL command? _____
    _____

34. Name three benefits of the DWF file format. What is one drawback to DWF files? _____
    _____

35. Do the following commands display extended entity data? List. Listurl. XDLIST. _____
    _____

36. Is it recommended to freeze layer URLLAYER? _____

37. Can DWG and DXF files be viewed by a Web browser? _____

38. How does the Whip plug-in differ from the Whip display driver? _____
    _____

39. Can you use the same DWF plug-in for Netscape Communicator and Internet Explorer? ____

40. How do you access DWF-related commands in the Web browser? _____
    _____

Chapter 21

Data Exchange, Object Linking and Embedding, Multilines, and Digitizing

Learning objectives

After completing this chapter, you will be able to:
- *Import and export files using the DXFOUT and DXFIN commands.*
- *Convert scanned drawings into the drawing editor using the DXB command.*
- *Export raster files using the SAVEIMG command.*
- *Understand the embedding and the linking functions of the OLE feature of Windows.*
- *Define multiline style and specify properties of multilines using the MLSTYLE command.*
- *Draw different types of multilines using the MLINE command.*
- *Edit multilines using the MLEDIT command.*
- *Configure the tablet to digitize drawings.*

DATA EXCHANGE IN AUTOCAD

Different companies have developed different software for applications such as CAD, desktop publishing, and rendering. This nonstandardization of software has led to the development of various data exchange formats that enable transfer (translation) of data from one data processing software to another. We will now discuss various data exchange formats provided in AutoCAD. AutoCAD uses the **.DWG** format to store drawing files. This format is not recognized by most other CAD software, such as Intergraph, CADKEY, and MicroStation. To solve this problem so that files created in AutoCAD can be transferred to other CAD software for further use, AutoCAD provides various data exchange formats, such as **DXF (data interchange file)** and **DXB (binary drawing interchange)**.

DXF FILE FORMAT (DATA INTERCHANGE FILE)

The DXF file format generates a text file in ASCII code from the original drawing. This allows any computer system to manipulate (read/write) data in a DXF file. Usually, DXF format is used for CAD packages based on microcomputers. For example, packages like SmartCAM use DXF files. Some desktop publishing packages, such as Pagemaker and Ventura Publisher, also use DXF files.

Creating a Data Interchange File (DXFOUT Command)

| **Command:** DXFOUT |

The **DXFOUT** command is used to create an ASCII file with a **.DXF** extension from an AutoCAD drawing file. Once you invoke the DXFOUT command, the **Create DXF File** dialog box (Figure 21-1) is displayed. By default, the DXF file to be created assumes the name of the drawing file from which it will be created. However, you can specify a file name of your choice for the DXF file by typing the desired file name in the **File Name:** edit box. AutoCAD automatically sets the extension of DXF files as **.DXF**. This can be observed in the Save as type: drop-down list where you can select the output file format.

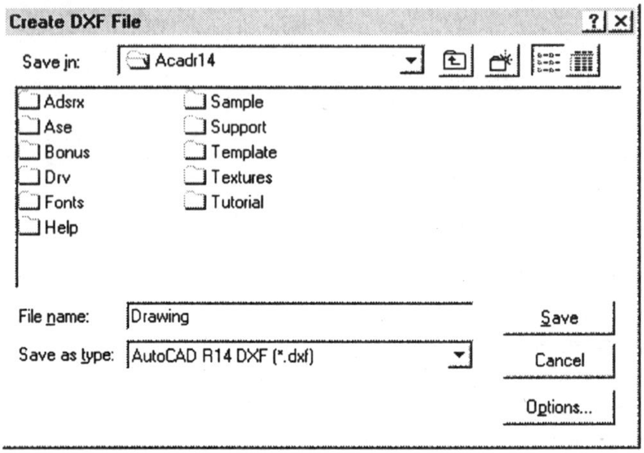

Figure 21-1 Create DXF File dialog box

Choose the Options button to display the **Export Options** dialog box (Figure 21-2). In this dialog box, enter the degree of accuracy for the numeric values. The default value for the degree of accuracy is six decimal places, however, this results in less accuracy than the original drawing, which is accurate to 16 places. You can enter a value between 0 and 16 decimal places.

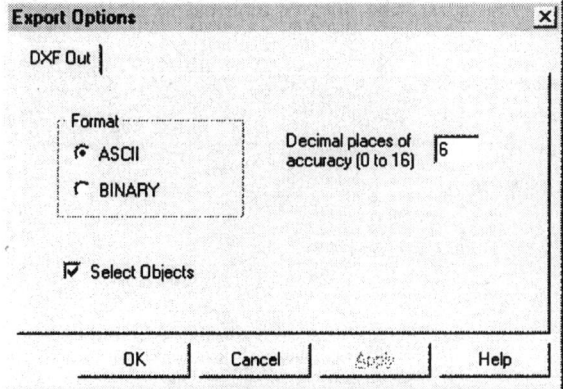

Figure 21-2 Export Options dialog box

In this dialog box, you can also activate the **Select Objects** check box, which allows you to specify objects you want to include in the DXF file. In this case, the definitions of named objects such as block definitions, text styles, etc., are not exported.

Now an ASCII file with a **.DXF** extension has been created, and this file can be accessed by other CAD systems. This file contains data on the objects specified in the DXFOUT command. By default, with the DXFOUT command, files are created in ASCII format. However, you can also create binary format files by activating the **Binary** radio button in the Export Options dialog box. Binary DXF files are more efficient and occupy only 75 percent of the ASCII DXF file. You can access a file in binary format more quickly than the same file in ASCII format.

Information in a DXF File

The DXF file contains data on the objects specified in the DXFOUT command. You can change the data in this file to your requirement. To examine the data in this file, load the ASCII file in word processing software. A DXF file is composed of four parts.

Header. In this part of the drawing database, all the variables in the drawing and their values are displayed.

Tables. All the named objects, such as linetypes, layers, blocks, text styles, dimension styles, and views, are listed in this part.

Blocks. The objects that define blocks and their respective values are displayed in this part.

Objects. Objects in the drawing are listed in this part.

Converting DXF Files into a Drawing File (DXFIN Command)

| Command: DXFIN |

You can import a DXF file into a new AutoCAD drawing file with the **DXFIN** command. Start a new drawing and then invoke the DXFIN command.

21-4 Customizing AutoCAD

Command: **DXFIN**

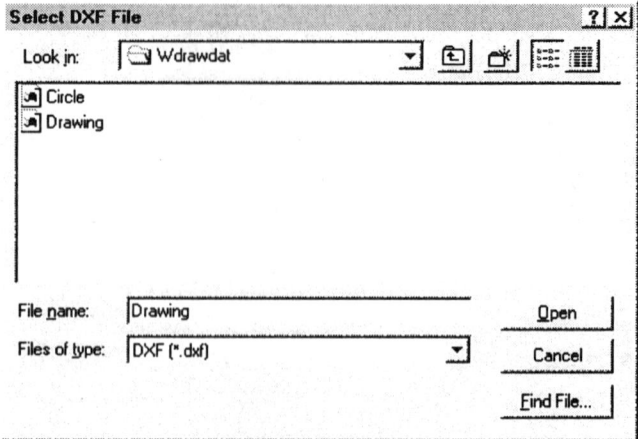

Figure 21-3 Select DXF File dialog box

After you invoke the DXFIN command, the **Select DXF File** dialog box (Figure 21-3) is displayed. In the File Name: edit box enter the name of the file you want to import into AutoCAD or select the file from the list. Choose the **Open** button. Once this is done, the specified DXF file is converted into a standard DWG file, regeneration is carried out, and the file is inserted into the new drawing. Now you can perform different operations on this file just as with other drawing files.

Importing Scanned Files into the Drawing Editor
DXB File Format

Pull-down:	Insert, Drawing Exchange Binary
Command:	DXBIN

AutoCAD offers another file format for data interchange: DXB. This format is much more compressed than the binary DXF format and is used when you want to translate large amounts of data from one CAD system to another. For example, when a drawing is scanned, a DXB file is created. This file has huge amounts of data in it. The **DXBIN** command is used to create drawing files out of DXB format files.

Command: **DXBIN**

After you invoke the DXBIN command, the **Select DXB File** dialog box is displayed. In the File Name: edit box, enter the name of the file (in DXB format) you want to import into AutoCAD. Choose the **Open** button. Once this is done, the specified DXB file is converted into a standard DWG file and is inserted into the current drawing.

> **Note**
> Before importing a DXF or DXB file, you must create a new drawing file. No editing or drawing setup (limits, units, etc.) can be performed in this file. This is because if you import a DFX or DXB file into an old drawing, the settings (definitions) of layers, blocks, etc., of the file being imported are overruled by the settings of the file into which you are importing the DXF or DXB file.

DATA INTERCHANGE THROUGH RASTER FILES

Until now, we have discussed importing and exporting files in the DXF file format. To uphold the accuracy of the drawing, the DXF file includes almost all the information about the original drawing file. The accuracy is maintained at the expense of DXF file size and degree of complexity of these files. There are many applications in which accuracy is not very important, like desktop publishing. In such applications, you are concerned primarily with image presentation. A very simple and effective way of storing an image for import/export is in the form of **raster files**. In a raster file, information is stored in the form of a dot pattern on the screen. This bit pattern is also known as a **bit map**. For example, in a raster file a picture is stored in the form of information about the position and color of the screen pixels. We will be discussing three types of raster files: TIFF files, TGA files, and BMP files. These formats make it possible to transfer a file from AutoCAD to other software.

TIFF (tagged image file format)

TGA (targa format)

BMP (bitmap format)

Exporting the Raster Files (SAVEIMG Command)

Pull-down:	Tools, Display Image, Save
Command:	SAVEIMG

If there is a picture on the screen that you want to save as a raster file, use the SAVEIMG command. With this command, you can save a rendered image to TIFF, TGA, or BMP types of raster files. These raster files can be used by most software. Sometimes, the raster files need to be converted into some other format before they can be used by certain software. This operation is performed with a file conversion program such as Hijaak or Pizazz Plus.

Command: **SAVEIMG**

Once you invoke this command, the **Save Image** dialog box is displayed. Depending on the Destination option you have selected in the **Rendering Preferences** dialog box, you will get one of two Save Image dialog boxes (Figures 21-4). The only difference lies in the Portion area.

21-6 Customizing AutoCAD

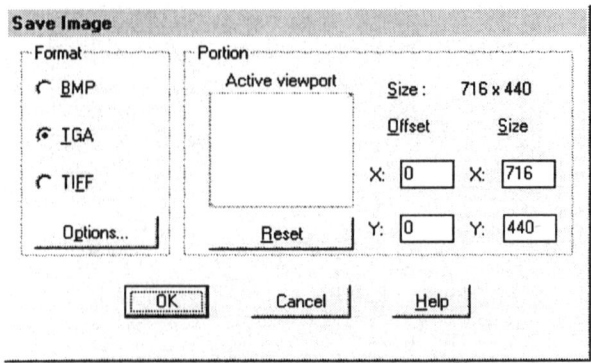

Figure 21-4 Save Image dialog box

Format. This area of the **Save Image** dialog box provides three raster formats: TGA, TIFF, and BMP. You can select any one of the formats by selecting the radio button next to the desired format. A brief description of each format follows.

TGA. (default format). With this format, you can create compressed or uncompressed 32-bit RGBA Truevision v2.0 format files. These files have the extension **.tga**.

TIFF. (tagged image file format). This file format has been developed by Aldus and Microsoft. With this format, you can create compressed or uncompressed 32-bit RGBA tagged image file format. These files have the extension **.tif**.

BMP. (bitmap format). These files have the extension **.bmp**.

Portion. In the **Portion** area of the **Save Image** dialog box, you can specify the portion of the screen you want to save in the specified file format. In case AutoCAD is configured for **rendering to a viewport** (in the Render Preferences dialog box), the Portion area provides options to save either the active viewport, the drawing area, or the full screen. If you are **rendering to a separate window**, the Portion area contains an image tile that graphically represents the section of the screen to be saved. You will notice that by default the entire screen is selected for saving. If you want to save only a section of the screen, you need to specify the desired section. This can be achieved by specifying two diagonally opposite points on the image tile. Specify the lower left corner of the desired section first, then specify the upper right corner. After specifying the two corner points, AutoCAD creates a box whose lower left corner point and upper right corner point coincide with the respective points you specified. Also, the values for the X and Y Offsets and Sizes are automatically updated.

Offset. This includes X and Y edit boxes. The values you enter in these edit boxes are treated as the X and Y offsets. The start point (lower left corner) of the area to be saved as a raster file can be specified by X and Y offset values. The default value for the offset is (0,0). The values entered are taken as the pixel values. The offset values should be specified within the screen size value. If you are rendering to a different window, then when you change the offset values, the change in the start point of the image selection area is reflected by a margin formed by two intersecting lines in the image tile. The point of intersection of these two lines is the start point of the image selection area. Values of X and Y offsets greater than the screen size are not accepted by AutoCAD. If you have selected the RND format, the Offset area is not available.

Size. With the Offset option, you can specify the lower left corner of the area to be selected for conversion into a raster file format. To specify the area completely, however, you need to specify the upper right corner of this area as well. This can be achieved with the **Size** option. In the case of the Offset option, the values entered in the X and Y edit boxes are the pixel values. The default for the Size option is the upper right corner of the display area. If you are rendering to a different window, the change in location of the upper right corner of the image selection area is reflected by two intersecting lines. If you have selected the RND format, the Offset area is not available.

Options. The display provided by selecting the **Options** option varies from one file format to another. The image compression options are displayed only for TGA and TIFF formats. For the BMP format, the image file compression is not supported, hence, the Options button is disabled.

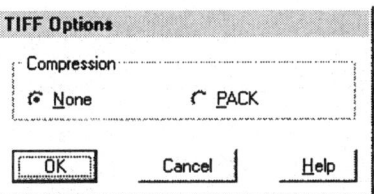   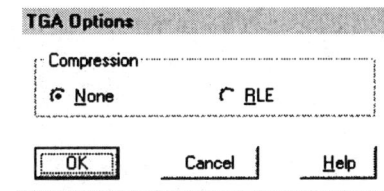

Figure 21-5 TIFF Options dialog box   **Figure 21-6** TGA Options dialog box

None. This is the default option. If this option is selected, compression does not take place.

PACK. With this option, run-length encoded image compression can be realized only for TIFF files (Figure 21-5). It uses Macintosh packbits.

RLE. This option works only for TGA format files (Figure 21-6). If you select this option, run-length encoded image compression can be realized.

Reset. When you select this option in the Save Image dialog box, the Offset values and the Size values are set to their default values.

Example 1

In this example, you will load a drawing or draw a figure and then use the SAVEIMG command to save the image.

Creating an image using the SAVEIMG command involves the following steps. Assume that the drawing is already loaded or drawn on the screen.

1. Invoke the SAVEIMG command by any one of the methods explained to display the **Save Image** dialog box.
2. Specify the file type in the **Format** area by selecting the radio button with the desired file type format.

3. Specify the area you want to save in the image file to be created by selecting the desired offset and size.
4. Choose the OK button.
5. The **Image File** dialog box is displayed. Enter the name you want to give the image file.

After this, AutoCAD starts saving the selected area to the file whose name you have specified. The saving operation is acknowledged by the following message: Writing file.

RASTER IMAGES*

A raster image consists of small square-shaped dots known as pixels. In a colored image, the color is determined by the color of pixels. The raster images can be moved, copied, or clipped. They can also be modified by using grips. You can also control the image contrast, transparency, and quality of the image.

The images can be 8-bit gray, 8-bit color, 24-bit color, or bitonal. The transparent images can be in color or gray scale. AutoCAD supports the following file formats.

Image Type	File Extension	Description
BMP	.bmp, .dib, .rle	Windows and OS/2 Bitmap Format
CAL-I	.gp4, .mil, .rst	Mil-R-Raster I
GIF	.gif	CompuServe Graphics Exchange Format
JFIF	.jpg	
PCX	.pcx	
PICT	.pct	
PNG	.png	
TARGA	.tga	
TIFF/LZW	.tif	Taffed image file format

Attaching Raster Images

Toolbar:	Insert, Image
Pull-down:	Insert, Image
Command:	IMAGE

Figure 21-7 Invoking the IMAGE command from the Insert toolbar

When you invoke the IMAGE command, AutoCAD displays the **Image** dialog box (Figure 21-8). To select the name of the file, choose the Attach button. If it is the first image you are inserting in the drawing, AutoCAD displays the **Attach Image File** dialog box; otherwise, the **Attach Image** dialog box is displayed. Select the file that you want to attach to the drawing; the selected image is displayed in the preview box. Choose the Open button to open the file and exit the dialog box. After selecting the file, the name of the file and its path and extension are displayed in the Attach Image dialog box. You can use the Browse button to select another file. Choose the OK button in the Attach Image dialog box and specify a point where you want to insert the image. Next, AutoCAD will prompt you to enter the scale factor. You can enter the scale factor or specify a point on the screen.

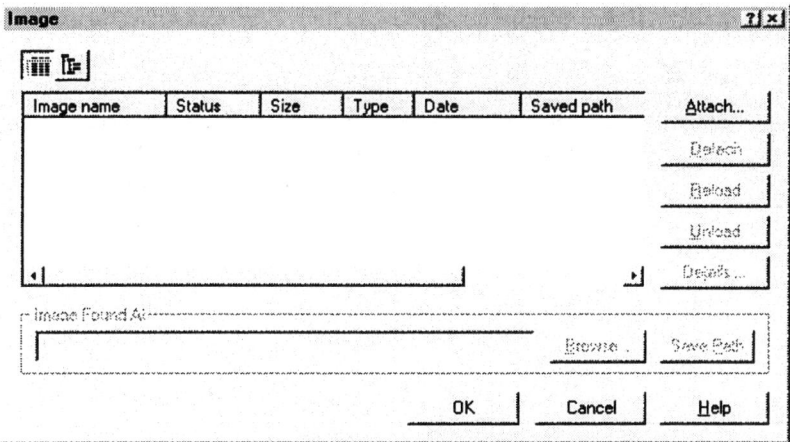

Figure 21-8 Image dialog box

The different options in the Image dialog box are:

Attach. Opens the Attach Image File dialog box, if it is the first image being inserted in the drawing; otherwise, AutoCAD opens the Attach Image dialog box.

Detach. Detaches the selected and all associated raster images from the current drawing. Once the image is detached, no information about the image is retained in the drawing.

Reload. Reloads an image. The changes made to the image since the last insert will be loaded on the screen.

Unload. Unloads an image. When you unload an image, AutoCAD retains information about the location and size of the image. If you reload the image, the image will appear at the same point and in the same size as the image was before unloading. Unloading the raster images enhances AutoCAD performance. Also, the unloaded images are not plotted.

Detail. Displays the details about the image like file name, saved path, file creation date, file size, file type, color and color path, pixel width and height, resolution, and default size. It also displays the image in the preview box.

Save Path. Saves the current path of the image.

Browse. Displays the Attach Image File dialog box.

Attach Image Dialog Box

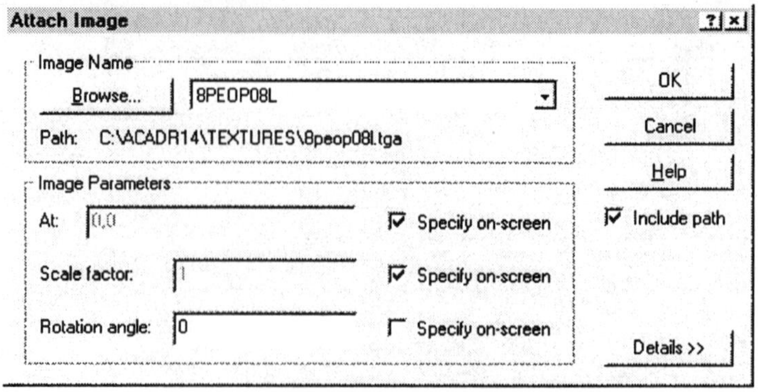

Figure 21-9 Attach Image dialog box

Browse Button. This button lets you browse through the files.

Image Parameters. You can specify the image parameters like the insertion point, scale factor, and rotation angle on the screen or by entering their values in the dialog box (Figure 21-9). If you want to specify these values on the screen, choose the corresponding check boxes.

Details. The Details button lists the information about the image, like horizontal and vertical resolution, image size in pixels, and image size in units.

EDITING RASTER IMAGE FILES*
Clipping Raster Images

Pull-down:	Modify, Clip
Command:	IMAGECLIP

You can clip a raster image by invoking the IMAGECLIP command. When you invoke the command, AutoCAD will prompt you to select the image to clip. Select the raster image by selecting the boundary edge of the image. The image boundary must be visible to select the image. At the next prompt, select the New boundary option and then define the clip boundary either by defining a rectangular or polygonal boundary. The clipping boundary must be specified in the same plane as the image or a plane that is parallel to it (Figure 21-10). The following is the command prompt sequence for the IMAGECLIP command:

Figure 21-10 Object before and after clipping

Command: **IMAGECLIP**
Select image to clip: *Select the image at the edge.*
ON/OFF/Delete/<New boundary>: *Enter to select New boundary.*
Polygonal/<Rectangular>: *Type P for Polygonal and then define the polygon shape.*

ON/OF. The ON/OFF option allows you to turn the clipping boundary on or off. When the clipping boundary is off, you can see the complete image; when it is on, the image that is within the clipping polygon is displayed.

Delete. The Delete option deletes the clipping boundary.

Adjusting Raster Image

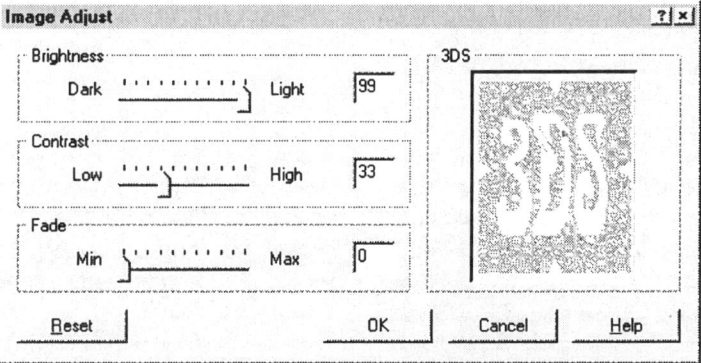

Figure 21-11 Image Adjust dialog box

The **IMAGEADJUST** command allows you to adjust the brightness, contrast, and fade of the raster image. When you invoke this command, AutoCAD displays the **Image Adjust** dialog box (Figure 21-11) which you can use to adjust the image. As you adjust the brightness, contrast, or fade, the image box displays the effect on the image. If you choose on the Reset button, the values are returned to default values (Brightness=50, Contrast=50, and Fade=0).

Image Quality
The **IMAGEQUALITY** command allows you to control the quality of the image that affects the display performance. A high-quality image takes longer time to display. When you change the quality, the display changes immediately without causing a REGEN. The images are always plotted using high-quality display.

Transparency
The **TRANSPARENCY** command is used with bitonal images to turn the transparency of the image background on or off. A bitonal image is one that consists of only a foreground and a background color. When you attach a bitonal image, the image assumes the color of the current layer. Also, the bitonal image and the bitonal image boundaries are always in the same color.

Image Frame
The **IMAGEFRAME** command is used to turn the image boundary on or off. If the image boundary is off, the image cannot be selected.

Other Editing Commands
You can use other editing commands like copy, move, and stretch to edit the raster image. You can also use the image as the trimming edge for trimming objects. However, you cannot trim an image. You can insert the raster image several times or make multiple copies of it. Each copy could have a different clipping boundary. You can also edit the image using grips.

Scaling Raster Images
The scale of the inserted image is determined by the actual size of the image and the unit of measurement (inches, feet, etc.). For example, if the image is 1"x1.26" and you insert this image with a scale factor of 1, the size of the image on the screen will be 1 AutoCAD unit by 1.26 AutoCAD units. If the scale factor is 5, the image will be five times larger. The image that you want to insert must contain the resolution information (DPI). If the image does not contain this information, AutoCAD treats the width of the image as one unit.

POSTSCRIPT FILES
PostScript is a page description language developed by Adobe Systems. It is used mostly in DTP (desktop publishing) applications. AutoCAD allows you to work with PostScript files. You can create and export PostScript files as well as convert PostScript files into regular AutoCAD drawing files (import). PostScript images have higher resolution than raster images. The extension for these files is **.EPS** (encapsulated PostScript).

PSOUT Command

| Command: | PSOUT |

As previously mentioned, any AutoCAD drawing file can be converted into a PostScript file. This can be accomplished with the PSOUT command. Once the PSOUT command is invoked, AutoCAD displays the **Create PostScript File** dialog box (Figure 21-12).

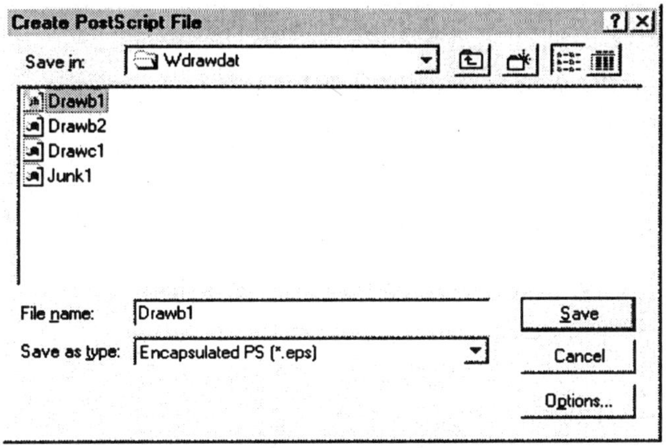

Data Exchange, Object Linking and Embedding, Multilines, and Digitizing 21-13

Figure 21-12 Create PostScript File dialog box

In the **File Name:** edit box, enter the name of the PostScript (EPS) file you want to create. Then you can choose the **Save** button to accept the default setting and create the PostScript file. You can also choose the **Options** button to change the settings through the **Export Options** dialog box (Figure 21-13) and then save the file. The Export Options dialog box has the following options:

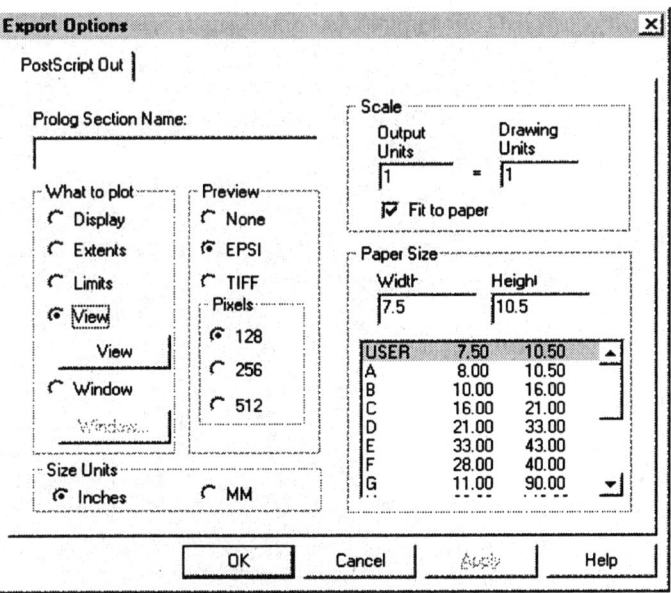

Figure 21-13 Export Options dialog box

Prolog Section Name
In this edit box, you can assign a name for a prolog section to be read from the acad.psf file.

What to plot
The What to plot area of the dialog box has the following options:

Display. If you specify this option when you are in model space, the image in the current viewport is saved in the specified EPS file. Similarly, if you are in paper space, the current view is saved in the specified EPS file.

Extents. If you use this option, the PostScript file created will contain the section of the AutoCAD drawing that currently holds objects. In this way, this option resembles the ZOOM Extents option. If you add objects to the drawing, they are also included in the PostScript file to be created because the extents of the drawing are also altered. If you reduce the drawing extents by erasing, moving, or scaling objects, then you must use the ZOOM Extents or ZOOM All option. Only then does the Extents option of the PSOUT command understand the extents of the drawing to be exported. If you are in model space, the PostScript file is created in relation to the model space extents; if you are in paper space, the PostScript file is created in relation to the paper space extents. If you invoke the PSOUT command Extents option when, perspective view is on and the position of camera is not out of the drawing extents, the following message is displayed:

> **PLOT Extents incalculable, using display**

In such cases, the EPS file is created as it would be created with the Display option.

Limits. With this option, you can export the whole area specified by the drawing limits. If the current view is not the plan view [viewpoint (0,0,1)], the Limits option exports the area just as the Extents option would.

View. Any view created with the VIEW command can be exported with this option. When this radio button is activated, the View button is also activated. Choose the View button to display the **View Name** dialog box from where you can select the view.

Window. In this option, you need to specify the area to be exported with the help of a window. When this radio button is activated, the Window button is also activated. Choose the Window button to display the **Window Selection** dialog box where you can select the Pick button and then specify the two corners of the window on the screen. You can also enter the coordinates of the two corners in the Window Selection dialog box.

Preview
The Preview area of the dialog box has two types of formats for preview images: EPSI and TIFF. If you want a preview image with no format, select the None radio button. If you select TIFF or EPSI, you are required to enter the pixel resolution of the screen preview in the **Pixel** area. You can select a preview image size of 128x128, 256x256, or 512x512.

Size Units
In this area, you can set the paper size units to Inches or Millimeters by selecting their corresponding radio buttons.

Data Exchange, Object Linking and Embedding, Multilines, and Digitizing 21-15

Scale
In this area, you can set an explicit scale by specifying how many drawing units are to be output per unit. You can check the **Fit to paper** check box so that the view to be exported is made as large as possible for the specified paper size.

Paper Size
You can select a size from the list or enter a new size in the Width and Height edit boxes to specify a paper size for the exported PostScript image.

PSIN Command

Pull-down:	Insert, Encapsulated PostScript
Command:	PSIN

AutoCAD allows you to import PostScript files. The PSIN (PostScript IN) command can be used to import a PostScript file. The PostScript file is imported into an AutoCAD drawing as a block.

Command: **PSIN**

The **Select PostScript File** dialog box (Figure 21-14) is displayed.

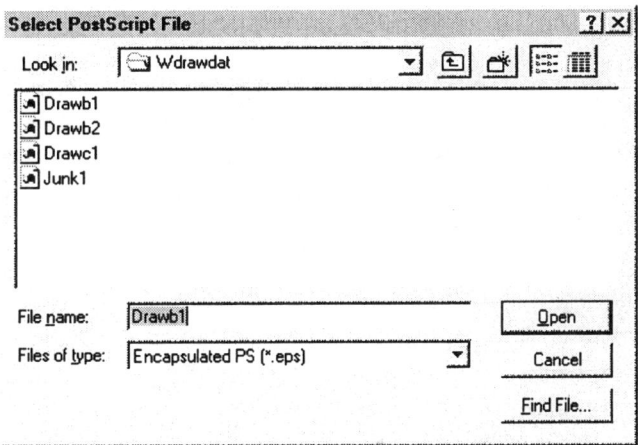

Figure 21-14 Select PostScript File dialog box

In this dialog box, all the EPS files in the current directory are listed. Select the file you want to import into AutoCAD, and then choose the **Open** button. The dialog box is cleared from the screen, and a rectangular box is displayed. You can move this rectangular box anywhere on the screen. The next prompts ask for the insertion point and the scale factor for the rectangular box (imported image), respectively:

Insertion point <0,0,0>: *Specify the location for the insertion point.*
Scale factor: *Specify the scale factor.*

The scale factor can also be specified by dynamically dragging the box. After you specify the scale factor, whatever is in the specified EPS file is inserted into the AutoCAD screen. From the prompts associated with PSIN command, it is obvious that the image is inserted into the AutoCAD drawing file as a block. By default, a box appears on the screen when you import an EPS file. This box contains the imported image, which you are unable to see until you specify the insertion point and the scale factor. If you want to see the image immediately instead of the box, set the **PSDRAG** variable to 1. By default, the value of this variable is 0; that is why, by default, you get the box. The quality of the rendering of the imported PostScript file depends on the **PSQUALITY** variable.

Command: **PSQUALITY**
New value for PSQUALITY <current>:

Different settings for **PSQUALITY** and their effects are as follows:

PSQUALITY = 0. When you insert an EPS file, the rectangular box representing the imported image is displayed. The file name of the imported EPS file is displayed in this rectangular box, and the size of this box corresponds to the size of the imported file.

PSQUALITY = A Positive Value. In this case, the imported image is displayed with the **PSQUALITY** value number of pixels per AutoCAD drawing unit. In other words, this is the resolution of the imported image. By default, the value of PSQUALITY is 75; hence, the imported image is displayed with 75 pixels per AutoCAD drawing unit. PostScript objects are filled.

PSQUALITY = A Negative Value. In such situations, the absolute value of the **PSQUALITY** variable controls the resolution of the imported image. The difference lies in the fact that the PostScript objects are not filled, but are displayed as outlines.

PostScript Fill Patterns (PSFILL Command)

| Command: PSFILL |

With the **PSFILL** command, you can fill closed 2D polylines with PostScript fill patterns. These fill patterns are declared in the **acad.psf** file. The declarations for a PostScript fill pattern are various parameters and arguments needed to influence the definition of a fill pattern on the screen. Some of the PostScript fill patterns are:

GRAYSCALE	LINEARGRAY	RADIALGRAY	SQUARE
WAFFLE	BRICK	ZIGZAG	STARS
AILOGO	SPECKS	RGBCOLOR	

You can also define your own fill patterns in the **acad.psf** file with the PostScript procedures. When you assign a fill pattern to a closed polyline, the pattern is not displayed on the screen. However, the fill pattern is recognized by the **PSOUT** command. If you import the PostScript file with a fill pattern using the **PSIN** command, the fill pattern is displayed on the screen as determined by the setting of **PSQUALITY, PSDRAG,** and **FILL** mode. Also, if you print such a file on a PostScript device, the fill pattern is printed. Depending on the fill pattern you specify,

Data Exchange, Object Linking and Embedding, Multilines, and Digitizing 21-17

AutoCAD issues prompts concerning various parameters and arguments needed by the fill pattern. For example, if you want to use the STARS fill pattern, the prompt sequence is:

Command: **PSFILL**
Select polyline: *Select the polyline to be filled.*
PostScript fill pattern (. = none) <current>/?: **STARS**
Scale <1.0000>: *Specify the scale factor for the fill pattern.*
LineWidth <1>: *Specify the line width for the fill pattern.*
ForegroundGray <100>: *Press Enter.*
BackgroundGray <0>: *Press Enter.*

Example 2

Draw a rectangle, and use the **PSFILL** command to assign the waffle pattern to the rectangle [Figure 21-15(a)]. Convert the rectangle to a PostScript file by using the **PSOUT** command, and then import the PostScript file using the **PSIN** command [Figure 21-15(b)].

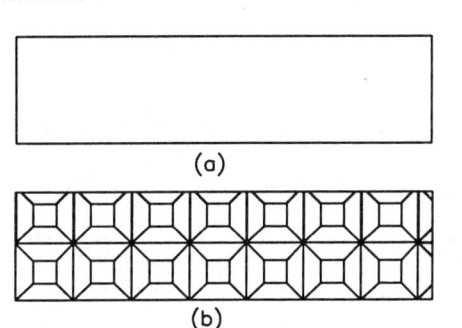

1. Set **PLINEWID** to 0 and use the **RECTANG** command to draw a rectangle [Figure 21-15(a)].

2. Use the **PSFILL** command to assign the waffle pattern to the rectangle.

Figure 21-15 Assigning the waffle pattern to a polyline

Command: **PSFILL**
Select polyline: *Select the rectangle.*
PostScript fill pattern (. = none) <current>/?: **Waffle**
Scale <1.0000>: ↵
Proportion <30>: ↵
LineWidth <1>: ↵
UpLeftGray <100>: ↵
BotRightGray <50>: ↵
TopGray <0>: ↵

3. Use the **PSOUT** command to convert the drawing into a PostScript file. Once the **PSOUT** command is invoked, AutoCAD displays the **Create PostScript File** dialog box. In the File Name: edit box, enter the name of the PostScript (EPS) file you want to create (Waffle). Then choose the **Options** button. AutoCAD displays **Export Options** dialog box.

4. In the **Preview** area of the Export Options dialog box, choose the **EPSI** radio button; then, choose the **128** radio button in the Pixels area.

5. In the **Scale** area, check the **Fit to paper** check box.

6. In the **Paper Size** area, enter 7.5 in the **Width** edit box and 10.2 in the **Height** edit box.

7. Choose the **Window** radio button in the **What to plot** area.

8. Choose OK in the Export Options dialog box, and then choose Save in the Create PostScript File dialog box.

9. In the command line area, the following prompts are displayed to select the window.

    First corner: *Specify a point close to lower left corner of the rectangle.*
    Second corner: *Specify a point close to upper right corner of the rectangle.*

10. Set **PSQUALITY** to -75 and **PSDRAG** to 0. Use the **PSIN** command to import and insert the PostScript file.

    Command: **PSIN**   *(Select the file Waffle from the dialog box.)*
    Insertion point <0,0,0>: *Specify the location for the insertion point.*
    Scale factor: *Specify the scale factor.*

OBJECT LINKING AND EMBEDDING

With Windows, it is possible to work with different Windows-based applications by transferring information between them. You can edit and modify the information in the original Windows application, and then update this information in other applications. This is made possible by creating links between the different applications and then updating those links, which in turn updates or modifies the information in the corresponding applications. This linking is a function of the OLE feature of Microsoft Windows. The OLE feature can also join together separate pieces of information from different applications into a single document. AutoCAD and other Windows-based applications, such as Microsoft Word, Notepad, and Windows WordPad support the Windows OLE feature.

For the OLE feature, you should have a **source document** where the actual object is created in the form of a drawing or a document. This document is created in an application called a **server application**. AutoCAD for Windows and Paintbrush can be used as server applications. Now this source document is to be linked to (or **embedded** in) the **destination document**, which is created in a different application, known as the **client application**. AutoCAD for Windows, Microsoft Word, and Windows WordPad can be used as client applications.

Clipboard

The transfer of a drawing from one Windows application to another is performed by copying the drawing or the document from the server application to the Clipboard. The drawing or document is then pasted in the client application from the Clipboard; hence, a Clipboard is used as a medium for storing the documents while transferring them from one Windows application to another. The drawing or the document on the Clipboard stays there until you copy a new drawing, which overwrites the previous one, or until you exit Windows. You can save the information present on the Clipboard with the **.CLP** extension.

Object Embedding

You can use the embedding function of the OLE feature when you want to ensure that there is no effect on the source document even if the destination document has been changed through the server application. Once a document is embedded, it has no connection with the source. Although editing is always done in the server application, the source document remains unchanged. Embedding can be accomplished by means of the following steps. In this example, AutoCAD for Windows is the server application and Windows WordPad is the client application.

1. Create a drawing in the server application (AutoCAD).

2. Open Windows WordPad (client application) from the **Accessories** group in the Program.

3. It is preferable to arrange both the client and the server windows so that both are visible (Figure 21-16).

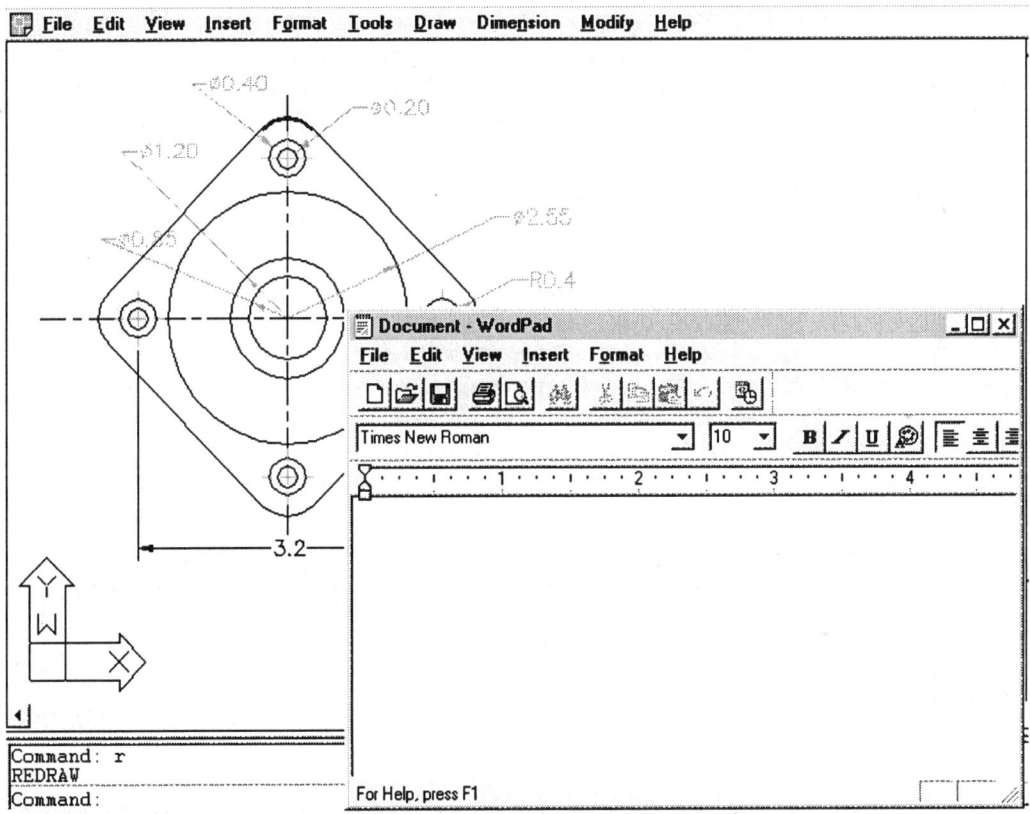

Figure 21-16 AutoCAD graphics screen with the WordPad window

4. In the AutoCAD graphics screen, use the **COPYCLIP** command. This command can be used in AutoCAD for embedding the drawings. This command can be invoked from the Standard toolbar by selecting the Copy to Clipboard icon, from the pull-down menu (Select Edit, Copy), or by entering **COPYCLIP** in the command line.

   Command: **COPYCLIP**

   The next prompt, **Select objects:**, allows you to select the entities you want to transfer. You can either select the full drawing by entering **All** or select some of the entities by selecting them. You can use any of the selection set options for selecting the objects. With this command the selected objects are automatically copied to the Windows Clipboard.

5. After the objects are copied to the Clipboard, make the WordPad window active. To get the drawing from the Clipboard to the WordPad application (client), select the **PASTE** command in the WordPad application. Invoke Paste from the **Edit** pull-down menu in Windows WordPad (Figure 21-17). You can also use **Paste Special** from the Edit pull-down menu, which will display a dialog box (Figure 21-18). In this dialog box, select the Paste radio button (default) for embedding, and then choose OK. The drawing is now embedded in the WordPad window.

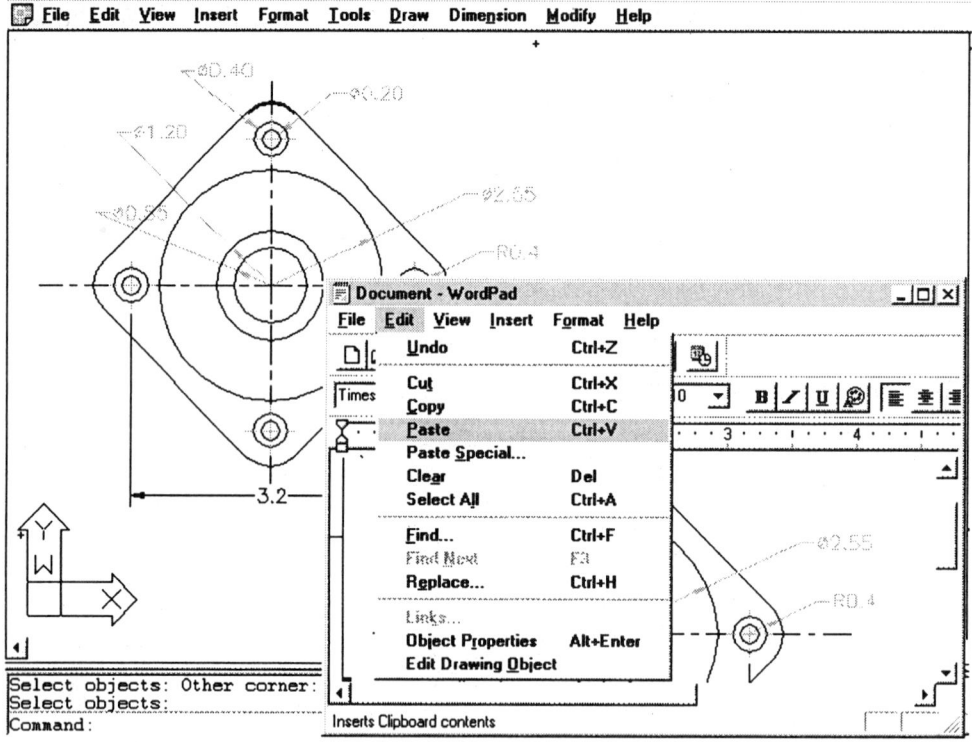

Figure 21-17 Pasting a drawing to the WordPad application by selecting Paste from the pull-down menu

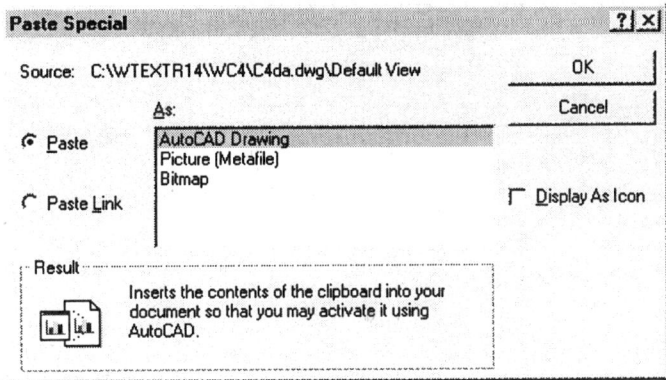

Figure 21-18 Paste Special dialog box

6. Your drawing is now displayed in the Write window, but it may not be displayed at the proper position. You can get the drawing in the current viewport by moving the scroll button up or down in the WordPad window. You can also save your embedded drawing by selecting Save from the File pull-down menu. It displays a **Save As** dialog box where you can enter a file name. You can also exit AutoCAD.

7. You can now edit your embedded drawing. Editing is performed in the server application, which in this case is AutoCAD for Windows. You can get the embedded drawing into the server application (AutoCAD) directly from the client application (WordPad) by double-clicking on the drawing in WordPad. The other method is by selecting **Edit Drawing Object** in the Edit pull-down menu. (This menu item has replaced **Object**, which was present before pasting the drawing.)

8. Now you are in AutoCAD, with your embedded drawing displayed on the screen, but as a temporary file with a file name, such as [Drawing in Document]. Here you can edit the drawing by changing the color and linetype or by adding and deleting text, entities, etc. In Figure 21-19 the two upper circles and their dimensions (diameter 20 and 40) have been erased, and a hexagon (dia 40) has been drawn in its place.

9. After you have finished modifying your drawing, select **Update WordPad** from the File pull-down menu in the server (AutoCAD). This menu item has replaced the previous Save menu item. When you select **Update**, AutoCAD automatically updates the drawing in Write (client application). Now you can exit this temporary file in AutoCAD.

10. This completes the embedding function so you can exit the Write application. While exiting, a dialog box which asks whether or not to save changes in WordPad is displayed.

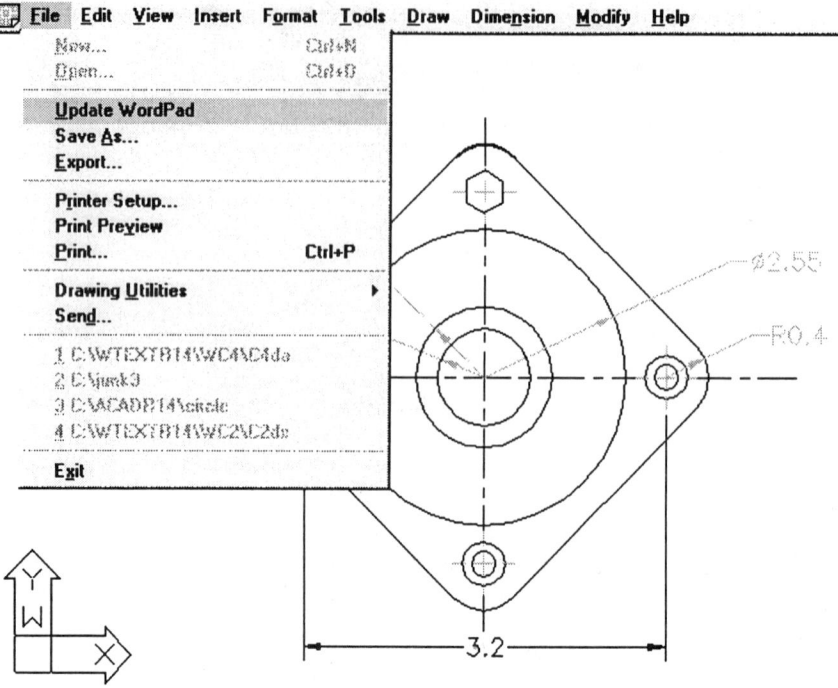

Figure 21-19 Selecting Update WordPad from the pull-down menu

Linking Objects

The linking function of OLE is similar to the embedding function. The only difference is that here a link is created between the source document and the destination document. If you edit the source, you can simply update the link, which automatically updates the client. This allows you to place the same document in a number of applications, and if you make a change in the source document, the clients will also change by simply updating the corresponding links. Consider AutoCAD for Windows to be the server application and Windows WordPad to be the client application. Linking can be performed by means of the following.

1. Open a drawing in the server application (AutoCAD). If you have created a new drawing, then you must save the drawing before you can link it with the client application.

2. Open Windows WordPad (the client application) from Accessories in the Program.

3. It is preferable to arrange both the client and the server windows so that both are visible.

4. In the AutoCAD graphics screen, use the **COPYLINK** command. This command can be used in AutoCAD for linking the drawing. This command can be invoked from the pull-down menu (Select Edit, Copy Link) or by entering **COPYLINK** at the Command: prompt. The prompt sequence is:

Command: **COPYLINK**

Data Exchange, Object Linking and Embedding, Multilines, and Digitizing

The COPYLINK copies the whole drawing in the current viewport directly to the Clipboard. Here you cannot select the objects for linking. If you want only a portion of the drawing to be linked, you can zoom into that view so that it is displayed in the current viewport prior to invoking the COPYLINK command. This command also creates a new view of the drawing having a name OLE1. Now you can exit AutoCAD.

5. Make the WordPad window active. To get the drawing from the Clipboard to the Write (client) application, select the **Paste Special** from the Edit pull-down menu, which will display the Paste Special dialog box. In this dialog box, select the Paste Link radio button for linking. The drawing is now linked to the WordPad window.

6. Your drawing is now displayed in the WordPad window. You can also save your linked drawing by selecting Save from the File pull-down menu. It displays a **Save As** dialog box where you can enter a file name.

7. You can now edit your linked drawing. Editing can be performed in the server application, which in this case is AutoCAD for Windows. You can get the linked drawing in the server (AutoCAD) directly from the client (WordPad) by double-clicking on the drawing in WordPad. The other method is by selecting **Edit Linked Drawing Object** in the Edit pull-down menu. (This menu item has replaced **Object**, which was present before pasting the drawing.)

8. Now you are in AutoCAD, with the original drawing displayed on the screen. You can edit the drawing by changing the color and linetype or by adding and deleting text, entities, etc. Then save your drawing in AutoCAD by using the SAVE command. You can now exit AutoCAD.

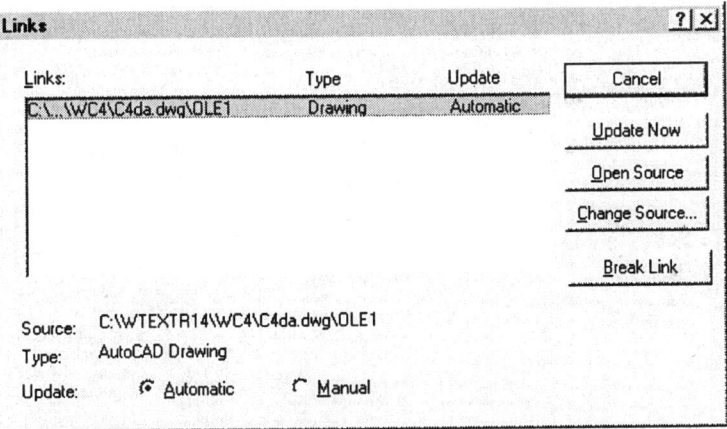

Figure 21-20 Links dialog box

9. You will notice that the drawing is automatically updated, and the changes made in the source drawing are present in the destination drawing also. This automatic updating is dependent on the selection of the **Automatic** radio button (default) in the **Links** dialog box (Figure 21-20). The Links dialog box can be invoked by selecting Links from the Edit pull-down menu. For updating manually, you can activate the **Manual** radio button in the dialog box. In the manual case, after making changes in the source document and saving it, you need to invoke the Links dialog box and then select the **Update Now** button; then, select Cancel. This will update the drawing in the client application and display the updated drawing on the WordPad.

10. Exit the Write application after saving the updated file.

CREATING MULTILINES

The AutoCAD Multiline feature allows you to draw composite lines that consist of multiple parallel lines. You can draw these lines with the **MLINE** command. Before drawing multilines, you need to set the multiline style. This can be accomplished using the **MLSTYLE** command. Also, editing of the multilines is made possible by the **MLEDIT** command.

DEFINING MULTILINE STYLE
(MLSTYLE COMMAND)

| **Pull-down:** | Format, Multiline Style... |
| **Command:** | MLSTYLE |

The **MLSTYLE** command allows you to set the style of multilines. You can specify the number of elements in the multiline and the properties of each element. The style also controls the end caps, the end lines, and the color of multilines and fill. When you enter this command, AutoCAD displays the **Multiline Styles** dialog box (Figure 21-21). With this dialog box, you can set the spacing between the parallel lines, linetype pattern, colors, solid fill, and capping arrangements. By default, the multiline style (STANDARD) has two lines that are offset at 0.5 and -0.5.

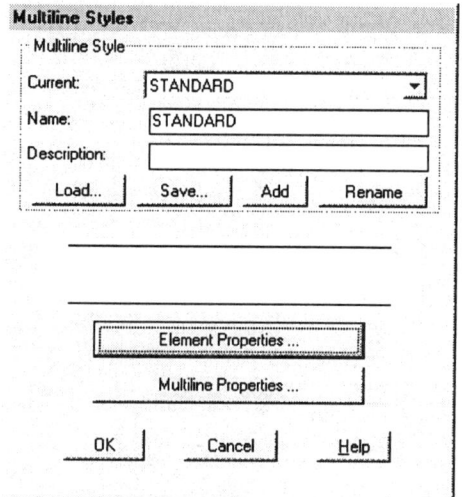

Figure 21-21 Multiline Styles dialog box

Multiline Style Area

The Multiline Style area of the Multiline Styles dialog box has the following options.

Data Exchange, Object Linking and Embedding, Multilines, and Digitizing   21-25

Current. This edit box displays and sets the current multiline style. If several styles have been defined, the name of the current style is displayed in the Current: edit box. You can use the arrow button to display the drop-down list which has all the predefined styles. You can choose any style to make it current. The list of multiline styles can include the multiline styles that have been defined in an externally referenced drawing (xref drawing).

Name. This edit box lets you enter the name of the multiline style you are defining. The STANDARD style is the default style. You can also use it to rename a style.

Description. This edit box allows you to enter the description of the multiline style. The description can be up to 255 characters long, including spaces.

Load.... This button allows you to load a multiline style from an external multiline library file (acad.mln). When you select this button, AutoCAD displays the **Load Multiline Style** dialog box. With this dialog box, you can select the style you want to make current. You can also use this dialog box to load a predefined multiline file (**.mln** file) by selecting the **File...** button and then selecting the file you want to load. Once the file is loaded, you can select a style that is defined in the **.mln** file.

Save.... This button lets you save or copy the current multiline style to an external file (**.mln** file). When you select this button, AutoCAD displays the **Save Multiline Styles** dialog box listing the names of the predefined multiline style (**.mln**) files. From the file listing, select, or enter the name of the file where you want to save the current multiline style.

Add. This button allows you to add the multiline style name (the style name displayed in the Name: edit box) to the current multiline file (**.mln** file).

Rename. This button allows you to rename the current multiline style. It will rename the multiline style that is displayed in the Current: edit box with the name displayed in the Name: edit box. You cannot rename the **Standard** multiline style.

Line Display Panel. The **Multiline Styles** dialog box also displays the multiline configuration in the display panel. The panel will display the color, linetype, and relative spacing of the lines.

Element Properties

If you select the **Element Properties...** button in the **Multiline Styles** dialog box, AutoCAD will display the **Element Properties** dialog box (Figure 21-22). Before making any changes in this dialog box, you should enter a new multiline name in the Name: edit box of the Multiline Style dialog box, and then make it current by selecting the Add button. The buttons in the Element Properties dialog box are activated only when the current style is not STANDARD. This dialog box gives you the following options for

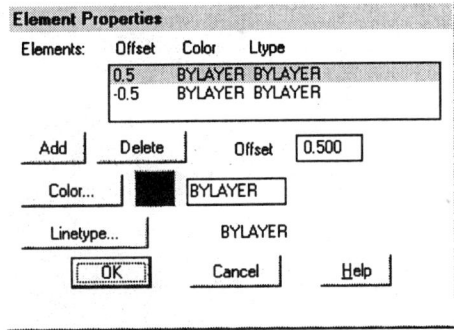

Figure 21-22 Element Properties dialog box

setting the properties of the individual lines (elements) that constitute the multiline.

Elements. This box displays the offset, color, and linetype of each line that constitutes the current multiline style. The lines are always listed in descending order based on the offset distance. For example, a line with 0.5 offset will be listed first and a line with 0.25 offset will be listed next.

Add. This button lets you add new lines to the current line style. The maximum number of lines you can add is 16. When you select the Add button, AutoCAD inserts a line with the offset distance of 0.00. After the line is added, you can change its offset distance, color, or linetype by selecting the Offset edit box or the Color... or Linetype... button.

Delete. This button allows you to delete the line highlighted in the Elements: list box.

Offset. This edit box allows you to change the offset distance of the selected line in the Elements: list box. The offset distance is defined with respect to the origin (0,0). The offset distance can be a positive or a negative value, which enables you to center the lines.

Color.... This button allows you to assign a color to the selected line. When you select this button, AutoCAD displays the standard color dialog box (Select Color dialog box). You can select a color from the dialog box or enter a color number or name in the edit box located to the right of the color swatch box in the Element Properties dialog box.

Linetype.... This button allows you to assign a linetype to the selected line. When you select this button, AutoCAD displays the standard linetype dialog box (Select Linetypes dialog box). After selecting the linetype, select the OK button to exit the dialog box.

Multiline Properties

If you select the Multiline Properties button from the Multiline Styles dialog box, AutoCAD will display the **Multiline Properties** dialog box (Figure 21-23). You can use this dialog box to define multiline properties such as display joints, end caps, and background fill. The buttons in this dialog box are activated only when the current style is not STANDARD. The dialog box provides the following options.

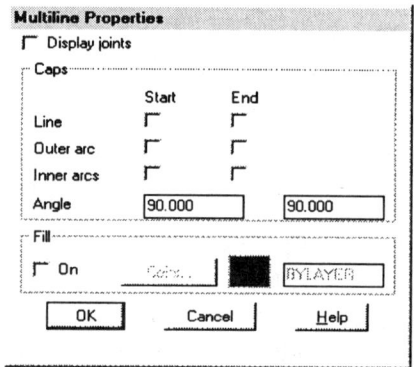

Figure 21-23 Multiline Properties dialog box

Display Joints. If you select the Display joints check box, AutoCAD will display a **miter** line (joint) across all elements of the multiline at the point where the two multilines meet. If you draw only one multiline segment, no miter line is drawn (because there is no intersection point).

Line. This option draws a line cap at the start and end of each multiline. It has two check boxes that control the start and end caps (Figure 21-24).

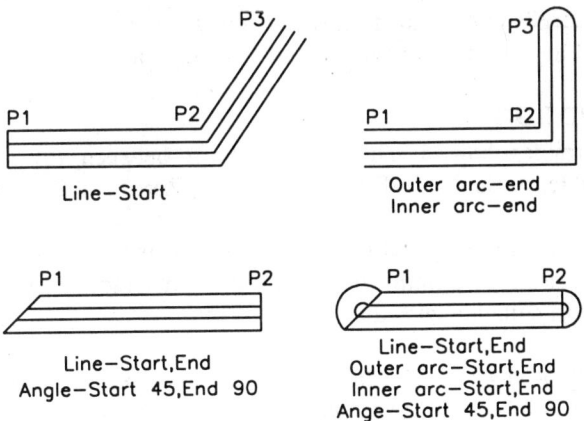

Figure 21-24 Drawing multilines with different end cap specifications

Outer Arc. This option draws an arc (semicircle) between the endpoints of the outermost lines.

Inner Arc. This option controls the inner arcs at the start and end of a multiline. The arc is drawn between the even-numbered inner lines. For example, if there are two inner lines, an arc will be drawn at the ends of these lines. However, if there are three inner lines, the middle line is not capped with an arc.

Angle. This option controls the cap angle at the start and end of a multiline. This angle can be from 10 degrees to 170 degrees.

Fill. This option toggles the fill on and off. If the fill is ON, you can select the color of the fill by selecting the color button. When you select this button, AutoCAD displays the standard color dialog box. You can also set the color by entering the color name or color number in the edit box located to the right of the color swatch box.

DRAWING MULTILINES (MLINE COMMAND)

Toolbar:	Draw, Multiline
Pull-down:	Draw, Multiline
Command:	MLINE

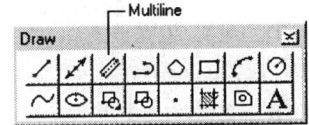

Figure 21-25 Selecting the Multiline icon from the Draw toolbar

The **MLINE** command can be used to draw multilines. The following is the command prompt sequence for the **MLINE** command:

Command: **MLINE**
Justification=Top, Scale=1.00, Style=STANDARD
Justification/Scale/STyle/<From point>: *Select a point.*
<To point>: *Select the second point.*
Undo/<To point>: *Select next point or enter U for undo.*
Close/Undo/<To point>: *Select next point, enter U, or enter C for close.*

When you enter the **MLINE** command, it always displays the status of the multiline justification, scale, and style name. The command provides the following options.

Justification Option

The justification determines how a multiline is drawn between the specified points. Three justifications are available for the MLINE command: Top, Zero, and Bottom (Figure 21-26).

Top. This justification produces a multiline in which the top line coincides with the selected points. Since the line offsets in a multiline are arranged in descending order, the line with the largest positive offset will coincide with the selected points.

Zero. This option will produce a multiline in which the zero offset position of the multiline coincides with selected points. Multilines will be centered if positive and negative offsets are equal.

Bottom. This option will produce a multiline in which the bottom line (the line with the least offset distance) coincides with the selected point when the line is drawn from left to right.

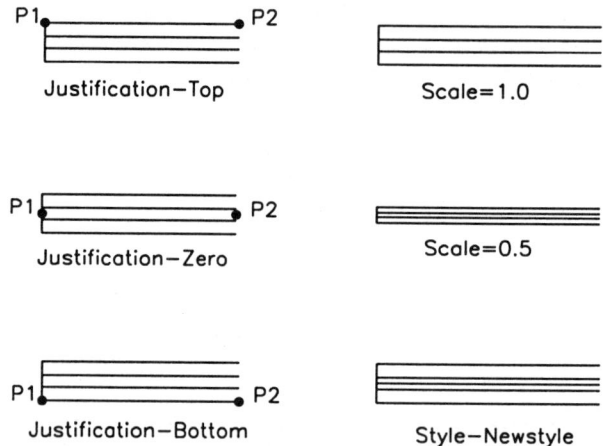

Figure 21-26 Drawing multilines with different justifications

Scale Option

The **Scale** option allows you to change the scale of the multiline. For example, if the scale factor is 0.5, the distance between the lines (offset distance) will be reduced to half. Therefore, the width of the multiline will be half of what was defined in the multiline style. A negative scale factor will reverse the order of the offset lines. Multilines are drawn so that the line with the maximum offset distance is at the top and the line with the least offset distance is at the bottom. If you enter a scale factor of -0.5, the order in which the lines are drawn will be reversed, and the offset distances will be reduced by half. (The line with the least offset will be drawn at the top.) Here it is assumed that the lines are drawn from left to right. If the lines are drawn from right to left, the offsets are reversed. Also, if the scale factor is 0, AutoCAD forces the multiline into a single line. The line still possesses the properties of a multiline. The scale does not affect the linetype scale (LTSCALE).

STyle Option

The **STyle** option allows you to change the current multiline style. The style must be defined before using the STyle option to change the style.

EDITING MULTILINES (USING GRIPS)

Multilines can be edited using grips (Figure 21-27). When you select a multiline, the grips appear at the endpoints based on the justification used when drawing multilines. For example, if the multilines are top-justified, the grips will be displayed at the endpoint of the first (top) line segment. Similarly, for zero- and bottom-justified multilines, the grips are displayed on the centerline and bottom line, respectively.

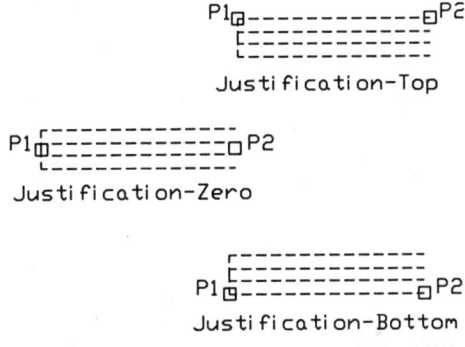

Figure 21-27 Using grips to edit multilines

> **Note**
> *Multilines do not support certain editing commands, such as BREAK, CHAMFER, FILLET, TRIM, and EXTEND. However, commands such as COPY, MOVE, MIRROR, STRETCH, and EXPLODE, and certain object snap modes can be used with multilines. You must use the MLEDIT command to edit multilines. The MLEDIT command has several options that make it easier to edit these lines.*

EDITING MULTILINES (USING THE MLEDIT COMMAND)

Toolbar:	ModifyII, Edit Multiline
Pull-down:	Modify, Object, Multiline
Command:	MLEDIT

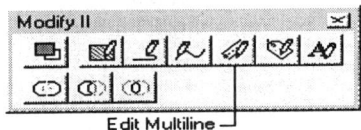

Figure 21-28 Edit Multiline icon in the ModifyII toolbar

When you enter the **MLEDIT** command, AutoCAD displays the **Multiline Edit Tools** dialog box (Figure 21-29). This dialog box contains five basic editing tools. To edit a multiline, first select the editing operation you want to perform by double-clicking on the image tile or by selecting the image tile; then select the OK button. Once you have selected the editing option, AutoCAD will prompt you to select the object or the points, depending on the option you have selected. If you press Enter after you are done editing, the dialog box will return, and you can continue editing. The following is the list of options for editing multilines:

    Cross Intersection
        Closed Cross
        Open Cross
        Merged Cross

21-30 Customizing AutoCAD

    Tee Intersection
        Closed Tee
        Open Tee
        Merged Tee
    Corner Joint
    Adding and Deleting Vertices
        Add Vertex
        Delete Vertex
    Cutting and Welding Multilines
        Cut Single
        Cut All
        Weld All

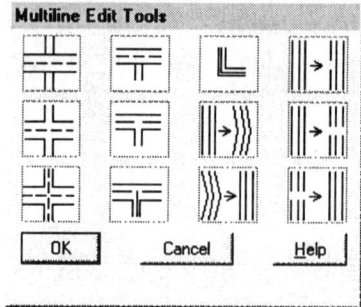

Figure 21-29 Multiline Edit Tools dialog box

Cross Intersection

With the **MLEDIT** command options, you can create three types of cross intersections: closed, open, and merged. You must be careful how you select the objects because the order in which you select them determines the edited shape of a multiline (Figure 21-30). The multilines can belong to the same multiline or to two completely different multilines. The following is the prompt sequence for creating the cross intersection:

    Command: **MLEDIT**

The **Multiline Edit Tools** dialog box is displayed. Select the Closed Cross image tile and then choose the **OK** button. The dialog box is removed from the screen, and the following prompts appear in the command line.

    Select first mline: *Select the first multiline.*
    Select second mline: *Select the second multiline.*
    Select first mline (or Undo):

If you select Undo, AutoCAD undoes the operation and prompts you to select the first multiline. However, if you select another multiline, AutoCAD will prompt you to select the second multiline. If you press Enter twice, the Multiline Edit Tools dialog box is returned.

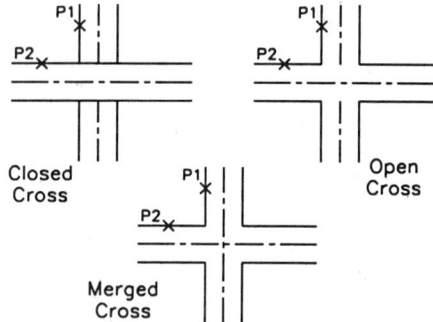

Figure 21-30 Using MLEDIT to edit multilines (Cross Intersection)

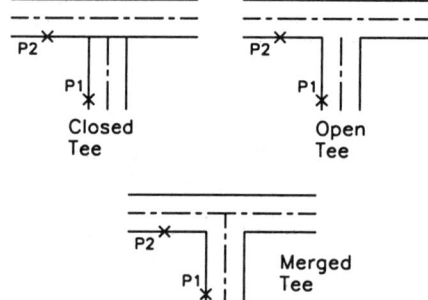

Figure 21-31 Using MLEDIT to edit multilines (Tee Intersection)

Tee Intersection

With the **MLEDIT** command options, you can create three types of tee-shaped intersections: closed, open, and merged. As with the cross intersection, you must be careful how you select the objects because the order in which you select them determines the edited shape of a multiline (Figure 21-31). The prompt sequence for the tee intersection is same as for the cross intersection.

Command: **MLEDIT**

The **Multiline Edit Tools** dialog box is displayed. Select the Closed Tee image tile and then choose the **OK** button. The dialog box is removed from the screen, and the following prompts appear in the command line.

Select first mline: *Select the first multiline.*
Select second mline: *Select the intersecting multiline.*
Select first mline (or Undo): *Select another multiline, enter Undo, or press Enter.*

Corner Joint

The **Corner Joint** option creates a corner joint between the two selected multilines. The multilines must be two separate objects (multilines) (Figure 21-32). When you specify the two multilines, AutoCAD trims or extends the first multiline to intersect with the second one. The following is the command prompt sequence for creating a corner joint:

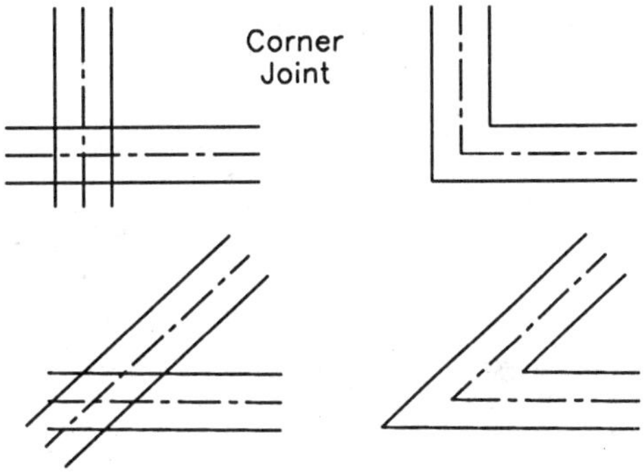

Command: **MLEDIT**

The **Multiline Edit Tools** dialog box is displayed. Select the Corner Joint image tile and then choose the **OK** button. The dialog box is removed from the screen, and the following prompts appear in the command line.

Figure 21-32 Using MLEDIT to edit multilines (Corner Joint)

Select first mline: *Select the multiline to trim or extend.*
Select second mline: *Select the intersecting multiline.*
Select first mline (or Undo): *Select another multiline, enter Undo, or press Enter.*

Adding and Deleting Vertices

You can use the **MLEDIT** command to add or delete the vertices of a multiline (Figure 21-33). When you select a multiline for adding a vertex, AutoCAD inserts a vertex point at the point where the object was selected. If you want to move the vertex, use grips. Similarly, you can use the **MLEDIT** command to delete the vertices by selecting the object whose vertex point you want to delete. AutoCAD removes the vertex that is in the positive direction of the selected multiline segment.

Command: **MLEDIT**

The **Multiline Edit Tools** dialog box is displayed. Select the Add Vertex image tile and then choose the **OK** button. The dialog box is removed from the screen, and the following prompts appear in the command line.

> Select mline: *Select the multiline for adding vertex.*
> Select mline (or Undo): *Select another multiline, enter undo, or press Enter.*

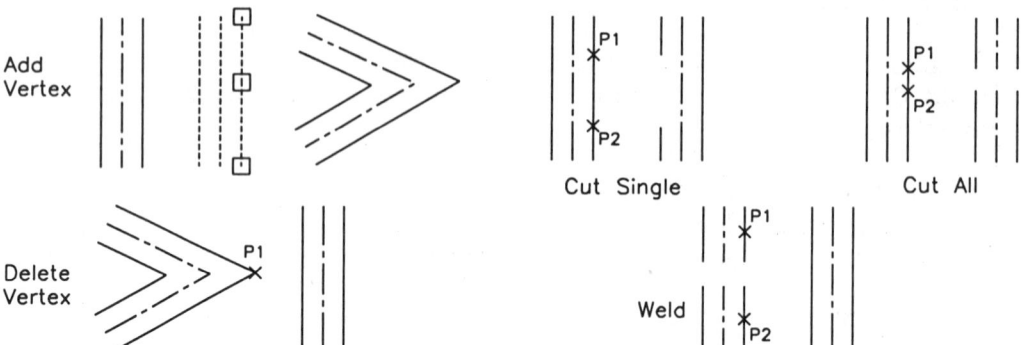

Figure 21-33 Using MLEDIT to edit multilines (adding and deleting vertices)

Figure 21-34 Using MLEDIT to edit multilines (cutting and welding multilines)

Cutting and Welding Multilines

You can use the **MLEDIT** command to cut or weld the lines. When you cut a multiline, it does not create two separate multilines. They are still a part of the same object (multiline). Also, the points selected for cutting the multiline do not have to be on the same element of the multiline (Figure 21-34).

> Command: **MLEDIT**

The **Multiline Edit Tools** dialog box is displayed. Select the Cut Single image tile and then choose the **OK** button. The dialog box is removed from the screen, and the following prompts appear in the command line.

> Select mline: *Select the multiline. (The point where you select the multiline specifies the first cut point.)*
> Select second point: *Select the second cut point.*
> Select mline (or Undo): *Select another multiline, enter Undo, or press Enter.*

The Weld option welds the multilines that have been cut by using MLEDIT command options (Figure 21-34). You cannot weld or join separate multilines.

> Command: **MLEDIT**

The **Multiline Edit Tools** dialog box is displayed. Select the Weld All image tile and then choose the **OK** button. The dialog box is removed from the screen, and the following prompts appear in the command line.

Select mline: *Select the multiline.*
Select second point: *Select the second multiline.*
Select mline (or Undo): *Select another multiline, enter Undo, or press Enter.*

-MLEDIT COMMAND

You can also edit the multilines using the command line and not the Multiline Edit Tools dialog box. To accomplish this, enter **-MLEDIT** at the Command: prompt.

Command: **-MLEDIT**
Mline editing options AV/DV/CC/OC/MC/CT/OT/MT/CJ/CS/CA/WA:

After you select an option, AutoCAD will prompt you to select the multilines or specify the points. The prompt sequence will depend on the option you select, as discussed earlier under the MLEDIT command. The following are the command line options:

AV	Add Vertex	OT	Open Tee
DV	Delete Vertex	MT	Merged Tee
CC	Closed Cross	CJ	Corner Joint
OC	Open Cross	CS	Cut Single
MC	Merged Cross	CA	Cut All
CT	Closed Tee	WA	Weld All

SYSTEM VARIABLES FOR MLINE

CMLJUST   Stores the justification of the current multiline (0-Top, 1-Middle, 2-Bottom)
CMLSCALE   Stores the scale of the current multiline
CMLSTYLE   Stores the name of the current multiline style

Example 3

In the following example, you will create a multiline style that represents a wood-frame wall system. The wall system consists of 1/2" wallboard, 3 1/2" 2x4" wood stud, and 1/2" wallboard.

Step 1. Use the **MLSTYLE** command to display the Multiline Styles dialog box. The current style, STANDARD, will be edited to create the new multiline style.

Step 2. Select the **Name:** edit box and replace the word **STANDARD** with **2x4_Wood**.

Step 3. Select the **Description:** edit box and enter **Wallboard Wood Framed 2x4 Partition**.

Step 4. Select the **Add** button to add the new style to the current multilines.

Step 5. Select the **Element Properties...** button to display the Element Properties dialog box. The element properties of the STANDARD multiline style remain.

Step 6. Select the **0.5** line definition in the **Elements** display box. Select the **Offset** edit box and replace **0.500** with **1.75**. This redefines the first line as being 1.75" above the centerline of the wall.

Step 7. Select the **-0.5** line definition in the **Elements** display box. Select the **Offset** edit box and replace **0.500** with **-1.75**. This redefines the second line as being 1.75" below the centerline of the wall.

Step 8. Select the **Add** button to add a new line to the current line style.

Step 9. The new **0.0** line definition in the **Elements** display box is already highlighted. Select the **Offset** edit box and replace **0.000** with **2.25**.

Step 10. Select the **Color** edit box and replace **BYLAYER** with **YELLOW**.

Step 11. Repeat steps 7 through 9, this time using the value -2.25 in step 8 to add another line to the current line style.

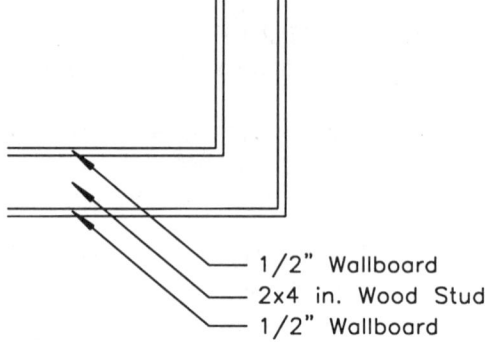

Figure 21-35 Creating a multiline style for wood-frame wall system

Step 12. Select the **OK** button to accept the changes to the **Element Properties** and return to the **Multiline Styles** dialog box. The new multiline style will be displayed (Figure 21-35).

Step 13. Select the **Save...** button to save the current multiline style to an external file for use in later drawing sessions. From the file listing, select the file (or enter the name of the file) where you want to save the current multiline style.

Step 14. Select the **OK** button to return to the drawing editor. To test the new multiline style, use the **MLINE** command and draw a series of lines.

DIGITIZING DRAWINGS

In digitizing, the information in the drawing is transferred to a CAD system by locating the points on the drawing that is placed on the top surface of the digitizing tablet. Sometimes it is easier to digitize an existing drawing than to create the drawing again on a CAD system. The digitizing of drawings is an important application in industry. The advantages of a drawing based on a CAD system have prompted companies to digitize drawings.

If you want to digitize a drawing, you should have a digitizer as large as the size of the largest drawing to be digitized so that you do not have to realign the drawing sheet. But situations often arise in which the digitizer is not as large as the drawing to be digitized. In such cases, after digitizing a part of the drawing (as large as the digitizer can accommodate), digitize the other parts until the entire drawing is digitized. Digitizing can be carried out in the Tablet mode. In the Tablet mode, AutoCAD uses the digitizing tablet as a digitizer, not as a screen pointing device. In this mode, the coordinate system of the paper drawings can be mapped directly into AutoCAD. The Tablet mode can be turned on or off with the **TABLET** command or by pressing function key F4.

    Command: **TABLET**
    Option (ON/OFF/CAL/CFG): **ON**   (or OFF)

TABLET COMMAND

| **Pull-down:** | Tools, Tablet |
| **Command:** | TABLET |

The first step in digitizing a drawing is to configure the tablet so that the maximum possible area on the tablet can be used. You can do this with the **TABLET** command.

    Command: **TABLET**

At the next prompt, enter CFG (Configuration option) to configure the tablet.

    Option (ON/OFF/CAL/CFG): **CFG**

After this, you are prompted to specify the number of tablet menus you want. When digitizing drawings, the whole tablet area is used for digitizing the drawing; hence, there are no tablet menus.

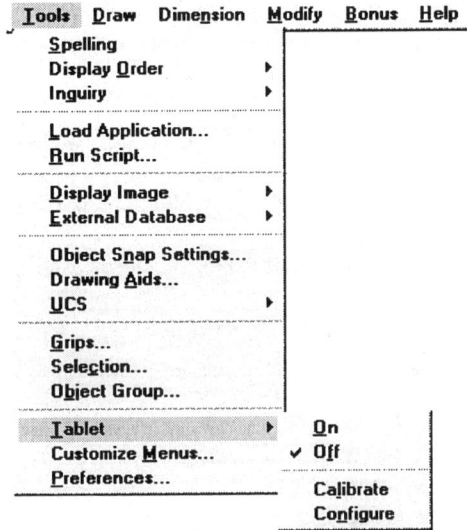

Figure 21-36 Tablet options in the Tools pull-down menu

    Enter the number of tablet menus desired (0-4)< >: **0**

At the next prompt, enter **Y** so that reconfiguration of the tablet is possible:

    Do you want to respecify the Fixed Screen Pointing area? <N> **Y**
    Digitize lower left corner of Fixed Screen pointing area: *Specify the lower left corner of the pointing area.*
    Digitize upper right corner of Fixed Screen pointing area: *Specify the upper right corner of the pointing area.*
    Do you want to specify the Floating Screen pointing area? <N>: **N**

The next step in digitizing a drawing is to calibrate the tablet. For this, position the drawing to be digitized to the tablet with the help of adhesive tape. Calibration is possible in model space and paper space.

    Command: **TABLET**
    Option (ON/OFF/CAL/CFG): **CAL**

After you select or enter CAL, the Tablet mode is turned on and the screen crosshairs cursor disappears. The process of calibration involves digitizing two or more points on the drawing and then entering their coordinate values. The points to be selected can be anywhere on the drawing. Once you have specified the position of the points and the coordinate values of those positions, AutoCAD automatically calibrates the digitizing area.

> Digitize point #1: *Select the first point.*
> Enter coordinates for point #1: *Enter the coordinates of the first point selected.*
> Digitize point #2: *Select the second point.*
> Enter coordinates for point #2: *Enter the coordinates of the second point selected.*
>
> Digitize point #3 (or RETURN to end): *Select the third point.*
> Enter coordinates for point #3: *Enter the coordinates of the second point selected.*
>
> Digitize point #4 (or RETURN to end): *Press Enter to end.*

AutoCAD will display Text Window displaying the transformation information (Figure 21-37).

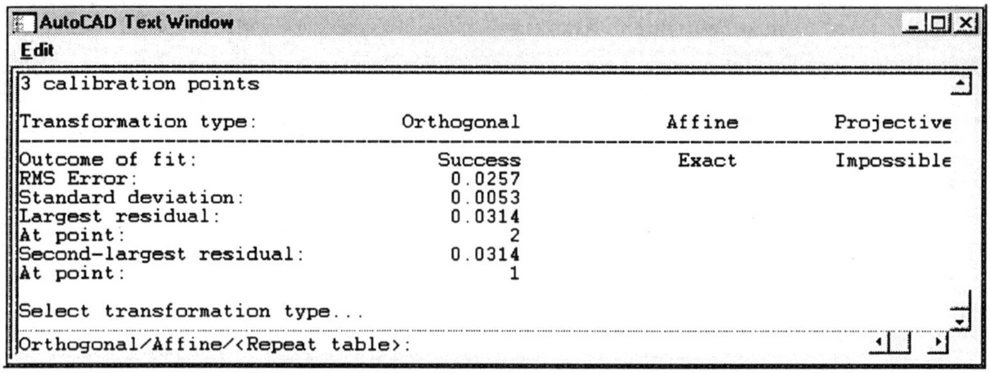

Figure 21-37 AutoCAD Text Window

> Select transformation type...
> Orthogonal/Affine/Projective/<Repeat table>: **A**

You can enter any number of points. The more points you enter, the more accurate is the digitizing. If you have entered only two points, AutoCAD will automatically compute orthogonal transformation. If you have entered three or more points, AutoCAD will compute orthogonal, affine, and projective transformations and determine which best fits the configuration of the selected points. After configuring and calibrating the tablet, the crosshairs cursor is redisplayed on the screen. Now, you can use AutoCAD commands to digitize the drawing.

Floating Screen Pointing Area*

You can configure the tablet for two pointing areas: fixed pointing area and floating pointing area. Before you configure the tablet, the entire tablet surface is the fixed screen pointing area. At this point the digitizer tablet performs like a pointing device. You can use the CFG option of the TABLET command to specify the new fixed screen pointing area. When you enter this command,

Data Exchange, Object Linking and Embedding, Multilines, and Digitizing 21-37

AutoCAD will prompt you to specify the two points: lower left corner and upper right corner of the new screen pointing area. Now, if you move the digitizer puck within the fixed pointing area, the cursor moves on entire screen because there is one-to-one correspondence between the fixed pointing area on the digitizer and the screen. You can also define a floating screen pointing area that is the same as the screen pointing area or you can specify a different area. The following is the prompt sequence to specify a floating screen pointing area:

Command: **TABLET**
Option (ON/OFF/CAL/CFG): **CFG**
Enter the number of tablet menus desired (0-4) < > : **0**

Do you want to respecify the Fixed Screen Pointing area? <N> N
Do you want to specify the Floating Screen pointing area? <N> : **Y**
Do you want the Floating Screen Pointing Area to be the same size as the Fixed Pointing Area? <Y> : **N**
Digitize lower left corner of Floating Screen pointing area: *Specify the lower left corner of the pointing area.*
Digitize upper right corner of Floating Screen pointing area: *Specify the upper right corner of the pointing area.*

Next, AutoCAD will prompt you to specify the toggle key. By default, you can use the function key F12 to toggle between the floating and fixed screen pointing area or you can define a button of the digitizer puck as the toggle key. You can still use the function key F4 to turn the Tablet mode on or off to digitize a drawing.

REVIEW QUESTIONS

Data Exchange and Object Linking and Embedding

1. You can import a DXF file into an AutoCAD drawing file with the _____ command.

2. The _____ command is used to create drawing files out of DXB format files.

3. If you import a DFX or DXB file into an old drawing, the settings of layers, blocks, etc., of the file being imported are _____ by the settings of the file into which you are importing the DXF or DXB file.

4. If you import a PostScript file with a fill pattern using the PSIN command, the fill pattern is not displayed on the screen. (T/F)

5. The _____ is used as a medium for storing the documents while transferring them from one Windows application to another.

21-38   Customizing AutoCAD

6. The _____ command can be used in AutoCAD for embedding drawings.

7. You can edit your embedded drawing in the _____ application.

8. You can get an embedded drawing into the server application directly from the client application by _____ _____ on the drawing.

9. The _____ command can be used in AutoCAD for linking a drawing.

10. The COPYLINK command copies a drawing in the _____ to the Clipboard.

11. The _____ command is used to create an ASCII format file with the **.DXF** extension from AutoCAD drawing files.

12. With the Binary option of the DXFOUT command you can also create binary format files. Binary DXF files are _____ efficient and occupy only 75 percent of the ASCII DXF file. File access for files in binary format is _____ than for the same file in ASCII format.

13. When importing full DXF files, you should import the DXF file into a _____ file so that valuable information is not lost.

14. In a _____ file, information is stored in the form of a dot pattern on the screen. This bit pattern is also known as _____ .

15. If there is a picture on the screen that you want to save as a raster file, use the _____ command.

16. With the SAVEIMG command you can create _____, _____, _____, or _____ type of raster files from the current drawing.

17. The section of screen to be saved can be specified by specifying _____ on the image tile.

18. If you are rendering to a viewport and you select the _____ option, then whatever is in the current viewport is saved to the raster file of the selected format.

19. If you are rendering to a viewport and you select the _____ option, the entire screen, including the pull-down menu area and the command line area, is saved to the raster file.

Mline and Mledit

20. You can draw multilines by using the _____ command.

21. The _____ command allows you to set the style of multilines.

22. The **Description** edit box allows you to enter the description of the _____.

23. The **Save...** button lets you save or copy the current multiline style to an external file (**.mln** file). (T/F)

24. The justification does not determine how a multiline is drawn between the specified points. (T/F)

25. The _____ option allows you to change the scale of a multiline.

26. The style must be _____ before using the STyle option to change the style.

27. Using the _____ command options, you can create three types of cross intersections: closed, open, and merged.

28. The **Zero** option will produce a multiline so that the zero offset position of the multiline coincides with the selected points. (T/F)

29. Multilines cannot be edited using grips. (T/F)

30. You can also edit the multilines without using the Multiline Edit Tools dialog box. To accomplish this, enter _____ at the Command: prompt.

31. _____ can be used to edit multilines. This command edits only multilines.

Digitizing

32. The Tablet mode can be turned on and off with the help of the _____ command.

33. The process of calibration involves digitizing _____ on the drawing and then entering their _____.

34. The digitizing process can be used to convert a drawing on paper to a CAD system. (T/F)

EXERCISES

Exercise 1

Draw two rectangles, and use the **PSFILL** command to assign stars and brick patterns to these rectangles. Convert the rectangles to the PostScript file using the **PSOUT** command, and then import the PostScript file using the **PSIN** command.

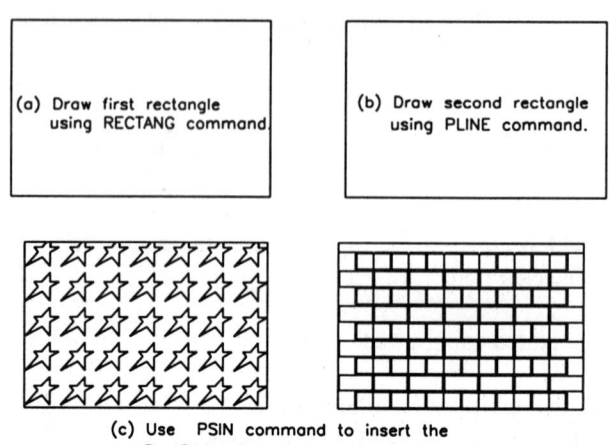

Figure 21-38 Drawing for Exercise 1

Appendix A

System Requirements and AutoCAD Installation

SYSTEM REQUIREMENTS

The following are the minimum system requirements for running AutoCAD Release 14:

1. Operating systems: Windows 95, Windows NT 3.51, and Windows NT 4.0.
2. 32 MB of RAM
3. Intel 486, Pentium 90, or better/compatible processor
4. 50MB of hard disk space
5. 64MB of disk swap space (minimum)
6. 10MB of additional memory (RAM) for each concurrent session of AutoCAD
7. 2.5 MB of free disk space required at the time of installation for temporary files that are removed when installation is complete.
8. 4X speed or faster CD-ROM for initial installation
9. 1024 by 768 SVGA display
10. Mouse or other pointing device
11. IBM compatible parallel port and hardware lock for international single user and educational versions only.
12. Service Pack 4 or 5 for Windows NT 3.51 (If you install internet utilities)

The following hardware is optional:

1. Printer
2. Plotter
3. Digitizing tablet
4. Serial port
5. CD-ROM drive for using Learning Assistant

INSTALLING AUTOCAD

The installation process has been simplified and it is now is easier and faster to install AutoCAD. Also, the installation procedure is more consistent with other Windows installations. The following steps describe the installation process for AutoCAD Release 14:

1. The first step is to insert AutoCAD Release 14 CD in the CD-ROM drive of your computer.

2. If AutoPlay is enabled, the **Setup** dialog box is automatically displayed and the InstallShield Wizard is installed on your system. (AutoPlay is a Windows 95 and Windows NT 4.0 feature that automatically runs an executable file. The AutoPlay feature can be turned off by holding down the Shift key when you insert the CD in the CD ROM drive, or by changing the play setting of CD ROM.) If your system matches the minimum hardware requirements, the Welcome dialog box is displayed on the screen.

3. Select Next button in the **Welcome** dialog box; the Software License Agreement dialog box is displayed on the screen.

4. After reading the Software License Agreement select **Accept** to continue installation; the **Serial Number** dialog box appears on the screen.

5. Enter the serial number and CD Key and then select the Next button. (The Serial Number and the CD Key can be found at the back of the box containing AutoCAD Release 14 CD.) If the numbers you entered are correct; the **Personal Information** dialog box is displayed.

6. Enter the requested information and then select the Next button (each field must contain at least one character) to display the next screen that shows the personal information you just entered. You can edit the information or select the Next button to confirm the data. The **Destination Location** dialog box is displayed on the screen.

7. In the Destination Location dialog box use the Browse button to select the directory where you want to install AutoCAD and then select the Next button to display the **Setup Type** dialog box.

8. In the Setup Type dialog box select the type of setup you prefer (Typical, Full, Compact, or Custom). If you select Custom, the **Custom Components** dialog box is displayed. From this dialog box clear the check boxes for the components that you do not want to install and then select the Next button. If your computer has enough hard disk space, the **Folder Name** dialog box is displayed.

9. In the Folder Name dialog box accept the AutoCAD R14 (default) folder or select any other folder. Select the Next button; the **Setup Confirmation** dialog box is displayed.

10. If the selection you made is OK, select the Next button in the Setup Confirmation dialog box. The installation program installs the program files on your system and when the installation is complete the **Setup Complete** dialog box is displayed on the screen.

Appendix B

Bonus Tools

BONUS TOOLS

Bonus tools consist of programs, sample drawings, driver files, and font files which enhance the existing AutoCAD. The **CAD tools** are productivity tools and are in the form of AutoLISP and ARX applications which help the user to increase the speed in drafting. The CAD Tools are Layers, Text, Modify, and Draw Tools.

LAYER TOOLS

Toolbar:	Bonus Layer Tools, Layer tool
Pull-down:	Bonus, Layers, Layer tool
Command:	Layer tool

The different layer tools have been added to help in the manipulation of layers.

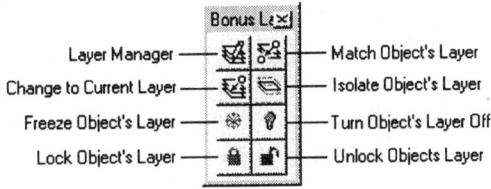

Figure B-1 Bonus Layer Tools toolbar

LMAN

When you invoke this command the **Layer Manager: Save and Restore Layer Settings** dialog box is displayed. Through this dialog box you can save, edit and restore the layers. The layers can also be exported to or imported from a LAY file.

LAYCUR

This command changes the layer of the selected objects to the current layer. The prompt sequence is:

Command: **LAYCUR**
Select objects to be CHANGED to the current layer
Select objects: *Select the objects whose layer you want to change.*
Select objects: **Enter**

LAYFRZ

  This command freezes the layer of the selected objects. The prompt sequence is:

Command: **LAYFRZ**
Options/Undo/<Pick an object on the layer to be FROZEN>: *Select the objects whose layer you want to freeze or enter O for Options.*
If you enter O then the next prompt issued is:
No nesting/Entity level nesting/<Block level nesting>: *Select an option.*

The **No nesting** option freezes the layer of the object selected. If a block or an xref is selected, the layer in which they are inserted are also frozen. The **Entity level nesting** freezes the layers of the objects selected even if the selected object is nested in xref or block. In the **Block level nesting**, if a block is selected then the layer in which the block is inserted is frozen. But if an xref is selected then the layer of the object is frozen and the xref layer is not frozen.

LAYISO

  With this command the layer of the selected objects is isolated by turning all other layers off. The prompt sequence is:

Command: **LAYISO**
Select objects on the layer(s) to be ISOLATED.
Select objects: *Select the objects whose layer you want to isolate.*

LAYLCK

  With this command the layer of the selected objects is locked. The prompt sequence is:

Command: **LAYLCK**
Pick an object on the layer to be LOCKED: *Select the object whose layer you want to lock.*

LAYMCH

  With this command the layer(s) of the selected object(s) is changed such that it matches the layer of selected destination object. The prompt sequence is:

Command: **LAYMCH**
Select objects to be changed.
Select objects: *Select the object whose layer you want to change.*
Type Name/Select entity on destination layer:

LAYOFF

  The layer of the selected object is turned off. The prompt sequence is:

Command: **LAYOFF**
Options/Undo/<Pick an object on the layer to be turned OFF>: *Select an object or specify an option.*

If you enter O for Options the next prompt issued is:
No nesting/Entity level nesting/<Block level nesting>:
If you select an object on the current layer you are prompted:
Really want layer <layer name>(the CURRENT layer) off?<N>:

TEXT TOOLS

Toolbar:	Bonus Text Tools, Text tool
Pull-down:	Bonus, Text, Text tool
Command:	Text command

The bonus text commands are used to manipulate the text and attributes.

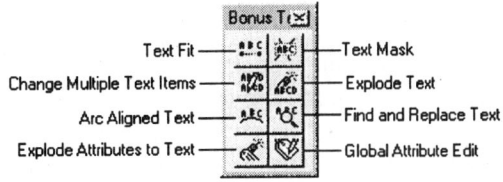

Figure B-2 Bonus Text Tools toolbar

ARCTEXT

  This command is used to place the text along an arc. The prompt sequence is:

Command: **ARCTEXT**
Select an Arc or an ArcAligned Text: *Select the arc or a text.*

The **ArcAlignedText Workshop - Create** dialog box is displayed. You can enter the text in the **Text** edit box and also choose a text style and a font for the text from the respective drop-down lists in the dialog box. You can also edit an arc aligned text by selecting it and then changing the different properties through the dialog box. In the Properties area of the dialog box you can change the height, width, and offset from the arc. After making the changes choose the OK button.

BURST

  This command explodes a block so that the attribute values are changed to text entities. The prompt sequence is:

Command: **BURST**
Select objects: *Select the block containing attributes.*

CHT

  This command is used to edit text strings. The text can be edited individually or globally. The prompt sequence is:

Command: **CHT**
Select objects: *Select a text string.*
Height/Justification/Rotation/Style/Text/Undo/Width: *Select an option.*

The **Height** option allows you to change the height of the text entity and also you can list the objects current height. With the **Justification** option you can use the ? suboption to list the different alignment options. With the **Location** option, you can specify a new location to place the selected text. Using the **Rotation** option you can specify a new rotation angle for all text. With the **Style** option you can change the text style of all the text by specifying a new style name. The style should already exist. The **Text** option displays the **Edit Text** dialog box to edit individual text lines. The **Individual** option is used to specify selected option change for individual text line. The **List** option lists all the information about the selected option.

FIND

 When you invoke this command, the **Find and Replace** dialog box is displayed which can be used to perform the find and replace functions.

You can enter the text string you want to change in the **Find** edit box. The text string which you want should replace the found text string and should be entered in the **Replace With** edit box. If you want that find and replace should be sensitive to upper and lower case usage then activate the **Case Sensitive** check box. If you want that find should replace all text or text strings then check the **Global Change** check box.

GATTE

 With this command the attribute values are changed globally for all insertions of a specific block. The prompt sequence is:

Command: **GATTE**
Block name/<Select block or attribute>: *Enter B to use a block name or select the block.*
Known tag names for block:MDLN
Select attribute or type attribute name: *Select attribute or enter the tag name.*
Number of inserts in drawing=2
Process all of them? <Yes>/No: *Select Yes to process all and No for processing the selected attributes.*

TEXTFIT

 This command shrinks or stretches text objects such that they fit between the specified start and end points. The prompt sequence is:

Command: **TEXTFIT**
Select text to stretch/shrink: *Select the text.*
Starting point/<Pick new ending point>: *Enter S for new start point or select an ending point.*
If you enter S for a new starting point then the next prompt is:
Pick new starting point: *Specify a new starting point.*
Ending point: *Specify the ending point.*

TEXTMASK

 This command hides the objects behind the selected text. TEXTMASK works together with WIPEOUT bonus routine. The prompt sequence is:

Command: **TEXTMASK**
Enter offset factor relative to text height <0.35>: *Specify a new value or pick points for the mask size.*

Select text to Mask
Select objects: *Select the text.*

TEXTEXP

 This command explodes the selected text into lines and arcs. These objects can then be assigned a thickness. The prompt sequence is:

Command: **TEXTEXP**
Select text to be EXPLODED: *Select the text.*

MODIFY TOOLS

Toolbar:	Bonus Standard Tools, Modify tool
Pull-down:	Bonus, Modify, Modify tool
Command:	Modify command

The bonus modify commands are based on the standard AutoCAD modify commands.

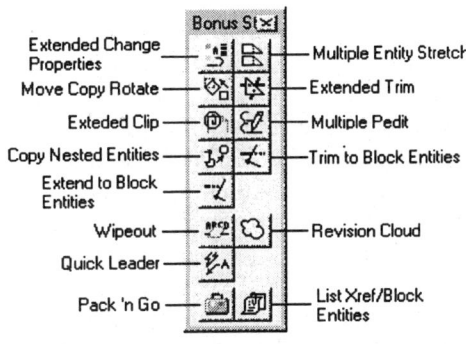

Figure B-3 Bonus Standard Tools toolbar

BEXTEND

 With this command the selected objects are extended to a selected boundary. The boundary can be a block or an xref object. The prompt sequence is:

Command: **BEXTEND**
Select edges for extend: *Select objects for boundary edge.*
<Select object to extend>/Project/Edge/Undo: *Select the objects to extend or select an option.*

BTRIM

 With this command the selected objects are trimmed to a selected cutting edge. The cutting edge can be a block or an xref object. The prompt sequence is:

Command: **BTRIM**
Select cutting edges: *Select objects for cutting edge.*
<Select object to trim>/Project/Edge/Undo: *Select the objects to trim or select an option.*

CLIPIT

This command allows you to separate some portion of block, xref drawing, or image by removing the display of the remaining objects using a polyline, circle, or arc. The prompt sequence is:

Command: **CLIPIT**
Pick a POLYLINE/CIRCLE/ARC for clipping edge

Select objects: *Select a clipping edge.*
Pick an IMAGE, a WIPEOUT, or an XREF/BLOCK to clip: *Select the object.*
Enter max error distance for resolution of arcs <0.0200>: *Select a resolution value.*

The curved boundary edge is traversed with a series of short straight segments, and the resolution of arc is the distance between the midpoint of the given segment and the arc. Hence a large error value creates less segments and faster speed but the arcs become less smooth.

EXCHPROP

This command allows you to change the properties of multiple objects. The prompt sequence is:

Command: **EXCHPROP**
Select objects: *Select the objects whose properties you want to change.*

The **Change Properties** dialog box is displayed on the screen. You can change the color of the objects by selecting the **Color** button to display the **Set Color** dialog box and then choose the new color. Similarly selecting the Layer and Linetype buttons opens the **Select Layer** dialog box and the **Select Linetype** dialog box respectively. You can select the desired layer or the linetype from the dialog box and then select the OK button to make the changes to the selected objects. In the Change Properties dialog box you can enter a new value for the linetype scale and the thickness in their corresponding edit boxes. In the **Polyline** area of the dialog box enter new values in the **Width** and **Elevation** edit boxes for the selected polylines. Specify new values in the Height and Style edit boxes in the **Text/Mtext/Attdef** area to make the changes to the selected text or attribute definition.

EXTRIM

This command allows you to trim the objects by specifying a cutting edge which can be a polyline, line, circle, or arc. The prompt sequence is:

Command: **EXTRIM**
Pick a Polyline, Line, Circle, or Arc for cutting edge..
Select objects: *Select a cutting edge.*
Pick the side to trim on: *Select a point to specify the side of trimming.*

The portion of the objects in the selected side are trimmed.

MOCORO

This single command can be used to move, copy, rotate, and scale the selected object or objects. The prompt sequence is:

Command: **MOCORO**
Select objects: *Select objects for modifying.*
Base point: *Select a base point.*
Move/Copy/Rotate/Scale/Base pt/<eXit>: *Specify an option.*

MPEDIT

 This command edits multiple polylines and also converts lines and arcs to polylines. The prompt sequence is:

Command: **MPEDIT**
Select objects: *Select objects.*
Convert Lines and Arcs to polylines? <Yes>: *Press Enter*
Open/Close/Width/Fit/Spline/Decurve/Ltype gen/eXit <X>: *Specify an option.*

MSTRETCH

 This command is used to stretch multiple objects. If the complete object is selected, this command works like the MOVE command. The prompt sequence is:

Command: **MSTRETCH**
Define crossing windows or crossing polygons...
CP[crossing polygon]/<Crossing first point>: *Specify first point of a crossing window.*
CP/Undo/<Crossing First point>: *Specify second point of a crossing window.*
CP/Undo/<Crossing First point>: *Press Enter.*
Remove objects/<Base point>: *Select a base point.*
Second base point: *Select a new base point.*

NCOPY

 This command is used to copy objects that are nested in an xref or block. The prompt sequence is:

Command: **NCOPY**
Select nested objects to copy: *Select objects.*
<Base point or displacement>/Multiple: *Select the base point or M for multiple copies.*
Second point of displacement: *Select the second point.*

XPLODE

This command is used to explode objects and controls all the properties of the component entities of a block or a group of blocks. The prompt sequence is:

Command: **XPLODE**
Select objects to Xplode.
Select objects: *Select objects.*
All/Color/LAyer/LType/Inherit from parent block/<Explode>: *Specify an option.*

The **All** option specifies color, linetype, and layer for the new entities. The **Color**, **LType**, and **LAyer** options also specify a new color, linetype, and the layer for the selected objects. The **Inherit** option verifies that the attributes of the selected block will be the attributes of the component entities. The **Explode** option explodes the selected object into its individual parts.

DRAW TOOLS

Toolbar:	Bonus Standard Tools, Draw tool
Pull-down:	Bonus, Draw, Draw tool
Command:	Draw command

The bonus draw commands are used to create entities.

QLEADER

  This command allows you to create the type of leader you specify by providing the prompts that are relevant for that kind of leader. The prompt sequence is:

Command: **QLEADER**
First Leader point or press Enter to set Options: *Select the starting point of leader.*
Next Leader point: *Select the second point of leader.*
Next Leader point: *Select the third point of leader.*
Next Leader point: *Press Enter.*
Enter leader text: *Enter the text.*

You can also enter press Enter at the **First Leader point or press Enter to set Options:** prompt to set options for the leader. When you press Enter the **Quick Leader Options** dialog box is displayed on the screen. The **Annotation/Format** tab displays the options for setting the Annotation, Annotation Memory, and Format for the leader. The **Points** tab displays the options for setting the number of points to draw the leader line. The **Angles** tab displays the options for setting the first and second segment angles. The **Attachment** tab displays the options for setting the left and right side Attachment method.

QLATTACH

This command is used to attach the leader line to an annotation object such as Mtext, Tolerance, or Block Reference Object. The prompt sequence is:

Command: **QLATTACH**
Select Leader: *Select the leader.*
Select Annotation: *Select the annotation object.*

QLDETACHSET

This command is used to detach the leader line from an annotation object such as Mtext, Tolerance, or Block Reference Object. The prompt sequence is:

Command: **QLDETACHSET**
Select objects: *Select the leader and the annotation objects.*
Select objects: *Press Enter*
Number of Leaders=n
Number of annotations detached=n

QLATTACHSET

This command is used to globally attach the leader line to an annotation object such as Mtext, Tolerance, or Block Reference Object. The prompt sequence is:

Command: **QLATTACHSET**
Select objects: *Select the leader and the annotation objects.*
Select objects: *Press Enter*
Number of Leaders=n
Number with annotations detached=n

REVCLOUD

This command is used to create a revision cloud shape from a polyline made up of sequential arcs. The prompt sequence is:

Command: **REVCLOUD**
Type REVCLOUD to draw Revision Cloud. To close, return to start point.
Arc length set at 0.375
Arc length/<Pick cloud starting point>: *Enter A to set a new arc length or enter the start of the cloud.*

As you move the crosshairs, continuous arcs are created having the set arc lengths. When the crosshairs are close to the start point, the last arc joins the start point to close the Revcloud and you come out of the command.

WIPEOUT

With this command the selected area is covered with the background color so as to hide it. The area you want to select must be defined by a closed polyline. The prompt sequence is:

Command: **WIPEOUT**
Frame/New <New>: *Press Enter to create a new wipeout.*
Select a polyline: *Select the closed polyline.*
Erase polyline? Yes/No <No>: *Enter you choice.*

The area inside the polyline boundary is wiped out. You can also enter F at the **Frame/New** <New>: prompt to display the following prompt:

On/Off <On>:

The **On** option displays the wipeout frame on all viewport entities. The **Off** options turns off the wipeout for all wipeout entities.

TOOLS COMMANDS

Toolbar:	Bonus Standard Tools, Tools
Pull-down:	Bonus, Tools, Tools command
Command:	Tool command

The bonus tool commands help you to manage the AutoCAD environment.

ALIASEDIT

This command allows you to edit the acad.pgp file by creating, modifying, and deleting AutoCAD command aliases on-the-fly. The prompt sequence is:

Command: **ALIASEDIT**

The **acad.pgp - AutoCAD Alias Editor** dialog box is displayed. When you choose the **Command Aliases** tab, the AutoCAD commands and their alias are displayed in the list area. When you choose the **Shell Commands** tab, the Shell commands and their alias with their prompts are displayed in the list area. The options in the dialog box are common to both the tabs. The **Add** button displays a dialog box to add a new AutoCAD alias and command (Command Aliases tab), or displays a dialog box to add a DOS alias, command, and prompt (Shell command tab). The **Delete** button allows you to delete an AutoCAD or DOS command alias from the Acad.pgp file through a dialog box. The **Edit** button allows you to edit the selected AutoCAD alias and command or DOS alias, command, and prompt. If you want a Confirmation dialog box to be displayed when creating or modifying aliases, activate the **Confirm changes** check box. If you want to save the changes and exit the Alias Editor, choose the **OK** button. If you want to exit without saving then choose the **Close** button. The **Apply** button allows you to save the changes and then continue editing.

BONUSPOPUP

This command loads and unloads the bonus pop-up application. Use the Ctrl key and right-click your pointing device to pop-up the menu. Use Alt key and right-click to select the drop-down to be used.

CONVERTPLINES

With this command all the pre-Release 14 polylines are converted to lightweight polylines by removing any attached xdata. The prompt sequence is:

Command: **CONVERTPLINES**
Type "YES" <all caps> to proceed: *Enter YES to convert.*

DIMEX

With this command the named dimension styles and all their settings are exported to an exported file.

Command: **DIMEX**

The **Dimension Style Export** dialog box is displayed. You can enter the desired DIM filename in the **Export filename** edit box. The list of dimension styles available in the current drawing are listed in the **Available Dimension Style** list box. You can select the dimension style you wish to write to the ASCII file. In the **Text Style Options** area you can choose the **Full Text Style Information** radio button or **Text Style Name Only** radio button.

DIMIM

With this command the named dimension styles are imported from a DIM file into the current drawing.

Command: **DIMIM**

The **Dimension Style Import** dialog box is displayed. You can enter the desired dimension style name in the **Import filename** edit box, or from the Open dialog box which can be invoked by selecting the **Browse** button. You can retain the dimension style in the current drawing or overwrite it by selecting the appropriate radio button in the **Import Options** area.

PACK

 With this command all the files associated with a particular drawing are copied to a specified location.

Command: **PACK**

The **Pack & Go** dialog box is displayed. All the filenames with their size, date, and drawing version (if any) is listed in the dialog box. You can choose the **Tree View** tab to display the files as a directory tree. If you choose the **Browse** button the **Choose Directory** dialog box is displayed where you can specify a path to copy files. The **Copy to** button copies the files to the specified path. The **Print** button allows you to print a report of information displayed in the dialog box. The **Report** button displays the Report dialog box specifying information about associated files.

SYSVDLG

This command allows you to edit the system variables and then save them on-the-fly.

Command: **SYSVDLG**

The **System Variable** dialog box is displayed and lists all the system variables. You can select the system variable you want to edit so that it is highlighted. Its current value and description are displayed in the **Value** and **Description** boxes respectively. You can change its value and then press Enter to update it. The **Save Custom** and **Read Custom** buttons display the corresponding dialog boxes so as to save the system variable settings to a SVF file, or restore the saved settings from a SVF file. The **On** toggle check box is used for the switch-type system variables to put them on or off.

XDATA

This command allows you to attach extended entity data (xdata) to a selected entity.

Command: **XDATA**
Select object: *Select the objects.*
Application name: *Enter the name of application.*
3Real/DIR/DISP/DIST/Hand/Int/LAyer/LOng/Pos/Real/SCale/STr/ < eXit >: *Enter an option.*

XDLIST
This command lists all the xdata associated with an object.

Command: **XDLIST**
Select object: *Select the objects.*
Application name < * >: *Enter the name of application.*

XLIST

This command lists the different properties such as object, block name, layer, color, and linetype of a nested object in a block.

Command: **XLIST**
Select nested xref or block object to list: *Select the object.*

An **Xref/Block Nested Object List** dialog box is displayed which lists all the information about the selected object.

ADDITIONAL COMMANDS
These commands are accessed through the command line only.

ASCPOINT
This command reads coordinate data from an ASCII file. The data is then generated in the form of a string of lines, a Polyline, a 3DPolyline, multiple copies of a selected group of objects, or Point entities.

Command: **ASCPOINT**
File to read: *Enter an ASCII filename.*
Comma/Space delimited < Comma >: *Enter a format.*
Generate Copies/Lines/Nodes/3Dpoly/ < Pline >: *Enter an option.*

BLOCK?
This command lists the different entities in a block definition.

Command: **BLOCK?**
Block name/ < Return to select >: *Press Enter.*
Pick a block: *Select the block.*
An entity type/ < Return for all >: *Press Enter for all entities or enter a type of entity.*

Appendix B: Bonus Tools   B-13

COUNT
This command counts the number of insertions of each block in the drawing or in the selected objects. It then displays it in a tabular form.

> Command: **COUNT**
> Press ENTER to select entire drawing or, Select objects: *Press Enter or select objects.*
> Counting block insertions...Block Count

CROSSREF
This command lists the name of the block that contains reference to a specified object which can be a layer, linetype, style, dimstyle, mlinestyle, or block.

> Command: **CROSSREF**
> Cross reference Block/LType/Style/Dimstyle/Mlinestyle/<Layer>: *Select type.*
> Name of <type> to cross reference: *Select object name.*

DBTRANS
This command translates the textual data contained in a drawing between various formats. This application needs to be loaded with APPLOAD or ARX. This command is used for translating drawings created inside AutoCAD R11 or previous versions to Release 14.

DOSLIB
This command lists the DOS related AutoLISP functions. This application needs to be loaded with APPLOAD or ARX.

GETSEL
This command creates a temporary selection set.

> Command: **GETSEL**
> Select Object on layer to Select from <*>: *Select object or press Enter for all layers.*
> Select type of entity you want <*>: *Select entity type or press Enter for all.*

JULIAN
This command contains the AutoLISP functions for conversion of Julian date.

PQCHECK
This command checks AutoLISP programs for errors or missing quotes or mismatched parenthesis.

> Command: **PQCHECK**
> Enter filename: *Enter the name of the file to be checked.*

SSX
This command creates a selection set.

> Command: **SSX**

Select objects/<None>: *Select a template object.*
>>Block name/Color/Entity/Flag/LAyer/LType/Pick/Style/Thickness/Vector: *Enter an option.*

It creates a selection set exactly like the selected entity or very similar to the filter list which you have adjusted.

Appendix C

AutoCAD Linetypes

The following are the linetypes that are defined in the ACAD.LIN file.

ISO02W100	ACAD_ISO03W100
ACAD_ISO04W100	ACAD_ISO05W100
ACAD_ISO06W100	ACAD_ISO07W100
ACAD_ISO08W100	ACAD_ISO09W100

C-2 Customizing AutoCAD

ACAD_ISO10W100	ACAD_ISO11W100
ACAD_ISO12W100	ACAD_ISO13W100
ACAD_ISO14W100	ACAD_ISO15W100
BORDER	BORDER2
BORDERX2	CENTER
CENTER2	CENTERX2
CONTINUOUS	DASHDOT
DASHDOT2	DASHDOTX2

Linetype		Linetype
DASHED		DASHED2
DASHEDX2		DIVIDE
DIVIDE2		DIVIDEX2
DOT		DOT2
DOTX2		HIDDEN
HIDDEN2		HIDDENX2
PHANTOM		PHANTOM2
PHANTOMX2		

Appendix D

AutoCAD Hatch Patterns

Following are the hatch patterns that are defined in the ACAD.PAT file.

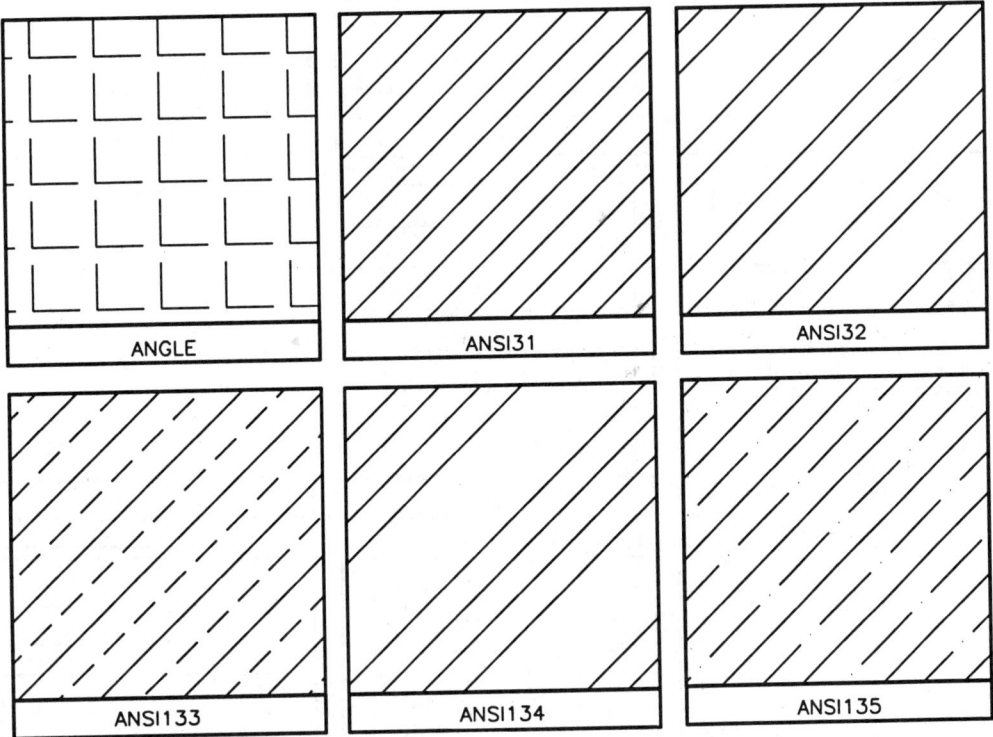

D-2 Customizing AutoCAD

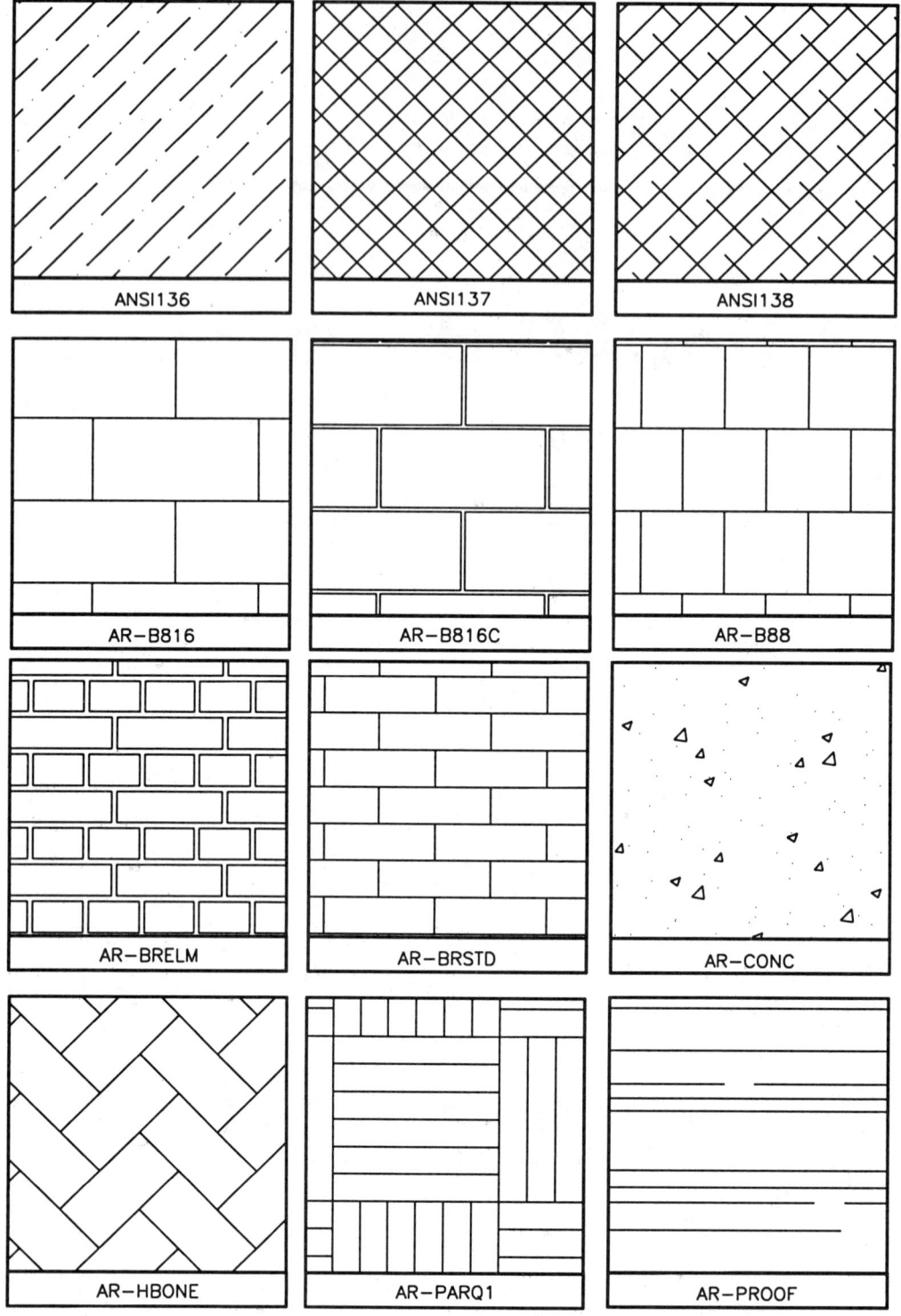

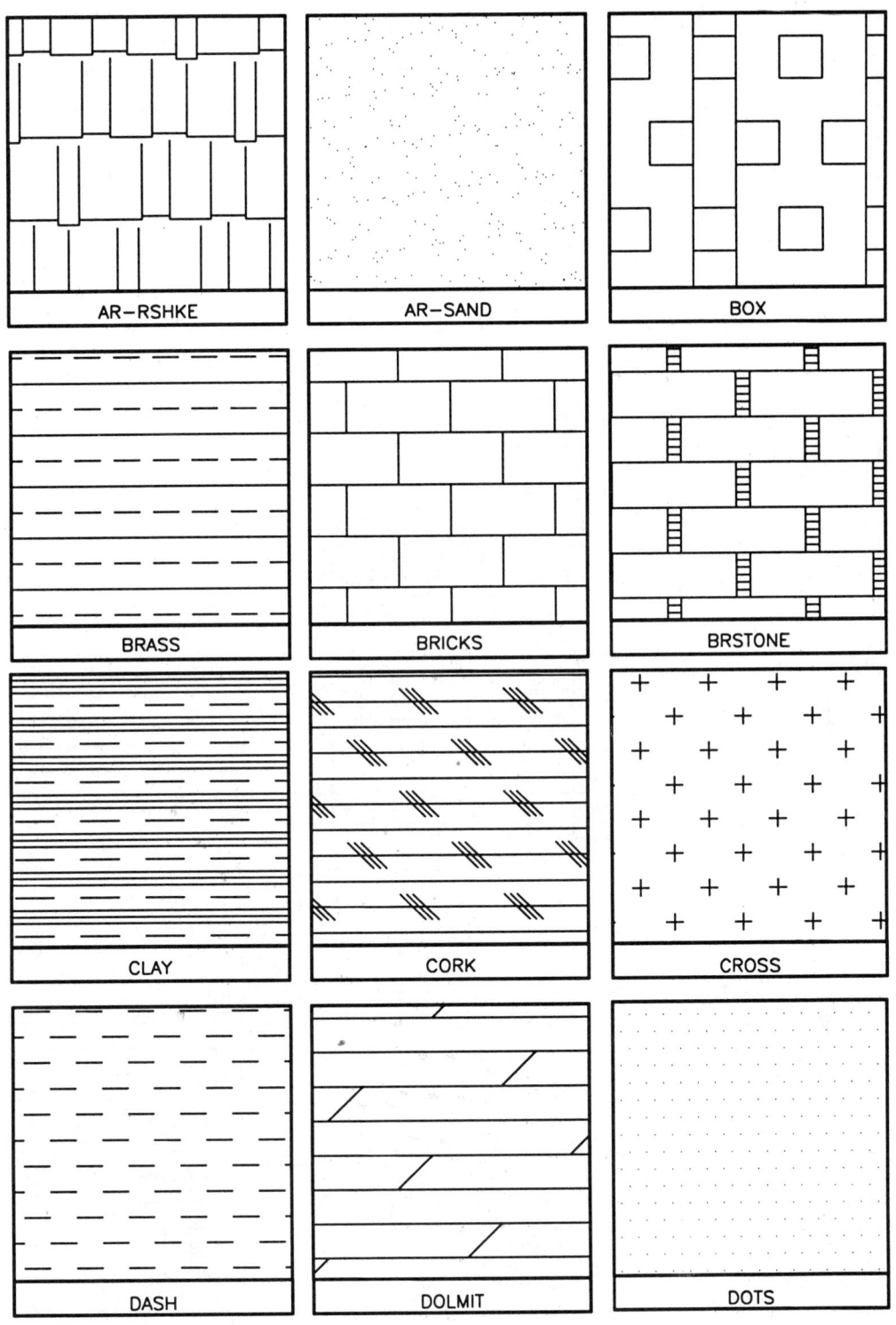

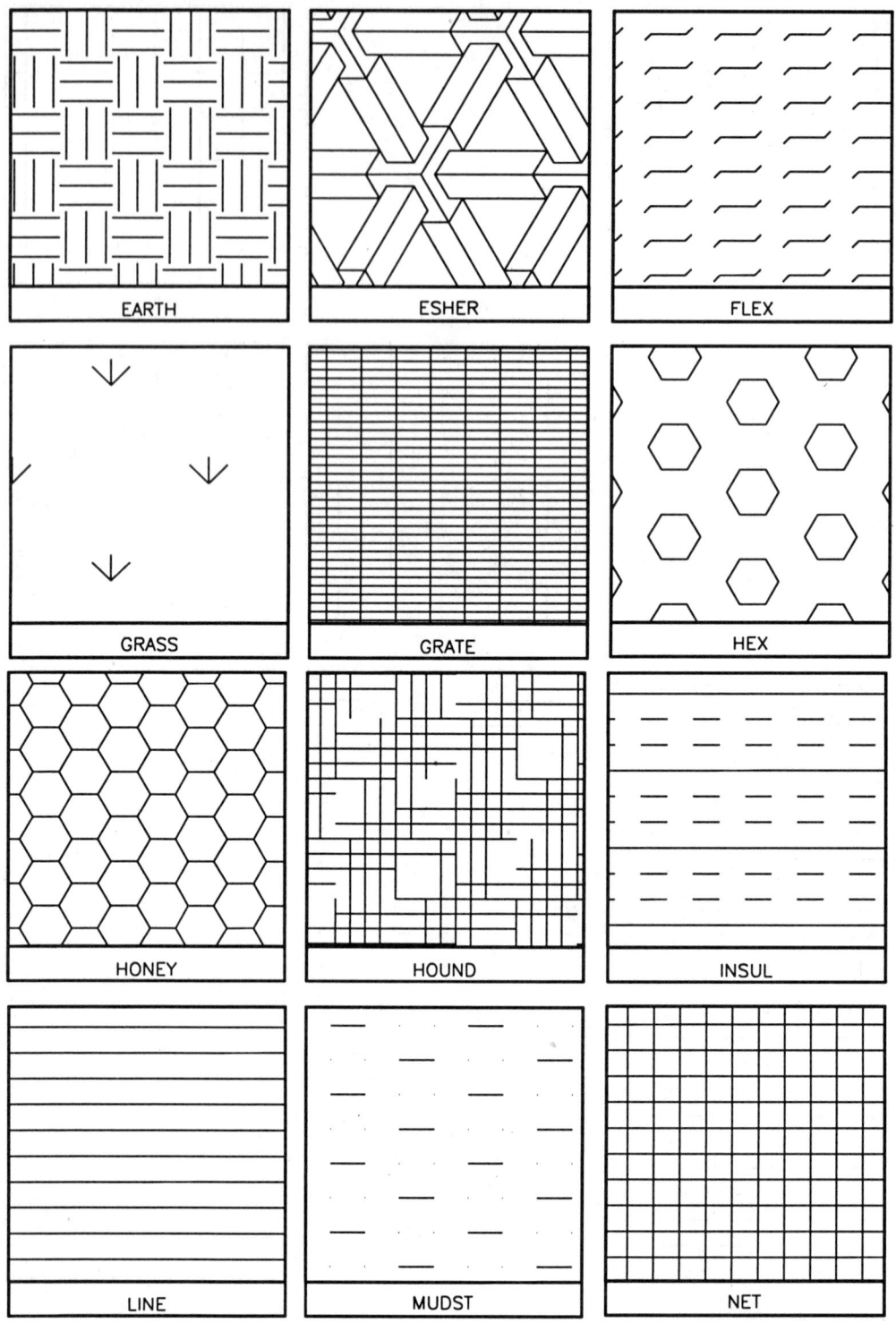

Appendix D: AutoCAD Hatch Patterns D-5

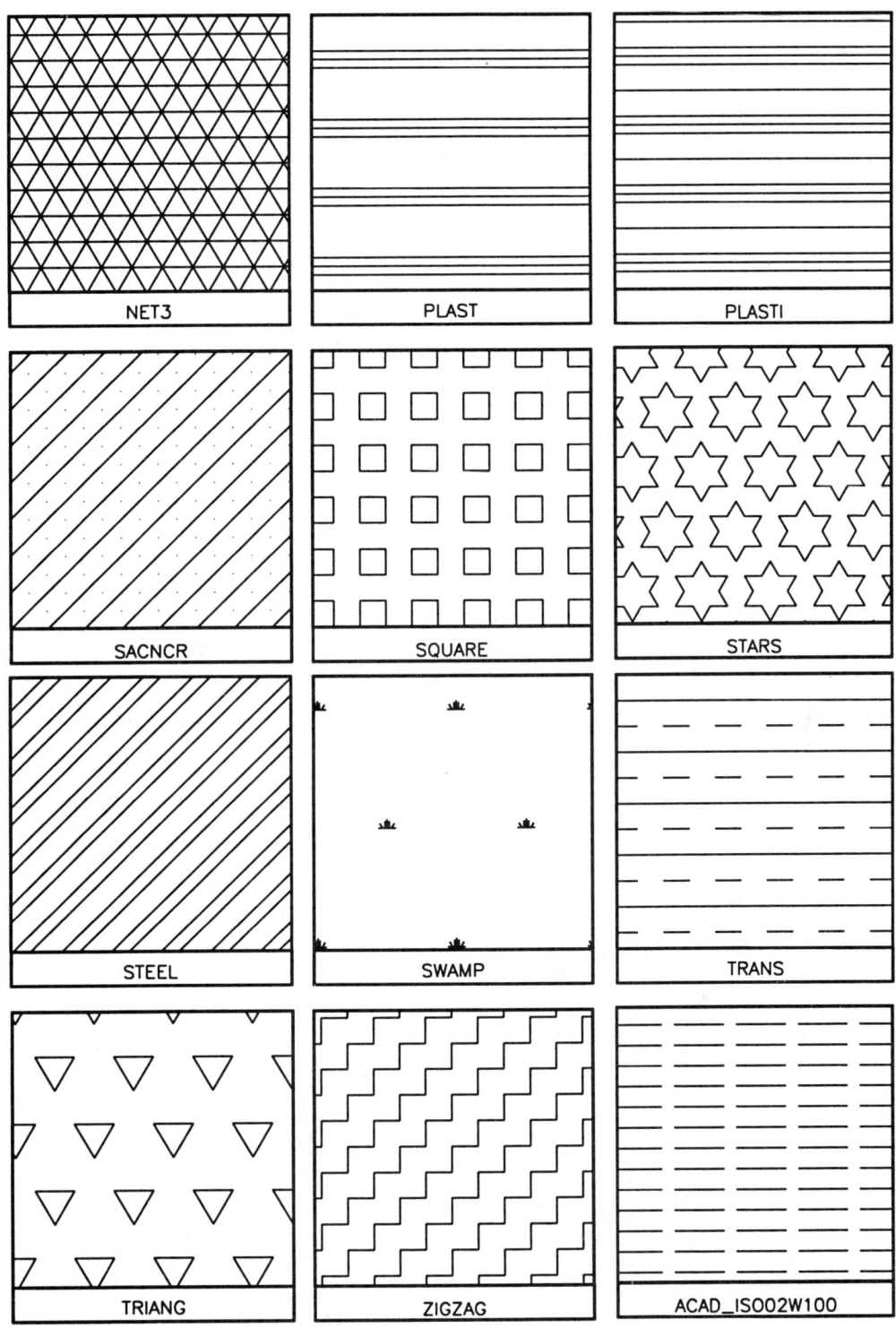

ACAD_ISO03W100	ACAD_ISO04W100	ACAD_ISO05W100
ACAD_ISO06W100	ACAD_ISO06W100	ACAD_ISO08W100
ACAD_ISO09W100	ACAD_ISO10W100	ACAD_ISO11W100
ACAD_ISO12W100	ACAD_ISO13W100	ACAD_ISO14W100

Appendix E

AutoCAD Text Fonts

The following are the Standard and True Type text fonts supported by AutoCAD Release 14.

Standard Fonts

```
ABCDEFGHIJKLMNOPQRS
TUVWXYZ
123456789

This is a sample text
to demonstrate the
effect of MONOTXT
Text font
```

```
ABCDEFGHIJKLMNOPQRS
TUVWXYZ
123456789

This is a sample text to
demonstrate the effect of
ROMANS Text font
```

```
ABCDEFGHIJKLMNOPQRS
TUVWXYZ
123456789

This is a sample text
to demonstrate the
effect of TXT Text
font
```

```
ABCDEFGHIJKLMNOPQRS
TUVWXYZ
123456789

This is a sample text to
demonstrate the effect of
ROMAND Text font
```

ABCDEFGHIJKLMNOPQRS
TUVWXYZ
123456789

This is a sample text to
demonstrate the effect
of ROMANC Text font

ABCDEFGHIJKLMNOPQRS
TUVWXYZ
123456789

This is a sample text to
demonstrate the effect
of ITALICT Text font

ABCDEFGHIJKLMNOPQRS
TUVWXYZ
123456789

This is a sample text to
demonstrate the effect
of ROMANT Text font

ABCDEFGHIJKLMNOPQRS
TUVWXYZ
123456789

This is a sample text to
demonstrate the effect of
SCRIPTC Text font

ABCDEFGHIJKLMNOPQRS
TUVWXYZ
123456789

This is a sample text
to demonstrate the effect
of ITALIC Text font

ABCDEFGHIJKLMNOPQRS
TUVWXYZ
123456789

This is a sample text to
demonstrate the effect of
SCRIPTS Text font

ABCDEFGHIJKLMNOPQRS
TUVWXYZ
123456789

This is a sample text to
demonstrate the effect
of ITALICC Text font

𝔄𝔅ℭ𝔇𝔈𝔉𝔊ℌ𝔍𝔍𝔎𝔏𝔐𝔑𝔒𝔓𝔔𝔕𝔖
𝔗𝔘𝔙𝔚𝔛𝔜𝔝
123456789

This is a sample text to
demonstrate the effect of
GOTHICE Text font

Appendix E: AutoCAD Text Fonts   E-3

𝔄𝔅ℭ𝔇𝔈𝔉𝔊𝔥𝔍𝔎𝔏𝔐𝔑𝔒𝔓
𝔔𝔕𝔖𝔗𝔘𝔙𝔚𝔛𝔜𝔷
123456789

This is a sample text to
demonstrate the effect of
𝔊𝔒𝔗𝔥𝔍ℭℭ𝔊 Text font

ABCDEFGHIJKLMNOP
QRSTUVWXYZ
123456789

This is a sample text to
demonstrate the effect of
GOTHICI Text font

АБВГДЕЖЗИЙКЛМНОПРСТ
УФХЦЧШЩ
123456789

Узит ит а тамплд удчу
уо гдмонтусауд узд
деедву ое ВШСИЛЛИВ
Удчу еону

АБЧДЕФГХИЩКЛМНОПЦРС
ТУВШЖЙЗ
123456789

Тхис ис а сампле тежт
то демонстрате тхе
еффечт оф ЧЙРИЛТЛЧ
Тежт фонт

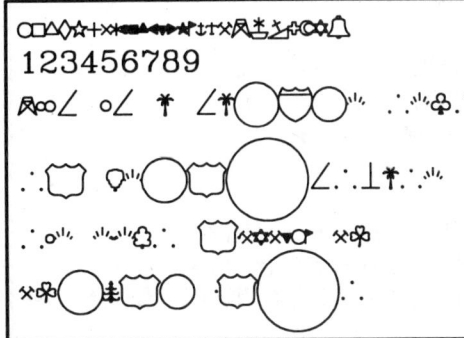

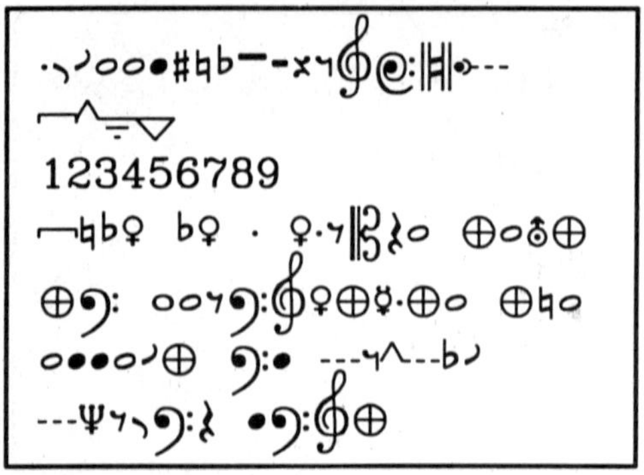

TrueType Fonts

```
·.⌐♪○○●#♭♭━━×⧫𝄞@:|H|♭---
⌐∿=∇
123456789
━♭♭♀  ♭♀ · ♀·⧫|♭♭♂○ ⊕○♂⊕
⊕♀): ○○⧫♀):⧫♀⊕♀·⊕○  ⊕♭○
○●●○⧫⊕  ♀):●  ---⧫∧--♭⧫
---Ψ⧫⧫♀):⧫ •♀):⧫⊕
```

ABCDEFGHIJKLMNOPQR
STUVWXYZ
123456789

This is a sample text to
demonstrate the effect
of COMPLEX Text font

ABCDEFGHIJKLMNOPQRS
TUVWXYZ
123456789

This is a sample text to
demonstrate the effect of
SIMPLEX Text font

ABCDEFGHIJKLMNOPQRSTUVWXYZ
123456789

This is a sample text to
demonstrate the effect of
ISOCP Text font

ABCDEFGHIJKLMNOPQRS
TUVWXYZ
123456789

This is a sample text to
demonstrate the effect of SWISS
True Type font

ABCDEFGHIJKLMNOPQRS
TUVWXYZ
123456789

This is a sample text to demonstrate
the effect of SWISSL True Type font

ABCDEFGHIJKLMNOPQRS
TUVWXYZ
123456789

This is a sample text to demonstrate
the effect of SWISSLI True Type font

ABCDEFGHIJKLMNOPQRS
TUVWXYZ
123456789

This is a sample text to
demonstrate the effect of
SWISSI True Type font

Appendix E: AutoCAD Text Fonts   E-5

ABCDEFGHIJKLMNOPQRS TUVWXYZ
123456789

This is a sample text to demonstrate the effect of SWISSB True Type font

ABCDEFGHIJKLMNOPQRSTUVWXYZ
123456789

This is a sample text to demonstrate the effect of SWISSC True Type font

ABCDEFGHIJKLMNOPQRS TUVWXYZ
123456789

This is a sample text to demonstrate the effect of SWISSBI True Type font

ABCDEFGHIJKLMNOPQRSTUVWXYZ
123456789

This is a sample text to demonstrate the effect of SWISSCL True Type font

ABCDEFGHIJKLMNOPQRS TUVWXYZ
123456789

This is a sample text to demonstrate the effect of SWISSK True Type font

ABCDEFGHIJKLMNOPQRSTUVWXYZ
123456789

This is a sample text to demonstrate the effect of SWISSCLI True Type font

ABCDEFGHIJKLMNOPQRS TUVWXYZ
123456789

This is a sample text to demonstrate the effect of SWISSKI True Type font

ABCDEFGHIJKLMNOPQRSTUVWXYZ
123456789

This is a sample text to demonstrate the effect of SWISSCI True Type font

ABCDEFGHIJKLMNOPQRSTUVWXYZ
123456789

This is a sample text to demonstrate the effect of SWISSCB True Type font

ABCDEFGHIJKLMNOPQRSTUVWXYZ
123456789

This is a sample text to demonstrate the effect of SWISSE True Type font

ABCDEFGHIJKLMNOPQRSTUVWXYZ
123456789

This is a sample text to demonstrate the effect of SWISSCBI True Type font

ABCDEFGHIJKLMNOPQRSTUVWXYZ
123456789

This is a sample text to demonstrate the effect of SWISSEL True Type font

ABCDEFGHIJKLMNOPQRSTUVWXYZ
123456789

This is a sample text to demonstrate the effect of SWISSCK True Type font

ABCDEFGHIJKLMNOPQRSTUVWXYZ
123456789

This is a sample text to demonstrate the effect of SWISSEB True Type font

ABCDEFGHIJKLMNOPQRSTUVWXYZ
123456789

This is a sample text to demonstrate the effect of SWISSCKI True Type font

ABCDEFGHIJKLMNOPQRSTUVWXYZ
123456789

This is a sample text to demonstrate the effect of SWISSEK True Type font

Appendix E: AutoCAD Text Fonts

ABCDEFGHIJKLMNOPQRS
TUVWXYZ
123456789

This is a sample text to demonstrate the effect of SWISSBO True Type font

---

ABCDEFGHIJKLMNOPQRS
TUVWXYZ
123456789

This is a sample text to demonstrate the effect of MONOSI True Type font

---

ABCDEFGHIJKLMNOPQRS
TUVWXYZ
123456789

This is a sample text to demonstrate the effect of SWISSKO True Type font

---

ABCDEFGHIJKLMNOPQRS
TUVWXYZ
123456789

This is a sample text to demonstrate the effect of MONOSB True Type font

---

ABCDEFGHIJKLMNOPQRSTUVWXYZ
123456789

This is a sample text to demonstrate the effect of SWISSCBO True Type font

---

ABCDEFGHIJKLMNOPQRS
TUVWXYZ
123456789

This is a sample text to demonstrate the effect of MONOSBI True Type font

---

ABCDEFGHIJKLMNOPQRS
TUVWXYZ
123456789

This is a sample text to demonstrate the effect of MONOS True Type font

---

ABCDEFGHIJKLMNOPQRS
TUVWXYZ
123456789

This is a sample text to demonstrate the effect of DUTCHI True Type font

ABCDEFGHIJKLMNOPQR
STUVWXYZ
123456789

This is a sample text to
demonstrate the effect of
DUTCHB True Type font

ABCDEFGHI
JKLMNOPQR
STUVWXYZ
123456789

THIS IS A SAMPLE TEXT
TO DEMONSTRATE THE
EFFECT OF BGOTHM
TRUE TYPE FONT

ABCDEFGHIJKLMNOPQRS
TUVWXYZ
123456789

This is a sample text to demonstrate
the effect of DUTCHBI True Type
font

ABCDEFGHIJKLMNOPQR
STUVWXYZ
123456789

This is a sample text to demonstrate
the effect of COMIC True Type
font

ABCDEFGHIJKLMNOPQRS
TUVWXYZ
123456789

This is a sample text to
demonstrate the effect of
DUTCHEB True Type font

ABCDEFGHIJKLMNO
PQRSTUVWXYZ
123456789

This is a sample text
to demonstrate the
effect of VINET True
Type font

ABCDEFGHI
JKLMNOPQR
STUVWXYZ
123456789

THIS IS A SAMPLE TEXT
TO DEMONSTRATE THE
EFFECT OF BGOTHL
TRUE TYPE FONT

±°′″Ø+−×÷=±°′″⌀....
...℞♀♂•◦○◦▫◻▫○▫
.%%® %® © ®©%©@®®®×®
®‖ ®®©‖%®®©©®® ®%©
®™ ®®©® ‖™ ′_○ .©™®
. ÷ ©® ™‖%©

Appendix F

Dialog Boxes

The following are the frequently used dialog boxes in AutoCAD Release 14.

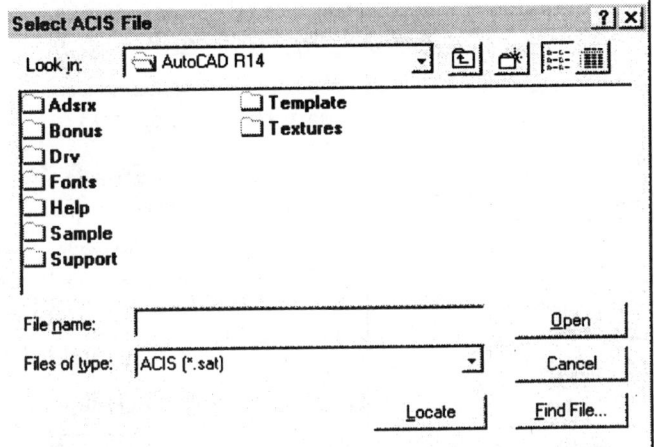

ACISIN (Select ACIS File dialog box)

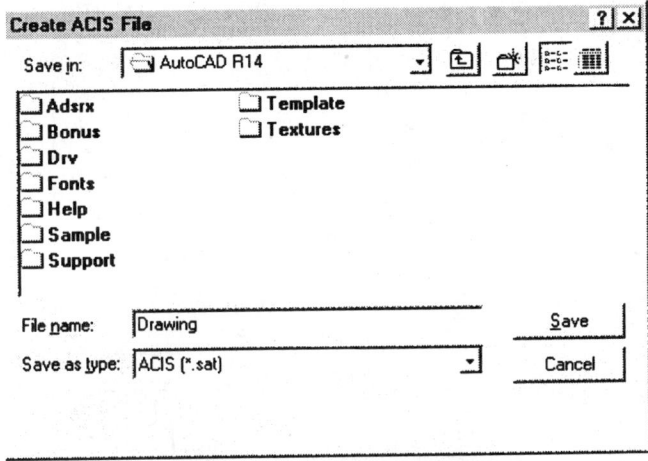

ACISOUT (Create ACIS File dialog box)

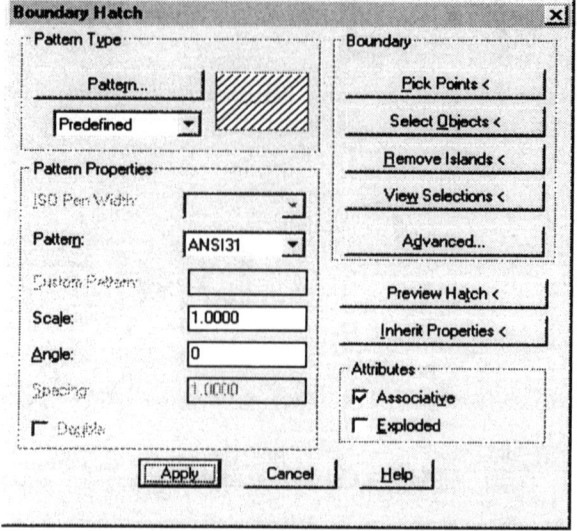

BHATCH (Boundary Hatch dialog box)

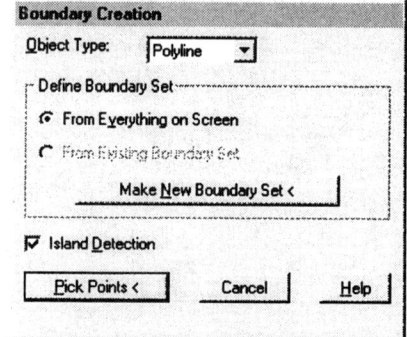

BMAKE (Block Definition dialog box)   **BOUNDARY** (Boundary Creation)

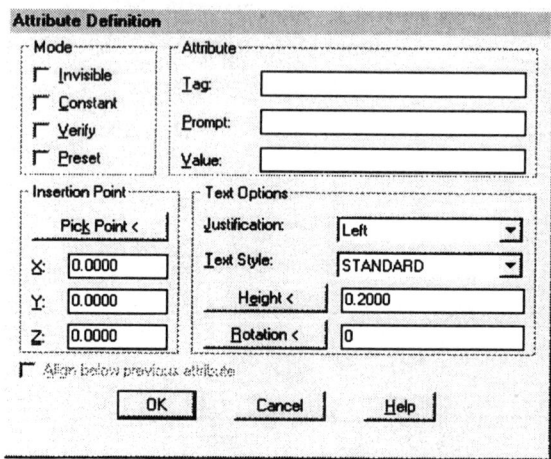

DDATTDEF (Attribute Definition dialog box)

Appendix F: Dialog Boxes F-3

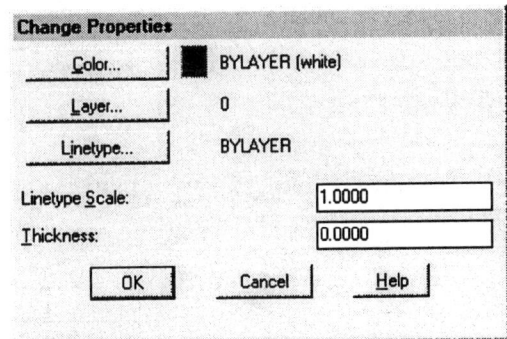

DDATTE (Edit Attributes dialog box)

DDATTEXT (Attribute Extraction dialog box)

DDCHPROP (Change Properties dialog box)

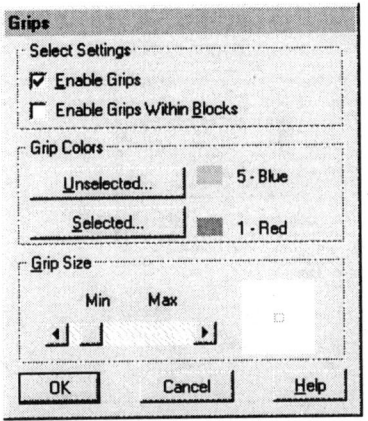

DDGRIPS (Grips dialog box)

DDIM (Dimension Styles dialog box)

F-4 Customizing AutoCAD

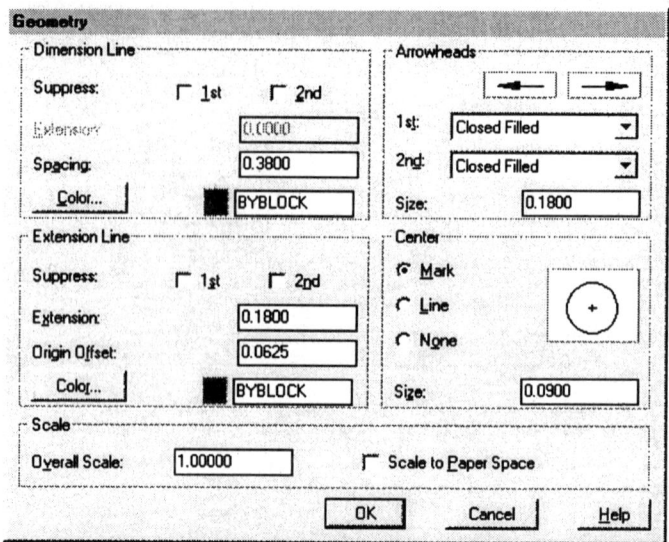

DDIM (Geometry dialog box)

DDIM (Annotation dialog box)

Appendix F: Dialog Boxes   F-5

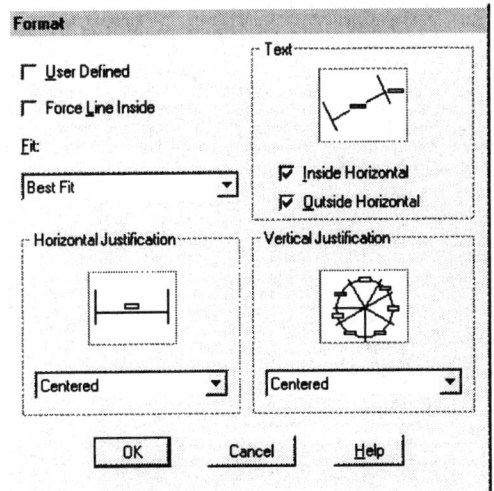

DDIM (Format dialog box)

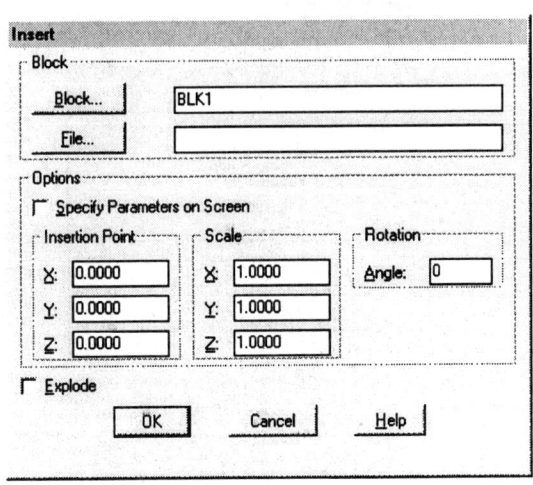

DDINSERT (Insert dialog box)

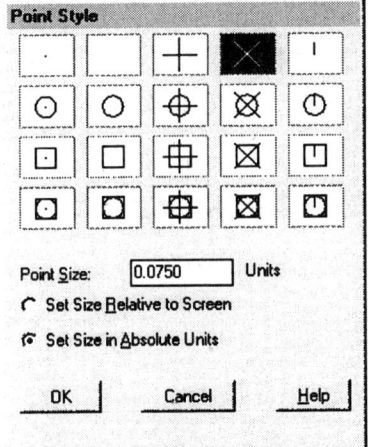

DDPTYPE (Point Style dialog box)

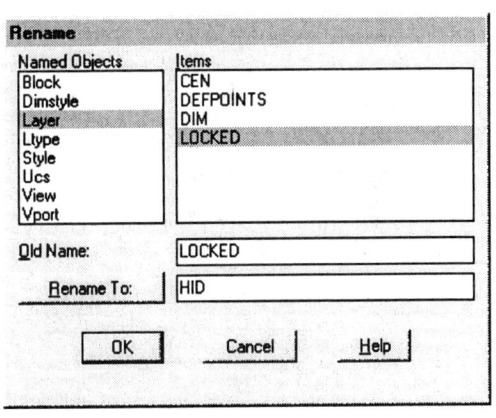

DDRENAME (Rename dialog box)

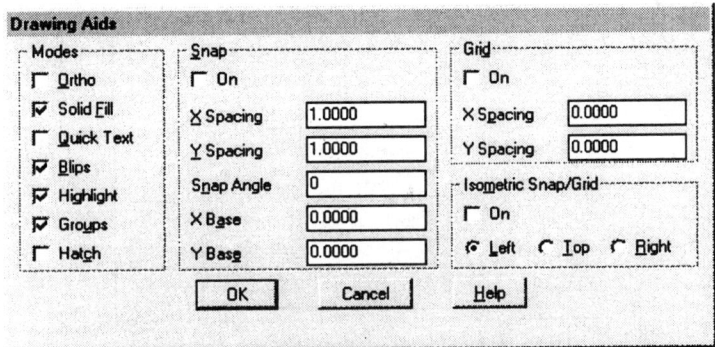

DDRMODES (Drawing Aids dialog box)

F-6 Customizing AutoCAD

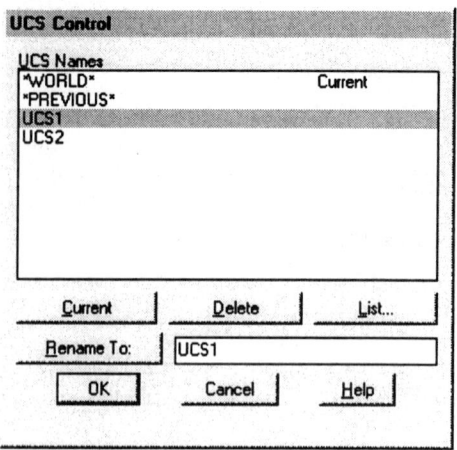

DDSELECT (Object Selection Settings)

DDUCS (UCS Control dialog box)

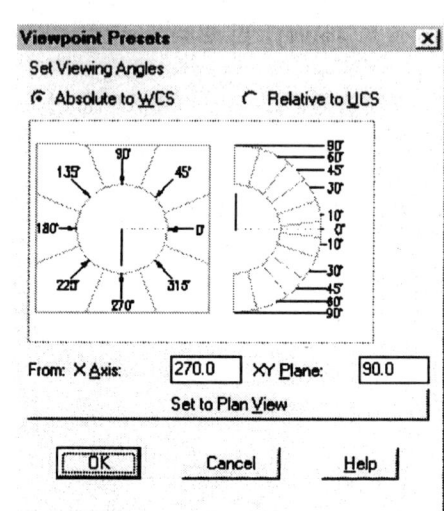

DDUNITS (Units Control dialog box)

DDVPOINT (Viewpoint Presets)

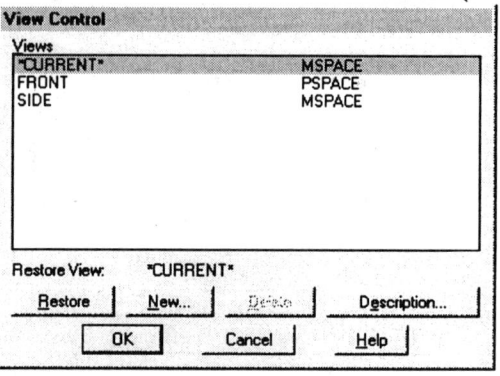

DDVIEW (View Control dialog box)

Appendix F: Dialog Boxes F-7

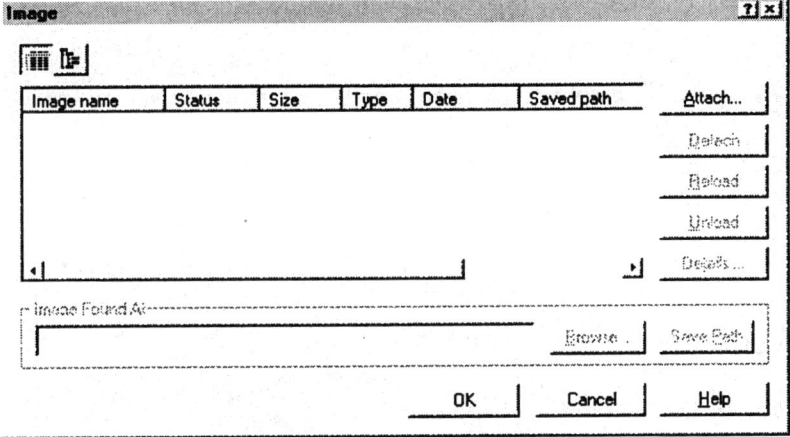

DXFIN (Select DXF File dialog box)

DXFOUT (Create DXF File dialog box)

IMAGE (Image dialog box)

GROUP (Object Grouping dialog box)

HELP (Help Topics: AutoCAD Help dialog box (Index tab)

Appendix F: Dialog Boxes   F-9

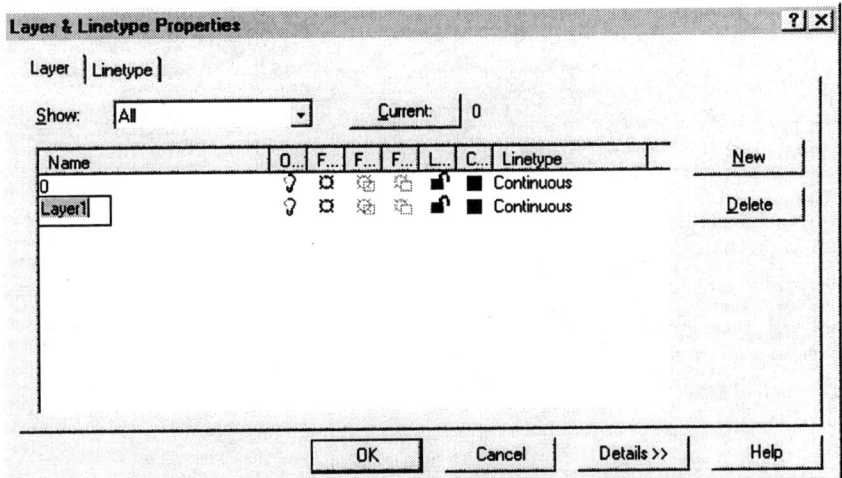

LAYER (Layer & Linetype Properties dialog box)

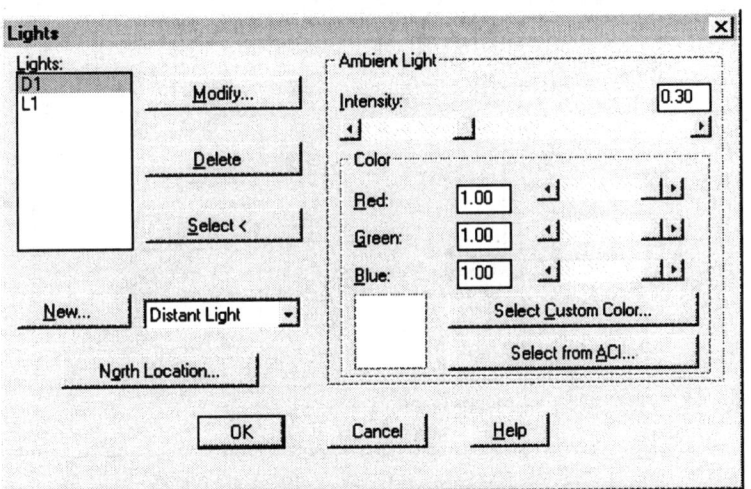

LIGHT (Lights dialog box)

MTEXT (Multiline Text Editor dialog box)

F-10 Customizing AutoCAD

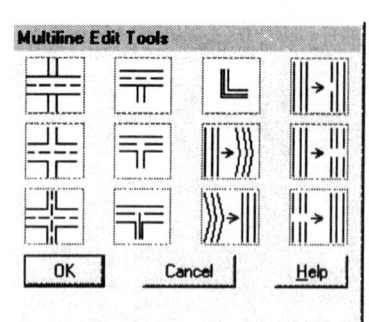

MLEDIT (Multiline Edit Tools dialog box)

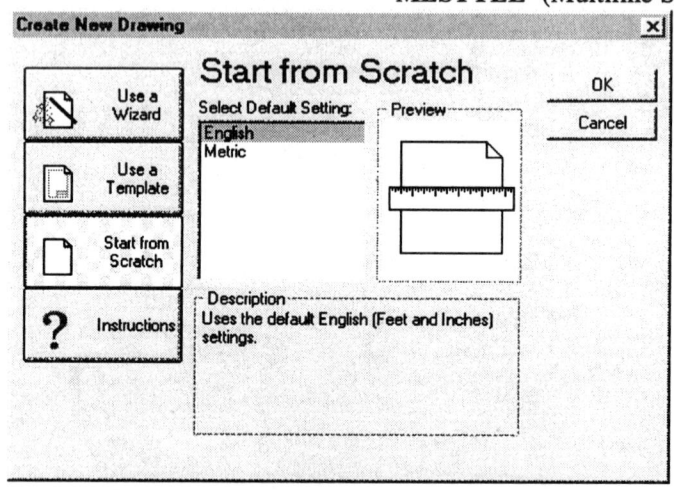

MLSTYLE (Multiline Styles dialog box)

NEW (Create New Drawing dialog box)

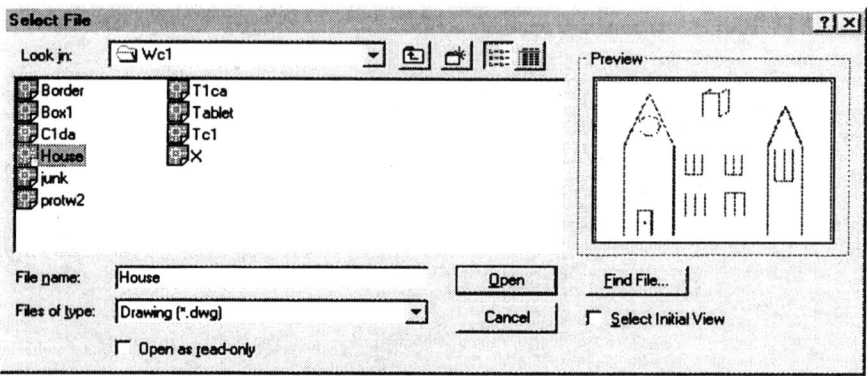

OPEN (Select File dialog box)

Appendix F: Dialog Boxes   F-11

OSNAP (Osnap Settings dialog box)

SPELL (Check Spelling dialog box)

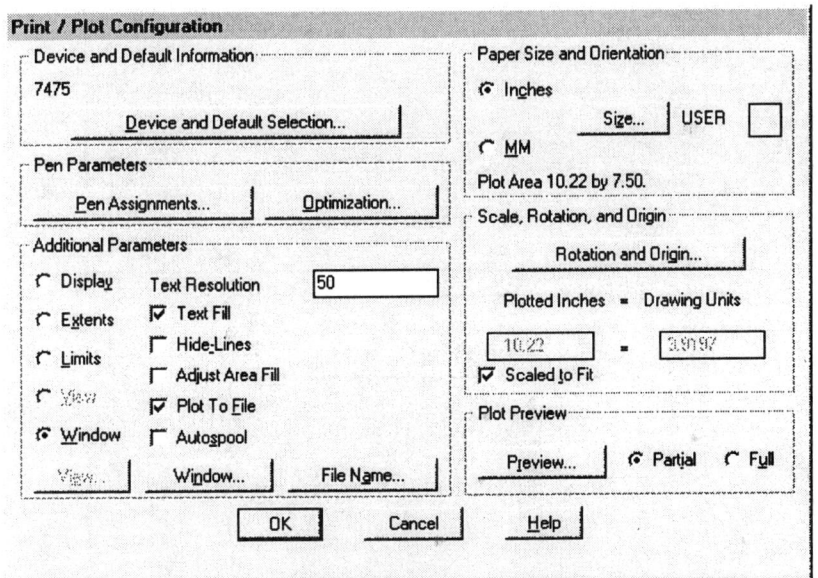

PLOT (Print/Plot Configuration dialog box)

F-12 Customizing AutoCAD

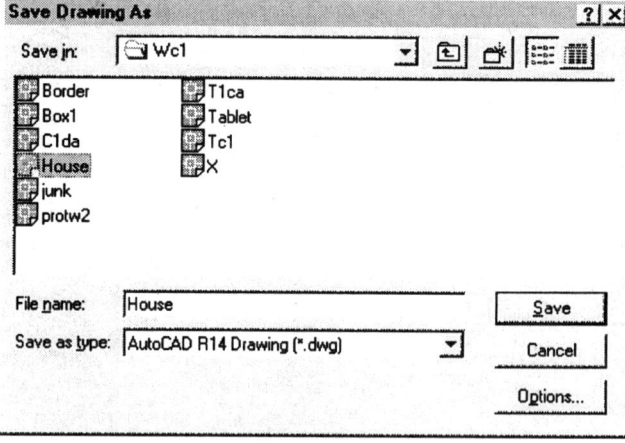

PSIN (Select PostScript File dialog box)

PSOUT (Create PostScript File dialog box)

SAVE (Save Drawing As dialog box)

Appendix F: Dialog Boxes F-13

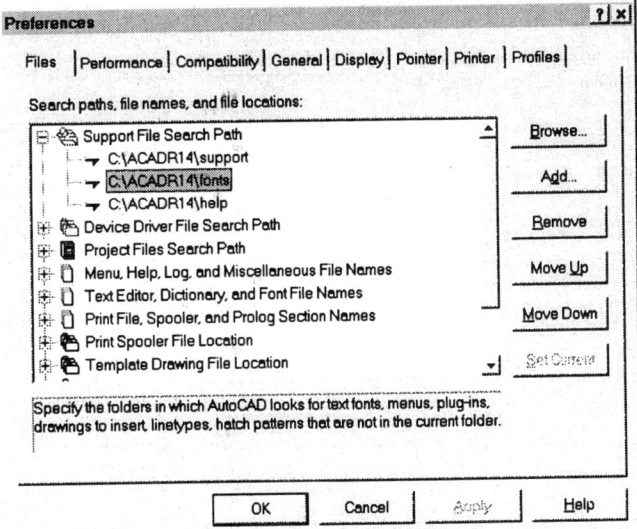

PREFERENCES (Preferences dialog box)

RENDER (Render dialog box)

F-14 Customizing AutoCAD

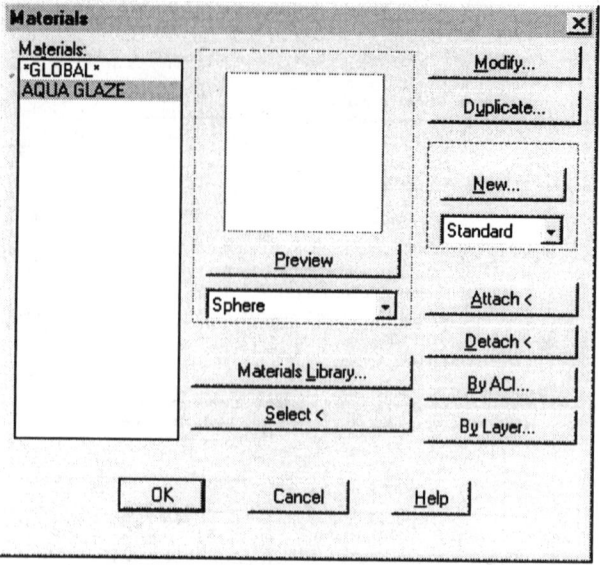

RMAT (Materials dialog box)

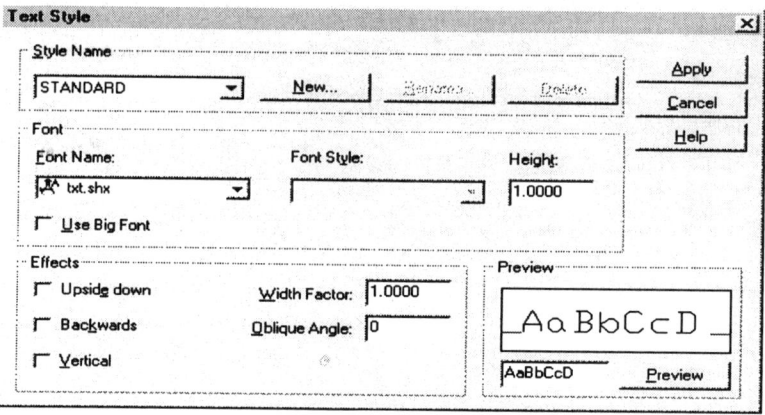

SAVEIMG (Save Image dialog box)

STYLE (Text Style dialog box)

WBLOCK (Create Drawing File dialog box)

XREF (External Reference dialog box)

Appendix G

Pull-down Menus

The following are the pull-down menus in AutoCAD Release 14.

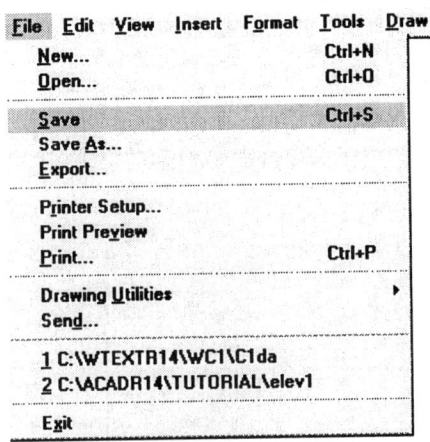

File pull-down menu

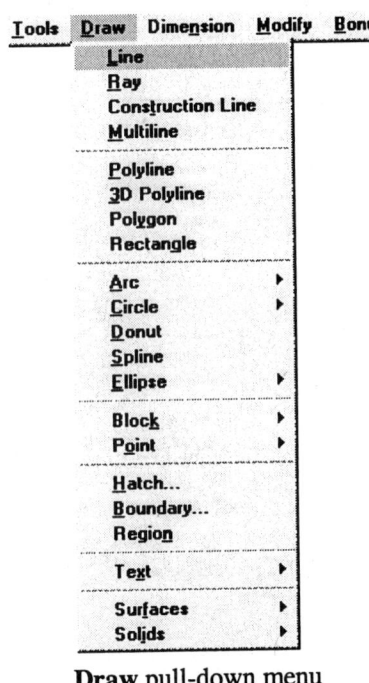

Draw pull-down menu

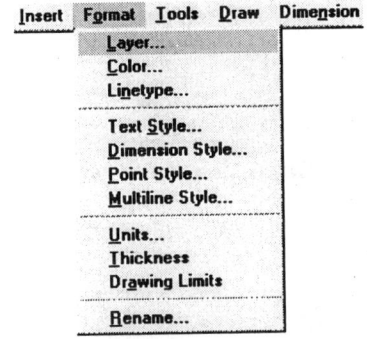

Format pull-down menu

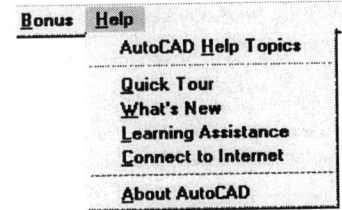

Help pull-down menu

G-1

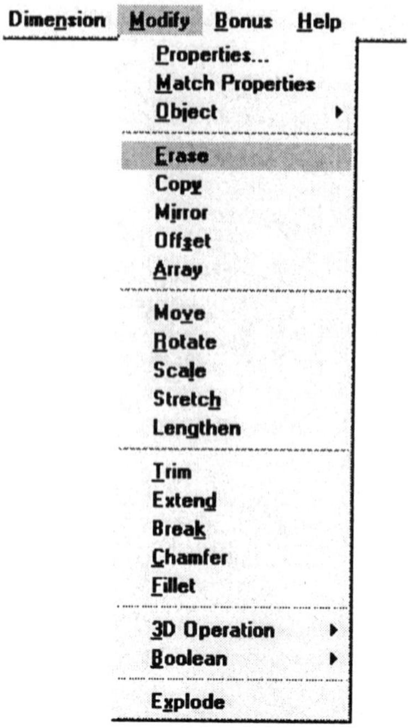

Modify pull-down menu

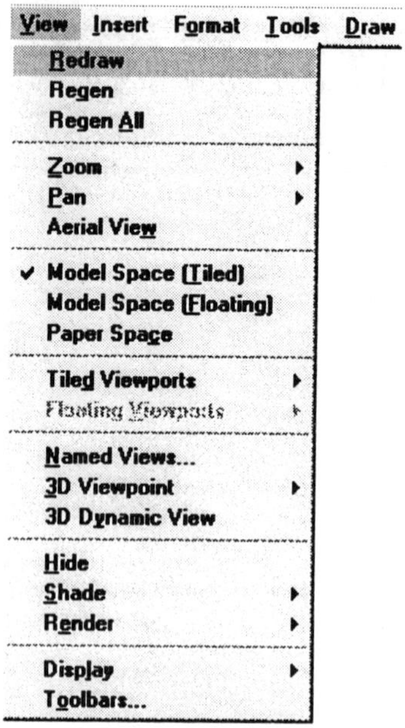

View pull-down menu

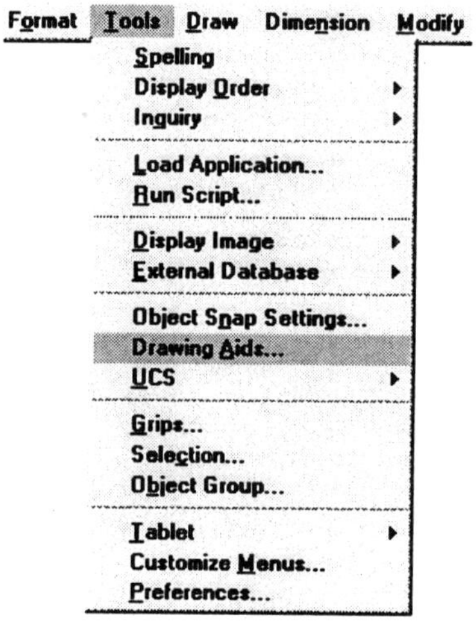

Tools pull-down menu

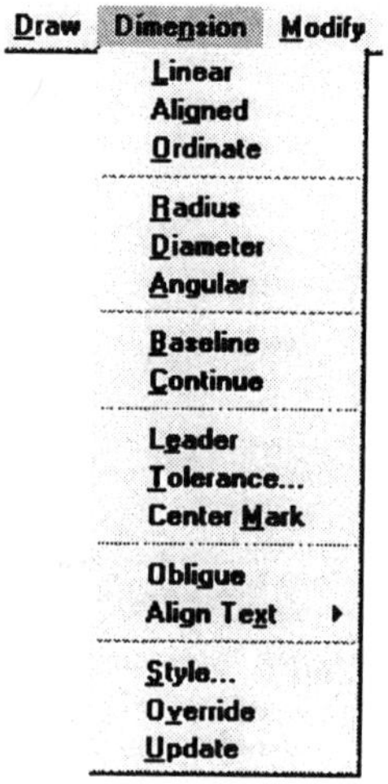

Dimension pull-down menu

Appendix G: Pull-down Menus   G-3

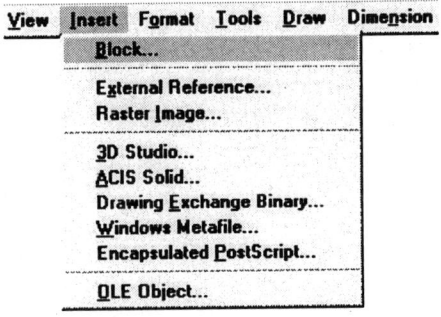

Insert pull-down menu

Appendix H

Toolbars

The following are the frequently used toolbars in AutoCAD Release 14.

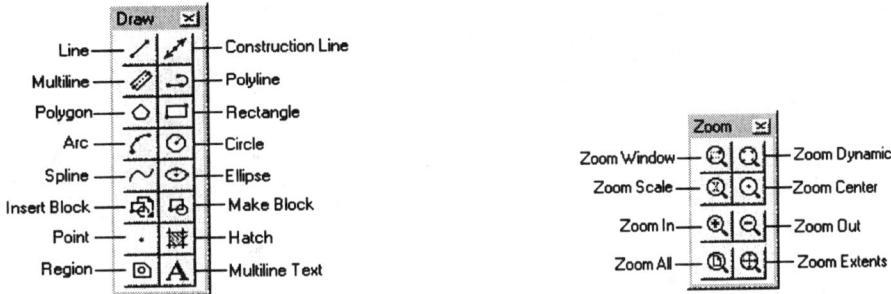

Figure H-1 Draw toolbar

Figure H-2 Zoom toolbar

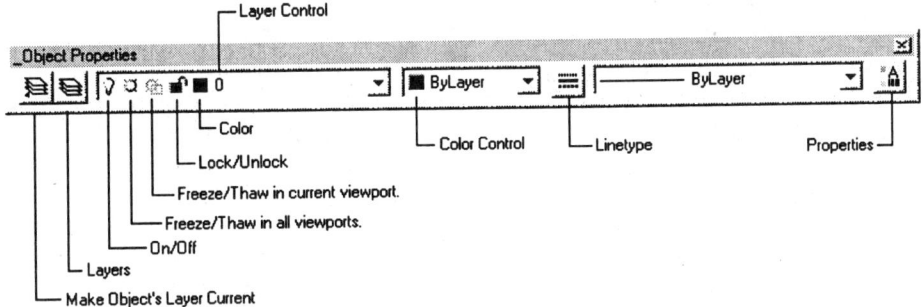

Figure H-3 Object Properties toolbar

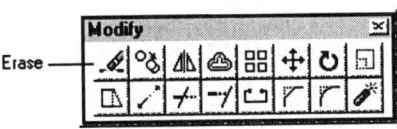

Figure H-4 Modify toolbar

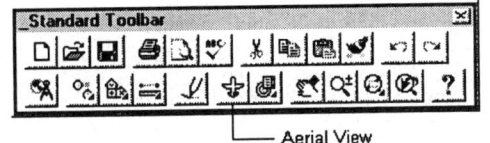

Figure H-5 Standard toolbar.

H-1

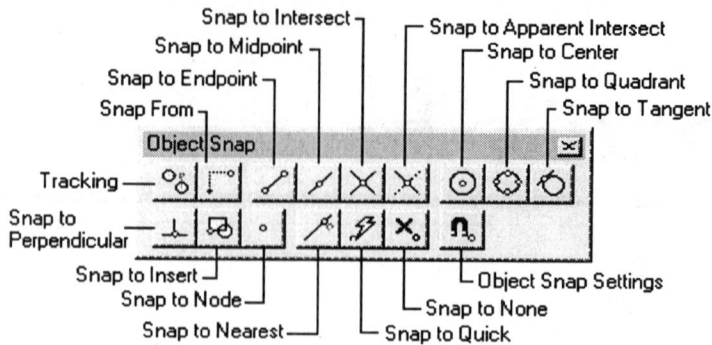

Figure H-6 Object Snap toolbar

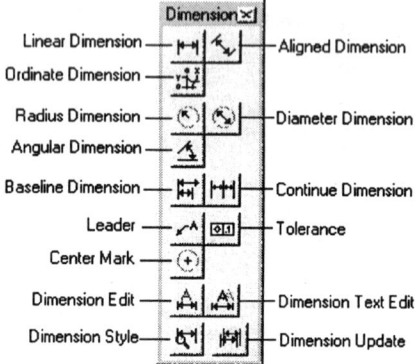

Figure H-7 Dimensioning toolbar

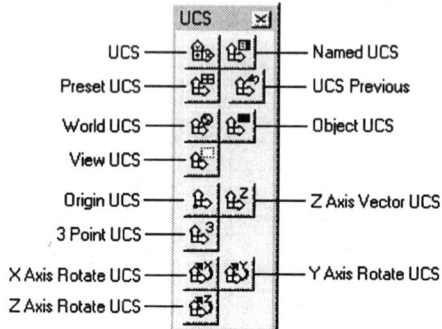

Figure H-8 UCS toolbar

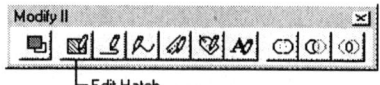

Figure H-9 Modify II toolbar

Figure H-10 Insert toolbar

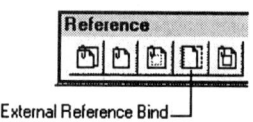

Figure H-11 Reference toolbar

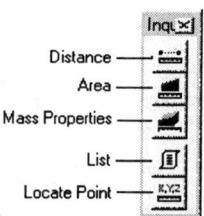

Figure H-12 Inquiry toolbar

Appendix H: Toolbars H-3

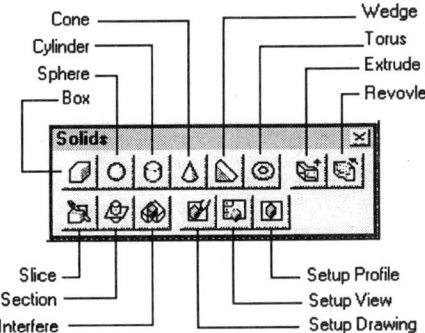

Figure H-13 Solids toolbar

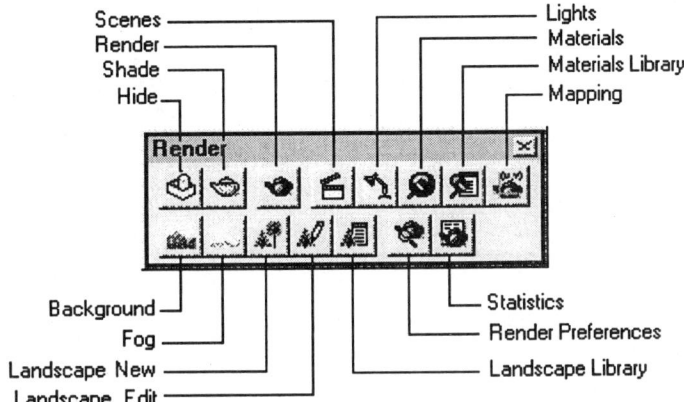

Figure H-14 Render toolbar

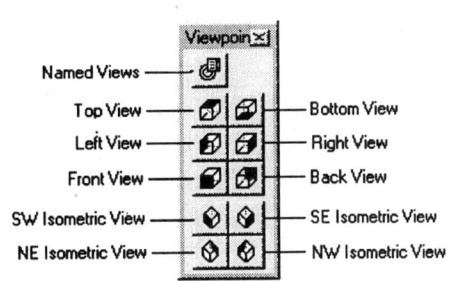

Figure H-15 Viewpoint toolbar

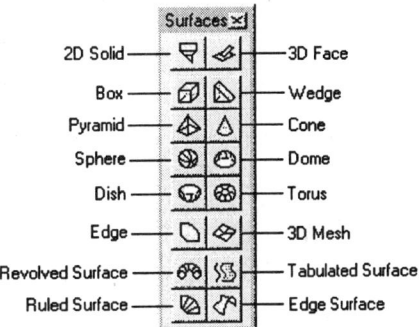

Figure H-16 Surfaces toolbar

Appendix I

AutoCAD Commands

Command	Description and Options

3D Draws a 3D polygon entities with surfaces. Options:
B - Draws a 3D box, C - Draws a wire frame having a cone shape, DI - Draws a 3D dish shaped (lower half of a sphere) polygon mesh by specifying the center and then the diameter or radius, DO - Draws the 3D upper half of the spherical polygon mesh by specifying the center and then the diameter or radius, M - Draws a polygon mesh by specifying the corners and the M and N sizes, P - Draws a 3D tetrahedron or a pyramid by specifying the relevant number of base points and the apex point or the top points, S - Draws a spherical polygon mesh by specifying the center and then the diameter or radius, T - Draws a polygon mesh having a toroidal shape. It is drawn by specifying the center and then the diameter or the radius, W - Draws a polygon wire frame having the shape of a wedge. It is drawn by specifying a corner, height, length, and width.

3DARRAY Draws a 3D rectangular or polar array. Options:
R - Rectangular 3D array, P - Polar array

3DFACE Draws a 3D surface with three or four sides.

3DMESH Draws a polygon mesh by specifying the size of M and N and the location of the vertices.

3DPOLY Draws a 3D polyline having segments of straight line. Options:
point - Draws the 3D polyline to the specified point, C - Closes the 3D polyline by joining the last point with the first point, U - Deletes the last segment.

3DSIN Displays a dialog box that allows to import the specified objects in a 3D Studio file.

3DSOUT Displays a dialog box that exports the AutoCAD objects with surface characteristics to a 3D studio file.

'ABOUT Displays AutoCAD version and serial numbers and other information.

ACISIN Imports an ASCII ACIS file into the AutoCAD drawing.

ACISOUT AutoCAD exports the selected 3D solid objects to an ASCII ACIS file.

ALIGN Allows specified objects to align with other objects by moving and rotating them.

AMECONVERT Changes the regions and solids of AME to region and solids of AutoCAD.

'APERTURE Controls the size of the object snap target box.

'APPLOAD Displays a dialog box that loads certain applications such as AutoLISP, ADS, and ARX.

ARC Draws an arc of any size. The default method is to specify two endpoints and a point along the arc. Options:
A - Included angle, C - Center point, D - Starting direction, E - Endpoint, L - Length of chord, R - Radius

AREA Computes the area and perimeter of different objects and of a region formed by specifying a sequence of points. Options:
A - Add mode, F - First point (Area by specifying points), O - Area of the object, S - Subtract mode

ARRAY Creates specified number of copies of a selected object. Options:
P - Polar array, R - Rectangular array

ARX Loads, unloads, and provides information about ARX applications.

.ASEADMIN Makes other programs attain access to database objects and sets up environment for external database commands through a dialog box.

ASEEXPORT The information about the link for selected object is exported to the external database file. The link information is stored in the text files and the link path name in the export table through a dialog box.

ASELINKS The link information is manipulated (edited or deleted) in the drawing and its blocks through a dialog box.

ASEROWS It edits the data in the database. Links and selection sets are also created through a dialog box.

ASESELECT Creates a selection set of objects, selects objects, and highlights them. Graphic and nongraphic data can be combined to create the selection set.

ASESQLED Displays a dialog box through which the external databases can be queried directly by executing the SQL statements.

ATTDEF Creates an attribute definition (characteristics of an attribute). Options:
I - Invisible mode: Attribute remains invisible, C - Constant mode: Constant value of attribute, V - Verify mode: Verifies attribute value is correct, P - Preset mode: Default value to attribute.

'ATTDISP Controls the visibility of the attribute globally. Options:
ON - Attributes made visible, OFF - Attributes made invisible, N - Current visibility kept.

ATTEDIT Edits attributes within a block or blocks.

ATTEXT Attribute information is extracted from the drawing. Options:
C - CDF: Comma-Delimited File, D - DXF: Drawing Interchange File, S - SDF: Space-Delimited File

ATTREDEF An existing block is redefined and its attributes are updated.

AUDIT Identifies errors in a drawing. Options:
Y - Corrects the errors, N - Informs about the error without correcting it

BACKGROUND Creates a background for your scene.

'BASE Sets the point of origin for inserting a drawing into another drawing.

BHATCH A specified enclosed area is filled with an associative hatch pattern through a dialog box. Previewing a hatch and adjusting the boundary is also possible.

'BLIPMODE Controls the appearance of marker blip that is displayed on the screen when a point is picked. Options:
ON - Marker blip displayed, OFF - Marker blip not displayed.

BLOCK Creates a compound object as a block definition from a set of entities. Options:
? - Lists names of previously defined blocks.

BMAKE Creates a block using a dialog box.

BMPOUT Creates a bitmap image of the drawing and saves the screen to a file having a .bmp extension.

BOUNDARY Creates a polyline or region of a boundary which defines an enclosed area.

BOX Creates a solid box which is 3D in nature with its base parallel to the XY plane.

BREAK Removes specified portions of an object or splits the object. Options:
  F - Respecifies first point.

BROWSER Launches the default Web browser defined in the registry of the system.

'CAL Calculates expressions which can be mathematical as well as geometrical.

CHAMFER Connects two nonparallel objects with a beveled line. Options:
  A - Chamfer distance is set using angle and distance, D - Sets chamfer distance, P - Chamfers entire polyline, T - Controls the trimming of the edges to chamfer line endpoints.

CHANGE Alters the properties of selected objects. Options:
  C - Change point - Changes lines, circles, Text, Attribute Definitions, Blocks, P - Changes properties like Color, Elev, LAyer, LType, and Thickness.

CHPROP Alters the drawing properties of selected objects. Options:
  C - Changes color, A - Changes layer, LT - Changes linetype, S - Changes linetype scale factor, T - Changes thickness.

CIRCLE Draws a circle using any of the four methods available. Options:
  C - Circle drawn on the basis of center point and diameter or radius, 3P - Drawn on the basis of 3 points on circumference, 2P - Drawn on the basis of 2 endpoints of the diameter, TTR - Circle drawn tangent to two objects with a specified radius.

'COLOR Sets color for subsequent objects drawn. Options:
  value - Sets color by number (1-255), name - Sets color by name, Byblock - The current color setting is inherited by the block at the time of insertion, Bylayer - Objects inherit the color of the layer in which they are drawn.

COMPILE Shape files and PostScript font files are compiled.

CONE Draws a 3D solid cone. Options:
  Center - Cone having circular base, E - Cone having elliptical base.

COPY Draws a copy of the selected object leaving the original object intact. The default method is to specify the base point. Options:
  M - Multiple copies of object in single COPY command.

COPYCLIP Copies the selected objects to the Clipboard.

COPYHIST Copies the text in the command line history to the Clipboard.

COPYLINK Copies the current view to the Clipboard.

CUTCLIP Copies objects to the Clipboard but erases them from the drawing.

CYLINDER Draws a 3D solid cylinder. Options:
  center - Specifies the center of the circular base, E - Forms an elliptical base.

DBLIST The database information about each entity in the drawing is listed.

DDATTDEF Creates an attribute definition (characteristics of an attribute) through a dialog box.

DDATTE Variable attribute values of a block are edited through a dialog box.

DDATTEXT Attribute information is extracted from the drawing using a dialog box.

DDCHPROP Alters the drawing properties (color, layer, linetype, and thickness) of selected objects via a dialog box.

DDCOLOR Displays a dialog box that sets color for subsequently drawn objects.

DDEDIT Displays a dialog box that allows the user to edit text and attribute definitions.

'DDGRIPS A dialog box is displayed through which grips are enabled and their color and size is set.

DDIM Using a series of dialog boxes, different dimension styles are created and modified.

DDINSERT Using a dialog box, a block or a drawing file is inserted into a drawing. This dialog box also allows the setting of different parameters like insertion point, scale, rotation, and explode.

DDMODIFY Displays the appropriate dialog box that controls the properties (color, layer, linetype, linetype scale, and thickness) of the existing objects.

'DDPTYPE Displays a dialog box that sets the point style and also the size of the point object.

DDRENAME Changes the names of different types of objects via a dialog box. The objects that can be renamed are Blocks, Dimension style, Layer, Linetype, Style, Ucs, and View.

'DDRMODES Displays a dialog box that sets drawing aids such as Ortho, Fill, Qtext, Blipmode, Highlight, Group, Snap, Grid, and Isoplane.

'DDSELECT Displays a dialog box that sets the object selection modes. It also sets the pickbox size and the object sort method.

DDUCS Controls the defined User Coordinate Systems via a dialog box.

DDUCSP Displays a dialog box that selects a preset User Coordinate System.

'DDUNITS Sets distance and angle display formats and precision through a dialog box. It also controls the direction of the angle.

DDVIEW Displays and restores the existing named views via a dialog box. It also creates new views.

DDVPOINT Controls the direction of 3D views through a dialog box.

'DELAY The execution of the next command is postponed for a specified time duration. In other words, a specified pause is provided within the script.

DIM Dimensioning mode is invoked and permits the use of dimension subcommands from AutoCAD's previous releases.

DIMALIGNED Creates a linear dimension aligned to the specified points or object.

DIMANGULAR Creates an angular dimension.

DIMBASELINE Starts drawing from the baseline of the previous dimension. The new dimension can be linear, angular, or ordinate.

DIMCENTER Draws center mark or center lines in circles and arcs.

DIMCONTINUE Starts drawing a new dimension from the second extension line of the previous or selected dimension. The new dimension can be a linear, angular, or ordinate dimension.

DIMDIAMETER Draws diameter dimensions for different circles and arcs.

DIMEDIT Edits dimension text and extension lines. Options:
  H - Dimension text moved back to default position, N - Dimension text is replaced, R - Dimension text is rotated, O - Extension lines placed at obliquing angle.

DIMLINEAR Draws linear dimensions.

DIMORDINATE Creates ordinate dimensions.

DIMOVERRIDE The settings of the dimensioning system variables concerning the dimension object are overridden. The current dimension style is not affected.

DIMRADIUS Draws radial dimensions for different circles and arcs.

DIMSTYLE New dimension styles are created and the existing ones are modified. Options:
R - Dimensioning system variable setting changed, S - Current settings of dimensioning variables saved, ST - Current values of dimensioning variables displayed, V - Dimensioning variable setting of a style is listed, A - Selected dimension objects are updated, ? - Named dimension styles are listed.

DIMTEDIT Dimension text is moved and rotated. Options:
A - Angle of dimension text changed, H - Dimension text moved to default position, L - Dimension text left justified, R - Dimension text right justified.

'DIST Distance and angle between two points is measured.

DIVIDE Places blocks or points as markers at equal distance along the length or perimeter of an entity, thus dividing it into a specified number of equal parts. Options:
B - Places blocks as markers.

DONUT Draws wide polyline with specified inside and outside diameters, thus forming a ring.

'DRAGMODE Controls the dragging feature for appropriate commands. Options:
ON - Permits dragging, OFF - Ignores dragging, A - Permits dragging wherever possible.

DRAWORDER Changes the display order of images and other objects.

DSVIEWER Displays the Arial View window.

DTEXT Writes text and displays it as it is entered. Text is written from a specified start point. Options:
J - Alignment of text is controlled by several options, A - Text aligned between two points, F - Fits text of specified height between two points, C - Text is centered horizontally, M - Text is centered horizontally and vertically, R - Text is justified right, BL - Bottom left, BC - Bottom center, BR - Bottom right, ML - Middle left, MC - Middle center, MR - Middle right, TL - Top left, TC - Top center, TR - Top right, S - Sets the text style.

DVIEW Parallel projection or perspective views are defined. Options:
CA - Sets camera position by rotating about the target, TA - Sets target position by rotating about the camera, D - Camera to target distance is set, PO - Locates target and camera points, PA - Pans image, Z - Zooms In/Out, TW - Tilts view around line of sight, CL - The view is clipped in front and back, H - Hidden lines removed on selected objects, OFF - Perspective viewing turned off, U - Last DVIEW operation reversed, X - Exits DVIEW.

DDWFOUT Exports a drawing web format file.

DXBIN Specially coded binary files are imported into a drawing.

DXFIN A drawing interchange file is imported.

DXFOUT A drawing interchange file of the current drawing is created.

EDGE The visibility of 3D sides is altered. By default the selected edge is hidden. Options:
D - Display mode invisible edges are highlighted.

EDGESURF A 3D polygon mesh is created which has four adjoining edges that define a Coons surface patch.

ELEV The elevation and extrusion thickness is set for the new objects.

ELLIPSE Draws ellipses or elliptical arcs using different options. Specifying the axis endpoint is the default method. Options:
A - Draws elliptical arc, C - Specifies the center point of the ellipse, I - Draws isometric circle in current isometric plane.

ERASE Erases the selected objects from the drawing.

EXPLODE Compound objects (blocks, dimensions, polylines, 3D solids, regions, polygon meshes, and multilines) are broken into their constituent parts.

EXPORT Objects are saved to other file formats via a dialog box.

EXTEND Lengthens a selected entity to meet another entity. Options:
P - Specifies projection mode like UCS and View, E - Controls the extension to implied or actual edge, U - Latest extension is undone.

EXTRUDE Solids are created by extruding 2D entities along a selected path. By default height of extrusion is to be specified which extrudes the object along the positive Z axis. Options:
P - Extrusion path is selected.

'FILL Controls whether multilines, traces, solids, or wide polylines are filled or not filled. Options:
ON - Fill mode is enabled, OFF - Fill mode is disabled.

FILLET The edges of two specified lines, arcs, or circles are filleted by construction of an arc of specified radius. The default method is to specify the two objects. Options:
P - Entire polyline is filleted, R - The radius of the fillet arc is specified, T - Controls the trimming of the edges to fillet arc endpoints.

'FILTER Creates a list of properties on the basis of which the objects are selected.

'FOG Provides visual clues for the apparent distance of objects.

'GRAPHSCR Flips to the graphics window from the text window.

'GRID A grid of dots at specified spacing is displayed. Options:
Grid spacing(X) - Grid set to specified value, ON - Grid turned on at current spacing, OFF - Grid turned off, S - Grid spacing set to current Snap interval, A - Grid set to different spacing in X and Y.

GROUP Creates and changes object selection groups, which are sets of objects having specific names.

HATCH A specified area is filled with a selected pattern. Options:

? - Lists the hatch patterns in acad.pat file, name - A pattern name as defined in acad.pat file is specified, U - User defined hatch pattern is specified. U can be followed by a comma and a hatch style. The different hatch styles are:, n - Normal or standard style. Hatching is performed from the inside of the outermost boundary. Moreover areas having odd number of boundaries around them are hatched, o - Hatches outermost area only, i - Hatches the complete area thus ignoring the internal structure.

HATCHEDIT Edits a hatch block through a dialog box. It sets the pattern type and properties and then applies it to a block.

'HELP Displays help for a specific command and also lists the commands and data entry options.

HIDE Regenerating of a 3D object is performed with the removal of hidden lines.

'ID The UCS coordinates of a specified point are displayed.

IMAGE Inserts images into an AutoCAD drawing file.

IMAGEADJUST Controls the brightness, contrast and fade values of the selected image.

IMAGEATTACH Attaches a new image object and definition.

IMAGECLIP Creates new clipping boundaries for single image objects.

IMAGEFRAME Controls the display of image frame on the screen.

IMAGEQUALITY Controls the display quality of images.

IMPORT Imports the different file formats into AutoCAD drawing via a dialog box.

INSERT Places a previously defined named block or drawing into the current drawing.
Options:
  X scale - Inserts copy of block with basepoint at insertion point, C - Insertion point and another point as the corners, XYZ - Scaling in all three dimensions.

INSERTOBJ Inserts a previously linked or embedded object.

INTERFERE Highlights all of the interfering solids and then creates new solids from the intersections of the interfering pairs of solids.

INTERSECT A new composite solid is created from the intersecting region of two or more solids.

'ISOPLANE An isometric plane is selected to be the current plane for an orthogonal drawing. Options:
  T - Switches to the next plane, L - Left-hand plane, T - Top plane, R - Right-hand plane.

'LAYER Creates layers and sets different properties for the specified layers. Options:
  ? - Lists defined layers, M - Creates a layer and makes it the current layer, S - Makes a specified already existing layer current, N - Creates one or more new layers, ON - Turns on the specified layers, OFF - Turns off the specified layers, C - Sets the color of the specified layer, L - Sets the linetype of the specified layer, F - Makes a layer invisible by freezing it, T - The frozen layer is thawed, LO - Locks layers, thus prevents editing on them, U - Unlocks specified locked layers.

LEADER Creates a line segment with an arrowhead that connects the text to a feature. The leader is created from a specified point to another point depending upon the options.
Options:
  A - Annotation is inserted at the end of leader line, F - Controls the type of leader (Spline, Straight, and Arrow), U - The last vertex point is removed.

LENGTHEN Alters the length of specified entities and the included angle of arcs.
Options:
  DE - Lengthens the object by a specified incremental distance, P - Alters the length by a specified percentage of its total length, T - Alters the length by specified total absolute length, DY - The object is lengthened to where its endpoint is dragged.

LIGHT Controls the lighting effects in the model space via a dialog box. It creates, modifies, deletes the lights, and controls the color system in a drawing. It manages different lights (Point light, Distant light, and Spotlight) through a series of dialog boxes.

'LIMITS Sets the drawing boundaries (and WCS grid extents) for the current space.
Options:
  Lower - Specifies 2 points-lower left corner and the left-upper right corner, ON - Limits checking is enabled, OFF - Limits checking is disabled.

LINE Draws straight line segments of any length by specifying the endpoints. Options:
  Enter (◂┘) - Continues from end of previous line or arc, U - Removes the most recent segment, C - Closes polygon.

'LINETYPE Defines line characteristics, loads linetypes, and sets them for new entities. It also creates new linetype, definitions to a library file. Options:
  ? - Lists linetypes in a file, C - Creates new linetype definition, L - Loads an already existing linetype definition, Sets linetype for new entities. Set suboptions:

name - Sets the specified linetype, Bylayer - Sets to inherit the linetype associated with layer, Byblock - The objects inherit the linetype of the block after it is inserted.

LIST Lists database information (type, layer, X,Y,Z position, thickness, etc.) about the specified entity.

LOAD Loads the shapes from the shape file to be used by the SHAPE command.

LOGFILEOFF The log file already opened is closed by this command.

LOGFILEON The subsequent contents of the text window are recorded into the log file.

LSEDIT Lets you edit a landscape object.

LSLIB Lets you maintain libraries of landscape objects.

LSNEW Lets you add realistic landscape items, such as trees and bushes, to your drawing.

'LTSCALE Sets the global scale factor of the linetype so as to alter the relative length of dashes and dots.

MASSPROP Calculates and lists the mass characteristics of 2D and 3D objects. The properties displayed are Area, Perimeter, Bounding box, Centroid. For Coplanar regions, additional properties displayed are Moments of Inertia, Products of Inertia, Radii of Gyration, and Principal Moments. The properties displayed for solids are Mass, Volume, and the properties of Coplanar region.

'MATCHPROP Copies the properties from one object to oneor more objects.

MATLIB Displays a dialog box that lists all the predefined materials (material list) and lists the materials in the selected library (library list). It also imports and exports materials between those two lists.

MEASURE Places blocks or points as markers at measured interval along the length or perimeter of an entity. Options:
B - Places blocks as markers.

MENU Loads a customized menu file into the menu area. The menu file contains the command strings and menu syntax.

MENULOAD Displays a dialog box that loads and permits you to add partial menu files to an already present base menu file.

MENUUNLOAD Displays the same dialog box as in the case MENULOAD command that can also be used to unload the partial menu files.

MINSERT Places multiple copies of a previously drawn named block or drawing into the current drawing in a rectangular array. Options:
? - Lists the defined block definitions, ~ - Displays a dialog box. After specifying the insertion point you get the following options: X scale - Inserts copy of block with basepoint at insertion point, C - Insertion point and another point as the corners, XYZ - Scaling in all three dimensions.

MIRROR Reflects objects so as to create their mirror images about a specified line.

MIRROR3D Reflects objects so as to create their mirror images about a specified plane. Options:
3points - 3points specify the mirroring plane, O - The plane of a planar object specifies the mirroring plane, L - The previous mirroring plane is taken as the present one, Z - The point on the plane and another point on the Z axis (normal) of the plane specifies the mirroring plane, V - A point on the viewing plane specifies the mirroring plane, XY/YZ/ZX - The mirroring plane is aligned to any one of the standard planes.

MLEDIT Displays a dialog box that controls intersection between multiple parallel lines and edits them. Different types of cross, tee, corner joints, and vertices can be created between multilines via the dialog box. It is also possible to cut and weld segments of a single multiline.

MLINE   Draws multiple parallel lines between two points. Options:
  J - Justification- How multiline is drawn between two points, S - Scale- Sets the width of the multiline, ST - Sets the multiline style.

MLSTYLE   Displays a dialog box that creates a multiline style, makes a specific style current, saves, adds a style to the current list, renames a style, adds a description to a style, and loads a style from the library file. It also controls the element properties (number, offset, color, and linetype) and the multiline properties (start and end caps, angle, and background color).

MOVE   Moves objects from one location to another by specifying a displacement.

MSLIDE   Creates a slide file from the current display.

MSPACE   Switches to model space in a floating viewport from paper space.

MTEXT   Creates paragraph text within a specified text boundary. Displays a dialog box where you can specify the different options for the multiline text.

MULTIPLE   Causes the repetition of the next command until it is cancelled.

MVIEW   Creates viewports and controls the number and layout of paper space viewports. You specify diagonal corners of new viewport as the default option. Options:
  ON - Viewport is turned on, OFF - Viewport is turned off, H - Hideplot- Hidden lines removed during plotting, F - Fit- Single viewport created which fills the display area completely, 2 - The specified area is divided into two viewports either horizontally or vertically, 3 - The specified area is divided into 3 viewports, 4 - The specified area is divided into 4 viewports, R - Restore- Viewport configurations changed into individual viewports.

MVSETUP   The specifications of a drawing are set. Depending upon the system variable TILEMODE, the working of MVSETUP is different. When TILEMODE is On, drawing scale factor, units type, and paper size is set and lastly a bounding box is drawn. When TILEMODE is Off a set of floating viewports is created. Options (TILEMODE Off):
  A - Aligns the view in a viewport with another viewport. The view can be panned in a specified direction, align it horizontally, vertically, or rotate it, C - The viewport can be created, S - Sets the scale factor of objects in the viewport, O - Options- The layer can be set, reset limits, set units, Xref attach, T - Creates a title block and drawing border, U - Reverses the previous operation.

NEW   Displays a dialog box that creates a new drawing.

OFFSET   Creates offset curves, concentric circles, and parallel lines at a specified distance from the original object. Options:
  value - specify the offset distance, T - Through- The offset object passes through the specified point.

OLELINKS   Updates, changes, and cancels existing OLE links.

OOPS   Restores those entities which have been erased by the last ERASE command.

OPEN   Displays a dialog box through which an existing drawing can be opened. The dialog box also displays the directory, files, preview, name of the file, and the pattern.

'ORTHO   The movement of the cursor is restrained to only vertical or horizontal directions and aligned with the grid. Options:
  ON - Constrains cursor movement, OFF - Does not constrain cursor movement.

'OSNAP   Specifies a point at an exact location on an object by setting the Object Snap modes. Options:
  END - Closest endpoint of arc (arcs and lines include polyline segments), elliptical arc, ray, mline, line and closest corner of trace, solid, 3D face, MID - Midpoint of

arc, elliptical arc, spline, ellipse, ray, solid, xline, mline, or line, INT - Intersection of line, arc, spline, elliptical arc, ellipse, ray, xline, mline, or circle, APPINT - Apparent or extended (projected) intersection (which may not actually intersect in 3D space) of line, arc, spline, elliptical arc, ellipse, ray, xline, mline, or circle, CEN - Center of arc, elliptical arc, ellipse, or circle, QUA - Quadrant point of arc, elliptical arc, ellipse, solid, or circle, PER - Point perpendicular to arc, elliptical arc, ellipse, spline, ray, xline, mline, line, solid, or arc, TAN - Tangent to arc, elliptical arc, ellipse, or circle, NOD - Point object, INS - Insertion point of text, block, shape, or attribute, NEA - Nearest point of arc, elliptical arc, ellipse, spline, ray, xline, mline, line, circle, or point, QUI - First snap point, NON - Turns Object Snap mode off.

'PAN Moves the drawing display by a specified displacement.

PASTECLIP Inserts data from the Clipboard.

PASTESPEC Inserts data from the Clipboard and controls the format of the data.

PEDIT Editing of 2D polyline, 3D polyline, or 3D mesh. Options:
  2D polyline C - Closes polyline segment, O - Closing segment removed, J - Joins to polyline, W - Specifies uniform width, E - Edits the vertices. The first vertex is marked by placing a X. Editing includes moving the X to next or previous vertex, adding a new vertex, setting the first vertex for break, moving the vertex, regenerating, straightening and attaching a tangent direction to the current vertex, F - Fits arc curves smoothly to the polyline, replacing each line segment with a pair of arcs, S - Vertices are used as a frame for spline curve, L - Linetype generation in a continuous pattern, U - Reverses the previous operation, X - Exits PEDIT. 3D polyline C - Closes polyline segment, O - Closing segment removed, E - Edits the vertices. Same suboptions as in 2D Edit except the tangent suboption, S - Vertices are used as a frame for spline curve, D - Removes a spline curve to its control frame, U - reverses the previous option, X - Exits PEDIT. 3D polygon mesh E - Edits vertices. The first vertex is marked by placing a X. Editing includes moving the X to next or previous vertex, moving the X marker to the next vertex or the previous vertex in the N direction, moving the marker to the next or previous vertex in the M direction, regenerating the mesh, S - Fits a smooth surface, D - The control point polygon mesh is restored, Mclose - M-direction polylines are closed, Mopen - M-direction polylines are opened, Nclose - N-direction polylines are closed, Nopen - N-direction polylines are opened.

PFACE A 3D polyface mesh is created.

PLAN Allows you to view the drawing from plan view of a User Coordinate System.
Options:
  C - Plan view of the current UCS, U - Plan view of the specified UCS, W - Plan view of the World Coordinate System.

PLINE Draws 2D polylines. The default is to draw a polyline between two specified points.
Options:
  A - Arc mode - Arc segments can be added to polyline. The arc segment starts from the endpoint of the previous polyline segment and can be drawn by specifying the endpoint of the arc, the included angle, center of the arc, starting direction of the arc, halfwidth of the arc, radius of the arc, and width. You can also close the polyline with the arc segment, or reverse the previous operation or you can shift to the Line mode, C - Closes the polyline, H - Sets the halfwidth, L - Draws polyline of specified length, U - Last polyline segment is removed, W - The width of the next segment is specified.

PLOT Displays a dialog box that allows you to plot the drawing to the plotting device or file. Through a series of dialog boxes you can set the different parameters, device information, drawing extents and limits, plot size, paper size, orientation, plot scale, rotation, and origin. You can also plot a view or a specific portion of the drawing and preview the plot.

POINT Draws a point object at a specified location.

POLYGON Draws a polygon (closed polyline object) having specified number of sides. Options:
C - Specifies the center of polygon. Suboptions:
I - Inscribed in the circle, C - Circumscribed about the circle, E - Defines one edge of the polygon.

PREFERENCES Displays a dialog box that permits you to customize the AutoCAD settings. Controls the units of measurement and sets the environment.

PREVIEW Shows how the drawing will look when it is printed or plotted.

PSDRAG An imported PostScript file is dragged into place by the PSIN command and as it is dragged, the PSDRAG command controls its appearance. Options:
0 - Only the bounding box and the file name of the image is displayed as the image is being dragged, 1 - The rendered PostScript image is displayed.

PSFILL A 2D polyline boundary is filled with a PostScript fill pattern. Options:
name - Fills the polyline with the specified pattern, ? - Lists all the previously defined PostScript fill patterns.

PSIN A PostScript file is inserted into a drawing.

PSOUT Creates an Encapsulated PostScript file into which the current view of the drawing is exported. Options:
D - Exports the current view. It can also include the EPSI or TIFF screen preview image, E - Only the portion of the current space which contains the entities is exported, L - The area defined by the limits is exported, V - The previous saved is exported, W - The portion you specify within a window is exported.

PSPACE Switches from a model space viewport to paper space.

PURGE Removes those references from the database which are not being used. Options:
B - Removes unused blocks, D - Removes unused dimstyles, LA - Removes unused layers, LT - Removes unused linetypes, SH - Removes unused shape files, ST - Removes unused text styles, AP - Removes unused APPID table, M - Removes unused mline styles, A - Removes all unused objects.

QSAVE Saves and backs up the drawing without asking for a filename.

QTEXT Sets the text and the attribute objects to be displayed without drawing the text detail. Options:
ON - Text displayed as a bounding box, OFF - Quick text mode off.

QUIT Exits AutoCAD without saving.

RAY Draws a semi-infinite line used as a construction line.

RECOVER Recovers a damaged and corrupted drawing.

RECTANG Creates a polyline rectangle by specifying the diagonally opposite corners.

REDEFINE Restores an AutoCAD built-in command which has been previously overridden by UNDEFINE.

REDO The effect of the previous command if it was UNDO is reversed.

'REDRAW Cleans up the current viewport by removing the blip marks and other stray pixels and redrawing missing portions of objects.

'REDRAWALL Refreshes or cleans up all the viewports.

REGEN Regenerates the current viewport.

REGENALL Regenerates all the viewports.

'REGENAUTO Controls automatic regeneration of the drawing. Options:

ON - Permits automatic regeneration, **OFF** - Does not permit automatic regeneration.

REGION   Region entities (2D enclosed areas) are created from a selection set.

REINIT   Reinitializes the I/O ports, digitizer, display, or parameters file.

RENAME   Alters the name of entities.
Options:
> B - Renames block, D - Renames dimstyle, LA - Renames layers, LT - Renames linetype, S - Renames style, U - Renames UCS, VI - Renames view, VP - Renames viewport configuration.

RENDER   Displays a dialog box that shades a 3D wireframe or solid, so that a realistically shaded image is created. It is possible to render the current scene or just the specified objects. You can also control the color map and the shading of different materials.

REPLAY   The BMP, TGA, or TIFF images are displayed via a dialog box.

'RESUME   Resumes an interrupted script.

REVOLVE   By revolving a 2D entity (polygon, closed polyline, circle, ellipse, donuts, etc.), a solid is formed. Options:
> point - The axis of revolution is specified by two points, O - The axis of revolution is specified by selecting an existing line or a segment polyline, X - The positive X axis used as the axis direction, Y - The positive Y axis used as the axis direction.

REVSURF   A polygon mesh is constructed by rotating a curve or profile around a specified axis.

RMAT   Displays a dialog box that manages the materials used for rendering. A new material can be created or the existing ones can be modified through a series of dialog boxes. It is possible to adjust the value and color of the materials. AutoCAD's color index can also be attached by layers or by using a color wheel.

ROTATE   Rotates specified entities about a base point. Options:
> angle - Rotates object through a specified angle, R - Rotates object with respect to the reference angle.

ROTATE3D   Rotates object about a 3D axis. Options:
> 2points - The axis of rotation is given by specifying 2 points, A - Axis by object- The axis of rotation is aligned with an object, L - The previous rotation axis is considered, V - The axis of rotation is aligned with the viewing direction, X/Y/Z - The axis of rotation is aligned with any one of the axes (X-axis, Y-axis, Z-axis).

RPREF   Displays a dialog box that controls the rendering preferences. It controls the color map, the behavior of the RENDER command by default, rendering display, and the image output setting. Through a series of subdialog boxes, the type of shading used and 3D solid faces can be controlled. You can also set the color and the aspect ratio of the output file.

RSCRIPT   Repeats a script continuously.

RULESURF   Creates a polygon mesh representing a ruled surface between two curves.

SAVE   A name is requested under which the drawing is saved. If the drawing is already named, then it is saved under the current filename.

SAVEAS   An unnamed drawing is saved with a filename or the current drawing is renamed.

SAVEIMG   Displays a dialog box that saves a rendered image to a file. Through the subdialog boxes, image compression for TGA and TIFF formats is possible.

SCALE   The size of the existing objects is changed. the default is to specify a scale factor. Options:
> R - The object is scaled according to the reference length and a new length.

SCENE Controls different scenes (particular view) in model space. Through a series of dialog boxes all the scenes in the current drawing are listed, new scenes can be added, scene names can be modified, and the lights can be controlled in the scene.

'SCRIPT Executes a command script.

SECTION Creates regions. from the intersection of a plane and solids. Options:
3points - Specifying 3 points on sectioning plane, O - Sectioning plane is aligned with the object, Z - Sectioning plane is aligned with the plane's normal direction, V - Sectioning plane is aligned with the viewing plane of current viewport, XY - Sectioning plane aligned with XY plane of UCS, YZ - Sectioning plane aligned with YZ plane of UCS, ZX - Sectioning plane aligned with ZX plane of UCS.

SELECT Creates a selection set of specified group of objects. Options:
AU - Automatic selection, A - Add mode - Objects are added to the selection set, ALL - Selects all objects, BOX - Objects inside or crossing a rectangle are selected, C - Objects are selected which lie inside and crossing an area specified by two points, CP - Those objects are selected which lie inside and crossing the polygon created by specifying points around the objects, F - Those objects are selected which are crossing the specified fence, G - Objects within a group are selected, L - Recently created object is selected, M - Objects are picked without highlighting them, P - Recent selection set is selected, R - Remove mode- Objects can be removed from the selection set, SI - Selects first object or a set of objects, U - Removes the most recently added object from the selection set, W - Selects those objects which lie completely inside an area specified by two points, WP - Selects those objects which lie completely inside an area specified by picking points around the objects.

SETUV Lets you map materials onto geometry.

'SETVAR Sets the values of the system variables. Options:
? - Lists the variables with their current values.

SHADE Displays a shaded picture of the drawing in the current viewport.

SHAPE Predefined shapes are inserted. Options:
? - Lists the shape names.

SHELL Permits the access to the commands in the operating system while in AutoCAD.

SHOWMAT Lists the material type and attachment method for a selected object.

SKETCH Allows you to draw freehand drawings. Options:
P - Pen- sketching pen raised and lowered, X - Reports the number of temporary lines drawn and then exits SKETCH Q - Temporary lines discarded and then exits SKETCH R - Temporary lines recorded as permanent, E - Removes portion of the temporary line, C - Pen lowered for sketching, . - Draws a straight line from endpoint of sketched line to current position of pen.

SLICE Solid is cut with a plane. Options:
3points - Cutting plane specified by defining 3 points, O - Cutting plane aligned with an object (Circle, ellipse, elliptical arc, 2D spline, or polyline), Z - Cutting plane specified by locating a point on Z-axis, V - Cutting plane aligned to the viewing plane of the current viewport, XY - Cutting plane aligned with the XY plane, YZ - Cutting plane aligned with the YZ plane, ZX - Cutting plane aligned with the ZX plane.

'SNAP The movement of the cursor is constrained to the snap spacing. Options:
ON - Snap mode is turned on, OFF - Snap mode is turned off, A - Sets different X and Y spacings, R - Snap grid is rotated, S - Sets the style (Standard or Isometric) of the snap grid.

SOLDRAW Generates profiles and sections in viewports created with SOLVIEW.

SOLID Draws polygons which are solid-filled.

SOLPROF Creates profile images of three-dimensional solids.

SOLVIEW Creates floating viewports using orthographic projection to lay out multi and sectional view drawings of 3D solid and body objects.

'SPELL Allows spell check of text objects in a drawing. If an ambiguous word is found then the dialog box is displayed that lists the alternatives for the word, or permits you to replace the current word with another one, or add the word to the dictionary.

SPHERE A 3D solid sphere is drawn.
Options:
R - Radius of the sphere, D - Diameter of the sphere.

SPLINE Draws smooth spline curves between points. Options:
Point - Specify points to define the spline curve. Suboptions:
Point - Adds spline curve segments by specifying points, C - Spline curve is closed, F - Fit Tolerance - The tolerance for fitting is changed, O - 2D or 3D spline - fit polylines are changed to splines.

SPLINEDIT Allows you to edit a spline entity. Options:
F - Fit data is edited. Suboptions:
A - Fit points are added, C - An open spline is closed, O - A closed spline is opened, D - Fit points are removed, M - Fit points are moved, P - A spline fit data is removed from database, T - Beginning and end tangents are edited, L - Tolerance value for spline fit are changed, X - Exits fit data option, C - An open spline is closed, O - A closed spline is opened, M - Move Vertex-The position of the control vertices is changed, R - Refines a spline by adding control points, or by increasing its order, or by changing the weight, E - Spline direction is reversed, U - Reverses the previous operation of SPLINEDIT, X - Exits SPLINEDIT command.

STATS Displays a dialog box that provides the rendering statistics. It also saves the statistics to a file.

'STATUS Lists the drawing statistics, modes, and extents.

STLOUT Creates a binary or ASCII file and stores the solid in the specified file.

STRETCH Stretches lines, arcs, and polylines by moving the endpoints to another specified location, and moves the objects.

STYLE Creates new text styles or modifies the existing ones. Options:
? - Lists the text styles.

SUBTRACT Subtracts the area of one set of regions from another and subtracts the volume of one set of solids from another, thus creating a new composite region or solid.

SYSWINDOWS Arranges windows and is equivalent to standard Window menu options in Windows applications.

TABLET Aligns the tablet with the coordinate system of a paper drawing.
Options:
ON - Tablet mode is turned on, OFF - Tablet mode is turned off, CAL - Calibrates the tablet, CFG - Configures tablet menu area and screen pointing area.

TABSURF Creates a polygon mesh which represents a tabulated surface formed from a path curve and direction vector.

TEXT Writes text using a variety of character pattern. The text prompt is displayed only once. Options: See DTEXT command for options.

'TEXTSCR Flips to the text window from the graphics window.

'TIME The date and time of drawing creation is displayed. It also displays the time and the date when the current drawing was last updated and controls an elapsed timer.
Options:

Appendix I, AutoCAD Commands I-15

D - Displays the updated times, O - Elapsed timer is turned on, OFF - Elapsed timer is turned off, R - Resets the user elapsed timer.

TOLERANCE   Creates and adds geometric tolerances to a drawing.

TOOLBAR   Displays, hides, and customizes toolbars.

TORUS   Draws a solid having the shape of a donut. Options:
R - Radius of the tube, D - Diameter of the tube.

TRACE   Draws filled lines having a specified width.

TRANSPARENCY   Controls whether background pixels in an image are transparent or opaque.

TREESTAT   Displays the current spatial index (position of objects in space) of a drawing. The information includes the number of nodes, number of objects, depth of branch, etc.

TRIM   Removes the extra portion of an entity which extends beyond a specified boundary. Options:
P - Sets projection mode, E - Controls trimming of objects till the implied edge, U - Reverses the previous operation of TRIM command, **U** - Reverses the effect of previous operation.

U   Reverses the most recent operation.

UCS   Sets and modifies user coordinate system. Options:
W - Current UCS set to World Coordinate System, O - Allows the shifting of the UCS origin, ZA - UCS defined with the positive Z axis, 3 - Sets new UCS origin and a new X and Y axes direction, OB - A new UCS is defined aligned to a specified object, V - A new UCS is defined whose XY plane is perpendicular to the viewing direction, X/Y/Z - Rotates the current UCS around X axis, or Y axis, or Z axis, P - Previous UCS is restored, R - A saved UCS is restored, S - Current UCS is saved to a name, D - Deletes the specified UCS, ? - Lists the saved coordinate systems.

UCSICON   Manages the location and the visibility of the UCS icon. Options:
ON - Coordinate system icon is enabled, OFF - Coordinate system icon is disabled, A - Icon is changed in all active viewports, N - Icon displayed at the lower left corner, OR - Icon displayed at the origin of current coordinate system.

UNDEFINE   A built-in AutoCAD command is disabled.

UNDO   Reverses the effect of commands. Options:
N - The effect of specified number of previous commands used is reversed, A - The effect of the menu items is reversed by a single U command, C - The UNDO command is limited or is turned off, BE - A number of operations are grouped together and are treated as a single operation, E - The group is terminated, M - Mark - A marker is placed in the undo information, B - Back- Undoes all work till the marker is encountered.

UNION   Combines the area of two or more regions, or the volume of two or more solids to create a composite region or solid.

'UNITS   Sets the coordinate and angle display formats and precision.

'VIEW   The graphics display is saved and restored as a view with a specified name. Options:
? - Lists the named views, D - Deletes specified views, R - Restores a specified view, S - Saves the display as a named view, W - Saves a portion of the display as a named view.

VIEWRES   Controls the appearance of objects by setting their resolution in the current viewport.

VPLAYER   Controls the visibility of layers in different viewports. Options:

? - Lists the frozen layers in a specified viewport, F - Layers are frozen in current, or all, or specified viewport, T - Layers are thawed in current, or all, or specified viewport, R - Rests the layers default visibility, N - New layers which are frozen in all viewports are created, V - Viewport Visibility Default- Controls thawing and freezing of layers.

VPOINT The viewing direction for 3D visualization. Options:
◄┘ - Displays compass and axis tripod for controlling viewing direction, V - Specify a point from which drawing can be viewed, R - New direction using two angles is specified.

VPORTS Divides the graphics display into a number of viewports. Options:
S - Current viewport is saved under a specified name, R - Restores previously saved viewport configuration, D - Removes a viewport configuration, J - Joins two viewports into one, SI - Displays a single viewport view, ? - Lists active viewport configuration, 2 - Divides the current viewport into two, 3 - Divides the current viewport into three, 4 - Divides the current viewport into four.

VSLIDE Displays an existing raster image slide file in the current viewport.

WBLOCK Writes a block definition or specified objects to a new disk file. Options:
name - Writes specified block to file, * - Writes the drawing to a new file, = - Same name for the block and the file, ◄┘ - Writes selected entities to a file.

WEDGE Creates a 3D solid in the shape of a wedge having its one of the faces as tapered and sloping. Options:
point - Specifies the first corner of the wedge. Suboptions:
point - Specifies the other corner of the wedge, C - Wedge having sides of equal length, L - Wedge with specified length, width, and height.
C - Creates wedge with specified center point. Suboptions:
point - Specifies the other corner of the wedge, C - Creates wedge having all sides equal, L - Creates wedge with specified length, width, and height.

WMFIN Imports a Windows metafile.

WMFOPTS Sets options for WMFIN.

WMFOUT Saves objects to a Windows metafile.

XATTACH Attaches an external reference to the current drawing.

XBIND Adds Xref's dependent symbols to a drawing. Options:
B - Binds a block to the current drawing, D - Binds a dimstyle to the current drawing, LA - Binds a layer to the current drawing, LT - Binds a linetype to the current drawing, S - Binds a style to the current drawing.

XCLIP Defines an xref clipping boundary and sets the front or back clipping planes.

XLINE Creates a line of infinite length. Options:
point - Specifies the point through the xline passes, H - Creates a horizontal xline, V - Creates a vertical xline, A - Creates a xline at an angle, B - Creates an xline through the vertex of two lines so that it bisects the angle between those two lines, O - Creates an xline parallel to another linear object.

XPLODE Breaks a compound object into its individual objects. Options:
G - Changes selected objects. Suboptions:
E - Explodes the entire compound object, A - Sets color, linetype, layer of the component entities, C - Sets the color, LA - Sets the layer, LT - Sets the linetype, I - Sets all the properties to that of the original compound object, I - Changes selected objects one by one.

XREF Manages external references to a drawing. Options:
A - Attaches an xref, ? - Lists xrefs in the drawing, B - Binds an xref permanently to a drawing, D - Detaches xrefs from the

drawing, P - Allows to edit the path name with a xref, R - Reloads one or a number of xrefs, O - Overlays an xref.

'ZOOM Changes the display of the entities in the current drawing. Options:

value - Scale(X/XP)- Changes the display by a specified scale factor, Scale X - Zoom relative to current scale, Scale XP - Scale relative to paper space, A - Zooms the entire drawing in current viewport, C - Displays at a specified center point, D - Displays the portion of the drawing with a view box, E - Displays the drawing extents, L - Displays a window by specifying the lower left corner and a magnification, P - Displays the previous view, V - Zooms out on virtual screen of current viewport, W - Displays an area specified by two corners of the window.

Appendix J

AutoCAD System Variables

Variable Name	Type and Description

ACADPREFIX　　　String
The ACADPREFIX variable contains the direction path for support files specified by the ACAD environment variable. Path separators are attached if needed. This is a read-only variable.

ACADVER　　　String
The ACADVER variable contains the AutoCAD version number, which can have values such as "14" or "14a". This variable is different from the DXF file $ACADVER header variable, which stores the drawing database level number. This is a read-only variable.

ACISOUTVER　　　Integer
The ACISOUTVER variable controls the ACIS version of SAT files created using the ACISOUT command. Initial value is 16.

AFLAGS　　　Integer
The AFLAGS variable establishes the attribute flags for the ATTDEF command bit-code. The initial value for this variable is 0. Basically, the value of this variable is the addition of the following:
0 - No attribute mode selected, 1 - Invisible, 2 - Constant, 4 - Verify, 8 - Preset.

ANGBASE　　　Real
The ANGBASE variable establishes the base angle 0 in relation to the current UCS. This variable is saved in the drawing and has an initial value of 0.0000.

ANGDIR　　　Integer
The ANGDIR variable establishes the angle from angle 0 in relation to the prevailing UCS. This variable is saved in the drawing and has an initial value of 0.
0 - Direction is counterclockwise, 1 - Direction is clockwise.

APBOX　　　Integer
The APBOX variable turns the AutoSnap aperture box on or off. Initial value is 1.
0 - AutoSnap aperture box is not displayed,
1 - AutoSnap aperture box is displayed.

APERTURE　　　Integer
The APERTURE variable defines the object snap target height in pixels. This variable is saved in registry and has an initial value of 10.

AREA　　　Real
The most recently calculated area with commands such as AREA, LIST, or DBLIST is stored in this variable. You can

examine this variable through the SETVAR command.

ATTDIA Integer
With the ATTDIA variable you can specify whether you want to enter the attribute value through the INSERT dialog box or from the command line. This variable is saved in the drawing and has an initial value of 0.
0 - Attribute values can be specified on the command line, 1 - Attribute values can be specified in the dialog box.

ATTMODE Integer
The ATTMODE variable controls the Attribute Display mode and is saved in the drawing and its initial value is 1.
0 - Attribute Display mode is off, 1 - Normal, 2 - On.

ATTREQ Integer
The value contained in the ATTREQ variable determines whether INSERT command uses the default attribute settings when the blocks are being inserted.
0 - The default values for all the attributes are used, 1 - This is also the initial value and enables prompts or dialog box for attribute values (depending on the value of ATTDIA variable).

AUDITCTL Integer
The AUDITCTL variable determines whether an .adt file (audit report file) will be created by AutoCAD. This variable is saved in registry.
0 - Does not allow writing of .adt files. This is also the initial value, 1 - Allows writing of .adt files.

AUNITS Integer
The AUNITS variable establishes the Angular Units mode and is saved in the drawing.
0 - Decimal degrees (initial value), 1 - Degrees/ minutes/ seconds, 2 - Gradians, 3 - Radians, 4 - Surveyor's units.

AUPREC Integer
The AUPREC variable establishes the angular units decimal places. This variable is saved in the drawing and has an initial value of 0.

AUTOSNAP Integer
The AUTOSNAP variable controls the display of AutoSnap marker, and Snap Tips and turns the AutoSnap magnet on or off. Initial value is 7.
0 - Turns off marker, Snap Tip and magnet, 1 - Turns on marker, 2 - Turns on Snap Tip, 4 - Turns on magnet.

BACKZ Real
The BACKZ variable contains the back clipping plane offset (in current drawing units) from the target plane for the current viewport. You can determine the distance between the back clipping plane and the camera point by subtracting the BACKZ value from the camera to target distance. This variable is saved in the drawing and is read-only.

BLIPMODE Integer
The visibility of the blip marks is controlled by BLIPMODE variable. This variable is saved in the drawing and its initial value is 0.
0 - Blip marks are not visible, 1 - Blip marks are visible.

CDATE Real
The calendar date and time is stored in this variable. This is a read-only variable.

CECOLOR String
The CECOLOR variable defines the color of new objects. This variable is saved in the drawing and its initial value is "BYLAYER" (256).

CELTSCALE Real
The CELTSCALE variable defines the current global linetype scale factor for objects. This variable is saved in the drawing and its initial value is 1.0000.

CELTYPE String
The CELTYPE variable defines the linetype that will be used in the new objects. This variable is saved in the drawing and its initial value is "BYLAYER."

CHAMFERA　　Real
The CHAMFERA variable defines the first chamfer distance. This variable is saved in the drawing and its initial value is 0.5000.

CHAMFERB　　Real
The CHAMFERB variable defines the second chamfer distance. This variable is saved in the drawing and its initial value is 0.5000.

CHAMFERC　　Real
The CHAMFERC variable sets the chamfer length. This variable is saved in the drawing and its initial value is 1.0000.

CHAMFERD　　Real
The CHAMFERD variable sets the chamfer angle. This variable is saved in the drawing and its initial value is 0.0000.

CHAMMODE　　Integer
With the CHAMMODE variable you can specify the method that will be used to create chamfers.
0 - This is the initial value and in this case two chamfer distances are required, 1 - One chamfer length and an angle are required.

CIRCLERAD　　Real
The CIRCLERAD variable defines the default circle radius. The initial value of this variable is 0.0000.

CLAYER　　String
The CLAYER variable sets the current layer. This variable is saved in the drawing and its initial value is "0".

CMDACTIVE　　Integer
The CMDACTIVE variable contains the bit-code that signifies whether an ordinary command, transparent command, dialog box, or script is active. Basically, the value of this variable is the addition of the following:
1 - Only ordinary command is active, 2 - Ordinary command as well as transparent command are active, 4 - Script is active, 8 - If this bit is set then Dialog box is active.

CMDDIA　　Integer
The CMDDIA variable determines whether the dialog boxes are enabled for only PLOT and external database commands. This variable is saved in registry and its initial value is 1.
0 - Dialog boxes are disabled, 1 - Dialog boxes are enabled.

CMDECHO　　Integer
The CMDECHO variable determines whether the prompts and input of a AutoLISP (command) function are echoed. This variable is saved in registry and its initial value is 1.
0 - Echoing is disabled, 1 - Echoing enabled.

CMDNAMES　　String
The CMDNAMES variable displays the name of the presently active command and transparent command. This variable is read-only.

CMLJUST　　Integer
The CMLJUST variable determines the justification of a multiline. This variable is saved in registry and its initial value is 1.
0 - Sets Top justification, 1 - Sets Middle justification, 2 - Sets Bottom justification.

CMLSCALE　　Real
The CMLSCALE variable determines the width of a multiline. For example, a scale factor of 3.0 generates a multiline that is thrice as wide as specified in the style definition. If the value is set to 0, the multiline takes the form of a single line. By specifying a negative scale factor, the order of offset lines is flipped. This variable is saved in registry and has an initial value of 1.0000.

CMLSTYLE　　String
The CMLSTYLE variable sets the name of the multiline style that is used to draw multilines. This variable is saved in registry and has an initial value "".

COORDS　　Integer
The COORDS variable determines when the coordinates are updated. This variable is

saved in the drawing and its initial value is 1.
0 - Coordinates are updated only upon picking points, 1 - Absolute coordinates are continuously updated, 2 - Absolute continuously plus, when a distance or angle are requested, then the distance and angle from the last point are displayed.

CURSORSIZE Integer
This variable determines the size of the crosshairs as a percentage of the screen size. Initial value is 5.

CVPORT Integer
The CVPORT variable establishes the identification number of the current viewport. When this value is changed, the current viewport is also changed in case the following conditions hold good:
1 - The specified identification number belongs to an active viewport, 2 - The cursor movement to the specified viewport is not locked by the command being executed, 3 - Tablet mode is off. The variable is saved in the drawing and its initial value is 2.

DATE Real
The DATE variable contains the current date and time as a Julian date and fraction in a real number. This variable is read-only.

DBMOD Integer
The DBMOD variable expresses the drawing modification status using bit-code. This variable is read-only. Basically the value of this variable is the addition of the following:
0 - No changes, 1 - The object database is changed, 2 - The symbol table is changed, 4 - The database variable is changed, 8 - The window is changed, 16 - The view is changed.

DCTCUST String
The DCTCUST variable shows the current custom spelling dictionary path and filename. This variable is saved in registry and its initial value is "".

DCTMAIN String
The DCTMAIN variable shows the current main spelling dictionary filename. Normally this file is located in the \support directory. The default main spelling dictionary can be specified using the SETVAR command. This variable is saved in registry and its initial value is "".

DELOBJ Integer
Thw DELOBJ variable determines whether objects used to draw other objects are kept or deleted from the drawing database. This variable is saved in the drawing and its initial value is 1.
1 - Objects are deleted from the drawing database, 0 - Objects are kept in the drawing database.

DEMANDLOAD Integer
This variable specifies if and when AutoCAD demand loads a third-party application if a drawing contains custom objects created in that application. Initial value is 3.

DIASTAT Integer
The method of exiting from the most recently used dialog box is held in the DIASTAT variable. This variable is read-only.
0 - Cancel, 1 - OK.

DIMALT Switch
The DIMALT variable controls the dimensioning in alternate units system. If the DIMALT variable is on (1), alternate unit dimensioning is facilitated. This variable is saved in the drawing and its initial value is Off (0).

DIMALTD Integer
The DIMALTD (DIMension ALTernate units Decimal places) variable controls the number of decimal places (decimal precision) of the dimension text in the alternate units if DIMALT variable is on. This variable is saved in the drawing and its initial value is 2.

DIMALTF Real
The DIMALTF variable (DIMension ALTernate units scale Factor) controls

alternate units scale factor. In case DIMALT variable is enabled, all the linear dimensions will be multiplied with this factor to generate a value in an alternate units system. The initial value for DIMALTF is 25.4. This variable is saved in the drawing.

DIMALTTD Integer

The DIMALTTD variable establishes the number of decimal places for the tolerance values of an alternate units dimension. This variable is saved in the drawing and has an initial value of 2.

DIMALTTZ Integer

The DIMALTTZ variable controls the suppression of zeros for alternate tolerance values. With this variable, the real-to-string transformation carried out by AutoLISP functions **rtos** and **angtos** is also influenced. This variable is saved in the drawing and has an initial value of 0.
0 - Suppresses zero feet and precisely zero inches, 1 - Includes zero feet and precisely zero inches, 2 - Includes zero feet and suppresses zero inches, 3 - Includes zero inches and suppresses zero feet. Value in the range of 0 and 3 influence only the feet and inches dimensions. However, you can add 4 to the above values to omit the leading zeroes in all decimal dimensions. If you add 8, the trailing zeroes are omitted. If 12 (both 4 and 8) is added, the leading and the trailing zeroes are omitted.

DIMALTU Integer

The DIMALTU variable establishes the units format for alternate units of all dimensions except angular. This variable is saved in the drawing and has an initial value of 2.
1 - Scientific, 2 - Decimal, 3 - Engineering, 4 - Architectural, 5 - Fractional.

DIMALTZ Integer

The DIMALTZ variable controls the suppression of zeros for alternate units dimension values. With this variable, the real-to-string transformation carried out by AutoLISP functions **rtos** and **angtos** is also influenced. This variable is saved in the drawing and has an initial value of 0.
0 - Suppresses zero feet and precisely zero inches, 1 - Includes zero feet and precisely zero inches, 2 - Includes zero feet and suppresses zero inches, 3 - Includes zero inches and suppresses zero feet. Value in the range of 0 and 3 influence only the feet and inches dimensions. However, you can add 4 to the above values to omit the leading zeroes in all decimal dimensions. If you add 8, the trailing zeroes are omitted. If 12 (both 4 and 8) is added, the leading and the trailing zeroes are omitted.

DIMAPOST String

With the help of DIMAPOST variable, you can append a text prefix, suffix, or both to an alternate dimensioning measurement. This can be done in case of all the dimensions except angular dimensions. The variable is saved in the drawing and has an initial value of "". In order to disable an existing suffix or prefix, set the value of this variable to a single period.

DIMASO Switch

The DIMASO variable governs the creation of associative dimensions. This variable is saved in the drawing (not in the dimension style) and its initial value is set to on.
Off (0) - The dimension created are not associative in nature and hence in such dimensions no association exists between the dimension and the points on the object. All the dimensioning entities such as arrowheads, dimension lines, extension lines, dimension text, etc. are drawn as separate entities. On (1) - The dimension created are associative in nature and hence in such dimensions there exists an association between the dimension and the definition points. If you edit the object, (editing like trimming or stretching) the dimensions associated with that object also change. Also, the appearance of associative dimensions can be preserved when they are edited by commands such as STRETCH or TEDIT. For example, a vertical associative dimension is retained as a vertical dimension even after an editing operation. The associative dimension is always generated with the same dimension variable settings as defined in the dimension style.

DIMASZ   Real

The DIMASZ (Dimension arrowhead size) variable specifies the size of dimension line and leader line arrowheads when DIMTSZ is set to zero. The size of arrowhead blocks set by DIMBLK is also controlled by DIMASZ variable. Multiples of this variable determine whether the dimension line and text will be located between the extension lines. This variable is saved in the drawing and has an initial value of 0.18 units.

DIMAUNIT   Integer

The DIMAUNIT variable establishes the angle format for angular dimensions. This variable is saved in the drawing and its initial value is 0. 0 - Decimal degrees format, 1 - Degrees/minutes/seconds format, 2 - Gradians format, 3 - Radians format, 4 - Surveyor's units format.

DIMBLK   String

DIMBLK variable replaces the default arrowheads at the end of the dimension lines with a user defined block. The user defined block that may replace the standard arrowhead can be a custom designed arrow or some other symbol. DIMBLK (DIMension BLocK) takes the name of the block as its string value. This variable is saved in the drawing and its initial value is no block (""). To discard an existing block name, set its value to a single period (.).

DIMBLK1   String

DIMBLK1 variable designates user defined arrow block for the first end of the dimension line. This option can be used only if the DIMSAH (DIMension Separate Arrow blocks) variable is on. The value of this variable is the name of earlier formulated block as in the case of DIMBLK. You can discard an existing block name by setting its value to a single period (.). This variable is saved in the drawing and its initial value is no block ("").

DIMBLK2   String

DIMBLK2 variable designates a user defined arrow block for the second end of the dimension line. This option can be used only if DIMSAH (DIMension Separate Arrow blocks) variable is on. The value of this variable is the name of earlier formulated block as in the case of DIMBLK. You can discard an existing block name, by setting its value to a single period (.). This variable is saved in the drawing and its initial value is no block ("").

DIMCEN   Real

The DIMCEN (DIMension CENter) variable governs the drawing of center marks and the center lines of circles and the arcs by the DIMCENTER, DIMDIAMETER, and DIMRADIUS commands. DIMCEN takes a distance as its argument. The value of the DIMCEN variable determines the result. This variable is saved in the drawing and its initial value is 0.0900.

0 - Center marks or center lines are not drawn, >0 - Center marks are drawn and their size is governed by the value of the DIMCEN. For example, a value of 0.250 displays center dashes which are 0.2500 units long, <0 - Center lines in addition to center marks are drawn and again the size of the mark portion is governed by the absolute value of the DIMCEN. The center lines extend beyond the circle or arc by the value entered. For example a value of -0.2500 for DIMCEN variable will draw a center dashes 0.25 units long ant also the center lines will be extended beyond the circle/arc by a distance of 0.25 units. With the DIMRADIUS and DIMDIAMETER commands, center mark or center line is generated only when the dimension line is located outside the circle or arc.

DIMCLRD   Integer

The DIMCLRD variable is used to assign colors to dimension lines, arrowheads, and the dimension leader lines. This variable can take any permissible color number or the special color labels BYBLOCK (0) or BYLAYER (256) as its value. If you use the SETVAR command, then you have to enter the integer number of the color you want to assign to the DIMCLRD variable. This variable is saved in the drawing and its initial value is 0.

DIMCLRE Integer
DIMCLRE variable is used to assign color to the dimension extension lines. Just as DIMCLRD, DIMCLRE (DIMension CoLOr Extension) can take any permissible color number or the special color labels BYBLOCK or BYLAYER. This variable is saved in the drawing and its initial value is 0.

DIMCLRT Integer
The DIMCLRT (DIMension CoLoR Text) variable is used to assign a color to the dimension text. DIMCLRT can take any permissible color number or the special color labels BYBLOCK (0) or BYLAYER (256). This variable is saved in the drawing and its initial value is 0.

DIMDEC Integer
The DIMDEC variable establishes the number for decimal places of a primary units dimension. This variable is saved in the drawing and its initial value is 4.

DIMDLE Real
By default the dimension lines meet the extension lines. But if you want that the dimension line to continue past the extension lines, DIMDLE (Dimension Line Extension) variable can be used for this function. DIMDLE is used only when DIMTSZ variable is nonzero (when DIMTSZ variable is nonzero, ticks are drawn instead of arrows). The dimension line will extend past the extension line by the value of DIMDLE. This variable is saved in the drawing and its initial value is 0.0000.

DIMDLI Real
The DIMDLI variable governs the spacing between the successive dimension lines when dimensions are created with the DIMCONTINUE and DIMBASELINE commands. Successive dimension lines are offset by the DIMDLI value, if needed, to avert drawing over the previous dimension. This variable is saved in the drawing and its initial value is 0.38 units.

DIMEXE Real
The extension of the extension line past the dimension line is governed by the DIMEXE (Dimension EXtension line Extension) variable. This variable is saved in the drawing and has an initial value of 0.18 units.

DIMEXO Real
There exists a small space between the origin points you specify and the start of the extension lines. The size of this gap is controlled by the DIMEXO (DIMension EXtension line Offset) variable. The offset distance is equal to the value of the DIMEXO variable. This variable is saved in the drawing and has an initial value of 0.0625 units.

DIMFIT Integer
The DIMFIT variable governs the placement of text and arrowheads inside or outside extension lines depending on the space available between the extension lines. This variable is saved in the drawing and its initial value is 3.
0 - In this case the text and arrowheads are positioned between the extension lines if enough space is available. Otherwise, the text and arrowheads are positioned outside the extension lines, 1 - In this case the text and arrowheads are positioned between the extension lines if enough space is available. Otherwise, if enough space is available for the text, it is positioned between the extension lines and the arrowheads are positioned outside the extension lines. If enough space is not found between the extension lines for the placement of the text, then both the text and arrowheads are positioned outside the extension lines, 2 - In this case the text and arrowheads are positioned between the extension lines if enough space is available. Otherwise, if enough space is available for the arrowheads only, they are positioned between the extension lines and the text is positioned outside the extension lines. If enough space is not found between the extension lines for the placement of arrowheads, then both the text and arrowheads are positioned outside the extension lines, 3 - In this case the text and arrowheads are positioned between the

extension lines if enough space is available. Otherwise, if enough space is available for the text only, it is positioned between the extension lines and the arrowheads are positioned outside the extension lines. If enough space is available for the arrowheads only, they are positioned between the extension lines and the text is positioned outside the extension lines. If enough space is not found between the extension lines for the placement of arrowheads and text, then both the text and arrowheads are positioned outside the extension lines, 4 - For this value leader lines are created when enough space is not available between the extension lines. Whether the text will be placed on the right or left of the leader depends on the horizontal justification, 5 - No Leader.

DIMGAP Real
The DIMGAP variable controls the space between the dimension line and the dimension text (distance maintained around the dimension text), when the dimension line is split into two for the placement of dimension text. The gap between the leader and annotation created with the LEADER command is also governed by DIMGAP variable. This variable is saved in the drawing and its initial value for DIMGAP is 0.0900 units. By entering a negative DIMGAP value, you can create a reference dimension, in which case you get the dimension text with a box drawn around it. DIMGAP value is also used by AutoCAD as the measure of minimum length needed for the segments of the dimension line. AutoCAD places the dimension text inside the extension lines only if the dimension line is split into two segments each of which is at least as long as DIMGAP. In case the text is positioned over or under the dimension line, it is placed inside the dimension line only if there is space for the arrows, dimension text, and a margin between them has a minimum value at least as much as DIMGAP: 2*(DIMGAP + DIMASZ).

DIMJUST Integer
The DIMJUST variable governs the horizontal dimension text position. This variable is saved in the drawing and its initial value is 0.
0 - The text is center justified between the extension lines, 1 - The text is placed next to the first extension line, 2 - The text is placed next to the second extension line, 3 - The text is placed above and aligned with the first extension line, 4 - The text is placed above and aligned with the second extension line.

DIMLFAC Real
The DIMLFAC (DIMension Length FACtor) variable acts as a global scale factor for all linear dimensioning measurements. The linear distances measured by dimensioning include coordinates, diameter, and radii. These linear distances are multiplied by the prevailing DIMLFAC value before they are projected as dimension text. In this manner DIMLFAC scales the contents of the default text. The angular dimensions are not scaled. Also DIMLFAC does not apply to the values held in DIMTM, DIMTP, or DIMRND. For example, if you want to scale the default dimension measurement by a value of 2, set the value of DIMLFAC to 2. When dimensioning in the paper space, if the value of DIMLFAC variable is not zero, then the distance measured is multiplied by the absolute value of DIMLFAC. In case of dimensioning in the model space, values less than zero are neglected, instead the value of DIMLFAC is taken as 1.0. If in paper space you select the Viewport option and try to change DIMLFAC from the Dim: prompt, AutoCAD will compute a value for the DIMLFAC for you. This is illustrated as follows:
Dim: **DIMLFAC**, Current value <1.0000> New value (Viewport): **V**
Select viewport to set scale: The scaling of model space to paper space is computed by AutoCAD and the negative of the computed value is assigned to DIMLFAC. This variable is saved in the drawing and its initial value is 1.0000.

DIMLIM Switch
The DIMLIM (DIMension LIMits) variable acts as a switch and creates the dimension limits as the default text if it is on (1). Also

DIMTOL is forced to be off. This variable is saved in the drawing and its initial value is off.

DIMPOST String
The DIMPOST variable is used to define prefix or suffix to the dimension measurement. The variable is saved in the drawing and has an initial value "" (empty string). DIMPOST takes a string value as its argument. For example if you want to have a suffix for centimeters, set DIMPOST to "cm". A distance of 4.0 units will be displayed as 4.0cm. In case tolerances are enabled, the suffix you have defined gets applied to the tolerances as well as to the main dimension. To establish a prefix to a dimension text, type " < > " and then the prefix at the same prompt.

DIMRND Real
The DIMRND (DIMension RouND) variable is used for rounding all the dimension measurements to the specified value. For example if the DIMRND is set to 0.10, then all the measurements are rounded to the nearest 0.10 unit. Likewise, a value of 1 for this variable will result in the rounding of all the measurements to the nearest integer. The angular measurements cannot be rounded. The variable is saved in the drawing and has an initial value of 0.0000.

DIMSAH Switch
The DIMSAH (DIMension Separate custom Arrow Head) variable governs the placement of user-defined arrow blocks instead of the standard arrows at the end of the dimension line. As explained earlier, DIMBLK1 variable places a user defined arrow block at the first end of the dimension line and DIMBLK2 places a user defined arrow block at the other end of the dimension line. This variable is saved in the drawing and its initial value is off.
On - DIMBLK1 and DIMBLK2 specify different user-defined arrow blocks to be drawn at the two ends of the dimension line, Off - Ordinary arrowheads or user-defined arrowhead block defined by the DIMBLK variable is used.

DIMSCALE Real
The DIMSCALE variable controls the scale factor for all the size-related dimension variables such as those that affect text size, center mark size, arrow size, leader objects, etc. The DIMSCALE is not applied to the measured lengths, coordinates, angles, or tolerances. The default value for this variable is 1.0000; and in this case the dimensioning variables assume their preset values and the drawing is plotted at full scale. If the drawing is to be plotted at half the size, then the scale factor is the reciprocal of the drawing size. Hence, the scale factor or the DIMSCALE value will be reciprocal of 1/2 which is 2/1 = 2.
0.0 - A default value based on the scaling between the current model space viewport and paper space is calculated. In case you are not using the paper space feature, then the scale factor is 1.0, >0 - A scale factor is computed that makes the text sizes, arrowhead sizes, and scaled distances to plot at their face value.

DIMSD1 Switch
The DIMSD1 (DIMension Suppress Dimension line 1) variable suppresses the drawing of the first dimension line when it is on. This variable is saved in the drawing and its initial value is off.

DIMSD2 Switch
The DIMSD1 (DIMension Suppress Dimension line 2) variable suppresses the drawing of second dimension line when it is on. This variable is saved in the drawing and its initial value is off.

DIMSE1 Switch
The DIMSE1 variable is used to suppress drawing of the first extension line. When DIMSE1 (DIMension Suppress Extension line 1) is on, the first extension line is not drawn. This variable is saved in the drawing and its initial value is off.

DIMSE2 Switch
The DIMSE2 variable is used to suppress drawing of the second extension line. When DIMSE2 (DIMension Suppress Extension line 2) is on, the second extension line is

not drawn. This variable is saved in the drawing and its initial value is off.

DIMSHO　　　　Switch
DIMSHO variable governs the redefinition of dimension entities while dragging into some position. If DIMSHO (DIMension SHOw dragged dimensions) is on, associative dimensions will be computed dynamically as they are dragged. The DIMSHO value is saved in the drawing (not in a dimension style) and its initial value is on (1). Dynamic dragging reduces the speed of some computers and hence in such situations DIMSHO should be set off (0). However, when you are using the pointing device to specify the length of the leader in Radius and Diameter dimensioning, the DIMSHO setting is neglected and dynamic dragging is used.

DIMSOXD　　　　Switch
If you want to place text inside the extension lines, you will have to set the DIMTIX variable on. And if you want to suppress the dimension lines and the arrow heads you will have to set the DIMDSOXD (DIMension Suppress Outside eXtension Dimension lines) variable on. DIMSOXD suppresses the drawing of dimension lines and the arrow heads when they are placed outside the extension lines. If DIMTIX is on and DIMSOXD is off and there is not enough space inside the extension lines for drawing the dimension lines, then dimension lines will be drawn outside the extension lines. In such a situation, if both DIMTIX and DIMSOXD are on, then the dimension line will be totally suppressed. DIMSOXD works only when DIMTIX is on. The DIMSOXD variable is saved in the drawing and its initial value is off.

DIMSTYLE　　　　String
DIMSTYLE variable is used for displaying the name of the present dimension style. DIMSTYLE is a read-only variable and is saved in the drawing. You can change the dimension style using the DDIM or DIMSTYLE command.

DIMTAD　　　　Integer
The DIMTAD (DIMension Text Above Dimension line) variable governs the vertical placement of the dimension text with respect to the dimension line. DIMTAD gets actuated when dimension text is drawn between the extension lines and is aligned with the dimension line, or when the dimension text is placed outside the extension lines. This variable is saved in the drawing and its initial value is 0.
0 - For this value the dimension text is placed at the center between the extension lines, 1 - The dimension text is placed above the dimension line and a single (unsplit) dimension line is drawn under it spanning between the extension lines. The exceptions to this arise when the dimension line is not horizontal and text inside the extension line is forced to be horizontal by making DIMTIH = 1. The space between the dimension line and the baseline of the lowest line of text is nothing but the prevailing DIMGAP value, 2 - The dimension text is placed on the side of the dimension line most remote from the defining points, 3 - The dimension text is placed to tune to a JIS representation.

DIMTDEC　　　　Integer
The DIMTDEC variable establishes the number of decimal places for the tolerance values for the primary units dimension. This variable is saved in the drawing and its initial value is 4.

DIMTFAC　　　　Real
With the DIMTFAC (DIMension Tolerance scale FACtor) variable you can control the scaling factor of the text height of the tolerance values in relation to the dimension text height set by DIMTXT. Suppose DIMTFAC is set to 1.0 (the default value for DIMTFAC variable), then the text height of the tolerance text will be equal to the dimension text height. If DIMTFAC is set to a value of 0.50, the text height of the tolerance is half of the dimension text height. This variable is saved in the drawing and its initial value is 1.0000. It is important to remember that the scaling of tolerance text to any requirement is possible only when DIMTOL is on and DIMTM and

DIMTP variable values are not identical, or when DIMLIM is on.

DIMTIH Switch
The DIMTIH (DIMension Text Inside Horizontal) variable controls the placement of the dimension text inside the extension lines for Linear, Radius, Angular, and Diameter dimensioning. DIMTIH is effective only when the dimension text fits between the extension lines.
On - If DIMTIH is on (the default setting), it forces the dimension text inside the extension lines to be placed horizontally, rather than aligned, Off - In case DIMTIH is off, the dimension text is aligned with the dimension line.

DIMTIX Switch
The DIMTIX variable draws the text between the extension lines. This variable is saved in the drawing and its initial value is Off.
On - When DIMTIX is set to on, the dimension text is placed amidst the extension lines even if it would normally be placed outside the extension lines, Off - If DIMTIX is off, the placement of the dimension text depends on the type of dimension. For example, if the dimensions are Linear or Angular, the text will be placed inside the extension lines by AutoCAD if there is enough space available. While as for the Radius and Diameter dimensions, the text is placed outside the object being dimensioned.

DIMTM Real
The DIMTM variable establishes the lower (minimum) tolerance limit for the dimension text. Tolerance is defined as the total amount by which a particular dimension is permitted to vary. The tolerance or limit values are drawn only if DIMTOL or DIMLIM variable is on. DIMTM (DImension Tolerance Minus) identifies the lower tolerance and DIMTP (DIMension Tolerance Plus) identifies the upper tolerance. You can specify signed values for DIMTM and DIMTP variables. If DIMTOL is on and both DIMTM and DIMTP have same value, AutoCAD draws the "±" symbol followed by the tolerance value. If DIMTM and DIMTP hold different values, the upper tolerance is drawn above the lower tolerance. Also a positive (+) sign is appended to the DIMTP value if it is positive. For minus tolerance value (DIMTM), the negative of the value you enter (negative sign if you enter positive value and positive sign if you enter negative value) is displayed. Signs are not appended with zero. This variable is saved in the drawing and its initial value is 0.0000.

DIMTOFL Switch
If DIMTOFL variable is turned on, a dimension line is drawn between the extension lines even if the text is located outside the extension lines. When DIMTOFL is off, for radius and diameter dimensions, the dimension line and the arrowheads are drawn inside the arc or circle, while the text and the leader are placed outside. This variable is saved in the drawing and its initial value is Off.

DIMTOH Switch
The DIMTOH (DIMension Text Outside Horizontal) variable controls the orientation of the dimension text outside the extension lines. If DIMTOH is on, it forces the dimension text outside the extension lines to be placed horizontally, rather than aligned. In case DIMTOH is off, the dimension text is aligned with the dimension line. You must have noticed that the variable DIMTOH is same as DIMTIH variable except it controls text drawn outside the extension lines. This variable is saved in the drawing and its initial value is On.

DIMTOL Switch
DIMTOL (DIMension with TOLerance) variable is used for controlling the appending of dimension tolerances to the dimension text. With DIMTM and DIMTP you can define the values of the lower and upper tolerances. If the DIMTOL variable is set on, the tolerances are appended to the default text. When DIMTOL is set on, DIMLIM variable is set off. This variable is saved in the drawing and its initial value is Off.

DIMTOLJ Integer
DIMTOLJ variable establishes the vertical justification for the tolerance values with respect to the normal dimension text. This variable is saved in the drawing and its initial value is 1.
0 - Bottom, 1 - Middle, 2 - Top.

DIMTP Real
The DIMTP (DIMension Tolerance Plus) variable establishes the upper (maximum) tolerance limit for the dimension text. Tolerance is defined as the total amount by which a particular dimension is permitted to vary. The tolerance or limit values are drawn only if DIMTOL or DIMLIM variable is on. If DIMTOL is on and both DIMTM and DIMTP have same value, AutoCAD draws the "±" symbol followed by the tolerance value. If DIMTM and DIMTP hold different values, the upper tolerance is drawn above the lower tolerance. Also a positive (+) sign is appended to the DIMTP value if it is positive. This variable is saved in the drawing and its initial value is 0.0000.

DIMTSZ Real
The DIMTSZ variable defines the size of oblique strokes (ticks) instead of arrowheads at the end of the dimension lines (just as in architectural drafting), for Linear, Radius, and Diameter dimensioning. This variable is saved in the drawing and its initial value is 0.0000.
0 - Arrows are drawn, >0 - Oblique strokes instead of arrows are drawn. The size of the ticks are computed as DIMTSZ * DIMSCALE. Hence if DIMSCALE factor is one then the size of the tick is equal to the DIMTSZ value. This variable is also used to determine whether dimension line and dimension text will get accommodated between the extension lines.

DIMTVP Real
The DIMTVP (DIMension Text Vertical Position) variable, controls the vertical placement of the dimension text over or under the dimension line. In certain cases DIMTVP is used as DIMTAD to control the vertical position of the dimension text. DIMTVP value holds good only when DIMTAD is off. The vertical placing of the text is done by offsetting the dimension text. The amount of the vertical offset of dimension text is a product of text height and DIMTVP value. If the value of DIMTVP is 1.0, DIMTVP acts as DIMTAD. However, if the value of the DIMTVP is less than 0.70, the dimension line is broken into two segments to accommodate the dimension text. This variable is saved in the drawing and its initial value is 0.0000.

DIMTXSTY String
The DIMTXTSTY variable specifies the text style of the dimension. This variable is saved in the drawing and its initial value is "STANDARD".

DIMTXT Real
The DIMTXT variable is used to control the height of the dimension text except if the current text style has a fixed height. This variable is saved in the drawing and its initial value is 0.1800.

DIMTZIN Integer
With the DIMZIN variable you can control the suppression of the zeros for tolerance values. The variable is saved in the drawing and its initial value is 0.
0 - Suppresses zero feet and precisely zero inches, 1 - Includes zero feet and precisely zero inches, 2 - Includes zero feet and suppresses zero inches, 3 - Includes zero inches and suppresses zero feet. You can add 4 to the above values to omit the leading zeroes in all decimal dimensions. If you add 8, the trailing zeroes are omitted. If 12 (both 4 and 8) is added, the leading and the trailing zeroes are omitted.

DIMUNIT Integer
The DIMUNIT variable establishes the linear units format for all dimension styles.
1 - Scientific units format, 2 - Decimal units format, 3 - Engineering units format, 4 - Architectural (stacked) units format, 5 - Fractional (stacked) units format, 6 - Architectural, 7 - Fractional, 8 - Window Desktop.

DIMUPT Switch
This variable governs the cursor functionality for User Positioned Text. This variable is saved in the drawing and its initial value is Off.
0 - The cursor controls the location of the dimension line only, 1 - The cursor controls the location of both the dimension text and the dimension line.

DIMZIN Integer
The DIMZIN (DIMension Zero INch) controls the suppression of the inches part of a feet-inches dimension when the distance is integral number of feet or the suppression of the feet portion when the distance is less than one foot. This variable is saved in the drawing and its initial value is 0.
0 - Suppress zero feet and exactly zero inches, 1 - Include zero feet and, exactly zero inches, 2 - Include zero feet, suppress zero inches, 3 - Include zero inches, suppress zero feet. If the dimension has feet and a fractional inch part, the number of inches is included even if it is zero. This is independent of the DIMZIN setting. For example a dimension such as 1'-2/3" never exist. It will be in the form 1'-0 2/3". The integer values 0-3 of the DIMZIN variable control the feet and inch dimension only, while as you can add 4 to omit the leading zeroes in all decimal dimensions. For example 0.2600 becomes .2600. If you add 8, the trailing zeroes are omitted. For example, 4.9600 becomes 4.96. If 12 (both 4 and 8) is added, the leading and the trailing zeroes are omitted. For example, 0.2300 becomes .23.

DISPSILH Integer
The DISPSILH variable governs the display of silhouette curves of body objects in a wireframe model. The variable is saved in the drawing and its initial value is 0.
0 - Silhouette curves of body objects not displayed, 1 - Silhouette curves of body objects displayed.

DISTANCE Real
The DISTANCE variable holds the distance value determined by the DIST command. This command is read-only.

DONUTID Real
The DONUTID variable establishes the default inside diameter of a donut. The initial value for this variable is 0.5000.

DONUTOD Real
The DONUTOD variable establishes the default outside diameter of a donut. It is important that the value of this variable be greater than zero. In case the value of DONUTID is greater than that of DONUTOD, then the two values are interchanged by the next command. The initial value for this variable is 1.0000.

DRAGMODE Integer
The DRAGMODE variable establishes the Object Drag mode while carrying out editing operations.
0 - Dragging disabled, 1 - Dragging enabled if invoked, 2 - Auto. This variable is saved in the drawing and is initially set to 2.

DRAGP1 Integer
The DRAGP1 variable establishes the regen-drag input sampling rate. This variable is saved in registry and is initially set to a value of 10.

DRAGP2 Integer
The DRAGP2 variable establishes the fast-drag input sampling rate. This variable is saved in registry and is initially set to a value of 25.

DWGCODEPAGE String
The DWGCODEPAGE variable holds the drawing code page. When you create a new drawing, this variable is set to the system code page. Otherwise, it is not maintained by AutoCAD. This variable describes the code page of the drawing. You can set this variable to any value by using the SYSCODEPAGE system variable or set it as undefined. It is a read-only variable and is saved in the drawing.

DWGNAME String
The DWGNAME variable holds the name of the drawing as specified by the user. In case the drawing has not been assigned a name, the DWGNAME variable conveys that the drawing is unnamed. The drive and

DWGPREFIX   String
The DWGPREFIX variable holds the drive and directory prefix for the drawing. This variable is a read-only variable.

DWGTITLED   Integer
The DWGTITLED variable reflects whether the present drawing has been named.
0 - Indicates that the drawing has not been named, 1 - Indicates that the drawing has been named.

EDGEMODE   Integer
With the EDGEMODE variable you can control how the EXTEND and TRIM commands determine boundary and cutting edges.
0 - In this case the selected edge is used without an extension, 1 - The object is trimmed or extended to an imaginary extension of the cutting or boundary edges. This is the initial value for this variable.

ELEVATION   Real
The ELEVATION variable holds the current 3D elevation associated to the current UCS for the current space. This variable is saved in the drawing and has an initial value of 0.0000.

EXPERT   Integer
The issuance of some prompts is controlled with the EXPERT variable. The initial value for this variable is 0.
0 - All the prompts are issued, 1 - The "About to regen, proceed?" prompt and "Really want to turn the current layer off?" prompts are suppressed, 2 - The preceding prompts and "Block already defined. Redefine it?" (BLOCK command) and "A drawing with this name already exists. Overwrite it?" (SAVE or WBLOCK commands) are suppressed, 3 - The preceding prompts and the ones issued by LINETYPE if you try to load a linetype that is already loaded or create a new linetype in a file that already defines it are suppressed, 4 - The preceding prompts and the ones issued by UCS Save and VPORTS Save in case the name you provide already exists are suppressed, 5 - The preceding prompts and the ones issued by the DIMSTYLE Save option, and DIMOVERRIDE in case the dimension style name you provide already exists, are suppressed. Whenever the EXPERT command suppresses a prompt, the corresponding operation is carried out as if you have entered Y as the response to the prompt. The EXPERT command can influence menu macros, scripts, AutoLISP, and the command functions.

EXPLMODE   Integer
The EXPLMODE variable govern whether the EXPLODE command can explode nonuniformly scaled blocks. This variable is saved in the drawing and its initial value is 1.
0 - Nonuniformly scaled blocks cannot be exploded, 1 - Nonuniformly scaled blocks can be exploded.

EXTMAX   3D Point
The EXTMAX variable holds the upper-right point of the drawing extents and is saved in the drawing. The drawing extents increase outward when new objects are drawn and reduce only when ZOOM All or ZOOM Extents is used. The variable is reported in the World coordinates for the current space.

EXTMIN   3D Point
The EXTMIN variable holds the lower-left point of the drawing extents and is saved in the drawing. The drawing extents increase outward when new objects are drawn and reduce only when ZOOM All or ZOOM Extents is used. The variable is reported in the World coordinates for the current space.

FACETRES   Real
The FACETRES variable adjusts the smoothness of shaded and objects whose hidden lines have been removed. This variable can be assigned values in the range of 0.010 to 10.0. The variable is saved in the drawing and has an initial value of 0.5.

FILEDIA   Integer
The FILEDIA variable suppresses the display of file dialog boxes. This variable is

saved in registry and has an initial value of 1.
0 - The file dialog boxes are disabled. However you can make AutoCAD to display the file dialog box by entering a tilde (~) as the response to the prompt. This applied for AutoLISP and ADS functions also, 1 - The file dialog boxes are enabled except when a script or AutoLISP/ADS program is active in which case only a prompt appears.

FILLETRAD　　　　Real
The FILLETRAD variable holds the current fillet radius and is saved in the drawing and its initial value is 0.5000.

FILLMODE　　　　Integer
FILLMODE variable indicates whether objects drawn with SOLID command are filled in. This variable is saved in the drawing and its initial value is 1.
0 - Objects are not filled, 0 - Objects are filled.

FONTALT　　　　String
The FONTALT variable specifies the alternate font to be used in case the specified font file cannot be found. In case you have not specified an alternate font, AutoCAD issues a warning. This variable is saved in registry and its initial value is "simplex.shx".

FONTMAP　　　　String
The FONTMAP variable specifies the font mapping file to be used in case the specified font file cannot be found. This file holds one font mapping per line. The original font and the substitute font are separated by a semicolon (;). This variable is saved in registry and its initial value is "Acad.fmp".

FRONTZ　　　　Real
The FRONTZ variable contains the front clipping plane offset (in current drawing units) from the target plane for the current viewport. You can determine the distance between the front clipping plane and the camera point by subtracting the FRONTZ value from the camera to target distance. This variable is saved in the drawing and is read-only.

GRIDMODE　　　　Integer
The GRIDMODE variable specifies whether the grid is turned on or off. This variable is saved in the drawing and its initial value is 0.
0 - The grid is turned off, 1 - The grid is turned on.

GRIDUNIT　　　　2D point
The GRIDUNIT variable specifies the X and Y grid spacing for the current viewport. The changes made to the grid spacing are manifested only after using the REDRAW or REGEN command. This variable is saved in the drawing and its initial value is 0.5000,0.5000.

GRIPBLOCK　　　　Integer
The GRIPBLOCK variable controls the assignment of grips in blocks. This variable is saved in registry and its initial value is 0.
0 - The grip is assigned only to the insertion point of the block, 1 - Grips are assigned to objects within the block.

GRIPCOLOR　　　　Integer
The GRIPCOLOR variable controls the color of nonselected grips. It can take a value in the range of 1 to 255. This variable is saved in registry and its initial value is 5.

GRIPHOT　　　　Integer
The GRIPHOT variable controls the color of selected grips. It can take a value in the range of 1 to 255. This variable is saved in registry and its initial value is 1.

GRIPS　　　　Integer
With the GRIPS variable you can make use of selection set grips for the Stretch, Move, Rotate, Scale, and Mirror grip modes. This variable is saved in registry and its initial value is 1.
0 - Grips are disabled, 1 - Grips are enabled.

GRIPSIZE　　　　Integer
The GRIPSIZE variable allows you to assign a size to the box drawn to show the grip. It sets its half height in pixels. This variable can be assigned a value in the range of 1 to 255. The variable is saved in registry and its initial value is 3.

HANDLES Integer
The HANDLES variable is always on (1), which states that object handles are enabled and can be accessed by applications. This variable is saved in the drawing and is read-only.

HIGHLIGHT Integer
The HIGHLIGHT variable governs object highlighting. Objects selected with grips are not influenced.
0 - Object selection highlighting is disabled, 1 - Object selection highlighting is enabled. This is the initial value for the variable.

HPANG Real
The HPANG variable specifies the angle of the hatch pattern. The initial value for this variable is 0.

HPBOUND Real
The HPBOUND variable governs the object type created by the BHATCH and BOUNDARY commands. This variable is saved in the drawing and its initial value is 1.
0 - A region is created, 1 - A polyline is created.

HPDOUBLE Integer
The HPDOUBLE variable governs the hatch pattern doubling for user-defined patterns. The initial value of this variable is 0.
0 - Hatch pattern doubling disabled, 1 - Hatch pattern doubling enabled.

HPNAME String
The default hatch pattern name is established with HPNAME variable. The name can be up to 34 characters and spaces are not allowed. Empty string ("") is returned if no default exists. To set no default enter a period (.). The initial value of this variable is "ANSI31".

HPSCALE Real
The hatch pattern scale factor is specified with HPSCALE variable. This variable cannot assume zero value. The initial value of this variable is 1.0000.

HPSPACE Real
The hatch pattern line spacing for user-defined simple patterns is specified by HPSPACE variable. This variable cannot assume zero value. The initial value of this variable is 1.0000.

INDEXCTL Integer
Controls whether layer and spatial indexes are created and saved in drawing. Initial value is 0.
0 - No indexes created, 1 - Layer index created, 2 - Spatial index is created, 3 - Layer and spatial are created.

INETLOCATION Real
Stores the Internet location used by BROWSER. Initial value is "www autodesk.com/acaduser".

INSBASE 3D point
The insertion base point established by the BASE command is stored in this variable. This point is defined in UCS coordinates for the current space. The variable is saved in the drawing and its initial value is 0.0000,0.0000,0.0000.

INSNAME String
The INSNAME variable establishes the default block name for DDINSERT or INSERT commands. To set no default enter a period (.). The initial value of this variable is "".

ISAVEBAK Integer
Improves the speed of incremental saves, especially for large drawings on Windows. Initial value is 1.
0 - No BAK file is created, 1 - A BAK file is created.

ISAVEPERCENT Integer
Determines the amount of wasted space tolerated in a drawing file. Initial value is 50.

ISOLINES Integer
The ISOLINES variable specifies the number of isolines per surface on objects. The variable can accept a value in the range of 0 to 2047. This variable is saved in the drawing and its initial value is 4.

LASTANGLE Real
The LASTANGLE variable holds the end angle of the last arc entered, with respect to the XY plane of the current UCS for the current space. This variable is a read-only variable.

LASTPOINT 3D point
The LASTPOINT variable holds the UCS coordinates for the current space of the most recently entered point. This variable is saved in the drawing and its initial value is 0.0000,0.0000,0.0000.

LASTPROMPT String
The LASTPROMPT variable stores the last string echoed to the command line. Initial value is "".

LENSLENGTH Real
The LENSLENGTH variable holds the length of the lens (in mm) used in perspective viewing for the current viewport. This variable is saved in the drawing.

LIMCHECK Integer
This variable governs the drawing of objects outside the specified drawing limits. This variable is saved in the drawing and its initial value is 0.
0 - Object can be drawn outside the drawing limits, 1 - Object cannot be drawn outside the drawing limits.

LIMMAX 2D point
The upper-right drawing limits stated in World coordinates (for the current space) are held in the LIMMAX variable. This variable is saved in the drawing and its initial value is 12.0000,9.0000.

LIMMIN 2D point
The lower-left drawing limits stated in World coordinates (for the current space) are held in the LIMMIN variable. This variable is saved in the drawing and its initial value is 0.0000,0.0000.

LISPINIT Integer
Specifies whether AutoLISP defined functions and variables are preserved when you open new drawing. it is saved in registry and the initial value is 1.
0 - AutoLISP functions and variables are preserved, 1 - AutoLISP functions and variables are valid in current drawing only.

LOCALE String
The LOCALE variable shows the ISO language code of the current AutoCAD version in use. The initial value of this variable is "enu" (varies by country).

LOGFILEMODE Integer
Specifies whether the contents of the text window are written to a log file. Its initial value is 0.
0 - Log file is not maintained, 1 - Log file is maintained.

LOGFILENAME String
Specifies the path for the log file. Initial value is "C:\ACADR14\acad.log".

LOGINNAME String
The LOGINNAME variable shows the user's name as specified while configuring when AutoCAD is loaded.

LTSCALE Real
The LTSCALE variable establishes global linetype scale factor. The variable is saved in the drawing and its initial value is 1.0000.

LUNITS Integer
The LUNITS variable establishes the Linear Units mode. The variable is saved in the drawing and its initial value is 2.
1 - Scientific units mode, 2 - Decimal units mode, 3 - Engineering units mode, 4 - Architectural units mode, 5 - Fractional units mode.

LUPREC Integer
The LUPREC variable establishes the linear units decimal places or denominator. The variable is saved in the drawing and its initial value is 4.

MAXACTVP Integer
The MAXACTVP variable specifies the maximum number of viewports to

regenerate at one time. The initial value of this variable is 48.

MAXOBJMEM Integer
Controls the object pager. Initial value is 0.

MAXSORT Integer
The MAXSORT variable sets the maximum number of symbol names of file names that are to be sorted by listing commands. In case the total number of items is greater than this number, then no items are sorted. This variable is saved in registry and its initial value is 200.

MEASUREMENT Integer
The MEASUREMENT variable sets the drawing units as English or metric. It is saved in drawing and the initial value is 0.
0 - English, 1 - Metric.

MENUCTL Integer
The MENUCTL variable governs the page switching of the screen menu. This variable is saved in registry and its initial value is 1.
0 - For this value the screen menu does not switch pages in response to a keyboard command entry, 1 - For this value the screen menu switches pages in response to a keyboard command entry.

MENUECHO Integer
The MENUECHO variable sets menu echo and prompt control bits. The initial value for this variable is 0. The variable is the addition of the following:
1 - The echo of menu items is suppressed, 2 - The display of system prompts during menu is suppressed, 4 - The ^P toggle of menu echoing is disabled, 8 - The input/output strings and debugging aid for DIESEL macros is displayed.

MENUNAME String
The MENUNAME variable contains the MENUGROUP name. In case the prevailing primary menu has no MENUGROUP name, the menu file includes the path if the location of the file is not defined in the AutoCAD environment setting. This variable is saved in the application leader and is a read-only variable.

MIRRTEXT Integer
The MIRRTEXT variable governs how the MIRROR command mirrors text. This variable is saved in the drawing and its initial value is 1.
0 - The text direction is retained, 1 - The text is mirrored.

MODEMACRO String
The MODEMACRO variable shows a text string or DIESEL expression on the status line. This string reveals information like the name of the current drawing, time/date stamp, or special modes. The initial value of this variable is "".

MTEXTED String
The MTEXTED variable sets the name of the program to be used for the editing of mtext objects. This variable is saved in registry and its initial value is "Internal".

OFFSETDIST Real
This variable sets the default offset distance. If the value of this variable is less than zero then the offset distance can be specified with the through mode. If the value of this variable is greater than zero then the default offset distance is established. The initial value of this variable is 1.0000.

OLEHIDE Integer
Controls the display of OLE objects in AutoCAD. It is saved in registry and its initial value is 0.
0 - All OLE objects are visible, 1 - OLE objects are visible in paper space, 2 - OLE objects are visible in paper space, 3 - No OLE objects are visible.

ORTHOMODE Integer
The ORTHOMODE variable governs the orthogonal display of lines or polylines. This variable is saved in the drawing and its initial value is 0.
0 - The Ortho mode is turned off, 1 - The Ortho mode is turned on.

OSMODE Integer
The OSMODE variable sets the running Object Snap modes using the following bit-codes:

0 - NONe object snap, 1 - ENDpoint object snap, 2 - MIDpoint object snap, 4 - CENter object snap, 8 - NODe object snap, 16 - QUAdrant object snap, 32 - INTersection object snap, 64 - INSertion object snap, 128 - PERpendicular object snap, 256 - TANgent object snap, 512 - NEArest object snap, 1024 - QUIck object snap, 2048 - APPint object snap. If you want to specify more than one object snap, enter the sum of their values. For example, if you want to specify the node and center object snaps, enter 4+8 = 12 as the value for the OSMODE variable. This variable is saved in the drawing and its initial value is 0.

OSNAPCOORD   Integer
Controls whether coordinates entered on the command line override running object snaps.
0 - Running object snap settings override keyboard entry, 1 - Keyboard entry overrides object snap settings, 2 - (Initial value) Keyboard entry overrides object snap setting except in scripts.

PDMODE   Integer
The PDMODE variable sets Point Object Display mode. This variable is saved in the drawing and its initial value is 0.

PDSIZE   Real
The PDSIZE variable sets the display size of the point object. This variable is saved in the drawing and its initial value is 0.0000.
0 - For this value, point is created at 5 percent of the graphics height, >0 - In this case the value entered specifies the absolute size, <0 - In this case the value entered specifies the percentage of the viewport size.

PELLIPSE   Integer
The PELLIPSE controls the type of ellipse created with the ELLIPSE command. This variable is saved in the drawing and its initial value is 0.
0 - A true ellipse object is drawn, 1 - A polyline representation of an ellipse is drawn.

PERIMETER   Real
The PERIMETER variable holds the most recently perimeter value computed by AREA, DBLIST, or LIST commands. This variable is a read-only variable.

PFACEVMAX   Integer
The PFACEVMAX variable sets the maximum number of vertices per face. This variable is a read-only variable.

PICKADD   Integer
The PICKADD variable controls additive selection of objects. This variable is saved in registry and its initial value is 1.
0 - PICKADD variable is disabled, 1 - PICKADD variable is enabled. All the objects selected by any method are added to the selection set. If you want to remove objects from the selection set, hold down the Shift key and select the objects.

PICKAUTO   Integer
PICKAUTO variable controls the automatic windowing feature when the "Select object" prompt appears. This variable is saved in registry and its initial value is 1.
0 - PICKAUTO variable is disabled, 1 - PICKAUTO variable is enabled and a selection window is automatically drawn at the Select objects prompt.

PICKBOX   Integer
The PICKBOX variable sets the object selection target half height (in pixels). This variable is saved in registry and its initial value is 3.

PICKDRAG   Integer
The PICKDRAG variable governs the method of drawing a selection window:
0 - For this value the selection window is drawn by clicking the pointing device at one corner and then clicking again at the other corner of the window, 1 - For this value the selection window is drawn by clicking the pointing device at one corner, holding down the pick button, dragging the cursor, and finally releasing the pick button of the pointing device at the other corner of the window. This variable is saved in registry and its initial value is 0.

PICKFIRST Integer
The PICKFIRST variable governs the method of object selection in such a manner that you can first select the object and then specify the desired edit or inquiry command. Its initial value is 1.
0 - PICKFIRST variable disabled, 1 - PICKFIRST variable enabled.

PICKSTYLE Integer
The PICKSTYLE variable controls the associative hatch selection and group selection. This variable is saved in the drawing and its initial value is 1.
0 - Associative hatch selection and group selection not possible, 1 - Group selection possible, 2 - Associative hatch selection possible, 3 - Associative hatch selection and group selection possible.

PLATFORM String
The PLATFORM variable specifies the platform of AutoCAD that is in use. This a read-only variable. Some of the platforms are:
Microsoft Windows - Sun/SPARCstation, 386 DOS Extender - DECstation, Apple Macintosh - Silicon Graphics Iris Indigo.

PLINEGEN Integer
The PLINEGEN variable sets the linetype pattern generation around the vertices of a 2D polyline. This variable does not affect polylines with tapered segments. This variable is saved in the drawing and its initial value is 0.
0 - Polylines are generated with a dash at each vertex, 1 - Linetype is created in a continuous pattern around the vertices of the polyline.

PLINETYPE Integer
Specifies whether AutoCAD uses optimized 2D polylines.
0 - PLINE creates old format polylines,
1 - PLINE creates optimized polylines,
2 - PLINE creates optimized polylines and the polylines in older drawings are converted on open.

PLINEWID Real
The default polyline width is stored in this variable. This variable is saved in the drawing and its initial value is 0.0000

PLOTID String
The PLOTID variable stores the current plotter's description. The plotter configuration can be changed by entering the plotter's full or partial description. This variable is saved in registry and its initial value is "".

PLOTROTMODE Integer
The PLOTROTMODE variable controls the orientation of plots. This variable is saved in the drawing and its initial value is 1.
0 - The effective plotting area is rotated in order to align the corner with the Rotation icon with the paper at the lower-left for a rotation of 0, top-left for a rotation of 90, top-right for a rotation of 180, and lower left for a rotation of 270, 1 - The lower-left corner of the effective plotting area is aligned with the lower-left corner of the paper.

PLOTTER Integer
The PLOTTER variable stores an integer number assigned for configured plotter. This integer number can be in the range of 0 to the number of configured plotters. You can change to some other configured plotter by entering the integer number assigned to the plotter. If you have 6 plotters, the valid numbers are 0, 1, 2, 3, 4, 5. This variable is saved in registry and its initial value is 0.

POLYSIDES Integer
The POLYSIDES variable establishes the default number of sides for a polygon. This variable can take values in the range of 3 to 1024. The initial value of this variable is 4.

POPUPS Integer
The POPUP variable shows the status of the presently configured display driver. This is a read-only variable.
0 - The dialog boxes, menu bar, pull-down menus, and icon menus are not supported, 1 - The dialog boxes, menu bar, pull-down menus, and icon menus are supported.

PROJECTNAME String
Stores the current project neame. Initial value is " ".

PROJMODE Integer
The PROJMODE variable establishes the current Projection mode for Extend or Trim operations. Its initial value is 1.
0 - True 3D mode established (no projection), 1 - Projection to XY plane of the current UCS, 2 - Projection to current view plane.

PROXYGRAPHICS Integer
Specifies whether images of proxy objects are saved in the drawing. The initial value is 1. 0 - Image is not saved, 1 - Image is saved.

PROXYNOTICE Integer
Displays a notice when you open a drawing containig custom objects created by an application that is not present. The initial value is 1.
0 - No proxy warning displayed, 1 - Proxy warning displayed.

PROXYSHOW Integer
Controls the display of proxy objects in a drawing. The initial value is 1.
0 - Proxy objects are not displayed, 1 - Graphic images are displayed for all proxy objects, 2 - Only bounding box is displayed for all proxy objects.

PSLTSCALE Integer
The PSLTSCALE variable governs the paper space linetype scaling. This variable is saved in the drawing and its initial value is 1.
0 - Special linetype scaling not allowed. Linetype dash lengths depend on the drawing units of the space in which the objects were drawn, 1 - Linetype scaling governed by viewport scaling. In case TILEMODE is set to 0, dash lengths depend on the paper space drawing units, even if objects are in model space.

PSPROLOG String
The PSPROLOG variable assigns a name for a prologue section which is to be read from the acad.psf file when PSOUT command is being used. This variable is saved in registry and its initial value is "".

PSQUALITY Integer
The PSQUALITY variable governs the quality of rendering of PostScript images and also whether these images are drawn as filled objects or as outlines. This variable is saved in the drawing and its initial value is 75.
0 - The PostScript image generation is disabled, >0 - Any value greater than zero specifies the number of pixels per AutoCAD drawing unit for the PostScript resolution and fills outlines, <0 - Value less than zero specifies the number of pixels per AutoCAD drawing unit, but uses the absolute value. This causes the AutoCAD to display PostScript paths as non-filled outlines.

QTEXTMODE Integer
The QTEXTMODE controls the Quick Text mode. This variable is saved in the drawing and its initial value is 0.
0 - The Quick Text mode is turned off and characters are displayed, 1 - The Quick Text mode is turned on and a box instead of text is displayed.

RASTERPREVIEW Integer
The RASTERPREVIEW variable determines whether the drawing preview images are saved with the drawing and in which format they will be saved. This variable is saved in the registry and its initial value is 1.
0 - No preview image created, 1 - Preview image is created.

REGENMODE Integer
The REGENMODE variable controls the automatic regeneration of the drawing. This variable is saved in the drawing and its initial value is 1.
0 - REGENAUTO is turned off, 1 - REGENAUTO is turned on.

RE-INIT Integer
The RE-INIT variable reinitializes the I/O ports, plotter, digitizer, display, and acad.pgp file. The following bit-codes are used for this process:

...value), 1 - Reinitialization of digitizer port, 2 - Reinitialization of plotter port, 4 - Reinitialization of digitizer, 8 - Reinitialization of display, 16 - Reinitialization of PGP file. You can specify more than one reinitialization by entering the sum of the values of the desired reinitializations.

RTDISPLAY Integer
Controls the display of raster images during realtime zoom or pan.
0 - Displays raster image content, 1 - Displays raster image outline only (initial value).

SAVEFILE String
The present auto-save filename is held in the SAVEFILE variable. This variable is a read-only variable. The initial value is "auto.sv$".

SAVENAME String
You can save the current drawing to a different name and this name is held in the SAVENAME variable. This variable is a read-only variable.

SAVETIME Integer
AutoCAD has provided the facility of automatically saving your work at specific intervals. You can specify the automatic save time intervals (in minutes) with the SAVETIME variable. This variable is set to an initial value of 120.
0 - Automatic save facility is disabled, 1 - The drawing is saved according to the intervals specified. Once you make changes to the drawing, the SAVETIME timer starts. SAVE, SAVEAS, or QSAVE commands reset and restart this timer. AutoCAD saves the drawing under the filename auto.sv$.

SCREENBOXES Integer
The SCREENBOXES variable stores the number of boxes in the screen menu area of the graphics area. In case the screen menu is disabled, this variable is set to zero. The value of this variable is susceptible to change during an editing session on platforms that allow the AutoCAD graphics window to be resized or the screen menu to be reconfigured while you are in an editing session. This is a read-only variable.

SCREENMODE Integer
The SCREENMODE variable holds a bit-code specifying the graphics/text state of the AutoCAD display. This is a read-only variable. Following are the bit values:
0 - Text screen is displayed, 1 - Graphics mode is displayed, 2 - Dual-screen display (text and graphics) is displayed.

SCREENSIZE 2D point
This variable holds the current viewport size in pixels. This is a read-only variable.

SHADEDGE Integer
The SHADEDGE variable governs the shading of edges in rendering. This variable is saved in the drawing and its initial value is 3.
0 - Faces are shaded and edges are not highlighted, 1 - Faces are shaded and edges are drawn in background color, 2 - Faces are not filled and edges are in object color, 3 - Faces are in object color and edges are drawn in background color.

SHADEDIF Integer
The SHADEIF variable establishes the ratio of diffuse reflective light to ambient light (in percent of diffuse reflective light). This variable is saved in the drawing and its initial value is 70.

SHPNAME String
The SHPNAME variable establishes the name of the default shape. The initial value for this variable is "". To set no default enter a period (.).

SKETCHINC Real
The SKETCHINC specifies the record increment for the SKETCH command. This variable is saved in the drawing and its initial value is 0.1000.

SKPOLY Integer
The SKPOLY variable decides whether SKETCH command generates lines or polylines. This variable is saved in the drawing and its initial value is 0.

0 - Lines are generated, 1 - polylines are generated.

SNAPANG   Real
The SNAPANG variable specifies the snap/grid rotation angle relative to the UCS for the current viewport. This variable is saved in the drawing and its initial value is 0. Changes to this variable are manifested only after a redraw is performed.

SNAPBASE   2D point
The SNAPBASE variable specifies the snap/grid origin point (in UCS X, Y coordinates) for the current viewport. This variable is saved in the drawing and its initial value is 0.0000,0.0000. Changes to this variable are manifested only after a redraw is performed.

SNAPISOPAIR   Integer
The SNAPISOPAIR variable controls the current isometric plane for the current viewport. This variable is saved in the drawing and its initial value is 0.
0 - Left, 1 - Top, 2 - Right.

SNAPMODE   Integer
The SNAPMODE variable controls the Snap mode. This variable is saved in the drawing and its initial value is 0.
0 - Snap disabled, 1 - Snap enabled for the current viewport.

SNAPSTYL   Integer
The SNAPSTYL variable establishes the snap style for the current viewport. This variable is saved in the drawing and its initial value is 0.
0 - Standard, 1 - Isometric.

SNAPUNIT   2D point
The SNAPUNIT variable specifies the X and Y snap spacing for the current viewport. This variable is saved in the drawing and its initial value is 0.5000,0.5000. The changes to this variable are manifested only after a redraw is performed.

SORTENTS   Integer
The SORTENTS variable governs the display of object sort order operations using the following values:
0 - SORTENTS is disabled, 1 - Sorts for object selection, 2 - Sorts for object snap, 4 - Sorts for redraw, 8 - Sorts for MSLIDE slide creation, 16 - Sorts for regens, 32 - Sorts for plotting, 64 - Sorts for PostScript output. More than one options can be selected by specifying the sum of the values of these options. Its initial value is 96. This value specifies sort operations for plotting and PostScript output.

SPLFRAME   Integer
The SPLFRAME variable governs the display of spline-fit polylines. This variable is saved in the drawing and its initial value is 0.
0 - The control polygon for spline fit polylines is not displayed. The fit surface of a polygon mesh is displayed while as the defining mesh is not displayed. Also invisible edges of 3D faces or polyface meshes are not displayed, 1 - The control polygon for spline fit polylines is displayed. The fit surface of a polygon mesh is not displayed while as the defining mesh is displayed. Invisible edges of 3D faces or polyface meshes are also displayed.

SPLINESEGS   Integer
The SPLINESEGS variable governs the number of line segments used to construct each spline. Hence, with this variable you can control the smoothness of the curve. This variable is saved in the drawing and its initial value is 8. With this value a reasonably smooth curve is generated which does not need a much regeneration time. The greater the value of this variable, the smoother the curve and greater the regeneration time and the space occupied by the drawing file.

SPLINETYPE   Integer
The SPLINETYPE variable specifies the type of spline curve that will be generated by Spline option of PEDIT command. This variable is saved in the drawing and its initial value is 6.

5 - Quadratic B-spline is generated, 6 - Cubic B-spline is generated.

SURFTAB1 Integer
The SURFTAB1 variable governs the number of intervals (tabulated surfaces) to be generated for TABSURF and RULESURF commands along the path curve. This variable also defines the mesh density in the M direction for REVSURF and EDGESURF commands. This variable is saved in the drawing and its initial value is 6. In case the path curve is a line, arc, circle, spline-fit polyline, or an ellipse, the path curve is divided into intervals equal to the value of SURFTAB1 by the tabulation lines. Else, if the path curve is a polyline (not spline-fit), the tabulation lines are generated at the ends of the polyline segments and if there are any arc segments, each segment is divided into intervals equal to the value of SURFTAB1 by the tabulation lines.

SURFTAB2 Integer
The SURFTAB2 variable defines the mesh density in the N direction for REVSURF and EDGESURF commands. This variable is saved in the drawing and its initial value is 6.

SURFTYPE Integer
The SURFTYPE variable governs the type of surface-fitting to be performed by the Smooth option of PEDIT command. This variable is saved in the drawing and its initial value is 6.
5 - Quadratic B-spline surface, 6 - Cubic B-spline surface, 8 - Bezier surface.

SURFU Integer
The SURFU variable specifies the surface density of polygon meshes in the M direction. This variable is saved in the drawing and its initial value is 6.

SURFV Integer
The SURFV variable specifies the surface density of polygon meshes in the N direction. This variable is saved in the drawing and its initial value is 6.

SYSCODEPAGE String
The SYSCODEPAGE variable expresses the system code page specified in the acad.xmf file. This variable is a read-only variable and is saved in the drawing.

TABMODE Integer
The TABMODE variable governs the use of Tablet mode.
0 - Tablet mode disabled (initial value), 1 - Tablet mode enabled.

TARGET 3D point
The TARGET variable holds the position of the target point (in UCS coordinates) for the current viewport. This is a read-only variable and is saved in the drawing.

TDCREATE Real
The TDCREATE variable holds the creation time and date of a drawing. This is a read-only variable and is saved in the drawing.

TDINDWG Real
The TDINDWG variable holds the total editing time. This is a read-only variable and is saved in the drawing.

TDUPDATE Real
The TDUPDATE variable holds the time and date of most recent update/save. This is a read-only variable and is saved in the drawing.

TDUSRTIMER Real
The TDUSRTIMER variable stores the user elapsed timer. This is a read-only variable and is saved in the drawing.

TEMPPREFIX String
The TEMPPREFIX variable stores the directory name configured for the placement of temporary files. The path separator is included. This is a read-only variable.

TEXTEVAL Integer
The TEXTEVAL variable determines the procedure of evaluation of text strings. The initial value for this variable is 0.
0 - All the responses to prompts for text strings and attributes are accepted as literals, 1 - In case the starting character of the text string is "(" or "!", it is treated as an

AutoLISP expression. The TEXTEVAL setting does not affect the DTEXT command. DTEXT command accepts all input as literals.

TEXTFILL   Integer
The TEXTFILL variable governs the filling of TrueType fonts. This variable is saved in the registry and has an initial value of 1.
0 - The text is displayed as outlines, 1 - The text is displayed as filled images.

TEXTQLTY   Real
The TEXTQLTY variable defines the resolution of TrueType, Bitstream, and Abode Type 1 fonts. The higher the value of this variable, the higher the resolution and lower the display and plotting speed. On the other hand the lower the value of this variable, the lower the resolution and higher the display and plotting speed. This variable is saved in the drawing and can take values in the range of 0 to 100.0, its initial value is 50.

TEXTSIZE   Real
The TEXTSIZE variable controls the text height of the text drawn with the current text style. But this is possible only if the style does not have a fixed height. This variable is saved in the drawing and its initial value is 0.2000.

TEXTSTYLE   String
The TEXTSTYLE variable stores the name of the current text style. This variable is saved in the drawing and its initial value is STANDARD.

THICKNESS   Real
The THICKNESS variable defines the current 3D thickness. This variable is saved in the drawing and its initial value is 0.0000.

TILEMODE   Integer
The TILEMODE variable governs entry into paper space and also how the AutoCAD viewports act. This variable is saved in the drawing and its initial value is 1.
0 - The paper space and viewport objects are enabled. The graphics area is cleared and you are prompted to use the MVIEW command to define viewports, 1 - Release 10 Compatibility mode is enabled. Automatically you are taken into Tiled Viewport mode and previously active viewport configuration is restored on the screen. Paper space objects including viewport objects are not displayed. MSPACE, PSPACE, VPLAYER, and MVIEW commands are disabled.

TOOLTIPS   Integer
The TOOLTIPS variable is concerned with the Windows version of AutoCAD and determines the display of ToolTips. Its initial value is 1.
0 - The display of ToolTips is turned off, 1 - The display of ToolTips is turned on.

TRACEWID   Real
The TRACEWID variable establishes default value for the width of the trace. This variable is saved in the drawing and its initial value is 0.0500.

TREEDEPTH   Integer
The TREEDEPTH variable specifies how many times the tree-structured spatial index may divide into branches. This variable is saved in the drawing and its initial value is 3020.
0 - The spatial index is totally suppressed. In this case the objects are processed in database order and hence it is not necessary to set the SORTENTS variable, >0 - TREEDEPTH variable is enabled. You can enter an integer of up to four digits. The first two digits indicate the depth of model space nodes and the second two digits indicate the depth of paper space nodes, <0 - If the value is negative then the model space objects are treated as 2D objects. Negative values are relevant for 2D drawings. This way memory is more efficiently utilized and there is no trade-off with the performance.

TREEMAX   Integer
The TREEMAX variable sets the limit to the maximum number of nodes in the spatial index. This way the memory use during regeneration of drawing is limited. Its initial value is 10000000.

TRIMMODE Integer
The TRIMMODE variable determines whether selected edges for chamfers and fillets will be trimmed.
0 - Selected edges are not trimmed after chamfering and filleting, 1 - Selected edges are trimmed after chamfering and filleting (initial value).

UCSFOLLOW Integer
The UCSFOLLOW variable controls the automatic displaying of a plan view when you switch from one UCS to another. All the viewports have the UCSFOLLOW facility and hence you need to specify the UCSFOLLOW setting separately for each viewport. This variable is saved in the drawing and its initial value is 0.
0 - Switch from one UCS to another, does not alter the view, 1 - Plan view of the new UCS is automatically displayed when you switch from one UCS to another.

UCSICON Integer
The UCSICON variable displays the present UCS icon using bit-code for the current viewport. The value of this variable is the sum of the following:
1 - Icon display is enabled, 2 - The icon moves to the UCS origin if the icon display is enabled. In case more than one viewport is active, each of the viewport can have a different value for the UCSICON variable. If you are in paper space, the UCSICON variable will contain the setting for the UCS icon of the paper space. This variable is saved in the drawing and its initial value is 1.

UCSNAME String
The UCSNAME variable contains the name of the current UCS. This is a read-only variable and is saved in the drawing. In case the current UCS is unnamed, then a null string is returned.

UCSORG 3D point
The coordinate value of the origin of the current UCS is held in the UCSORG variable. This is a read-only variable and is saved in the drawing.

UCSXDIR 3D point
The X axis direction of the current UCS for the current space is held in UCSXDIR variable. This is a read-only variable and is saved in the drawing.

UCSYDIR 3D point
The Y axis direction of the current UCS for the current space is held in UCSYDIR variable. This is a read-only variable and is saved in the drawing.

UNDOCTL Integer
The UNDOCTL variable holds a bit-code expressing the state of the UNDO command. This is a read-only variable. The value of this value is the addition of following values:
0 - UNDO command is disabled, 1 - UNDO command is enabled, 2 - Just one command can be undone, 4 - Auto-group mode is enabled, 8 - Some group is presently active.

UNDOMARKS Integer
The UNDOMARKS variable contains the number of marks that have been put in the UNDO command's control stream by the Mark option. In case a group is presently active, the Mark and Back options cannot be accessed. This variable is a read-only variable.

UNITMODE Integer
The UNITMODE variable governs the units display format. This variable is saved in the drawing and its initial value is 0.
0 - The fractional, feet and inches, and surveyor's angles are displayed as previously defined, 1 - The fractional, feet and inches, and surveyor's angles are displayed in the input format. This variable is saved in the drawing and its initial value is 0.

USERI1-5 Integer
Stores and retrieves integer values. Initial value is 0.

USERR1-5 Integer
Stores and retrieves real numbers. Initial value is 0.0000.

USERS1-5 — Integer
Stores and retieves text string data. Initial value is " ".

VIEWCTR — 3D point
The VIEWCTR variable stores the center of view in the current viewport, defined in the UCS coordinates. This variable is a read-only variable and is saved in the drawing.

VIEWDIR — 3D vector
The VIEWDIR variable contains the viewing direction in the current viewport expressed in the UCS coordinates. The camera position is expressed as a 3D offset from the target position. This variable is a read-only variable and is saved in the drawing.

VIEWMODE — Integer
The VIEWMODE variable governs Viewing mode for the current viewport using bit-code. The value for this variable is the addition of the following bit values:
0 - Viewing mode disabled, 1 - Perspective view active, 2 - Front clipping on, 4 - Back clipping on, 8 - UCS Follow mode on, 16 - Front clip not at eye. In case it is on, the front clipping plane is determined by the front clip distance stored in the FRONTZ variable. If it is off, the front clipping plane passes through the camera point and in this case FRONTZ variable is not taken into consideration. If the front clipping bit (2) is off then this flag is neglected. This variable is a read-only variable and is saved in the drawing.

VIEWSIZE — Real
The VIEWSIZE variable contains the view height in the current viewport and is defined in the drawing units. This variable is a read-only variable and is saved in the drawing.

VIEWTWIST — Real
The VIEWTWIST variable contains the view twist angle for the current viewport. This variable is a read-only variable and is saved in the drawing.

VISRETAIN — Integer
The VISRETAIN variable determines whether changes to the visibility of layers in xref are saved in the current drawing.
0 - Changes to On/Off, Freeze/Thaw, color, and linetype settings for the xref-dependent layers are not saved in the current drawing, 1 - Changes to the xref layer definitions in the current drawing are saved with the current drawing.

VSMAX — 3D point
The VSMAX variable contains the upper-right corner of the virtual screen of the current viewport and is expressed in UCS coordinates. This variable is a read-only variable and is saved in the drawing.

VSMIN — 3D point
The VSMIN variable contains the lower-left corner of the virtual screen of the current viewport and is expressed in UCS coordinates. This variable is a read-only variable and is saved in the drawing.

WORLDUCS — Integer
The WORLDUCS variable expresses whether the UCS is the same as the WCS. This variable is a read-only variable.
0 - Current UCS and WCS are different, 1 - Current UCS and WCS are not different.

WORLDVIEW — Integer
The WORLDVIEW variable determines whether UCS changes to WCS during DVIEW or VPOINT commands. This variable is saved in the drawing and its initial value is 1.
0 - Current UCS is not changed, 1 - Current UCS is changed to WCS till the DVIEW or VPOINT command is in progress. The DVIEW and VPOINT command input is with respect to the current UCS.

XCLIPFRAME — Integer
Controls visibility of xref clipping boundaries and its initial value is 0.
0 - Clipping boundary is not visible, 1 - Clipping boundary is visible.

XLOADCTL Integer
Turns demand load on and off and controls whether it loads the original drawing or a copy. Initial value is 1.
0 - Turns off demand loading; entire drawing is loaded, 1 - Turns on demand loading; reference fle is kept open, 2 - Turns on demand loading; a copy of reference file is opened.

XLOADPATH String
Creates a path for storing temporary copies of demand-loaded xref files. Initial value is " ".

XREFCTL Integer
The XREFCTL variable determines whether AutoCAD writes .xlg files (external reference log files). This variable is saved in registry and its initial value is 0.
0 - Xref log files are not written, 1 - Xref log files are written.

Index

***AUX1, 10-35, 36
***BUTTONS, 10-35
***BUTTONS1, 10-33
***BUTTONS1, 8-3
***HELPSTRING, 5-38
***IMAGE, 10-49
***IMAGE, 7-3, 6, 13
***IMAGE, 9-60
***POP1, 7-5, 13
***POP1, 9-60
***SCREEN, 9-57
***TABLET1, 10-20, 23
**Aliasname, 5-35
-LAYER, 1-9, 18
-MLEDIT command, 21-33
-MLEDIT, 21-33
.EPS (encapsulated postscript), 21-12
3D drawing, 1-17

A

ACAD.LIN, 3-1, 13
ACAD.MNU file, 5-2, 10, 6-8, 7-9, 9-6, 10-1, 31
ACAD.PAT, 3-25, 28
ACAD.PGP, 2-1
Acad.psf, 21-16
Accelerator keys, 5-32
Adding and deleting vertices, 21-31
Adding hatch pattern slides to autocad slide library, 3-37
Advantages of a tablet menu, 6-3
Alias, 5-35
Alignment definition, 3-6
Alignment specification, 3-8
Alignment field specification, 3-2

Alternate linetypes, 3-12
ASCII control character, 9-36, 38
ASCII format, 21-3
Assigning commands to a tablet, 6-13
Attribute
   ATTDEF, 18-1, 4
   ATTDEF command, 18-4
   ATTDIA, 18-8
   ATTDISP, 18-1, 2, 15
   ATTEDIT, 18-1, 17, 22
   ATTEXT, 18-1, 14
   Attribute area, 18-3
   Attribute definition dialog box, 18-2
   Attribute extraction dialog box, 18-11
   Attribute value, 18-1
   ATTRIBUTES, 18-1
   BLOCK, 18-6
   CHANGE, 18-7
   Comma delimited file (cdf), 18-11
   Controlling attribute visibility, 18-15
   DDATTDEF, 18-1, 5, 6
   DDATTE, 18-1, 16
   DDATTEXT, 18-1, 11
   DDEDIT, 18-7
   Defining attributes, 18-2
   Drawing interchange file (dxf), 18-11
   Edit attribute definition dialog box, 18-7
   Editing all attribute, 18-18
   Editing attribute tags, 18-7

   Editing attributes (ATTEDIT command), 18-17
   Editing attributes (DDATTE command), 18-16
   Editing attributes with a specific attribute value, 18-19
   Editing attributes with specific attribute tag names, 18-18
   Editing specific blocks, 18-18
   Editing visible attributes only, 18-18
   Extracting attributes, 18-11
   Global editing of attributes, 18-17
   Individual editing of attributes, 18-22
   Inserting blocks with attributes, 18-7
   Inserting text files in the drawing, 18-25
   Multiline text editor, 18-25
   Space delimited file (sdf), 18-11
   Template file, 18-11
   Text options, 18-4
   Using the DDEDIT command, 18-7
Autocad menu file, 6-2
Autocad tablet template, 6-2
AutoLISP, 10-30, 42
AutoLISP
   About AutoLISP, 12-1

Absolute number, 12-4
Addition, 12-2
Algorithm, 12-38
ANGBASE, 12-25, 27
ANGDIR, 12-25, 27
Angtos, 12-6
Assoc, 13-6, 7, 9
Atan, 12-5
AutoLISP, 12-1, 13-1, 13
Cadr, 12-20, 24
Car, 12-19, 24
Car, cdr, and cadr functions, 12-19
CHAMFERA, 12-17, 18
CHAMFERB, 12-17, 18
Cmdecho, 12-23
Command, 12-11, 37, 13-3
Conditional functions, 12-38
Cons, 13-6, 7
Cos, 12-5
Decremented number, 12-4
Defun, 12-8, 23
Defun, 13-2
Degrees to radians, 12-49
Division, 12-3
Editing the drawing database, 13-1
Entget, 13-5, 7, 8
Entmod, 13-7, 9
Equal to, 12-7
Flowchart symbols, 12-39
Flowchart, 12-38, 13-12
FORTRAN, 12-1
Getangle, 12-18, 25, 26
Getcorner, getdist, and setvar functions, 12-15
Getcorner, 12-15
Getdist, 12-16, 18
Getint, 12-28
Getint, getreal, getstring, and getvar functions, 12-28
Getorient, 12-26
Getpoint, 12-10, 23

Getreal, 12-28
Getstring, 12-28
Getvar, 12-29, 13-3
Graphscr, textscr, princ, and terpri functions, 12-21
Graphscr, 12-21, 23
Greater than or equal to, 12-8
Greater than, 12-8
Group Codes for ssget "X", 13-4
How the database is retrieved and edited, 13-8
If, 12-38, 41, 42
Incremented number, 12-4
Incremented, decremented, and, 12-4
Itoa, 12-33
Itoa, rtos, strcase, and prompt functions, 12-33
Less than or equal to, 12-7
Less than, 12-7
Limmax, 13-3
Limmin, 13-3
LISP programming language, 12-1
LISP, 12-1
List function, 12-19
List processor, 12-1
Loading an autolisp program, 12-14
Mathematical operations, 12-2
Multiplication, 12-3
Not equal to, 12-7
Persistent AutoLISP, 12-47
Polar and sqrt functions, 12-29
Polar, 12-29, 37
Preferences, 12-47
Princ, 12-21
Progn, 12-42
PROMPT, 12-34

Radians to degrees, 12-49
Relational statements, 12-6
Repeat, 12-46, 13-11
Rtos, 12-33
Setq, 12-9
Setvar, 12-16, 23
Sin, 12-5
SMLayout, 12-2
Sqrt, 12-30, 48
Ssget "X", 13-3, 4, 11
Ssget, 13-1, 2, 3, 8
Sslength, 13-5, 11
Ssname, 13-5, 7, 8
Strcase, 12-34
Subst, 13-6, 7, 9
Subtraction, 12-3
Terpri, 12-21
Textscr, 12-21
Trigonometric functions, 12-5
UNITS, 12-25
While, 12-43, 44, 45
Automatic menu swapping, 6-14
Automatic menu swapping, 9-35

B

BASE, 1-2
Binary DXF files, 21-3
Bit map, 21-5
BLIPMODE, 1-2
BMP files, 21-5
Btnname, 5-34
Button menu, 8-5
Buttons and auxiliary, 10-31
BUTTONS, 10-31
Bylayer, 1-2

C

Cadr, 9-41
Calibrate the tablet, 21-35
Car, 9-41
CHAMFERA, 1-2
CHAMFERB, 1-2

Clipboard, 21-18
CMDDIA, 2-14
CMDECHO, 9-41
CMLJUST, 21-33
CMLSCALE, 21-33
CMLSTYLE, 21-33
COLOR, 1-2
Command definition, 5-8, 7-4
Command definition without enter or space, 9-38
Command aliases, 4-7
Comments, 4-6
Configure the tablet, 21-35
Control characters, 9-37
COPYCLIP, 21-20
Copying a tool icon, 5-41
COPYLINK, 21-23
Corner joint, 21-31
Create dxf file dialog box, 21-2
Create new drawing dialog box, 1-3
Create postscript file dialog box, 21-12, 17
Creating linetypes, 3-3
Creating a new image and tooltip for an icon, 5-39
Creating custom toolbars with flyout icons, 5-42
Creating linetype files, 3-8
Creating template drawings, 1-1
Cross intersection, 21-30
Custom hatch pattern file, 3-36
Customize the toolbars, 5-38
Customizing drawings according to, 1-10
Customizing drawings with layers, 1-6
Customizing drawings with viewports, 1-14
Customizing AutoCAD, 7-1
Customizing Buttons And Auxiliary Menus, 10-31

Customizing Tablet Area-1, 10-17
Customizing Tablet Area-2, 10-26
Customizing Tablet Area-3, 10-27
Customizing Tablet Area-4, 10-29
Customizing a drawing with paper space, 1-17
Customizing a tablet menu, 6-3
Customizing Image Tile Menus, 10-42
Customizing Pull-down And Cursor Menus, 10-36
Customizing the toolbars, 5-38
Customizing The Screen Menu, 10-47
Customizing the, 4-1
Cutting and welding multilines, 21-32

D

Data exchange, object, 21-1
DCL (Dialog Control Language)
$value, 14-41
Action_tile, 14-14
Alignment attribute, 14-8
Aspect_ratio and color attributes, 14-36
Aspect_ratio attribute, 14-36
Atof and rtos functions, 14-22
Autolisp functions, 14-13, 21, 39
Big_increment, 14-36
Boxed column tile, 14-17
Boxed radio column tile, 14-25
Boxed row tile, 14-16
Button and text tiles, 14-4
Button tile, 14-4

Color attribute, 14-36
Column, boxed column, and toggle tiles, 14-17
Column tile, 14-17
Components of a dialog box, 14-2
Dialog box, 14-2
Dialog box components, 14-2
Dialog boxes, 14-2
Dialog control language, 14-1
Dimx_tile and dimy_tile, 14-39
Displaying a new dialog box, 14-10
Done_dialog, 14-14
Edit box tile, 14-31
Edit_width attribute, 14-32
End_image, 14-41
Fill_image, 14-40
Fixed_width and alignment attributes, 14-8
Fixed_width attribute, 14-8
Get_tile and set_tile functions, 14-22
Graphical user interface, 14-2
Image tile, 14-35
Integer values, 14-5
Is_default, 14-7
Key attribute, 14-6
Key, label, and is_default attributes, 14-6
Label attribute, 14-7
Load_dialog, 14-13
Loading a dcl file, 14-10
Logand and logior, 14-21
Managing dialog boxes with autolisp, 14-14
Max_value, 14-35
Min_value, 14-35
Min_value and max_value attributes, 14-35

Min_value, max_value, small_increment,, 14-35
Mnemonic attribute, 14-17
New_dialog, 14-13
Predefined attributes, 14-6
Predefined radio button, radio column,, 14-25
Radio button tile, 14-25
Radio column tile, 14-25
Radio row tile, 14-26
Real values, 14-5
Reserved words, 14-5
Row and boxed row tiles, 14-16
Row tile, 14-16
Slider and image tiles, 14-35
Slider tile, 14-35
Small_increment, 14-36
Small_increment and big_increment attributes, 14-36
Start_dialog, 14-14
Start_image, 14-40
String values, 14-5
Text tile, 14-4
Tile attributes, 14-5
Toggle tile, 14-17
Unload_dialog, 14-13
Use of standard button subassemblies, 14-12
Use of the label attribute in a boxed column, 14-7
Use of the label attribute in a button, 14-7
Use of the label attribute in a dialog box, 14-7
Using autolisp function to load a dcl file, 14-12
Vector_image, 14-40
Width and edit_width attributes, 14-31
Width attribute, 14-31
Default setup values, 1-1
Definition of the line pattern, 3-2

Defun, 9-41
DELAY command, 2-9
Deleting a toolbar, 5-41
Deleting the icons from a toolbar, 5-41
Descriptive text, 3-5
Design of tablet template, 6-3
Design of the menu, 9-9
Design of the screen menu, 9-10
Device driver, 10-43
Dialog box, 7-2, 3, 10-43
Diesel expression in menus, 9-41
DIESEL
　Addition, 15-10
　Angtos, 15-13
　Customizing the status line, 15-3
　DIESEL, 15-1
　Diesel expressions in menus, 15-8
　Diesel string functions, 15-10
　Division, 15-11
　Edtime function, 15-16
　Eq function, 15-13
　Equal to, 15-11
　Eval function, 15-14
　Fix function, 15-14
　Getvar function, 15-14
　Greater than, 15-13
　Greater than or equal to, 15-13
　If function, 15-15
　Less than, 15-12
　Less than or equal to, 15-12
　Linelen function, 15-16
　Macro expressions using diesel, 15-4
　Macrotrace system variable, 15-9
　MODEMACRO, 15-1
　Modemacro system variable, 15-2
　Multiplication, 15-11
　Not equal to, 15-12
　Relational statements, 15-11
　RTOS function, 15-15
　Status line, 15-1
　Strlen function, 15-16
　Subtraction, 15-10
　Upper function, 15-16
　Using autolisp with MODEMACRO, 15-6
Digitizing tablet, 6-2
Digitizing drawings, 21-34
DIM, 1-9, 14
DIMALT, 1-2
DIMALTD, 1-2
DIMALTF, 1-2
DIMASO, 1-2
DIMASZ, 1-2
DIMPOST, 1-2
DIMSCALE, 1-11, 2-3
Displaying a submenu, 5-23, 7-3
DRAGMODE, 1-2
Dxb file format, 21-4
DXBIN, 21-4
DXF file format, 21-2
DXF file, 21-3
DXFIN command, 21-3
DXFOUT command, 21-2

E

Edit a slide, 2-24
EDIT, 2-2
Editing multilines (using grips), 21-29
Effect of angle and scale factor on hatch, 3-29
Element properties dialog box, 21-25
Element, 5-35
Elements of linetype specification, 3-3
ELEVATION, 1-2
EPS, 21-13, 17
EXPLODE command, 3-36

Exporting the raster files, 21-5
External command, 4-6

F
FILLETRAD, 1-2
FILLMODE, 1-2
Floating viewports, 2-19
Floating pointing area, 21-36
Floating screen pointing area, 21-36, 37
Flyname, 5-35

G
GETDIST, 9-41
GETPOINT, 9-41
GRID, 1-2, 2-2, 3, 14, 9-32
GRIDMODE, 1-2

H
Hatch description, 3-25
Hatch an area, 3-27
Hatch angle, 3-26
Hatch boundaries, 3-27
HATCH command, 3-29
Hatch name, 3-25
Hatch pattern definition, 3-25, 28, 29
Hatch pattern library file, 3-25
Hatch pattern specification, 3-28
Hatch pattern with dashes and dots, 3-30
Hatch pattern, 3-26
Hatch spacing, 3-29
Hatch with multiple descriptors, 3-31
Header line, 3-1, 25
How hatch works, 3-27

I
Icon, 5-35
Id_big, 5-35
Id_small, 5-34, 35

Image tile menu section, 7-2, 10-43
Image tile menus, 10-42
Image tile, 7-2
Image, 10-43
IMAGEADJUST command, 21-11
IMAGECLIP command, 21-10
IMAGEFRAME command, 21-12
IMAGEQUALITY command, 21-12
Initial drawing setup, 2-1
Internet
  Accessing a drawing on the internet, 20-17
  Attaching URLs to objects, 20-6
  ATTACHURL command, 20-6
  AutoCAD on the internet, 20-1
  BROWSER command, 20-1
  Create a DWF file, 20-11
  DETACHURL, 20-8
  Drag and drop, 20-15
  Drawback to using URLs, 20-5
  Drawing web format, 20-1
  DWF plug-in commands, 20-14
  Embedding a DWF file, 20-16
  File compression, 20-12
  FTP site, 20-18
  Inserting a block from the internet, 20-20
  INSERTURL, 20-20
  Internet, 20-1
  Launching the web browser, 20-1
  Listing a URL, 20-7
  LISTURL, 20-7
  Local file, 20-4
  Localhost, 20-4
  Microsoft internet explorer, 20-17
  Netscape navigator, 20-17
  OPENURL, 20-17
  Removing a URL, 20-8
  SAVEURL, 20-20
  Saving a drawing to the internet, 20-20
  Selecting a URL, 20-7
  SELECTURL, 20-7
  Servername, 20-4
  Drawing web format, 20-11
  Uniform resource locator, 20-3
  URL, 20-2, 3
  Viewing DWF files, 20-12
  Web site, 20-18
  Whip plug-in, 20-13
  Windows high performance, 20-12
  Worldwide web, 20-1
Invoking a script file when loading autocad, 2-11
ISOPLANE, 1-2

L
LAYER, 2-5
Library of standard linetypes, 3-1
LIMITS, 1-3, 7, 12, 14, 18, 2-2, 3, 9-33
LIMMAX, 1-2
LIMMIN, 1-2
Line pattern, 3-6
Line style, 3-5
Linetype definition, 3-1
Linetype description, 3-2
Linetype specification, 3-2
Linetype name, 3-2
LINETYPE, 3-6
Linking function, 21-22
Linking objects, 21-22
Links dialog box, 21-24
LIST, 9-41

Load multiline style dialog box, 21-25
Loading a template drawing, 1-5
Loading a pull-down menu, 8-9
Loading an image tile menu, 5-24
Loading an image menu, 8-9
Loading Image Tile Menus, 10-16
Loading menus, 5-10, 23, 6-8
Loading menus, 7-9, 8-9, 9-6
Loading screen menus, 5-23, 8-9, 10-16
Long menu definitions, 9-31
Ltscale and Dimscale, 1-12
LTSCALE commnad, 3-9
LTSCALE factor for plotting, 3-11
LTSCALE factor, 3-11
LTSCALE, 1-2, 8, 11, 14, 2-2, 3, 3-9

M

Macro, 5-35, 10-42
Macros, 8-1
Memory reserve., 4-6
Menu bar area, 5-3
Menu bar titles, 5-8
Menu command repetition, 9-33
MENU command, 5-11, 6-3, 9, 7-10, 9-7, 10-22
Menu item repetition, 7-8
Menu item label, 9-1
Menu items with single object selection mode, 9-40
Menu-specific help, 5-38
MENUCMD, 5-30
MENUECHO, 9-35
MENULOAD, 5-29, 36
MIRRTEXT, 1-2
Miter line, 21-26

MLEDIT, 21-24
MLINE, 21-24, 27
MLSTYLE, 21-24
Mnemonic key, 5-29
Model space, 1-17, 2-19
Modifying linetypes, 3-13
MSLIDE command, 2-18
MSLIDE, 2-19
MSPACE, 1-19
Multibutton pointing device, 8-1
Multiline properties, 21-26
Multiline edit tools dialog box (figure 21-29), 21-29
Multiline styles, 21-24
Multiline, 21-24
Multiple submenus, 9-16
MVIEW, 1-19

N

Nested submenus, 9-9, 10-48
New hatch pattern library file, 3-25

O

Object embedding, 21-19
OLE, 21-18
Online help, 5-38
Orient, 5-34
ORTHO, 2-2
ORTHOMODE, 1-2
OS command name., 4-6

P

Paper space, 1-17
Partial menu, 5-28
Partial menus, 5-27
PASTE command, 21-20
Paste special, 21-23
Pattern line, 3-1, 2, 10-32
Pick button, 8-2
Pline width, 1-12
PLINE, 1-8, 14, 19
PLINEWID, 1-8

Pointing device, 6-2, 8-1, 10-32
Polar, 9-41
POLYLINE, 9-32
POP1, 5-7
Postscript files, 21-12
Postscript fill patterns, 21-16
Preloading slides, 2-21
PRINC, 9-41
Prototype drawing, 1-14, 1-17
PSDRAG, 21-16
PSFILL command, 21-16
PSIN command, 21-15
PSOUT command, 21-12
PSPACE, 1-19
PSQUALITY, 21-16
Pull-down and cursor menus, 10-36
Pull-down menu or cursor menu, 5-13, 22
Pull-down menu, 10-42

R

Raster files, 21-5
Raster images, 21-8
REINIT command, 4-9
Render
   Ambient, 19-24
   Ambient light, 19-9
   ARX command, 19-2
   Assigning materials to layers, 19-22
   Assigning materials to the autocad color index (ACI), 19-21
   Attach by AutoCAD color index dialog box, 19-21
   Attach by layer dialog box, 19-22
   Attaching material to an object, 19-19
   Attaching materials, 19-19
   Attenuation, 19-10
   Attributes area, 19-24

Index 7

Autocad render light source, 19-9
AutoCAD's RENDER Command, 19-1
Back face, 19-2
Back Face Normal is Negative, 19-7
Bump Map, 19-25
Changing the parameters of a material, 19-23
Color pattern, 19-24
Colors area, 19-8
Defining and rendering a scene, 19-15
Defining new materials, 19-25
Destination, 19-7
Detaching materials, 19-23
Discard back faces, 19-7
Distant light, 19-11
Exporting a material from a drawing, 19-26
FACETRES variable, 19-2
File option, 19-7
File output configuration dialog box, 19-7, 27
File type area, 19-8
Geographic location, 19-14
Gouraud, 19-6
Image file dialog box, 19-27
Image specification dialog box, 19-28
Inserting and modifying lights, 19-11
Inserting distant light, 19-11
Inserting point light, 19-15
Interlace area, 19-8
Inverse linear, 19-10
Inverse square, 19-10
Library file dialog box, 19-20
Light icon scale, 19-7
Lights dialog box, 19-11

Loading and unloading autocad render, 19-2
Materials dialog box, 19-20
Materials library dialog box, 19-19
Modify distant light dialog box, 19-13
Modify point light dialog box, 19-15
Modify scene dialog box, 19-18
Modify standard material dialog box, 19-23
Modifying a scene, 19-18
Modifying distant light, 19-13
New distant light dialog box, 19-12, 16
New point light dialog box, 19-15, 16
New scene, 19-17
New scene dialog box, 19-17
New standard material dialog, 19-25
None, 19-10
Normal, 19-1
Obtaining rendering information, 19-18
Phong, 19-6
Point light, 19-9
Postscript options area, 19-8
Reflection, 19-24
Refraction, 19-25
RENDER command, 19-2
Render dialog box, 19-3
Render options dialog box, 19-6
Render quality area, 19-6
Render window option, 19-7
RENDER.MLI, 19-19
Rendering, 19-1

Rendering file dialog box, 19-27
Rendering options, 19-6
Rendering preferences dialog box, 19-5
Rendering procedures, 19-7
Rendering type, 19-6
REPLAY command, 19-28
Replay command to display the replay dialog box, 19-28
Replaying a rendered image, 19-28
Roughness, 19-25
Save image dialog box, 19-27
SAVEIMG command, 19-27
Saving a render-window rendered image, 19-27
Saving a rendering, 19-26
Saving a rendering to a file, 19-26
Saving a viewport rendering, 19-27
Selecting different properties, 19-5
Smooth shade, 19-6
Smoothing angle, 19-7
Spotlight, 19-9
Statistics dialog box, 19-18
Sun angle calculator, 19-13, 14
Tga options area, 19-8
Transparency, 19-25
Vector, 19-1
Viewport option, 19-7
VIEWRES command, 19-2
Restrictions, 7-10
RESUME command in transparent mode, 2-10
RESUME command, 2-10
Root menu, 9-14
Rows, 5-34
RSCRIPT command, 2-8

S

Save image dialog box, 21-5
Save multiline styles, 21-25
SAVEIMG command, 21-5
Saving hatch patterns in a separate file, 3-36
Screen menu, 9-1
Screen pixels, 21-5
SCREEN, 9-3
SCRIPT command, 2-3
Script file name, 2-4
Section label, 5-2, 7-3, 8-3
Sections of the ACAD.PGP file, 4-6
Select DXB file dialog box, 21-4
Select DXF file dialog box, 21-4
Select postscript file dialog box, 21-15
SETQ, 9-41
SETVAR, 1-3, 8, 14, 2-2, 3, 9-41
Shapes
   ASCII code, 11-19
   ASCII files, 11-1
   Bulge factor, 11-13
   Code 000
   End of shape definition, 11-6
   Code 001
   Activate draw mode, 11-6
   Code 002
   Deactivate draw mode, 11-6
   Code 003
   Divide vector lengths by next byte, 11-8
   Code 004
   Multiply vector lengths by next byte, 11-9
   Code 007
   Subshape, 11-10
   Code 008
   X-Y displacement, 11-10
   Code 009
   Multiple x-y displacements, 11-11
   Code 00a or 10
   Octant arc, 11-11
   Code 00b or 11
   Fractional arc, 11-12
   Code 00c or 12
   Arc definition by displacement and bulge, 11-13
   Code 00d or 13
   Multiple bulge-specified arcs, 11-14
   Code 00e or 14
   Flag vertical text, 11-14
   Codes 005 and 006
   Location save/restore, 11-9
   Compiling and loading shape/font files, 11-3
   Data byte, 11-2, 7, 16
   Defbytes, 11-2
   Direction code, 11-2
   Direction vector, 11-7, 8, 16
   Direction vectors, 11-6
   Draw mode, 11-6, 7
   End of a shape definition, 11-6
   End offset, 11-12
   Explanation, 11-7
   Header, 11-1
   Header line, 11-3
   Hexadecimal notation, 11-17
   Hexadecimal number, 11-2
   Highradius, 11-12
   Insert a shape, 11-4
   Length specification, 11-2
   Line feed, 11-19
   Load command, 11-4
   Lowradius, 11-12
   Maximum length of the vector, 11-2
   Maximum number of saves and restores, 11-10
   Negative displacement, 11-10
   Nonstandard fractional arc, 11-12
   Nonstandard vectors, 11-10, 11
   Octant, 11-11
   Octant arc, 11-11, 17
   Octant boundary, 11-11
   Position stack overflow in shape, 11-10
   Position stack underflow in shape, 11-10
   Positive displacement, 11-10
   Restrictions, 11-10
   Shape definition, 11-19
   Shape description, 11-1
   Shape files, 11-1, 18
   Shape name, 11-2, 18
   Shape number, 11-2, 7, 15, 18, 21
   Shape specification, 11-2, 3, 7, 15
   Special codes, 11-6
   Standard codes, 11-6
   Start offset, 11-12
   Subshape, 11-10
   Subshape code, 11-10
   Text font description, 11-18
   Text font files, 11-18
   Text fonts, 11-18
   Vector, 11-10, 16
   Vector length, 11-2
   Vector length and direction encoding, 11-2
   Vectors, 11-2
Simple hatch pattern, 3-28
Single, 9-40
Slide presentation, 2-18
Slide libraries, 2-23
Slide library file, 2-24
Slide show, 2-1
SLIDELIB, 2-23, 25, 3-38

Slides for image tile menus, 7-8
SNAP, 1-3, 14, 2-2, 3, 9-32
Special handling for button menus, 8-5
SQL
   Union, 17-21
   Accessing data in external databases, 17-9
   Administration dialog box, 17-6
   ASEADMIN command, 17-6
   ASEEXPORT command, 17-26
   ASELINKS command, 17-14, 18
   ASEROWS command, 17-9, 10, 18
   ASESQLED command, 17-21
   ASEUNLOAD command, 17-9
   AutoCAD SQL2 environment (ASE), 17-1
   Catalog, 17-2
   Column, 17-5
   Connect to environment dialog box, 17-7
   Creating displayable attributes, 17-16
   Current row, 17-10
   Database, 17-2
   Database management system, 17-3
   Defining keys, 17-5
   Deleting links, 17-15
   Dirty read transaction, 17-5
   Edit row dialog box, 17-12, 19
   Editing data in a Table, 17-10
   Editing links (ASELINKS), 17-14
   Editing rows, 17-18
   Environment, 17-2
   Establishing the database environment, 17-6
   Export Links dialog box, 17-26
   Forming selection sets, 17-19
   Generating reports from exported data, 17-26
   Isolation levels, 17-5
   Link path names dialog box, 17-8
   Linking a database with a drawing, 17-13
   Links dialog box, 17-14, 18
   Make displayable attribute dialog box, 17-16
   Nonrepeatable read transaction, 17-6
   Phantom read transaction, 17-6
   Relational database, 17-3
   Row, 17-5
   Rows dialog box, 17-9, 18
   Schema, 17-2
   SELECT command, 17-21
   Select objects dialog box, 17-20
   Select row by key values dialog box, 17-10
   Session, 17-2
   SQL cursor dialog box, 17-22
   SQL editor dialog box, 17-21
   SQL2, 17-2
   Subtract A-B, 17-21
   Subtract B-A, 17-21
   Transaction, 17-2
   Using SQL statements (ASESQLED), 17-21
Standard AutoCAD menu, 5-1, 10-47
Standard pull-down menus, 5-2
Standard tablet menu, 6-2
Standard template drawings, 1-1
Standard viewport configuration, 1-14
Submenu definition, 5-22
Submenu reference, 5-23
Submenu reference, 7-3
Submenu definition, 8-8
Submenu reference, 8-8
Submenu definition, 9-8
Submenu reference, 9-8
Submenu definition, 10-15
Submenu reference, 10-15
Submenus, 5-22, 22, 42, 43, 47, 7-2, 8-8, 11, 9-7, 67, 10-14
Swapping pull-down menus, 10-42
System variables for mline, 21-33

T

Tablet configuration, 6-7
Tablet command, 6-7
Tablet menu, 6-2
Tablet menus with different block sizes, 6-9
Tablet mode, 21-34
Tablet template, 6-2
TABLET, 21-35
TABLET1, 6-2
TABLET2, 6-2
TABLET3, 6-2
TABLET4, 6-2, 10-29
Tbarname, 5-34
Tee intersection, 21-31
Template drawings, 1-1, 6
Template designs, 6-4
Template overlay, 10-26
Text height, 1-12
TEXTSIZE, 2-3
TGA files, 21-5

The standard AutoCAD menu, 10-1
TIFF files, 21-5
TILEMODE, 1-2, 18, 19
Title of the image tile menu, 7-2
Title of the image tile menu, 10-43
Toggle functions, 9-37
Toolbar, 5-33, 34
TRACEWID, 1-2
TRANSPARENCY command, 21-12

U

UNITS, 1-4, 13
Use of AutoLISP in menus, 9-40
Use of control characters in menu items, 9-36

V

Virtual keys, 5-33
Visible, 5-34
Visual basic
   About visual basic, 16-1
   ActiveX automation, 16-1
   Add method, 16-3
   AddArc, 16-4
   AddCircle, 16-3
   Additional VBA examples, 16-20
   AddLine method, 16-3, 13
   AddText, 16-4
   ANGBASE, 16-11
   ANGDIR, 16-11
   AngleFromXAxis method, 16-15
   AutoCAD object library, 16-2
   BASIC, 16-1
   Dot notation, 16-3
   Event driven, 16-7
   GetAngle method, 16-10
   GetDistance method, 16-10
   GetOrientation, 16-11
   GetPoint method, 16-10
   Installing VBA, 16-2
   Integrated development environment (IDE), 16-5
   Methods, 16-3
   More VBA examples, 16-20
   Object browser, 16-5
   Objects, 16-2
   PolarPoint method, 16-14
   Project, 16-7
   Properties, 16-3
   VBA specific help, 16-4
   Visual basic (VB), 16-1
   Visual basic for applications (VBA), 16-1
VPOINT, 1-15, 19
VSLIDE command, 2-19

W

What are script files?, 2-1
What are slides, 2-18
What is a slide show, 2-18
What is the ACAD.PGP file, 4-1
Windows wordpad, 21-19
Writing a pull-down menu, 5-3
Writing a tablet menu, 6-4
Writing an image tile menu, 7-3
Writing button and auxiliary menus, 8-2

X

Xval, 5-34

Y

Yval, 5-34

Z

ZOOM, 1-18, 19, 2-2, 3, 9-33